High Resolution Archaeology and Neanderthal Behavior

Vertebrate Paleobiology and Paleoanthropology Series

Edited by

Eric Delson
Vertebrate Paleontology, American Museum of Natural History,
New York, NY 10024, USA
delson@amnh.org

Eric J. Sargis
Anthropology, Yale University
New Haven, CT 06520, USA
eric.sargis@yale.edu

Focal topics for volumes in the series will include systematic paleontology of all vertebrates (from agnathans to humans), phylogeny reconstruction, functional morphology, Paleolithic archaeology, taphonomy, geochronology, historical biogeography, and biostratigraphy. Other fields (e.g., paleoclimatology, paleoecology, ancient DNA, total organismal community structure) may be considered if the volume theme emphasizes paleobiology (or archaeology). Fields such as modeling of physical processes, genetic methodology, nonvertebrates or neontology are out of our scope.

Volumes in the series may either be monographic treatments (including unpublished but fully revised dissertations) or edited collections, especially those focusing on problem-oriented issues, with multidisciplinary coverage where possible.

For other titles published in this series, go to
www.springer.com/series/6978

High Resolution Archaeology and Neanderthal Behavior

Time and Space in Level J of Abric Romaní (Capellades, Spain)

Edited by

Eudald Carbonell i Roura

Institut Català de Paleoecologia Humana i Evolució Social (IPHES), Area de Prehistoria, Universitat Rovira i Virgili (URV), C/Escorxador s/n, 43003 Tarragona, Spain

Springer

Editor
Eudald Carbonell i Roura
Institut Català de Paleoecologia Humana i Evolució Social (IPHES)
Area de Prehistoria
Universitat Rovira i Virgili (URV)
C/Escorxador s/n
43003 Tarragona
Spain

ISSN 1877-9077 e-ISSN 1877-9085
ISBN 978-94-007-3921-5 e-ISBN 978-94-007-3922-2
DOI 10.1007/978-94-007-3922-2
Springer Dordrecht Heidelberg New York London

Library of Congress Control Number: 2012933099

Cover illustration: Photographs (from top to bottom) of worked bone, red deer mandible, lithic refitting, *Iberomys cabrerae* molar, and SEM image of butchery use wear, superimposed on a view of level J occupation surface. Photos IPHES.

Printed on acid-free paper

Springer is part of Springer Science+Business Media (www.springer.com)

Preface

The behavioral strategies of Neanderthals are currently one of the key questions in archeological and paleoanthropological research. There are different reasons for this interest in Neanderthal behavior, related both to the development of empirical studies and the introduction of new theoretical paradigms that have changed the understanding of the material record of prehistoric groups. First, accessing behavior is the only way to approach one of the classic problems of archeological research: the variability of Middle Paleolithic archeological assemblages. The multifactor nature of this variability, closely linked to economic strategies and daily activities, necessarily implies examining the complexity of human behavior as an approach to explaining changes in the characteristics of archeological assemblages. Second, Neanderthal behavior has taken on special importance in the framework of the debate on the nature of the cultural transformations defining the transition from the Middle to the Upper Paleolithic. The behavioral capacities of Neanderthals are a key component in clarifying the scope of differences between this human species and anatomically modern humans, an essential question concerning the evolutionary role of Neanderthals and the way in which we understand their culture.

At the site of Abric Romaní, we have never found skeletal remains attributed to *Homo neanderthalensis*. However, the lithic assemblages from all the archeological levels (except level A, corresponding to the Upper Paleolithic) exhibit methods of flake production and tool manufacture consistent with the technological characteristics traditionally associated with this human species. In addition, the chronology of these layers fits perfectly into the temporal range of the last Neanderthals. Therefore, we have always worked with the hypothesis that the archeological evidence found at this site was abandoned by hominids belonging to this European human species.

The scientific and popular debates about *Homo neanderthalensis* began from the very moment the first Neanderthal remains were discovered in Feldhofer Cave in 1856. Soon after that discovery, the remains were studied and interpreted by renowned members of the scientific community. Some of them suggested that the remains from Feldhofer Cave actually corresponded to a *Homo sapiens* affected by a serious pathology. In the mid nineteenth century creationism was still a common explanation for the appearance of living beings and the existence of human species different from *Homo sapiens* was difficult to accept.

One of the more passionate debates that arose as the number of fossils increased concerned the interaction or hybridization between Neanderthals and modern humans. This debate particularly intensified when new radiometric data began to suggest that the two species coexisted in certain European regions for at least 8 ka. And the controversy was further fuelled by evidence suggesting that previously, about 90 ka ago, Neanderthals and modern humans coexisted in the Near East, coinciding with the first

Homo sapiens migration out of Africa. This is still a crucial scientific debate. In spite of numerous excavations undertaken in Europe, fossils of these two species have never been found together in the same archeological layer. However, the genetic analysis recently published by Svante Pääbo and the Max Planck Institute in Leipzig, in coordination with an extensive team of researchers, found that direct contact did indeed occur between the two species during the Upper Pleistocene. These studies show that non-African *Homo sapiens* share between 1 and 4% of their gene pool with the extinct species.

Another topic traditionally treated in scientific works on Neanderthals concerns the factors involved in their extinction. This debate started at the beginning of the twentieth century and is still alive at the beginning of the twenty-first, and probably constitutes one of the most controversial questions in the paleoanthropological and archeological research about our genus.

Some of the arguments used in these debates throughout the past century emerged from the erroneous interpretation of some of the first Neanderthal fossils, like that found at La Chapelle-aux-Saints. Boule's incorrect reconstruction of this fossil contributed to the distorted view of Neanderthals that was dominant in the scientific and mass-culture arenas during most of the twentieth century. The anatomical characteristics of Neanderthals and the lack of symbolic expressions in the archeological assemblages produced by these hominids were arguments used to suggest that their cognitive and organizational patterns were less complex than those exhibited by modern humans.

However, there is evidence to dispute these inferences. It is true that the skull of *Homo neanderthalensis* was different from that of modern humans, but it had a large cranial volume—larger than that of *Homo sapiens*—which seems at odds with the purported inability of Neanderthals to develop symbolic expression. In the same way, it has also been argued that Neanderthals had some impediments to speech, or at least were incapable of the same level of communication that modern humans are capable of. However, the discovery of several ear bones in the Sima de los Huesos of Atapuerca, dated to 500 ka, has allowed the structure of the auditory area of *Homo heidelbergensis* to be reconstructed. This area is similar to that exhibited by *Homo sapiens*, which indicates that human species older than modern humans were probably capable of speech. Although funerary practices are also controversial, intentional burials have been well documented among European and Near Eastern hominids and provide sound evidence supporting the behavioral complexity of Neanderthals.

We believe that the social complexity of *Homo neanderthalensis* is beyond question considering the growing amount of data derived from archeological inquiry. Well verified information is essential to solving the debates described above. This is the only valid method in scientific endeavor: fieldwork should be done after a consistent hypothesis about behavioral complexity has been posed. Only then can we avoid the speculative loop that has often characterized the scientific inquiry into the social and evolutionary complexity of Neanderthals.

The aim of this monograph is to share the scientific information gained from the large-surface excavations carried out in level J, one of the archeological levels forming the Abric Romaní sequence. We wish to present new information about the behavioral patterns of Neanderthals living in northeastern Iberia 50 ka ago. We would like to contribute to the debate on the degree of complexity and organization characterizing these hominids from the multidisciplinary study of this archeological level. In addition, we think that the data yielded by this level are relevant to some of the big issues related to the emergence, evolution and extinction of *Homo neanderthalensis*. Level J is one of the richest of the sequence, both in the quantity of archeological remains and in its occupation structures.

It has been almost thirty years since our team started the excavations at this site on the banks of the Anoia River. In Spanish archeology, it was already a classic site when our work began, having been discovered in 1909 and excavated at different times during the twentieth century. Abric Romaní is a rockshelter formed in a 50 m thick tufaceous formation. Sediments accumulated in this rockshelter throughout the Upper Pleistocene until it was totally filled in during MIS 2. At the beginning of the 1980s, we planned an excavation over a surface large enough to yield a paleoethnographic picture of the spatial strategies of Neanderthals. Since then, this large-surface strategy has been a fundamental component of our data recovery process. We are convinced that spatially oriented studies can lend a great deal towards understanding the level of behavioral complexity achieved by Neanderthals during Marine Isotope Stage 3.

The goal of our research is to discover the behaviors of these European hominids at a specific time and in a specific place in order to establish their social and organizational complexity. This is the first step towards a basis for comparison with the complexity and organization of modern humans arriving in Europe 40 ka ago. When fieldwork began we were convinced that through the recovery of reliable data we would be able to determine whether Neanderthals had a complex social structure.

An important characteristic of this study on level J is that most of the contributing researchers have been excavating the Abric Romaní for some time, some of them for over 20 years. This means that their experience is based on praxis and this close empirical knowledge is very useful for a reliable interpretation of the archeological record. They have been a part of the logical sequence made up of the starting hypothesis, excavation, data recovery, study, discussion and, finally, publication of the results.

More than ten archeological levels have been excavated over a surface equivalent to 90% of the total extension of the site. The excavation strategy followed since 1983 has been directed explicitly towards the reconstruction of the behavioral strategies of Neanderthal groups through the excavation of a large surface area, which includes most of the surface occupied originally. This has led to the excavation of an area measuring nearly 300 m^2, undertaken with careful attention to the spatial distribution of the archeological remains and the identification of structures. This has yielded a diachronic perspective on spatial patterns spanning over more than 10 ka, conditioned by the rapid sedimentation rate characterizing the tufa deposits, which increases the temporal resolution of the occupation layers. Thick sterile layers separate these levels, which considerably diminishes the temporal depth of the palimpsests. This is the case of level J, where the excavated surface is approximately 240 m^2.

In addition to tufa formation, other sedimentary processes have played an important role at Abric Romaní. The cyclic events of roof collapse conditioned the occupation of the site by Neanderthals, as the accumulation of blocks in some areas restricted the habitability of the rockshelter as a whole. As we will see in this book, level J is a good example of this.

Another goal of this work has been to place this far-reaching archeological record in a well-defined environmental context. Level J formed during MIS 3, a period characterized by a high climatic instability, during which cold phases alternated with wet and temperate interstadials. Some colleagues have suggested that these climatic conditions played an important role in the population dynamics of Neanderthals and even determined their extinction.

But the natural environment is not the only driving force that should be taken into account. The historical environment is also an essential key to interpreting the archeological record. This historical context is represented by the patterns defining the Middle Paleolithic as a developmental stage in material culture and social organization. Manufacturing of lithic and wood artifacts, provisioning of raw materials, food and fuel, processing and consumption of faunal and plant resources, and spatial organization are behavioral domains partly conditioned by long-term processes that appeared during Middle Pleistocene times.

Animal hunting and processing behaviors provide insight into the complexity of foraging strategies. Classic questions in faunal studies, like the opposition between specialization and diversification, can be clarified through the analysis of bone assemblages. The formation dynamics defined by the alternating occupations of humans and carnivores is another topic of interest that will be considered in the faunal analysis, although Abric Romaní is characterized by the dominant role of humans in the generation of the archeological record.

Among these behavioral patterns, microspatial interactions are particularly important. The role of hearths should be emphasized, as they are highly abundant in all the archeological levels excavated so far, and have allowed us to designate the formation of hearth-related activity areas as one of the essential features of Neanderthal spatial behavior. These areas can be interpreted as household spaces similar to those identified among contemporary hunter-gatherer groups, a comparison that makes it possible to approach the social dynamics of prehistoric groups. From this point of view, hearths were the basic points of reference in the formation of the archeological record. The dimensional and morphological variability of combustion structures shows the complexity of their functional patterns. These structures determined the spatial distribution of the archeological remains and were particularly important as evidence of social relations. From this perspective, the production and use of fire played a primary role in reinforcing human sociability. Level J can therefore be of great value in testing hypotheses concerning the social structure of Neanderthal groups.

The Abric Romaní rockshelter was a point of reference in the landscape that was visited repeatedly, possibly following a cyclical pattern. This temporal dimension is also an essential part of interpreting the archeological record. The central role of human behavior in the interpretation of the archeological record has been linked to a generalization of ethnoarcheological models as an essential referent in the reconstruction of the formation dynamics of assemblages. However, the use of these models to identify behavioral strategies gives rise to various problems, some of which are not always explicitly approached by researchers. One of the more pronounced of these problems is related to the different time scales that define archeological assemblages and ethnoarcheological contexts. Most archeological assemblages are palimpsests of one type or another, whose formation can span periods of hundreds or even thousands of years and to which many natural and cultural processes of very diverse character can have contributed. From this point of view, then, one must question the extent to which ethnographic models, defined by very different time scales, can provide suitable explanations for these assemblages, and whether misconceptions might occur in assemblage interpretations due to differences concerning formation time.

To approach these essential questions in current archeological research, it is necessary to study assemblages whose time scale is as close as possible to the ethnographic time scale, that is to say, to increase to the maximum the temporal resolution of our assemblages. Achieving this goal is not always easy, since the possibility of accessing increasingly higher temporal levels depends partly on the natural formation processes of the deposits and their stratigraphic resolution. Middle Paleolithic assemblages are often difficult to interpret in temporal terms. Most of them are deep palimpsests formed by the accumulation of archeological remains over long periods. Due to these formation processes, identifying spatial patterns is particularly challenging in many Pleistocene sites featuring low sedimentation rates. For this reason, deposits characterized by high-resolution geological formation processes are especially attractive, as they provide stratigraphic levels covering time periods that are considerably shorter compared to other contexts. These types of deposits are especially suitable for a behavioral reading of the archeological record, and the Abric Romaní is one of these deposits.

In spite of the high temporal resolution favored by geological formation dynamics, some data indicate that the level J assemblage corresponds to a palimpsest formed by an indeterminate number of occupation episodes. The study of level J will be directed to the temporal dissection of this palimpsest, identifying higher resolution assemblages of remains from which it will be possible to access behavioral patterns with a considerable degree of certainty. In this context, the spatial data will be fundamental, since they reveal the dynamics of mobility and artifact transport from which the formation sequence of the archaeological assemblage can be established. The temporal and spatial interpretations will therefore be closely linked.

This spatio-temporal interpretation will be achieved through the information yielded by a wide range of analytic fields that constitute the different chapters of the monograph, including aspects related both to natural formation dynamics (stratigraphy, palaeoenvironment, biostratigraphy, taphonomy, etc.) and human activities (lithic technology, faunal processing, habitat structures, spatial distribution, wood implements, etc.). The last section will discuss whether the spatio-temporal perspective that we propose opens up a new view of Neanderthal behavior, different from that derived from works that do not address the importance of time resolution in the formation of archeological assemblages.

We are very thankful to Eric Delson and Eric Sargis (editors of the Vertebrate Paleobiology and Paleoanthropology book series at Springer) for their help and advice during the completion of this book. Special thanks go to the staff of Springer for their patience.

Following the Series' requirements, each chapter was peer-reviewed by two or three referees in addition to the editorial team. We are very grateful to all the reviewers: Dan Adler, Rosa Maria Albert, Peter Andrews, Javier Baena, Mercé Bergadà, Jean-Phillipe Brugal, Dan Cabanes, José Carrión, Gloria Cuenca, Yolanda Fernández-Jalvo, Paul Goldberg, Naama Goren-Inbar, Donald Henry, Steven Kuhn, Carlos Lorenzo, Richard Macphail, Sally McBrearty, Jordi Nadal, Oriol Oms, Marco Peresani, and Gina Semprebon for their very helpful comments.

We also thank all the institutions and companies that have provided and continue to provide financial support to the Abric Romaní excavations: Generalitat de Catalunya, Diputació de Barcelona, Ajuntament de Capellades, Tallers Gràfics Romanyà-Valls S.A., Bercontrés-Centre de Gestió Mediambiental S.L., and Constructora de Calaf SAU.

We would like to express also our gratitude to Raül Bartrolí, our man in Capellades, for his assistance and logistical support during fieldwork.

Finally, we especially thank all the generous excavators that have taken part in the fieldwork at Abric Romaní since 1983.

Tarragona, November 2010 Eudald Carbonell

Contents

Contributors

Ethel Allué
Institut Català de Paleoecologia Humana i Evolució Social (IPHES), Universitat Rovira i Virgili (URV), C/Escorxador s/n, 43003 Tarragona, Spain
eallue@iphes.cat

Maria Bennàsar
Institut Català de Paleoecologia Humana i Evolució Social (IPHES), Universitat Rovira i Virgili (URV), C/Escorxador s/n, 43003 Tarragona, Spain
mlluc@prehistoria.urv.cat

James L. Bischoff
US Geological Survey, ms/470, 345 Middlefield Rd, Menlo Park, CA 94025, USA
jbischoff@usgs.gov

Hugues-Alexandre Blain
Institut Català de Paleoecologia Humana i Evolució Social (IPHES), Universitat Rovira i Virgili (URV), C/Escorxador s/n, 43003 Tarragona, Spain
hablain@iphes.cat

Ruth Blasco
Institut Català de Paleoecologia Humana i Evolució Social (IPHES), Universitat Rovira i Virgili (URV), C/Escorxador s/n, 43003 Tarragona, Spain
rblasco@iphes.cat

Pilar Bravo
Institut Català de Paleoecologia Humana i Evolució Social (IPHES), Universitat Rovira i Virgili (URV), C/Escorxador s/n, 43003 Tarragona, Spain
pbravo@codex.cat

Francesc Burjachs
Institució Catalana de Recerca i Estudis Avançats (ICREA), Universitat Rovira i Virgili (URV), C/Escorxador s/n, 43003 Tarragona, Spain
francesc.burjachs@icrea.cat

Dan Cabanes
Department of Structural Biology, Kimmel Center for Archaeological Science, Weizmann Institute of Science, 76100 Rehovot, Israel
dan.cabanes@weizmann.ac.il

Isabel Cáceres
Institut Català de Paleoecologia Humana i Evolució Social (IPHES), Universitat Rovira i Virgili (URV), C/Escorxador s/n, 43003 Tarragona, Spain
icaceres@iphes.cat

Maria Gerard Campeny
Institut Català de Paleoecologia Humana i Evolució Social (IPHES), Universitat Rovira i Virgili (URV), C/Escorxador s/n, 43003 Tarragona, Spain
gcampeny@iphes.cat

Eudald Carbonell i Roura
Institut Català de Paleoecologia Humana i Evolució Social (IPHES), Universitat Rovira i Virgili (URV), C/Escorxador s/n, 43003 Tarragona, Spain
eudald.carbonell@urv.cat

María Gema Chacón
Institut Català de Paleoecologia Humana i Evolució Social (IPHES), Universitat Rovira i Virgili (URV), C/Escorxador s/n, 43003 Tarragona, Spain; UMR7194 - Département de Préhistoire, Muséum national d'Histoire naturelle, 1, rue René Panhard, 75013 Paris, France
gchacon@iphes.cat

Marie-Agnès Courty
CNRS-MNHN UMR 5198, Institut Català de Paleoecologia Humana i Evolució Social (IPHES), C/Escorxador s/n, 43003 Tarragona, Spain
courty@mnh.fr

Felipe Cuartero
Institut Català de Paleoecologia Humana i Evolució Social (IPHES), Universitat Rovira i Virgili (URV), C/Escorxador s/n, 43003 Tarragona, Spain
felipecuartero@yahoo.es

Montserrat Esteban-Nadal
Institut Català de Paleoecologia Humana i Evolució Social (IPHES), Universitat Rovira i Virgili (URV), C/Escorxador s/n, 43003 Tarragona, Spain
mesteban@iphes.cat

Isabel Expósito
Institut Català de Paleoecologia Humana i Evolució Social (IPHES), Universitat Rovira i Virgili (URV), C/Escorxador s/n, 43003 Tarragona, Spain
iexposito@iphes.cat

Cristina Fernández-Laso
Institut Català de Paleoecologia Humana i Evolució Social (IPHES), Universitat Rovira i Virgili (URV), C/Escorxador s/n, 43003 Tarragona, Spain
cfernan@prehistoria.urv.cat

Maria Joana Gabucio
Institut Català de Paleoecologia Humana i Evolució Social (IPHES), Universitat Rovira i Virgili (URV), C/Escorxador s/n, 43003 Tarragona, Spain
mjgabucio@iphes.cat

Ana García
Institut Català de Paleoecologia Humana i Evolució Social (IPHES), Universitat Rovira i Virgili (URV), C/Escorxador s/n, 43003 Tarragona, Spain
anagarbo@gmail.com

Mª Dolores García-Antón
Institut Català de Paleoecologia Humana i Evolució Social (IPHES), Universitat Rovira i Virgili (URV), C/Escorxador s/n, 43003 Tarragona, Spain
madolores.garciaanton@urv.cat

Bruno Gómez de Soler
Institut Català de Paleoecologia Humana i Evolució Social (IPHES), Universitat Rovira i Virgili (URV), C/Escorxador s/n, 43003 Tarragona, Spain
bruno@prehistoria.urv.cat

Rosa Huguet
Institut Català de Paleoecologia Humana i Evolució Social (IPHES), Universitat Rovira i Virgili (URV), C/Escorxador s/n, 43003 Tarragona, Spain
rhuguet@iphes.cat

Nuria Ibáñez
Institut Català de Paleoecologia Humana i Evolució Social (IPHES), Universitat Rovira i Virgili (URV), C/Escorxador s/n, 43003 Tarragona, Spain
nibanez@iphes.cat

Juan Manuel López-García
Institut Català de Paleoecologia Humana i Evolució Social (IPHES), Universitat Rovira i Virgili (URV), C/Escorxador s/n, 43003 Tarragona, Spain
jmlopez@iphes.cat

Patricia Martín
Institut Català de Paleoecologia Humana i Evolució Social (IPHES), Universitat Rovira i Virgili (URV), C/Escorxador s/n, 43003 Tarragona, Spain
patricia.martin@estudiants.urv.cat

Jordi Martinell
Departament d'Estratigrafia, Paleontologia i Geociències Marines, Universitat de Barcelona Facultat de Geologia, Campus de Pedralbes, Barcelona, Spain
jmartinell@ub.edu

Kenneth Martínez
Institut Català de Paleoecologia Humana i Evolució Social (IPHES), Universitat Rovira i Virgili (URV), C/Escorxador s/n, 43003 Tarragona, Spain
kenneth@prehistoria.urv.cat

Laura Muñoz
Institut Català de Paleoecologia Humana i Evolució Social (IPHES), Universitat Rovira i Virgili (URV), C/Escorxador s/n, 43003 Tarragona, Spain
lencinar@prehistoria.urv.cat

Florent Rivals
Institució Catalana de Recerca i Estudis Avançats (ICREA), Universitat Rovira i Virgili (URV), C/Escorxador s/n, 43003 Tarragona, Spain
florent.rivals@icrea.es

Antonio Rodríguez-Hidalgo
Institut Català de Paleoecologia Humana i Evolució Social (IPHES), Universitat Rovira i Virgili (URV), C/Escorxador s/n, 43003 Tarragona, Spain
arodriguez@iphes.cat

Jordi Rosell
Institut Català de Paleoecologia Humana i Evolució Social (IPHES), Universitat Rovira i Virgili (URV), C/Escorxador s/n, 43003 Tarragona, Spain
jordi.rosell@urv.cat

Robert Sala
Institut Català de Paleoecologia Humana i Evolució Social (IPHES), Universitat Rovira i Virgili (URV), C/Escorxador s/n, 43003, Tarragona, Spain
robert.sala@icrea.es

Palmira Saladié
Institut Català de Paleoecologia Humana i Evolució Social (IPHES), Universitat Rovira i Virgili (URV), C/Escorxador s/n, 43003 Tarragona, Spain
psaladie@iphes.cat

Pablo Sañudo
Institut Català de Paleoecologia Humana i Evolució Social (IPHES), Universitat Rovira i Virgili (URV), C/Escorxador s/n, 43003 Tarragona, Spain
pablo.sanudo@bolomor.com

Alex Solé
Institut Català de Paleoecologia Humana i Evolució Social (IPHES), Universitat Rovira i Virgili (URV), C/Escorxador s/n, 43003 Tarragona, Spain
asole@prehistoria.urv.es

Josep Vallverdú
Institut Català de Paleoecologia Humana i Evolució Social (IPHES), Universitat Rovira i Virgili (URV), C/Escorxador s/n, 43003 Tarragona, Spain
josep@prehistoria.urv.cat

Manuel Vaquero
Institut Català de Paleoecologia Humana i Evolució Social (IPHES), Universitat Rovira i Virgili (URV), C/Escorxador s/n, 43003 Tarragona, Spain
manuel.vaquero@urv.cat

Chapter 1
Introduction: Neanderthal Behavior and Temporal Resolution of Archeological Assemblages

Manuel Vaquero

Abstract Neanderthal behavior is currently one of the main topics in paleoanthropological and archeological research. This interest is largely related to the debate about the emergence of modern humans and the extinction of Neanderthals. In this context, some researchers have considered Neanderthal behavior as archaic and essentially different to the "modern behavior" characteristic of *Homo sapiens*. We present in this chapter a general outline of this debate in different domains of the archeological research, from technology and subsistence to spatial patterns. Moreover, we point out that these behavioral issues should be approached by taking into account the temporal nature of the archeological assemblages, since temporal resolution may be a primary factor in interassemblage variability. Due to these time-dependent formation processes, Abric Romaní appears as a site particularly suitable to yield information on Neanderthal behavior.

Keywords Neanderthal behavior • Abric Romaní • Time scales • Settlement patterns

Neanderthal Behavioral Strategies and Middle Paleolithic Variability

The behavioral patterns of ancient hominids are among the main concerns of Paleolithic archeology. Although chronocultural issues also play an important role in research, especially in periods such as the Upper Paleolithic, there seems to be general agreement about the significance of behavioral strategies in the variability of archeological assemblages. Understanding behavior is one of the principal goals of Paleolithic archeologists, but at the same time, behavior has become a key factor in explaining the variability of the archeological record. This interest in behavioral patterns also has clear evolutionary implications, especially if we take into account two related issues. On the one hand, behavior is a basic form of adaptation to natural and social environments, and the Darwinian struggle for survival often takes place in the behavioral realm (Cronk 1991; Krebs and Davies 1991; Smith and Winterhalder 1992; Cronk et al. 2000). On the other hand, behavior is subject to evolutionary forces and the behavioral patterns of current human populations may be understood as the outcome of a long process during which these patterns progressively developed. Paleolithic archeology provides the essential database for insight into this second issue as it is the best means by which to approach the behavioral strategies of hominid populations prior to modern humans.

Among ancient hominids, Neanderthals play a central role in the ongoing debate on the evolution of behavioral capabilities. The behavioral strategies of Neanderthals are currently one of the key questions in paleoanthropological research. Neanderthals were the last of the ancient hominids prior to the worldwide expansion of modern humans, with whom they coexisted during several millennia. This interest in Neanderthal behavior has developed for different reasons related both to the publication of empirical works and to the introduction of new theoretical paradigms that have changed the understanding of the material record of prehistoric groups. First, it is clear that accessing behavior is the only surefire way of approaching one of the classic problems in archeological research: the variability of Middle Paleolithic archeological assemblages. The multifactorial character of this variability, closely tied to economic strategies and daily activities, necessarily implies an approach to the complexity of human behavior as a way of explaining the changes in the characteristics of archeological assemblages. In fact, recent

M. Vaquero (✉)
Institut Català de Paleoecologia Humana i Evolució Social (IPHES), Universitat Rovira i Virgili (URV), C/Escorxador s/n, 43003 Tarragona, Spain
e-mail: manuel.vaquero@urv.cat

E. Carbonell i Roura (ed.), *High Resolution Archaeology and Neanderthal Behavior: Time and Space in Level J of Abric Romaní (Capellades, Spain)*, Vertebrate Paleobiology and Paleoanthropology, DOI: 10.1007/978-94-007-3922-2_1, © Springer Science+Business Media B.V. 2012 explanations of Mousterian variability have emphasized behavioral factors as opposed to classical preconceptions based on a normative sense of culture and the definition of chronocultural entities. Second, Neanderthal behavior has taken on special importance in the framework of the debate on the nature of the cultural transformations that define the passage from the Middle Paleolithic to the Upper Paleolithic. In this context, the behavioral capacities of Neanderthals are a key issue in clarifying the scope of the differences between them and anatomically modern humans, an essential question concerning the evolutionary role of Neanderthals and the manner in which we understand their culture.

It is worth emphasizing, however, that this evolutionary scenario implies some negative consequences for Neanderthals due to the inevitable intertwining of biological and cultural issues. It seems that it is the Neanderthals' fate to be systematically compared with modern humans, or more precisely, with Upper Paleolithic modern humans. In this framework it is not surprising that Neanderthal behavior appears less complex or developed, although we should wonder if such an appraisal is at least partly the result of comparing two successive historical periods. In this respect, it would be an instructive exercise to consider what differences in behavioral capabilities could be inferred by comparing, for instance, Upper Paleolithic and Neolithic material cultures. The archeological differences between these two periods are in some respects more impressive than those found between the Middle and Upper Paleolithic. However, nobody has proposed an explanation in terms of behavioral capabilities for the Neolithic Revolution, surely due to the fact that both Upper Paleolithic and Neolithic populations look anatomically like us.

So, the Neanderthals versus modern humans debate has given rise to two opposing perspectives on Neanderthal behavior. According to some authors, Neanderthals did not have the capabilities required to develop fully modern behavior. This inability would have been related to Neanderthal biological and cognitive patterns, which determined that their mental and linguistic capacities were significantly different from those of modern humans (Klein 1995, 2000; Mellars 1996a, b; Mithen 1994, 1996a, b; Tattersall 1999; Wynn and Coolidge 2004). For example, Mithen argued that the Middle-Upper Paleolithic transition represented the emergence of modern human intelligence, based on the connectivity between different cognitive domains. In general, this conclusion emerges from the comparison between Middle and Upper Paleolithic archeological records and the alleged clear-cut differences that can be observed between them. The adaptive advantages provided by these new behavioral patterns would explain the replacement of Neanderthals by modern humans. Meanwhile, other researchers argue that the behavior inferred from Middle Paleolithic sites exhibits the basic features that characterize modern behavior. They point out that the differences between the Middle and Upper Paleolithic can be attributed to various social, economic and historical causes, but they do not imply any biologically determined behavioral inability of Neanderthals (Hayden 1993; Soffer 1994; d'Errico et al. 1998).

It is important to recognize, however, that from the very start this debate entails a problem that is difficult to resolve: to what extent can we define "modern human behavior"? The human population, both historic and of today, exhibits considerable behavioral variability; variability that makes discovering the features that can be accepted as common to all past and present modern humans no easy task. From an archeological point of view, an operational definition of behavioral modernity should avoid any essentialist concept about human nature and focus on what humans do. As a matter of fact, we are what we do. This is especially important in evaluating arguments based on intellectual or mental capabilities. These capabilities are useful in defining behavioral modernity only if they are expressed in how material activities are performed.

There are different versions of the Neanderthal inability hypothesis, depending on the behavioral domain, although this argument is normally associated with the idea that the emergence of the Upper Paleolithic represented a behavioral revolution—the *human revolution* (Mellars and Stringer 1989). It is considered the great breakthrough in the evolution of humankind, the beginning of culture as we understand it today (Binford 1985) or "the conquest of nature by the spirit" (Otte 1996). Bearing this complete set of behavioral innovations, modern human populations would have had decisive adaptive advantages over local Neanderthal populations, which would ultimately be doomed to extinction due to their inability to successfully compete with the newcomers. Although many Upper Paleolithic features were not exclusively associated with modern humans, as shown by some Chatelperronian assemblages, this has not changed the general terms of this assumption on the behavioral disadvantages of Neanderthal ways of life. Most of these theories are based more or less explicitly on the alleged inability of Neanderthals to develop symbolic behavior. Symbolic expression represented a huge step forward in human evolution and the capacity to use symbols in everyday life would have radical consequences on all human activities, especially on those associated with the production of material culture. The appearance of clear evidence of mobile and parietal art coinciding with the beginning of the Upper Paleolithic is the most conclusive manifestation of this symbolic revolution. As Henshilwood and Marean (2003) pointed out, modern human behavior is defined by the use of symbolism to organize behavior. Several works provide a full account of this approach (Klein 1995, 2000; Mellars 1996a; Noble and Davidson 1996), which we will only briefly summarize here.

In the first place, if Neanderthals lacked the ability for symbolic behavior, this would be evident in the technological realm; however it is clear that some knapping methods widely represented in Middle Paleolithic assemblages, such as Levallois technology, exhibit high levels of technical complexity, similar to those inferred from Upper Paleolithic technologies. In addition, blade reduction methods, once considered exclusive to Upper Paleolithic technologies, have been well documented in Middle Paleolithic assemblages. However, some clear-cut differences still remain in tool manufacture, and these differences are the main argument of those advocating the Neanderthal cognitive drawback perspective. For instance, it has been pointed out that the Middle Paleolithic toolkit is not standardized and shows little diversity, which some say points to a lack of competence in adapting tool morphology to the function at hand. Middle Paleolithic tools were versatile and therefore less efficient than the specialized artifacts typical of the Upper Paleolithic, which were designed for specific functions. This is especially clear in the manufacture of tools interpreted as comprising part of the hunting gear, such as the wide array of projectile points that appear from the start of the Upper Paleolithic. Among such tools, it is worth highlighting those made from raw materials other than stone, like bone and antler, which contrasts with the scant use of these materials during the Middle Paleolithic. This evidence suggests that hafting technology and the use of composite tools were more developed during the Upper Paleolithic, which may be related to the increased use of hunting techniques based on throwing weapons. Although the use of hafting and throwing spears was probably not unknown in the Middle Paleolithic, it seems clear that it was less systematic and widespread than in the Upper Paleolithic (Shea 2006).

Upper Paleolithic technology shows a higher level of complexity and capacity for innovation (Hoffecker 2005). This would explain the trend of temporal change exhibited by the Upper Paleolithic chronocultural sequence, as well as the clearer pattern of geographical variability in lithic assemblages as opposed to the comparatively diachronous monotony of Middle Paleolithic assemblages. The ultimate explanation for this difference rests on the inability of Neanderthals to give to their tools an imposed form according to well-defined mental templates (Chase 1991). Another consequence of such a cognitive feature would be the lack of symbolic significance in artifact shape and therefore the inability to transmit social and ethnic information through material culture. This temporal and spatial patterning in tool morphology has been interpreted as the first clear evidence of stylistic variability in technology. Stylistic expression in material culture would have been a method of marking social and ethnic boundaries.

Food-procurement strategies also show this specific characteristic of Neanderthal behavior. Two issues have focused the debate on this subject: the role of scavenging in the exploitation of animal resources and the degree of specialization in subsistence patterns. The former issue was put forward by Binford (1985), who suggested that Middle Paleolithic peoples were incapable of hunting large animals due to their low degree of long-range planning and cooperation. The remains of the large animals found at archeological sites would therefore correspond to resources obtained through scavenging. However, Binford's hypothesis has received little support, and many studies of Middle Paleolithic faunal assemblages indicate that hunting was the primary strategy for obtaining large animal carcasses (Chase 1988; Auguste et al. 1998; Marean 1998; Patou-Mathis 2006). In spite of this evidence, it has been suggested that Neanderthal hunting practices would have been less systematic and intensive, exhibiting a lower level of logistical organization (Mellars 1989), related to Neanderthals' inability to make long-term plans. In addition, their communicative flaws would have been a handicap in developing cooperative hunting techniques.

As for the degree of specialization in subsistence patterns, most Middle Paleolithic sites do not exhibit clear evidence of economic specialization, defined as the deliberate selection of a particular species from the animal resources available in the environment, although apparent examples of specialization are not unheard of (Gaudzinski and Roebroeks 2000). This pattern is more common in Upper Paleolithic faunal assemblages (Mellars 1973; Orquera 1984), although such specialization has basically been documented in Late Upper Paleolithic fauna (Chase 1989) and some studies have pointed out that Middle and Early Upper Paleolithic subsistence strategies were not significantly different (Grayson and Delpech 2002, 2003; Adler 2006). As a matter of fact, the faunal changes observed in some regions at the beginning of the Upper Paleolithic show a higher diversification of subsistence strategies, which were particularly characterized by the increasing exploitation of small resources (Aura et al. 2002). In addition, some authors (Kuhn and Stiner 2006; O'Connell 2006) have suggested that diversified patterns enabled more efficient exploitation of the environment, which would have been a key factor in the more successful adaptation of modern humans.

Finally, the lack of symbolic expression among Neanderthals would have also had critical consequences in the social realm (Pettitt 2000). Several different authors (Soffer 1994; Pettitt 2000; Kuhn and Stiner 2006) propose that Neanderthal society would have differed from modern patterns of social organization—as defined, for example, by Rodseth et al. (1992)—on several different levels, from gender relationships and the division of economic activities between the sexes to group structure and the formation of regional networks. Based on the spatial distribution of the remains at Combe Grenal, Binford suggested that Neanderthal social structure was characterized by a separation between males and females, who formed separate foraging units and got together only during mating periods (cited in Mellars 1992, pp. 357–359). This pattern would be radically different from that characteristic of present-day human groups, which is based on household units formed by the permanent relationship between pairs of males and females. Other researchers (Soffer 1994; Kuhn and Stiner 2006) have also pointed out a gender-related difference between Neanderthals and modern humans, but in the opposite direction. They argue that the Middle Paleolithic record does not show the sexual division of labor characterizing modern hunter-gatherer societies. Male and female roles would have been similar in Neanderthal groups and their economy would therefore be less efficient, since the division of activities according to sex allows the subsistence pattern to be more diversified and, as a consequence, less fragile in periods of resource stress.

Regardless of their cognitive significance, the virtual absence of personal ornaments and grave offerings associated with human burials during the Middle Paleolithic indicates that Neanderthal communities were characterized by scant social differentiation in terms of gender, age or status. According to Zilhao (2002), Middle Paleolithic burials do not show a differential treatment of individuals by age classes, which suggests that developmental stages did not have social significance in Neanderthal society, or at least such significance was not expressed in material cultural and funerary practices. One of the most extreme statements about the non-modern character of Neanderthal society can be found again in the work of Pettitt (2000), who points out that age thresholds were simply determined by the physical achievements of individuals, in the context of a lifecycle defined as "nasty, brutish and short". Under this assumption, biology would have been more important than culture in the constitution of Neanderthal society.

Large-scale social interactions and the formation of large social or ethnic units would have been another change caused by the enhanced communicative abilities of modern people. The density of Neanderthal populations would have been low and they would have organized themselves into small social units, both at local and regional levels. The lack of an extended network of information exchange would have had particularly negative effects during periods of environmental decline characterized by an increase in resource stress. According to Finlayson (2004), the fragmentation of Neanderthal populations in the context of the climatic instability of MIS 3 was the ultimate cause of their extinction. The appearance of symbolically mediated forms of communication would have allowed the formation of alliances and kinship relationships over large territories. This would have permitted a more effective adaptation to high-risk environments defined by an unpredictable and unreliable distribution of resources. Neanderthal bands would have been unable to establish social bonds and, as a consequence, band size would be smaller.

However, in order to discuss these questions related to social structure we first need an approach to spatial behavior. In this book we pay special attention to behavioral strategies related to settlement and spatial organization. Spatial behavior, both at the regional and the intrasite levels, is closely linked to social structure and is a good approach to understanding aspects like group size, occupation type, site function, specialization of activity areas, mobility patterns, social relationships, etc. Furthermore, these issues are central to any understanding of archeological assemblage variability. There is no doubt that Neanderthals regularly used caves and rockshelters as occupation or activity sites, although open-air settlements are certainly underrepresented in the Middle Paleolithic record due to the traditional focus of research on cave and rockshelter sites. Nevertheless, some data indicate that the use of these locations was somewhat different to that inferred from the Upper Paleolithic record. The alternating use of cave and rockshelters by humans and carnivores was common during the Middle Paleolithic. Mixed assemblages derived from this alternation form a well-documented archeological context (Brugal and Jaubert 1991). It seems that this scenario changed with the appearance of the Upper Paleolithic, in which human impact was clearly dominant, and carnivore activity—especially that associated with large carnivores—showed a significant decline. This change could be attributed to more intense or longer occupations during the Upper Paleolithic, which may indicate that caves and rockshelters were more commonly used for residential purposes. As mentioned earlier, it has also been argued that modern humans had an enhanced capacity to settle in harsh environments at high latitudes and high-mountain areas that were scarcely visited by Neanderthals (Mellars 1989; Gamble 1994). This would have also been related to the use of symbolism to construct social and regional networks for sharing information and resources, which would be essential to successful adaptation in such high-risk environments. In any case, settlement patterns are fundamental in reconstructing the cultural processes responsible for the formation of the archeological assemblage and many studies have recently suggested that these patterns are a primary factor in Middle Paleolithic interassemblage variability (Henry 1992; Kuhn 1995; Marks and Chabai 2001; Richter 2001; Bolus 2004; Depaepe 2004; Soressi 2004; Burke 2006; Wallace and Shea 2006).

When we talk about settlement patterns we are basically dealing with two interrelated issues: mobility patterns and types of sites. As for the former, it seems clear that mobility plays an essential role in the hunter-gatherer way of life and we can presume that it conditioned many aspects of Neanderthal behavior. In general, mobility patterns can be characterized by means of various distinct features, such as the distance traveled in each movement, the total distance covered during an annual cycle, the number of moves, and the character of the movement—residential or logistical (Kelly 1983). With regard to the latter, site variability can be analyzed on two levels: site type or function, and occupation type, that is, length of occupation and group size. Site function basically means differentiating between residential and non-residential sites. This is a particularly important issue, especially if we take into account the social relevance of concepts such as the residential campsite, because such campsites can be interpreted as social spaces based on interpersonal communication and food-sharing. The functional variability of sites is therefore a key issue to understanding the spatial behavior of Neanderthals, and the archeological criteria that might provide insight into that variability are particularly significant.

Residential sites have been defined as "…the locus out of which foraging parties originate and where most processing, manufacturing and maintenance activities take place" (Binford 1980, p. 9). They are where the family units that make up the band live and where resources gathered in the foraging territory are brought back to. The home base is the basic type of site in hunter-gatherer settlement patterns. Nevertheless, finding residential sites in archeological contexts poses serious challenges derived from the temporal dimension of archeological assemblages. Many of the criteria used to identify such sites are also employed to infer occupation length, which will be discussed in further detail later on in this work. However, now is a good time to emphasize one such criterion. In ethnographic contexts, residential sites are always characterized by the role played by hearths as the focal points of their spatial organization. Hearths are an essential characteristic of the household areas that define residential campsites, and most activities are carried out around them. From this point of view, identifying combustion structures in archeological sites seems to be a strong basis for characterizing them as residential campsites. The fact that hearths are commonly found in Middle Paleolithic sites suggests that residential locations were an important component of Neanderthal settlement patterns.

In addition to these residential sites, special activity sites complete the range of functional variability in settlement strategies. As we have seen, the specialization concept has frequently come up in discussions about the emergence of modern human behaviors. The manufacture of functionally specialized tools and subsistence strategies specializing in certain resources have been considered indicators of modern behavioral competences. The search for specialized sites is a common theme in research on settlement patterns. Among these sites, those devoted to lithic raw material provisioning (stone quarry sites) and the first processing of animal resources (kill sites or hunting camps) are the most commonly identified in the Middle Paleolithic record. The former were places in which lithic resources were extracted and initially worked. They are defined by the almost exclusive presence of lithic remains corresponding to the first stages of the reduction sequence and are normally located close to raw material outcrops (Geneste 1985; Turq 1992). In the latter, bone remains are overwhelmingly dominant and lithic remains are mainly represented by selected tools used in butchering activities (Chase 1989; Farizy and David 1992; Costamagno et al. 2006). When carcasses were obtained through hunting, the bones remaining at the kill site correspond mainly to the less useful parts of the animal, since the richest anatomical portions would have been brought back to the residential camp.

Beyond the functional differences between residential sites and special-purpose sites, a second level of variability concerns the differences than can be found between residential occupations. This topic is particularly difficult to approach, as the two features that define this variability—occupation length and group size—are especially elusive to archeological inquiry. In recent years, several Middle Paleolithic assemblages have been attributed to short-term occupations characterized by a limited number of activities (Roebroeks 1986; Geneste 1988; Deloze et al. 1994; Defleur and Cregut-Bonnoure 1995; Conard and Adler 1997; Martínez Moreno et al. 2004; Vallverdú et al. 2005). In some cases, intrasite spatial information indicates that occupation events took place in restricted and very limited areas, with no evidence of links between different activity areas. Based on these data, some authors (Mellars 1996a, b; Kolen 1999) argue that Neanderthal settlements as a whole are characterized by the ephemeral occupations of small groups, unlike modern humans who exhibit higher settlement variability, including long-term residential camps. However, we should wonder if this proliferation of short-term occupations is an artifact of the resolution problems affecting archeological interpretation. Small assemblages corresponding to short events are easier to interpret in settlement terms than the huge accumulations derived from long-term occupations, which are difficult to differentiate from occupational palimpsests. We will come back to this subject in the next section.

Differences in occupation length and/or group size have frequently been used to explain the variability of archeological assemblages. In general, studies on Middle

Paleolithic settlement patterns have been more or less explicitly influenced by hunter-gatherer ethnoarcheology. In this respect, it is important to bear in mind the role played by the forager/collector distinction proposed by Binford (1980) and the different mobility patterns—residential and logistical—associated with these settlement strategies. Research on modern hunter-gatherers shows that some campsites are characterized by long stays and are occupied for several weeks or months, while others are used for considerably shorter occupations of only a few days or even hours. In the annual cycle, these differences were associated with the aggregation/segregation dynamics typical of hunter-gatherer bands, which ultimately depended on changes in the distribution and availability of resources. In this context, several authors have argued that some Middle Paleolithic assemblages would correspond to long-term occupations. For example, a radial settlement pattern based on seasonal or multi-seasonal base camps was proposed for the Middle Paleolithic of the Near East (Marks and Freidel 1977; Coinman et al. 1986; Lieberman 1993). By the same token, Henry (1992) pointed out that settlement in Southern Jordan included two different types of sites: transitory camps occupied by small groups and longer encampments. Lithic assemblages would be different at these site types, since ephemeral camps show an emphasis on tool manufacture and maintenance, while the full range of lithic processing activities were carried out at long-term sites.

This kind of settlement patterning has also been suggested for the European Middle Paleolithic. For example, Richter (2006) suggested that Middle Paleolithic settlement in Germany was characterized by a seasonal pattern based on a dichotomy between mountain areas and plains. The plain sites would correspond to autumn–winter camps and would have been occupied by larger human groups for extended periods. On the other hand, spring and summer campsites would be located in mountain areas and correspond to ephemeral occupations by small groups. For the Middle Paleolithic of Crimea, Marks and Chabai (2001) also proposed that variability in settlement strategies was defined by different types of sites largely characterized according to occupation length: ephemeral kill/butchery loci, ephemeral camps, short-term camps and base camps. In general, several settlement models have incorporated the coexistence of short-term and long-term campsites in the European Middle Paleolithic (Peresani 2001; Rolland 2001; Tillet 2001; Depaepe 2004; Moncel 2004). Meanwhile, other studies have stressed the absence of long-term residential sites in some regions (Conard 2001).

Settlement patterns have particularly been used to explain the variability of Middle Paleolithic lithic assemblages. Dibble and Rolland suggested that denticulate-rich assemblages would have resulted from the lower occupation intensities characteristic of milder conditions, while the assemblages dominated by highly reduced sidescrapers would correspond to longer residences more common in colder periods (Rolland and Dibble 1990; Dibble and Rolland 1992). Settlement dynamics might also explain the differences between types A and B of Mousterian of Acheulian Tradition (Soressi 2004) or between the Mousterian and the Micoquian in Central Europe (Richter 2001). Other authors have also pointed out the effects of mobility patterns on retouched tool inventories (Geneste 1988; Meignen 1988) and core reduction strategies (Kuhn 1995; Wallace and Shea 2006).

However, these approaches to archeological variability pose a far-reaching problem: how to identify different occupation types through the archeological record. Too often settlement dynamics are used to explain assemblage variability, but rarely have independent criteria for determining occupation type been proposed. It has been pointed out (Burke 2006) that trying to correlate ethnographically derived site typologies to archeological contexts is unrealistic and that a distinction should only be made between residential and non-residential locations. As we will discuss more thoroughly in the next section, these problems are derived from the temporal resolution of archeological assemblages and, as a consequence, the difficulties in differentiating between single occupations and palimpsests (Roebroeks 1988; Vermeersch 2001). For example, it seems clear that the quantity of remains, once used to infer occupation length, cannot actually be considered a valid criterion, especially for identifying long-term campsites. Small assemblages can be confidently attributed to short-term occupations, but the opposite cannot be assumed, since large assemblages may also be the product of multiple short events that took place over a long time span. Therefore, it is worth discussing other criteria that have been proposed both to identify residential locations and to measure occupation length. Among such criteria we emphasize the following:

- Origin of lithic raw materials. Raw materials coming from the vicinity of the site would be more intensively exploited as occupation length increased. On the other hand, short-term camps would correspond to contexts of higher mobility and would show higher percentages of exotic raw materials (Richter 2006).
- Proportion of certain artifact classes. Richter (2006) has suggested that denticulates were tools for daily use and their absolute number would reflect occupation time. However, this is at odds with the hypothesis that longer occupations are characterized by more intense use of lithic raw materials, therefore leading to the presence of higher percentages of heavily reduced artifacts, such as some types of sidescrapers.
- Diversity of activities. Residential locations are defined by the performance of a wide range of activities, including the processing and cooking of animal and plant

resources, and the manufacture and maintenance of tools. As occupation length increased, it would become more and more likely that new activities would be carried out at the site.

- Evidence of fire making. As previously discussed, residential locations are defined by the occurrence of household areas that are organized around hearths. The size and thickness of burnt deposits can be considered evidence of combustion intensity and, therefore, occupation length. For example, the thick ashy deposits associated with the combustion structures of Kebara would suggest that this site functioned as a home base during the Middle Paleolithic (Meignen et al. 1998).
- Site structure. Ethnoarcheological evidence indicates that the structure of a hunter-gatherer campsite is characterized by the coexistence of different household areas, each of them associated with at least one hearth-related accumulation of remains. In addition, some spatial features, like secondary refuse areas, tend to be better defined as occupation length increases. Identifying dumping sites is therefore a good indicator of long-term occupations.
- Transport of animal resources. Food sharing is a basic component of the definition of a residential campsite. Carrying animal carcasses back to the site—especially the nutritionally richer parts—is characteristic of residential sites, as opposed to the special-purpose sites from which these resources are removed.
- Contribution of carnivores to the faunal assemblage. Evidence of carnivore activity would be scarce in assemblages derived from long-term occupation sites, in which most faunal remains would correspond to human agency. Moreover, extended occupation can protect bones against carnivore scavenging, since there is more time for bones to lose their appeal (Yellen 1991, p. 186). The percentage of bones showing carnivore damage would therefore decrease as occupation length increased.

These criteria will be examined very closely in this book. We will analyze whether they are free from the effects derived from palimpsest formation and can be considered good approaches to determining occupation type or if they are inevitably flawed by the low temporal resolution of archeological assemblages. One of our goals is to establish whether, beyond the identification of residential sites, differences in occupation length or group size can be realistically ascertained.

As stated above, the discussion of settlement patterns is closely related to spatial analyses at the intrasite level. Some authors have argued that specificity of Neanderthal behavior would also be evident at the intrasite level, since the spatial layout of individual campsites would not show the traits that define a "modern use of space". According to Mellars (1996), the spatial patterns documented in some Middle Paleolithic sites are simply a pragmatic response to functional requirements such as the quest for greater comfort or the need for fire in resource processing, but they do not exhibit a deeper cognitive or conceptual structure in the organization of activities. Deliberate living structures are clearer in the Upper Paleolithic, showing a preconceived form that Middle Paleolithic spatial structures lack. It has also been argued that Neanderthal bands would have been smaller and their sites would have therefore been less extensive and complex, although it should be stressed that camp size is a feature that is particularly difficult to determine using archeological data. Similarly, Wadley (2001, 2004) claimed that behaviorally modern humans used space symbolically for social purposes, arranging themselves in social groupings based on kinship, gender, age, status or skill. The distribution of remains in Middle Paleolithic sites exhibits an unstructured pattern, while Upper Paleolithic settlements are characterized by recurring patterns of spatial organization, including the appearance of specific activity areas, the clustering of activities around hearths and the segregation of refuse middens. The most extreme version of this position was expressed by Pettitt (1997), who argued that Middle Paleolithic humans were characterized by the repetition of a single model, showing a simple organization similar to that observed among non-human carnivores. Under this hypothesis, the spatial distributions observed in some sites would be explained by site constraints and biomechanics.

Kolen (1999) proposed the term "centrifugal living structure" for the spatial structures left behind by Middle Paleolithic humans. Although in some ways these structures share features with nest building in primate societies, they are different from primate nests in other aspects, and they are different from modern human dwellings. Middle Paleolithic structures would describe fluid life histories characterized by continuous use, reuse and centrifugal cleaning—changing continuously and demonstrating an emphasis on process and flexibility. They were never "finished" in the sense of modern structures, which were defined by well-planned and goal-oriented trajectories and showed well-demarcated episodes of production, use and abandonment. In addition, these "centrifugal living structures" would correspond to small-scale occupations by individuals or small groups. This would indicate an ephemeral use of space that would be at odds with the characterization of Neanderthal sites as home bases defined by long periods of occupation. In short, Kolen pointed out that Neanderthals' use of space was "situational" in character, since they "... didn't bother a lot about places and environmental conditions when they were not actually there" (Kolen 1999, p. 161). This would be far from the symbolic and mythical links established between modern humans and places.

However, some traits used to define "modern spatial behavior" do not seem to be supported by available ethnoarcheological information. In fact, terms such as "pragmatic" or "situational" used to characterize Neanderthal spatial behavior seem particularly suitable for defining the spatial patterns of modern hunter-gatherers. If specific activity areas are considered a hallmark of spatial modernity, most ethnographically documented hunter-gatherer groups should be characterized as behaviorally archaic. According to the data provided by several authors (Yellen 1977; O'Connell 1987; O'Connell et al. 1991; Fisher and Strickland 1991; Jones 1993) about the spatial patterning in hunter-gatherer camps, most activities are carried out around hearths in household areas. Food processing and consumption and manufacturing activities take place in these multifunctional areas. Only certain activities that require a lot of space or are particularly messy tend to be located outside the household area, in special activity areas. For example, O'Connell (1987, pp. 83–84) distinguished between four types of such areas in Alyawara campsites: shady spots, roasting pits, auto repair stations and defecation areas. It therefore seems clear that the spatial segregation of activities is not a systematic feature in modern hunter-gatherer camps, especially in those with a forager settlement system. In addition, the activities carried out in the special activity areas tend to produce a limited amount of debris and the archeological visibility of these areas is therefore highly dependent on occupation length. Evidence of special activity areas would hardly be apparent in short-term camps.

In addition, clustering activities around household hearths tends to create the undifferentiated 'smudge' of food and manufacturing remains considered by Wadley as typical of non-modern spatial patterning. Remains from different activities lie together in these multifunctional areas, forming the main archeological accumulations in the campsite layout. This patterning is modified by refuse disposal strategies, which tend to segregate remains according to size. Small remains tend to remain in the activity area, while large remains are most commonly discarded outwards (Binford 1978; Hayden and Cannon 1983; O'Connell 1987). This occurs because large remains are more likely to hinder future activities, whereas small items are less disturbing and tend to be quickly buried. Moreover, size sorting is strengthened by post-depositional processes, both of natural and human origin, which also tend to separate remains by size. Because of these intentional and unintentional processes, spatial associations depend more on the size of the remains than on their use context. Remains generated during the same activity may be spatially segregated, while items of similar size from different activities may be found together in the archeological record. If we take into account the ethnoarcheological evidence, it seems that a "pragmatic" consideration such as the hindrance factor plays a more important role in the spatial distribution of remains than the symbolic or mythical significance of places. Human behavior is pragmatic and situational in nature, so these concepts can hardly be used to characterize Neanderthal spatial patterns as archaic.

However, ethnoarcheological research also indicates that occupation length is one of the principal factors affecting archeological spatial patterning. In the first place, some patterns tend to be more archeologically visible as occupation length increases and there is a spatial redundancy in the use of certain areas. This is not only the case of the abovementioned special activity areas, but also of secondary refuse areas. Secondly, increasing occupation periods may lead to changes in camp structure, especially in contexts of a marked reduction in mobility or sedentism. A reduction in residential mobility brings about an increase in the number of site types, and campsites tend to be more complex (Hitchcock 1987). Specific activity areas are more common than in mobile situations and, more importantly, some activities previously carried out in the multifunctional hearth-related area, like cooking, may be spatially segregated. Secondary dumping areas are particularly well defined and refuse disposal is also more complex, with the appearance of specialized or organized dumps in which different types of remains are located separately. In addition, spatial differentiation according to age, gender or social status is more common, which tends to strengthen the symbolic significance of the space. As we have seen, some of the characteristics derived from the increase in occupation length, like the proliferation of special activity areas or the differentiation of space according to symbolic and social categories, are among those used to define "modern spatial behavior". The relationship of these features with occupation length suggests that they have no cognitive implications.

As in settlement patterns, time is again an essential factor in behavioral variability. Spatial behavior, both at the regional and intrasite levels, largely depends on temporal dynamics. Therefore, attempting to understand such behavior without taking time into account is a near to impossible undertaking. However, we need to ask ourselves whether these temporal dynamics are accessible through archeological inquiry. When we ask about short-term or long-term occupations, we are posing these questions in terms of a fine-grained ethnographic time scale, but the evidence available to answer them—the archeological assemblages—is normally constructed from a coarse-grained geological time scale: that associated with the formation of archeological layers. In this book we will try to provide some insight into the consequences of this temporal discrepancy on the knowledge of ancient behavioral strategies.

Temporal Resolution of Assemblages and Ethnographic Interpretations

Today, any approach to Neanderthal behavior must take into account the methodological questions associated with the interpretation of the archeological record in behavioral terms. As mentioned above, the central role of human behavior in Paleolithic archeology has been linked to a generalization of ethnoarcheological models as an essential referent in the reconstruction of the formation dynamics of assemblages. Hunter-gatherer ethnoarcheological research has provided most of the middle-range theories needed to achieve the systemic reconstruction of prehistoric ways of life. The archeological literature of the last 40 years is replete with hunter-gatherer studies that have played a central role in archeological debates about practically all the behavioral domains cited in the previous paragraph. Ethnoarcheological studies devoted to the most systemic levels of behavior, like settlement strategies or intrasite spatial patterns, have been especially influential among the archeological audience.

Nevertheless, the use of these models in identifying behavioral strategies gives rise to several problems, some of which are not always explicitly approached by researchers. Among these problems, of particular interest are the different time scales defining archeological assemblages and ethnoarcheological contexts. This lack of temporal correspondence is the central argument in the "Pompeii premise" debate (Binford 1981, 1986; Schiffer 1985). According to Binford, archeological time represents a different order of reality from that of a living community, and reconstructing phenomena pertaining to ethnographic time would be an illusory endeavor. Moreover, the temporal pace of change shown by the archeological record is much slower than that perceived in an ethnographic context. Binford's archeological time is basically characterized by processes occurring over long intervals. However, it has been pointed out (Lucas 2005, p. 47) that Binford's characterization of ethnographic time is misleading, since ethnographic context is multi-temporal and it is also enveloped in deep time scales. Regardless of the accuracy of this critique, it seems clear today that there is not simply one archeological time scale.

Archeology is particularly well-suited to approaching different temporal scales. Theoretical developments on this subject have recently been emphasized in connection with discussions about the validity in archeology of the time scale system proposed by the *Annales* school and the three scales distinguished by Braudel (1949) in historical time: the long term of environmentally determined dynamics, the medium term of social and economic structures, and the short term of events or individuals. According to the *Annales* theory, these time scales are characterized by different rates of change. Long-term processes have traditionally been a favorite subject of research due to the temporal depth of archeological evidence (Bailey 1983, 2007; Bintliff 1991; Knapp 1992; Smith 1992; Lucas 2005). The shortest temporal scale, that of the single action, is also easily accessible through archeological inquiry, since any artifact found in an assemblage can be interpreted as the outcome of a specific action or event. However, the temporal scale corresponding to "ethnographic time" is particularly difficult to isolate through the archeological method. Ethnoarcheological studies are normally carried out during short time spans made up of days, weeks or months. This "quick time" is especially evident in studies approaching spatial patterns, both at the regional and intrasite levels, which are based on the concept of "occupation", that is, the uninterrupted stay at the same site by a single human group. Although short-term events are visible in the archeological record, and long-term processes are also implicit in ethnographic contexts, there are differences between ethnographic and archeological time in terms of accessibility to the intermediate time-scale or "occupation" time.

This difference between ethnographic and archeological time is even more conclusive if we consider the criteria normally used to create the archeological assemblages from which behavioral patterns are inferred. In general, assemblages are defined according to a geological time scale. All the remains found in the same stratigraphical unit are included in the same assemblage. The slow sedimentation rates dominant in most archeological deposits, together with the reduction of the sedimentary volumes caused by some post-depositional processes (Brochier 1999), make the recovery of occupation floors, especially in cave and rockshelter sites, extremely unlikely. It seems clear that practically all archeological assemblages are palimpsests of one type or another, the formation of which can span periods of hundreds or even thousands of years and to which many natural and cultural processes of a very diverse character may have contributed (Bailey 2007). The succession of different events has even been documented in archeological assemblages traditionally characterized by their high temporal resolution (Julien et al. 1992; Ketterer et al. 2004). Although the discovery of archeological horizons formed by a single occupation cannot be ruled out, it seems highly unlikely in most contexts, particularly at sites that were repeatedly occupied over time. From this perspective, one cannot help but wonder to what point the use of ethnographic models, defined by very different time scales, are suitable for explaining these assemblages and what misconceptions might be introduced in assemblage interpretation by differences concerning formation time. According to some authors (Smith 1992; Lake 1996; Murray 2002), the disjunction between ethnographic models and the low temporal resolution of many archeological assemblages can make ethnographically derived interpretations problematic. According to Murray (1993), archeological theories must pay more attention to the nature of their own data, rather than relying on ethnographic information.

We should also ask ourselves which time scale provides the most suitable approach to understanding human behavior. There are reasons to believe that the shortest time scale—the event—is the best suited for making behavioral inferences (Brooks 1982). The deposition and characteristics of material remains depend on decisions made by individuals at specific times and places with the aim of solving specific needs. Stratigraphically defined archeological assemblages are simply the sum of an unknown number of such decisions. In fact, decision-making by individuals may be considered one of the main causes of human adaptation and, therefore, of cultural variability. An ideal explanation of an archeological assemblage would be one that accounts for each of the activity events that contributed to its formation.

Moreover, it seems clear that assemblage formation length can be an important factor in assemblage variability and it should therefore be considered in interassemblage comparisons. If an assemblage was formed over a long period, it would be more likely that different activities would be carried out at the site, including some relatively uncommon ones. The variability of an assemblage would therefore increase as the formation period of that assemblage increased. Interpreting archeological assemblages as the product of a sequence of different events raises other interesting implications. One of them concerns the search for internal coherence that characterizes many archeological explanations. Many times the whole assemblage is explained as the product of the same behavior, since it is assumed that the same constraints conditioned all the events represented in the assemblage. However, this is an unwarranted assumption as there may have been significant differences concerning the contexts, circumstances, needs, and constraints affecting those events. From this point of view, it seems unlikely that the entire assemblage could be explained by the same factors. In fact, it would be possible to find contradictory behaviors—for example, economizing and uneconomizing behaviors—represented in the same assemblage.

This problem becomes particularly evident if we analyze how the variability of an archeological assemblage can be interpreted. The behavioral variability attested to by an archeological assemblage can be considered as an expression of the different options available for humans during the period in which the assemblage was formed. In this sense, the variability of assemblages can be correlated to the variability of human behavior at a point in time. However, this same assemblage variability can be alternatively interpreted as the temporal succession of different behaviors during the assemblage formation period. In this case, there would be no correspondence between assemblage variability and behavioral variability, since the former would be the result of a pooling together of different behavioral moments. For example, we can imagine a faunal assemblage formed by different animal species. According to the synchronous interpretation, this diversity of taxa would show a wide range economy in which humans exploited the different resources available in the surroundings of the site. However, if each resource is associated with a specific temporal event, the behavioral interpretation might be completely different, since each one of these events would show a specialized behavior focused on a given resource. The sum of different specialized events can produce a diversified assemblage.

A palimpsest may be formed by the succession of natural and cultural events, sometimes creating particularly equivocal associations (c.f. Byers 2002). Recovery of faunal remains from carnivore activity or other natural processes together with evidence of human occupations is relatively common in Paleolithic sites. Taphonomic analysis can be particularly informative about the sequence of natural modifications that has affected an archeological assemblage (López González et al. 2006). However, in this book we will focus on cultural palimpsests. Some studies have pointed out that the interpretation of such palimpsests can be flawed by serious misconceptions derived from putting together events of a different character. These misconceptions can affect domains that are crucial to understanding the behavioral strategies of prehistoric hunter-gatherers, such as provisioning strategies, diet breadth or carcass transport. For example, Grayson and Delpech (1998) argued that diet breadths inferred from archeological assemblages reflect the addition of an uncontrolled number of collecting events distributed over an uncontrolled period of time, and they are therefore not comparable to diet breadths measured from ethnographic data. Certain variables, like the number of taxa, seem particularly sensitive to differences in the amount of time represented in the faunal assemblage. This disagreement between ethnographic and archeological data was also pointed out by Monahan (1998) in analyzing Hadza carcass transport. Variability in the patterns of carcass field processing and transport can be fully ascertained when each transport event is observed individually. On the other hand, this variability is masked when all the faunal remains from these events are aggregated in a single assemblage. It seems clear that time averaging could mask significant differences in resource exploitation patterns, as also shown by Lyman (2003), who argues that we should establish the temporal scale of archeological assemblages and ask questions relating to that scale.

The consequences of time averaging can also lead to serious misconceptions in the interpretation of lithic assemblages, since it tends to reduce the behavioral variability of the events responsible for their formation. If the assemblage is considered as a whole, without discerning each specific technical event, it may be characterized by the events producing the highest number of remains, although these events may have been the least common and therefore the least significant from a behavioral point of view. In addition, grouping all the activity episodes together gives rise to a mixed assemblage that may not correspond to any of the events actually performed. For example, cores can be introduced into the sites at different stages of the reduction sequence. If we put together the artifacts coming from those different stages, we will get an assemblage suggesting that the reduction sequences were entirely carried out at the site, since all the products are present—cortical products, small and large flakes, and cores. In fact, the case may be that no entire sequence was represented and what are left are only different parts of sequences. The entire reduction sequence picture would be an illusion created by grouping together different segments of sequences.

Therefore, approaches to subsistence and technological behavior can be seriously conditioned by time averaging, but its consequences are especially apparent in spatial analyses, both at the regional and the intrasite level. As discussed in the previous paragraph, research on these domains has been highly influenced by the use of ethnoarcheological data and, as a consequence, it is particularly sensitive to the differences between ethnographic time and archeological time. Ethnoarcheological approaches to settlement patterns are highly dependent on temporal dynamics, since occupation length is a main factor in site typologies and spatial organization. However, inferences about this ethnographic time can be seriously flawed by a lack of correspondence with archeological time. As for subsistence strategies and technical activities, we should consider the extent to which occupation type and the intrasite spatial patterns inferred from palimpsests can be biased by time averaging. A spatial analysis should not be exclusively spatial, it should be also temporal. Temporal relationships between remains must be analyzed in order to test whether distribution patterns really correspond to the organization strategies of ancient humans, or if they are conditioned by the temporal dynamics of assemblage formation. This link between space and time is strengthened by the fact that spatial data are basic to obtaining information about temporal dynamics. As we will see in level J, the spatial distribution and directionality of refits and the differential scatter of remains produced in the same activity episode can provide information about the temporal ordering of different events. We will discuss this subject at length in the conclusions of this book.

In order to approach these key questions through current archeological research it is necessary to rely on assemblages whose time scale is as close as possible to the ethnographic time scale, in order to increase the temporal resolution of assemblages as much as possible. Achieving this goal is not always easy, since the possibility of accessing increasingly higher temporal levels depends on several factors, some of them not easily accessible through archeological research. It depends partly on the natural formation processes of the deposits and their stratigraphic resolution. The temporal resolution of stratigraphic units is highly conditioned by sedimentation rates. This makes deposits characterized by rapid sedimentary rates especially attractive because they provide stratigraphic levels with time spans that are markedly shorter than in other contexts. Moreover, the temporal resolution of assemblages is also conditioned by the occupation redundancy of human groups. As the frequency of occupation increases, it becomes more difficult to isolate assemblages of high temporal resolution. Therefore, in deposits characterized by rapid sedimentary rates, a low occupation redundancy would be especially suitable for attempting a behavioral reading of the archeological record.

At first glance, Abric Romaní (Capellades, Spain) is one of these deposits. On the one hand, its stratigraphic sequence, dated between 70 and 40 ka, is characterized by the dominance of travertine formation dynamics. This is a particularly rapid formation process—the sedimentary rate has been calculated at 0.46 mm/yr (Bischoff et al. 1988), which has produced archeological levels of high temporal resolution. This can be seen, for example, in the paleoenvironmental sequence derived from pollen analysis (Burjachs and Julià 1994), which is more detailed than those obtained at most archeological sites. Because of the high sedimentary rate, the archeological levels took less time to form and include fewer occupation events than in other sedimentary contexts. Human occupations were notably discontinuous, since the rockshelter was not habitable during the travertine formation periods. Thick, sterile layers separate the levels, which considerably diminish the temporal depth of the palimpsests. As a matter of fact, the Abric Romaní sequence can be considered a natural sequence punctuated by short periods of human occupation. Archeological layers are vertically well delimited and the mixing of elements from different layers is unlikely.

On the other hand, some data suggest that the archeological levels are characterized by relatively limited occupation redundancy. Compared with the huge accumulations found at other sites, these levels have yielded a relatively low density of remains. For example, the density of lithic remains larger than 1 cm is less than 5 in most archeological levels: 1.3 artifacts/m^2 in level H. Even the levels with the highest densities—levels E and J—do not reach 25 artifacts/m^2. So, even if we take into account the irregular distribution of the remains (finds tend to be clustered in well-defined accumulations and are practically absent in other areas), this means that both the lithic and faunal assemblages are relatively small. The archeological horizons are normally very thin, less than 10 cm in most layers, which suggests a limited overlapping of activities in the same areas. Moreover, Abric Romaní meets other conditions particularly suitable for behavioral inquiry. The sedimentary context was especially favorable for the conservation of some archeological evidence, like hearths and wood remains. Burnt areas are perfectly visible on the yellowish travertine surfaces, allowing the location and size of hearths to be documented even when their sedimentary deposits were affected by post-depositional processes. This explains why so many hearths have been recorded in Abric Romaní and facilitates understanding into the role played by these structures in the spatial organization of the site. Furthermore, Capellades travertine has proved an excellent material for U-series dating, which has resulted in a reliable chronological control for the entire sequence.

In addition, the excavation strategy followed since the interventions began in 1983 has been directed explicitly toward the reconstruction of the behavioral strategies of the Neanderthal groups that occupied the shelter. The research project started under the theoretical influence of processual archeology, and one of its main goals is to achieve an ethnographic interpretation of the archeological record. A central aspect of this strategy has been the excavation of a large surface, which includes most of the originally occupied surface. This has led to the excavation of an area of nearly 300 m^2, paying special attention to the spatial distribution of the archeological remains and the identification of structures. Among these structures the hearths should be emphasized, as they are very abundant in all the archeological levels so far excavated, and have allowed us to characterize the formation of hearth-related activity areas as one of the essential features of Neanderthal spatial behavior. These areas can be interpreted as household spaces similar to those identified among contemporary hunter-gatherer groups; a comparison that makes it possible to approach the social dynamics of prehistoric groups.

The aim of this monograph is to provide some insight into Neanderthal behavior through the multidisciplinary study of an archeological level of the Abric Romaní site: level J. This is the third study of this kind published to date on the Abric Romaní levels, after the monographs devoted to levels H (Carbonell 1992) and I (Carbonell 2002). The spatial dimension has been fundamental in all these works, in keeping with the archeological paradigm that has guided the excavation of this site since the current project began in 1983. However, the approach to spatial patterning has undergone some changes over the course of the years. The study of level H was strictly spatial and was clearly under the influence of the "occupation floor" concept. Although the possibility that the archeological assemblage was the result of different occupations was recognized, the interpretation had essentially a synchronous character and the possible consequences of the temporal dynamics were not developed. Temporal issues played a major role in the level I monograph. The starting assumption was still essentially spatial, but the final conclusion was largely based on identifying different occupation events in different areas of the rockshelter. Level I was characterized as a horizontal palimpsest formed by several hearth-related accumulations corresponding to different occupation events (Vallverdú et al. 2005). One of the primary goals of this monograph on level J is to integrate spatial and temporal data with the aim of testing whether this temporal perspective can provide new insight into Neanderthal behavior.

This level, dated at 50 ka, is one of the richest of the sequence, both in the quantity of archeological remains uncovered there and in the number of occupation structures it holds (more than 50 hearths have been identified). From this point of view, comparing level J with levels I and H can provide insight into the effects of time on the formation of archeological assemblages. Are the differences observed between levels J and H–I the result of behavioral changes in settlement strategies or are they simply the product of differences in formation length? In spite of the high temporal resolution favored by the dynamics of the geological formation, some data indicate that the level J assemblage corresponds to a palimpsest formed by an indeterminate number of occupation episodes. The study of level J is directed toward the temporal dissection of this palimpsest, identifying assemblages of remains of higher resolution from which it is possible to access behavioral patterns with a certain degree of confidence. In this context, spatial data are fundamental as they reveal the dynamics of mobility and artifact transport from which the formation sequence of the archeological assemblage can be established. The temporal and spatial interpretations are therefore very closely linked.

This spatial–temporal interpretation is achieved by means of the information yielded by researchers from a wide range of analytic fields whose contributions constitute the different chapters of this monograph, including aspects related both to natural formation dynamics (stratigraphy, paleoenvironment, biostratigraphy, taphonomy, etc.) and human activities (lithic technology, faunal processing, habitat structures, spatial distribution, wood implements, etc.). Chapter 2 deals with the geological and geomorphological characteristics of the Capellades area, as well as the chronostratigraphic sequence of Abric Romaní, paying especial attention to the stratigraphic set in which level J is located. We also present in this chapter the history of the archeological excavations carried out in this site, in which different theoretic and methodological paradigms followed one another during the twentieth-century. Chapter 3 focuses on the spatial distribution of archeological remains and combustion structures, bringing a preliminary overview of the spatial patterns that will be further developed in the chapters devoted to lithic and bone remains. The natural and cultural processes that contributed to the formation of level J deposit are presented in Chap. 4 using the data provided by micromorphological analysis. This chapter emphasizes the paleoecological reconstruction, but also the anthropogenic processes, among which hearth construction played a primary role. Paleoecology is the central theme of Chap. 5, which summarizes the results from different domains: pollen and charcoal analysis, large and small vertebrate paleontology, microvertebrate taphonomy, and malacology. The natural processes affecting the archeological assemblages are also discussed in Chap. 6, in which the taphonomic study of bones and charcoals is presented.

The second section of this book, made up of Chaps. 7, 8 and 9, focuses on human activities by reviewing the provisioning and use of lithic, faunal and plant resources. The study of the lithic assemblage is presented in Chap. 7, in which two different analytical levels have been distinguished. The first one is based on the analysis of artifact attributes in order to identify the technical strategies used in the production of lithic implements. Raw material and usewear analyses are also included in this section. In the second level, we focus on spatial and refitting data for identifying single technical events and increasing the temporal resolution of the analysis. This temporal dimension is also a primary concern of Chap. 8, devoted to the zooarcheological study of the faunal assemblage. The anatomical representation of the different taxa and the traces of human activity on bones allow the primary agent of faunal assemblage formation to be identified. These data are also used for characterizing level J in terms of occupation length and site function. Chapter 9 focuses on the exploitation of plant resources, which will be examined using two kinds of evidence: charcoal remains derived from the use of wood as fuel and wood pseudomorphs that suggest the manufacture of wood artifacts. Finally, the last section of this monograph focuses on whether the spatial–temporal perspective that we propose gives rise to a new view of Neanderthal behavior, different from that derived from works that are unaware of the importance of time resolution in the formation of archeological assemblages.

References

Adler, D. S., Bar-Oz, G., Belfer-Cohen, A., & Bar-Yosef, O. (2006). Ahead of the game. Middle and Upper Palaeolithic hunting behaviors in the Southern Caucasus. *Current Anthropology, 47*(1), 89–118.

Auguste, P., Moncel, M.-H., & Patou-Mathis, M. (1998). Chasse ou "charognage": acquisition et traitement des rhinocéros au Paléolithique moyen en Europe occidentale. In J.-Ph. Brugal, L. Meignen & M. Patou Mathis (Eds.), *Economie préhistorique: les comportements de subsistance au Paléolithique* (pp. 133–151). Sophia Antipolis: Editions APDCA.

Aura, J. E., Villaverde, V., Pérez Ripoll, M., Martínez Valle, R., & Guillem, P. (2002). Big game and small prey: Paleolithic and Epipaleolithic economy from Valencia (Spain). *Journal of Archaeological Method and Theory, 9*(3), 215–268.

Bailey, G. N. (1983). Concepts of time in quaternary prehistory. *Annual Review of Anthropology, 12*, 165–192.

Bailey, G. (2007). Time perspectives, palimpsests and the archaeology of time. *Journal of Anthropological Archaeology, 26*, 198–223.

Binford, L. R. (1978). Dimensional analysis of behavior and site structure: Learning from an Eskimo hunting stand. *American Antiquity, 43*(3), 330–361.

Binford, L. R. (1980). Willow smoke and dogs' tails: Hunter-gatherer settlement systems and archaeological site formation. *American Antiquity, 45*, 4–20.

Binford, L. R. (1981). Behavioral archaeology and the "Pompeii premise". *Journal of Anthropological Research, 37*, 195–208.

Binford, L. R. (1985). Human ancestors: Changing views of their behavior. *Journal of Anthropological Archaeology, 4*, 292–327.

Binford, L. R. (1986). In pursuit of the future. In D. J. Meltzer, D. D. Folwer, & J. A. Sabloff (Eds.), *American archaeology past and future* (pp. 459–479). Washington, DC: Smithsonian Institution.

Bintliff, J. (1991). The contribution of an Annaliste/structural history approach to archaeology. In J. Bintliff (Ed.), *The Annales school and archaeology* (pp. 1–33). London: Leicester University Press.

Bischoff, J. L., Julià, R., & Mora, R. (1988). Uranium-series dating of the Mousterian occupation at Abric Romaní, Spain. *Nature, 332*, 68–70.

Bolus, M. (2004). Settlement analysis of sites of the Blattspitzen complex in Central Europe. In N. J. Conard (Ed.), *Settlement dynamics of the Middle Paleolithic and Middle Stone Age II* (pp. 201–226). Tübingen: Kerns Verlag.

Braudel, F. (1949). *La Méditerranée et le Monde Méditerranéen a l'époque de Philippe II*. Paris: Armand Colin.

Brochier, J. E. (1999). Couche archéologique, sol archéologique et distributions spatiales: quelques réflexions (géo)archéologiques sur un vieux problème. In V. M. Rosselló (Ed.), *Geoarqueologia i Quaternari litoral: Memorial Maria Pilar Fumanal* (pp. 91–95). València: Universitat de València.

Brooks, R. L. (1982). Events in the archaeological context and archaeological explanation. *Current Anthropology, 23*(1), 67–75.

Brugal, J.-Ph, & Jaubert, J. (1991). Les gisements paléontologiques pléistocènes à indices de frequentation humaine: un nouveau type de comportement de prédation? *Paléo, 3*, 15–41.

Burjachs, F., & Julià, R. (1994). Abrupt climatic changes during the last glaciation based on pollen analysis of the Abric Romani, Catalonia, Spain. *Quaternary Research, 42*, 308–315.

Burke, A. (2006). Neanderthal settlement patterns in Crimea: A landscape approach. *Journal of Anthropological Archaeology, 25*, 510–523.

Byers, D. A. (2002). Taphonomic analysis, associational integrity, and depositional history of the Fetterman Mammoth, Eastern Wyoming, U.S.A. *Geoarchaeology, 17*(5), 417–440.

Carbonell, E. (1992). Abric Romaní, nivell H: un model d'estratègia ocupacional al Plistocé Superior mediterrani. *Estrat, 5*, 157–308.

Carbonell, E. (2002). *Abric Romaní nivell I. Models d'ocupació de curta durada de fa 46.000 anys a la Cinglera del Capelló (Capellades. Anoia, Barcelona)*. Tarragona: Universitat Rovira i Virgili.

Chase, P. G. (1988). Scavenging and hunting in the Middle Paleolithic. The evidence from Europe. In H. L. Dibble & A. Montet-White (Eds.), *Upper Pleistocene prehistory of Western Eurasia* (pp. 225–232). Philadelphia, PA: The University Museum, University of Pennsylvania.

Chase, P. G. (1989). How different was Middle Palaeolithic subsistence? A zooarchaeological perspective on the Middle to Upper Palaeolithic transition. In P. Mellars & C. Stringer (Eds.), *The human revolution. Behavioural and biological perspectives in the origin of modern humans* (pp. 321–337). Edinburgh: Edinburgh University Press.

Chase, P. G. (1991). Symbols and Paleolithic artifacts: Style, standardization, and the imposition of arbitrary form. *Journal of Anthropological Archeology, 10*, 193–214.

Coinman, N., Clark, G. A., & Lindly, J. (1986). Prehistoric hunter-gatherer settlement in the Wadi el'Hasa, West-Central Jordan. In L. Straus (Ed.), The end of the Paleolithic in the Old World (pp. 129–169). Oxford, England: BAR International Series 284.

Conard, N. J. (2001). River terraces, volcanic craters and Middle Paleolithic settlement in the Rhineland. In N. J. Conard (Ed.), *Settlement dynamics of the Middle Paleolithic and Middle Stone Age* (pp. 221–250). Tübingen: Kerns Verlag.

Conard, N. J., & Adler, D. S. (1997). Lithic reduction and hominid behavior in the Middle Paleolithic of the Rhineland. *Journal of Anthropological Research, 53*, 147–175.

Costamagno, S., Meignen, L., Beauval, C., Vandermeersch, B., & Maureille, B. (2006). Les Pradelles (Marillac-le-Franc, France): A mousterian reindeer hunting camp. *Journal of Anthropological Archaeology, 25*, 466–484.

Cronk, L. (1991). Human behavioral ecology. *Annual Review of Anthropology, 20*, 25–53.

Cronk, L., Chagnon, N., & Irons, W. (2000). *Adaptation and human behavior: An anthropological perspective*. Hawthorne, NY: Aldine de Gruyter.

d'Errico, F., Zilhão, J., Julien, M., Baffier, D., & Pelegrin, J. (1998). Neanderthal acculturation in Western Europe? A critical review of the evidence and its interpretation. *Current Anthropology, 39*, 1–44.

Defleur, A., & Cregut-Bonnoure, E. (1995). *Le gisement paléolithique moyen de la grotte des Cèdres (Le Plan-d'Aups, Var). Documents d'Archéologie Française, 49*. Paris: Editions de la Maison des Sciences de l'Homme.

Deloze, V., Depaepe, P., Gouedo, J.-M., Krier, V., & Locht, J.-L. (1994). *Le Paléolithique moyen dans le nord du Sénonais (Yonne). Contexte géomorphologique, industries lithiques et chronostratigraphie. Documents d'Archéologie Française, 47*. Paris: Editions de la Maison des Sciences de l'Homme.

Depaepe, P. (2004). Fonctions des Sites et Gestion du Territoire: Le Paléolithique Moyen Récent de la Vallée de la Vanne (France). In N. J. Conard (Ed.), *Settlement dynamics of the Middle Paleolithic and Middle Stone Age II* (pp. 273–293). Tübingen: Kerns Verlag.

Dibble, H. L., & Rolland, N. (1992). On assemblage variability in the Middle Paleolithic of Western Europe. History, perspectives and a new synthesis. In H. L. Dibble & P. Mellars (Eds.), *The Middle Paleolithic: Adaptation, behavior and variability* (pp. 1–28). Philadelphia, PA: The University Museum, University of Pennsylvania.

Farizy, C., & David, F. (1992). Subsistence and behavioral patterns of some Middle Paleolithic local groups. In H. L. Dibble & P. Mellars (Eds.), *The Middle Paleolithic: Adaptation, behavior and variability* (pp. 87–96). Philadelphia, PA: The University Museum, University of Pennsylvania.

Finlayson, C. (2004). *Neanderthals and modern humans. An ecological and evolutionary perspective*. Cambridge, MA: Cambridge University Press.

Fisher, J. W., & Strickland, H. C. (1991). Dwellings and fireplaces: keys to Efe Pygmy campsite structure. In C. S. Gamble & W. Boismier (Eds.), *Ethnoarchaeological approaches to mobile campsites* (pp. 215–236). Michigan: International Monographs in Prehistory.

Gamble, C. (1994). *Timewalkers: The prehistory of global colonization*. Cambridge, MA: Harvard University Press.

Gaudzinski, S., & Roebroeks, W. (2000). Adults only Reindeer hunting at the Middle Palaeolithic site Salzgitter Lebenstedt, Northern Germany. *Journal of Human Evolution, 38*, 497–521.

Geneste, J.-M. (1985). *Analyse lithique d'industries moustériennes du Périgord: une approche technologique du comportement des groupes humains au Paléolithique moyen*. Unpublished Ph.D. Bordeaux: Université de Bordeaux I.

Geneste, J.-M. (1988). Les industries de la Grotte Vaufrey: technologie du débitage, économie et circulation de la matière première lithique. In J.-Ph Rigaud (Ed.), *La Grotte Vaufrey à Cenac et Saint-Julien (Dordogne). Paleoenvironnements chronologie et activités humaines* (pp. 441–517). Paris: Mémoires de la Société Préhistorique Française XIX.

Grayson, D. K., & Delpech, F. (1998). Changing diet breadth in the early Upper Palaeolithic of Southwestern France. *Journal of Archaeological Science, 25*, 1119–1129.

Grayson, D. K., & Delpech, F. (2002). Specialized early Upper Palaeolithic hunters in Southwestern France? *Journal of Archaeological Science, 29*, 1439–1449.

Grayson, D. K., & Delpech, F. (2003). Ungulates and the Middle-to-Upper Paleolithic transition at Grotte XVI (Dordogne, France). *Journal of Archaeological Science, 30*, 1633–1648.

Hayden, B. (1993). The cultural capacities of Neandertals: A review and re-evaluation. *Journal of Human Evolution, 24*, 113–146.

Hayden, B., & Cannon, A. (1983). Where the garbage goes: Refuse disposal in the Maya highlands. *Journal of Anthropological Archaeology, 2*, 117–163.

Henshilwood, C. S., & Marean, C. W. (2003). The origin of modern human behavior. Critique of the models and their test implications. *Current Anthropology, 44*(5), 627–651.

Henry, D. O. (1992). Transhumance during the Late Levantine Mousterian. In H. L. Dibble & P. Mellars (Eds.), *The Middle Paleolithic: Adaptation, behavior, and variability* (pp. 143–162). Philadelphia, PA: The University Museum, University of Pennsylvania.

Hitchcock, R. K. (1987). Sedentism and site structure: Organizational changes in Kalahari Basarwa residential locations. In S. Kent (Ed.), *Method and theory for activity area research. An ethnoarchaeological approach* (pp. 374–423). New York: Columbia University Press.

Hoffecker, J. F. (2005). Innovation and technological knowledge in the Upper Paleolithic of Northern Eurasia. *Evolutionary Anthropology, 14*(5), 186–198.

Jones, K. T. (1993). The archaeological structure of a short-term camp. In J. Hudson (Ed.), *From bones to behavior: Ethnoarchaeological and experimental contributions to the interpretation of faunal remains* (pp. 101–114). Center for Archaeological Investigations: Southern Illinois University.

Julien, M., Karlin, C., & Valentin, B. (1992). Déchets de silex, déchets de pierres chauffées. De l'intéret des remontages à Pincevent (France). In J. L. Hofman & J. G. Enloe (Eds.), *Piecing together the past: Applications of refitting studies in archaeology* (pp. 287–295). Oxford, England: BAR International Series 578.

Kelly, R. L. (1983). Hunter-gatherer mobility strategies. *Journal of Anthropological Research, 39*(3), 277–306.

Ketterer, I., Pigeot, N., & Serra, S. (2004). Le temps de l'occupation. Une histoire des activités et des comportements. In N. Pigeot (Ed.), *Les derniers magdaléniens d'Étiolles. Perspectives culturelles et paléohistoriques* (pp. 235–254). Paris: Éditions du CNRS.

Klein, R. G. (1995). Anatomy, behavior, and modern human origins. *Journal of World Prehistory, 9*(2), 167–198.

Klein, R. G. (2000). Archeology and the evolution of human behavior. *Evolutionary Anthropology, 9*(1), 17–36.

Knapp, A. B. (1992). Archaeology and Annales: Time, space, and change. In A. B. Knapp (Ed.), *Archaeology, Annales, and ethnohistory* (pp. 1–21). Cambridge, MA: Cambridge University Press.

Kolen, J. (1999). Hominids without homes: On the nature of Middle Palaeolithic settlement in Europe. In W. Roebroeks & C. S. Gamble (Eds.), *The Middle Palaeolithic occupation of Europe* (pp. 139–175). Leiden: Leiden University.

Krebs, J. R., & Davies, N. B. (1991). *Behavioural ecology: An evolutionary approach*. Oxford, England: Blackwell.

Kuhn, S. L. (1995). *Mousterian lithic technology. An ecological perspective*. Princeton, NJ: Princeton University Press.

Kuhn, S. L., & Stiner, M. C. (2006). What's a mother to do? The division of labor among Neandertals and modern humans in Eurasia. *Current Anthropology, 47*(6), 953–980.

Lake, M. (1996). Archaeological inference and the explanation of hominid evolution. In J. Steele & S. Shennan (Eds.), *The Archaeology of human ancestry: Power, sex, and tradition* (pp. 184–206). London: Routledge.

Lieberman, D. E. (1993). The rise and fall of seasonal mobility among hunter-gatherers. The case of Southern Levant. *Current Anthropology, 34*(5), 599–631.

López-González, F., Grandal-d'Anglade, A., & Vidal-Romaní, J. R. (2006). Deciphering bone depositional sequences in caves through the study of manganese coatings. *Journal of Archaeological Science, 33*(5), 707–717.

Lucas, G. (2005). *The archaeology of time*. London: Routledge.

Lyman, R. L. (2003). The influence of time averaging and space averaging on the application of foraging theory in zooarchaeology. *Journal of Archaeological Science, 30*, 596–610.

Marean, C. (1998). A critique of the evidence for scavenging by Neanderthals and early modern humans: New data from Kobeh Cave (Zagros Mountains, Iran) and Die Kelders Cave 1 Layer 10 (South Africa). *Journal of Human Evolution, 35*, 111–136.

Marks, A. E., & Chabai, V. P. (2001). Constructing Middle Paleolithic settlement systems in crimea: Potentials and limitations. In N. J. Conard (Ed.), *Settlement dynamics of the Middle Paleolithic and Middle Stone Age* (pp. 179–204). Tübingen: Kerns Verlag.

Marks, A. E., & Freidel, D. (1977). Prehistoric settlement patterns in the Avdat/Aqev area. In A. E. Marks (Ed.), *Prehistory and Paleoenvironments in the Central Negev, Israel. Volume 3. The Avdat/Aqev area* (pp. 131–159). Dallas, TX: Southern Methodist University.

Martínez Moreno, J., Mora, R., & de la Torre, I. (2004). Methodological approach for understanding Middle Palaeolithic settlement dynamics at La Roca dels Bous (Noguera, Catalunya, Northeast Spain). In N. J. Conard (Ed.), *Settlement dynamics of the Middle Paleolithic and Middle Stone Age II* (pp. 393–414). Tübingen: Kerns Verlag.

Meignen, L. (1988). Un exemple de comportement technologique differentiel selon les matières premières: Marillac, couches 9 et 10. In L. Binford & J. Ph. Rigaud (Eds.), L'Homme de Néandertal. Vol. 4. La technique (pp. 71–79). Liège: ERAUL.

Meignen, L., Beyries, S., Speth, J., & Bar-Yosef, O. (1998). Acquisition, traitement des matières animales et fonction du site au Paléolithique moyen dans la grotte de Kébara (Israël): approche interdisciplinaire. In J.-Ph. Brugal, L. Meignen & M. Patou Mathis (Eds.), *Economie préhistorique: les comportements de subsistance au Paléolithique* (pp. 227–241). Sophia Antipolis: Editions APDCA.

Mellars, P. (1973). The character of the middle-upper palaeolithic transition in south-west France. In C. Renfrew (Ed.), *The explanation of culture change: Models in prehistory* (pp. 255–276). London: Duckworth.

Mellars, P. (1989). Major issues in the emergence of modern humans. *Current Anthropology, 30*, 349–385.

Mellars, P. A. (1992). Technological change in the mousterian of Southwest France. In H. L. Dibble & P. Mellars (Eds.), *The middle paleolithic: Adaptation, behavior and variability* (pp. 29–43). Philadelphia: University of Pennsylvania.

Mellars, P. (1996a). *The Neanderthal legacy. An archaeological perspective from Western Europe*. Princeton, NJ: Princeton University Press.

Mellars, P. (1996b). Symbolism, language, and the Neanderthal mind. In P. Mellars & K. Gibson (Eds.), *Modelling the early human mind* (pp. 15–32). Cambridge, MA: The McDonald Institute for Archaeological Research.

Mellars, P., & Stringer, C. (1989). *The human revolution. Behavioural and biological perspectives in the origin of modern humans*. Edinburgh: Edinburgh University Press.

Mithen, S. (1994). From domain specific to generalized intelligence: A cognitive interpretation of the Middle/Upper Palaeolithic transition. In C. Renfrew & E. B. W. Zubrow (Eds.), *The ancient mind. Elements of a cognitive archaeology* (pp. 29–39). Cambridge, MA: Cambridge University Press.

Mithen, S. (1996a). *The prehistory of the mind: A search for the origins of art*. Religion and science. London: Thames & Hudson.

Mithen, S. (1996b). Domain-specific intelligence and the Neandertal mind. In P. Mellars & K. Gibson (Eds.), *Modelling the early human mind* (pp. 217–229). Cambridge, MA: The McDonald Institute for Archaeological Research.

Monahan, C. M. (1998). The Hadza Carcass transport debate revisited and its archaeological implications. *Journal of Archaeological Science, 25*, 405–424.

Moncel, M.-H. (2004). Continuité et variabilité des occupations humaines à la fin du Pléistocène moyen et au début du Pléistocène supérieur (stades isotopiques 9 à 3). L'exemple de la moyenne vallée du Rhône (France). In N. J. Conard (Ed.), *Settlement dynamics of the Middle Paleolithic and Middle Stone Age II* (pp. 295–315). Tübingen: Kerns Verlag.

Murray, T. (1993). Archaeology and the threat of the past: Sir Henry Rider Haggard and the acquisition of time. *World Archaeology, 25*, 175–186.

Murray, T. (2002). Evaluating evolutionary archaeology. *World Archaeology, 34*(1), 47–59.

Noble, W., & Davidson, I. (1996). *Human evolution. Language and mind. A Psychological and archaeological inquiry*. Cambridge, MA: Cambridge University Press.

O'Connell, J. F. (1987). Alyawara site structure and its archaeological implications. *American Antiquity, 52*(1), 74–108.

O'Connell, J. F. (2006). How did modern humans displace Neanderthals? Insights from hunter-gatherer ethnography and archaeology. In N. J. Conard (Ed.), *When Neanderthals and modern humans met* (pp. 43–64). Tübingen: Kerns Verlag.

O'Connell, J. F., Hawkes, K., & Blurton Jones, N. (1991). Distribution of refuse-producing activities at Hadza residential base camps: Implications for analyses of archaeological site structure. In E. M. Kroll & T. D. Price (Eds.), *The interpretation of archaeological spatial patterning* (pp. 61–76). New York: Plenum Press.

Orquera, L. A. (1984). Specialization and the Middle/Upper Paleolithic transition. *Current Anthropology, 25*(1), 73–98.

Otte, M. (1996). Le bouleversement de l'humanité en Eurasie vers 40.000 ans. In E. Carbonell & M. Vaquero (Eds.), The last Neanderthals, the first anatomically modern humans: A tale about the human diversity (pp. 95–106). Tarragona: Universitat Rovira i Virgili.

Patou-Mathis, M. (2006). Comportements de subsistance des Néandertaliens d'Europe. In B. Demarsin & M. Otte (Eds.), Neanderthals in Europe (pp. 67–76). Liège: ERAUL 117.

Peresani, M. (2001). An overview of the Middle Paleolithic settlement system in North-Eastern Italy. In N. J. Conard (Ed.), *Settlement dynamics of the Middle Paleolithic and Middle Stone Age* (pp. 485–506). Tübingen: Kerns Verlag.

Pettitt, P. B. (1997). High resolution Neanderthals? Interpreting Middle Palaeolithic intrasite spatial data. *World Archaeology, 29*(2), 208–224.

Pettitt, P. (2000). Neanderthal lifecycles: Developmental and social phases in the lives of the last archaics. *World Archaeology, 31*(3), 351–366.

Richter, J. (2001). For lack of a wise old man? Late Neanderthal land-use patterns in the Altmühl River Valley, Bavaria. In N. J. Conard (Ed.), *Settlement dynamics of the Middle Paleolithic and Middle Stone Age* (pp. 205–219). Tübingen: Kerns Verlag.

Richter, J. (2006). Neanderthals in their landscape. In B. Demarsin & M. Otte (Eds.), *Neanderthals in Europe* (pp. 51–66). Liège: ERAUL 117.

Rodseth, L., Wrangham, R. W., Harrigan, A. M., & Smuts, B. B. (1992). The human community as a primate society. *Current Anthropology, 32*, 221–254.

Roebroeks, W. (1986). Archaeology and Middle Pleistocene stratigraphy: The case of Maastricht-Belvédère (NL). In A. Tuffreau & J. Sommé (Eds.), *Chronostratigraphie et faciès culturels du Paléolithique Inférieur et Moyen dans l'Europe du Nord-ouest* (pp. 81–88). Paris: Supplément au bulletin de l'A.F.E.Q. 26.

Roebroeks, W. (1988). *From find scatters to early hominid behaviour: A study of Middle Paleolithic riverside settlements at Maastricht-Belvédère (The Netherlands). Analecta Praehistorica Leidensia 21*. Leiden: University of Leiden.

Rolland, N. (2001). Determinants of Middle Paleolithic settlement organization: A review of evidence, based on the record from Western Europe. In N. J. Conard (Ed.), *Settlement dynamics of the Middle Paleolithic and Middle Stone Age* (pp. 545–571). Tübingen: Kerns Verlag.

Rolland, N., & Dibble, H. L. (1990). A new synthesis of Middle Paleolithic variability. *American Antiquity, 55*(3), 480–499.

Smith, M. E. (1992). Braudel's temporal rhythms and chronology theory in archaeology. In A. B. Knapp (Ed.), *Archaeology, Annales, and ethnohistory* (pp. 23–34). Cambridge, MA: Cambridge University Press.

Smith, E. A., & Winterhalder, B. (1992). *Evolutionary ecology and human behavior*. New York: Aldine de Gruyter.

Schiffer, M. B. (1985). Is there a "Pompeii premise" in archaeology? *Journal of Anthropological Archaeology, 41*, 18–41.

Shea, J. J. (2006). The origins of lithic projectile point technology: Evidence from Africa, the Levant, and Europe. *Journal of Archaeological Science, 33*, 823–846.

Soffer, O. (1994). Ancestral lifeways in Eurasia. The Middle and Upper Paleolithic records. In M. H. Nitecki & D. V. Nitecki (Eds.), *Origins of anatomically modern humans* (pp. 101–119). New York: Plenum Press.

Soressi, M. (2004). From the Mousterian of Acheulian tradition type A to type B: A change in technical tradition, raw material, task, or settlement dynamics? In N. J. Conard (Ed.), *Settlement dynamics of the Middle Paleolithic and Middle Stone Age II* (pp. 343–366). Tübingen: Kerns Verlag.

Tattersall, I. (1999). *The last Neanderthal: The rise success, and mysterious extinction of our closest human relatives*. Boulder, CO: Westview Press.

Tillet, T. (2001). *Le Paléolithique Moyen dans les Alpes et le Jura: exploitation de milieux de contraintes d'altitude*. In N. J. Conard (Ed.), *Settlement dynamics of the Middle Paleolithic and Middle Stone Age* (pp. 421–446). Tübingen: Kerns Verlag.

Turq, A. (1992). *Le Paléolithique inférieur et moyen entre les vallées de la Dordogne et du Lot*. Ph.D. Dissertation, Université de Bordeaux I.

Vallverdú, J., Allué, E., Bischoff, J. L., Cáceres, I., Carbonell, E., Cebrià, A., et al. (2005). Short human occupations in the Middle Palaeolithic level i of the Abric Romaní rock-shelter (Capellades, Barcelona, Spain). *Journal of Human Evolution, 48*(2), 157–174.

Vermeersch, P. M. (2001). Middle Paleolithic settlement patterns in West European open-air sites: Possibilities and problems. In N. J. Conard (Ed.), *Settlement dynamics of the Middle Paleolithic and Middle Stone Age* (pp. 395–417). Tübingen: Kerns Verlag.

Wadley, L. (2001). What is cultural modernity? A general view and a South African perspective from Rose Cottage Cave. *Cambridge Archaeological Journal, 11*(2), 201–221.

Wadley, L. (2004). Late Middle Stone Age spatial patterns in Rose Cottage Cave, South Africa. In N. J. Conard (Ed.), *Settlement dynamics of the Middle Paleolithic and Middle Stone Age II* (pp. 23–36). Tübingen: Kerns Verlag.

Wallace, I. J., & Shea, J. J. (2006). Mobility patterns and core technologies in the Middle Paleolithic of the Levant. *Journal of Archaeological Science, 33*, 1293–1309.

Wynn, T., & Coolidge, F. L. (2004). The expert Neandertal mind. *Journal of Human Evolution, 46*, 467–487.

Yellen, J. E. (1977). *Archaeological approaches to the present*. New York: Academic Press.

Yellen, J. E. (1991). Small mammals: Post-discard patterning of !kung san faunal remains. *Journal of Anthropological Archaeology, 10*, 152–192.

Zilhão, J., & Trinkaus, E. (2002). Social implications. In J. Zilhão & E. Trinkaus (Eds.), *Portrait of the artist as a child. The Gravettian human skeleton from the Abrigo do Lagar Velho and its archeological context* (pp. 519–541). Lisboa: Instituto Português de Arqueologia.

Part I
The Abric Romaní and the Archeological Level J (ca. 50 ka)

Chapter 2
The Abric Romaní Site and the Capellades Region

Josep Vallverdú-Poch, Bruno Gómez de Soler, Manuel Vaquero and James L. Bischoff

Abstract The main goal of this chapter is to provide a general presentation of the Abric Romaní site, paying special attention to archeological level J. First, we will present the basic information concerning the geological and geomorphological characteristics of the Capellades region, which are fundamental to understand site formation. We will summarize the history of the archeological excavations carried out at the Abric Romaní, since the discovery of the site in 1909 to the current works. This history shows the different theoretical and methodological paradigms that dominated European Prehistory in the course of the twentieth century. Within this context, the excavation of level J will be exposed in detail. Finally, we will present the stratigraphy of the site and the chronological framework, which is largely based on U/Th dating of tufas.

Keywords Abric Romaní • Geology • Capellades • Research history • Stratigraphy • Chronology

In this chapter, we will provide a general presentation of the Abric Romaní site in general and archeological level J in particular. First, we describe the main geological and geographical characteristics of the Capellades area, in which the Abric Romaní is located. Second, we summarize the long history of archeological research at the site through the various different excavations carried out at the rockshelter since its discovery in 1909. We then describe the the history of the excavation of level J in detail. Finally, we present chronostratigraphical data related to the site, again with special attention to level J.

The Geology and Geography of the Capellades Region (NE Iberian Peninsula)

The Abric Romaní is a wide rockshelter (Abric) in a travertine cliff called *Cinglera del Capelló*, located in a karstic landscape near Capellades (Barcelona, Spain) on the west bank of the Anoia River, 50 km west of Barcelona. The Abric Romaní has an elevation of 265 m above sea level. The cliff escarpment is orientated NW–SE with the entrance on the NE side of the wall, facing the Capellades Gorge. Its coordinates are 1°41′30″ longitude E and 41°32′ latitude N. At this point, the Anoia valley forms a narrow gorge, which in historic and prehistoric times was one of the main natural passages between the inner regions of Catalonia and the coastal areas. This cliff harbors several rockshelters with evidence of prehistoric occupation.

The Capellades area in which the Abric Romaní lies opens towards the Conca d'Òdena (Ódena basin), an erosional marginal basin in the Eastern Ebro Basin created by the Anoia River in its course towards the Mediterranean Sea as a tributary of the Llobregat River. The Anoia shaped the Capellades Gorge that connects the Penedès Depression and the Ebro Basin (Fig. 2.1). The Vallès-Penedès normal fault marks the subsidence of the Penedès Depression and the uplift of the Prelitoral Range and Ebro Basin. This morphostructural frame allowed for a new hydrological network that was incised capturing the Ebro Basin, thereby reversing the direction of pre-existing Pleistocene drainage (Gallart 1981).

There are three structural units in the Capellades region (Fig. 2.1):

J. Vallverdú-Poch (✉) · B. Gómez de Soler · M. Vaquero
Institut Català de Paleoecologia Humana i Evolució Social (IPHES), Universitat Rovira i Virgili (URV), C/Escorxador s/n, 43003 Tarragona, Spain
e-mail: josep@prehistoria.urv.cat

B. Gómez de Soler
e-mail: bruno@prehistoria.urv.cat

M. Vaquero
e-mail: manuel.vaquero@urv.cat

J. L. Bischoff
US Geological Survey, ms/470, 345 Middlefield Rd, Menlo Park, CA 94025, USA
e-mail: jbischoff@usgs.gov

E. Carbonell i Roura (ed.), *High Resolution Archaeology and Neanderthal Behavior: Time and Space in Level J of Abric Romaní (Capellades, Spain)*, Vertebrate Paleobiology and Paleoanthropology,
DOI: 10.1007/978-94-007-3922-2_2, © Springer Science+Business Media B.V. 2012

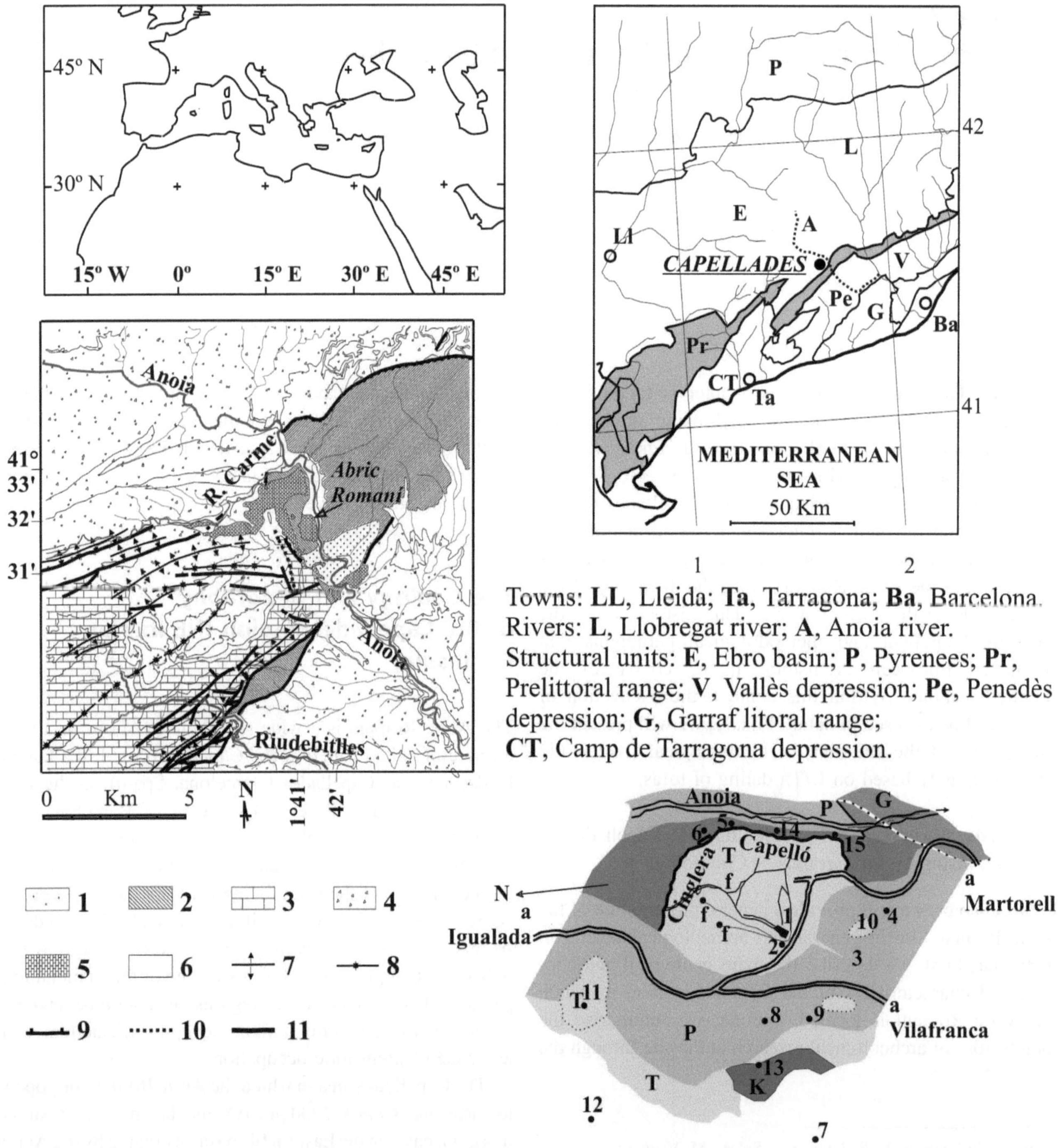

Fig. 2.1 Geographical location and main geochronological units of the Capellades region. Legend (IGME 1975): *1* Plutonic intrusions. *2* Paleozoic. *3* Mesozoic. *4* Cenozoic. *5* Quaternary travertines. *6* Quaternary. *7* Anticline. *8* Syncline. *9* Normal fault. *10* Inverse (thrust) fault. *11* Fault. At the bottom right, geological outline and water sources in the vicinity of Capellades (Vidal 1911, p. 111): *G* granites; *P* slates; *T* tuff; *K* travertines; *1* Bassa or Font Gran; *2* Font Petita; *3* Font Cuitora; *4* Font del Llargandaix; *5* Font de la Reina; *6* Abric Romaní; *7* Font de Riudacost; *8* Pou del Cardús; *9* Mina del Artigues; *10* Travertí del Cementeri; *11* Turó de Torre Nova; *12* Font de Frígols; *13* Terreres de la Garca; *14* Agut Station; *15* Barret del Capelló; *f* public springs

(1) The Ebro Basin
(2) The Prelitoral Range
(3) The Prelitoral Depression
(4) The travertine Capellades-Carme area

The Ebro Basin is an extensive triangular unit, bordered by the Pyrenees and the Basque-Cantabrian Ranges to the north, by the Iberian Ranges to the south and by the Catalan Coastal Ranges to the east (Fig. 2.2). This depression is

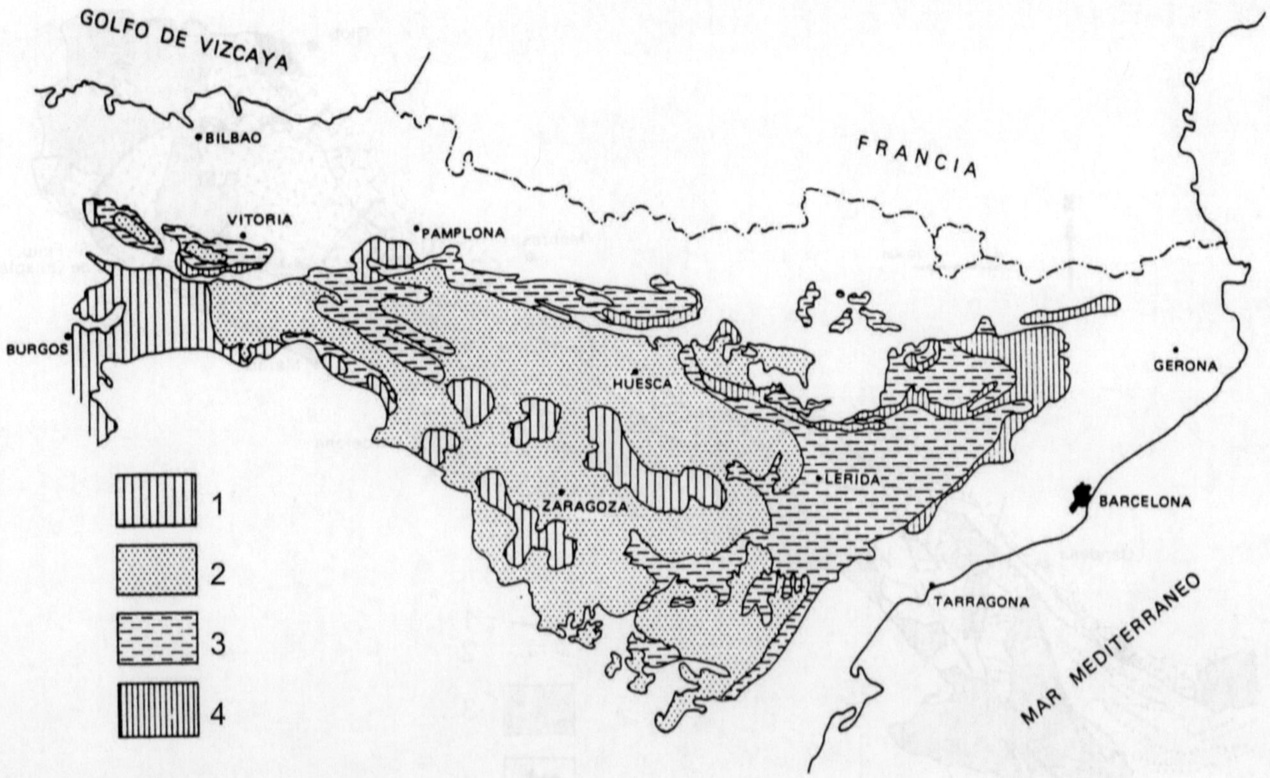

Fig. 2.2 Ebro Basin. From Riba et al. 1983. Legend: *1* Upper Miocene. *2* Lower Miocene. *3* Oligocene. *4* Eocene

mainly drained by the Ebro River and corresponds to a tertiary sedimentary basin with Eocene deposits formed by erosive processes affecting the bordering ranges (Solé Sabarís 1958–1964).

Two marine sedimentary cycles dominated the development of the Ebro Basin. The first marine transgression took place during the Ilerdian (Lower Eocene) and deposited platforms of neritic limestone and marls. The second cycle began in Catalonia during the Lutetian (Middle Eocene), reaching the Igualada area in the Bartonian (*Fm. Collbàs*, *Fm. Igualada* and *Fm. Tossa* from Upper Eocene) (Riba et al. 1983). A regression subsequently took place, causing the formation of conglomerates in the margins of the Catalan Coastal Ranges (Montserrat, St. Llorenç de Munt, etc.). During the transition between the Eocene and Oligocene, the basin was not covered by the sea and, therefore an endorheic system was established, forming a large interior lake by continuous continental contributions. This sedimentation was characterized by alluvial fans in the Oligocene. During the Neogene, sedimentation ended in the Catalan area (Gutiérrez and Peña 1994). This endorheic system concluded at the end of the Tertiary due to the elevation of the Iberian Peninsula and the sinking of the Mediterranean. Tertiary materials were subjected to great erosion and the basin was partially cast towards the sea by the river Ebro.

The Prelitoral Ranges are part of the Catalan Coastal Ranges. This name designates the reliefs extending along the Catalan coast between l'Empordà and the Iberian Range, which correspond to the Mediterranean System as defined by Solé Sabarís (1958–1964). It is range stretching north-east to south-west over 250 km that connects to the Pyrenees in the north. This system is formed by two parallel coastal ranges separated by an intermediate depression (Fig. 2.3).

One of the most outstanding features of the central part of the Prelitoral Range is the Capellades Strait, formed by the Anoia River. In the Capellades area, the Anoia divides the Prelitoral Range into two different lithological areas: the Paleozoic materials to the east and the Triassic materials to the west. The Paleozoic formations are composed of gray–blue Silurian slates, which are dark and glossy, and crossed by quartz ledges and some porphyry dikes (García Rodrigo 1957). Between Capellades and Cabrera d'Igualada, there is a plutonic outcrop formed by intrusive granitic materials, which has given rise to a phenomenon of regional metamorphism in contact with the Paleozoic. The Triassic formations west of the Anoia begin with the Bundsandstein materials of essentially fluvial origin (Anadón et al. 1979). The Muschelkalk is formed by dolomitic limestones separated by red clays. Finally, the Keuper is formed by yellowish clayish dolomites, gray marls, with occasional gypsum strata at the top of the formation (García Rodrigo 1957).

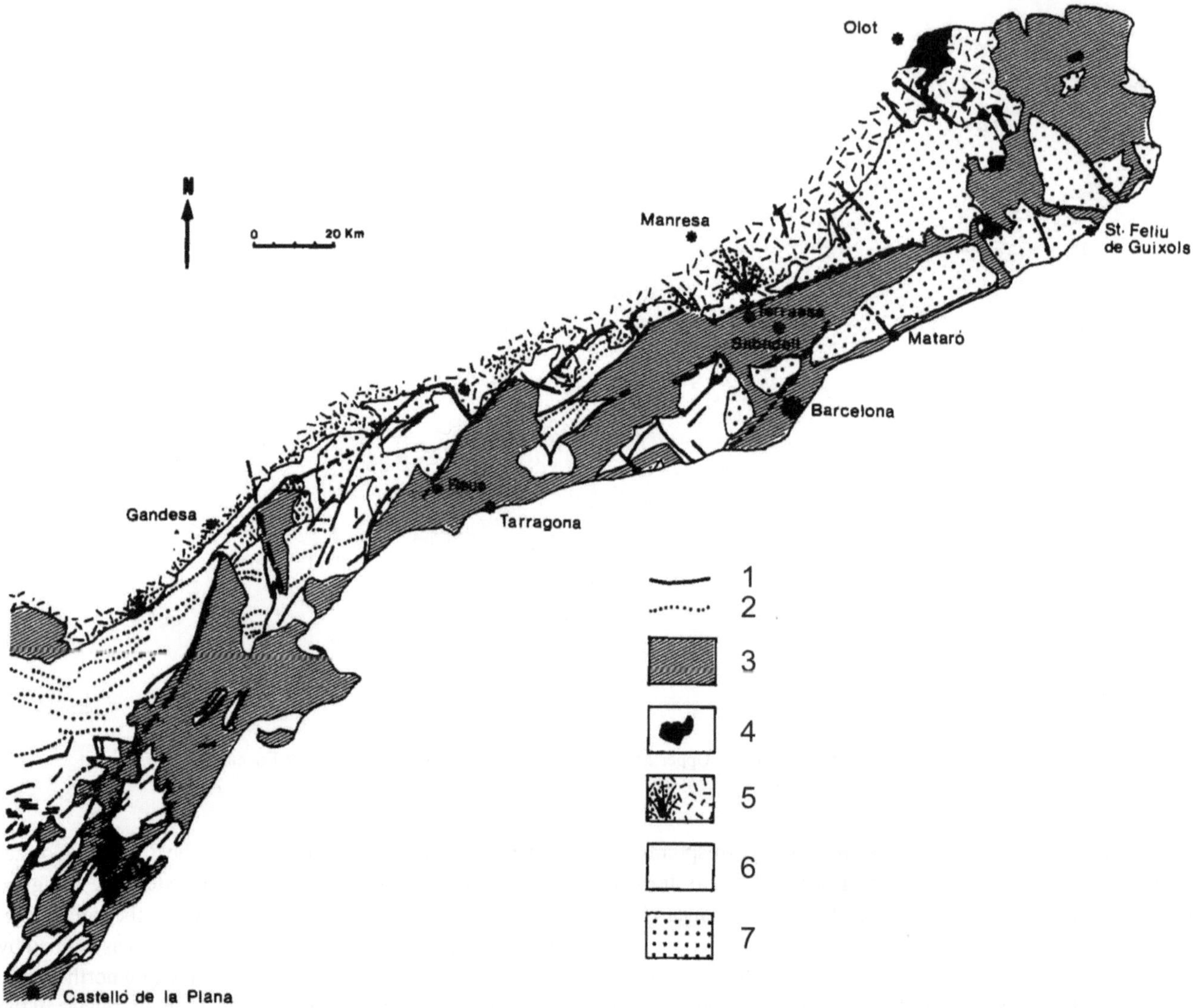

Fig. 2.3 Tectonic map of the Mediterranean system. Modified from Anadon et al. 1979 in Sala 1994. Legend: *1* Faults. *2* Folds. *3* Neogene from inner depressions. *4* Volcanic rocks from Neogene and Quaternary origin. *5* Paleogene from the Ebro basin and the Mora depression. *6* Mesozoic cover. *7* Paleozoic basement

The Prelitoral Depression, also known as the Vallès-Penedès rift, was formed by the sinking of a large block during the last movements in the Alpine orogeny. It is approximately 200 km long and 12 km wide, although its width tends to decrease in the north. During the Neogene, a group of normal faults related to distensive phenomena affected the western Mediterranean. The Llobregat fault divided the depression into two large rifts: the Vallès rift to the north-east and the Penedès rift to the south-west.

The Penedès rift has a sedimentary depth of 2,000 m and was filled by Triassic materials from the Prelitoral Range, marine Paleogene materials from the Ebro Basin and fluvial Quaternary sediments. At the end of the Lower Miocene, the sea penetrated through the Penedès Depression to the Llobregat valley when the fault became active and the rifts sank. During the Upper Miocene, a regression gave rise to a significant erosive phase. In the Pliocene, a transgressive phase fossilized the paleotopography developed during the previous phase (Sala 1994). According to Gallart (1981), the occurrence of more than one marine phase has not been proven and, with the exception of the first such phase during the Burdigalian–Serravallian age, the Pliocene deposits correspond to continental formations (Gallart 1981).

For this area García Rodrigo (1957) established three Miocene facies: the Piera Series, the Vallbona Series and the Guixera Series. However, Gallart (1981) later attributed the Guixera Series to the Pliocene. The Piera Series is of fluvial origin and can be found to the east of Piera creek. It extends to the south and south-west, forming the Badorc massif. It is a reddish-yellow formation composed of alternating conglomerate banks, silt and sandstones. The Vallbona Series is of local origin related to the erosion of the borders of the basin. It is located in the triangle formed by the Badorc massif and Piera creek, the Paleozoic of the Prelitoral Range to the north and north-west and Vallbona to the south-west. The Guixera Series is a detrital continental formation (Gallart, *Op. cit*), located in the triangle formed by the Anoia River to the north-east, the Riudebitlles River to the south-west, the confluence of these two rivers to the south-east and the Paleozoic materials of the Prelitoral Range to the north-west. In general, it is made up of alternating pudding stone banks, sandstones and clays. Chert nodules appear in the conglomerates of this formation.

The structural unit of the travertines of the *Cinglera del Capelló* cliff overlies the Paleozoic slates of the Prelitoral Range. In the Capellades Gorge the Anoia River runs along a tectonic fracture, often called the Anoia fault, which lies roughly perpendicular to the Vallès Penedès fault. The tectonic movements of these two faults produced the horst of the Prelitoral Range. Cenozoic materials from the Ebro Basin can also be found on the block near Carme. The post-Alpine subsidence of this block led to the lateral erosion of Paleozoic, Mesozoic and Cenozoic-aged materials and promoted springs and travertine formation in the fault of the Capellades and Carme region (Vidal 1911). These travertines fossilized the Anoia fault and therefore mark the end of its tectonic activity. However, the fault is laterally active through a small satellite fault which fractures the travertine deposits and forms the *Pla de Capellades* mesa (IGME 1973, 1975). This satellite fault runs parallel to the Anoia fault and caused this small part of the block to sink, resulting in a flat travertine surface, on which the town of Capellades has been built (Solé Sabaris et al. 1957).

The travertine platform, at a height of nearly 22 m above the river, is flooded by the multilayer aquifer of Carme and Capellades. Its hydrological model is described by Vidal (1911). This aquifer forms travertine deposits that outcrop at the Òdena and Penedès basins (García Rodrigo 1957). Springs originating from lithological discontinuities between fractured Paleozoic, Mesozoic, and Cenozoic sedimentary materials have led to differing aquifer elevations. In the area of Capellades, Romani and Vidal found different sedimentary ages for lacustrine and karstic spring travertine deposits, such as Pla de la Torre and the Pla de Capellades and the above-mentioned *Cinglera del Capelló* and the Abric Romaní lacustrine gorge (Muro et al. 1987; Bartrolí et al. 1995).

Quaternary Geology

Capellades is located on the the border between the Penedès and Òdena basins. Studies on the Quaternary of the Anoia valley unite these two regions as a single area (García Rodrigo 1957; Gallart 1980, p. 372; Josa 1985). Travertines from the Capellades area have been recognized as a stratigraphic marker, widely cited in regional studies on Quaternary-Neogene geology. García Rodrigo (1957) mentions two river terraces (+20 and +1) and interfluves near the town of Badorc, covered by two types of slope deposits composed of gravel and silt. The upper slope deposits are considerably widespread, while the reddish lower slope deposits are limited to the areas near escarpments. The position of these materials provides insight into some Quaternary erosional processes. The reddish sediments would have formed a ramp where they were deposited, probably synchronously to the Villafranquian travertine formation of Capellades, Cabrera and Badorc (García Rodrigo 1957). This deposition was then followed by a period of erosion, removing part of the reddish sedimentary material and reducing its extension to the current area. This new erosional relief hosts a second slope deposit, which in the Conca d'Òdena correlates with glacis 2 (G2), which laterally adjoins the +25 m Anoia terrace (T2) (Gallart 1991). The reddish slope deposits correspond to glacis 3 (G3) (Gallart 1991).

The geomorphology of the Anoia terraces in the Òdena basin is as follows (Gallart 1991):

Upper terrace (T4): 380 m high and 80 m above the level of the Anoia River. It locally contains bedded Eocene marine materials identical to those that constitute the G3 glacis. Furthermore, its elevation suggests a connection with the travertines of Capellades and Carme and, consequently, a Lower Pleistocene age.

High terrace (T3): 55 m above the level of the Anoia River. It contains a lot of lacustrine limestone and has a cemented relief, eroded/dissected into a discontinuous surface.

Middle terrace (T2): 25 m above the level of the Anoia River. This terrace is very extensive and contains two subunits: the more elevated T2b (+30 m) with coarser deposits and T2a, which is lower and more abundant. The lower T2a seems to link to glacis G2 and contains dark argillic soil.

Lower terrace (T1): 8–14 m above the level of the Anoia River. It consists of silt, gravel and sand deposits, with carbonate nodules and grey-colored soil on top.

Lower terrace (T0): 1–2 m above the level of the Anoia River. It consists of alluvial flood deposits.

In the area of the Penedès basin, Gallart indicates five Pleistocene levels between the Anoia and Riudebitlles fluvial valleys (Gallart 1981) (Fig. 2.4):

Level 5 (early Quaternary): Similar to conglomerates of Riudebitlles outcrops (Gallart 1985), though more torrential in character, this level is made up of deposits called *rañas* or *sasos* in other regions of the Iberian Peninsula. The morphometry of the distal segments of these deposits is

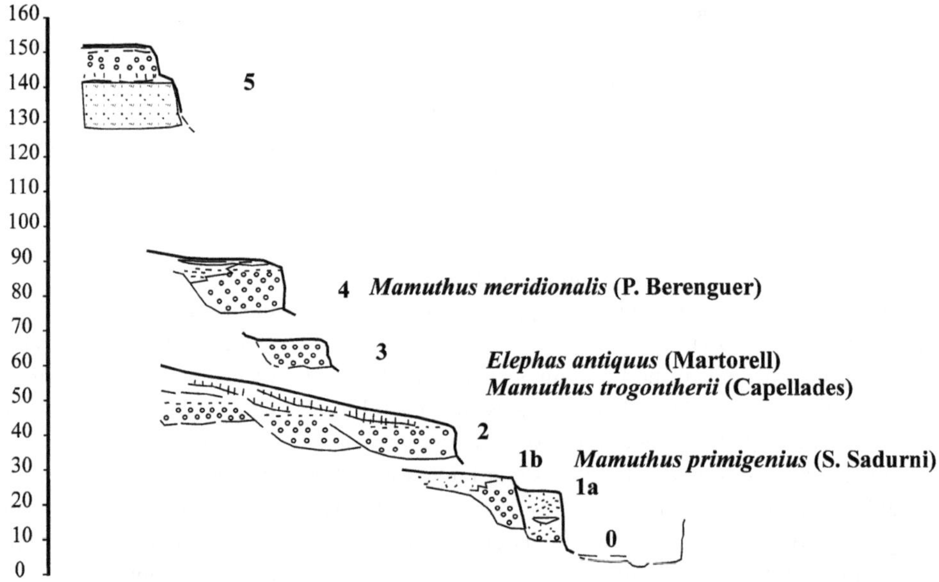

Fig. 2.4 Fluvial and slope surfaces with the proboscidean Quaternary biostratigraphy in the Llobregat and Anoia valleys (Gallart 1981)

similar to that of those located on the Anoia terraces. The possibility of two stages in these torrential deposits is suggested by the presence of a thick interbedded (1 m) fersialitic soil. This level may be related to Anoia River deposits measuring 140 m whose terrace morphology has not been preserved.

Level 4: This level is made up of accumulations of material and some alluvial fan deposits with poor morphological preservation, with an elevation of 90 m and a thickness of up to 15 m. The lower boundary surface of these accumulations is usually cemented and pedological processes mark the upper boundary surface capped with a caliche horizon. This level can be correlated to the 85–90-m Llobregat terrace (Manresa, Puig Berenguer), prior to the Cromer-Galerian biozone.

Level 3: A level comprised of cemented alluvial deposits, poorly preserved, at an elevation of 65 m.

Level 2: The most characteristic geomorphological forms in this level are the glacis, which formed in three phases, separated by periods of dissection. The fluvial deposits show the existence of two incision stages separating three morphological units. Their morphological preservation is poor due to colluvial and eolian erosion, which has limited the differentiation of forms. This level corresponds to deposits of the Barcelona city and Llobregat River terraces (Virgili 1960). The soils and paleosoils of these deposits contain argillic and carbonate-rich horizons.

Level 1: This level consists of well-preserved glacis-terraces with a relative height of 30 m. Soil forming processes produced brown soils (7.5 YR 5/4), with partial leaching of carbonates in the upper horizon. Pseudomicelia have accumulated in the lower horizons. In Sant Sadurní, *Elephas primigenius* remains suggest a Middle Pleistocene age for the lower Anoia terrace complex (Calzada 1975). A small level over 22 m is composed of silt and sulfates.

Level 0: This level consists of the Anoia floodplains and contains an incised and meandriform stream.

Morphogenetic periods in the Penedès Basin are characterized by stream efficiency, storm erosion, physical weathering and eolian erosion and sedimentation. This morphogenesis points to a relationship with cold Quaternary periods, but in the Penedès area no clear periglacial record below 800 m exists. There were no extremely cold periods because soil formation processes on colluvial slopes and in eolian silts were synchronous with their deposition (Gallart 1981).

One of the narrower sections of the Capellades Gorge is in the travertine cliff called the *Cinglera del Capelló*. Travertine stratigraphy shows facies associations originating from lacustrine and spring paleoenvironments. In Badorc, travertines are 20–50 m thick and are located on the hilltop. Below the travertines, a river terrace (+20 m) contains *Paraelephas trogontherii nesti* POHLIG (Solé Sabaris et al. 1957). Travertine U-series dating taken, from the Bofia sample of basal travertine units indicates an age of over 350 ka (Bischoff et al. 1988). Neotectonic activity caused faulting of the travertine complex between Pla de la Torre and Pla de Capellades during the Lower Pleistocene.

History of Excavations

The Abric Romaní has a long history of research. It was the first Middle Paleolithic site known in Catalonia and was excavated in different periods throughout the

Fig. 2.5 Amador Romaní (in the middle of the photograph) during the excavation of the Abric Romaní in 1909 (Photo: Arxiu Museu Molí Paperer de Capellades)

Fig. 2.6 Amador Romaní watching the excavation works. This image corresponds to the beginning of the excavation in the middle part of the shelter in 1909 (Photo: Arxiu Museu Molí Paperer de Capellades)

twentieth century. It has undergone three major periods of excavation: the Amador Romaní excavations (1909–1930), the Eduard Ripoll excavations (1956–1962) and the current excavations (1983–2009). This long research history pays testimony to the various different theoretical and methodological paradigms that have been dominant in the archeological thinking of this century. The Amador Romaní excavations exhibit the construction of the basic scientific procedures that have characterized Paleolithic archeology up to the present. The second period of excavations represents the dominance of the historic and cultural paradigm that marked the development of European archeology. Finally, the current period corresponds to the crisis of this paradigm and the spreading of the assumptions linked to processual archeology. In this sense, the history of the Abric Romaní is the history of the emergence, consolidation and transformation of archeological science.

Fig. 2.7 Sieving of sediments during the 1909 excavation. Amador Romaní is on the left (Photo: Arxiu Museu Molí Paperer de Capellades)

The Amador Romaní Excavations (1909–1930)

The site was discovered and the first archeological excavations were conducted by Amador Romaní i Guerra, an industrialist from Capellades who began prospecting in the *Cinglera del Capelló* in 1905 (Muro et al. 1989). Over several years, Romaní brought to light many rockshelters containing archeological deposits. Most of these were recent prehistory sites, but some of them yielded Pleistocene layers. Among these, the most significant was the Abric Romaní, previously known as the *Balma del Fossar Vell* (Old Grave Rockshelter), because it was used as a cemetery during the nineteenth century. After the discovery of the site, on 9th August 1909, the first excavations were immediately undertaken under the sponsorship of the Institute for Catalan Studies (IEC) (Figs. 2.5, 2.6, 2.7, 2.8). These works were nominally directed by Norbert Font i Sagué—who named the site Abric Romaní—and, after his sudden death in 1910, by Lluís Marià Vidal. However, in the field direction was effectively given by Amador Romaní. Vidal published the first results a short time after he took charge of the project (Vidal 1911–1912). That paper, along with Romaní's diary (the *Atlas*), are the primary sources of information about this first period of excavation (Bartrolí et al. 1995). Excavations continued until 1911 when IEC sponsorship came to an end, but Amador Romaní kept working intermittently in the *Cinglera del Capelló* almost until his death in 1930.

These excavations affected the upper part of the stratigraphy, from the top to level J. The uppermost layers, from A to D, were almost entirely removed, although these levels were severely damaged by the nineteenth century burials. The first works focused on the western part of the shelter, where a 4.5 m trench was dug perpendicular to the wall. This trench was subsequently enlarged, forming a 6 × 6 m pit, which is currently known as Pit 2. This pit was deepened to layer 12 of Romaní's stratigraphy, except in the area near the shelter wall, where it went down to layers 13–14 (level J of the current stratigraphy). In the middle of this pit a circular well, 1.6 m in diameter, was excavated with the aim of reaching the bottom of the deposit. This sounding—the Romaní well—did not reach the bottom of the sequence, but permitted the documentation of eight additional meters of the stratigraphy. At the base of the well, Romaní excavated a gallery 4 m in length that reached the shelter wall.

Romaní then extended the excavation towards the eastern part of the shelter, bringing the area affected by excavations to nearly 80 m^2. Nevertheless, east of Pit 2 Romaní's excavation only reached the top of layer 9—level E of the current excavations. The maps included in his *Atlas* indicate that the surface area Romaní excavated roughly corresponds to rows M–V and columns 43–60 of the current grid. However, he only occasionally touched the shelter wall and left several sedimentary remnants at the rear of the site. It should be stressed that the travertine layers forming the bulk of the sequence are considerably harder close to wall.

The first stratigraphic description was published by Vidal (1911–1912), although it was slightly different from that presented by Romaní in *Atlas*. Both Romaní and Vidal describe the sequence as a 12 m thick deposit of calcareous sand, interbedded with thin layers of black sediment—the archeological levels—and thick travertine strata. The stratigraphic sequence described by Romaní was made up of 31 layers, 12 of them containing evidence of human

Fig. 2.8 Amador Romaní sitting in the middle of the rockshelter in 1924 (Photo: Arxiu Museu Molí Paperer de Capellades)

occupation. However, the fourteen lowermost layers were only identified in the Romaní well, and Romaní provided very little information about them. The layers from the uppermost six meters, documented over a more or less extended surface, were thoroughly described from a geological and archeological point of view. This was the sequence used as a reference in subsequent works and basically corresponds to the profile described in Pit 2 (Fig. 2.9). From top to bottom, this sequence was as follows (Bartrolí et al. 1995, p. 118).

Layer 0: This was a disturbed layer containing some lithic implements and many human remains from the nineteenth century burials. This layer was the top of the sequence in most of the site, except near the shelter wall, where layer 1 was still present.

Layer 1: This was originally the top of the deposit, but was only preserved at the back of the shelter when Romaní started his excavations. It was between 0.5 and 2 m thick and completely filled the shelter. This layer was formed by red silts, characterized by Romaní and Vidal as *diluvium roig*, or red flood, and was archeologically virtually sterile.

Layer 2: This was an archeological layer of between 5 and 50 cm thick (Vidal 1911–1912) which corresponded to the only Upper Paleolithic assemblage documented in the Abric Romaní sequence. It was only preserved in three relatively small areas close to the shelter wall. It is called level A in the current excavations.

Layer 3: This was a travertine layer with an irregular thickness of between 15 and 40 cm. Amador Romaní designated this kind of hard travertine layer using different names such as *bancal estalagmític* (stalagmitic terrace), *bancal caliç* (calcareous terrace) or *calissa estalagmítica* (stalagmitic limestone), in order to differentiate it from the softer travertine deposits, which were called *terra sabulosa caliça* (calcareous sandy soil).

Layer 4: This was formed by black sediment with many charcoals and contained evidence of Mousterian industry. Thickness was also highly variable—between 20 and 50 cm. According to Romaní, this was the first archeological level preserved in the entire site and provided the richest assemblage. Vidal (1911–1912, p. 281) and Romaní mention the recovery of two human teeth, but they are intrusions from the modern burials. Layer 4 corresponds to level B of the current excavations.

Layer 5: This was a travertine layer with a maximum thickness of 10 cm.

Layer 6: This was a 10 cm thick archeological layer that was made up of black sediment with charcoals and Mousterian artifacts. In addition to the recovery of lithic and bone remains, Amador Romaní also documented a carbonized wood artifact. This is the first reference to the presence of wooden implements in the archeological levels of the Abric Romaní. This layer corresponds to level C of the current excavations.

Layer 7: This was a 10–15 cm thick deposit of white sediment found only in a small area near the shelter wall. It was absent in the rest of the site, in which layer 6 lay directly over layer 8.

Layer 8: This corresponds to a travertine layer.

Layer 9: This is a Mousterian archeological layer made up of red silts similar to those of layer 1. These characteristics have made this layer a point of reference in the stratigraphy and it has served to correlate the Amador Romaní sequence with those of subsequent works. It corresponds to level E of the current excavations.

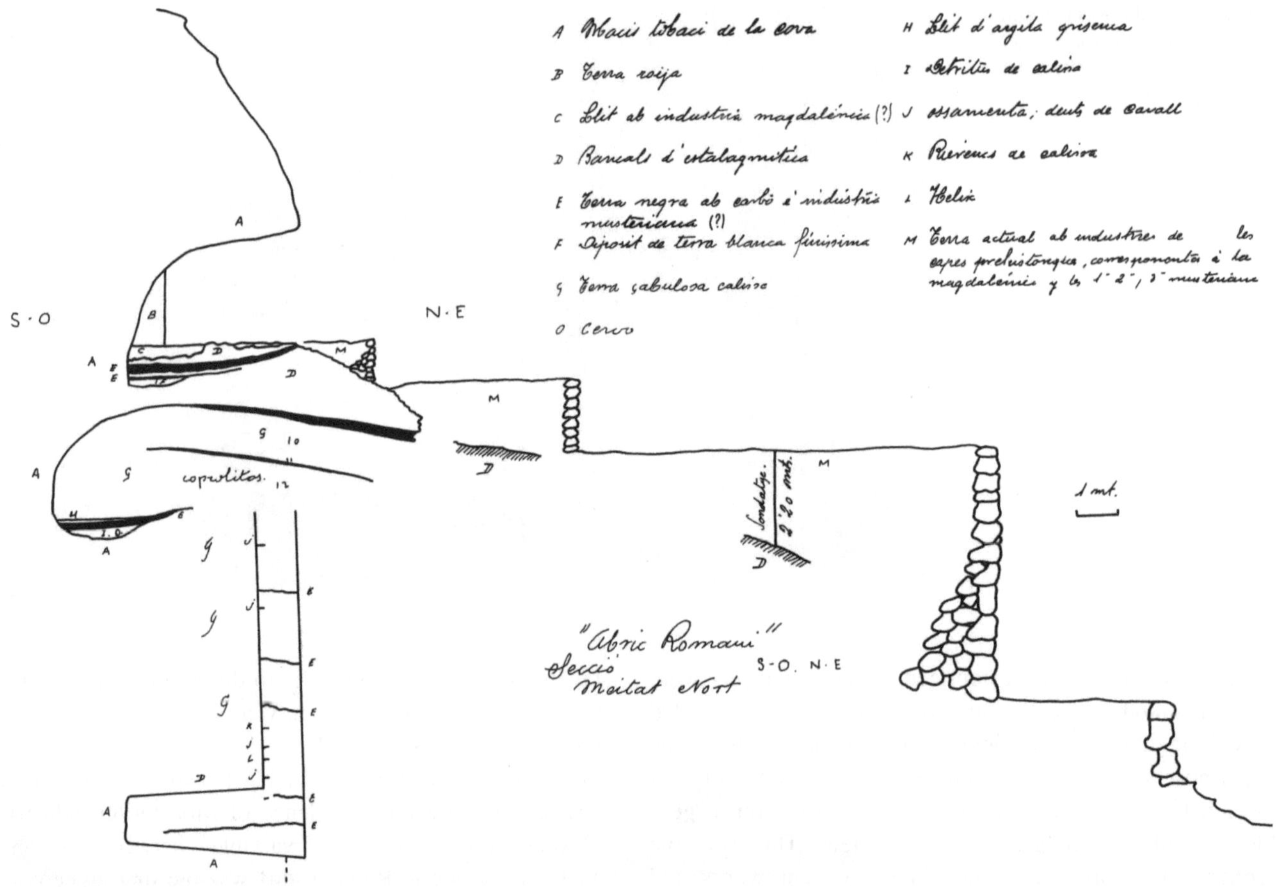

Fig. 2.9 Stratigraphic sequence documented by Amador Romaní at Pit 2 and the Romaní Well. Modified from the *Atlas* of Amador Romaní

Layer 10: This layer, with a thickness of 0.5–1.5 m, was characterized as a deposit of calcareous sandy sediment without archeological remains, but included many hyena coprolites.

Layer 11: This was an archeological layer with Mousterian artifacts. It appears in the profiles as a very thin and discontinuous layer. It can be correlated with level I of the current excavations.

Layer 12: This layer was also described by Romaní as a deposit of calcareous sand with many coprolites and malacological remains, but lacking other archeological evidence.

Layer 13: This was a 20 cm thick archeological layer formed by black sediment with Mousterian artifacts.

Layer 14: This layer was described by Amador Romaní as a deposit of *calissa estalagmítica en lámines*—blades of stalagmitic limestone. It seems clear that this corresponds to a layer of travertine blocks and clasts. Only charcoal and bone remains were found. Layers 13 and 14 can be correlated to level J of the current excavations. This layer and the one above it were only documented in the deepest area of Pit 1.

Layer 15: This travertine layer was the lowermost layer documented in Pit 1.

The rest of the sequence corresponds to the Romaní well, in which between five and nine additional archeological horizons were identified. This was the stratigraphic sequence depicted by Amador Romaní in the *Atlas*, although it was slightly different from that published by Lluís M. Vidal. The sequence presented by Vidal was made up of only 12 layers, as Romaní's layers 7, 14 and 15 were excluded. Due to the absence of layer 7, there was a change in the name of the units underlying layer 6; Romaní's layer 8 was Vidal's layer 7 and so on (Table 2.1). At any rate, this was the valid sequence used for the entire shelter, excepting the stratigraphic succession identified inside a small cavity located in the north-western end of the site, the *Coveta Nord* (Northern Small Cave). Due to the absence of travertine layers in this cavity, the archeological levels above layer 8 overlapped to form a thick palimpsest. The main features of Romaní's stratigraphy were confirmed in subsequent works, although the current excavations have shown that some archeological levels, like levels D, H and F–G, were Romaní. They were very thin and low-density levels that

Table 2.1 Correspondence between the different systems used to designate the archeological levels

Amador Romani	Vidal (1911–1912)	Lumley and Ripoll (1962)	C.R.P.E.S. (1983–1989)	Current name
Layer 2	Layer 2	Hearth 2	C.II 1.0.1	Level A
Layer 4	Layer 4	Hearth 4	C.II 1.1.0	Level B
Layer 6	Layer 6	Hearth 7	–	Level C
Layer 8A?	–	Hearth 8	C.II 1.4	Level D
Layer 9	Layer 8	Hearth 9	C.III 1.0.0	Level E
–	–	–	C.III 2.1.6	Level F
–	–	–	C.III 2.1.8	Level G
–	–	Layer 10	C.III 2.2.5	Level H
Layer 11	Layer 10	Hearth 11	C.III 2.2.7	Level I
Layer 13	Layer 12	Hearth 12	C.III 19.0	Sublevel Ja
Layer 14	–	Layer 13	–	Sublevel Jb

went unnoticed due to the excavation methods used at that time. Moreover, the Romaní excavations allowed the archeological layers to be characterized from the chrono-cultural point of view. At first, layer 2 was attributed to the Magdalenian, although its Aurignacian character was soon recognized. The remaining layers were assigned to the Mousterian. These Mousterian layers were considered as a whole, and no differences between levels were pointed out.

The Excavations of Dr. Eduard Ripoll (1956–1962)

Unfortunately, the line of research started by Amador Romaní and the Institute for Catalan Studies had no continuity. The Spanish Civil War (1936–1939) imposed a profound break in archeological activities. This "dark period" persisted until the mid-1950s, when the end of the international isolation of Franco's regime favored a reactivation of archeological works, frequently associated with the collaboration of foreign scholars. In this context, a new stage of excavations at the Abric Romaní began. The V INQUA Congress held in Spain in 1957 included a visit to the Abric Romaní leading the Archeological Research Service of the Barcelona Provincial Council to promote these excavations, which started in 1956 under the direction of Dr. Eduard Ripoll Perelló. The primary goal was to verify the stratigraphic sequence outlined by Amador Romaní (Ripoll 1958, p. 14). The fieldwork was undertaken in part by French researchers like Georges Laplace, who studied and published the description of the Upper Paleolithic assemblage recovered by Romaní in layer 2 (Laplace 1962), and Henry de Lumley.

Compared to the work led by Amador Romaní, relatively little is known about the excavations carried out between 1956 and 1962, especially in terms of fieldwork processes. We do not know whether there were field diaries, and the papers published as a result of these excavations (Ripoll 1958, 1959; Laplace 1962; Lumley and Ripoll 1962; Ripoll and Lumley 1964–1965) provide little information about the excavation methods used or the areas affected by these works. Apart from these publications, the only sources that give some insight into Ripoll's excavations are some imprecise and often contradictory oral accounts, short notes published by other authors (Pericot 1960–1961, 1964; Maluquer 1969) and occasional references accompanying the archeological remains recovered during that period and stored in the Capellades Museum. Methodology was essentially guided by stratigraphic criteria and consisted of digging various trenches and pits in different areas of the site. It is important to mention that the attribution of some of these actions to Ripoll's excavation is hypothetical and is based on the archeological methods used at the time in combination with the absence of references in Amador Romaní's works. Therefore, to the best of our knowledge, the actions carried out during this period seem to have been the following:

- A 4 × 4 m pit was dug in the central part of the site, corresponding to squares N–Q/52–55 of the current grid (Pit 1). The excavation of this area started at the level at which Amador Romaní (layer 9) stopped and continued to layer 14. At the bottom, the pit was smaller (3 × 3 m) and had a more irregular outline.
- A 4.5 × 3.5 m pit was excavated at the easternmost end of the site (Pit 3), in the area corresponding to squares G–J/67–71 of the current grid. This pit was 3.25 m deep and affected the upper part of the stratigraphic sequence to the lower part of layer 8.
- A 1-m wide L-shaped trench was also excavated in the middle of the site, starting from the south-eastern corner of Pit 1. At 3.5 m in length, the trench turned to the east at a right angle and extended three additional meters.

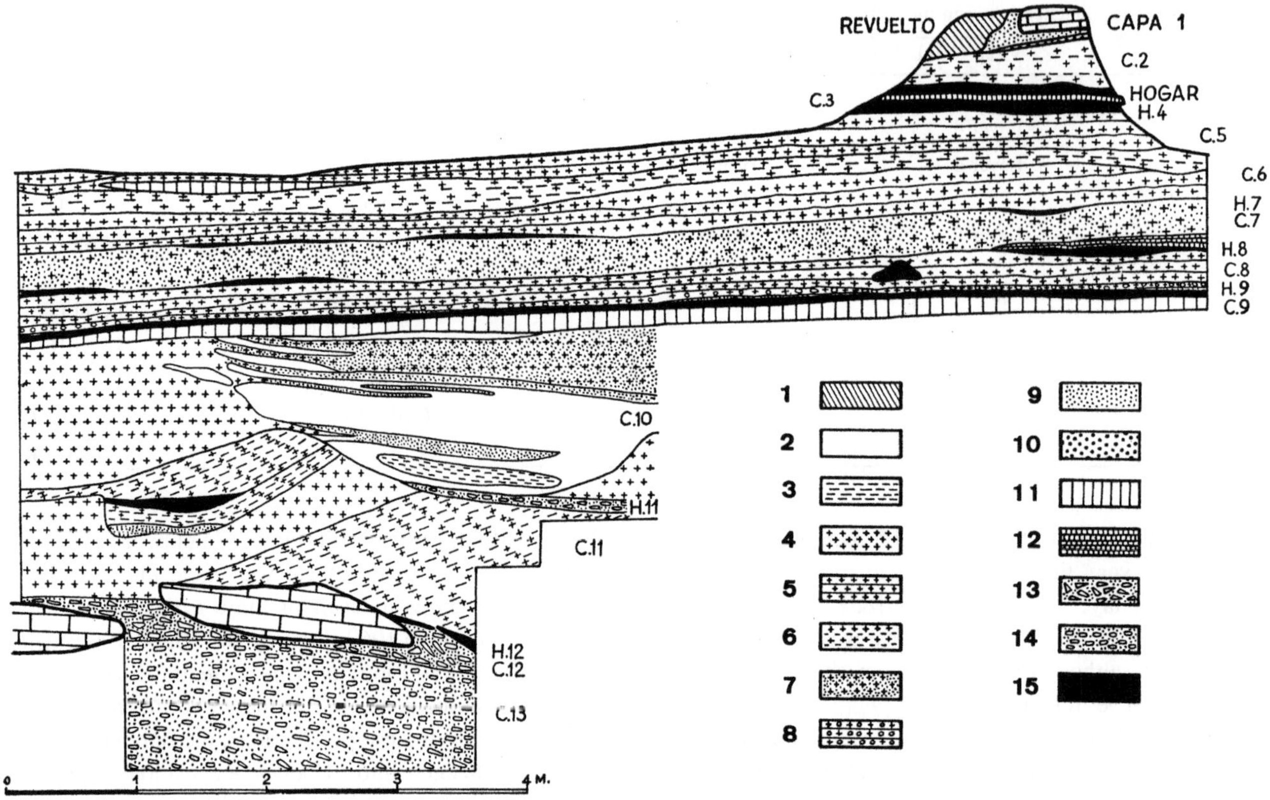

Fig. 2.10 Stratigraphic sequence documented during the Ripoll excavations. From Ripoll and Lumley 1965. Legend: *1* disturbed; *2* chalk; *3* lumpy chalk; *4* hardened tufa; *5* hardened and bedded tufa; *6* chalky bedded tufa; *7* hardened bedded sandy tufa; *8* tufa rich in plant remains; *9* silty sand; *10* calcareous spherical nodules; *11* sandy silts; *12* calcareous crust; *13* medium-sized clasts in yellow sandy matrix; *14* small clasts in yellow sandy matrix; *15* hearth

It had a maximum depth of 2 m and reached layer 12 of Romaní's sequence.

- It is possible that the *Coveta Ripoll,* or Ripoll's Small Cave, was also partially excavated during these years. This small cavity was formed in the middle of the shelter due to the growth of a stalagmitic column in the area of squares R–T/46–51. However, this excavation affected only the sequence made up of layers 9–12. Due to the hardness of the travertine layers above layer 9, the excavators penetrated the cavity leaving the travertine of layer 8 as a false roof. To prevent it from collapsing, this roof was supported by two vertical props that rested on two small planks. These planks were placed into two hollows excavated in layer 13.

Ripoll and de Lumley published the stratigraphic study derived from these excavations (Lumley and Ripoll 1962, p. 4–5). It describes only the six uppermost meters of the sequence because they were unaware of the existence of the Romaní well. This sequence was very similar to that described by Amador Romaní, although there are a few differences (Fig. 2.10). To refer to the archeological units they used the term *foyer* (hearth). From top to bottom, the stratigraphic sequence was described as follows:

Layer 1: Sandy silts.

Layer 2: Stratified calcareous travertine.

Hearth 2: Archeological level. Amador Romaní's layer 2.

Layer 3: Calcareous crust.

Hearth 4: Archeological level. Amador Romaní's layer 4.

Layer 5: Hardened stratified travertine that becomes calcareous travertine towards the bottom.

Layer 6: Hardened stratified travertine.

Hearth 7: Archeological level.

Layer 7: Hardened and stratified sandy travertine.

Hearth 8: Archeological level.

Layer 8: Hardened stratified travertine, rich in plant remains at the bottom.

Hearth 9: Archeological level. Amador Romaní's layer 9. Lumley and Ripoll highlighted the great abundance of faunal and lithic remains in hearth 9. This was the only level that yielded a statistically significant number of remains and was therefore presented separately.

Layer 9: Sandy silts.

Layer 10: This was a very complex unit formed by different depositional facies. At the western end of the section, there was a *gour* deposit basically formed by "calcareous dust", but also including some lenses of "lumpy clay" and

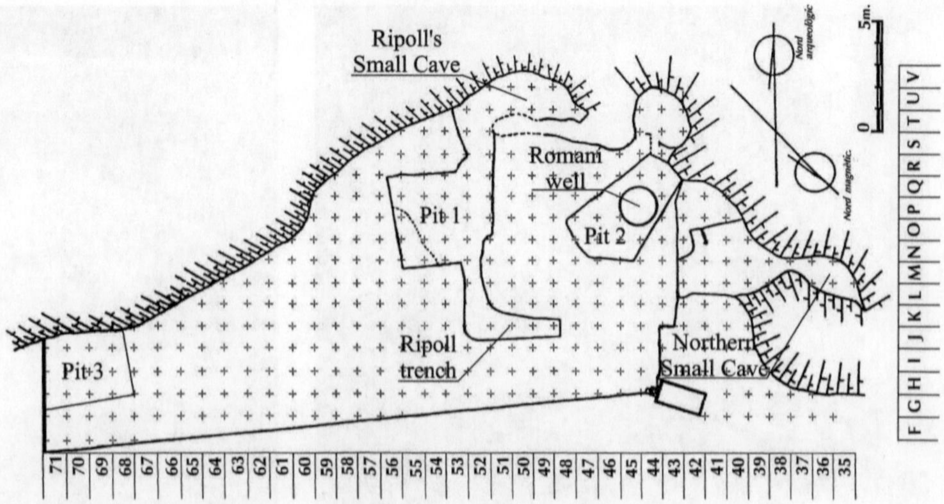

Fig. 2.11 Abric Romaní plan made in the beginning of the last excavation phase. The different pits and trenches from previous excavations are indicated (Drawing: Josep Maria Prats)

silty sand. There was a deposit of hardened and stratified sandy travertine at the top. To the east of the section, this layer was formed by a homogeneous hardened travertine. Although Lumley and Ripoll did not mention the recovery of archeological remains, some artifacts labeled as layer 10 have been found in the Capellades Museum.

Layer 11: This layer was formed by different travertine facies with some lenses of silty clays. A discontinuous archeological unit (hearth 11) was included in this layer. It can be correlated with layers 11 and 12 of the Amador Romaní sequence.

Hearth 12: Archeological level corresponding to layer 13 of Romaní's sequence.

Layer 12: This was a clastic deposit formed by medium-sized travertine fragments in a sandy matrix. Some large blocks from the roof collapse were also present.

Layer 13: Small clasts in a yellow sandy matrix. Layers 12 and 13 can be correlated to layer 14 of the Amador Romaní stratigraphy. This clastic deposit was considerably thicker in Pit 2 than in Pit 1, where it was excavated by Romaní. Like layer 10, this layer was not considered an archeological unit by Lumley and Ripoll, although remains labeled as coming from the level have been found in the Capellades Museum.

Lumley and Ripoll (1962, p. 3–7) organized this sequence into three stratigraphical sets that were interpreted in environmental terms.

1. Levels of fallen blocks of cryoclastic origin. This set was 1.5 m thick and was subdivided into two units. The upper unit (layer 12) was formed by large travertine blocks, while the lower one (layer 13) consisted of small angular clasts in a sandy matrix. The layer formed by small fragments was attributed to periods characterized by daily freeze–thaw cycles. Otherwise, the falling of large blocks would occur as the result of seasonal cycles. These deposits would indicate the cold and humid conditions associated with the end of Würm II.
2. Travertine layers (layers 11–2). This 4 m thick set was mainly formed by travertine of different facies with silt and sand lenses. It suggested the very wet conditions of the Würm II–III interstadial, although hearth 2 was attributed to the beginning of Würm III due to the presence of an Upper Paleolithic assemblage.
3. This set was comprised of the red silty sands of layer 1.

The principal outcome of Ripoll's excavation was the integration of the Abric Romaní into the framework of western prehistory, which was characterized at the time by the dominance of the historical-cultural paradigm. The lithic assemblage from layer 2 was studied by Laplace (1962), who attributed most of the artifacts to the *synthétotype aurignaco-gravettien* and highlighted its resemblance to a group of assemblages that were later included in the Protoaurignacian. The assemblages from the Mousterian layers were characterized by Lumley and Ripoll (1962) within the framework of Mousterian facies defined by Bordes. The artifacts were classified according to the Bordes type-list and all the Mousterian layers were attributed to the Denticulate Mousterian. This made it possible to compare the Abric Romaní assemblages with other European Mousterian sites. Similarly, correlating the stratigraphy with paleoenvironmental stages provided the first relative chronologies for the sedimentary deposit. The collaboration of foreign researchers and publications in international journals (Laplace 1962; Lumley and Ripoll 1962) made the site known among the European scientific community. As a result, references to the Abric Romaní appeared in many studies about the Middle Paleolithic of south-west Europe (Escalon de Fonton 1963; Lumley and Berard 1964; Freeman and González Echegaray 1967; Altuna 1972).

Fig. 2.12 Excavation of level J during the 1994 field season

Fig. 2.13 Excavation of level J during the 1994 field season

After the end of this second period of excavations, the Abric Romaní remained under the control of the Barcelona Provincial Council, but scientific activity was practically nonexistent. In the second half of the 1970s, there was a brief reactivation of archeological research in the *Cinglera del Capelló*. In 1976, Ripoll and Freeman conducted excavations at the nearby site of Abric Agut. However, these works scarcely affected the Abric Romaní. In 1978, the profile of Pit 2 was rejuvenated in order to take samples for several types of analyses, although in the end, only the pollen analyses were performed (Metter 1978; Deguillaume 1987). This lack of continuity in the excavations also gave rise to some negative consequences. Due to the absence of effective protective measures, the site was affected by the activity of amateur excavators who severely damaged the stratigraphic testimonies left at the back of the shelter. This plundering of the site persisted practically until the beginning of the third excavation period in 1983.

The Current Excavations (1983–2009)

Archeological excavations in the Abric Romaní restarted in 1983 under the direction of Eudald Carbonell, Artur Cebrià and Rafael Mora. The excavations were carried out by the Centre de Recerques Paleo-Eco-Socials (CRPES) of Girona between 1983 and 1988, and Rovira i Virgili University (URV) from 1989 to the present. Since 1989, Eudald Carbonell has been the only fieldwork director. These excavations have been conducted in the framework of a research project focusing on the entire *Cinglera del Capelló*, including works at other sites such as Abric Agut, Balma dels Pinyons and Balma de la Costa de Can Manel.

The first three years (1983–1985) were devoted to preparing the rockshelter for the excavation of a large surface. The site was protected with a roof and a 1-m grid system was set up. One of the primary undertakings was to delimit the old pits and trenches, and remove the sediments that filled them (Fig. 2.11) This revealed the Romaní

Fig. 2.14 Excavation of level J during the 1994 field season

well, which was deepened several times—in 1997 and 1992—adding four additional meters to the stratigraphic sequence originally documented by Romaní. However, the bedrock has not yet been reached. It was evident from the beginning that most of the uppermost levels had been removed in former excavations and a significant surface was preserved only from layer 9 downwards. At first, a numerical code was adopted to name the stratigraphic units, but this system was abandoned in 1990. Since then we have used an alphabetical system for the archeological levels. The correlation between this system and previous denomination systems can be seen in Table 2.1.

Apart from the excavation of the small remains of the uppermost levels (A–D) mainly preserved all along the shelter wall, our work has mostly been devoted to extensively excavating the Middle Paleolithic layers over large surfaces. Levels E to J were still significantly altered by the old pits and trenches, but from level K downwards the archeological units were virtually intact. The area affected by the excavations was progressively enlarged, reaching a maximum surface of 315 m^2, which represents most of the area occupied by prehistoric humans. Between 1983 and 1986, a first excavation area was defined, which was delimited by columns 43–60 and rows H–V. This area was enlarged in 1989–1990 to include the area defined by squares H–G/43–60. Finally, the excavation was extended to column 70 in 1991 and to column 40 in 1995.

The theoretical background of this research project was largely influenced by the principles of Processual Archeology. The excavation strategy has been directed towards achieving a paleoethnographical interpretation of the archeological record. Special attention has been paid to the spatial distribution of the remains. All artifacts and bones were three-dimensionally recorded and spatial structures were carefully documented. It was fairly clear from the beginning that the hearths constituted a basic feature in understanding the organization of the humans who occupied this site. Moreover, research has been conducted from an interdisciplinary perspective in order to approach a large array of behavioral, chronostratigraphical and paleoenvironmental questions. Great effort has been devoted to establishing the sedimentary dynamics and chronology of the deposit (Bischoff et al. 1988, 1994; Mora et al. 1988b; Giralt and Julià 1996; Arteaga et al. 2001b; Vallverdú et al. 2005b), as well as the paleoenvironmental sequence

Fig. 2.15 The Abric Romaní during the 1994 field season. The surface shown in the image corresponds to sublevel Ja. Some of the combustion structures found in this sublevel and the *Coveta Ripoll* in the back of the shelter can be seen

(Burjachs and Julià 1994, 1996). Behavioral issues have been approached from the perspective of different disciplines: bone taphonomy and zooarcheology (Aimene et al. 1996; Aimene 1998; Anconetani and Rosell 1998; Cáceres 1998, 2002; Cáceres et al. 1998; Saladié and Aimene 2000; Rosell 2001), the study of plant remains (Carbonell and Castro-Curel 1992; Castro-Curel and Carbonell 1995; Allué 2002; Cabanes et al. 2007), lithic technology (Mora 1988; Carbonell et al. 1997; Vaquero 1997, 1999b; Morant and García-Antón 2000; Martínez Molina 2002, 2005), and spatial organization (Vaquero et al. 1997, 1998, 2001a, 2004; Vaquero 1999a, 2005, 2008; Vaquero and Pastó 2001; Martínez Molina and Rando 2000; Pastó et al. 2000; Arteaga et al. 2001a). In addition, several general papers have been published throughout these years (Mora et al. 1988; Carbonell et al. 1994, 1996a, b; Vaquero et al. 2001b; Martínez Molina et al. 2005), as well as interdisciplinary studies of specific archeological levels (Carbonell 1992, 2002; Vallverdú et al. 2005a).

Level J was primarily excavated between 1993 and 1996, except for at the westernmost part of the site (columns 40–42), which was excavated between 1998 and 1999 (Figs. 2.12, 2.13, 2.14, 2.15). The excavated area was comprised of the space between columns 40–65 and rows G–V, which represents a total surface area of 250 m^2. As we have seen, level J was the last archeological unit significantly affected by the previous excavations, although less so than the overlying levels. The Ripoll trench did not reach level J and the surface affected by Pits 1 and 2 was smaller than in the upper levels. Pit 2 was only deepened to level J in a relatively small area (squares Q43–44, R43–44, S42–43, and T42–44). The main sector damaged by these previous works was that of Pit 1, corresponding to an irregular area of 10 m^2 (O–R/52–55). As described earlier, the *Coveta Ripoll* deposit was minimally damaged by the props used to sustain the travertine roof. Finally, a small circular pit located in K52 was also documented measuring a little more than 1 m in diameter. Although the authorship of this pit is unknown, its characteristics are reminiscent of those of the Romaní well.

The profiles of Pit 1 provided the first stratigraphic guidance for the excavation of level J. The examination of these sections revealed that there were two archeological units in level J, which were separated by a sterile layer. These units were designated sublevels Ja and Jb. The ongoing fieldwork was carried out taking into account

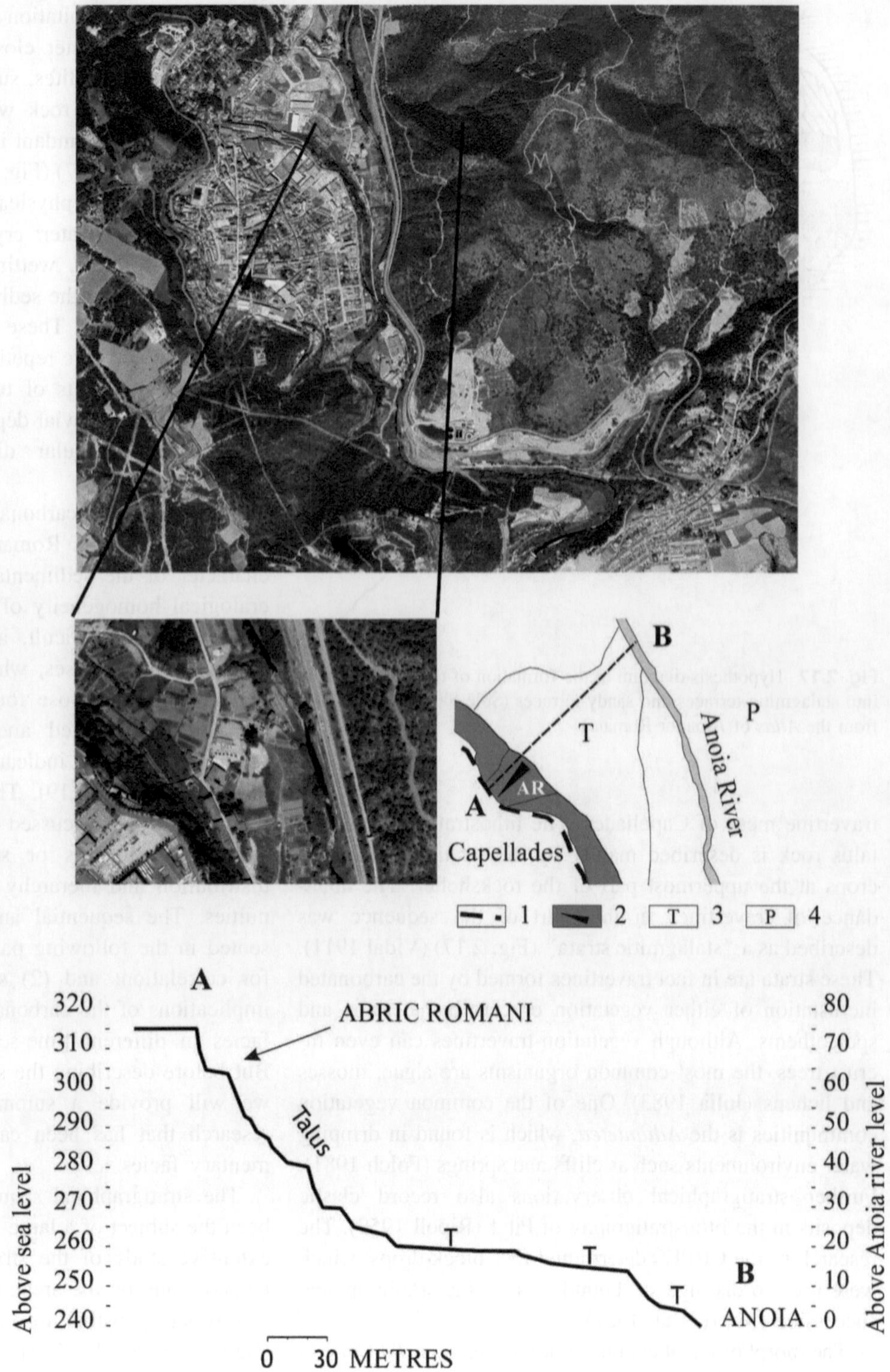

Fig. 2.16 Location of the Abric Romaní, Capellades. Orthophotomap of Capellades and the *Cinglera del Capelló*. Detail of the orthophotomap of the north-west section of the *Cinglera del Capelló* with the cartography of the lower terraces of the Anoia valley, the slope taluses, and the escarpment of the Capellades travertine mesa where the Abric Romaní is located. Topographic profile. Legend: *1* cliff escarpment; *2* talus slope; *3* Quaternary fluvial terraces; *4* Paleozoic slope; *AR* Abric Romaní; *A* and *B*, topographic transect

this archeostratigraphic pattern, although it was soon evident that it was not possible to document this sequence throughout the shelter. Only one archeological unit was recognized in many areas. In fact, sublevels Ja and Jb were clearly identified only in the middle of the shelter, in the area defined by columns 40–54 and rows L–O.

Stratigraphy and Sedimentary Processes

The stratigraphy of the Abric Romaní is considered within the talus sedimentary processes of the *Cinglera del Capelló* cliff (Fig. 2.16). The talus of the Abric Romaní occupies an NNE convex-shaped area exposed on the north side of the

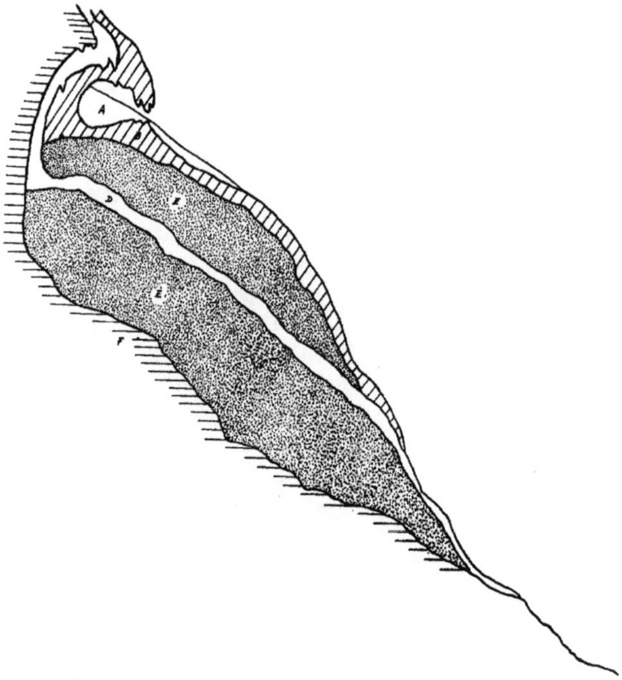

Fig. 2.17 Hypothesis-diagram of the formation of the Abric Romaní into stalagmitic terraces and sandy terraces (Solé Sabaris et al. 1957, from the *Atlas* of Amador Romaní)

travertine mesa of Capellades. The lithostratigraphy of the talus rock is described mainly by the stratigraphical outcrops at the uppermost part of the rockshelter. The abundance of travertines in this part of the sequence was described as a "stalagmitic strata" (Fig. 2.17) (Vidal 1911). These strata are in fact travertines formed by the carbonated incrustation of either vegetation or clastic materials, and speleothems. Although vegetation-travertines can even incrust trees, the most common organisms are algae, mosses and lichens (Julià 1983). One of the common vegetation communities is the *Adiantetea*, which is found in dripping water environments such as cliffs and springs (Folch 1981). Further stratigraphical observations also record clastic deposits in the lithostratigraphy of Pit 1 (Ripoll 1959). The research of the CRPES determined four block-drops, which were used to distinguish 'boundaries' in the stratigraphical succession (Muro et al. 1987).

The morphological evolution of the *Cinglera del Capelló* is characterized by the formation of alveoles and carbonate curtains (the *Capelló*). Most of the sedimentary materials deposited in the talus rock originated from the wall and roof of the Abric Romaní. Only in the upper section of the Abric Romaní is there evidence of a morphological change towards a cave-microenvironment where allochthonous aeolian sediment is trapped. The sedimentary processes in the alveoles, at the foot of the wall, are formed by clastic rhythms of gravel sand calcarenites more or less cemented by carbonate precipitation on dripping domes and flat surfaces. Dripping water close to alveoles can develop stalagmites and stalactites, such as the *covetas*. The alveoles are microforms of rock walls very similar to weathering forms in tafoni, abundant in coastal, arid and semiarid settings (Salomon 1997) (Fig. 2.18). These microforms are the product of several physical weathering processes that are concatenated by water: cryoclasty, haloclasty (salt crystal growth), insolation, wetting and drying. Underneath the carbonate cornice, the sedimentary processes are related to talus cone deposits. These sedimentary processes are characterized by a cyclic repetition of: (1) decimetric to metric biochemical deposits of tuffas and speleothem; (2) decimetric to metric fluvial deposit; (3) rockfall deposits originating from granular disaggregation, block fall and megablock fracture.

The variety of carbonate materials in the lithostratigraphy of the Abric Romaní illustrates the autochthonous character of the sedimentation. At first glance, the mineralogical homogeneity of the calcite makes granulometric estimations difficult, as do the abundant carbonated cementation processes, which can erase the discontinuity surfaces such as those found in dripping dome deposits. The highly stratified and cyclic nature of the stratigraphical sequence indicates numerous paleoenvironmental changes (Fig. 2.19). The outcrop of the *Coveta Nord* section will be discussed in order to highlight the allostratigraphical units or sequences and to identify the distribution and hierarchy of the stratigraphical discontinuities. The sequential analysis of the stratigraphy presented in the following paragraphs (1) indicates timelines for correlation; and (2) suggests the climato-geological implications of the carbonate and siliciclastic sedimentary facies in different time-scales of environmental change. But before describing the sequences of the Abric Romaní, we will provide a summary of the lithostratigraphical research that has been carried out in the field of sedimentary facies.

The stratigraphical sequence of the Abric Romaní has been the subject of a large volume of quality research. The extensive study of the stratigraphical sequence has been helped along by the archeological intervention of Romaní, who was responsible for the excavation of the Romaní well (Bartrolí et al. 1995) (Fig. 2.20). Subsequently, Lumley and Ripoll (1962) developed the first climatic interpretations. At this stage in the research, only the definition of the lithological character of the stratigraphical units was discussed.

The pioneering works of Romaní deal primarily with the top section of the rockshelter. Layers 1 to 13–14 (level J of the current stratigraphy) were described in detail (Bartrolí et al. 1995). In his *Atlas*, A. Romaní goes into the reconstruction of the shelter and its relationship to the cliff

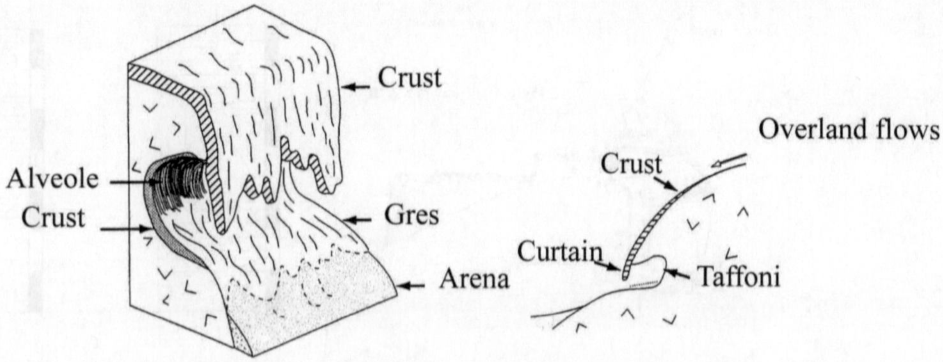

Fig. 2.18 Taffoni, a weathering microform typical of arid zones (Salomon 1997)

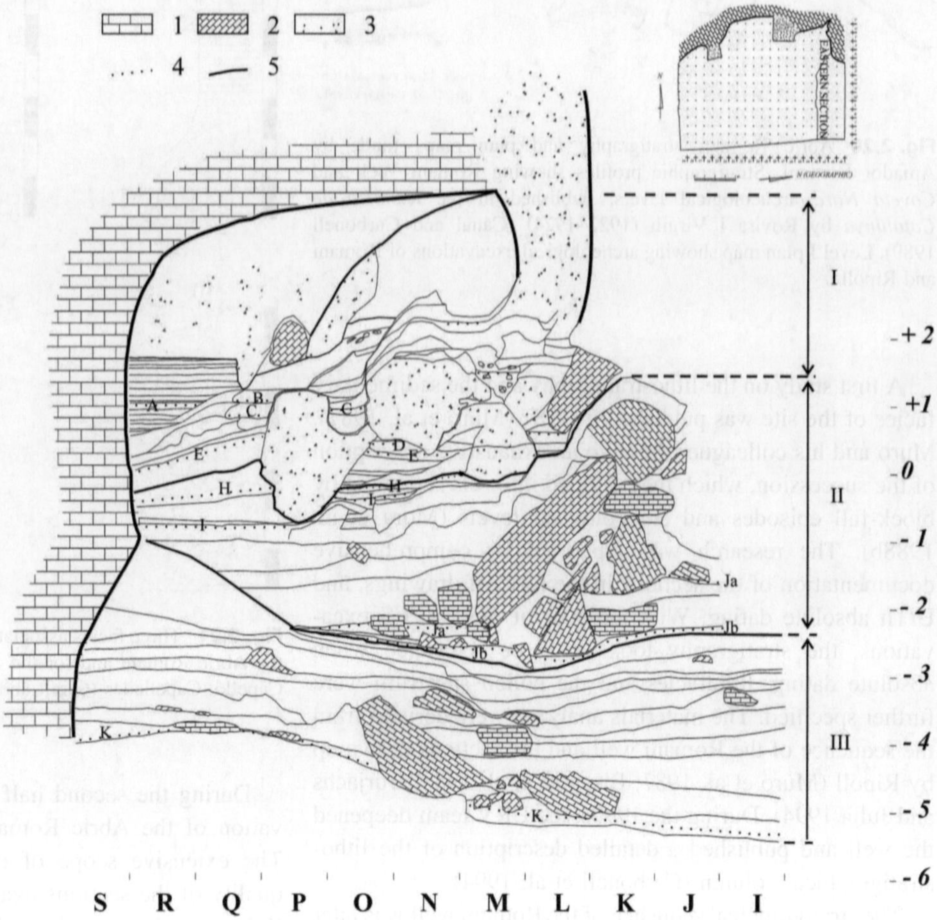

Fig. 2.19 Stratigraphy of the *Coveta Nord* section (SCN). Legend: *1 Cinglera del Capelló*, Abric Romaní wall; *2* blocks fallen form the cornice; *3* stalagmites and stalactites; *4* archeological levels; *5* major discontinuities

(Bartrolí et al. 1995). He also puts forward the hypothesis of three stalagmitic deposits stratified with sandy deposits (Fig. 2.17). Vidal also described the lithological character of the layers, following the outline established by Romaní (Vidal 1911; Vaquero 1992).

In 1957, the Abric Romaní was visited by the V INQUA International Conference (Solé Sabarís et al. 1957), although a detailed lithostratigraphical description and interpretation did not appear until some time later (Ripoll et al. 1965, p. 107). This publication deals with the Ripoll works in Pit 1 (Fig. 2.10), which was six meters deep and and describes the layers in Pit 1 that lay beneath Romaní's layer 13 (Level J in the current stratigraphy). Ripoll and Lumley (1965) divided and described these six meters as comprised of 1, basal cryoclastic gravel; 2, travertine stratum; and 3, superficial red silty sand. The basal cryoclastic gravels comprising strata 12 and 13 correspond to level J of the current stratigraphy. The proposed interpretation for this section of the sequence is characterized by a rapid rate of sedimentation and its correspondence to a cold, wet climate.

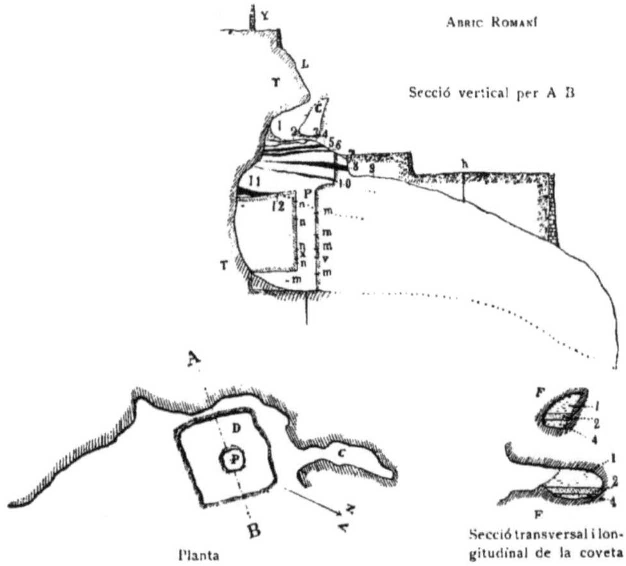

Fig. 2.20 Abric Romani stratigraphy and plan maps made by Amador Romaní. Stratigraphic profiles showing Romani Well and *Coveta Nord* archeological layers, published in the *Història de Catalunya* by Rovira i Virgili (1922–1924) (Canal and Carbonell 1989). Level J plan map showing archeological excavations of Romaní and Ripoll

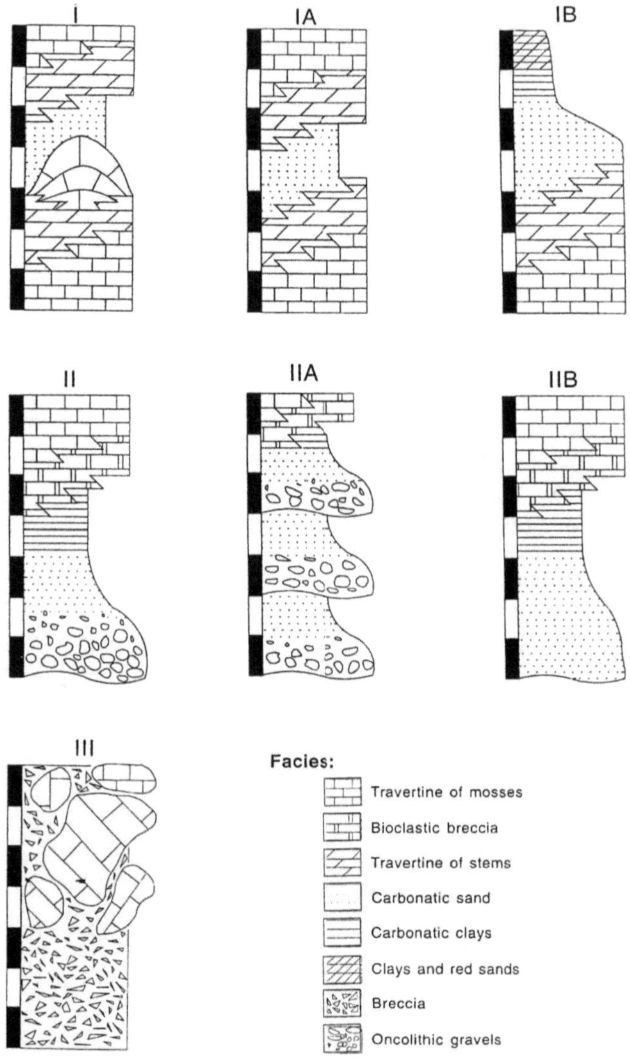

Fig. 2.21 Three facies associations established by Giralt and Julià for the Abric Romaní and for the Abric de la Consagració, Cinglera del Capelló, Capellades (Giralt and Julià 1996, p. 115)

A first study on the lithostratigraphy and the sedimentary facies of the site was published in 1987 (Muro et al. 1987). Muro and his colleagues drew up an exhaustive description of the succession, which they divided into sets separated by block-fall episodes and archeological levels (Mora et al. 1988b). The research was supported by comprehensive documentation of the sections in the form of drawings, and U/Th absolute dating. With the start of the CRPES excavations, the stratigraphy took on more relevance when absolute dating, lithofacies and the pollen spectrum were further specified. The materials analyzed were derived from the sequence of the Romaní well and the sections drawn up by Ripoll (Muro et al. 1987; Bischoff et al. 1988; Burjachs and Julià 1994). During the 1990s, the URV team deepened the well and published a detailed description of the lithostratigraphical column (Carbonell et al. 1994).

The stratigraphical sequence of the Romaní well was later deepened to 16.30 m (Carbonell et al. 1994). The U/Th dating of the bottom facies fixed a date of 70 ka BP (Burjachs and Julià 1994; Giralt et al. 1996). The most detailed lithostratigraphical sequence was obtained during the first years of archeological research conducted by Rovira i Virgili University. In 1994, Santiago Giralt (Carbonell et al. 1994) completed the *Cinglera del Capelló* lithostratigraphy and established four sedimentary series or facies associations. Later, Giralt (Giralt et al. 1996) established three facies associations through lithostratigraphy at the Abric de la Consagració and the Abric Romaní (Fig. 2.21).

During the second half of the 1990s, the surface excavation of the Abric Romaní rapidly transformed the site. The extensive scope of excavated surface improved the quality of the sections available for stratigraphical studies. Below we describe the sedimentary geology of the *Coveta Nord* section (SCN) (Fig. 2.22).

Coveta Nord Lithofacies

Bioconstructions (boundstones) and speleothems. The main sediments are tufa (moss-generated travertines) and stromatolites (calcareous microbial laminae) in laminar sets (Viles and Goudie 1990). Water dripping from the rock-shelter roof formed the dome-like beds.

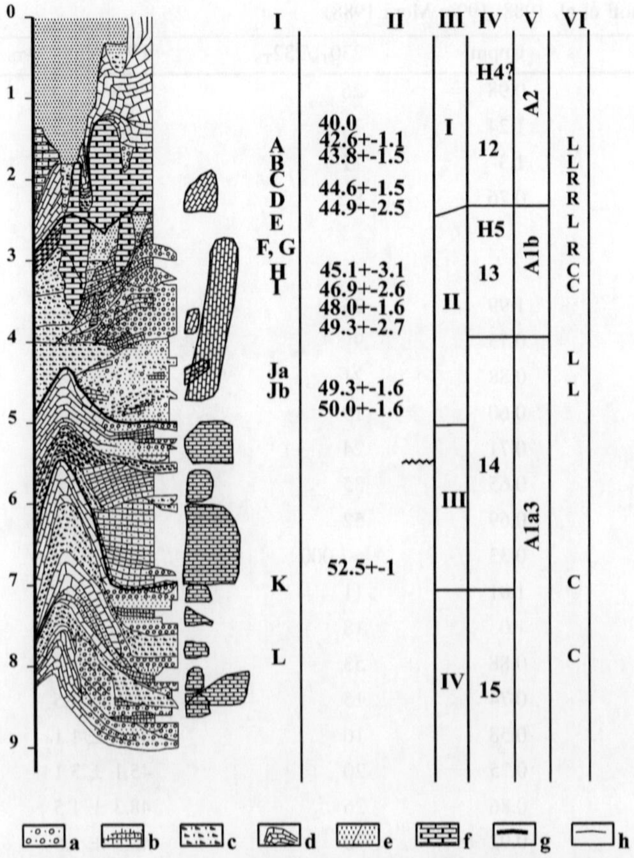

Fig. 2.22 Abric Romaní lithostratigraphy and sequence units in the *Coveta Nord* section. Lithology: **a** travertine and calcarenite conglomerates; **b** moss bioconstruction; **c** angular platy travertine gravels; **d** cemented and laminated algal bioconstructions; **e** siliciclastic calcarenite and calcilutite; **f** stalagmitic dome; **g** sequence limits; **h** boundary surfaces. Comments: *I* archaeological levels; *II* U/Th dates (Bischoff et al. 1988; Mora 1988); *III* sequence numbers; *IV* Groenland stadial and interstadial isotopic stages and Heinrich events within the GISP-2 age-model (Stuiver et al. 2000; Blunier et al. 2001); *V* polinic phases and zones; *VI* short (S) and long (L) time span of human occupation estimated for the archeological levels; *R* indication of relict archeological levels

Gravels, blocs and megablocks. Travertine megablocks, blocks, slabs and gravels found in poorly stratified wedge-shaped beds. Open-structure platy gravels show a typical fine-paired rhythm of scree or talus slope deposits formed by cryoclastism. Color change of the fine fraction from lithochromic grey (10 YR 7/2) to brown indicates decalcification, as can be seen in level J.

Conglomerates, calcarenites and tufas. Round to angular, gravel-sized porous or coated (pisoliths, oncoliths), calcarenites and calcilutites are well stratified beds of horizontal and graded discontinuous (stoneline) sedimentary structures. Conglomerates, sands and tufas fill ribbon-like channel and tabular bedforms. Calcarenite and calcilutite color is yellow to brownish (10 YR 6/4–10 YR 7/4).

Siliciclastic calcarenites and calcilutites. Silt and fine sand silicate minerals (quartz, feldspars, etc.), calcarenites and calcilutites make up irregular and massive to stratified beds. Commonly, this lithofacies has gradational lower contacts and abrupt upper contacts in massive beds of aeolian origin, but laminated and graded beds are also found, mixed with clast to matrix-supported platy gravels and nodules. Thin sections of siliciclastic calcarenite and calcilutite samples taken from the top of the Abric Romaní sequence show a platy microstructure caused by ice segregation (Arteaga et al. 2001b).

Abric Romaní Sequences

Two main lithological units characterize the 20 m thick stratigraphic section. The uppermost unit, composed of three meters of brown siliciclastic calcilutites, was interpreted as an aeolian deposit (Bartroli et al. 1995; Bischoff et al. 1988). The excavation carried out by Amador Romaní removed the greater part of this uppermost unit that originally covered the entire rockshelter. At present, only some small pockets remain adhered to the shelter wall. Below this unit, with a minimum thickness of 17 m, lies the base unit, composed of travertine gravels and blocks, calcarenites and calcilutites, with archeological levels interbedded throughout the entire column (Mora et al. 1988b; Carbonell et al. 1994; Giralt et al. 1996).

The *Coveta Nord* is the profile that has been used to define the stratigraphy of the Abric Romaní (Fig. 2.22). We describe the synthetic stratigraphy of the *Coveta Nord* beginning at the bottom in order to separate cycles of retreat and downward accretion of the carbonate curtain and sediment supply.

In sequence IV, close to the shelter wall, the facies are wedges of gravel alternating with domal bioconstructions. Below the cornice and outside the dripline, facies change to tabular conglomerates and sands, alternating with wedges of gravel.

The basal beds of sequence III onlap towards the shelter wall and consist of moss-generated tufa interbedded with very thin wedges of truncated gravels and blocks. A disconformity in the upper third part of the succession represents a period of erosion.

The bottom of sequence II consists of two wedge-shaped beds of calcarenite and moss-generated tufa interbedded with gravel and megablocks. Close to the shelter wall two dome-like relief forms are present, composed of cemented to open-structure platy fine gravels and calcarenites. A dripline in front the central alveole forms a long spelethemic dome. The upper succession, fining upward below

Table 2.2 U/Th dates of travertine samples from Abric Romani (Bischoff et al. 1988, 1994; Mora 1988)

Archeological location	USGS lab number	Depth	Uppm	$230_{Th}/232_{Th}$	U/Th date, kyrs
Above level A	87-88	30	0.98	26	40.8 ± 1.3[a]
Above level A	87-36f	−10	1.24	7	39.4 ± 1.5
Above level A	87-35f	−20	1.3	11	42.9 ± 1.6
Above level A	87-44	−30	0.76	8.5	39.1 ± 1.5
Above level A	90-AR4	−50			41.8 ± 0.8[a]
Below level A	90-AR3	−70			42.7 ± 1.3[a]
Below level A	87-37f	−60	1.99	25	43.8 ± 1.5
Below level B	87-52	−110	0.73	9	43.4 ± 1.5
Below level B	87-32f	−120	0.88	21	48.1 ± 3
Below level B	87-107	−120	0.60	19	43.1 ± 1.5
Below level B	86-33f	−100	0.71	24	44.0 ± 1.3
Below level B	86-35f	−100	0.65	35	42.4 ± 7.5
Below level C	87-41f	−150	0.69	52	44.2 ± 1.5
Below level C	87-133	−150	0.33	>1,000	44.6 ± 1.5
Below level C	87-126	−180	1.01	11	44.9 ± 2.5
Above level E[b]	03-64	−200	1.0	18	38.1 ± 0.9
Above level E[b]	02-18	−200	0.88	53	36.4 ± 1.4
Above level H[b]	02-22	−290	0.74	15	49.0 ± 1.5
Above level H[b]	03-65	−290	0.58	16	46.5 ± 1.1
Above level I	87-55	−340	0.75	20	45.1 ± 3.1
Above level I[b]	02-21	−340	0.86	26	48.3 ± 1.5
Below level I	87-54	−360	0.92	70	45.3 ± 1.5
Below level I	87-129	−360	0.79	105	47.7 ± 1.6
Below level I	87-66	−360			48.6 ± 2.3
Below level I	87-64	−360			46.9 ± 2.6
Below level I	87-56	−360			46.3 ± 2.4
Below level I	87-60	−380			49.2 ± 3.3
Below level I	87-123	−370	0.96	86	48.0 ± 1.6
Below level I	87-59	−380			47.4 ± 2.5
Below level I	87-57	−390			49.3 ± 2.7
Above level J	87-61	−480	0.55	>1,000	49.3 ± 1.6
Above level J	87-58	−490			49.2 ± 2.9
Above level J[b]	02-19	−500	0.80	34	46.6 ± 1.7
Below level J	87-3	−520	1.82	>1,000	50.0 ± 1.6
Below level J	87-16	−520	1.80	>1,000	50.8 ± 0.8
Above level K[b]	02-24	−600	0.45	13	50.0 ± 2.2
Above level K[b]	07-19	−600	0.68	>1,000	51.6 ± 0.3
Above level L	86-58f	−650	2.00	>1,000	52.0 ± 1.26
Above level L	87-10f	−650	1.84	>1,000	53.0 ± 0.8
Above level L	87-128	−650	1.00	>1,000	51.9 ± 1.6
Above level L	87-4f	−685	0.65	>1,000	52.2 ± 1.6
Above level L[b]	02-20	−700	0.77	146	50.6 ± 2.0
Above level M[b]	02-23	−800	0.98	79	51.8 ± 1.4
Above level M[b]	03-67	−800	0.64	164	61.7 ± 2.2
Above level N	87-17f	−820	1.34	27.5	54.9 ± 1.7
Above level N	87-5f	−820	1.22	27	54.1 ± 1.6

(continued)

Table 2.2 (continued)

Archeological location	USGS lab number	Depth	Uppm	$^{230}Th/^{232}Th$	U/Th date, kyrs
Above level N[b]	02-17	−850	0.82	50	55.8 ± 2.3
Above level O[b]	07-9	−950	0.71	330	54.6 ± 0.4
Above level P[b]	07-10	−980	0.76	149	54.24 ± 0.42
Romaní well	86-65f	−1010	1.56	13	55.0 ± 2.6
Romaní well	87-131	−1010	0.80	30	53.4 ± 1.6
Romaní well	87-129	−1155	0.92	29.5	54.5 ± 1.7
Romaní well	87-11f	−1155	1.46	26	55.5 ± 1.7
Romaní well	86-57f	−1155	1.60	32	60.6 ± 1.7
Romaní well	86-51f	−1180	1.93	40	57.2 ± 0.8
Romaní well	87-12f	−1180	1.79	29	59.6 ± 1.7
Romaní well	87-130	−1180	0.97	27.5	59.0 ± 1.7
Romaní well	86-67f	−1240	1.73	23	63.2 ± 0.9
Romaní well	87-132	−1240	0.91	43	59.0 ± 1.7
Romaní well	86-63f	−1240	1.75	14	60.1 ± 1.8
Romaní well	87-62	−1240	0.93	15	58.0 ± 2.6
Romaní well	87-63	−1240	0.89	16	59.6 ± 2.6
Romaní well[b]	93-12	−1425	0.85	8	60.0 ± 2.0
Romaní well[b]	93-13	−1480	0.65	20	57.0 ± 2.0
Romaní well[b]	93-14	−1580	1.10	121	61.0 ± 2.0
Romaní well[b]	93-15	−1640	0.98	18	70.0 ± 2.0

[a] Mean of several dates
[b] Previously unpublished dates

Table 2.3 ^{14}C (AMS) dates from the Abric Romaní (Bischoff et al. 1994; Carbonell et al. 1994). 2σ calibration has been made using Stuiver et al. (2000)

Level	Lab. Ref.	Radiocarbon age	Cal BP age	Material
A	AA-7395	37290 ± 990	43610–41250	Charcoal
A	AA-8037A	35400 ± 810	42690–38810	Charcoal
A	AA-8037B	37900 ± 1000	44180–41500	Charcoal
A	NZA-1817	28440 ± 650	35330–31010	Charcoal
A	NZA-1818	23160 ± 490	29030–26870	Charcoal
B	NZA-2312	43500 ± 1200	49630–44150	Charcoal
B	AA-7396	29230 ± 530	35760–32680	Charcoal
D	NZA-2313	40680 ± 940	46000–42720	Charcoal
E	NZA-2314	43200 ± 1100	49190–44070	Charcoal
H	NZA-2315	44500 ± 1200	50570–44770	Charcoal
H	NZA-3138	44140 ± 5930	59120 -37840	Charcoal
J	NZA-2316	47100 ± 2100	55910–45350	Charcoal

a second dripline in the shelter wall, shows conglomerates and sands interbedded with channelized calcarenites, stem bioconstructions and stromatolithic crust dipping towards the shelter wall in a talus cone asymmetrical form. These deposits are interbedded with fine beds of open-structure to partially filled platy gravels. The sequence II upper boundary surface contains the first occurrence of silici-clastic calcilutites in the Abric Romaní.

Sequence I contains a basal wedge of blocks and megablocks, capped by two domic biochemical deposits, one of which is rekarstified and the other consisting of stromatolithic laminations in cosets, dipping towards the

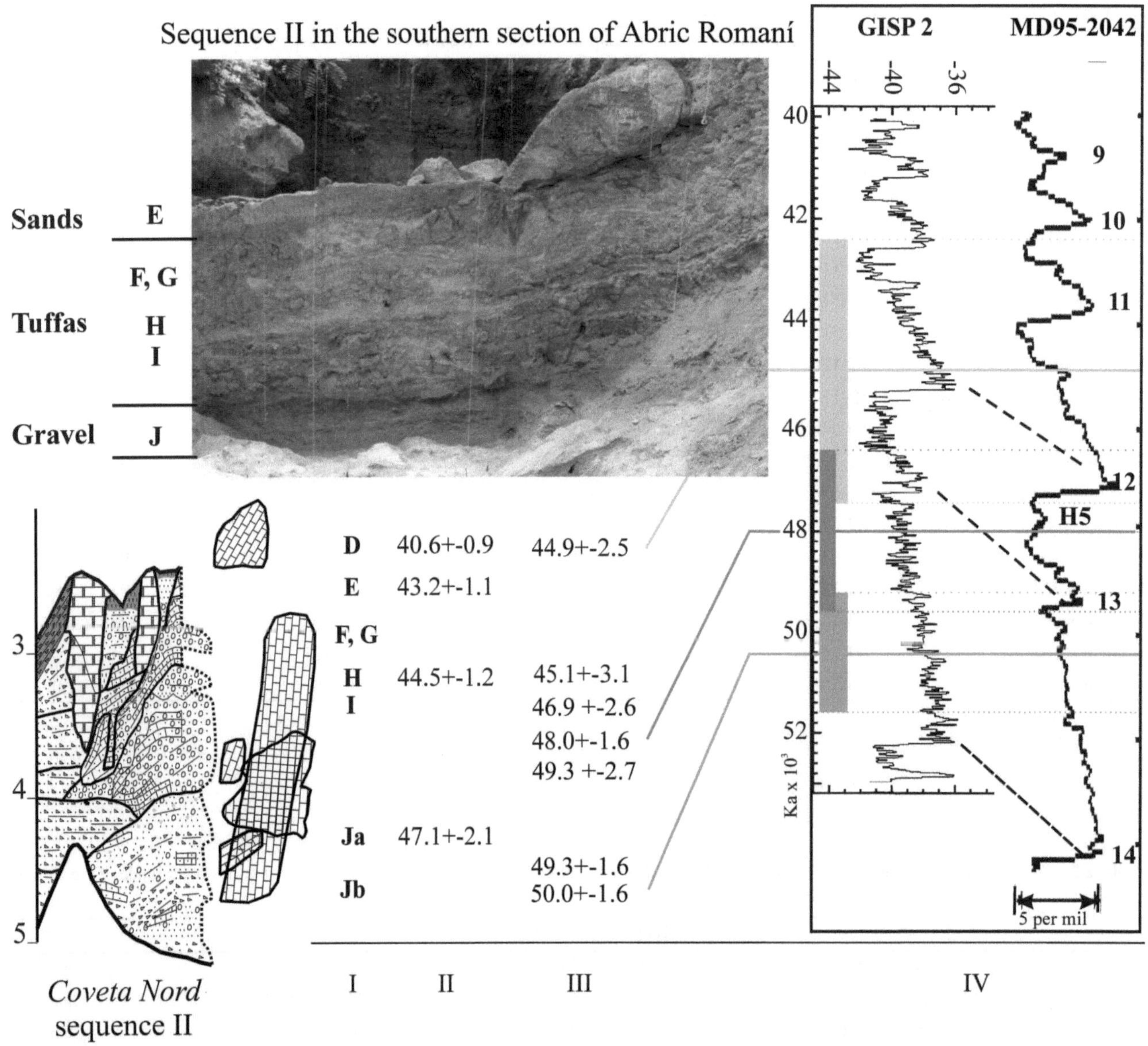

Fig. 2.23 *Coveta Nord* sequence II and level J complex in the radiometric and paleoclimatic time models of environmental change (Stuiver et al. 2000; Shackleton et al. 2004). Legend: *I* archeological levels; *II* radiocarbon (ka); *III* U-series calendar dates; *IV* Dansgaard–Oeschger 9–14 events (D–O) and Heinrich event 5 (HE) chronology in GISP2 and MD95-2042 time ages

shelter wall. Two superimposed channel fills are interstratified with the domic deposits. The upper half of the sequence consists of well stratified and cemented fine gravels, calcarenites and moss-generated tufa in horizontal beds. This sequence is capped by a second layer of siliciclastic calcilutites at the top of the stratigraphy.

Level J Chronology

The moss-generated tufa was found to be an excellent medium for coherent uranium-series dating (Bischoff et al. 1988). An ambitious program of dating has been carried out in the Abric Romaní in close collaboration with the US Geological Survey. Radiometric analysis yielded 66 dates in correct stratigraphic order that define the chronology of the sequence (Table 2.2). The dates indicate that the Romaní sequence covers the period between 70 and 40 ka BP. In addition, a series of ^{14}C AMS dates from the upper archeological levels (A to J) has been also obtained (Table 2.3).

The base of level J has been dated with samples USGS87-3 and USGS87-16 of the Romaní well at 50,000 ± 1600 to 50,800 ± 800 BP respectively (Bischoff et al. 1988). These samples, which come from below the cemented platy wall gravels, were found to significantly predate the fall of the large blocks.

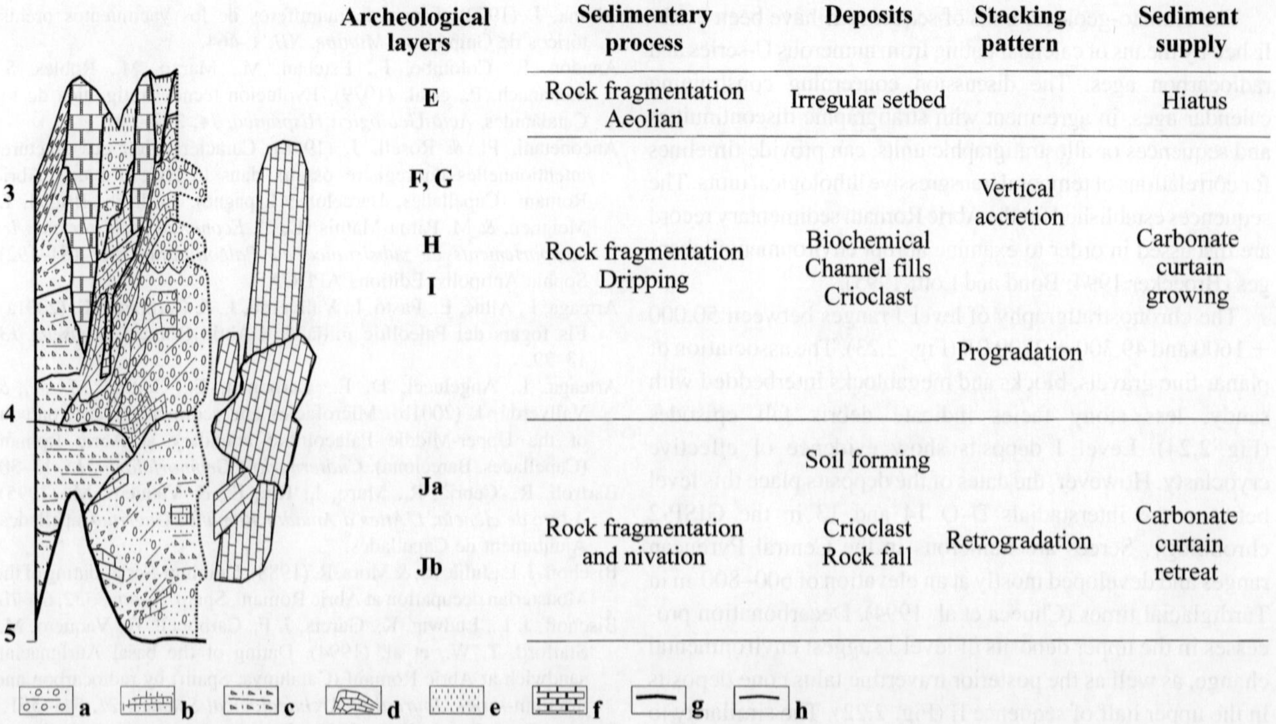

Fig. 2.24 Sedimentological summary of Abric Romaní sequence II. Lithology: **a** conglomerates of travertines and calcarenites; **b** moss bioconstruction; **c** angular, platy travertine gravels and granules; **d** cemented and laminated algal bioconstructions; **e** siliciclastic (*grey hacht*) calcarenite and calcilutite; **f** stalagmitic dome; **g** sequence limits; **h** boundary surfaces

The level J deposits are located underneath a thick dome-like structure of travertine tufa, which has an extended lateral continuity and is U/Th-dated at the base at around 49,300 ± 1600 BP (Fig. 2.23) (Bischoff et al. 1988). This talus cone contains gours (rimstone dams) and conglomerates, and sand tufa beds where archeological levels F–G, H and I have been excavated. Table 2.2 indicates the dates obtained from the first tufa bed of the talus cone, which has been dated at various points. Considerable horizontal coherence was found within the group of dates, indicating the strong reliability of the sampled and dated carbonated materials. Almost all of the dated samples are described as filiform tufa with ramifications (Mora 1988). Their U/Th age indicates a range of between 49,300 ± 1600 BP in the USGS87-61 sample, the only sample described by Mora as 'travertine over blocks' in section PM56, and 49,200 ± 2900 BP in the 'filiform with ramifications' sample (USGS87-58 in the M53-55 section). The sample of travertine over blocks (45a in the PM56 section by Mora) allows us to place this date at the tufa bed that separates Ja and Jb. The USGS87-57 sample, also in section PM56, seals two archeological levels, which can be interpreted as level Ja and level Jsup, dated at 49,300 ± 2700 BP (Mora 1988).

Radiocarbon dating of sublevel Ja has been published at 47,100 ± 210 ^{14}C BP (Carbonell et al. 1994). Archeological levels H and E have been radiocarbon-dated at 44,500 ± 1200 and 43,200 ± 1100 ^{14}C BP respectively (Fig. 2.23) (Table 2.3).

The chronostratigraphy of sublevels Ja, Jb and Jsup illustrates the statistical uncertainty of the calendar dates for correlation with the ice and marine age records (Blunier et al. 2001; Shackleton et al. 2004). A short span of Dansgaard–Oeschger (D–O) interstadials and Heinrich events makes it difficult to use lithofacies as a record of climatic change. Long-term cooling and Bond cycles seem to provide a more adequate time scale for age precision of level J and sequence II calendar dates (Fig. 2.23). Sequence II has a chronostratigraphy that is correlated with the paleoclimatic age models close to the Bond cycle of D–O 14 to Heinrich event 5. However, the sedimentary lithofacies of the Abric Romaní can guide the climato-geological correlation and provide dates that are more constrained. Thus, the information collected on travertines and speleothems has been largely assigned to interstadials and interglacials (Henning et al. 1983). Furthermore, the siliciclastic calcilutite facies indicate periods of aeolian sedimentation, which are characteristic for arid and semiarid seasonal humidity systems. These facies are located for the first time in the Abric Romaní at the top of sequence II, between 45,100 ± 3100 and 44,900 ± 2500 BP, very close to the GISP-2 chronology of Heinrich event 5 despite the range error in the dates (Blunier et al. 2001).

The climato-geologic units of sequence II have been established by means of calendar dating from numerous U-series and radiocarbon ages. The discussion concerning constraining calendar ages, in agreement with stratigraphic discontinuities and sequences or allostratigraphic units, can provide timelines for correlations of temporal transgressive lithological units. The sequences established for the Abric Romaní sedimentary record are discussed in order to examine abrupt environmental changes (Broecker 1994; Bond and Lotti 1995).

The chronostratigraphy of level J ranges between 50,000 ± 1600 and 49,300 ± 2700 BP (Fig. 2.23). The association of planar fine gravels, blocks and megablocks interbedded with sandy, less stony facies indicate debris fall episodes (Fig. 2.24). Level J deposits show evidence of effective cryoclasty. However, the dates of the deposits place this level between the interstadials D–O 14 and 13 in the GISP-2 chronology. Screes are numerous in the Central Pyrenean ranges and developed mostly at an elevation of 600–800 m in Tardiglacial times (Chueca et al. 1994). Decarbonation processes in the upper deposits of level J suggest environmental change, as well as the posterior travertine talus cone deposits in the upper half of sequence II (Fig. 2.22). The similarity to the last generation of taluses of the Central Pyrenees might suggest a climate change of a magnitude of stadial and interstadial scale. This climatic change in sequence II can be correlated, due to its proximity in time, to the D–O 13 chronostratigraphy, which is established between 48 and 50 ka with the MD95-2042 age model (Fig. 2.23). Our provisional conclusion is therefore that the deposits of level J lie within the central part of the Bond cycle made up of D–O 16 and Heinrich event 5, with climatic change similar to that which separates isotopic events 14 and 13 in Greenland stadial-interstadials (Walker et al. 1999).

Acknowledgments We are very grateful to the two anonymous reviewers for their helpful suggestions on the original draft. The authors thank the Museu Molí Paperer of Capellades and especially Raül Bartrolí for providing access to the photographs of the Amador Romaní excavations.

References

Aimene, M., Cáceres, I., Huguet, R., Ibáñez, N., Rosell, J., & Saladié, P. (1996). Procesos de aprovechamiento de la fauna en el Abric Romaní (Capellades, Barcelona). In G. Meléndez, M. F. Blasco, & I. Pérez (Eds.), *II Reunión de Tafonomía y fosilización* (pp. 19–26). Zaragoza: Institución Fernando el Católico.

Aimene, M. (1998). Les différents aspects de l'activité anthropique du niveau E de l'Abric Romaní (Barcelone, Espagne). In J.-Ph. Brugal, L. Meignen, & M. Patou-Mathis (Eds.), *Economie préhistorique: les comportements de subsistance au Paléolithique* (pp. 193–203). Sophia Antipolis: Editions APDCA.

Allué, E. (2002). *Dinámica de la vegetación y explotación del combustible leñoso durante el Pleistoceno Superior y el Holoceno del Noreste de la Península Ibérica a partir del análisis antracológico*. Ph.D. Dissertation, Universitat Rovira i Virgili.

Altuna, J. (1972). Fauna de mamíferos de los yacimientos prehistóricos de Guipúzcoa. *Munibe, XII*, 1–464.

Anadón, P., Colombo, F., Esteban, M., Marzo, M., Robles, S., Santanach, P., et al. (1979). Evolución tecnoestratigráfica de los Catalánides. *Acta Geológica Hispánica, 14*, 242–270.

Anconetani, P., & Rosell, J. (1998). Caractérisation des fractures intentionnelles du registre osseux dans le Niveau I de l'Abric Romaní (Capellades, Barcelone, Espagne). In J.-Ph. Brugal, L. Meignen, & M. Patou-Mathis (Eds.), *Economie préhistorique: les comportements de subsistance au Paléolithique* (pp. 181–192). Sophia Antipolis: Editions APDCA.

Arteaga, I., Allué, E., Pastó, I., Vallverdú, J., & Carbonell, E. (2001a). Els fogars del Paleolític mitjà de l'Abric Romaní. *Cypsela, 13*, 13–29.

Arteaga, I., Angelucci, D. E., Carbonell, E., Courty, M. A., & Vallverdú, J. (2001b). Microfacies and sedimentary environment of the Upper-Middle Palaeolithic boundary at Abric Romani (Capellades, Barcelona). *Cuaternario y Geomorfología, 15*, 15–30.

Bartrolí, R., Cebrià, R., Muro, I., Riu, E., & Vaquero, M. (1995). *A frec de ciència. L'Atles d'Amador Romani i Guerra*. Capellades: Ajuntament de Capellades.

Bischoff, J. L., Julià, R., & Mora, R. (1988). Uranium-Series dating of the Mousterian occupation at Abric Romani, Spain. *Nature, 332*, 68–70.

Bischoff, J. L., Ludwig, K., Garcia, J. F., Carbonell, E., Vaquero, M., Stafford, T. W., et al. (1994). Dating of the basal Aurignacian sandwich at Abric Romaní (Catalunya, Spain) by radiocarbon and uranium-series. *Journal of Archaeological Science, 21*, 541–551.

Blunier, T., & Brook, E. J. (2001). Timing of millenial-scale climate change in Antarctica and Greenland during the last glacial period. *Science, 291*, 109–112.

Bond, G. C., & Lotti, R. (1995). Iceberg discharges into the North Atlantic on millennial time scales during the last glaciation. *Science, 267*, 1005–1010.

Broecker, W. S. (1994). Ocean circulation: An unstable superconveyor. *Nature, 367*, 414–415.

Burjachs, F., & Julià, R. (1994). Abrupt climatic changes during the last glaciation based on pollen analysis of the Abric Romani, Catalonia, Spain. *Quaternary Research, 42*, 308–315.

Burjachs, F., & Julià, R. (1996). Palaeoenvironmental evolution during the Middle-Upper Palaeolithic transition in the NE of the Iberian Peninsula. In E. Carbonell & M. Vaquero (Eds.), *The last Neandertals, the first anatomically modern humans: A tale about the human diversity* (pp. 377–383). Tarragona: Universitat Rovira i Virgili.

Cabanes, D., Allué, E., Vallverdú, J., Cáceres, I., Vaquero, M., & Pastó, I. (2007). Hearth structure and function at level J (50kyr, bp) from Abric Romaní (Capellades, Spain): Phytolith, charcoal, bones and stone-tools. In M. Madella & D. Zurro (Eds.), *Plants, people and places: Recent studies in Phytolithic analysis* (pp. 98–106). Oxford, England: Oxbow Books.

Cáceres, I. (1998). Le niveau I de l'Abric Romaní (Barcelone, Espagne): séquence d'intervention des différents agents et processus taphonomiques. In J.-Ph. Brugal, L. Meignen, & M. Patou-Mathis (Eds.), *Economie préhistorique: les comportements de subsistance au Paléolithique* (pp. 173–180). Sophia Antipolis: Editions APDCA.

Cáceres, I. (2002). *Tafonomía de yacimientos antrópicos en karst. Complejo Galería (Sierra de Atapuerca, Burgos), Vanguard Cave (Gibraltar) y Abric Romaní (Capellades, Barcelona)*. Ph.D. Dissertation, Universitat Rovira i Virgili.

Cáceres, I., Rosell, J., & Huguet, R. (1998). Séquence d'utilisation de la biomasse animale dans le gisement de l'Abric Romaní (Barcelone, Espagne). *Quaternaire, 9*, 379–383.

Calzada, S. (1975). Desembocaba el río Anoia directamente en el mediterráneo. *Boletin de la Real Sociedad Española de Historia Natural*. Especial Centenario, I, 61–70.

Canal, J., & Carbonell, E. (1989). *Catalunya paleolítica*. Girona: Patronat Francesc Eiximenis.

Carbonell, E. (Ed.), (1992). Abric Romaní, nivell H: un model d'estratègia ocupacional al Plistocé Superior mediterrani. *Estrat, 5*, 157–308.

Carbonell, E. (2002). *Abric Romaní Nivell I. Models d'ocupació de curta durada de fa 46.000 anys a la Cinglera del Capelló (Capellades Anoia, Barcelona)*. Tarragona: Universitat Rovira i Virgili.

Carbonell, E., & Castro-Curel, Z. (1992). Palaeolithic wooden artefacts from the Abric Romani (Capellades, Barcelona, Spain). *Journal of Archaeological Science, 19*, 707–719.

Carbonell, E., Cebrià, A., Rosell, J., Sala, R., & Vaquero, M. (1996a). Els conjunts III i IV de l'Abric Romaní (Capellades, campanyes 1988–1994). Una seqüència d'ocupacions del paleolític mitjà. *Tribuna d'Arqueologia, 1994–1995*, 27–38.

Carbonell, E., Cebrià, A., Allué, E., Cáceres, I., Castro, Z., Díaz, R. et al. (1996b). Behavioural and organizational complexity in the Middle Palaeolithic from the Abric Romaní. In E. Carbonell & M. Vaquero, M. (Eds.), *The last Neandertals, the first anatomically modern humans: A tale about the human diversity* (pp. 385–434). Tarragona: Universitat Rovira i Virgili.

Carbonell, E., Giralt, S., & Vaquero, M. (1994). Abric Romaní (Capellades, Barcelone, Espagne): une importante séquence anthropisée du Pléistocène Supérieur. *Bulletin de la Société Préhistorique Française, 91*, 47–55.

Carbonell, E., Sala, R., & Vaquero, M. (1997). Technical system of configuration of denticulated artefacts at Abric Romaní (Capellades, Barcelona). In A. Ramos-Millán & M. A. Bustillo (Eds.), *Siliceous rocks and culture* (pp. 411–417). Granada: Universidad de Granada.

Castro-Curel, Z., & Carbonell, E. (1995). Wood Pseudomorphs from level I at Abric Romani, Barcelona, Spain. *Journal of Field Archaeology, 22*, 376–384.

Chueca, J., Gómez Ortiz, A., Lampre, F., & Peña Monné, J. L. (1994). El periglaciarismo heredado y actual de la Cordillera Pirenaica y del Sistema Costero Catalán. In A. Gómez Ortiz, M. Simón, & F. Salvador (Eds.), *Periglaciarismo en la Península Ibérica, Canarias y Baleares* (pp. 93–118). Granada: Universidad de Granada.

Deguillaume, S. (1987). Analyse palynologique du sommet du remplissage de l'Abri Romani (Catalogne, Spain). Memoire de D.E.A., Museum National d'Histoire Naturelle.

de Lumley, H., & Berard G. (1964). Les industries moustériennes du Bassin de Cabasse (Vallée de l'Issole, Var). *Bulletin du Musée d'Anthropologie préhistorique de Monaco, 11*, 81–119.

de Lumley, H., & Ripoll, E. (1962). Le remplissage et l'industrie moustérienne de l'Abri Romani. *L'Anthropologie, 66*, 1–35.

Escalon de Fonton, M. (1963). Beaucaire. Ioton. *Gallia Préhistoire, VI*, 271–273.

Folch, R. (1981). *La vegetació dels països catalans*. Barcelona: Ketres.

Freeman, L. G., & González Echegaray, J. (1967). La industria musteriense de la cueva de la Flecha. *Zephyrus, XVIII*, 43–61.

Gallart, F. (1980). *Estudi geomorfològic del Penedès, sector Anoia i Riudebitlles*. Barcelona: Universitat de Barcelona.

Gallart, F. (1981). Neógeno superior y Cuaternario del Penedès (Catalunya, España). *Acta geológica hispánica, 16*, 151–157.

Gallart, F. (1985). Nuevos datos sobre la terminación del relleno sedimentario en las fosas meridionales de los Catalánides: Depresiones de Reus-Valls y Penedès. *Acta Geológica Hispánica, 20*, 141–144.

Gallart, F. (1991). Estudi geomorfològic de la Conca d'Òdena. *Estrat, 3*, 9–45.

García Rodrigo, B. (1957). El valle del Anoia. *Memorias comisión Instituto Geológico Provincial, XVI*, 45–80.

Giralt, S., & Julià, R. (1996). The sedimentary record of the Middle-Upper Palaeolithic transition in the Capellades area (NE Spain). In E. Carbonell & M. Vaquero (Eds.), *The last Neandertals, the first anatomically modern humans: A tale about the human diversity* (pp. 365–376). Tarragona: Universitat Rovira i Virgili.

Gutiérrez, M., & Peña, J. L. (1994). Depresión del Ebro. In M. Gutiérrez (Ed.), *Geomorfología de España* (pp. 305–350). Madrid: Editorial Rueda.

Henning, G. J., Grün, R., & Brunnacker, K. (1983). Speleothems, travertines and paleoclimates. *Quaternary Research, 20*, 1–29.

IGME (1973). Mapa Geológico de España. E. 1:50.000. Villafranca del Panades. Madrid: Servicio de Publicaciones, Ministerio de Industria.

IGME (1975). Mapa Geológico de España. E. 1:50.000. Igualada. Madrid: Servicio de Publicaciones, Ministerio de Industria.

Josa, R. (1985). *Estudi cronoseqüencial dels sòls sobre les terrasses de l'Anoia*. Ph.D. Dissertation, Universitat de Barcelona.

Julià, R. (1983). Travertines. In P. Scholle, D. Bebout, & C. Moore (Eds.), *Carbonate depositional environments* (pp. 64–72). Tulsa, OK: The American Association of Petroleum Geologist.

Laplace, G. (1962). Le Paléolithique Supérieur de l'Abri Romani. *L'Anthropologie, 66*, 36–43.

Maluquer, J. (1969). Memoria resumen de las actividades arqueológicas en la zona del distrito universitario de Barcelona durante el año último. *Noticiario Arqueológico Hispánico, X–XI–XII*, 253–257.

Martínez Molina, K. (2002). Actividades concretas y su organización espacial en el interior del yacimiento del paleolítico medio del Abric Romaní (Capellades, Barcelona): análisis funcional de objetos remontados. In I. Clemente, R. Risch, & J. F. Gibaja (Eds.), *Análisis funcional. Su aplicación al estudio de las sociedades prehistóricas* (pp. 111–120). Oxford, England: BAR International Series.

Martínez Molina, K. (2005). *Análisis funcional de las industrias líticas del Pleistoceno Superior. El Paleolítico Medio del Abric Romaní (Capellades, Barcelona) y el Paleolítico Superior de Üçagizli (Hatay, Turquía) y del Molí del Salt (Vimbodí, Tarragona). Cambios en los patrones funcionales entre el Paleolítico Medio y el Superior*. Ph.D. Dissertation, Universitat Rovira i Virgili.

Martínez Molina, K., García, J., Chacón, M. G., & Fernández-Laso, M. C. (2005). Le Paléolithique moyen de l'Abric Romaní. Comportements écosociaux des groupes néandertaliens. *L'Anthropologie, 109*, 815–839.

Martínez Molina, K., & Rando, J. M. (2000). Organización espacial y de la producción lítica en el desarrollo de las actividades durante las ocupaciones del Paleolítico Medio. Nivel Ja del Abric Romaní (Capellades, Barcelona). In *Actas do 3° Congresso de Arqueologia Peninsular* vol. II (pp. 215–229). Porto: ADECAP.

Metter, E. (1978). *Contribution à l'etude palynologique de l'Abri Romani (Catalogne-Espagne)*. Université de Provence: Rapport de D.E.A.

Mora, R. (1988). *El Paleolítico Medio en Catalunya*. Ph.D. Dissertation, Universitat Autònoma de Barcelona.

Mora, R., Carbonell, E., Cebrià, A., & Martínez, J. (1988a). Els sòls d'ocupació a l'abric Romaní (Capellades, Anoia). *Tribuna d'Arqueologia, 1987–1988*, 115–123.

Mora, R., Muro, I., Carbonell, E., Cebrià, A., & Martínez, J. (1988b). Chronostratigraphy of "Abric Romaní". In M. Otte (Ed.), *L'Homme de Néandertal. La Chronologie* (pp. 53–59). Liège: ERAUL.

Morant, N., & García-Antón, M. D. (2000). Estudio de las materias primas líticas del nivel I del Abric Romaní. In *Actas do 3° Congresso de Arqueologia Peninsular* vol. II (pp. 203–212). Porto: ADECAP.

Muro, I., Mora, R., Carbonell, E., & Cebria, A. (1987). Ensayo de interpretación del marco geomorfológico de un yacimiento del Paleolítico Medio Catalán: Abric Romaní (Capellades, Anoia). *Cypsela, VI*, 125–131.

Muro, J. I., Mora, R., Martínez, J., Bergadà, M., Bartrolí, R., & Bartrolí, R. (1989). Aproximació històrica al paleolític de

Catalunya: Amador Romaní i Guerra i els seus treballs en el camp de l'arqueologia (1873–1930). *Estrat, 1*, 7–14.

Pastó, I., Allué, E., & Vallverdú, J. (2000). Mousterian Hearths at Abric Romaní, Catalonia (Spain). In C. B. Stringer, R. N. E. Barton, & J. C. Finlayson (Eds.), *Neanderthals on the edge* (pp. 59–67). Oxford, England: Oxbow Books.

Pericot, L. (1960–1961). Memoria de las actividades del Servicio de Investigaciones Arqueológicas en el año 1960. *Ampurias, XXII–XXIII*, 369–375.

Pericot, L. (1964). Delegación de zona del Servicio Nacional de Excavaciones arqueológicas (Barcelona). Memoria resumen de la labor realizada en 1960–61. *Noticiario Arqueológico Hispánico, VI*, 278–295.

Riba, O., Regunat, S., & Villena, J. (1983). Ensayo de síntesis estratigráfica y evolutiva de la cuenca del Ebro. In J. A. Comba (Ed.), *Geología de España* (pp. 131–159). Madrid: Libro Jubilar J.M. Rios.

Ripoll, E. (1958). Excavaciones en el Abrigo Romaní. *San Jorge. Revista Trimestral de la Diputación de Barcelona, 30*, 14–15.

Ripoll, E. (1959). Excavaciones en el Abrigo Romani (Capellades, Barcelona). *Ampurias, XXI*, 47–248.

Ripoll, E., & de Lumley, H. (1965). El Paleolítico Medio en Catalunya. *Ampurias, XXVI–XXVII*, 1–70.

Rosell, J. (2001). *Patrons d'aprofitament de les biomasses animals durant el Pleistocè inferior i mig (Sierra de Atapuerca, Burgos) i superior (Abric Romaní, Barcelona)*. Ph.D. Dissertation, Universitat Rovira i Virgili.

Sala, M. (1994). Sistema Costero Catalán. In M. Gutiérrez (Ed.), *Geomorfología de España* (pp. 287–304). Madrid: Editorial Rueda.

Saladié, P., & Aimene, M. (2000). Análisis zooarqueológico de los niveles superiores del Abric Romaní (Cataluña): actividad antrópica. In *Actas do 3° Congresso de Arqueologia Peninsular* vol. II (pp. 189–201). Porto: ADECAP.

Salomon, J. N. (1997). Les influences climatiques sur la géomorphologie karstique: exemple des milieux tropicaux et arides. *Quaternaire, 8*, 107–117.

Shackleton, N. J., Fairbanks, R. G., Tzu-chien, C., & Parrenin, F. (2004). Absolute calibration of the Greenland time scale: Implication for Antarctic time scales and for dC^{14}. *Quaternary Science Reviews, 23*, 1513–1522.

Solé Sabaris, L., Virgili, C., & Ripoll Perelló, E. (1957). Les matériaux quaternaires de la région pré-littorale aux environs de Capellades. In *V Congrès International INQUA. Livret guide de l'excursion B1* (pp. 5–38). Madrid-Barcelone: INQUA.

Solé Sabarís, L. (1958–1964). *Geografia de Catalunya*. Barcelona: Aedos.

Stuiver, M., Reimer, P. J., & Reimer, R. W. (2000). CALIB 4.3. [WWW program and documentation]URL:<http://www.calib.org>, using the data set given by M. Stuiver, P. J. Reimer, E. Bard, W. E. Beck, G. S. Burr, K. A. Hughen, B. Kromer, F. G. McCormac, J. v.d. Plicht & M. Spurk (1998): INTCAL98 radiocarbon age calibration 0-24,000 BP, *Radiocarbon, 40*, 1041–1083.

Vallverdú, J., Allué, E., Bischoff, J. L., Cáceres, I., Carbonell, E., Cebrià, A., et al. (2005a). Short human occupations in the Middle Palaeolithic level i of the Abric Romaní rock-shelter (Capellades, Barcelona, Spain). *Journal of Human Evolution, 48*(2), 157–174.

Vallverdú, J., Vaquero, M., Courty, M.-A., & Carbonell, E. (2005b). Procesos sedimentarios rápidos y ocupaciones humanas poco frecuentes. La sección del Pleistoceno Superior de la Coveta Nord del Abric Romaní (Capellades, comarca de l'Anoia, Barcelona). In M. Santonja, A. Pérez-González, & M. J. Machado (Eds.), *Geoarqueología y patrimonio en la Península Ibérica y el entorno mediterráneo* (pp. 319–331). Soria: ADEMA.

Vaquero, M. (1992). Processos de canvi tecnològic al voltant del 40.000 BP. *Continuïtat o ruptura. Estrat, 5*, 9–156.

Vaquero, M. (1997). *Tecnología lítica y comportamiento humano: organización de las actividades técnicas y cambio diacrónico en el Paleolítico Medio del Abric Romaní (Capellades, Barcelona)*. Ph.D. Dissertation, Universitat Rovira i Virgili.

Vaquero, M. (1999a). Intrasite spatial organization of lithic production in the Middle Palaeolithic: The evidence of the Abric Romaní (Capellades, Spain). *Antiquity, 73*, 493–504.

Vaquero, M. (1999b). Variabilidad de las estrategias de talla y cambio tecnológico en el Paleolítico Medio del Abric Romaní (Capellades, Barcelona). *Trabajos de Prehistoria, 56*, 37–58.

Vaquero, M. (2005). Les stratégies de transport d'outils dans un contexte résidentiel: un exemple du Paléolithique moyen. In D. Vialou, J. Renault-Miskovsky, & M. Patou-Mathis (Eds.), *Comportements des hommes du Paléolithique moyen et supérieur en Europe: territoires et milieux* (pp. 121–132). Liège: ERAUL.

Vaquero, M. (2008). The history of stones: Behavioural inferences and temporal resolution of an archaeological assemblage from the Middle Palaeolithic. *Journal of Archaeological Science, 35*, 3178–3185.

Vaquero, M., & Pastó, I. (2001). The definition of spatial units in Middle Palaeolithic sites: The Hearth-related assemblages. *Journal of Archaeological Science, 28*(11), 1209–1220.

Vaquero, M., Alegre, P., Galindo, E., Martínez, K., Martorell, S., Plana, X., et al. (1997). Organización espacial de la producción lítica en un yacimiento del Paleolítico Medio: niveles I y J del Abric Romaní (Capellades, Barcelona). In R. de Balbín & P. Bueno (Eds.), *II Congreso de Arqueología Peninsular. Tomo I.- Paleolítico y Epipaleolítico* (pp. 35–49). Zamora: Fundación Rei Afonso Henriques.

Vaquero, M., García-Antón, D., Mallol, C., & Morant, N. (1998). L'organisation spatiale de la production lithique dans un gisement du Paléolithique moyen: le niveau Ja de l'Abric Romaní (Capellades, Barcelona, Espagne). In *XIII U.I.S.P.P. Congress Proceedings - Forli, 8–14 September 1996* (pp. 777–782). Forli: ABACO Editions.

Vaquero, M., Rando, J. M., & Chacón, G. (2004). Neanderthal spatial behavior and social structure: Hearth-related assemblages from the Abric Romaní Middle Palaeolithic site. In N. J. Conard (Ed.), *Settlement dynamics of the Middle Paleolithic and Middle Stone Age* (Vol. 2, pp. 367–392). Tübingen: Kerns Verlag.

Vaquero, M., Chacon, G., Fernández, C., Martínez, K., & Rando, J. M. (2001a). Intrasite spatial patterning and transport in the Abric Romaní Middle Paleolithic site (Capellades, Barcelona, Spain). In N. J. Conard (Ed.), *Settlement dynamics of the Middle Paleolithic and Middle Stone Age* (pp. 573–595). Tübingen: Kerns Verlag.

Vaquero, M., Vallverdú, J., Rosell, J., Pastó, I., & Allue, E. (2001b). Neandertal Behavior at the Middle Palaeolithic site of Abric Romaní, Capellades, Spain. *Journal of Field Archaeology, 28*(1–2), 93–114.

Vidal, L. M. (1911). Abric Romaní, Estació Agut, Cova de l'Or o dels Encantats. Estacions Prehistòriques de les èpoques musteriana, magdaleniana i neolítica a Capellades i Sta. Creu d'Olorde. *Ann. Ins. d'Estudis Catalans, IV*, 267–302.

Viles, H. A., & Goudie, A. S. (1990). Tufas, travertines and allied carbonate deposits. *Progress in Physical Geography, 14*(1), 19–41.

Virgili, C. (1960). Estudio de los minerales pesados del cuaternario continental del litoral catalán. *Anales de Edafología y Agrobiología, XIX*, 229–240.

Walker, M. J. C., Björck, S., Lowe, J. J., Cwynar, L. C., Johnsen, S., Knudsen, K. L., et al. (1999). Isotopic 'events' in the GRIP ice core: A stratotype for the Late Pleistocene. *Quaternary Science Reviews, 18*, 1143–1150.

Chapter 3
Spatial Patterns in Level J

Pablo Sañudo, Josep Vallverdú-Poch and Antoni Canals

Abstract The excavation strategy at the Abric Romaní has been largely guided by spatial criteria. Moreover, spatial data are fundamental for explaining the archeological assemblages and this spatial perspective is present in most chapters of this book, especially in those dealing with lithic and faunal remains. The primary goal of this chapter is to show a preliminary outline of spatial patterns in level J, advancing some issues that will be further developed in the following chapters. The spatial distribution of both natural and cultural features is presented, paying especial attention to hearths, which played a central role in the spatial organization of Neanderthals. An archeostratigraphic analysis is also carried out, allowing the two main stratigraphic units of level J—sublevels Ja and Jb—to be distinguished.

Keywords Spatial patterns • Hearths • Archeostratigraphy • Size sorting

Introduction

Research conducted in the Abric Romaní in recent years has yielded important data that has furthered our ability to reconstruct the spatial patterning and behavioral organization of Middle Paleolithic human groups. The occupational pattern of the Abric Romaní consists of spatial units featuring the typical characteristics of hearth-related assemblages, around which most daily activities were carried out (Carbonell 1992, 2002; Carbonell et al. 1996; Vaquero and Pastó 2001; Vaquero et al. 2001; Vallverdú et al. 2005).

In this chapter we present the spatial analysis of level J—a level with a degree of occupational reiteration higher than the other levels of the shelter studied so far and one which comprises an undetermined number of occupational events. The spatial patterning of level J is well preserved due to the accumulative character of the cultural features (Vallverdú and Courty 2012).

The distribution of the combustion structures and scatters of archeological remains indicate that there are at least three general types of spatial units located in the internal, central and external topographical zones. The spatial patterning of archeological items such as combustion structures, blocks and slabs and other natural features, as well as the distances to the rockshelter wall or the dripline, might be a fundamental requisite for site structure research (Kent 1991). Combustion structures of Middle Paleolithic age in extensive excavations are scarce and may provide empirical evidence for recognizing the organizational properties of the archeological record (Binford 1983, 1996). This empirical combination of structures (cultural and topographical) and accumulations of lithics and bones provides two preliminary topics of discussion worth exploring: (1) the archeological visibility of spatial units and activity episodes; and (2) the analogies with spatial patterns and activity areas of modern foragers as a guideline for site structure research. The spatial patterning and temporal analysis of activity episodes will be revisited and further developed in subsequent chapters.

P. Sañudo (✉) · J. Vallverdú-Poch · A. Canals
Institut Català de Paleoecologia Humana i Evolució Social (IPHES), Universitat Rovira i Virgili (URV), C/Escorxador s/n, 43003 Tarragona, Spain
e-mail: pablo.sanudo@bolomor.com

J. Vallverdú-Poch
e-mail: josep@prehistoria.urv.cat

A. Canals
e-mail: acanals@prehistoria.urv.cat

Primary Characteristics of the Surfaces of Level J and Sublevels Ja and Jb

The occupied floors of level J contain various different natural features (blocks, megablocks and the rockshelter wall) that influence the occupation pattern. The main sedimentary form

E. Carbonell i Roura (ed.), *High Resolution Archaeology and Neanderthal Behavior: Time and Space in Level J of Abric Romaní (Capellades, Spain)*, Vertebrate Paleobiology and Paleoanthropology,
DOI: 10.1007/978-94-007-3922-2_3, © Springer Science+Business Media B.V. 2012

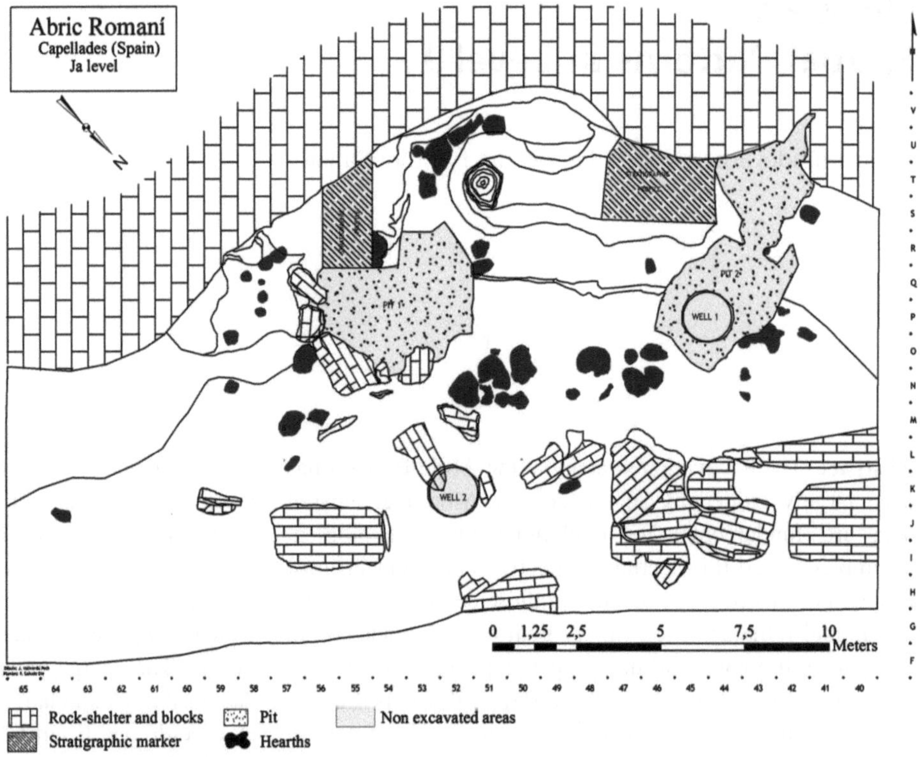

Fig. 3.1 The excavated surface of level J showing the main previously excavated pits and stratigraphic outcrops

of level J is made up of outcrops of megablocks (up to 3 m along the major axis) fallen prior to the human occupation and aligned with the dripline of the rockshelter. Other relevant morphologic features are the Coveta Ripoll (Ripoll's Small Cave) and its speleothem in the center of the rockshelter concavity. This small cave is a karstic cavity with a stalactite at the front. The cavity is separated from the rest of the level J surface area and is only accessible by means of a ramp near Pit 1.

The travertine megablocks, which have fallen from the ceiling of the shelter, rise up at different heights in level J. These megablocks create a natural division between the central and external zones only in the northern area of the excavated surfaces (Fig. 3.1). The inner and central zones are covered by the roof of the shelter, creating a protected area suitable for inhabitation. The internal zone has a limited surface area available for human activities. Carbonate ramps and dripping domes border the three available flat surfaces in the interior zone. A large section of the archeological record on the northern side of the inner zone was removed during previous excavations. The central zone offers a wide surface area that dips towards the southeast. This area is bordered by the dripline and is constrained by the travertine megablocks on the northern side of the excavated surface, where the richness of the archeological record decreases. The central zone was affected by previous excavations in front of the carbonate ramp of the *Coveta Ripoll* (Pit 1) and in the north of the excavated areas (around Romaní well 1). The exterior zone has irregular stone surfaces made up of large travertine fragments and has a moderate NNE slope (Fig. 3.2).

Sublevel Ja was detected in the middle and northern areas of this central zone, whereas J and sublevels Ja–Jb were found in the inner zone, the southern area of the central zone and the external zone. Level J can therefore be defined as a set of high-resolution archeostratigraphic units in front of the *Coveta Ripoll*. The total area of sublevel Ja is 331.28 m^2, although the excavated surface was reduced to 289.85 m^2 due to the former excavations and the stratigraphic columns in the inner zone (Fig. 3.1).

Figure 3.3 shows a representation of the travertine blocks recorded in sublevels Ja and Jsup during fieldwork. Aligned medium-sized blocks were documented below the dripline of the rockshelter. The blocks are aligned in overlapping bow shapes which have been described in rock cliffs stabilized by vegetation (Bertran and Texier 1999). The high density of small blocks in the northern sector of the excavated area indicates that the travertine blocks naturally fell along the dripline. Therefore, evidence of modifications to the location of the travertine blocks below the dripline of the rockshelter is difficult to pinpoint. In the south, the band from column 52 on continues to be assigned to sublevel Ja, though it is possible that the record of the blocks in this area is over-represented in the plots. Some of the blocks in the

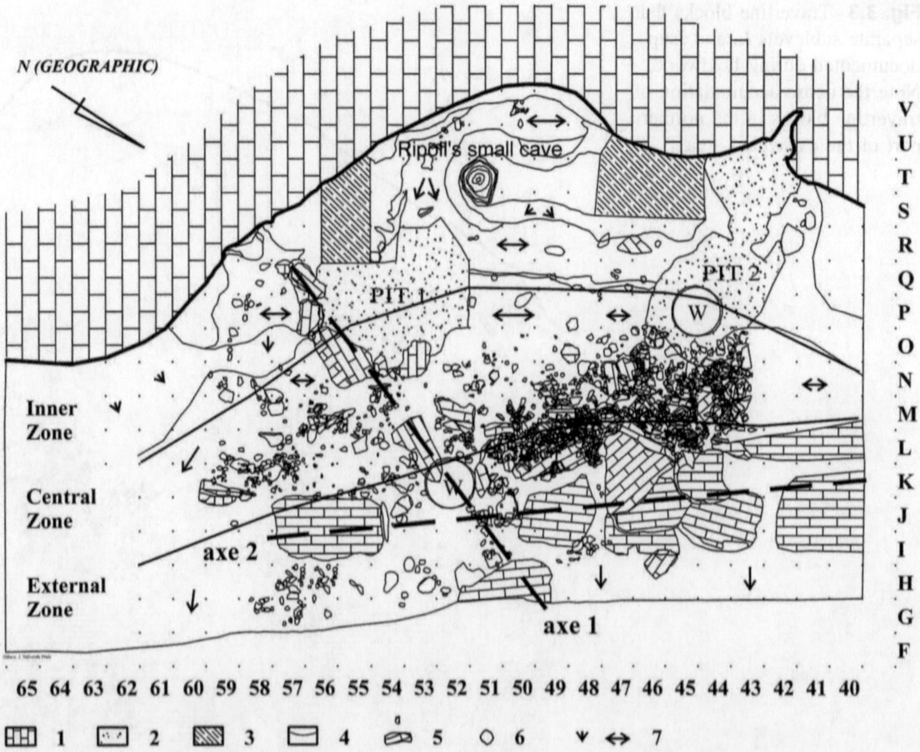

Fig. 3.2 Topography of the level and geomorphical units. *1* Abric Romaní cliff. *2* Previous archeological fieldwork pits and wells. *3* Stratigraphic marker. *4* Main topographic discontinuities. *5* Boulders. *6* Compact travertine slab. *7* Abrupt to flat surfaces

southern area may belong to the Ja–Jb separation. The travertine blocks in the central and inner zone may have been moved by humans. However, some small block alignments documented in the inner zone of sublevel J–Ja seem to suggest the development of fissures following dissolution planes in the travertine cliff (Fig. 3.2).

The Archeostratigraphic Definition of Level J

During fieldwork, level J was subdivided according to geological criteria into three sublevels: Jsup, Ja and Jb. The division between Jsup and Ja was established by the presence of a thin sterile layer, only distinguishable in some areas of the shelter, while the Ja–Jb subdivision was established by a thin travertine layer present across the entire surface of the central space. In order to corroborate these fieldwork definitions an archeostratigraphic analysis was performed which focused on defining the occupational events discernible in level J. Archeostratigraphic analysis establishes the diachronic relationship between cultural deposits by delimitating continuous sterile layers (Canals 1993). Cultural deposits may present a vertical stratification of archeological items, formed by the sum of occupational events in the same space. These events are difficult to separate in homogeneous geological deposits. Archeostratigraphy allows occupational events in homogeneous deposits to be separated through the identification of sterile layers.

The entire surface of level J was analyzed using profiles with a thickness of 25 cm in order to identify archeostratigraphic features throughout the level. After a first observation, 6 cross-sectional, 10 longitudinal and 12 short profiles were chosen for a deeper and more detailed analysis (Fig. 3.4). Short sections were used to examine certain processes in some characteristic areas.

The analysis of the profiles reveals the existence of two archeostratigraphic units, separated by a continuous sterile layer (Fig. 3.5). This sterile layer is only identifiable in the central zone (rows L–P and columns 40–53), corresponding to the thin travertine layer identified during fieldwork. These archeostratigraphic units are not distinguishable outside of the central area, forming an occupational palimpsest in the peripheral areas. Two units have been defined from the results of the archeostratigraphic analysis: Ja and Jb. These units are clearly separated in the central zone and joined at the periphery, with good temporal resolution found only in the central zone. In the peripheral areas it is not possible to separate the two units. Consequently, each of these units has been defined as the corresponding part of the record of the central area and the total record of the peripheral areas.

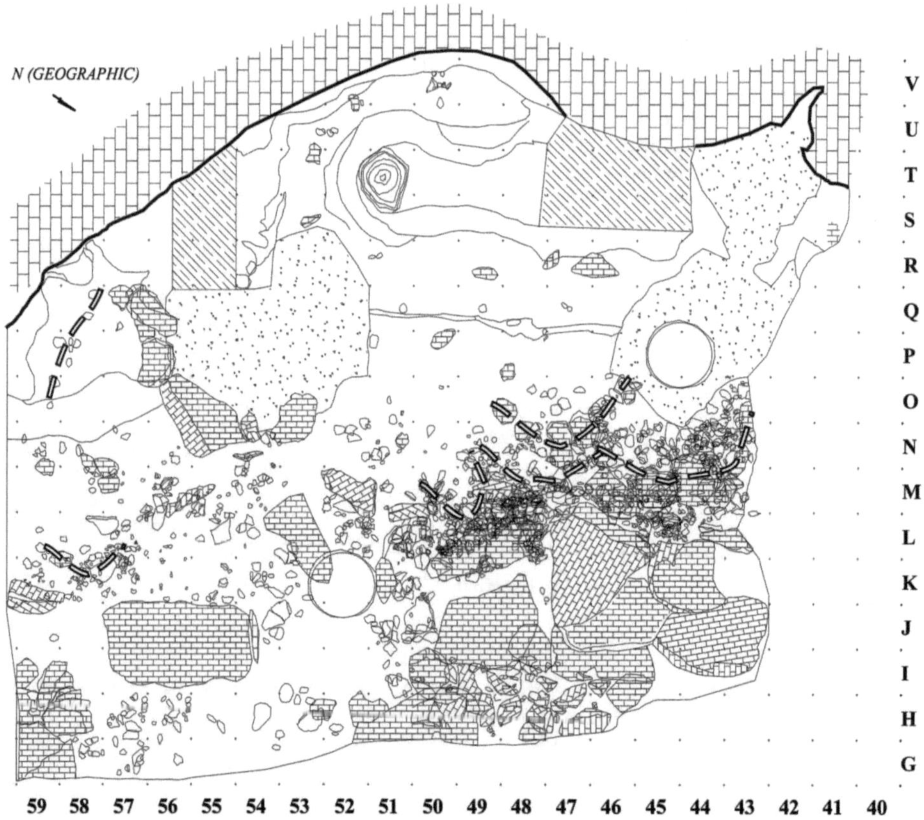

Fig. 3.3 Travertine blocks that separate sublevels Ja and Jsup documented during fieldwork. Note the dense accumulation of travertine blocks in the northern part of the excavated area

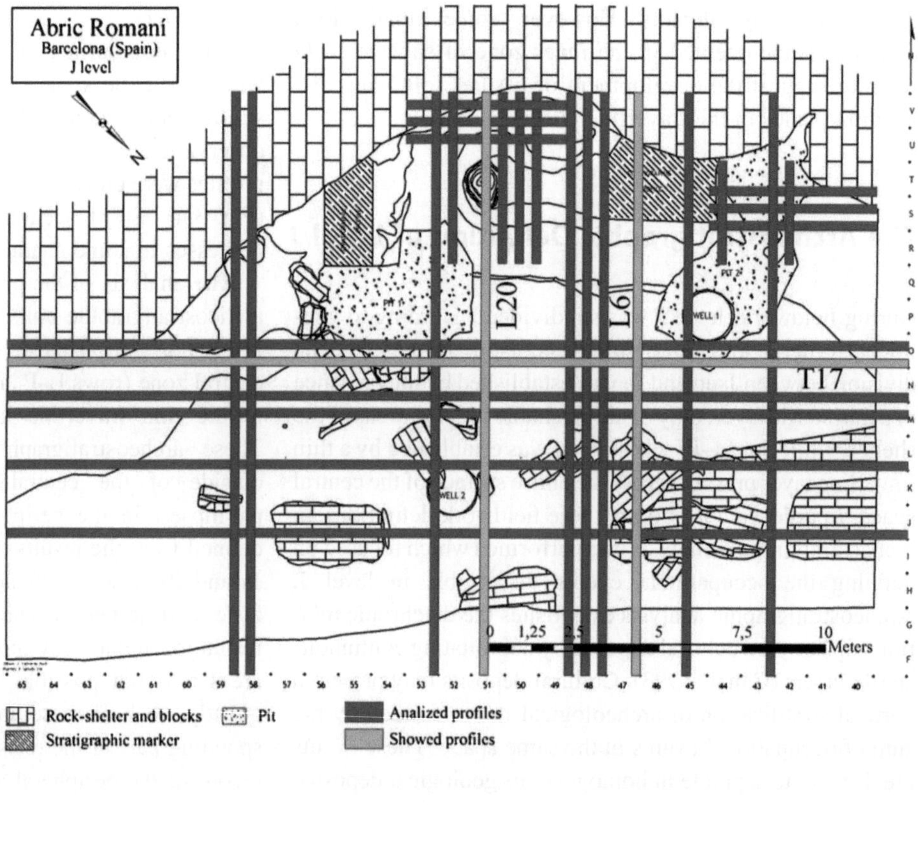

Fig. 3.4 Location of the archaeostratigraphic profiles observed in the level J complex. Selected archaeostratigraphic profiles T17, L6 and L20

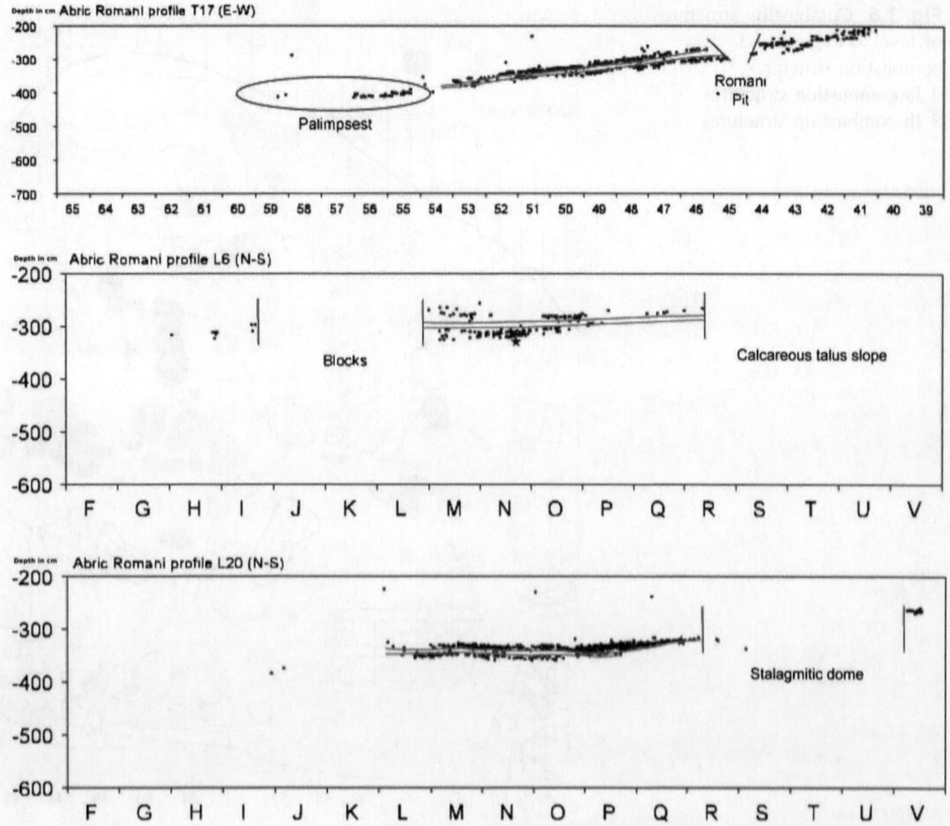

Fig. 3.5 Archaeo-stratigraphic profiles T17, L6 and L20 showed in the Fig. 2.4

Spatial Patterns of the Cultural Remains in Sublevel Ja

Groups of Combustion Structures and Burned Travertine Blocks

We have recorded a large number of combustion structures (n = 61) in level J, which demonstrates that fire use is a generalized behavior in the occupational patterns of the Abric Romaní (Fig. 3.6). The thermal alterations identified were verified in different ways, and are notably present in faunal remains, lithic artifacts and sediments. The evidence of fire use in sediments consists of the presence of microstratified carbonaceous and burned facies called combustion structures (Fig. 3.7). These thermal modifications and carbonaceous (charcoal-sediment-rich) incorporations found in the sedimentary record and archeological remains are well differentiated from natural oxidation and wild fires.

Combustion structures have well delimited forms and carbonaceous compositions and are always related to archeological remains (Fig. 3.8). The combustion structures of level J can be classified as flat and with or without stones (Perlès 1976).

However, the burned sedimentary materials of sublevel Ja consist of (1) travertine blocks and slabs (Fig. 3.9), and (2) the combustion structures of the internal and central zones (Fig. 3.10). These combustion structures are characterized by (a) rectangular or circular shapes of different sizes or (b) small, irregular shapes.

The combustion structures are stratified in the inner zone because levels Ja and Jb are indistinguishable, forming a micro-palimpsest (Fig. 3.11). In the central zone these combustion structures feature burned blocks and are microstratified, indicating an occupational palimpsest (Fig. 3.10). Some combustion structures of the central zone are small and irregularly shaped, which suggests the existence of secondary waste deposits whose microfacies are detailed in Vallverdú and Courty (2012).

The burned travertine blocks of sublevel Ja are found in groups within, around and outside combustion structures (Fig. 3.9). Planimetry shows the location of these burned blocks in sectors K–L/57–58 (Bb1, burned block 1); M56–57 (Bb2); N–O/50–51 (Bb3); N–O/46–47 (Bb4); and L/44–45 (Bb5). Burned blocks and combustion structures logically show the same spatial location in sublevel Ja, except for those located in Bb1 and Bb5 (Fig. 3.9). These groups suggest that combustion structure waste was carried to the boundary between the central and external zone and discarded there.

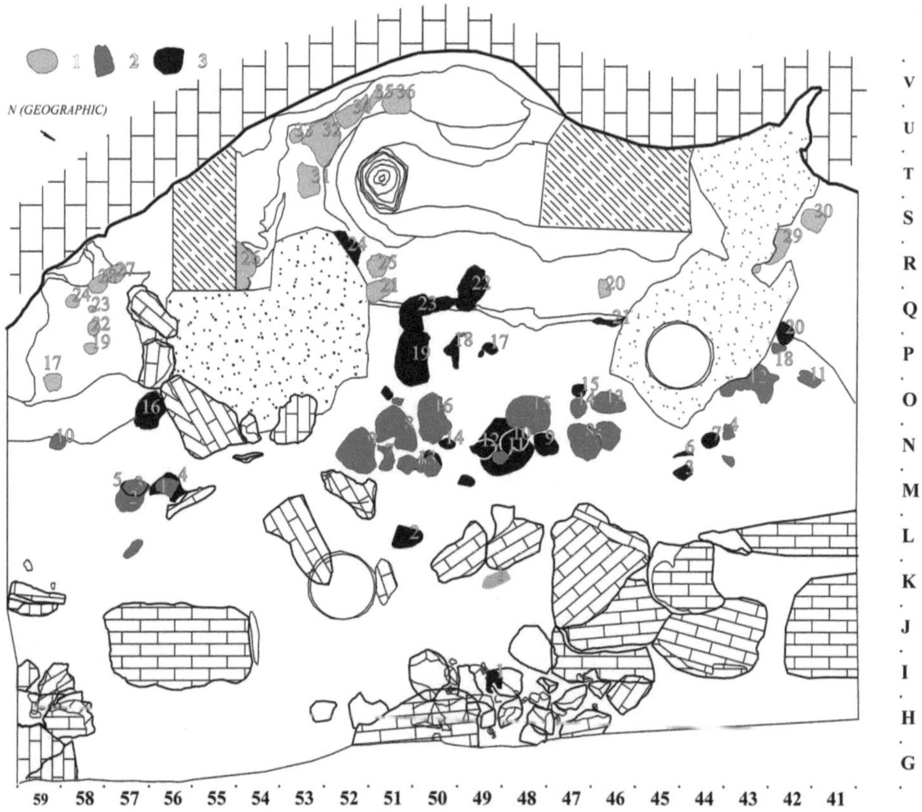

Fig. 3.6 Combustion structures of level J. Legend: *1* J combustion structures; *2* Ja combustion structures; *3* Jb combustion structures

Fig. 3.7 Combustion structures viewed from south to north. In the center, the combustion structure in square N52

The distances between the groups of combustion structures and the shelter wall may represent a productive approach to finding an analogy with behaviors observed in caves and rockshelters inhabited by modern human groups (Galanidou 2000; Henry 2003; Henry et al. 2004). The hearth location pattern in level J shows a bimodal distribution (Fig. 3.11). The asymmetry, with one primary median near the rockshelter wall, is caused by the over-representation of combustion activity areas in the level J geological palimpsest in the internal zone. Figure 3.11

Fig. 3.8 Vertical view of combustion structure N52

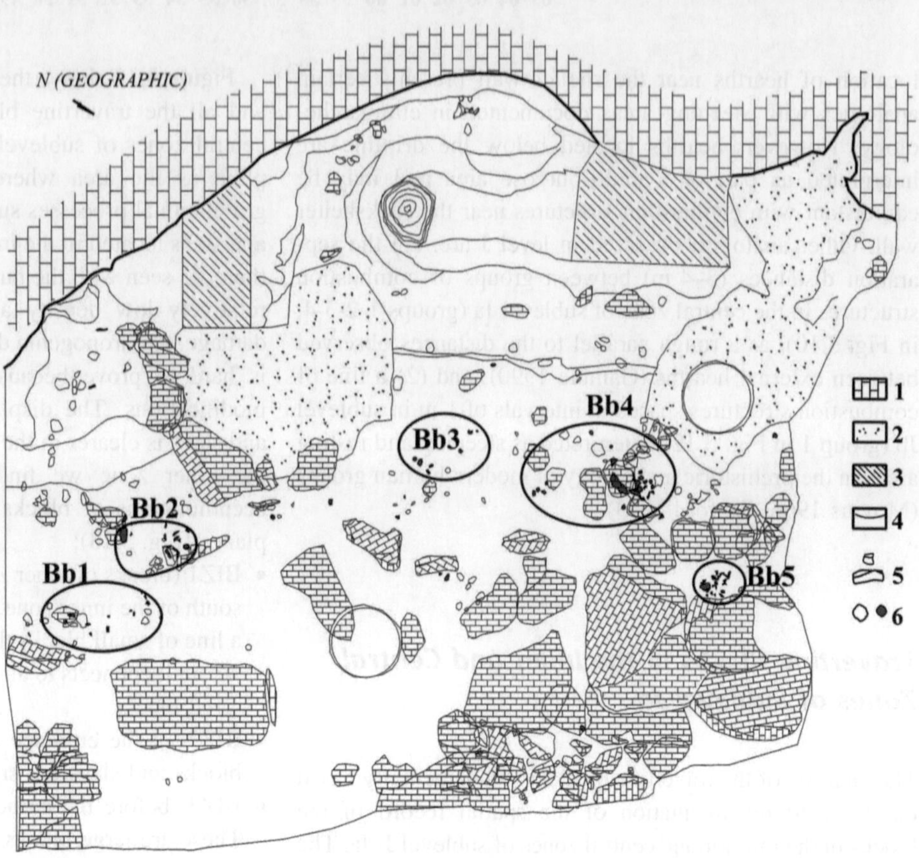

Fig. 3.9 Groups of burned travertine blocks. Burned block numbers (B1) shaded in *grey*. Legend: *1* Romaní cliff; *2* previous archeological fieldwork; *3* stratigraphic profiles; *4* main topographic discontinuities; *5* boulders; *6* compact travertine slab

shows the distribution near the shelter wall in distance classes for level J overall and for sublevels Ja and Jb, considering the same internal zone combustion structures for each sublevel. The hearths are located 0.5–2 m from the rockshelter wall. Other groups of hearths are located in the central zone, 4–8 m from the rockshelter wall, below the dripline. This bimodal pattern in hearth location with respect to the rockshelter wall has also been observed at Tor Faraj (Jordan) and compared with a large sample of modern cases (Henry et al. 2004). Henry et al. suggest that the

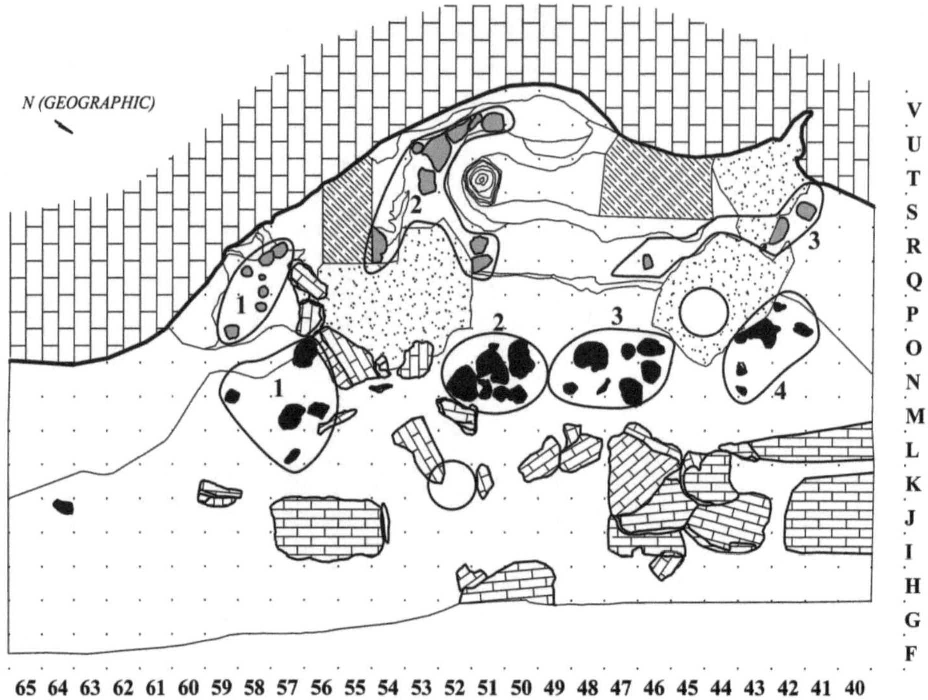

Fig. 3.10 *Grey* and *black* shaded areas are groups of combustion structures in sublevel J–Ja

location of hearths near the shelter wall presents certain analogies with sleeping areas documented in ethnoarcheology. However, hearths located below the dripline are interpreted as part of a multi-purpose area and may be equidistant with combustion structures near the rockshelter wall. Other analogies observed in level J are: (1) the separation distances (3–4 m) between groups of combustion structures in the central zone of sublevel Ja (groups 1-2-3-4, in Fig. 3.10), as a rough parallel to the distances observed between external hearths (Gamble 1990); and (2) a line of combustion structures spaced at intervals of 1 m in sublevel Jb (group 1 in Fig. 3.12), interpreted as sleeping and resting areas in the prehistoric archeology of modern human groups (Movius 1966; Binford 1983).

Travertine Blocks in the Inner and Central Zones of Sublevel Ja

The analysis of the travertine blocks is fundamentally based on the field documentation of the spatial record of the blocks in the interior and central zones of sublevel J–Ja. The blocks recorded for this study have more than 10 cm of longitudinal axis. Travertine slabs have a crystalline fabric and were found among the small-sized blocks during fieldwork. They present scarce thermal and/or volumetric modifications and are small (between 10 and 20 cm^2), but stand out for their thickness of less than 5 cm.

Figure 3.13 shows the plot of the combustion structures and all the travertine blocks and slabs in the inner and central zones of sublevel Ja. The identification of the dripline as the area where the blocks were deposited by gravitational processes suggests that some travertine blocks and slabs located in the inner and central zones were carried there, as seen with the burned blocks. Slabs are present in a relatively low density and some show signs of thermal damage. Anthropogenic displacement of the travertine slabs is hard to prove because many of them show no clear modifications. The displacement of the travertine blocks and slabs is clearer in the inner zone. On the flat surfaces of the inner zone we find three low-density and aligned accumulations of blocks and slabs not linked to fracture planes (Fig. 3.13):

- BIZ1 (blocks of inner zone 1): located in a flat area in the south of the inner zone. To the south of this group there is a line of small blocks that continues into the central zone, where it connects to an accumulation of small to medium-sized blocks.
- BIZ2: at the entrance and inside the *Coveta Ripoll*; the blocks and slabs are small inside the cave.
- BIZ3: before the carbonated talus of the *Coveta Ripoll*. These are large blocks placed in a line. A low density of small blocks and slabs can also be observed.

The central zone contains:

- BCZ1 (blocks of central zone 1), BCZ2, BCZ4: small blocks and plaques around the fireplaces (some rubefied) along with some medium-sized blocks. In BCZ4 large blocks are present but these are outcrops according to the

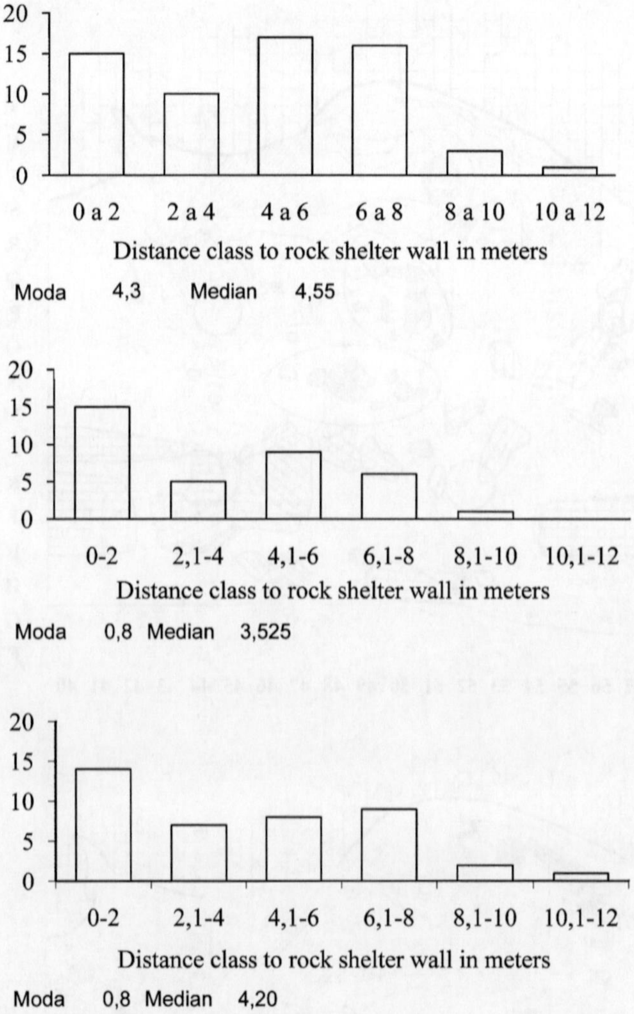

Fig. 3.11 Distance classes of combustion structures to the rockshelter wall. *Top* distances of all combustion structures in level J to the rockshelter wall. *Middle* distance classes of the combustion structures of sublevel J–Ja. *Bottom* distance classes of the combustion structures of sublevel J–Jb

field logs and belong to the sedimentation that separates Ja and Jb.

- BCZ3: small and medium-sized blocks at the foot of the step of the inner zone.

We might put forward the hypothesis that some blocks were moved to the boundary of the external zone in order to clear the central area. However, identifying blocks that may have been moved to this side from the central zone was difficult during field surveying. Figure 3.13 does not show the accumulation of small blocks below the dripline. This representation is an approach to the Ja surface with blocks and slabs, burned and unburned, plus the groups of combustion structures. This figure also shows: (1) empirical evidence of two groups of burned blocks transferred to the dripline; (2) a plot with the hypothetical transfer of blocks and slabs towards the inner and central zones.

Raw Materials and Artifact Classes in Lithic Assemblages

There are slight differences in the distribution of the lithic remains according to the most common raw materials (chert, limestone and quartz). The distribution of chert artifacts does not differ appreciably from the global distribution of all lithic artifacts (Figs. 3.14, 3.15). Limestone artifacts are more focused in the central area, with little dispersion in other areas (Fig. 3.16). The quartz remains are the most clustered and are concentrated within the central area, with scarce presence in the rest of site (Fig. 3.17).

The distribution of the lithics by artifact classes shows diverse groupings, which provide interesting data for the spatial analysis. The distribution of flakes and flake fragments, the most common artifacts, matches the distribution of the whole industry (Figs. 3.18 and 3.19). However, the distribution of other categories, such as retouched artifacts and cores, is interesting (Figs. 3.20 and 3.21). Retouched artifacts are located in greater or lesser densities around the hearth-related activity areas, except for a very few which are dispersed in the outer zone. These elements are very large, therefore making it very likely that they were discarded outside of the activity areas, towards this outer zone. Core distribution is more consistent with the zones containing a greater density of small lithic artifacts, in the context of the defined spatial units, although some of them appear in boundary areas and even dispersed in the outer zone (Fig. 3.21).

The distribution of the lithic remains shows an almost total use of the surface area of the shelter, and an adaptation to the topographical features. The study of the lithic assemblage suggests the existence of diachronic occupational events (see Vaquero et al. 2012), which seems to indicate that sublevel Ja corresponds to a period of frequentation and not to a single occupational event.

Faunal Remains

Faunal processing has been identified by the location of impact flakes resulting from bone breakage in the last stage of the faunal consumption sequence. The main locations of the impact flakes indicate the existence of faunal processing activity areas in front of the *Coveta Ripoll*, at the center of the central zone, and in the internal areas (Fig. 3.22). Impact flakes are almost totally absent in the external zone. Bones with cut marks, although they are not as good indicators of activity location as impact flakes, are distributed in

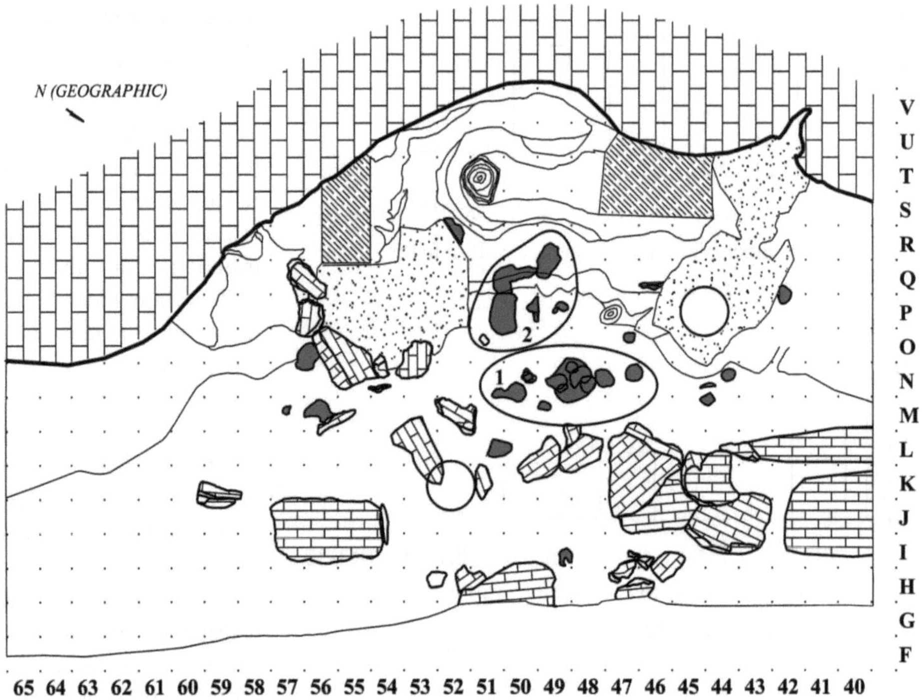

Fig. 3.12 *Grey shaded areas* are groups of combustion structures recorded only in sublevel Jb

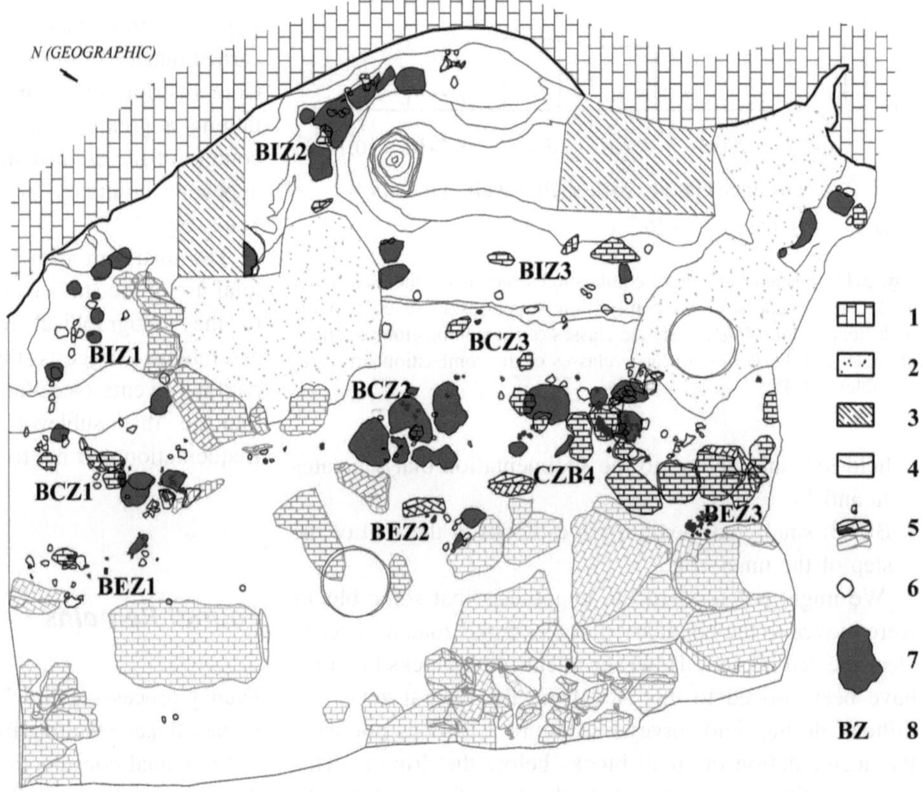

Fig. 3.13 Combustion structures and groups of travertine blocks of sublevel J–Ja indicating spatial, volumetric or thermal modification. The burned blocks are represented in shaded *grey* and the rest of the modified blocks are represented with a *black* outline. Legend: *1* Romaní cliff; *2* previous archeological fieldwork pits; *3* stratigraphic profiles; *4* main topographic discontinuities; *5* modified blocks in *black* contours, and structural blocks and megablocks in *grey* contours; *6* travertine slabs; *7* combustion structures; *8* groups of travertine blocks related to site zones: *BIZ* blocks of the internal zone; *BCZ* blocks of the central zone; *BEZ* blocks of the external zone

a manner consistent with that of impact flakes (Fig. 3.23). The good spatial preservation of these remains, which coincides with the zones where the impact flakes are located, is due to the fact that most of them are very small.

Size Sorting in Ja Distribution Patterns

Remain size is calculated by means of the area of the object, considered as the product of the length of the artifact

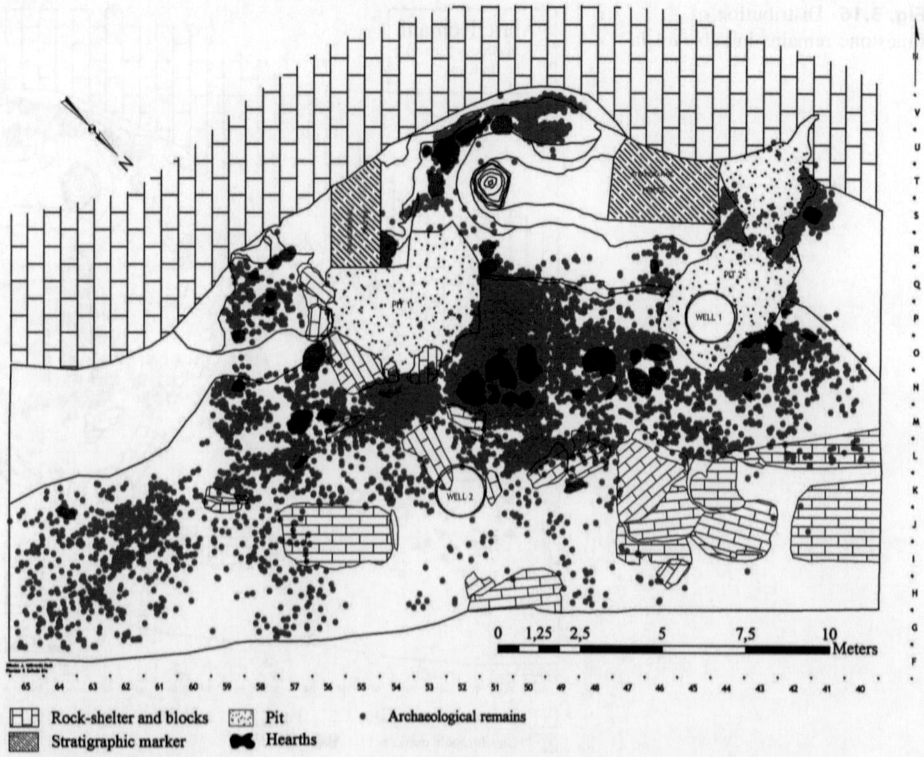

Fig. 3.14 Distribution of all three-dimensionally recorded archeological remains in sublevel Ja

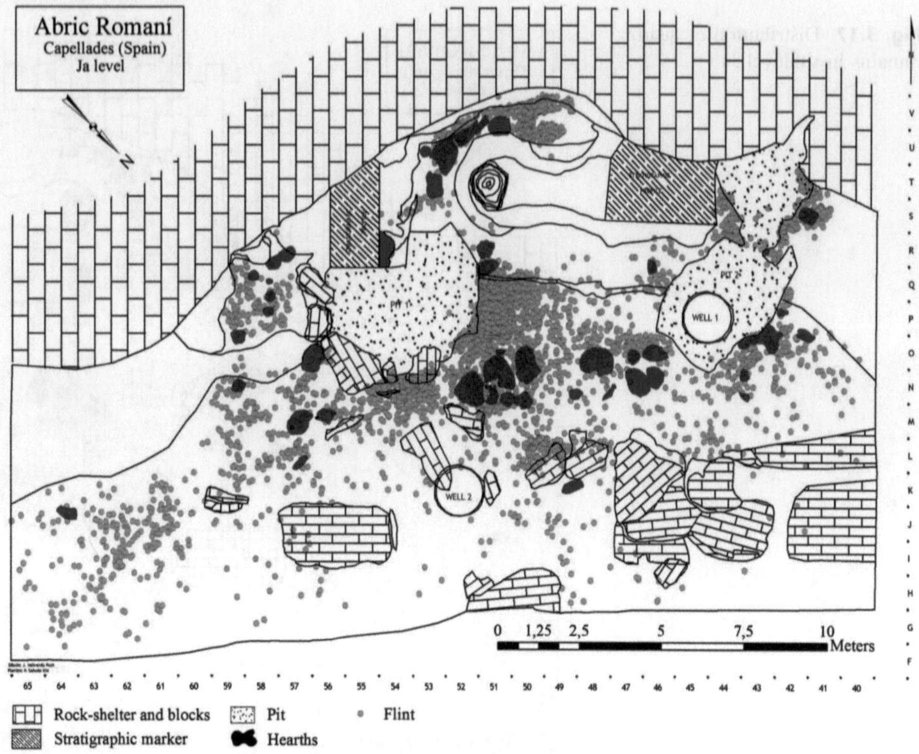

Fig. 3.15 Distribution of chert remains in sublevel Ja

multiplied by its width. The result is an area measurement in square millimeters that allows each object to be classified into one of five categories: very small (0–500 mm^2), small (501–1000 mm^2), medium (1001–1500 mm^2), large (1501–2000 mm^2) and very large (>2000 mm^2).

Small remains are considered to be the products that best indicate the location of knapping activities, according to the drop zone formation model of external hearths (Binford 1978). Small remains tend to be quickly buried and are therefore less affected by postdepositional processes of

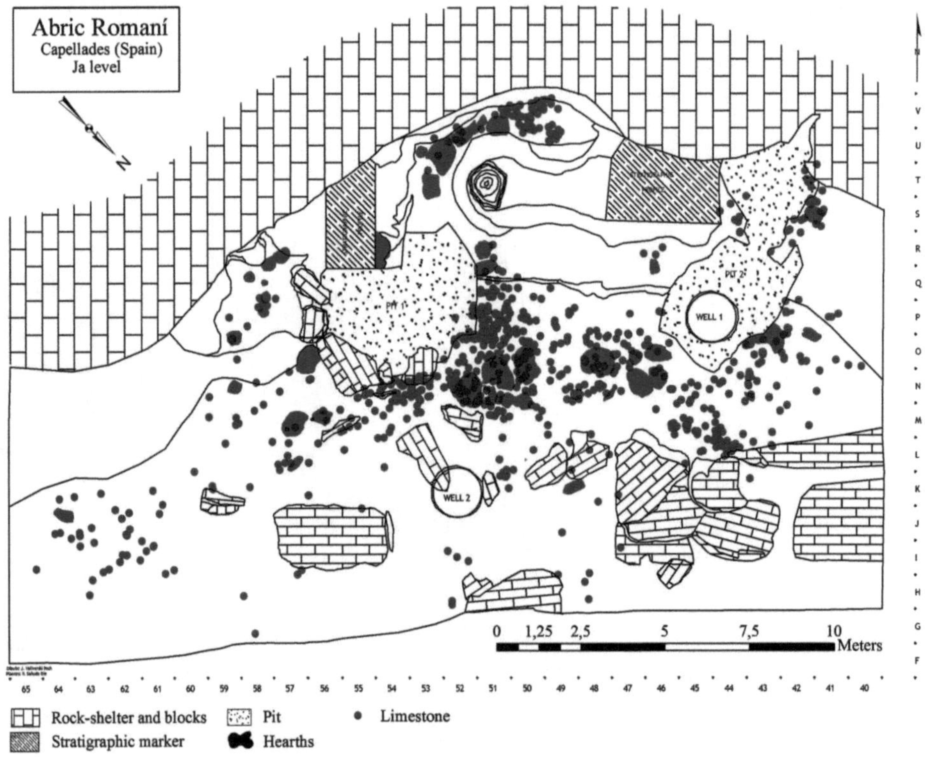

Fig. 3.16 Distribution of limestone remains in sublevel Ja

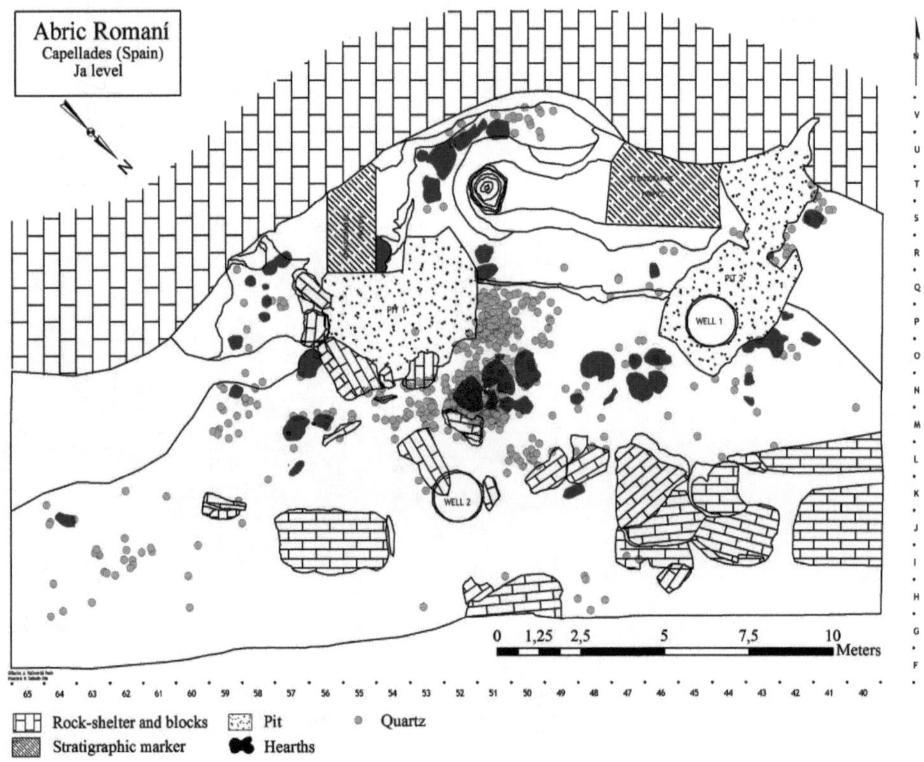

Fig. 3.17 Distribution of quartz remains in sublevel Ja

natural or human origin, allowing the activity area location to be identified (Stevenson 1991).

Very small remains are distributed in different concentrations, always linked to the combustion structures, and are found at variable densities depending on the zone (Fig. 3.24). The highest densities correspond to accumulations in the central zone in front of and inside the *Coveta Ripoll*. These seem to be related to the edges of the

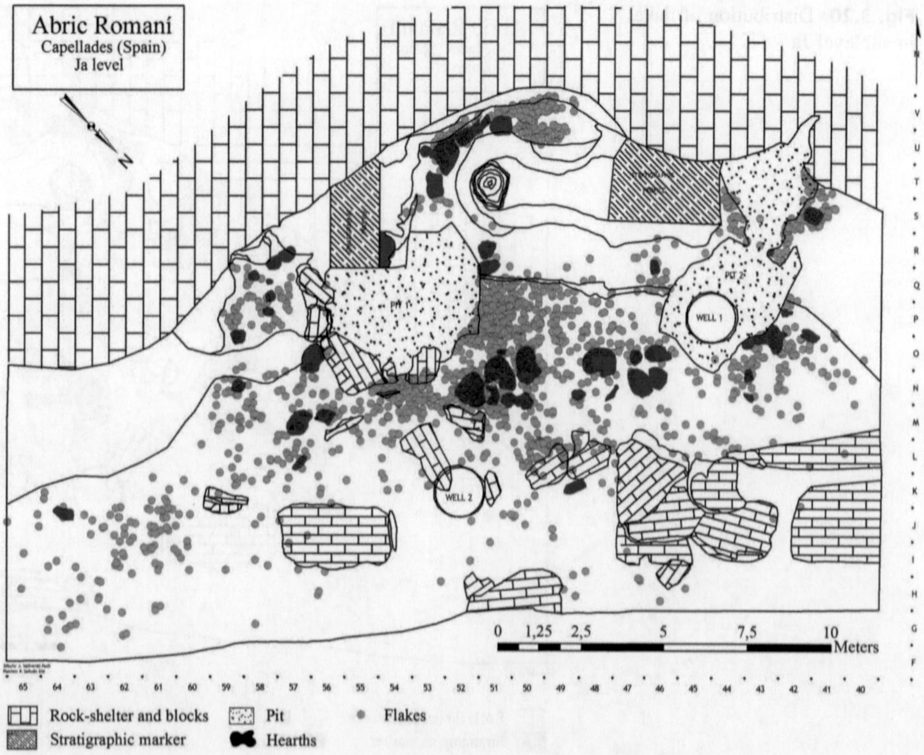

Fig. 3.18 Distribution of flakes in sublevel Ja

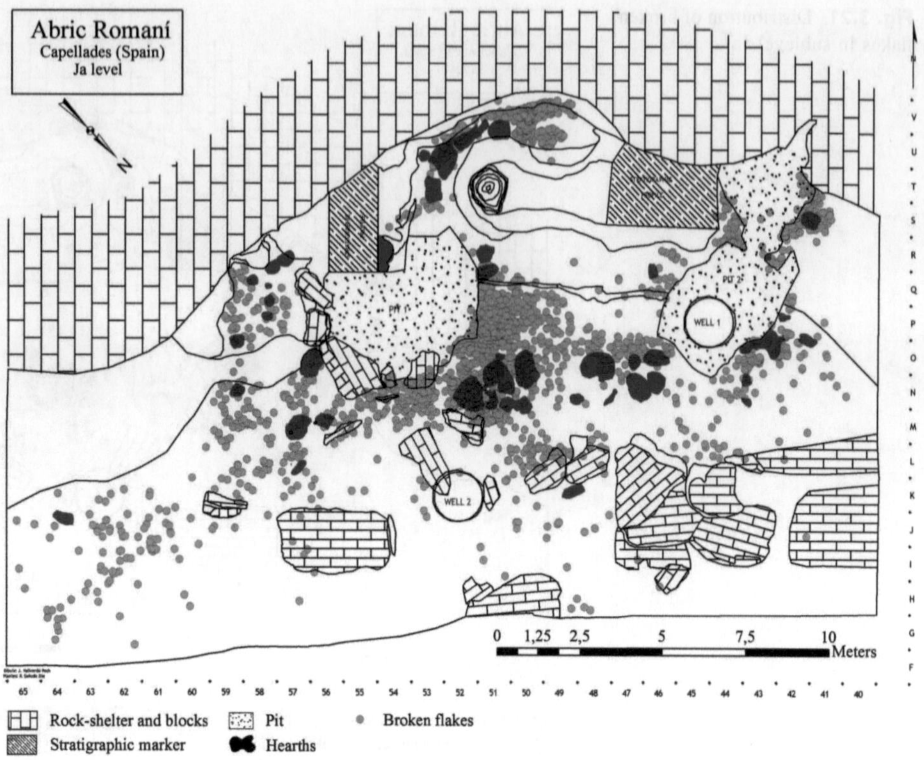

Fig. 3.19 Distribution of broken flakes in sublevel Ja

combustion structures, generally on one of their sides. The density of small remains reinforces the location of the activity areas and the division of sublevel Ja into different spatial units and allows for greater differentiation among them. The NW area of the external zone presents an almost total absence of very small remains, which corroborates the practical absence of activity episodes in this area. There are slight differences in the distribution of very small lithic and

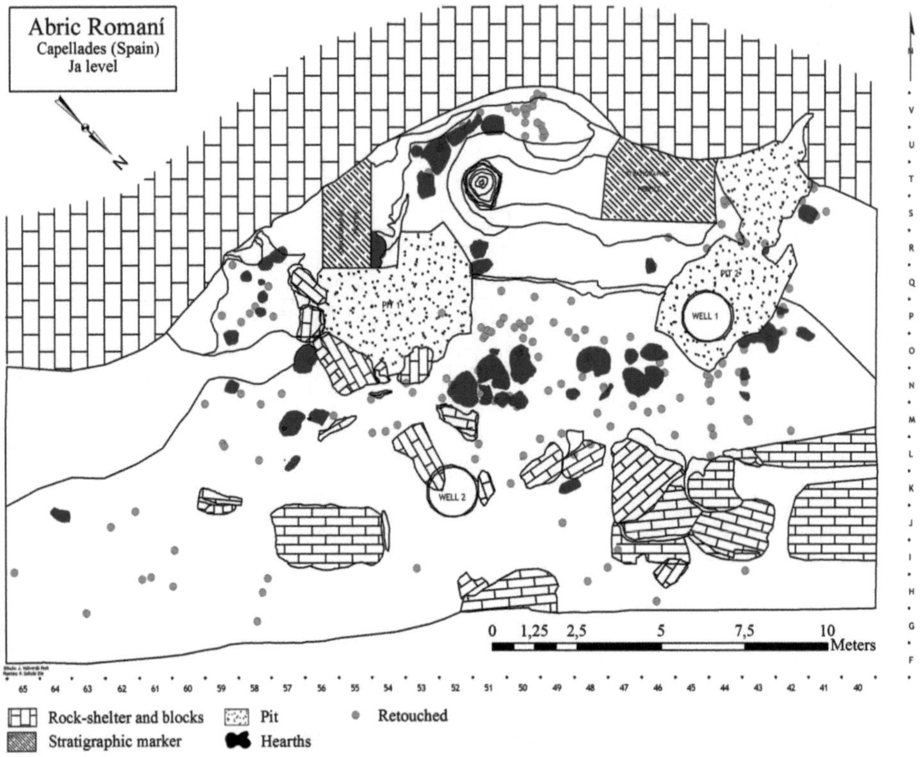

Fig. 3.20 Distribution of flakes in sublevel Ja

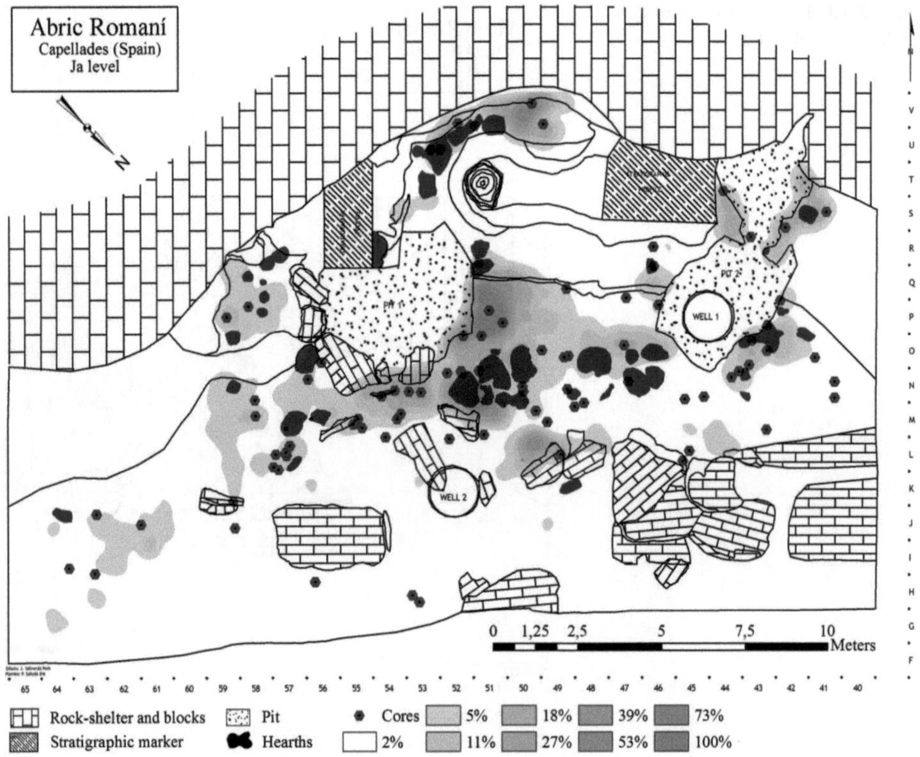

Fig. 3.21 Distribution of broken flakes in sublevel Ja

faunal remains (Figs. 3.25, 3.26). The very small lithic artifacts (Fig. 3.24) are more clearly concentrated at the western end of the external zone and in the combustion structures located towards the south. There is a high density of faunal remains between the central area and the external area, where few lithic artifacts were recovered, but no faunal remains were found in the combustion structures in the western part of the central zone.

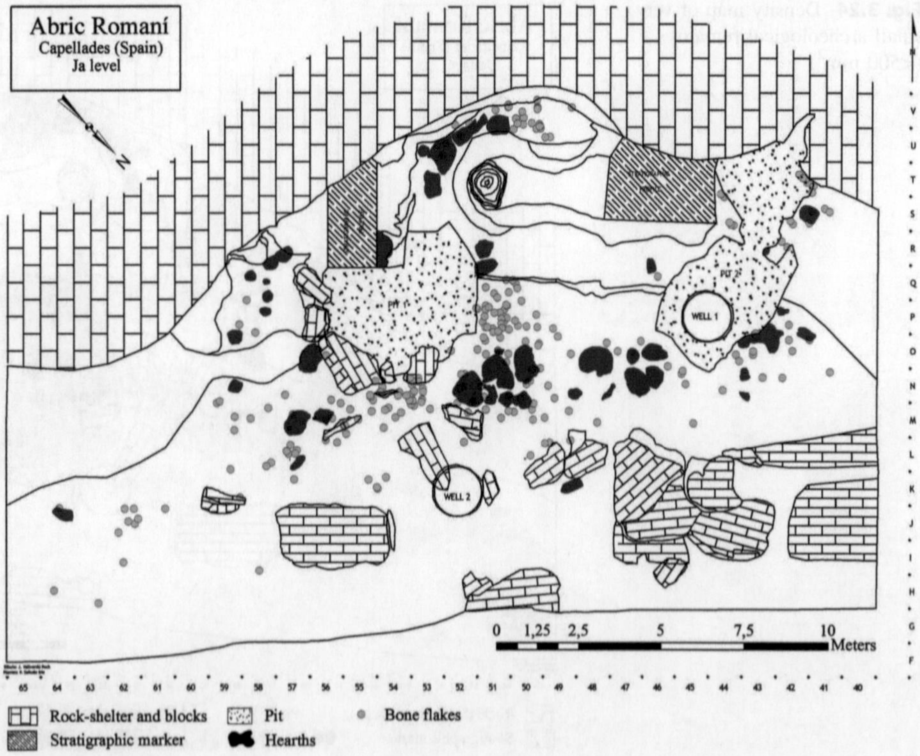

Fig. 3.22 Distribution of impact flakes in sublevel Ja

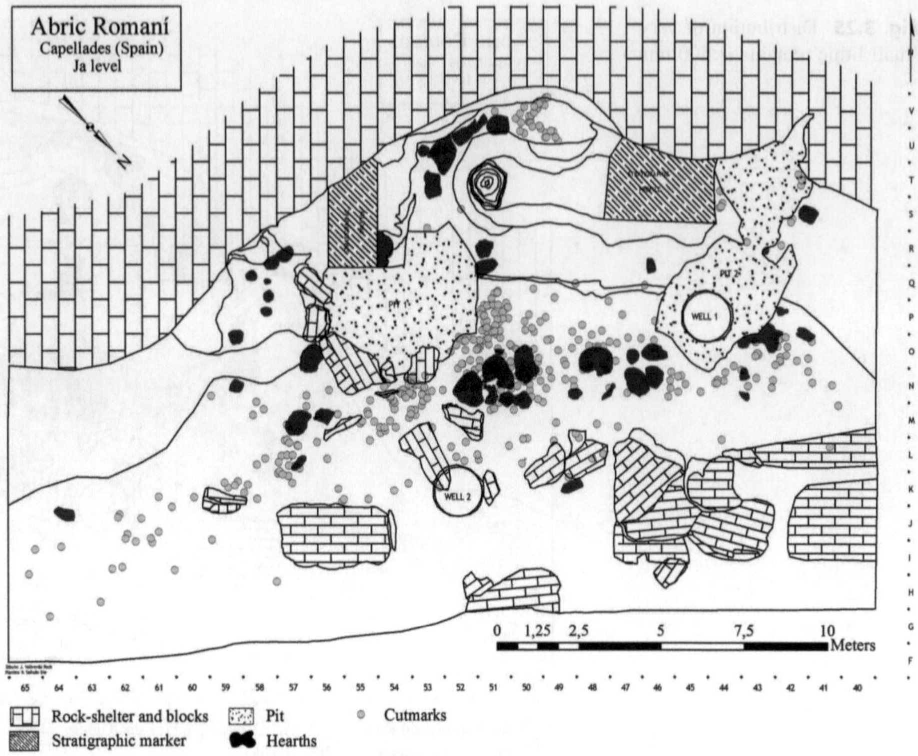

Fig. 3.23 Distribution of bone remains with cut marks

The small remains show a distribution similar to that of the very small remains, but with a higher homogeneity in their dispersion (Fig. 3.27). The zones with the highest densities of small remains are slightly blurred, although they are located in the same areas mentioned previously. The absence of small elements in the external peripheral zones may be related to the absence of activity episodes.

The remaining size categories become progressively less concentrated as size increases, resulting in homogenous

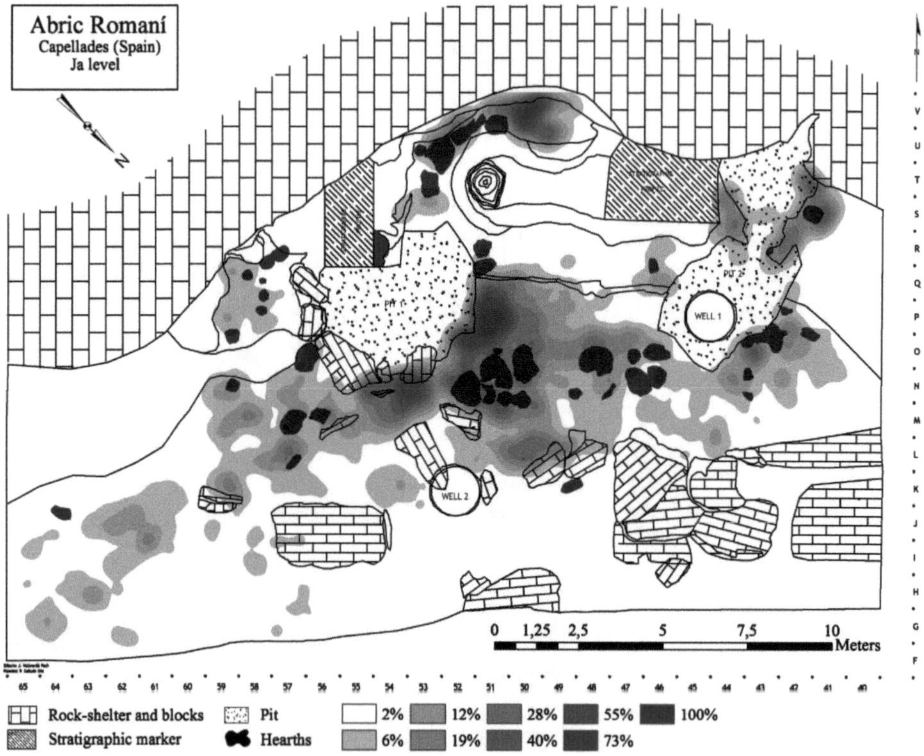

Fig. 3.24 Density map of very small archeological remains (<500 mm^2)

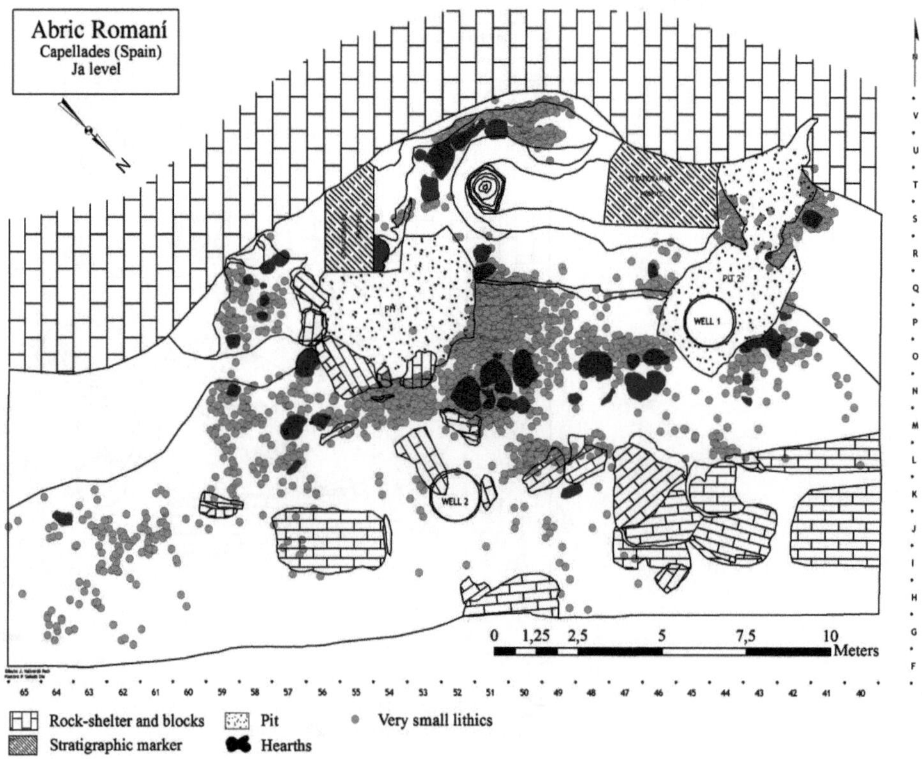

Fig. 3.25 Distribution of very small lithic remains (<500 mm^2)

distributions coinciding with the general distribution of the remains (Figs. 3.28, 3.29). The accumulations in the marginal areas with higher proportions of very small and small remains lack concentrations of large remains. They are evenly spread out (Fig. 3.29), perhaps due to intense circulation (scuffage) in the activity area, which led to their unintentional displacement (Stockton 1973; Stevenson 1985, 1991; Theunissen 1998).

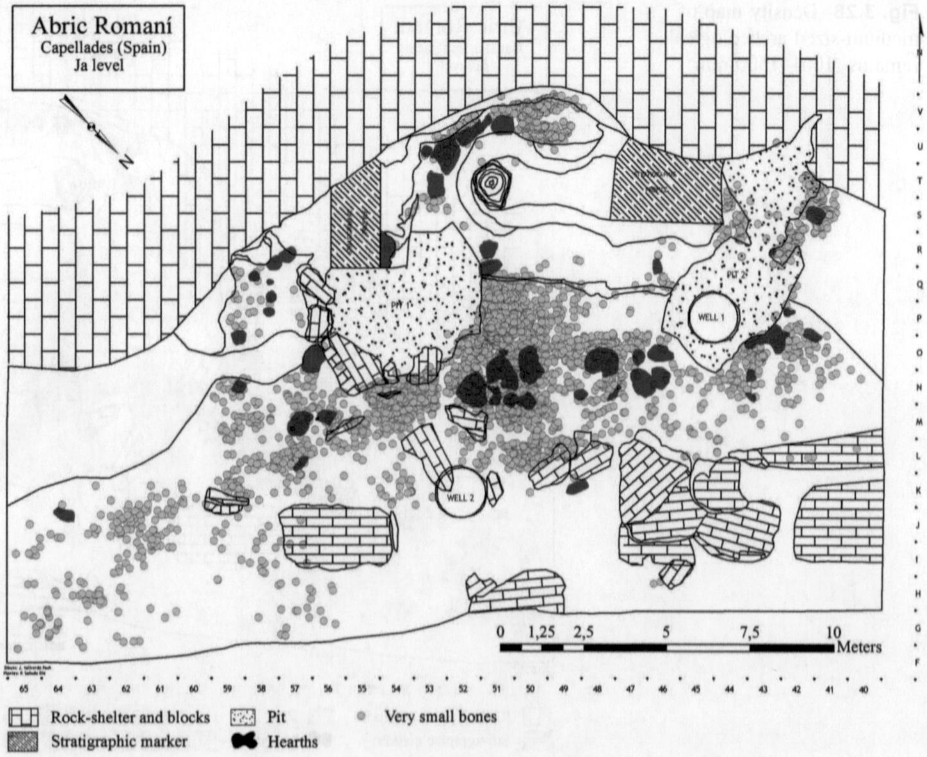

Fig. 3.26 Distribution of very small bone remains (<500 mm^2)

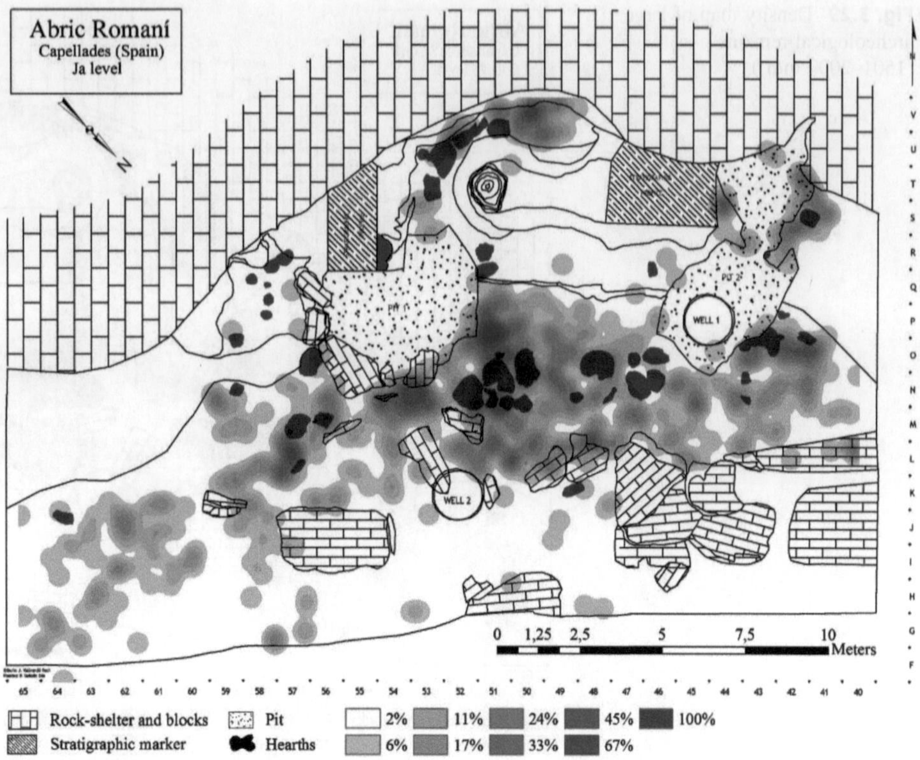

Fig. 3.27 Density map of small archeological remains (501–1000 mm^2)

Burned Faunal and Lithic Elements

Burned remains are present in all combustion structure groups, albeit at varying degrees, with greater densities recovered from inside and in front of the *Coveta Ripoll* (Fig. 3.30). The density of burned remains decreases progressively the further they are from the combustion structures. Furthermore, the elements with a greater degree of

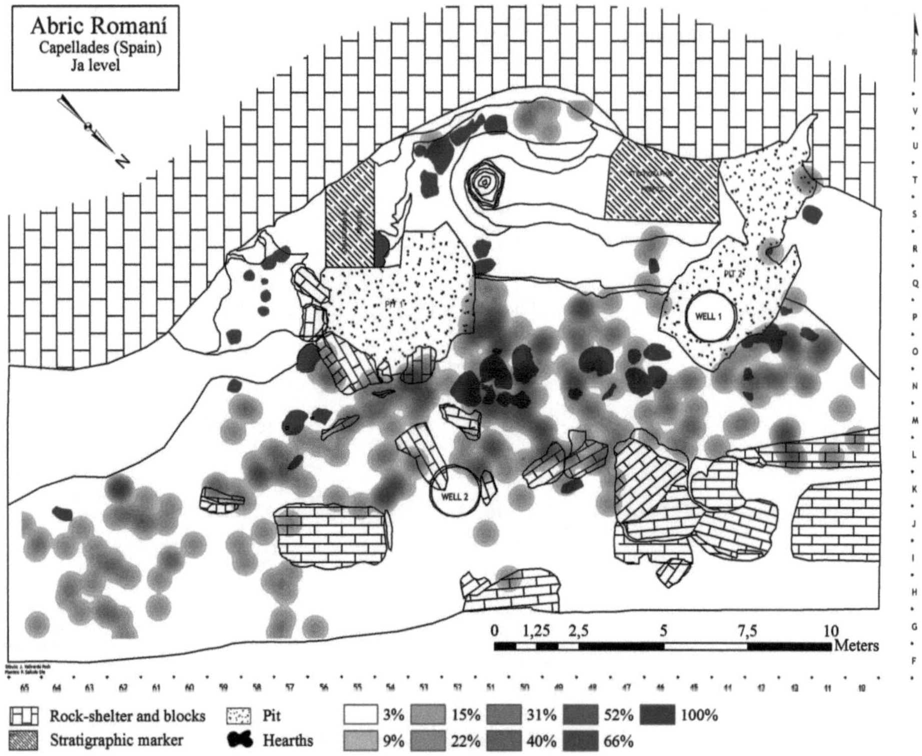

Fig. 3.28 Density map of medium-sized archeological remains (1001–1500 mm^2)

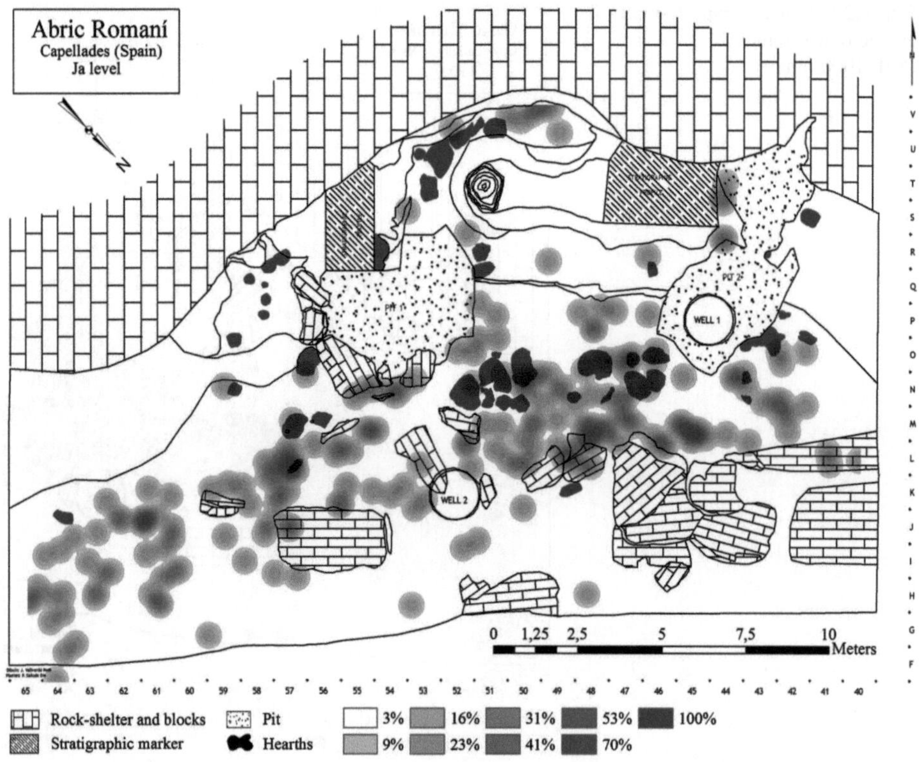

Fig. 3.29 Density map of large archeological remains (1501–2000 mm^2)

alteration are located near the center of the combustion structures, again with the degree of cremation decreasing progressively the further away they get. The degree of cremation of the faunal remains recovered has allowed us to make a density map indicating the points at which the most burned elements and the elements with the highest degrees of cremation were found (Fig. 3.31). The location of the burned elements confirms the good general preservation of

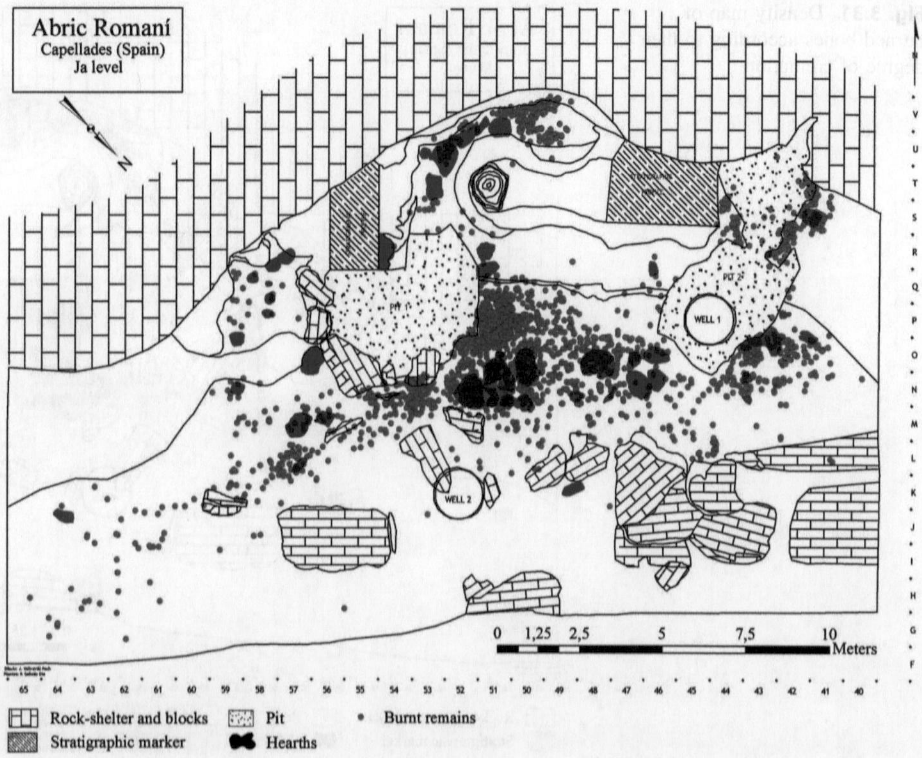

Fig. 3.30 Distribution of archeological remains altered by fire in sublevel Ja

the archeological record and in particular of the hearth-related areas.

Some differences were detected between the distribution of the burned lithic artifacts and that of burned faunal remains, not in their general spatial distribution, but in their location with respect to the combustion structures (Figs. 3.32, 3.33). Burned lithics are related to the combustion structures, but tend to be located outside of them, while the highest densities of burned bones are within the combustion structures. There are two high-density peaks that do not match the central point of the combustion structures close to them. The burned remains of the *Coveta Ripoll* appear slightly displaced towards the back of this cavity due to the topography of this zone, which presents a small depression where material tends to accumulate. Another high concentration of burned elements is located in front of and at the entrance to the *Coveta Ripoll.*

Spatial Units and Activity Episodes in Sublevel Ja

As noted above, the occupation pattern of level J is organized into three distinct zones: external, central and internal. In each one of these zones, we have identified a variable number of spatial units, called external spatial units (ESU), central spatial units (CSU) and internal spatial units (ISU) followed by an identifying number (Fig. 3.34).

These spatial units attempt to go beyond the diachronic study of faunal and lithic remains in order to isolate activity episodes and establish temporal relationships between them. This is done using the information provided by the refits. We describe spatial units through the spatial analysis of archeological items, taking into account: (1) the surface available according to the topographic zones of the rock-shelter; (2) the condition of the hearth-related assemblage and the combustion structure groups described above.

External Spatial Units

The external zone is comprised of spatial units that are not protected by the rockshelter roof. We suggest two main spatial units with an unlimited available surface area. Unit ESU-1 is below the dripline on a ramp parallel to the south wall of the rockshelter, and ESU-2 is on a moderate slope towards the NNE where the travertine megablocks may constrain the surfaces available for occupation.

External Spatial Unit 1 (ESU-1). The archeological remains are widely dispersed, conditioned by the natural slope and other elements that may have been intentionally displaced to this zone. This superposition of archeological elements pertaining to diverse occupational events includes evidence of specific activity episodes carried out in situ and

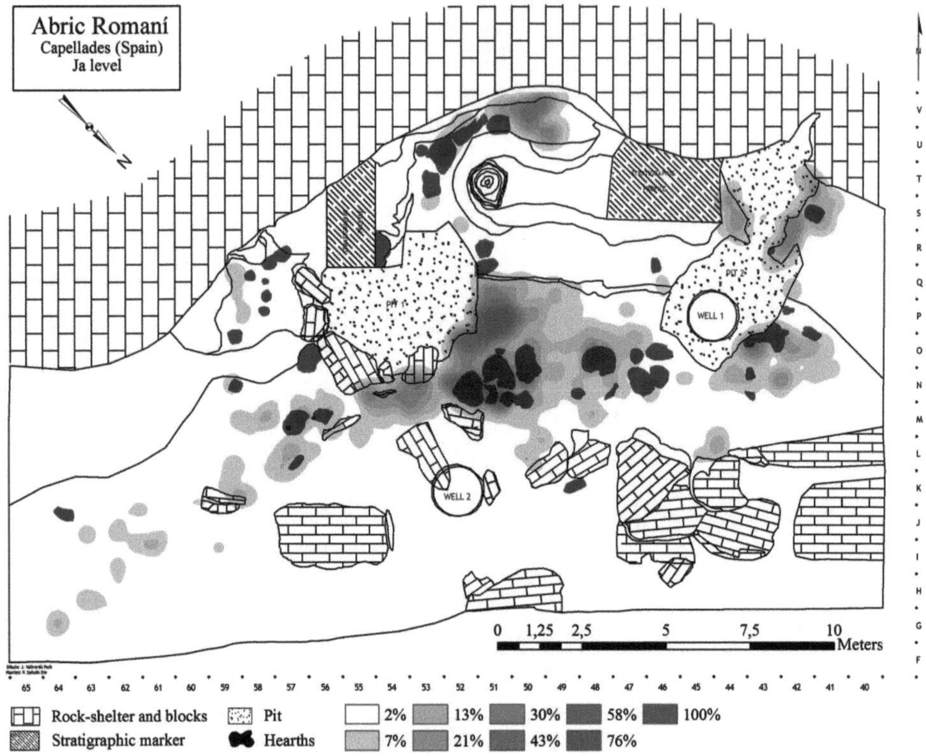

Fig. 3.31 Density map of burned bones according to their degree of cremation

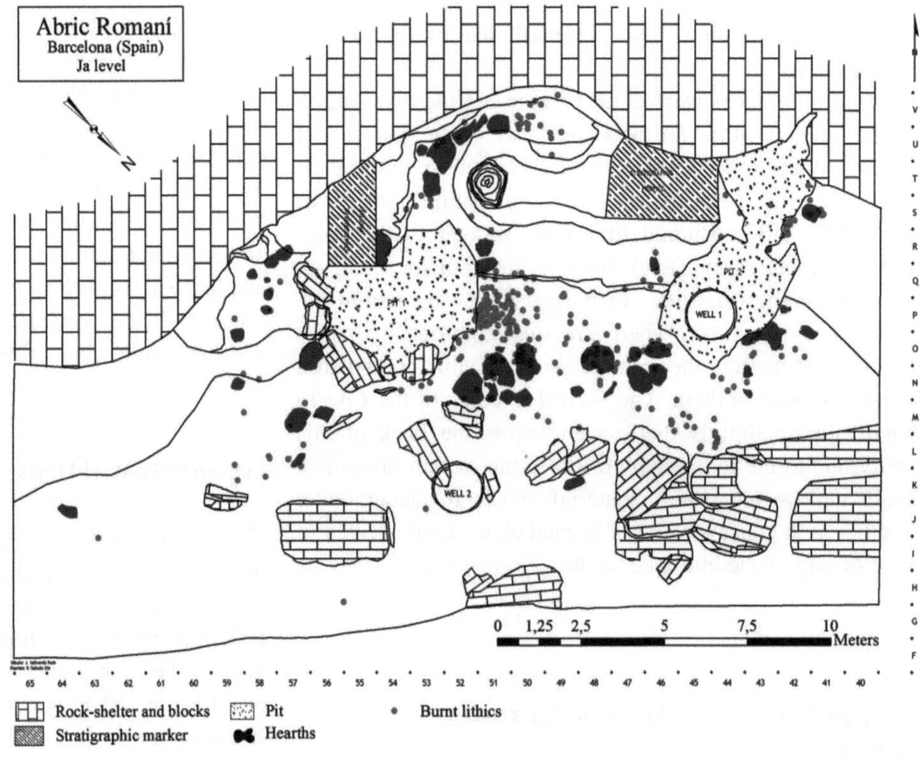

Fig. 3.32 Distribution of lithic remains altered by fire

linked to a specific combustion structure. Lithic artifacts in this unit consist of a concentration of small artifacts in J–I/61–63, made up mainly of chert remains. In this concentration there is a set of quartz artifacts that might correspond to an isolated activity episode. A faunal processing episode, characterized by a concentration of small bones and impact flakes, was also detected in this area.

Three artifacts with use-wear traces have been found in this spatial unit. They are denticulates used to cut bones and meat, and one of them was also used in hideworking.

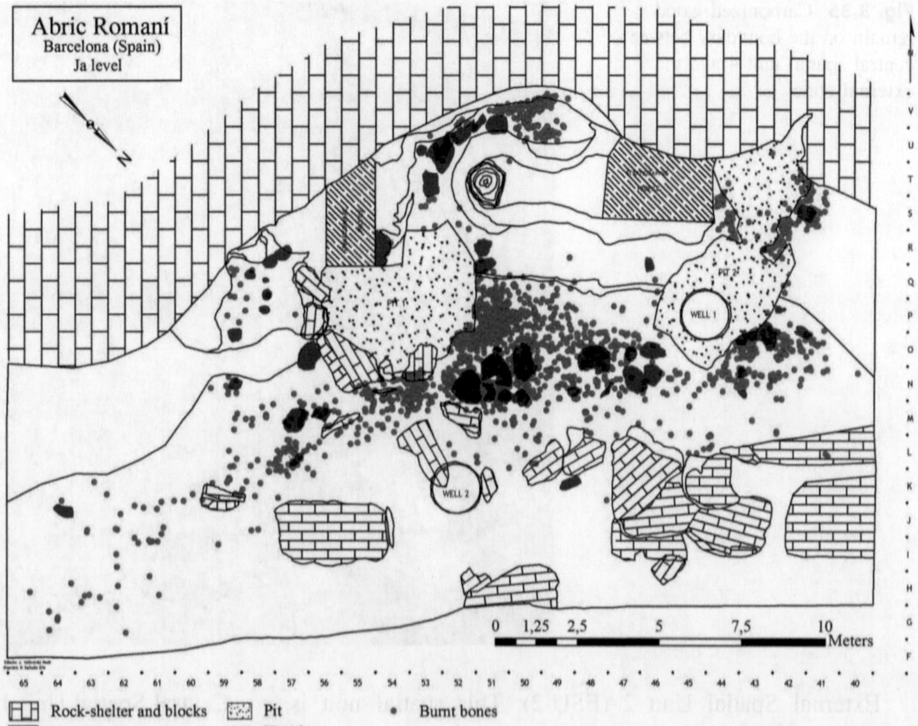

Fig. 3.33 Distribution of bone remains altered by fire

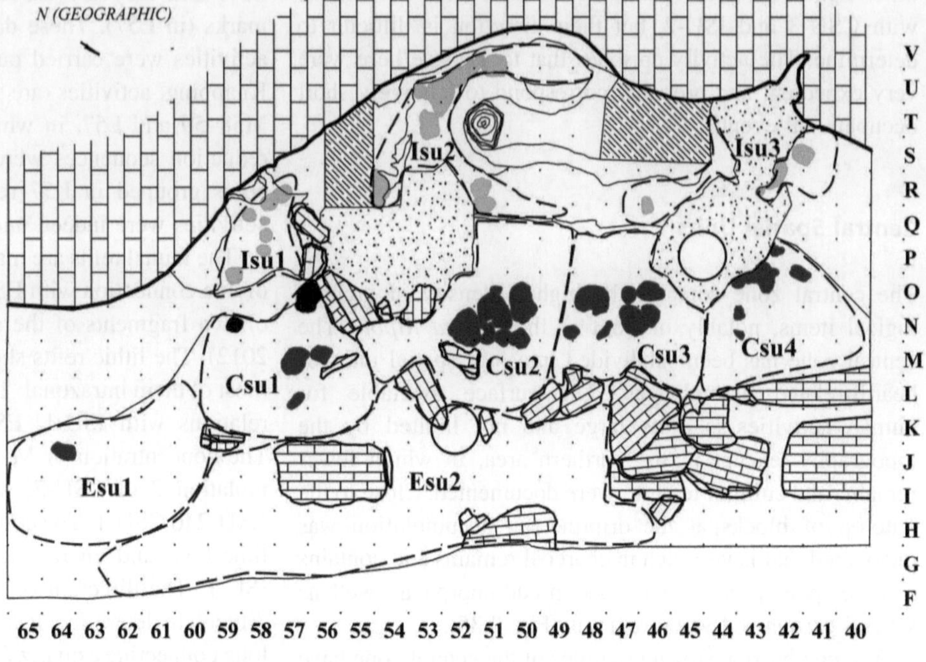

Fig. 3.34 Spatial units of sublevel Ja. *ESU* External spatial units, *CSU* Central spatial units, *ISU* internal spatial units

Bone refits are scarce and almost all are intrazonal, except one connection with CSU-1. The lithic refits provide different long-distance connections with ISU-1, CSU-1, CSU-2 and CSU-3 (see Vaquero et al. 2012). The direction of these connections is difficult to determine in most cases, but some refits indicate a predominant movement towards ESU-1, favored by the slope. The relation of this unit to the rest of the shelter is difficult to determine, due to the dispersion of the remains in this zone and the sporadic character of the activity episodes.

Fig. 3.35 Carbonized wood remain on the boundary between central spatial unit 4 and the external zone

External Spatial Unit 2 (ESU-2). This spatial unit is located in the external area furthest from the protection of the rockshelter and yielded the lowest density of remains in sublevel Ja. The activity episodes here are not well defined. Only one knapping sequence has been identified. Refits with other zones are scarce, although connections were made with CSU-3 and ISU-2, but their direction is difficult to determine. The activity episodes that took place here were very expedient and probably correspond to relatively short occupational events.

Central Spatial Units

The central zone contains the highest density of archeological items, notably in front of the *Coveta Ripoll*. The central zone has been subdivided into four spatial units of hearth-related assemblages. The surface available for human activities is quite large and not limited by the topography, except in the northern area, in which fewer remains and cultural features were documented. Close to the outcrop of blocks at the dripline, an accumulation was uncovered that is very rich in charcoal remains and contains a large, poorly preserved wood pseudomorph as well as several carbonized wood remains (Fig. 3.35).

The combustion structure groups of the central zone have an aggregate pattern with the exception of the southern accumulation. Combustion structures tend to be large and rectangular, but some are small and irregular. The scattered pattern of the southern combustion structure group, especially those located on the boundary with the internal zone, indicate an undifferentiated record between sublevels Ja and Jb.

Central Spatial Unit 1 (CSU-1). Different concentrations of small remains have been discovered in M58–59, M55–56, L57 and K59, which are linked to the location of the different combustion structures. There are abundant bone impact flakes, especially concentrated in M55–56 and L57, and they are associated with small bones with cut marks (in L57). These data suggest that faunal processing activities were carried out on the northern side of CSU-1. Knapping activities are related to the accumulations of M58–59 and L57, in which different complete and partial reduction sequences were performed. The presence of six cores grouped in L57 reinforces the deduction that such activities were indeed undertaken here.

The faunal refits are mainly intrazonal, with the exception of one connection with ESU-1 corresponding to the refitting of two fragments of the same long bone (see Rosell et al. 2012). The lithic refits show a high number of connections, most of them intrazonal. The interzonal connections indicate relations with ISU-1, ESU-1, CSU-2, ISU-2 and ISU-3. The concentration of M–N/58–59 is connected with accumulation 2 of CSU-2, with a preferential direction from CSU-2 to CSU-1. The connections with ISU-1, all of them in limestone and chert, suggest a direction from CSU-1 to ISU-1. The direction of the connection with ISU SW 2 is difficult to determine. It corresponds to a refit more than 11 m long connecting a quartz fragment with percussion marks and a flake detached from it.

Central Spatial Unit 2 (CSU-2). This unit was partly affected by previous excavations, during which the three-dimensional locations of the material recovered were not documented (Fig. 3.36). The main activity zone is located on the east side of the combustion structures, with three

Fig. 3.36 Photograph of the central zone viewed from north to south with Romaní Pit 1 in the background and central spatial units 2 and 3 in the *left*

large concentrations of materials. We have identified remains from different reduction sequences (see Vaquero et al. 2012), limestone fragments with percussion marks, and bone impact flakes.

All the indicators of activity appear concentrated on the south side of CSU-2, in contrast to the large and very large elements, which appear mainly concentrated on the north side, between the blocks. This distribution fits the drop/toss area model proposed by Binford (1983), commonly observed in hunter-gatherer societies, in which the activity is focused on one side of the hearth and different refuse areas are located on the periphery. In the case of CSU-2, there is a significant toss area on the north side, where a concentration of bones with carnivore damage has also been documented. Carnivores would have been attracted to this zone by the high concentration of bone refuse.

Lithic and bone refits resulted in a high number of both intrazonal and interzonal connections. Abundant internal (intrazonal) refits are found in CSU-2, which connect the three main accumulations described above. CSU-2 is connected by refits to CSU-1-3, ISU-1-2-3 and ESU-1 (see Vaquero et al. 2012; Rosell et al. 2012). Most connections originate in CSU-2 and go out towards the other areas, although there are also some connections in the opposite direction.

Three artifacts with use-wear marks have been found: two denticulates and one flake used for defleshing. They were located in the accumulation contiguous to the external area and the accumulation located in the hearth-related area. The location of these artifacts coincides with the higher densities of very small bones and impact flakes. All these elements seem to indicate that fauna processing occurred in this unit.

All the elements described in this spatial unit suggest an important position in the organization of the shelter, at least judging by the density of archeological remains that would be associated with higher frequentation. This interpretation is reinforced by the high number of connections that relate this unit to other areas and the predominant directionality from CSU-2 towards the other units.

Central Spatial Unit 3 (CSU-3). There are seven non-superimposed combustion structures in this unit, around which a considerable concentration of archeological items were deposited, although in lower densities than in CSU-2. In the part affected by the previous excavations, the three-dimensional location of the archeological items is not available. The main focus of activity is located on the southern side, with a secondary zone on the north-western side, partially coinciding with ISU-3. This unit displays a very low density of small bones and impact flakes, which contrasts with the density of the archeological elements in general. These data indicate that faunal processing was not the preferential activity performed here. The lithic industry points to a significant knapping episode next to the combustion structures close to Pit 2. These combustion structures are located beneath the majority of this lithic concentration and contain few burned artifacts. This suggests an inner diachrony for this unit, since these knapping episodes would correspond to an occupational event posterior to these combustion structures and associated with other combustion structures in this unit.

In addition to intrazonal connections, bone refits show interzonal connections with the nearby concentrations in units CSU-4 and ISU-2. It is difficult to relate these connections to intentional displacements of faunal remains, although this hypothesis should not be entirely ruled out. The lithic connections reveal a relationship with units CSU-2, ISU-2, CSU-4 and ESU-1 and -2. The predominant direction is towards other units, except in the case of the connections with CSU-2. This spatial unit displays a lower level of occupational redundancy than CSU-2, and it was likely a peripheral unit judging by the high number of connections from CSU-2 to CSU-3. Some activity episodes that began in CSU-2 (defleshing, processing, knapping, etc.) were displaced to CSU-3, where the processes continued.

Central Spatial Unit 4 (CSU-4). This unit was considerably affected by Pit 2. The main activity episodes took place around a concentration of small remains where some impact flakes were recovered. A considerable accumulation of remains, slightly separated from the combustion structures, was found around L–M/44–45. Most lithic and bone refits show intrazonal connection lines. The few extrazonal connection lines indicate associations with CSU-2, CSU-3 and ISU-3. The direction of some of these has been defined, indicating displacements from CSU-4 towards ISU-3 and CSU-3. The connections with CSU-2 were found to have directionality from this area towards CSU-4.

The accumulation in squares L–M/44–45 exhibits very specific characteristics; it corresponds to an activity episode isolated from the nearest combustion structures and performed in the area protected by the large blocks fallen from the roof. There are no refits among the artifacts found in this concentration except a refit of two chert flakes connecting with square O42 and one faunal refit with square L54, which cannot easily be interpreted as an intentional displacement. This concentration may correspond to an isolated episode related to a lithic production activity.

Internal Spatial Units

Three spatial units have been determined in the internal zone, distinguished by empty adjacent areas of topographic origin. The accumulations are well delimitated by empty areas featuring notable topographic discontinuities such as sharp slopes like the dripping dome in the *Coveta Ripoll* or the gentle slopes of the ramps close to the northern and southern areas of rockshelter wall. Accumulations and cultural structures are next to three small cavities which make up the entire concavity of the rockshelter. These constrained topographical accumulations are clearly present in the southern section (ISU-1) and partially present in the central part of the internal zone (ISU-2 or the *Coveta Ripoll*). ISU-3 is located in the northern part of the zone and is separated by Pit 1 into two parts, reducing the reliability of those parts as a single archeological accumulation. The combustion structures are smaller than in the central area. In both wide (ISU-1) and narrow (ISU-2) available surfaces combustion structures display an aligned pattern (ISU-2).

Internal Spatial Unit 1 (ISU-1). The unit is made up of a set of activity episodes carried out around seven small and medium-sized combustion structures (Fig. 3.37). The archeological elements found in this unit are small and very small, with an absence of large and very large remains. This indicates that the activity events that occurred in this spatial unit generated small remains and were probably followed by cleaning activities that moved the largest and most bothersome remains to secondary deposits. Only three impact flakes were found and bone remains are scarce, which indicates that faunal processing was a very uncommon activity, in contrast with lithic production, with evidence of a knapping area found in P-Q58. No faunal refits have been identified in this area and the connection lines provided by the lithic industry are scarce. These lines connect ISU-1 with CSU-1 and -3, although it is very difficult to determine the direction of movement.

Internal Spatial Unit 2 or the *Coveta Ripoll* Spatial Unit (ISU-2). The location of this spatial unit benefits from the semi-enclosed area created by the speleothem and the rockshelter wall. In this narrow space there is a limited surface available for human occupation. This spatial unit was affected by Ripoll's excavations and the three-dimensional locations of the materials recovered on the entrance ramp and inside the narrow passage of the *Coveta Ripoll* are unknown.

Two combustion structures with a small number of archeological items were found at the mouth of the *Coveta Ripoll*. Six combustion structures are located in the narrow inner passage of CSU-2, where high densities of archeological items coincide with the location of the depression. All the data indicate that several activity episodes took place in this area, linked to both faunal processing and the production and maintenance of lithic artifacts. The activities related to faunal processing are corroborated by the high density of impact flakes and the existence of a denticulate used in defleshing activities. In addition, two connection lines have been documented: one intrazonal and the other interzonal in the direction of square P51 of CSU-2, which contains a considerably high density of material.

Some lithic reduction sequences were entirely carried out in this area, but many knapping episodes correspond to the exploitation of cores previously reduced in other locations and transported to this spatial unit. The connections obtained from the refitting of the lithic material allow a relationship between this area and the CSU-1-2-3-4, ISU-3 and ESU-1 areas to be established. This relationship seems more reliable with CSU-1-2-4, since the intentionality of the transport seems clear. The directionality of these connections occurs generally from zones CSU-1 and CSU-2 towards the *Coveta Ripoll* (ISU-2) and from the *Coveta Ripoll* towards CSU-2.

Fig. 3.37 Photograph of internal spatial unit 1 and combustion structures on a flat surface

A denticulate with use-wear marks was recovered in this unit, indicating that hides were worked here (see Vaquero et al. 2012). This tool refits with an object located in CSU-2, which allows us to establish a relation between the two areas with a high degree of reliability. The direction of this connection is difficult to determine, but it is interesting to note that this hide-working activity was carried out in an area peripheral to the central zone.

Internal Spatial Unit 3 (ISU-3). This spatial unit is not well preserved because its central part was removed during previous excavations (Fig. 3.38). This makes it difficult to interpret this area and establish its relationship with the rest of the shelter. Only one hearth has been preserved intact, around which the archeological items are concentrated. This allows the definition of an activity area, proven by the presence of very small faunal and lithic remains and some impact flakes.

Several different connections were drawn by bone and lithic refits. All of them are intrazonal except one lithic refit that connects with area CSU-4, which we will discuss in more detail later. The bone refits connect fragments located within the area and mostly correspond to impact flakes derived from long bone breakage. The refitted fragments are very close to one another, both horizontally and vertically, which suggests that they did not undergo postdepositional movements. These connections show at least one activity episode related to faunal processing. Only three lithic refits have been identified (two intrazonal and one extrazonal), making it the unit with the fewest refits in the entire level. This extrazonal refit connects with area CSU-4 and is a flake from a small-artifact production sequence. This sequence is widely scattered and it is difficult to associate this connection with intentional transport.

These data seem to indicate that ISU-3 functioned separately from the rest of the shelter, and it is difficult to ascertain whether it was occupied at the same time as the other spatial units or whether it corresponds to an isolated and very expedient occupational event.

The archeological accumulation on the western surface of this spatial unit points to a single activity episode. Its location near the stalactite of the *Coveta Ripoll* makes this area unsuitable for performing most types of activities. A concentration of small bones was found around a small combustion structure, which could indicate that an expedient processing event occurred here. Another interesting find in this area was an upper D4 from a horse that belongs to the same dental series as an upper D3 located in CSU-3 (square P48).

Conclusion and Discussion

The existence of several occupational events defines the occupational pattern and the formation process in sublevel Ja. This sublevel can be considered a semi-palimpsest or occupational micro-palimpsest, characterized by high temporal resolution. The formation of the spatial units is due to the superposition of domestic activities in the same space, repeatedly selected over different occupational events, resulting in a repetitive occupation pattern. This phenomenon is most clearly observed in CSU-2 and -3, where there are higher densities of remains and cultural structures,

Fig. 3.38 Photograph of internal zone spatial unit 3 showing the area excavated and removed during previous excavations (Pit 2)

suggesting the existence of several activity episodes. These two units occupy the central space of the concavity of the shelter.

Around these two central areas, seven spatial units have been identified, which present a lower degree of occupational redundancy and in some cases are linked to more expedient activity episodes. These units display a higher temporal resolution than CSU-2 and -3, due to the short length of human occupations or because they correspond to more specific events that were not necessarily synchronous. Lithic and bone refits indicate that most peripheral spatial units were related to CSU-2 and -3, which suggests that these areas and the central space were connected. However, bi-directional connection lines between these areas, which would corroborate their contemporaneity, have not been identified. There is a directional flow in the refits, whose central focus is CSU-2 and -3.

The spatial distribution of cultural and natural features suggests a large available surface beyond the dripline. The NE of the excavated surface provides effective lateral protection enhanced by a prominent outcrop of megablocks. These topographic features suggest analogies with other protected and defensive sites (Rose and Marshall 1996; Rolland 2004). However, considerable densities of archeological items are not found here and exogenous elements and refits indicate that the large and open CSU-2 serves as the backbone of the occupational organization of sublevel Ja (Fig. 3.39). The remaining spatial units are peripheral to CSU-2 as the primary diachronic spatial unit. The spatial units exhibit characteristics typical of hearth-related domestic units (Vaquero and Pastó 2001). The occupation of these peripheral spatial units varies according to the different occupational events, with greater or smaller occupational reiteration depending on each case. The central space seems to be where the primary activities were carried out, although the high superposition of activity episodes makes them difficult to individualize. This is the main occupational pattern of sublevel Ja, although there were also other much more expedient occupational events, which are characterized by a more basic occupational strategy. These occupational events correspond to short visits to the shelter, probably by a few individuals, during which less significant activity episodes occurred.

The temporal span of the occupations is an aspect that cannot be overlooked. Although such time spans are difficult to determine with exactitude, they can be approximated. The occupational events of sublevel Ja exhibit characteristics typical of short-term occupations, characterized by low densities of remains and little reuse of resources, but with the organization of a large camp divided into different interrelated spatial units. The density of remains in certain areas of sublevel Ja must be related more to the superposition of different occupational events than to a single long occupation.

Sublevel Ja provides a structured record that is extremely useful in improving our understanding of the occupational strategies and spatial organization of Neanderthals. The archeostratigraphic definition of this sublevel over most of the surface of the rockshelter, the extensive excavated area and the good general conservation of the remains make a high spatial resolution approach possible. The occupation of sublevel Ja presents, therefore, a complex organizational pattern, based on extensive organization around different domestic units. The most important activities were carried out around the hearths, which act as the social and economic centers of the occupations.

The heterogeneous dispersion of bones and lithic remains is mainly adapted to the spatial units in front of the *Coveta Ripoll*. The spatial distributions of archeological elements adjacent to cultural and natural features provide scales that allow empirical observations to be made about the size of the occupied surfaces. The size of the occupied

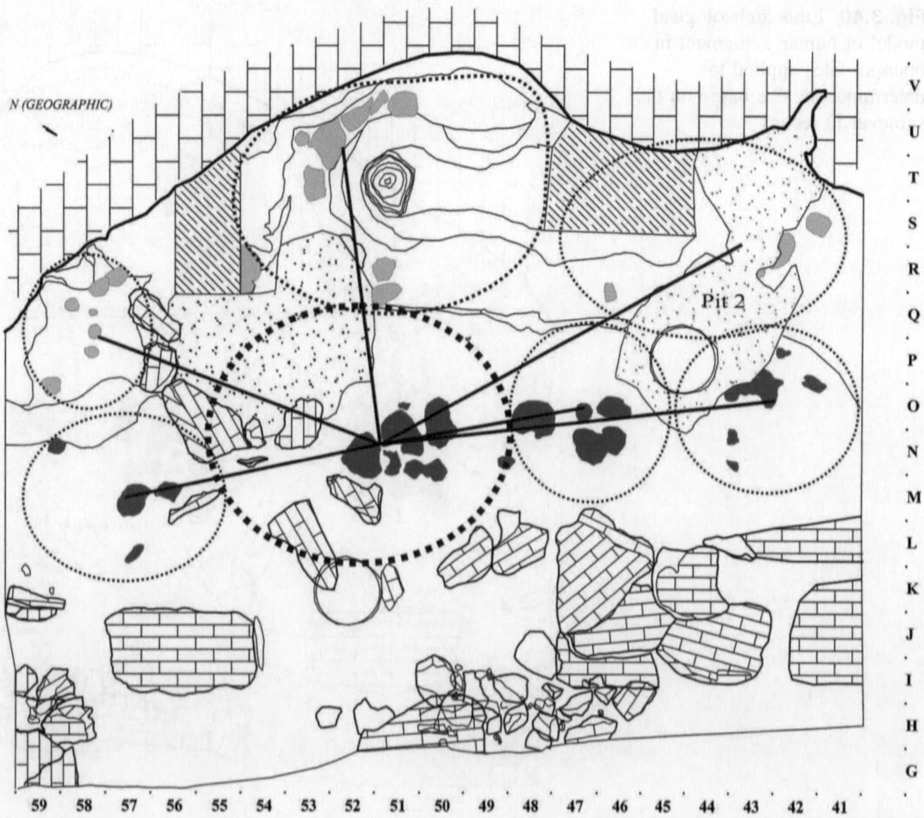

Fig. 3.39 Empirical evidence of spatial units which exhibit characteristics typical of hearth-related domestic units where CSU-2 serves as the backbone of the occupational organization of sublevel Ja

surfaces has been established using estimated areas around combustion structures (Vallverdú et al. 2005). But sublevel Ja is made up of a large number of combustion structures and it is not possible to identify well delimitated accumulations around a single combustion structure. Grouped combustion structures have been documented in other archeological levels (Carbonell et al. 2007), but the combustion structure groups in Ja contain a number of burned blocks and combustion structures that differ from those in other archeological levels. In the combustion structure groups of the interior zone the geological palimpsest makes measuring site size difficult. Furthermore, in Jb group 1 the size of the site cannot be clearly established due to the overlap of combustion structures (Fig. 3.12). In this Jb group, for instance, the variability between combustion structure size and linear spacing suggests an incongruent use or reoccupation (Brooks and Yellen 1987). In short, the size of the occupied floors of the interior zone of sublevel Ja and the central zone of Jb cannot be measured according to spatial patterns through archeological features like combustion structures and accumulations of blocks and slabs. However, empirical evidence of site size can possibly be found in the combustion structure groups of the central zone of sublevel Ja. The spacing of these combustion structure groups suggests an analogy with the external hearths described in prehistoric archeology and ethnoarcheology (Gamble 1990).

In sublevel Ja deposits of burned blocks were transferred to the boundaries between the central and external zones. Regular cleaning activities have been well documented in ethnoarcheology as part of the maintenance of domestic spaces in residential units (Schiffer 1987; Binford 1983). This evidence of cleaning in the combustion structures can orient the estimation of the integrity of the palimpsest and increase the archeological visibility of combustion structure reuse. Only in the central zone have burned travertine blocks and slabs been documented and their appearance in related waste deposits verify the congruent use (reuse) of these combustion structures. Finally, another analogy can be drawn with the bimodal spatial distribution of the combustion structures located in the central zone and inner zone of sublevel Ja.

The bimodal pattern of fire use in sublevel Ja points to the threshold of its archeological visibility when analyzed by means of ethnoarcheological analogy. The delimitation of analytical units is hard to prove when we observe the stratified use of fire in the inner zone of the rockshelter. However, more complex spatial units of analysis, like the distance between the combustion structures of the internal and central zones, constitute a scale of observation for testing and exploring the spatial patterns of residential units and site size determination (Henry et al. 2004; Binford 1987).

The spatial pattern of fire use, with combustion structures in the inner zone equidistant to those of the central zone, suggests an analogy based on observations of the residential

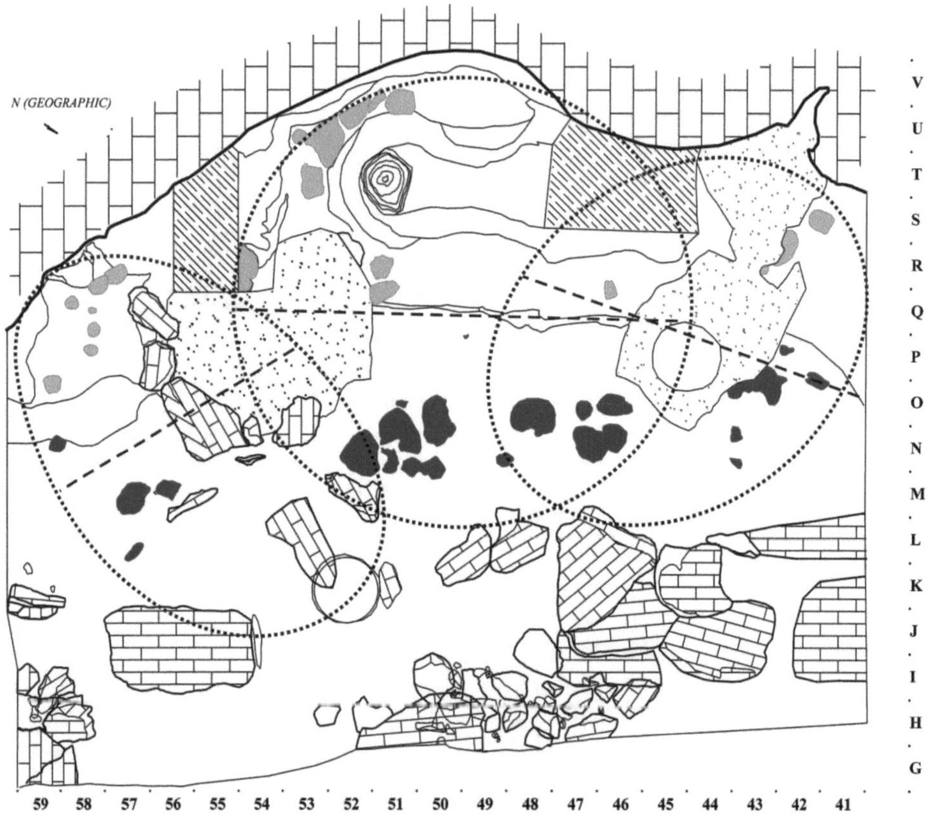

Fig. 3.40 Ethnoarcheological model of human settlement in open-air sites applied to determine site size based on the sublevel Ja record

units of open-air sites (Fig. 3.40) (Nicholson and Cane 1991). The few interzonal and very few bidirectional refits between the inner and central spatial units also suggests this spatial pattern, which can be used to estimate the size of the occupied floors. The scarce archeological visibility of the synchronic use of the two zones may be significant considering the geological palimpsest of the inner zone and the occupational palimpsest of the central zone. A high degree of reliability in the association between these spatial units, suggested by the direction of the denticulate tool connection, indicates inconsistencies in the interpretation of ISU-2 as a peripheral area of CSU-2 and the ethnological analogy with residential units of open-air sites (Gould and Watson 1982). In fact, this empirical evidence based on the direction of the refit may be interpreted as suggesting that reoccupation is more archeologically visible in ISU-2 than in CSU-2 (Stevenson 1991; Brooks and Yellen 1987).

The definition of residential camps has been summarized as a space with two zones, one of which is used as a sleeping area and the other as a cooking area (Binford 1978). Sleeping and resting areas are located along the internal band of this domestic space. Repetitive tasks, such as faunal processing and consumption, are performed around hearths in the external band. These domestic spaces are hard to identify in the spatial patterns of sublevel Ja. But other special purpose activity areas, suggested by the dumped accumulations of burned travertine blocks or the faunal remains in CSU-2 at the mouth of the *Coveta Ripoll*, have also been taken into account in order to identify the expected behavior in the generic structure of the residential occupations (Binford 1978). The burned rocks and large accumulations of bones and lithics in front of the *Coveta Ripoll* would correspond to refuse areas used to clear the limited available surfaces for inhabitation, as in the exogenous model of small caves presented by Thomas (1983) (Fig. 3.41). Finally, there are empty areas between the spatial units of the internal zones, which add an intriguing element in the attempt to determine the number of individuals in each occupation. In summary, the occupation floors of level J feature occupation dynamics characteristic of both open-air sites and small shelters and caves at the threshold of their archeological visibility.

The recognition of these organizational properties of the archeological remains cannot be based only on inductive inferences. Ethnological guidelines are required in order to suggest relevant observations about the cultural repertoires of human groups. Empirical evidence on the domestic spaces of archaic human groups has received little attention in archeological excavation strategies and research projects (Henry et al. 2004; Julien 2002; Kroll 1994). Therefore, the need to identify activity areas within the site structure continues to be one of the greatest challenges in elucidating the unknown cultural history of archaic human groups. The evidence presented in this monograph with regard to that

Fig. 3.41 Exogenous model of small caves for the cultural repertoires discovered in sublevel Ja (Thomas 1983)

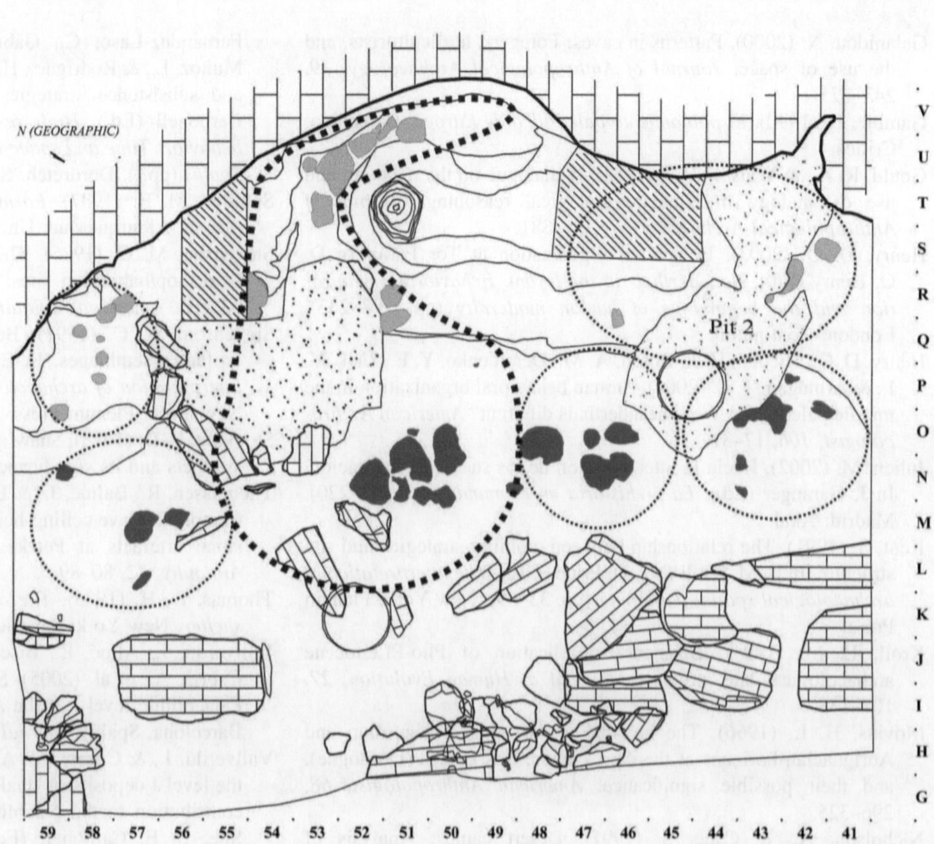

elusive cultural history provides insight into the territorial systems of modern human groups.

Acknowledgments We sincerely thank all the excavators who carefully recorded the spatial coordinates of the archeological remains recovered in level J. We would also like to thank two anonymous reviewers by their very helpful comments. Our thanks also go to the editor, Eudald Carbonell, for inviting us to participate in this book.

References

Bertran, P., & Texier, J. P. (1999). Sedimentation processes and facies on a semivegetated talus, Lousteau, Southwestern France. *Earth Surface Processes and Landforms, 24*, 177–187.

Binford, L. R. (1978). *Nunamiut ethnoarchaeology*. New York: Academic Press.

Binford, L. R. (1983). *In pursuit of the past. Decoding the archaeological record*. New York: Thames & Hudson.

Binford, L. R. (1987). Researching ambiguity : Frames of reference and site structure. In S. Kent (Ed.), *Method and theory for activity area research* (pp. 449–512). New York: Columbia University Press.

Binford, L. R. (1996). Hearth and home: The spatial analysis of ethnographically documented rock shelter occupations as a template for distinguishing between human and hominid use of sheltered space. In N. J. Conard & F. Wendorf (Eds.), *Middle Palaeolithic and Middle Stone Age settlement systems. Proceedings of the XIII UISPP Congress* (pp. 229–239). Forli: ABACO.

Brooks, A. S., & Yellen, J. E. (1987). The preservation of activity areas in the archaeological record: Ethnoarchaeological and archaeological work in northwest Ngamiland, Botswana. In S. Kent (Ed.), *Method and theory for activity area research. An ethnoarchaeological approach* (pp. 63–106). New York: Columbia University Press.

Canals, A. (1993). *Méthode et téchniques archéostratigraphiques pour l'étude des gisements archéologiques en sédiment homogène: application au complexe CIII de la Grotte de Lazaret, Nice (Alpes Maritimes)*. Ph.D. dissertation.

Carbonell, E. (Ed.). (1992). Abric Romaní, Nivell H: Un Model D'estratègia Ocupacional Al Plistocè Superior Mediterrani. *Estrat, 5*, 159–308.

Carbonell, E. (Ed). (2002). *Abric Romaní Nivell I. Models d'ocupació de curta durada de fa 46.000 anys a la Cinglera del Capelló (Capellades, Anoia, Barcelona)*. Tarragona: Universitat Rovira i Virgili.

Carbonell, E., Cebrià, A., Allué, E., Cáceres, I., Castro, Z., Díaz, R., Esteban, M., Ollé, A., Pastó, I., Rodríguez, X. P., Rosell, J., Sala, R., Vallverdú, J., Vaquero, M., & Vergès, J. M. (1996). Behavioural and organizational complexity in the middle palaeolithic from the Abric Romani. In E. Carbonell & M. Vaquero (Eds.), *The last neandertals the first anatomically modern humans: A tale about the human diversity. Cultural change and human evolution: The crisis at 40 Ka BP* (pp. 385–434). Tarragona: Universitat Rovira i Virgili.

Carbonell, E., Lorenzo, C., & Vallverdú, J. (2007). Centralidad espacial y operativa de los neanderthales. Análisis espacial diacrónico de las actividades de combustión en el Abric Romaní (Capellades, Anoia, Barcelona). In E. Baquedano (Ed.), *Universo Neanderthal 1* (pp. 197–219). Soria: Fundación Duques de Soria.

Galanidou, N. (2000). Patterns in caves: Foragers, horticulturists, and the use of space. *Journal of Anthropological Archaeology, 19*, 243–275.

Gamble, C. (1990). *El poblamiento paleolítico de Europa*. Barcelona: Critica.

Gould, R. A., & Watson, P. J. (1982). A dialogue on the meaning and use of analogy in ethnoarchaeological reasoning. *Journal of Anthropological Archaeology, 1*, 355–381.

Henry, D. O. (2003). Behavioral organization at Tor Faraj. In D. O. Henry (Ed.), *Neanderthals in the levant. Behavioral organization and the beginnings of human modernity* (pp. 198–233). London: Continuum.

Henry, D. O., Hietala, H. J., Rosen, A. M., Demidenko, Y. E., Usik, V. I., & Armagan, T. L. (2004). Human behavioral organization in the middle paleolithic: Were neanderthals different? *American Anthropologist, 106*, 17–31.

Julien, M. (2002). Hacia la interpretación de los suelos de habitación. In J. Garanger (Ed.), *La prehistoria en el mundo* (pp. 195–220). Madrid: Akal.

Kent, S. (1991). The relationship between mobility strategies and site structure. In E. M. Kroll & T. D. Price (Eds.), *The interpretation of archaeological spatial patterning* (pp. 33–59). New York: Plenum Press.

Kroll, E. M. (1994). Behavioral implication of Plio-PLeitocene archaeological site structure. *Journal of Human Evolution, 27*, 107–138.

Movius, H. L. (1966). The hearths of the upper Perigordian and Aurignacian horizons at the Abri Pataud, Les Eyzies (Dordogne), and their possible significance. *American Anthropologist, 68*, 296–325.

Nicholson, A., & Cane, S. (1991). Desert camps: Analysis of Australian aboriginal proto-historic campsites. In C. S. Gamble & W. A. Boismier (Eds.), *Ethnoarchaelogical approaches to mobile campsites* (pp. 263–354). Ann Arbor, MI: International Monographs in Prehistory.

Perlès, C. (1976). Le feu. In H. de Lumley (Ed.), *La Préhistoire française. Tome I: Les civilisations paléolithiques et mésolithiques de la France* (pp. 679–683). Paris: CNRS.

Rolland, N. (2004). Was the emergence of home bases and domestic fire a punctuated event? A review of the middle pleistocene record in Eurasia. *Asian Perspectives, 43*, 248–280.

Rose, L., & Marshall, F. (1996). Meat eating, hominid sociality, and home bases revisited. *Current Anthropology, 37*, 307–338.

Rosell, J., Blasco, R., Huguet, R., Cáceres, I., Saladié, P., Rivals, F., Bennàsar, M., Bravo, P., Campeny, G., Esteban-Nadal, M., Fernández-Laso, C., Gabucio, M. J., Ibáñez, N., Martín, P., Muñoz, L., & Rodríguez-Hidalgo, A. (2012). Occupational patterns and subsistence strategies in level J of Abric Romaní. In E. Carbonell (Ed.), *High resolution archaeology and neanderthal behavior: Time and space in level J of Abric Romaní (Capellades, Spain)* (pp.). Dordretch: Springer.

Schiffer, M. B. (1987). *Formation processes of the archaeological record*. Alburquerque: University of the New Mexico Press.

Stevenson, M. C. (1985). The formation of artefact assemblages at workshop/habitation sites: Models from peace point in northern Alberta. *American Antiquity, 50*, 63–81.

Stevenson, M. C. (1991). Beyond the formation hearth-associated artifact assemblages. In E. M. Kroll & T. D. Price (Eds.), *The interpretation of archaeological spatial patterning* (pp. 269–299). New York: Plenum Press.

Stockton, E. D. (1973). Shaw's creek shelter: Hhuman displacement of artefacts and its significance. *Mankind, 9*, 112–117.

Theunissen, R., Balme, J., & Beck, W. (1998). Headroom and human trampling: Cave ceiling-height determines the spatial patterning of stone artefacts at Petzkes Cave, Northern New South Wales. *Antiquity, 72*, 80–89.

Thomas, D. H. (1983). *The archaeology of monitor valley: Gatcliff shelter*. New York: American museum of natural history.

Vallverdú, J., Allué, E., Bischoff, J. L., Cáceres, I., Carbonell, E., Cebrià, A., et al. (2005). Short human occupations in the middle Palaeolithic level i of the Abric Romaní rock-shelter (Capellades, Barcelona, Spain). *Journal of Human Evolution, 48*, 157–174.

Vallverdú, J., & Courty, M.-A. (2012). Microstratigraphic analysis of the level J deposits. A dual paleoenvironmental-paleoethnographic contribution to the paleolithic archaeology at the Abric Romani Site. In E. Carbonell (Ed.), *High resolution archaeology and neanderthal behavior: Time and space in level J of Abric Romaní (Capellades, Spain)* (pp.). Dordretch: Springer.

Vaquero, M., Chacón, M.G., Cuartero, F., García-Antón, M[a] D., Gómez de Soler, B., & Martínez, K. (2012). The lithic assemblage of level J. In E. Carbonell (Ed.), *High resolution archaeology and neanderthal behavior: Time and space in level J of Abric Romaní (Capellades, Spain)* (pp.). Dordretch: Springer.

Vaquero, M., & Pastó, I. (2001). The definition of spatial units in middle Palaeolithic sites: The hearth-related assemblages. *Journal of Archaeological Science, 28*, 1209–1220.

Vaquero, M., Vallverdú, J., Rosell, J., Pastó, I., & Allué, E. (2001). Neandertal behaviour in the middle Palaeolithic site of Abric Romani, Capellades, Spain. *Journal of Field Archaeology, 28*, 93–114.

Chapter 4
Microstratigraphic Analysis of Level J Deposits: A Dual Paleoenvironmental-Paleoethnographic Contribution to Paleolithic Archeology at the Abric Romaní

Josep Vallverdú-Poch and Marie-Agnès Courty

Abstract Our microstratigraphic study, based on microfacies analyses of large thin sections, aims to improve our perception of level J deposits obtained by field observations during its extensive excavation. The refined understanding of pedo-sedimentary processes is at first expected to decipher the mineralogical homogeneity of carbonate sedimentary facies as a suite of depositional episodes controlled by distinctive changes in climate-linked factors (hydrology, thermal regime, vegetation) in a karstic setting. Estimating the magnitude, timing and duration of these processes at the finest level of resolution helps to sketch an image of the changing environmental setting of human occupation in this rockshelter. It provides a paleoecological reconstruction, which offers a framework for the paleoethnographic analysis of microfacies and contributes to the multidisciplinary prehistoric archaeological project undertaken at the Abric Romaní.

Keywords Microstratigraphy • Paleoenvironment • Formation processes • Palimpsest • Hearth

The Abric Romaní Archeological Complex J: The Field Perception

The most relevant features of the level J surface observed in the field are related to the use of fire, the presence of travertine blocks and significant accumulations of anthropic remains and artifacts (Fig. 4.1). Quantitatively, the anthropogenic imprint is obvious, characterized by large quantities of remains and carbonaceous (charcoal-rich) sediments, which hinder a clear delimitation of the accumulations. Lithic and bone densities are comparable with those of level E. In his doctoral thesis, Vaquero provides numerous examples of refits throughout the entire excavated surface (Vaquero 1997).

It is especially interesting to look at the initial hypotheses set out in the excavation logbook to explain the formation processes of level J. He records a surface covered with an accumulation of small to medium-sized blocks located behind a series of large blocks. This area of block accumulation yielded large bone remains and dispersed accumulations of dark sediments (Fig. 4.2). This was described as either an activity area or a refuse zone resulting from cleaning activities (Martínez Molina and Rando Moral 2000). Although less systematically, the presence of small-sized material inside the rockshelter was also initially noted. At the top of archeological level J, yellow organo-mineral accumulations were documented, interpreted as coprolites when spherical. Other accumulations with the shape of well-defined thin lenses, which contained bones, were interpreted as carnivore regurgitation deposits. Another component set out in the logbook are the poorly-defined spatial patterns of the combustion structures and the archeological artifacts, or the hearth related assemblages (Vaquero and Pasto 2001; Fig. 4.3). Vaquero's log also emphasizes the presence of a scour in the central area of the shelter, in the boundary between the central zone and the internal zone, a feature of considerable importance in the formation processes of level Ja. The scour area occupies a significant area between P46 and P51. Water dripping and flowing over the talus slope and the rockshelter wall, or from the stalagmite of the *Coveta Ripoll*, might also explain the low density of archeological remains in this empty area of level J. Scouring episodes or erosion might also explain the poorly-defined contours observed in the combustion structures located in these sectors (Fig. 4.4). The logbook explains the difficulties encountered by the excavation team

J. Vallverdú-Poch (✉)
Institut Català de Paleoecologia Humana i Evolució Social (IPHES), Universitat Rovira i Virgili (URV), C/Escorxador s/n 43003, Tarragona, Spain
e-mail: josep@prehistoria.urv.cat

M.-A. Courty
CNRS-MNHN UMR 5198, Institut Català de Paleoecologia Humana i Evolució Social (IPHES), C/Escorxador s/n 43003 Tarragona, Spain
e-mail: courty@mnh.fr

E. Carbonell i Roura (ed.), *High Resolution Archaeology and Neanderthal Behavior: Time and Space in Level J of Abric Romaní (Capellades, Spain)*, Vertebrate Paleobiology and Paleoanthropology,
DOI: 10.1007/978-94-007-3922-2_4, © Springer Science+Business Media B.V. 2012

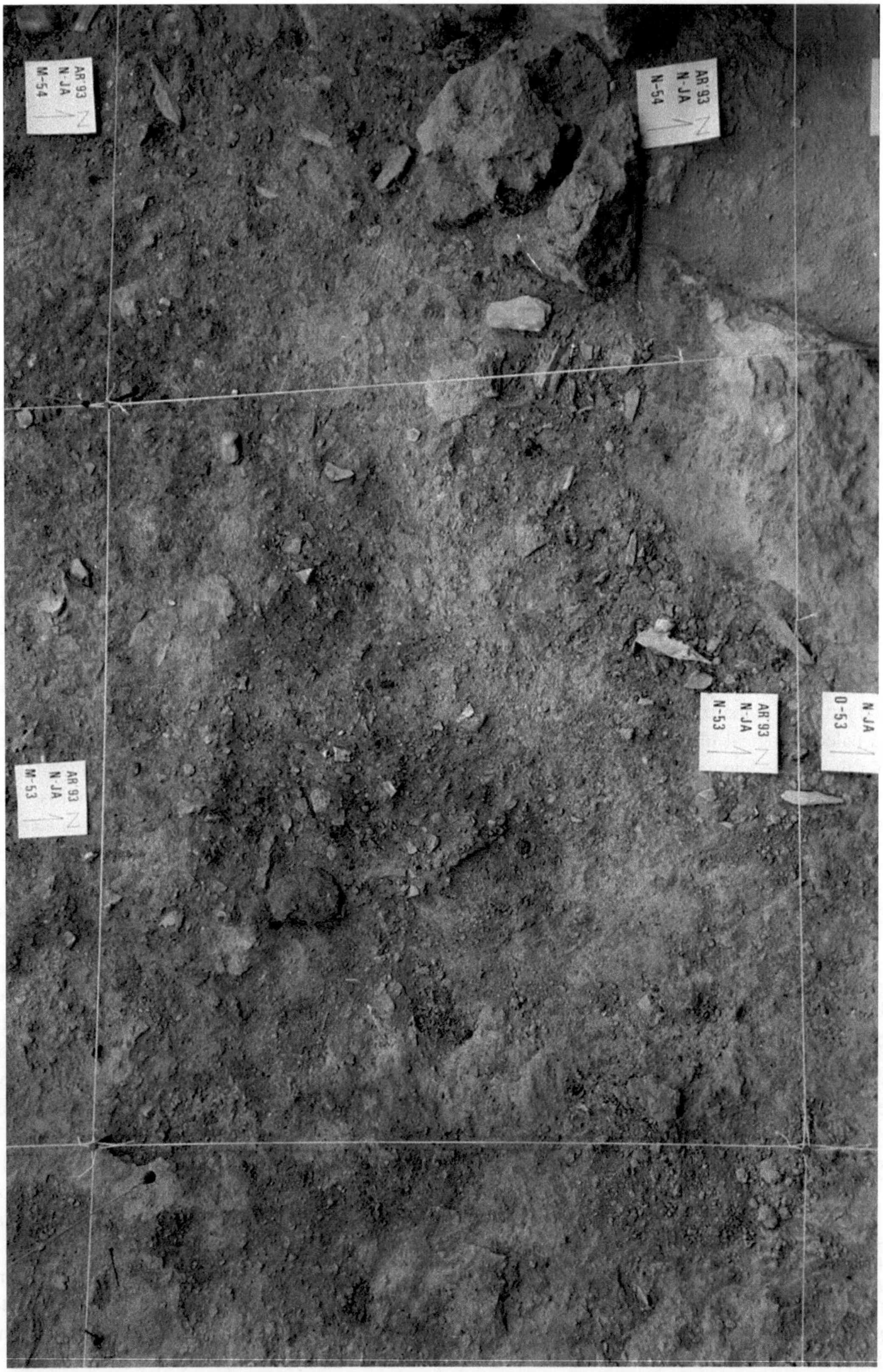

Fig. 4.1 Photograph showing a small area in the central zone of sublevel Ja made by the accumulations of stones and bones with carbonaceous sediments

Fig. 4.2 Arrangement of large bone remains accumulated between flat travertine blocks between the central and external zones of sublevel Ja

to propose a coherent model, and describes the presence of small objects, burned stones and a heterogeneous sedimentological composition. Rando Moral (1999) attributes the scarcity of bones and the absence of modifications to a fluvial current (Rando Moral 1999). Other evidence of fluvial activity in level J is supported by field observations of the excavated

Fig. 4.3 Combustion structure group 1 showing travertine blocks and slab in the central zone of sublevel J. A slab of travertine, which is darker due to thermal impact, can be seen in the *center* to the *left* of the photograph. Also noticeable is the good delimitation of the each combustion structure

lithological areas. The base of level Ja in the central zone is composed of sands and dripping tufa that make up a carbonate incrustation (travertine), the upper surface of which is irregular. The artifacts of level Ja were primarily documented between sands and gravel in almost all the field annotations. Lastly, a layer of yellowish-brown sands and tufa was also detected in level Jsup. Level Jsup, which is archeostratigraphically clearly distinct from level Ja, and is only documented in the northern-central zone of the excavated area.

The stratification of levels Ja and Jb is found between rows L and O and columns 43 and 53 (Vaquero 1997), and other sublevels were delimited in the field. The areas in which level Ja cannot be distinguished from Jb are categorized as level "J", specifically in the *Coveta Ripoll* and on part of the carbonate wall. In the eastern zone, almost all the materials recovered were assigned to Ja. In the external zone, elements between the blocks were assigned to "Ja/Jb"

Level J Lithostratigraphic Units (Informal)

At the outcrop below the *Coveta Nord* in the central zone, the stratigraphy of level J appears in the field as a distinctive succession of well-stratified units or beds, in which sublevels Ja and Jb can be distinguished. From bottom to top, the lithostratigraphic units are described as follows (Fig. 4.5):

1. Lowermost level J deposits. Fine to medium-size crystalline platy gravels, open-structured to clast-supported filled with a bright yellowish brown sand matrix. The gravels contain concretions. Towards the wall slope, clast-supported gravels filled up light yellow silt matrix.
2. Yellowish-brown massive sands, micro-stratified with discontinuous and horizontal fine platy crystalline gravels, and large meter-sized fallen blocks.
3. Archeological level Jb. Fine platy crystalline gravels and blocks with open-structure to clast-supported bright gray sand matrix. Platy gravels are slightly cemented into concretions.
4. Light brown massive sand and platy fine crystalline gravels. Platy gravels showing slight concretion, and subrounded tufa-clast gravels in the very coarse fraction. Against the wall, platy fine to medium-sized gravels have an open structure to clast-supported filled with a sand matrix.
5. Archeological level Ja. Platy and medium crystalline gravels with open structure to clast-supported filled with light brown sands. Meter-sized blocks with bright yellow sands and gravels are present up to the rockshelter wall.
6. Archeological level Jsup. Weakly cemented graded sand in microstratified beds made up of very light brown to bright gray sands, tufa gravel-sized and decimeter-sized blocks. Towards the foot of the rockshelter wall, these sands become fine platy crystalline gravel microbreccia. There are also superficial pisoids and ooids, and gravel-sized subrounded tufa fragments.

Level J Sedimentary Forms and Lithofacies

The level J deposits have been related to rock fall sedimentary processes in the Abric Romaní cliff. Talus deposits, which occur on very steep bedrock cliffs and slopes, are a

Fig. 4.4 Photograph of sublevel Ja as seen from the north, and the monoliths of undisturbed samples to make thin sections. At the foreground, there is the well-delimited combustion structure group 3 of the central zone. Towards the *right* of the photograph, the talus slope deposits are lighter in color and form a gentle slope. In the *center*, combustion structures with irregular and diffuse contours appear in the *right half* of the photograph

very common sedimentary form in prehistoric caves and rockshelters. The deposits of level J are a stratification between the talus slope sediments derived from the travertine wall of the rockshelter, and the talus cone sediments, fallen from the dripline of the roof of the Abric Romaní (Figs. 4.6 and 4.7). The units of level J display marked changes in the lateral facies, which are related to these two possible sediment sources. These two sedimentary sub-environments delineate the internal zone, the central zone and the external zone.

In the central zone the sedimentary lithofacies forms are lenticular and wedge-shaped (Fig. 4.5). The vertical profile form a rhythmic sedimentary succession made up of: (i) facies 1 of fine to medium-size platy gravels in poorly microstratified beds, open-structured (clast-supported) to totally filled with sands, and including blocks and mega-blocks; (ii) facies 2 of sands and granules in massive to graded beds with blocks. The thickness of these sedimentary facies coarsens upward to facies 2 and fines upward to facies 1 (Fig. 4.5). In Fig. 4.5, the column lists at least four episodes of block-fall that have been identified inside the archeological level J complex. Relevant mega-blocks are observed at the base of level J, which indicates the initial retreat of the rockshelter cliff. Meanwhile, the thickness of facies 2 at the top of the sedimentary succession suggests a volumetric increase of the carbonate curtain in the roof of the rockshelter cliff, and an areal extension of the related sedimentary products, and hence a lateral accretion of the talus cone deposits onlapping over the internal zone of the rockshelter fill.

Materials and Methods

A sedimentary geological study is always required in depositional contexts occupied by past humans. In the context of level J, this study makes an original contribution to the identification of the effects of anthropogenic activities on sedimentation processes. Meanwhile, the study of sedimentary and pedological processes in reaction to the effects of natural factors contributes to broadening our knowledge of the depositional history in rockshelters and caves in karstic environments affected by past environmental changes. In addition to biochemical sedimentation, cryoclastism and granular disaggregation appear to have played a major role in the sedimentary process of level J. The combined actions of runoff, cryoturbation, and biological crust formation have contributed to the modification of the initial clastic deposits during their accumulation on the ground surface.

Sampling Strategies and Objectives

The initial horizontal sampling strategy adopted for level J comprised the collection of sediment samples at different points in the combustion structures. Complementary samples were collected along two microstratigraphic transects in order to complement the stratigraphic description between the central and internal zones performed in the field. Additional elements from particular locations were

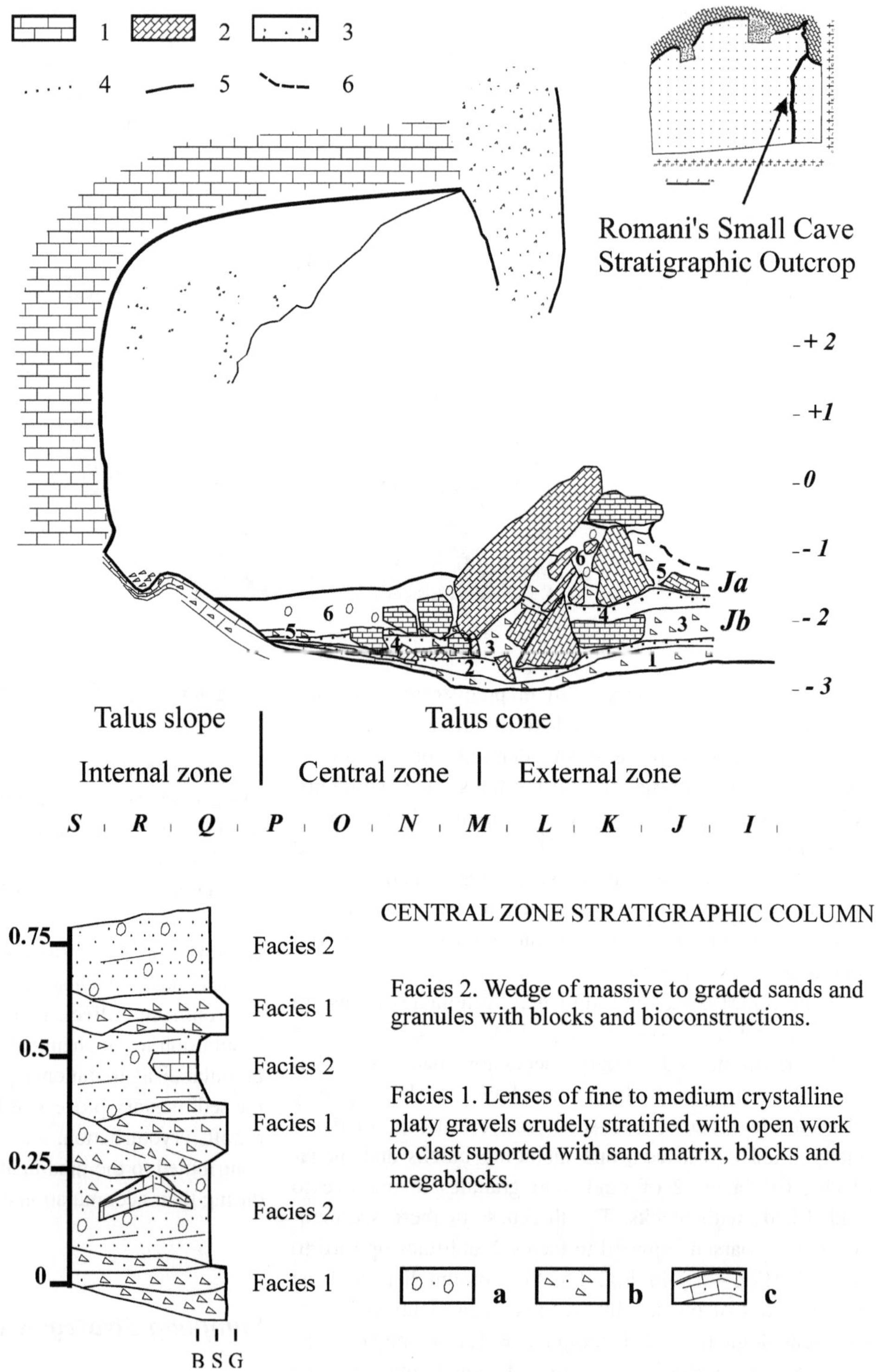

Fig. 4.5 Lithostratigraphic panel of central and internal zones of J level described in the *Coveta Nord* outcrop. Below, synthetic stratigraphic column of the central zone. *1* shelter wall; *2* megablocks; *3* carbonated curtain; *4* archeological level; *5* boundary surfaces; *6* disturbed surface. B, boundstone; S, sand (calcarenites); G, gravel; **a**, subrounded gravel and sand; **b**, fine to medium platy crystalline gravel and sand; **c**, boundstone

sampled, and other components, such as coprolites, were gathered to answer specific questions.

Basically, two types of samples were taken: samples of loose sediment, mainly from combustion structures; and undisturbed samples, in order to make large (125 mm × 65 mm) thin sections. The large thin section sediment samples come from the excavated surfaces of Ja and Jb (Fig. 4.6). Some of the samples are also indicated on the maps of sublevels Ja and Jb because they could not be distinguished during sampling. In the tables (see Appendix 2) we also indicate field recordings and the number of thin sections obtained for each sample, grouped according to the field strategy. Sampling of combustion structures was performed during the fieldwork in order to

Fig. 4.6 Photograph showing level J talus cone deposits in the central zone. The outcrop of column 43 in the Romaní grid. The combustion structure at the *top* of the photograph corresponds to sublevel Ja. The sedimentary succession shows the lithofacies rhythm of gravels and sands with blocks overlapping above the talus slope deposits on the *bottom left* of the photograph

clarify the anthropic formation processes, as well as lateral and vertical changes of facies in the zones where it was not possible to establish a distinction between sublevels Ja and Jb.

The decision to carry out microstratigraphic sediment sampling and spatial strategy sampling was initially made in 1994 in order to characterize sublevel Ja. But during field operations and because it was difficult to separate sublevel Ja from Jb in some areas of the excavated surface, the scope of field sampling was expanded to include sublevel Jb during 1995 fieldworks. Most of the samples from sublevel Ja were taken from from the central zone limited by columns 48 and 51 and rows M and R (Fig. 4.8). The sampling strategy for sublevel Jb also focuses on the central zone, where an axis of combustion structures was found in N46–50 (Fig. 4.8).

Description and Classification of Microfacies

Thin sections were initially described by developing a processed-based approach, derived from the guidelines provided by soil micromorphology and sedimentary petrography handbooks (Leighton and Pendexter 1962; Elf-Aquitaine 1975; Bullock et al. 1985; Douglas and Thompson 1985; Tucker 1988; Freytet and Verrecchia 1989; Verrecchia 1989; FitzPatrick 1993; Fédoroff and Courty 1994; Adams et al. 1997; Gómez Gras 1997; Stoops 2003). Thin section description was directed toward the analysis of the pedo-sedimentary characteristics of each microstratigraphic unit, considered as the result of the succession and interaction of distinctive depositional and postdepositional processes that are controlled by the interplay between natural and human agents. In this procedure, a detailed classification of pedo-sedimentary microfacies has been established to facilitate the analysis of a large number of thin sections sampled from a single depositional setting and showing subtle facies changes. Therefore, this classification is based on the distinctive categories of the pedo-sedimentary facies.

The classification of the microfacies of level J was also enhanced by estimating the weathering intensity undergone by carbonate rock fragments. Weathering intensity was identified by referring to a previous study that provided a calibration between carbonate crystalline and cryptocrystalline features and micro-environmental conditions in cave and rockshelter deposits using modern analogues (Courty 1986). This method also bridges the investigative gap between the travertine cliff and the rockshelter fill subsystems. The parameters used for relating these two subsystems are set out in Fig. 4.9. The description of the thin sections covers the following general aspects:

1. Texture of the coarse fraction (CF) and the fine fraction (FF). Frequency estimation, nature (biocrystalline, impregnative, crystalline) of the very coarse fraction (VCF). Description of the CF/FF ratio and the related distribution of both. Characterization of the shape of the Coarse Fraction. Determination of the abundance and frequency of sedimentary components of mineral and biological origin (Tables 4.1, 4.2). Estimation of the abundance of pedofeatures and anthropic components.
2. Degree of weathering intensity (weak, strong and late) of the carbonate rock fragments. Carbonate weathering is estimated from the corrosion features of the crystalline calcium carbonate components, their loss of birefringence and/or partial disaggregation, and the degree of amorphization of the calcitic fine mass (Courty et al. 1994b). For micritic and microsparitic components, corrosion develops from the periphery, resulting in the micritic fine mass indicated by a brown color. Corrosion of the sparitic components develops from cleavage planes, which show irregular outlines and iridescence with small cavities or fenestral porosity, along with a reduced transparency. More specifically, the identification and description of crystalline calcitic pedofeatures and sedimentary processes in the vadose and epiphreatic zones have also been established employing micromorphological and

Fig. 4.7 Detail of the talus slope stratigraphy near the rockshelter wall in the internal zone of level J made from the dripping dome which closes the *Coveta Ripoll*. Photograph taken during fieldwork in level K. Level J is located in the *upper* surface boundary of this dome-shaped sedimentary succession

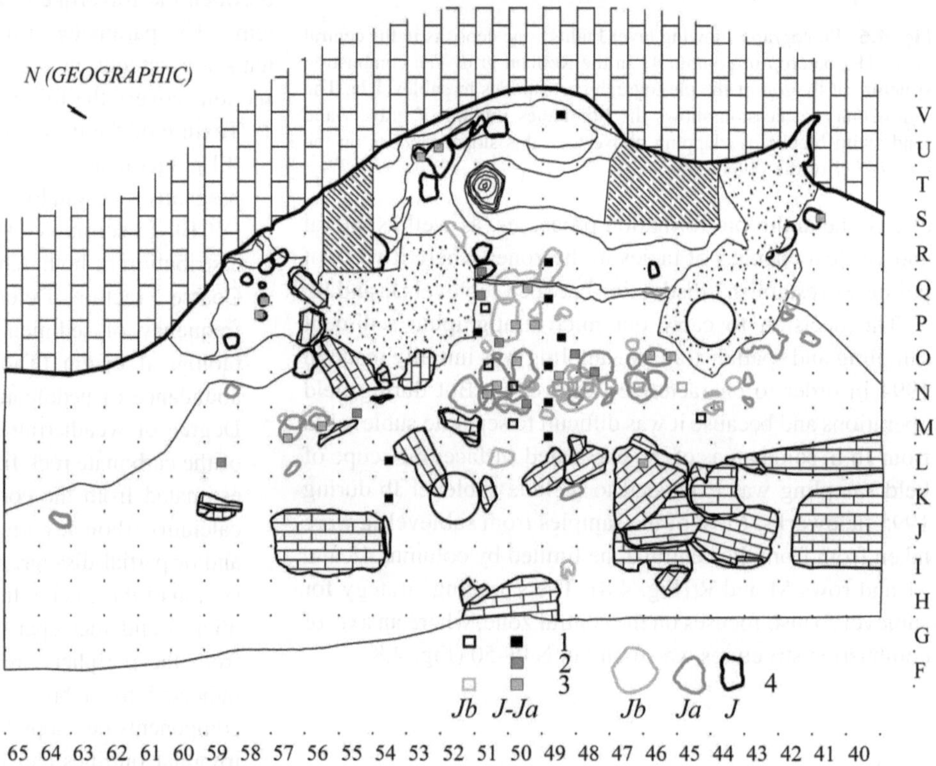

Fig. 4.8 Undisturbed samples (monoliths) and combustion structures located at the surface of level J. Legend: *1* microstratigraphic transect monoliths; *2* other sampled monoliths; *3* combustion structure monoliths; *4*, combustion structures shapes

petrographic terminology (Kahle 1977; Freytet and Verrecchia 1989). These criteria have allowed three stages of weathering intensity to be distinguished that are specific to environmental conditions (Table 4.3).

3. Sedimentary structure. Naked-eye examination and scan of thin sections provide complementary information on the sedimentary bedding, its geometry, bed boundary (boundary type and form) and its overall geometry

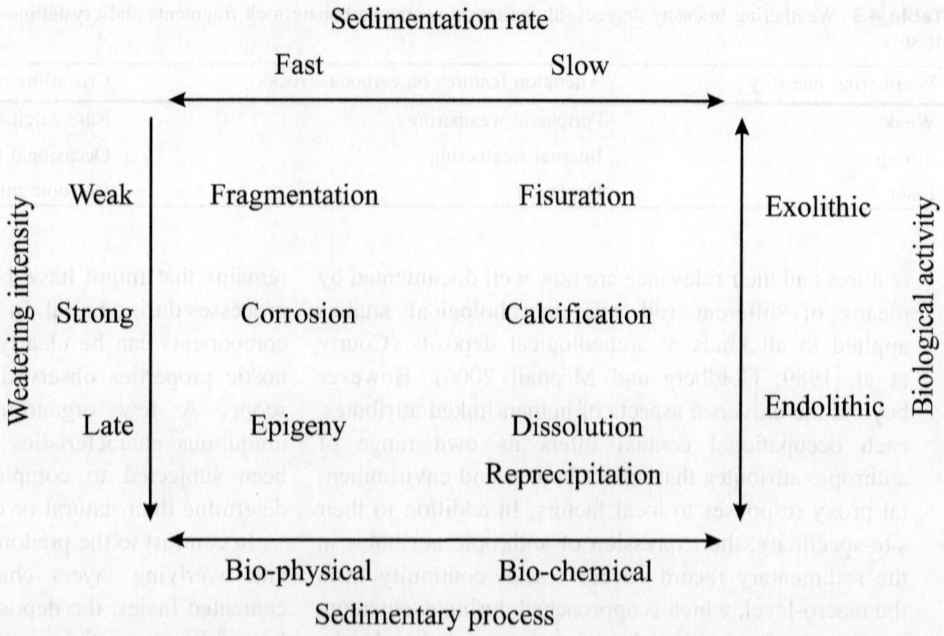

Fig. 4.9 Weathering intensity of carbonate rock fragments and related sedimentary processes, sedimentation rate and biological activity observed on microscopic features at the level J microfacies (based on Courty 1986)

Table 4.1 Frequency class for particular components (Bullock et al. 1985)

Frequency	%
Very dominant	>70
Dominant	50–70
Common	30–50
Frequent	15–30
Few	5–15
Very few	<5

Table 4.2 Abundance of anthropic components and pedofeatures (Bullock et al. 1985)

Abundance	%
Very abundant	>20
Abundant	10–20
Many	5–10
Occasional	2–5
Rare	<2

observed in the field (Campbell 1967). For the anthropic facies of level J, we considered more specifically:

- (i) laminar sedimentary structures, about 1 cm thick, which are continuously parallel with a gradual boundary ranging from straight to undulated.
- (ii) microlaminar sedimentary structures, between 2 and 5 mm thick, which are discontinuous and non-parallel with a straight to undulated boundary.
- (iii) stratified sedimentary structures, over 2 cm thick, which are non-parallel and non-continuous with an undulated to gradual boundary.

In addition, the internal fabric of the sedimentary structures has been described combining the descriptive criteria proposed by soil micromorphology and sedimentary petrography specialists (Humbert 1976; Tucker 1988, 2001; Courty 1990).

4. Microstructure. Microstructure refers to the geometry of solid components and related voids at mesoscopic scales. The soil's capacity to organize into aggregates is called pedality. Heterogeneous materials might be made up of more than one type of microstructure. Pedological and sedimentary structure description helps in identifying the polarity of the surfaces of the microstratigraphic units, their structural evolution, hydrological regime and anthropic disturbances caused during burial (Boiffin and Bresson 1987; Gé et al. 1993; Chartres et al. 1994).
5. Pedological features. There are systematic guides to pedological features described in thin sections of soils and regoliths, some with an emphasis on examining carbonate crystalline and cryptocrystalline features (Fig. 4.8; Bullock et al. 1985; Verrecchia 1989; Rabenhorst et al. 1991; Stoops 2003). Soil micromorphology contributes to the microscopic identification and classification of pedofeatures, and the estimation of the maturity of relict pedological processes (Douglas and Thompson 1985; Rabenhorst et al. 1991; Magaldi and Tallini 2000).
6. Anthropic features. Traces of the effects of humans within host sediments comprise both distinctive anthropic microvestiges, which are closely linked to material culture, and specific assemblages produced by human-linked mechanisms, i.e. burning, trampling, dumping or material-processing (Courty et al. 1989, 1994a; Wattez 1990; Gé et al. 1993; Chartres et al. 1994; Macphail et al. 1997; Stoops 2003). The key characteristics of anthropic

Table 4.3 Weathering intensity degrees, alteration on coarse carbonate rock fragments and crystalline-criptocrystalline pedofeatures (Courty 1986)

Weathering intensity	Alteration features on carbonate rocks	Cristalline and cryptocrystalline pedofeatures
Weak	Peripheric weathering	Rare calcitic pedofeatures
Strong	Internal weathering	Occasional to abundant calcitic pedofeatures
Later	Depletion	Isotropic micromass with carbonate punctuations

features and their relevance are now well documented by means of different soil micromorphological studies applied to all kinds of archeological deposits (Courty et al. 1989; Goldberg and Mcphail 2006). However beyond the universal aspects of human-linked attributes, each occupational context offers its own range of anthropic attributes that reflect human and environmental proxy responses to local factors. In addition to their site-specificity, the expression of anthropic attributes in the sedimentary record covers a scale-continuity from the macro-level, which is approached during excavation, down to the finest scales that are only revealed under the microscope. Therefore, analyzing the anthropic attributes that are recorded in the host sedimentary matrix remains a delicate task. Following the guiding principles now commonly used in archeological microfacies analysis, we intend to build here, from the characteristics of level J, the first database of anthropic features specific to the Abric Romaní.

Lastly, the organization of these parameters is expressed by defining sedimentary microfacies according to their textural characteristics, color and microstructure. The anthropic microfacies of level J are classified and organized according to sedimentary structure type, and other parameters, such as the estimation of the frequency of burned sedimentary materials and microaggregation, are systematically added. In this classification we generate subtypes that are primarily related to the nature of basic components and postdepositional modifications of these anthropic microfacies, by describing their porosity, the development of aggregation, the microstructure and the CF/FF related distribution.

Results: Basic Components and Relevant Microfacies

Clastic and Anthropic Basic Components

The basic components are made up of various types of mineral and organic elements that are either of natural origin (although potentially affected by human activities as long as they remain on the ground surface) or cultural remains that might have been altered by pedosedimentary processes during burial. In level J, most natural and cultural components can be clearly separated based on their diagnostic properties observed under the petrographic microscope. A few organo-mineral elements that display ubiquitous characteristics at the microscopic level have been subjected to complementary analyses in order to determine their natural or cultural origin.

In contrast to the predominance of tufa in the underlying and overlying layers characterized by their distinctive cemented facies, the deposits of level J are mostly derived from fallen travertinic wall materials that were then disaggregated and transported short distances by means of flowing water. Rare in situ tufas display the great diversity of the crystalline assemblage typical of travertine produced from the symbiotic growth of mosses, algae and bacteria (Verrecchia et al. 1995). The diverse carbonate components of biological origin can therefore be recognized in most cases due to to their crystalline fabric, although they occur as a continuum of clastic elements ranging from gravels to silt. Carbonate constituents modified by pedological processes during the formation of level J show a wide range of weathering stages that have undergone a variable loss of their original characteristics. The travertine rock fragments with nearly intact morphology and a distinctive dissolution pattern along cleavage planes are typical of the rapid circulation of water unsaturated with $CaCO_3$, especially water derived from snowmelt. In contrast, carbonate rock fragments that display diffuse contours and darkened micrite reflect more intense weathering by slow percolation of solutions highly enriched in organic acids. Pale yellow to yellowish-brown carbonate rock fragments with an angular to sub-angular morphology and a homogeneous internal fabric represent the late stage of carbonate weathering. Their amorphous fine mass consists of weakly crystallized hydroxyapatite that replaced the original calcitic fabric resulting from biochemical alteration due to the growth of a cryptogamic plant association on the cliff travertine walls or on the ground surface of the rockshelter. The hardness of this cryptogamic weathering crust contrasts with the friability of the dissolved carbonate rock fragments, which explains why they are commonly found in the coarse fraction.

Ooids and pisoliths occur repeatedly throughout the level J deposits expressing the rhythmic precipitation of carbonates controlled by bacterial activity. In the context of the

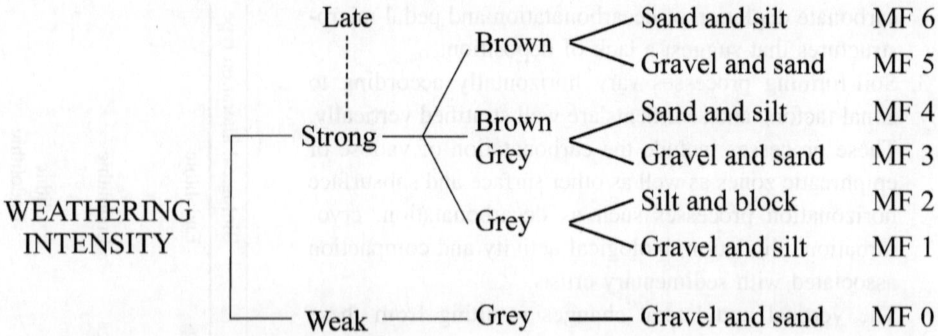

Fig. 4.10 Particle sizes, colors and intensity of weathering on the carbonate rock fragments exemplifying the microfacies classes established

deposits of level J, these ooids and pisoliths formed locally in areas where intense dripping of carbonate-saturated water occurred, and may have later been transported by a higher energy flow.

The low detrital content of the travertine precursor has resulted in an overall petrological homogeneity, although schist fragments derived from the surroundings are occasionally found. This provides an ideal context in which to fully explore the environmental relevance of the microfacies made up of natural elements in relation to the two linked conditions controlling the rate and type of clastic production: (1) meteorological factors affecting the travertine rockshelter wall just before block detachment; (2) local environmental factors at the soil surface scale of the rockshelter and the postdepositional evolution of the soil and occupational surfaces during burial processes.

Lithics, bones, charred plant fragments and burned clasts found in thin section are the most obvious cultural materials and are consistent with the larger remains retrieved during the excavation and sieving of level J deposits. The ashy components, as complex deposits with significantly variable consistency and color, were not clearly visible in the field. At the microscopic level, they display a range of distinctive crystalline and crypto-crystalline assemblages and morphologies that allow their history in the occupation area and their transformation during burial to be traced. The micro-artifact assemblage presents variable degrees of heating, although all to a limited extent, repeated in different locations across the excavated surface, which provides complementary information for identifying human activities on the finest scale, especially combustion parameters (burning processes, intensity, duration, combustible source, dispersion mode), and the occupational context.

The organo-mineral micro-aggregates occur as angular to subangular elements, with colors ranging from yellow to reddish-yellow, brown or pale brown, either randomly distributed or locally concentrated bearing evidence of in situ fragmentation. Their isotropic fine mass shows pure crypto-crystalline microaggregates, with microcontrasted particles occurring as rare silicate minerals or crystal intergrowths of druse microsparite typical of the ashes of burned deciduous leaves (Courty et al. 1989). Although the exact origin of the organo-mineral components still requires further research, their properties seem to match human-processed materials originating from numerous sources, including local plants, which may have been used as a 'floor' covering (Macphail et al. 1997; Nadel et al. 2004).

Clastic Microfacies of the Level J Deposits

These sedimentary and soil processes form a well delimited conceptual paleoenvironmental framework for the deposits and microfacies of level J (Fig. 4.10). The microfacies of the level J deposits were classified according to the degree of weathering of the carbonate rock fragments and the soil micromorphology of carbonate crystalline and cryptocrystalline pedofeatures based on sedimentary petrographic examination (Fig. 4.9). The micromorphological description and microfacies characteristics and inferences are presented in Appendix 1 and summarized in Tables 4.4 and 4.5. Our main paleoenvironmental interpretations are also presented in Table 4.6. These results indicate that:

1. The sediments originated in the travertine cliff and contain angular fine to coarse silt-size silicate components. They are rare in lithofacies 1, and occur occasionally in lithofacies 2, indicating an aeolian sediment input.
2. The primary sedimentary process is the cryoclastic fall of gravel and boulders. Some gray and brownish-gray microfacies with pisolitic-oncolithic components suggest water laid deposits originating from dripping and flowing currents. Other sedimentary processes have led to the formation of sedimentary structures in surface deposits, such as graded beds and biological or sedimentary crusts. Some characteristic layers of graded gravels and sands were deposited by dense flows of aggregates (lithoclasts). Other sedimentary crusts with gradation in the brown rhythmic lithofacies are described as the results of soil horizon formation, characterized by pedofeatures of

carbonate depletion or decarbonatation and pedal microstructures that suggest a lack of deposition.

3. Soil-forming processes vary horizontally according to zonal factors, and sediments are well stratified vertically. These processes include the carbonatation of vadose or epiphreatic zones as well as other surface and subsurface horizonation processes such as decarbonatation, cryoturbation, eluviation, biological activity and compaction associated with sedimentary crusts.

The vertical and lateral changes resulting from these sedimentary and soil-forming processes, detailed in the microfacies, allow us to distinguish the descriptive parameters that are most significant to a paleoecological interpretation (Table 3.6). The precipitation and seasonal regime are derived from the number of wet months per year (Cecil 2002).

Clastic Microfacies of Weak Weathering Intensity

Some sediments were only weakly affected by weathering, and they are located at the talus slope near the rockshelter wall. Weathering of weak intensity is characterized by the presence of loose carbonate cappings of variable thickness on at least one of the faces of the surface-coated rock fragments. This capping is produced by the activity of exolithic lichens. This has been observed in cave and shelter walls which are protected from weathering, but which, more importantly, are not permanently affected by moisture (Courty 1986). The development of traces of weak weathering on rock fragments suggests a high frequency of wetting and drying cycles, with an order of magnitude that prevents the development of high intensity weathering at the rockshelter wall.

The scarcity of pedofeatures is consistent with the open structure and loose packing suggests the rapid burial of these microfacies.

Microfacies featuring traces of weak weathering mainly contain the lithological and microstructural characteristics described for microfacies 0 (see Appendix 1): gravels and sandy silt located along the talus slope near the rockshelter wall.

Clastic Microfacies of High Weathering Intensity

Microfacies displaying evidence of high intensity weathering are much more abundant and are mainly located at the talus cone deposits. Intense weathering induces modifications in the size characteristics of the rock fragments, leading to a loss of cohesion in the deposit. This acute weathering has been documented at walls that are not well

Table 4.4 Soil micromorphology of level J microfacies

Microfacies	Microstructure	Porosity	Pedality	Calcitic features	Textural features	Biological alteration type
MF 6	Microaggregate	Channels Fisures	Granular Crumb	Biocrystallithics Pseudomorphics Criptocrystallines	Incomplete infillings	Endolithic
MF 5	Microaggregate	Channels Fisures	Crumb Granular	Criptocrystallines	Incomplete infillings	Endolithic
MF 4	Vughy	Vughs	Fisural	Biocrystallithics Pseudomorphics Criptocrystallines	Dense infillings Loss capping	Exolithic Endolithic
MF 3	Spongy	Vughs	Crumb	Typic Pseudomorphics Criptocrystallines	Classified infillings Packed capping	Exolithic Endolithic
MF 2	Vesicular	Fisures Vesicles	Laminar	Biocristalithics Pseudomorphics Criptocrystallines	Classified infillings Compact capping	Exolithic
MF 1	Vesicular	Fisures Vesicles	Laminar	Criptocrystallines	Classified infillings Compact capping	Exolithic
MF 0	Compound packing	Simple to compound packing	Crumb	Criptocrystalines	Incomplete infilling Loss capping	Exolithic

Table 4.5 Particle-sizes nature and ratios, calcitic pedofeatures and observed sedimentary processes in order to elucidate the sedimentary origin of the microfacies

Microfacies	VCF and CF	FF	CF/FF ratio	Calcitic features	Sedimentary microfacies origin
MF 6	Subrounded microsparitic coarse sand	Brown microspar	2/8 to 8/2	Biocrystallithics Pseudomorphics Criptocrystallines	Endolithic sandy desagregation from carbonate curtain
MF 5	Angular crystalline gravel and subrounded microsparitic sand	Brown microspar	6/4	Criptocrystallines	Endolithic gravel fragmentation from shelter wall
MF 4	Subrounded microsparitic coarse sand	Grey brownish microspar	8/2	Biocristallithics Pseudomorphics Criptocrystallines	Endolithic sandy desagregation from carbonate curtain
MF 3	Angular crystalline gravel and subrounded microsparitic sand	Grey brownish microspar	9/1	Typic Pseudomorphics Criptocrystallines	Exolithic gravel fragmentation from shelter wall
MF 2	Microsparitic coarse silt	Grey microspar	8/2	Biocristalithics Pseudomorphics Criptocrystallines	Exolithic silt fragmentation (?) from carbonate curtain
MF 1	Angular crystalline gravels and subangular esparitic sands	Grey microspar	8/2	Criptocrystallines	Exolithic gravel fragmentation from shelter wall
MF 0	Angular crystalline gravels	Brownish grey microspar	8/2	Criptocrystallines	Exolithic gravel fragmentation from shelter wall

Table 4.6 Paleoenvironmental and climate-geologic implications suggested in the micromorphological and petrographical observations based on level J microfacies

Microfacies	Sedimentation rate	Weathering intensity	Pedological process	Precipitation regime	Degree of seasonality	Hidryc regime
MF 6: Brown sand and silt with blocks	Very slow	Strong and late	Humification Pedoturbation Decarbonation	Humid	Minimal	Drained
MF 5: Brown gravel and sand	Fast	Strong	Pedoturbation Decarbonatation Silt iluvation	Subhumid	Medial	Drained
MF 4: Grey-brown sand and silt with blocks	Slow	Strong	Carbonatation Cryoturbation Eluviation	Dry-subhumid	Maximal	Drained
MF 3: Grey-brown gravel and sand	Very fast	Strong	Pedoturbation Cryoturbation Eluviation	Dry-subhumid	Maximal	Drained
MF 2: Grey silt and sand with blocks	Slow	Strong	Carbonatation Cryoturbation Silty iluviation	Semiarid	Minimal	Confined
MF 1: Grey gravel and silt	Fast	Strong	Carbonation Cryoturbation	Dry-subhumid	Maximal	Confined
MF 0: Grey gravel and sand	Very fast	Weak	Silt iluviation	Dry	Aseasonal	Drained

protected and are colonized by algae, lichens and mosses. The microscopic examination of weathered fragments reveals the endolithic penetration of mycelic hyphae through pre-existing dissolution hollows or through new fissures caused by the production of organic acid.

In addition to rock fragments fallen on to the ground surface of the rockshelter, the crystalline and cryptocrystalline carbonate pedofeatures are related to a variety of soil formation processes: microaggregation of surficial horizons, textural illuviation-eluviation, and carbonatation and decarbonatation. Therefore, we suggest that acutely weathered microfacies occurred in at least two ways (Courty 1986):

- In weakly weathered microfacies, carbonate rock fragments close to the wall originated in the cliff. During phases of intense weathering the rock fragments fallen on the ground surface inside the rockshelter were acutely altered even after their fall.
- When the wall supplied carbonate rock fragments due to high intensity weathering, endolithic weathering continued inside the rockshelter infilling producing the gray to brownish-gray color of these strongly weathered microfacies.

The very coarse fraction commonly preserves fragmented calcitic coatings. The crystalline fabrics show considerable fissural-vacuolar porosity and contain abundant black dots located on the cleavage planes of the sparitic cements. These cements have a weaker birefringence compared to unweathered calcite. Travertine rock fragments are yellowish in natural light. The fine fraction is micritic-microsparitic and is gray to yellowish-brown. It is distributed mainly in dusty coatings around grains and in crystallitic low-density ground-masses.

We interpret the intensely weathered gray microfacies as the expression of the postdepositional evolution of weakly weathered microfacies. The chemical processes are related to calcite solubility and its concomitant dissolution-precipitation. This solubility is influenced by the amount of soluble CO_2 in the bicarbonated water. The amount of soluble CO_2 in the solution depends on temperature as well as other factors, such as evaporation, which causes variations in the quantity and partial pressure of CO_2 leading to calcite precipitation or dissolution. The biochemical processes of precipitation are of the same nature as the crystalline features. Therefore, the abundance of biocrystals indicates the presence of organisms, which contribute to slightly increasing the quantity of organic acids in water. We identified several species including *Microcodium, Rivuliaria,* and *Phormidium* (Freytet 1990; Verrecchia et al. 1995). The contribution of algae and bacteria to the physical and chemical processes of dissolution and concomitant precipitation of sparite to micrite has been described by Caudwell (1987). In thin sections, this can be observed in the faint form of isolated black dots documented in the sparitic cements of carbonate rock fragments and biocrystals.

Depending on the load of organic acids in the sediments, the processes of biochemical precipitation contribute to the greater or lesser weathering of the carbonate fragments on the shelter ground, expressed by the three microstructural arrangements of acutely weathered microfacies: gray vesicular; brownish-gray vughy and brown granular.

Rhythm of Gray Vesicular Microstructure Lithofacies

The weakly weathered carbonate rock fragments fallen on the shelter ground present typical surficial corrosion by micro-organisms due to minor percolation by water loaded with organic acids. This weak endolithic weathering is related to cycles of sedimentary material sporadically falling from the cornice. This would have likely occurred during periods in which dry conditions inhibited cliff fragmentation by exfoliation. Silts are abundant in tufa fragmented by superficial freezing. When the cornice froze, there would have been very little water and preferential circulation may have occurred in preexisting fissures during thawing, which would thereby have favored the fall of the mega-blocks associated with these microfacies at the bottom of sedimentary sequence of level J. Near the rockshelter wall, conditions may have been wetter or less frozen, inducing a higher frequency of cycles, and resulting in the cryoclastic exfoliations which were filled with carbonate silts.

The biochemical cementation responsible for the formation of the tufa especially affected these gray vesicular microfacies near the cliff wall. Humidity on the shelter ground favored its preservation by impregnative incrustation coming from the roof dripping lines. Buried superficial biological crusts were also observed in these microfacies. Therefore, we suspect that rapid sedimentation was interrupted by dissolution phases when temperature and humidity increased on the shelter floor.

In summary, this intensely weathered vesicular gray microfacies was formed on the rockshelter ground, as indicated by the sedimentary record composed of (see Appendix 1): microfacies 1, gray sandy silts and gravels at the talus slope; and microfacies 2, gray sandy silts and blocks at the talus cone.

Rhythm of Brownish-Gray Vughy Microstructure Lithofacies

An increase in the organic acid content of the shelter fill was documented in the weakly weathered carbonate rock fragments. The postdepositional evolution of these fragments resulted in a characteristic brownish-gray vughy microstructure. The increase in organic acid content corresponds to an increase in the endolithic weathering in the more humid clastic fillings on the surface of the shelter floor. This increase in humidity is also expressed in secondary carbonates, and contributed to the formation of tufa and biocrystals. Therefore, the increased humidity in the shelter fill reflects an increase in the frequency of wetting and drying cycles on the cornice, leading to increased exfoliation of the shelter wall. Consequently, marked fragmentation gave rise to a high sedimentation rate. These facies show a high content of sub-rounded gravel-sized tufa, with abundant sparitic pseudomorphic moss fragments and microsparitic-micritic algae fabrics. During the postdepositional evolution of these facies, solutions weakly loaded with organic acids provoked dissolution, and developed a slight carbonate corrosion of the shelter floor deposits. Biological activity must have been limited to a short wet period, possibly during thawing. These microfacies lack the pedofeatures of carbonate corrosion, and when drained conditions were moderate, they were accompanied by eluvial translocation, as witnessed in the fillings of carbonate brownish silts and in the vughy microstructure. Eluviation is recorded through the textural pedofeatures of the active layer, similar to those of a frozen soil surface affected by significant percolation. The lixiviation process is indicated by slightly corroded carbonate crystalline pedofeatures under poorly drained conditions.

The sandiest microfacies, which are found in the talus slope of the wall, indicate increasing humidity, which reduced the number of wetting–drying cycles. Little fragmentation took place in the presence of carbonatation and decarbonatation processes, which is suggested by the formation of buried biological crusts. The fragmentation of blocks and coarse gravels also suggests increasing humidity and fissure development at the shelter wall. The weathering intensity of the shelter fill indicates slow sedimentation, which favored the growth of the carbonate curtain in the roof of the shelter.

Moreover, in this context of increasing humidity in a moderately drained hydrological regime, the microfacies of brownish-gray vughy microstructure also show the result of abundant percolations, as reflected by corroded textural coatings and fillings in tabular vesicles below granules and gravels (piprakes). Bicarbonate saturation can be high in meltwaters and leaching processes can be halted by precipitation. Therefore, the most highly developed carbonate impregnative crystalline and cryptocrystalline pedofeatures of all the level J microfacies are found in areas where conditions were wetter and more poorly drained, which is also where their sedimentary accretion developed.

The lithological characteristics of these strongly weathered brownish-gray vughy microfacies are (see Appendix 1): microfacies 3, brownish-gray silts with sands and gravels at the talus slope; and microfacies 4, brownish-gray sands and silts with blocks at the talus cone.

Rhythm of Brown Granular Microstructure Lithofacies

The carbonate rock fragments affected by high intensity weathering have an irregular boundary and rarely show weakly weathered carbonate coatings. Their external surface is commonly dented and features low-birefringence crystals. The structural porosity of the fabric of the coarse and very coarse fraction also presents these edge characteristics. Where high intensity weathering was well developed, brown areas were documented in sparite and microsparite. When weathering intensity was not very high, sparite displayed micrite-microsparite gray aggregates, which can be interpreted as the area where rizomes caused incrustation and corrosion (Callot et al. 1985). When the zones of brown micrite and microsparite translocate, they corrode and disaggregate into subrounded coarse sands with rough boundaries. In crystalline carbonate rock fragments, weathering is indicated by the yellowing of sparites, which develop a brown color linked to a pronounced loss of birefringence, proportional to the increase in weathering. These tufa fragments were more resistant to endolithic corrosion. The break-up of these carbonate rock crystalline fabrics produced subrounded to subangular sparite sands with abundant depletion features, which generally show gray secondary microsparite-micrite coatings and hypocoatings. The fine fraction of acutely weathered tufa rock fragments is brown in color. Lastly, when the shelter wall produced a deposit of intensely weathered carbonate fragments, two acutely weathered brown microfacies were distinguished: 5 and 6 (see Appendix 1).

Under less aggressive or rapid burial conditions, the travertine rock fragments and fine constituents do not show signs of weathering after falling (Courty 1986). In this case, a high sedimentation rate is expected because rock fragmentation is enhanced by the high intensity weathering of the continually wet cornice. Biological activity is poorly developed at the talus slope, which is manifested by microaggregated crumbs set in open-structured platy gravels. The rapid burial of these facies emphasizes the dry character of the sedimentation. These facies thus generally display the lithological characteristics of microfacies 5, gravels and sands with complex packing at the talus slope of the rockshelter wall (see Appendix 1).

Under slower burial conditions, the weathering of fallen rock fragments continues as granular disaggregation on the ground surface of the rockshelter, which results in a characteristic intensely weathered brown microfacies. The sedimentation rate is lower, since the cornice and the sedimentary fillings of the shelter ground are stabilized by biological activity, which limits mechanical fragmentation. The marked carbonate incrustation of vegetation or tufa in the carbonate curtain in the roof of the rockshelter means indicates continuous high humidity. A relevant redistribution of carbonates is marked by the abundant biocrystalline and impregnative crystalline pedofeatures in the sandiest microfacies of the level J complex. The detachment of microfacies is followed by the preferential dissolution of fissures and zonal corrosion at the cornice, resulting in the sedimentation of intensely weathered brown blocks and coarse gravel. Corrosion in the microfacies is manifested by the abundance of subrounded sands, fragments of biominerals, highly weathered plant pseudomorphs and granular aggregates of microsparite. This postdepositional evolution caused a slower sedimentary rate than that of the microfacies made up of gravels and crumby brown sands. The increase in biological activity in a context of increasing humidity stimulated then granular disaggregation in the filling, leading to the lithological characteristics of microfacies 6, coarse sands and silts with blocks, located in the talus cone of the shelter filling (see Appendix 1).

Clastic Microfacies with Late Weathering

Late weathering did not affect the cohesion of the deposits but it has been documented in the microfacies by the near total disappearance of secondary carbonates. This type of weathering is seldom seen in the sedimentary facies of level J. The few microstratigraphic units that show this type of weathering correspond to sublevel Jsup.

Thin sections allow the the initial texture of the rock to be identified. Micrite accumulations appear as brownish-gray. The carbonate rock fragments of crystalline cement show isolated patches of low-birefringence sparite with irregular boundaries and isotropic pale yellow masses covering only one face of the fragment. This sort of weathering has been recognized in limestone walls in contact with layers of humus and mosses, which show a juxtaposition of crystalline cements and isotropic yellow zones. Three zones have been distinguished in these crystalline limestones: an isotropic yellow external layer with moss incrusts a central zone with isolated and much corroded calcite crystals in a clear yellow amorphous groundmass, and an external part with dissolution cavities and low-birefringence calcite (Courty 1986). The weakened wall of the rockshelter disaggregates very easily. The epigenesis of limestone rock by phosphates seems to be responsible for these late weathering microfacies. This organo-mineral epigenesis is related to the activity of micelial and bacterial hyphae, since they generate chelates from humic and fulvic acids derived from their decomposition.

In level J, this type of weathering is specifically expressed by rare subrounded yellow aggregates, with a rough boundary, black dots and an isotropic micromass. Some of them contain silicate mineral particles probably trapped at the cornice, which suggests aeolian transport, indicating slow granular disaggregation, sedimentation and limited mechanical fragmentation. They are associated with the intensely weathered facies with abundant crystalline features, which indicates that humid conditions prevailed during their postdepositional evolution. This is consistent with a low sedimentation rate on the ground surface of the rockshelter. This leads us to deduce that late weathering in these microfacies developed in intensely weathered carbonate rock fragments. The abundance of brown silts in these microfacies correlates with the absence of coarse fractions, which were literally dissolved and disaggregated by processes such as corrosion and phosphatisation, thereby eliminating the sedimentary record observable at a microscopic scale.

These microfacies are composed of yellowish-brown sandy silt and are distributed over the entire shelter floor surface, including a sample from the *Coveta Ripoll*.

The Sedimentary View of Cultural Activities

The microscopic characterization of anthropic microfacies is summarized in Appendix 1. Their hierarchical characteristics and interpretations are presented in Table 4.7. In contrast with the clastic microfacies previously described, which contained a minor proportion of micro-vestiges which are not associated with a distinctive human-linked micro-pattern, anthropic microfacies are characterized by a comprehensive set of features mainly derived from human activities. Extensively recognized throughout level J, they display a significant diversity that is generally similar to that typically encountered in archeological contexts where cultural factors are shown to have played a dominant part in site formation processes (Gé et al. 1993). In general, the anthropic microfacies of Ja display a high degree of preservation of human-linked attributes, due to minimal alteration by natural mechanisms. Anthropic signatures have obviously been altered by pedo-sedimentary processes throughout the course of human occupation and later during burial, but never resulted in a complete blurring of the cultural information. In addition, microscopic observation of sedimentary samples reveals a vertical succession formed of infra-millimetric to pluri-centimetric individual anthropic microfacies without significant input of natural sediments during the time interval between the formation of Jb and that of Ja, except in some places. Thus, the finely stratified sedimentary succession studied from key locations throughout the level J complex offers a series of high resolution records of the occupation episodes.

Table 4.7 Anthropic microfacies description in order to interpret cultural sedimentary process in occupied floors of level J

	Micromass of the Anthropic remains	Burnt frequency	Burnt fraction	Microstructure	Relative distribution CF/FF	Microfacies types	Interpreation of cultural sedimentary process
Laminar (L)	Brown silt	F to Fr	FF and CF	Intergrain microaggregate	Enaulic and porphyric	L.1.1	Trampled occupied floor
				Vesicular	Chitonic and gefuric	L.1.2	
	Brown silt	F to C	FF and CF	Intergrain microaggregate	Enaulic	L.2.1	Trampled fill deposits
				Briged grain	Chitonic	L.2.2	
	Burned brown organomineral silt	D	FF	Intergrain microaggregate	Enaulic	L.3.1	Trampled burnt organomineral sediments
				Vesicular	Gefuric and enaulic	L.3.2	
	Burned black organomineral silt	VD	FF and CF	Intergrain microaggregate	Porphyric and enaulic	L.4.1	Fire-use
					Gefuric and chitonic	L.4.2	
Microlaminar (M)	Brown silt	F to Fr	FF and CF	Channel and massive	Enaulic and porphyric	M.1.1	Covered occupied floor
				Vesicular to fine platy	Gefuric and chitonic	M.1.2	
	Burned brown organomineral silt	F to C	FF	Intergrain microaggregate	Enaulic and porphyric	M.2	Covered organomineral surface
	Burned black organomineral silt	D to VD	FF	Intergrain microaggregate	Enaulic and porphyric	M.3.1	Covered combustion structure
				Briged grain	Gefuric	M.3.2	
Stratified (S)	Brown yellowish silt and sands aggregates	F to Fr	FF and CF	Intergrain microaggregate	Enaulic	S.1.1	Organomineral sediment fill
				Briged grain	Gefuric and chitonic	S.1.2	
	Brown silt	F to C	FF and CF	Intergrain microaggregate	Enaulic	S.2.1	Fill of heterogeneous burnt sediments
				Briged grain	Gefuric and chitonic	S.2.2	
	Burned black organomineral silt	D to VD	FF and CF	Intergrain microaggregate	Enaulic	S.3.1	Fill of burnt sediments
				Briged grain	Gefuric and chitonic	S.3.2	

Trampling Sedimentary Process and Anthropic Microfacies

Diagnostic signatures of trampling are shown to have affected every depositional context occupied by humans. Trampling signatures result from the combined effects of two ranges of processes: (1) structural modifications under pressure that are closely related to the type of pre-existing surface and local conditions; (2) incorporation of microdebris and fine mass components that have been transported by accident across activity areas by the occupant's feet when walking around. The structural modifications range from dense compaction documented best in the trampling of wet clay-rich materials to loss disaggregation by the trampling of dry friable surfaces. Trampled facies thus provide a relevant record of the local conditions at the exact time of human activity. The provenance of the incorporated microdebris and their intrinsic properties allow the circulation of human occupants between different activity areas to be traced. Trampling facies are inherently well-integrated after a prolonged period, which is also generally well expressed by their high degree of microstructural maturity.

The formation of micro- and meso-sized granular or platy compact fragmented aggregates (clods), made up mainly of brown silts and fine sands with very fragmented integrated anthropic microvestiges, is considered to be a product of the degradation of the living floor surface that had been trampled by humans in the circulation zones of the occupation surfaces. This hypothesis is based on the examination of these aggregates in experimental and ethnoarcheological contexts (Davidson et al. 1992; Goldberg and Whitbread 1993; Courty et al. 1994a; Chartres et al. 1994; Macphail et al. 2004). Structural degradation by compaction due to trampling is explained by analogy with the formation process of structural crusts (slacking crusts and others forms such as sieving crusts) produced by the impact of raindrops disintegrating and compacting soil surfaces (Boiffin and Bresson 1987; Chartres et al. 1994).

When this structural crust develops, structural porosity closes and develops into a sedimentary crust, thereby compacting the soil surface and enhancing lateral infiltration, especially when the infiltration capacity is not sufficient during intense rains (Boiffin and Bresson 1987). Other types of surficial crusts have been characterized as products of the fragmentation of sedimentary components through the application of mechanical forces, for example agricultural machinery, which provokes the explosion of trapped air and the fragmentation of soil aggregates which subsequently seals the structural porosity of the soil surface (Jonguerius 1970).

In level J, human compaction modified by natural processes has been established by means of sedimentary microfacies. We interpret the postdepositional modifications of sublevel Jb as the result of cryoturbation and those of sublevel Ja as of the result of soil-formation processes. The modifications caused by cryoturbation in sublevel Jb have produced a considerable redistribution of the sedimentary components, partially originating from fragmentation caused by trampling. In sublevel Ja, the microstructural characteristics indicate a low to medium degree of incidence of surface horizon processes such as biological activity, which partially destroys natural compaction during burial because pedofauna regenerate structural porosity. The sealing of porosity as expressed by compaction and the development of lateral infiltration has been best observed in examples of sedimentary and biological crusts weakly affected by bioturbation. These sedimentary crusts are a sedimentary figure of polarity for strata because they result from a sedimentary process taking place on subaerial surfaces. They result from graded laminar currents or aggregate flows, as discussed earlier in the analysis and interpretation of the sedimentary microfacies. In level J, the sedimentary crusts have been used as a basis for the classification of laminar anthropic microfacies. The stratification of these sedimentary crusts illustrates how the cumulative character of compacted surfaces has favored the preservation of occupation surfaces. In sedimentary petrography these sedimentary forms of subaerial surfaces, though they are characteristic of slope deposits, are seldom considered, while they have been explored with great success in agronomy by means of soil micromorphology (Mücher et al. 1988; Miedema 1997; Vliet-Lanoë 1997).

Trampling sedimentary processes involve the redistribution of sedimentary particles, clods and finely fragmented anthropic components on occupation surfaces. The occurrence frequency of anthropic components and burned sediments, and the nature and aggregation of the sedimentary components in these microfacies have been estimated for the purpose of classifying the results of sedimentary trampling processes into the following microfacies types:

- L.1. Trampled occupation surface at the periphery of combustion structures and bare surfaces of anthropic vestiges. Coarse and fine fractions have few to frequent burned components. Aggregation is rare to very abundant.
- L.2. Trampled occupation surface on dumped or scattered sediments. Frequent to common burned sedimentary components. Pedal aggregation is rare.
- L.3. Trampled combustion structure and organo-mineral or/and charcoal-rich sediments. Dominant burned components in the fine fraction and the coarse fraction, and frequent to common burned constituents in the very coarse fraction. Aggregates of organo-mineral nature are abundant.
- L.4. Use of fire. Burned constituents are dominant and organo-mineral aggregates are rare.

Microlaminar anthropic microfacies, with a straight to curved boundary and an abrupt upper contact, have been interpreted as being the sedimentary record of occupation surfaces covered with mats (Gé et al. 1993; Matthews et al. 1997). Covered occupation surfaces explain the conservation of fine laminae or sets of laminae with very fine anthropic micro-remains, produced during compaction. A covered living floor, as an instantaneously buried surface, prevents the integration and redistribution of the compacted layers by means of sedimentary and soil surface processes. In short, the mechanical effects of repeated compaction on a covered living floor produce anthropic microlaminar microfacies.

The rare incorporation of microvestiges into anthropic microlaminar microfacies, as observed in level J, is indicative of relatively impervious mats. Consequently, the cryptocrystalline organo-mineral components stand out. Postdepositional transformations are required for recording these microlaminar microfacies, with the occurrence of mechanical soil processes, the appearance of fissuration and a high degree of modification by cryoturbation, and the incidence of soil surface processes like low to medium intense biological activity, which is assessed by documenting the porosity of channels and excremental fabrics. Porous channels with compact walls were detected, showing the distinctive morphology of worm burrowing. Lastly, we described pure isotrope yellowish microaggregates, which characterize the components of Ja anthropic microfacies of impervious covers, and were probably formed by means of wetting and sealing soil by depositional crusts.

The thermal impact, nature and aggregation of the constituents have also been used to interpret the following microlaminar anthropic microfacies:

M.1. Covered occupation surfaces at the periphery of combustion structures and bare occupation surfaces. Few to very few burned sedimentary constituents. Abundant fine platy aggregates limited by subvertical and subhorizontal fissures.

M.2. Covered occupied organo-mineral floors. Frequent to common burned constituents of the fine and coarse fractions. Rare organo-mineral aggregates.

M.3. Covered combustion structures. Burned coarse and fine fractions are dominant. Pedal aggregates are rare to occasional, bordered by fissures.

The Fill Sedimentary Process and Anthropic Microfacies

The marked contrast between anthropic fill facies and all other types of anthropic facies was at first perceived during the excavation, but was not fully understood at the time due to their original geometry. Consequently, sampling was at first designed to better determine their true nature, while their extent and spatial links to the surrounding facies were not so closely examined. Therefore, we focus here on their overall characteristics at the scale of individual thin sections.

The anthropic fill facies occur as a succession of distinctive stratigraphic units of over than 2 cm in thickness. Their boundaries are rarely concordant; they often appear as inclined surfaces that are easily visible in naked-eye observations of the thin section, although the sharp contact cannot be followed in the field due to the weak cohesion.

The anthropic fill facies are characterized by a simple to complex packing compound microstructure. The fine fraction appears as local concentrations of loosely packed fine aggregates, whereas the coarse fraction often shows angular clasts affected by in situ fragmentation. Anthropic features are rare to very abundant with a variable proportion of burned components. Pedofeatures are always rare, mainly represented by occasional excremental fabrics, channels with compact hypocoatings, loose incomplete infillings of sands and silts, loose incomplete infillings of anhedral calcite and acicular calcite, which also appear in the form of crystal intergrowths. Pedofeatures most often appear as randomly distributed and not part of a coherent microstructural pattern as observed in all the other anthropic and natural facies.

The estimation of the constituents bearing thermal impact, and the nature and aggregation of the components in these stratified anthropic facies allowed us to distinguish the following categories:

S.1. Filling of sediments collected from peripheral and bare occupation surfaces. Few to very few burned constituents and very rare aggregation. Filling of sediment collected from land occupied in the suburbs.

S.2. Filling of sediments collected from scattered combustion structures. Frequent to dominant burned constituents and rare to occasional aggregates.

S.3. Filling of sediments collected from combustion structures. Burned constituents are very dominant and aggregation is very rare.

Fire-Use and Anthropic Microfacies

Considering the expected focal point of combustion structures in the spatial patterning of activity areas, the elementary types of anthropic microfacies are at first defined on the basis of the assemblage of combustion attributes. They basically comprise (i) burned components of human (cultural vestiges and microdebris) and natural origin (clastic debris); (ii) fire residues (ash, charcoal) included in the fine mass; (iii) the fine mass which has been modified to variable degrees by fire-induced thermal processes. Such spreads of sediments in relation to the use of the fire are often called 'combustion structures'.

The elementary types of microfacies exhibit the diagnostic assemblage and basic components generated by in situ combustion. Their prime constitutive criteria are represented by their twofold to threefold micro-stratified module, which was caused by a series of interdependent burning episodes: (i) an upper combustion sub-unit, and (ii) a lower combustion unit often showing two distinctive sub-units.

The upper combustion sub-unit comprises the subsurfaces that are directly exposed to high temperature heating. Above all its basic components exhibit significant thermal modifications characterized by distinctive morphological, structural and compositional changes. The fine mass is dominantly made up of burned organo-mineral microaggregates with high-birefringence microsparite and fine carbonaceous components, including micro-charcoals. Sparite and high-birefringence microsparite are byproducts of the thermal microfragmentation of the clastic sparite into microsparite and of the ashy components generated by means of the complete calcination of wood fuel (pseudomorphic microsparite). Their original distinctive morphology can no longer be clearly recognized in thin sections due to dissolution at their edges and along cleavage planes and to humic staining. The upper combustion sub-unit contained a minor amount of cultural microvestiges, mostly represented by burned bone micro-fragments and organo-mineral aggregates. The bone remains commonly show a distinct coloration ranging from yellow, yellowish-red, and red to whitish-gray. Occasionally they present distinct peripheral coloration or double coloring, especially yellow and red. The organo-mineral aggregates are subrounded to subangular with sharp edges, and isotropic or micro-contrasted, with a yellow to yellowish-red homogeneous fine mass that can contain silicate minerals or crystal intergrowths of druse microsparite, which has been interpreted as calcium oxalates from leaves (Courty et al. 1989; see Appendix 1).

The various types of burned carbonate rock fragments encountered in the upper combustion sub-unit suggest a wide range of burning conditions. Long duration burning at low temperature (<300°C) is traced by the darkening of carbonate rocks without marked changes in crystallinity. Neoformation of acicular calcite in spherulites, together with fine cracking, microsparitisation and increased birefringence are transformations related to burning temperatures at medium to high intensities (400–600°C). The most extreme stage of thermal alteration is represented by intensively cracked travertine rock fragments that show the pale to dark gray isotropic fine mass typical of quicklime transformation known to occur at temperatures above 750°C (Fig. 4.15, microphoto 1).

Carbonaceous micro-remains are comprised of black vesicular carbon, clusters of translucent carbon fibers and whitish blue ribbon-shaped carbon. The black carbon elements contrast with charred remains because they lack anatomical structures and have a vitreous appearance with a cracked pattern, occasionally with degassing vesicles (Fig. 4.18, c1). None of these carbonaceous micro-remains is common in archeological contexts, and so far, they have only been identified in high-temperature combustion byproducts from wild fires. In level J, their close association with combustion structures is clearly related to cultural practices although they may result from the human processing of natural products.

The lower combustion unit corresponds to the overlying materials of the intensely burned surface that has been exposed to temperatures ranging from less than a hundred degrees to a few ten degrees but with a rapid decreasing temperature gradient according to depth. The marked color changes result mostly from the oxidation of iron components present in both the fine fraction and the coarse fraction, together with low temperature carbonization of the organic components included in the fine mass. This heat-induced micro-stratification is relatively well expressed when the fuels are directly lying on the natural deposits originally made of carbonate rock fragments embedded in an homogeneous microsparite fine fraction. In contrast, heat-induced micro-stratification is less detectable, or even undetectable, when burned microfacies are superimposed over other types of anthropic facies, simply due to their heterogeneity and/or lack of humic layer when burned.

Two sub-types of microfacies are distinguished based on the nature of their burned components (L4, L3, M3 and S3 in Appendix 1): (i) a homogeneous burned microfacies with an upper combustion sub-unit displaying similar thermal modification components and rare micro-vestiges; (ii) a heterogeneous burned microfacies with an upper combustion sub-unit made up of rock fragments and more frequent micro-vestiges heated at various temperatures.

Two sub-types of burned microfacies are distinguished based on their frost-linked assemblage: (i) dense platy burned microfacies with fissure porosity and a loose packing of platy dense aggregates rich in micro-divided charcoals; link cappings rich in highly birefringent microsparite are present in the coarse components; (ii) loose platy burned microfacies with open packing of randomly distributed platy aggregates in the upper combustion sub-unit and coarse elements embedded within finely disaggregated link cappings. This suggests combustion affecting a formerly frozen ground surface. In contrast, the former type was originally a combustion facies that was later affected by surface freezing.

Fragmentation and color change are the two dominant structural modifications seen on burned carbonate rock fragments. Thermal impact is expressed as described for heterogeneous limestone (Wattez 1990): an increase in the

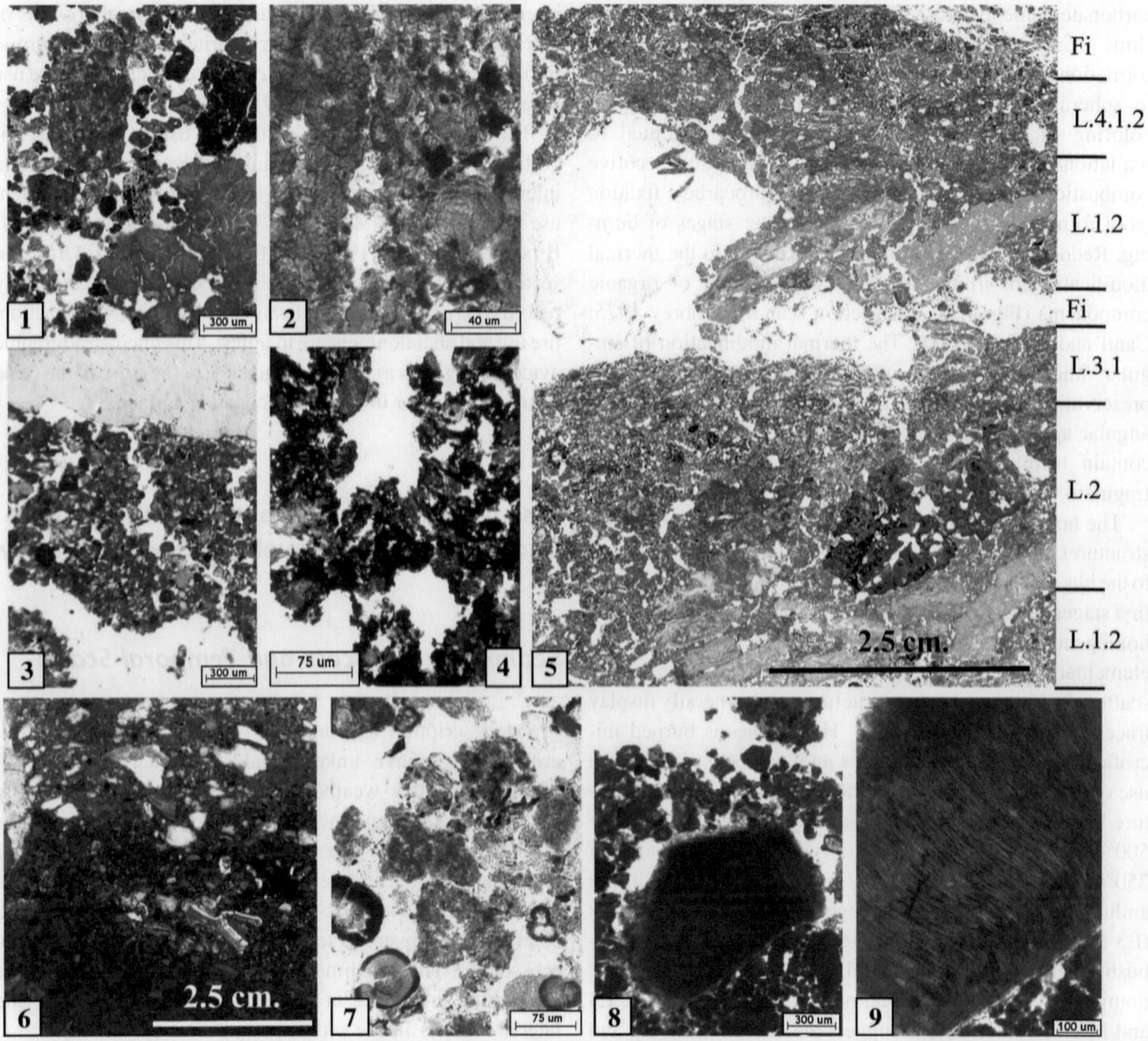

Fig. 4.11 Sedimentary components and microscopic vestiges of human origin showing thermal transformations included in the anthropic microfacies. *1* opacification and fissuration of carbonated materials. *2 reddening* (rubefaction) and particle-fragmentation detachment of sparite and microsparite of high birefringence. *3* compact aggregates and microdivided cryptocrystalline organo-mineral components. *4* charred charcoal and *ash* formation (100X). *5* thin section scan of a combustion structure sample (square M56). Two microstratified burned surfaces are evident, separated by a layer of clastic microfacies 5 suggesting occupational reuse and abandon of the combustion structure. Note a white fragment of chert without thermal modification in the *left corner* of the photograph. Legend at *right side* of the thin section: *L* laminar and its numeric subtype code; *Fi* artifact fissure produced during production of the thin section. *6* impregnated block scan showing anthropic microfacies M4 (heterogeneous burned carbonate rock fragments) and an erosional truncation in the *bottom* part of the sample. *7* rounded and coarse spherulites. *8* fragment of burned skin. *9* detail of the burned skin showing a packed fibrous shapes similar to a histological structure

fissure network and disintegration of its constituents, microdivision of sparitic and microsparitic cements, and opacification of micritic zones (Fig. 4.11, microphotos 1 and 2). The break-up of carbonate rocks by means of thermal impact is manifested in the formation of abundant angular high-birefringency microsparite (Fig. 4.11, microphoto 2). Such fragmentation has also been described for surface soil aggregates during wildfires (Greene et al. 1990). The release of sedimentary rock components, especially in tufa rock fragments, can be seen in the presence of highly birefringent spherical fine sands (Carozzi 1960; Fig. 4.11, microphoto 3). Complete (pseudomorphic) or incomplete burning on the charcoal from wood fuel was uncommon (Fig. 4.11, microphotography 4). There are also few carbonate rock fragments in the anthropic microfacies which show cryptocrystalline transformation or microdivision of carbonate cements caused by burning at high temperatures (lime—CaO) (Fig. 4.11, thin section scan 6). Crystal neoformations caused by thermal impact have been attributed to spherulites (Fig. 4.11, microphoto 7). Changes in the coloring of these spherulites are basically attributed to oxidation/reduction and opacification. Differing consecutive combustion modes are the result of hydrocarbon fixation (soot formation), probably during the later stages of burning. Reddening of soil-sediment is attributed to the thermal modification of iron oxides and the presence of organic components (Fig. 4.11, thin section scan 5; Limbrey 1975; Canti and Linford 2000). The thermal modification of singular microvestiges of anthropic origin enhances their preservation. In Fig. 4.11, microphotographs 8 and 9 of an angular aggregate of black-brown fibrous shapes appear to contain histological structures, which we attribute to a fragment of burned hide (Wattez 1990).

The homogeneous burned microfacies of the combustion structures of level J (L4, M3, S3) have components analogous to the black and brown residues which were formed during the first stages of the transformation of charcoal into ashes. These homogeneous burned microfacies are related to the preserved elementary hearths, or combustion activity zones more or less scattered within combustion structures, and generally display traces of low-intensity burning. Homogeneous burned microfacies are of limited thickness and may be a result of the use of fire over a small surface. Residues of higher temperature burning, like gray ashes from combustion intensities of 500°C and white lime residues from combustion intensities of 750°C, are very rare and have only been observed in the anthropic S3 microfacies. Heterogeneous burned microfacies (L3 and M2) are abundant in samples localized in the combustion structures of the central area. The internal zone is dominated by homogenous burned microfacies L4. The L3 and M2 microfacies types suggest the lateral scattering and covering of burned microfacies, corresponding to the repeated use of fire within the same combustion structure. In contrast, anthropic microfacies S3 and S2 suggest episodes of maintenance, cleaning or other specific activities which took place in the combustion structures of the occupation surfaces.

Thermal modifications recorded in the travertine rock fragments appear consistent, under taphonomic conditions favorable to their preservation, with calcitic and burned microfabrics incorporated during fuel consumption (Wattez 1990). We were able to distinguish a considerable number of microstratigraphic units of L3 and M3 microfacies, containing a high percentage of heterogeneous burned sedimentary components, sampled as activity areas implying the use of fire. These have been interpreted as the product of repeated fire use and help us to determine scattered fire-use residues. As observed in microfacies, the fiber-radiated spherulitic neoformations observed along the rims of corroded brown microsparites on heated carbonate rock fragments (high intensity weathering in MF 5 and 6), and crystal growths of acicular fiber-radiated acicular sparite caused by thermal impact were also interpreted as evidence of repetitive fire use (Fig. 4.15, microphotography 7).

The sedimentary record of the fire use in anthropic microfacies does not only contain the record of combustion intensities but also evidence of the congruent or incongruent use of combustion structures (Hughes and Lampert 1977; Brooks and Yellen 1987; Schiffer 1987). In Fig. 4.15, thin section scan 5 is shown to contain an exceptional sedimentary record where the cyclic stratification of dispersion-compaction, fire use and abandonment are manifest. This microstratigraphic evidence suggests a spatial redundancy, or the reuse of the same area as a place for fire on the occupation surface.

Discussion: Paleoenvironments and Human Occupation in the Level J Microstratigraphy

Sedimentary Process and Temporal Scales

In the description and analysis of clastic sedimentary microfacies, we have linked local sedimentary micro-environments with weathering intensities. We thereby established two entrance points from which materials originate, in accordance with the arrangement of the paleosurfaces of level J: the internal zones over the talus slope at the foot of the rockshelter wall, and the central and external zones over the talus cone below the dripline of the rockshelter. These sedimentary micro-environments and the weathering intensities in these zones enable us to describe lateral changes in the sedimentary facies (Fig. 4.12):

1. In the internal zone, lithofacies 1 shows an alternating rhythm of weak to intense weathering.
2. In the central zone, the rhythms of lithofacies 1 and 2 show high intensity weathering and endolithic biological activity that increases upwards.
3. In the external zone, lithofacies 2 shows high weathering intensity.

These changes in lateral facies indicate that the most detailed sedimentary record is located in the central zone where lithofacies 1 and 2 are interbedded. The stratification in the central zone points to sediment deposition by intermittent events. Sedimentation in the internal zone occurs when cryoclastism is effective or when the travertine wall is wet. In this area cryoclastic fragmentation may have limited the environmental and temporal threshold of weathering intensity. Consistently high intensity weathering of the carbonate curtain continuously supplies sedimentary materials to the central and external zones. In this area

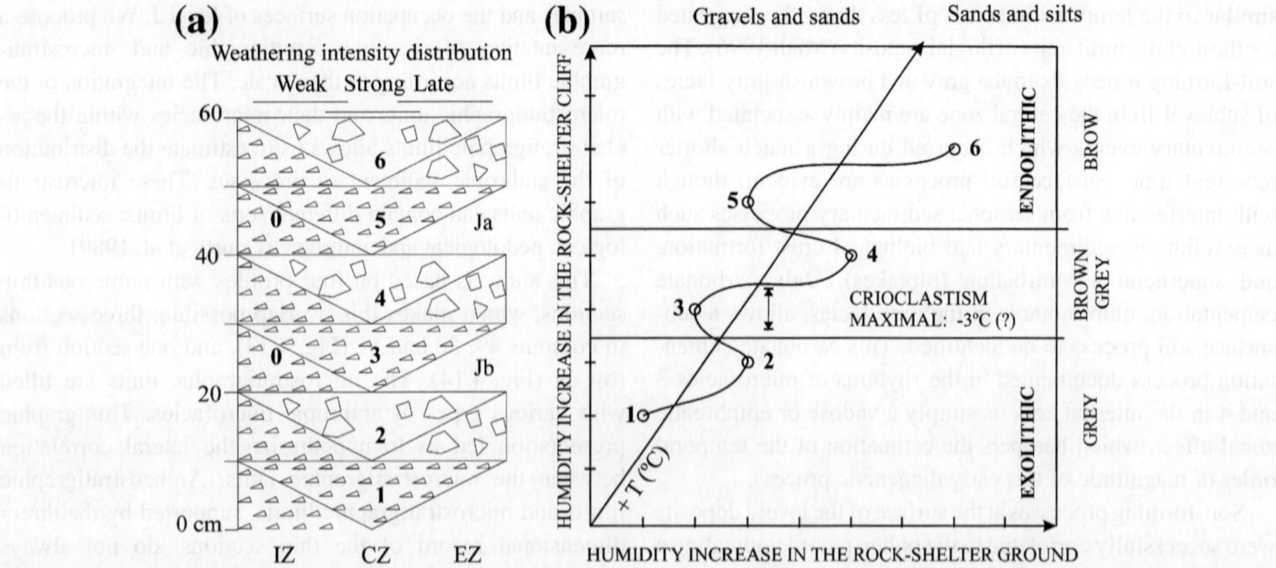

Fig. 4.12 Synthesis of the lateral and vertical clastic microfacies changes in the lithofacies rhythms. *A*. Synthetic lithostratigraphy of level J and its lateral variations according to the geomorphological zones examined: *IZ* internal zone; *CZ* central zone *EZ* external zone. Microfacies numbers: *0 gray* gravels and sands; *1 gray* gravels and silts; *2 gray* silts and blocks; *3 grayish-brown* gravels and sands (Jb); *4 grayish-brown* sands and silts with blocks; *5 brown* gravels and sands (Ja); *6 brown* sands and silts with blocks. *B*. Model for the interpretation of clastic microfacies 0–6 in good agreement with humidity on the carbonate wall and the clastic fill of the rockshelter

cryoclastic fragmentation may restrict the environmental and temporal threshold of late degrees of weathering.

The open-structure fabric of lithofacies 1, which is composed of clast-supported fine platy gravel, indicates the fall of individual rock fragments. Frost activity is a dominant process in most cold and humid alpine zones and cryoclastism is commonly deduced from sedimentary petrography studies of scree, or *talus d'éboulis*, in a periglacial geomorphologic context. The fall of slabs and blocks takes longer to develop because pressure fissures have to develop first. Sedimentary petrography and rock weathering studies also suggest that biochemical mechanisms are responsible for the transport of medium and fine fractions in the formation processes of rhythmic facies 1 and 2. Sandy and granular disaggregation and exfoliation have been observed in taffoni microforms of carbonate rocky escarpments. Rock fall processes through hydration, hydrolysis, cryoclastism, and more frequently through haloclasty (a physical weathering caused by the growth of salt crystals) take place in taffoni of crystalline and sedimentary rocks. Haloclasty, which affects carbonate and sulfate rocks, is more frequent in coastal, desert and semi-desert regions. The cumulative processes of rock fall and the weathering of rocky escarpments by means of water suggest fragmentation by granular disaggregation and exfoliation.

The transport of debris-fall leads to stratification. On flat slopes, the stratification of lithofacies 1 and lithofacies 2 was most often found to be the result of debris flow, solifluction or wash (Selby 1993). Talus stratification in level J has created a gently sloping surface, suggesting low gravity-induced movement. *Grèze litées* or *groise litée* is a French term for common gravitational deposits in mid-latitude geographical areas. Talus or scree formation processes need effective cryoclastism at about 0°C and current flows, in order to sort sedimentary materials. Stratification and particle selection indicate nivation: the thawing of periodic snow patches, and the freezing and thawing of the soil moisture which comes from meltwater, cause downslope movement by geliflution, soil or snow creep, and sheet flow (Bertran 2004). Other stratified slope deposits and forms have been described as dry grain flows and aggregate flows (Steijn 2002). Aggregate flows are initiated by the overland flow of water-laid deposits occurring during high intensity rainstorms. Dry grain flow activity characterizes sedimentary processes during longer spells of dry weather.

A lithostratigraphical synthesis of the level J complex and its sedimentary rhythms is presented in Fig. 4.12. It indicates that endolithic weathering develops upwards in the sedimentary succession of level J. In Fig. 4.12 we present an estimate of temperatures, based on the expressed cryoclastism and the degree of humidity at the travertine wall and the cornice of the rockshelter. The rhythmicity in the central zone characterizes an alternating sedimentation rate. An approximated estimate of about 60 cm ky^{-1} has been proposed by Bischoff et al. (1988) from the geochronological U/Th study of the total stratigraphic succession. We propose a rate that emphasizes the hierarchy of biological processes within the sedimentary record and their development through time or maturity. The formation of an immature soil in the rhythmic brownish lithofacies, composed of microfacies 5 and 6 of sublevel Ja on the talus cone of the central zone, may be similar to the temporal intervals of less than 3 ky suggested for the architectural units of fluvial deposits (Miall 1996). The soil-forming processes in the gray and brownish-gray facies of sublevel Jb in the central zone are mainly associated with sedimentary events which occurred during a much shorter recurrent time. Surface soil processes are evident, though with interference from seasonal sedimentary processes such as exfoliation, sedimentary and biological crust formation, and superficial cryoturbation (piprakes). Only carbonate cementation, more notable in the gray facies, allows a subsurface soil process to be identified. This carbonate cementation process documented in the rhythms of microfacies 3 and 4 in the internal zone is simply a vadose or epiphreatic zonal effect, which hampers the estimation of the temporal order of magnitude of this early diagenetic process.

Soil-forming processes at the surface of the level J deposits were successfully correlated to the pollen record, indicating a slight increase in tree taxa in the last peak of pollen zone 3. This environmental change has a millennial order of magnitude for a sedimentation rate of 0.6 m ky^{-1}. The temporal recurrence of level J deposits is on a millennial order of magnitude, as documented in the boundary surface made up of the rhythmic microfacies 5 and 6 lithofacies.

The sedimentary microfacies suggest a sub-millennial to seasonal temporal order of magnitude for the sedimentary processes. These processes can be interpreted as events with a seasonal to centennial order of magnitude. Cryoclastism, granular disaggregation, travertine incrustation and sedimentary crusts resulting from aggregate-rich flood sediments find their origin in cyclical processes of less than a hundred years each. This temporal estimation of the cyclicity observed in sedimentary processes suggests that the moisture record corresponds to the hydrological variations in the Capellades plain. Therefore, the insolation deficit at the wall of the Abric Romaní is expressed by the fact that cryoclasty was one of the main sedimentary processes at work in level J. However, these cryoclastic microfacies can also be attributed to a decennial climatic change, characteristic of the variability related to solar activity and the North Atlantic Oscillation (NAO) (Weedon 2003).

Microstratigraphy of the Level J Anthropic Microfacies

The study of the microstratigraphy of anthropic microfacies attempts to integrate anthropic sedimentary processes within the archeostratigraphic units, which are defined using the X, Y and Z position of archeological remains. To this end, we positioned microstratigraphic units and anthropic microfacies in order to establish the continuity between the sedimentary samples and the occupation surfaces of level J. We propose a representation of the archeostratigraphic and microstratigraphic limits according to their scale. The integration of the microstratigraphic units and their microfacies within the archeostratigraphic limits allows us to estimate the distribution of the anthropic sedimentary processes. These microstratigraphic units can contain different sorts of limits: sedimentological, pedological and anthropic (Courty et al. 1989).

This study is based on four profiles with numerous thin sections, which makes this overlap possible: three sections in columns 49, 50 and 51 (Fig. 4.13), and one section from row N (Fig. 4.14). The microstratigraphic units are filled with various types of anthropic microfacies. This graphic presentation led us to hypothesize the lateral correlation between the microstratigraphic units. Archeostratigraphic limits and microstratigraphic limits, supported by the three-dimensional record of the thin sections, do not always overlap and coincide exactly. Thus, the limits of the microstratigraphic units, and especially of the units without anthropic facies, have been adjusted to the archeostratigraphic limits delimited by the two-dimensional projections of the remains. These archeostratigraphic profiles contain the anthropic microfacies of a large part of the archeostratigraphic units that separate Ja, Jb and Jsup.

In the field stratigraphy, sublevels Ja and Jb were documented when sedimentary processes generated gravels. Level J beds with no recorded anthropic activity are characterized by sands, silts and blocks. The microscopic observation of thin sections confirms this association between the recording/non-recording of anthropic activity, and the granulometry observed in the field. Figures 4.17 and 4.18 show:

1. The hypothetical correlation between the microstratigraphic and archeostratigraphic limits.
2. The association of anthropic microfacies and microstratigraphic units and their hypothetical lateral changes.

Microstratigraphic Units with Unclear Delimitation

The sedimentary samples of the profiles which were not well delimited are localized in P–S/49–51. One of the difficulties encountered during our fieldwork, which contributed to the decision to carry out sedimentary sampling, was the presence of combustion structures we assigned to sublevel Ja and possibly to Ja–b, in R49, P49–50, P50 and Q50 (Fig. 4.13, columns 49 and 50). Their microstratigraphic assignation to sublevel Ja or Jb is a key question because the context of these sublevels is an indicator of the spatial analysis of these archeological structures. This complicated microstratigraphy is an example of an occupational palimpsest.

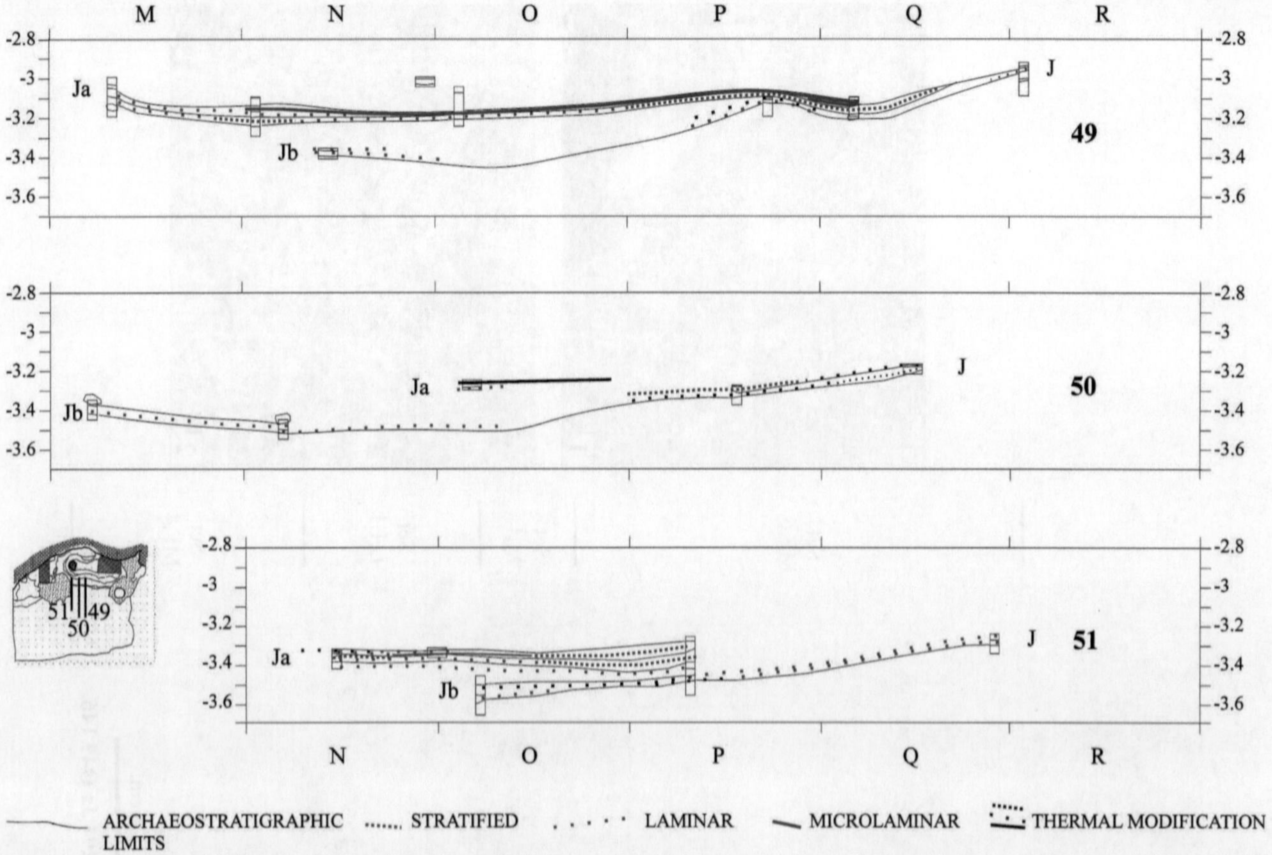

Fig. 4.13 Archeostratigraphic units and anthropic microfacies correlation at rows 49, 50 and 51, corresponding to the central and internal zones of level J

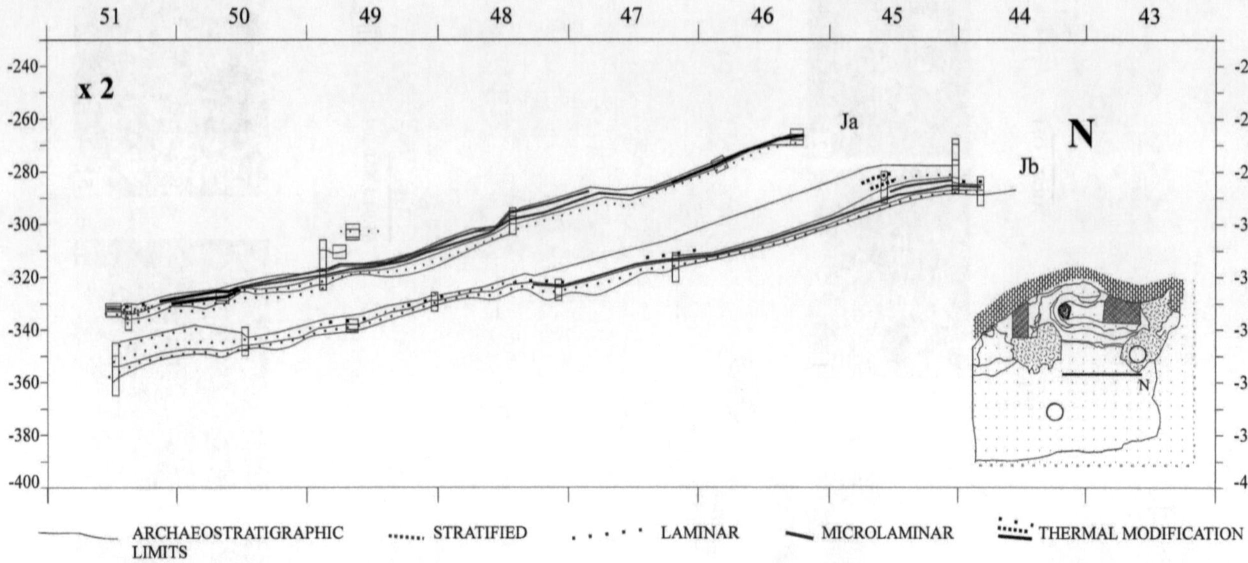

Fig. 4.14 Archeostratigraphic units and anthropic microfacies at row N. These units cover the northern half of the central zone

The archeostratigraphic profiles in Figs. 4.17 and 4.18 show that all the samples of the combustion structures in this area belong to sublevel Jb. In summary, in this palimpsest of combustion structures archeostratigraphic data suggest the loss of delimitation between Ja and Jb. We propose that these samples from combustion structures belong to sublevel Jb on the basis of the cryogenic modifications recorded in sedimentary microfacies 3,

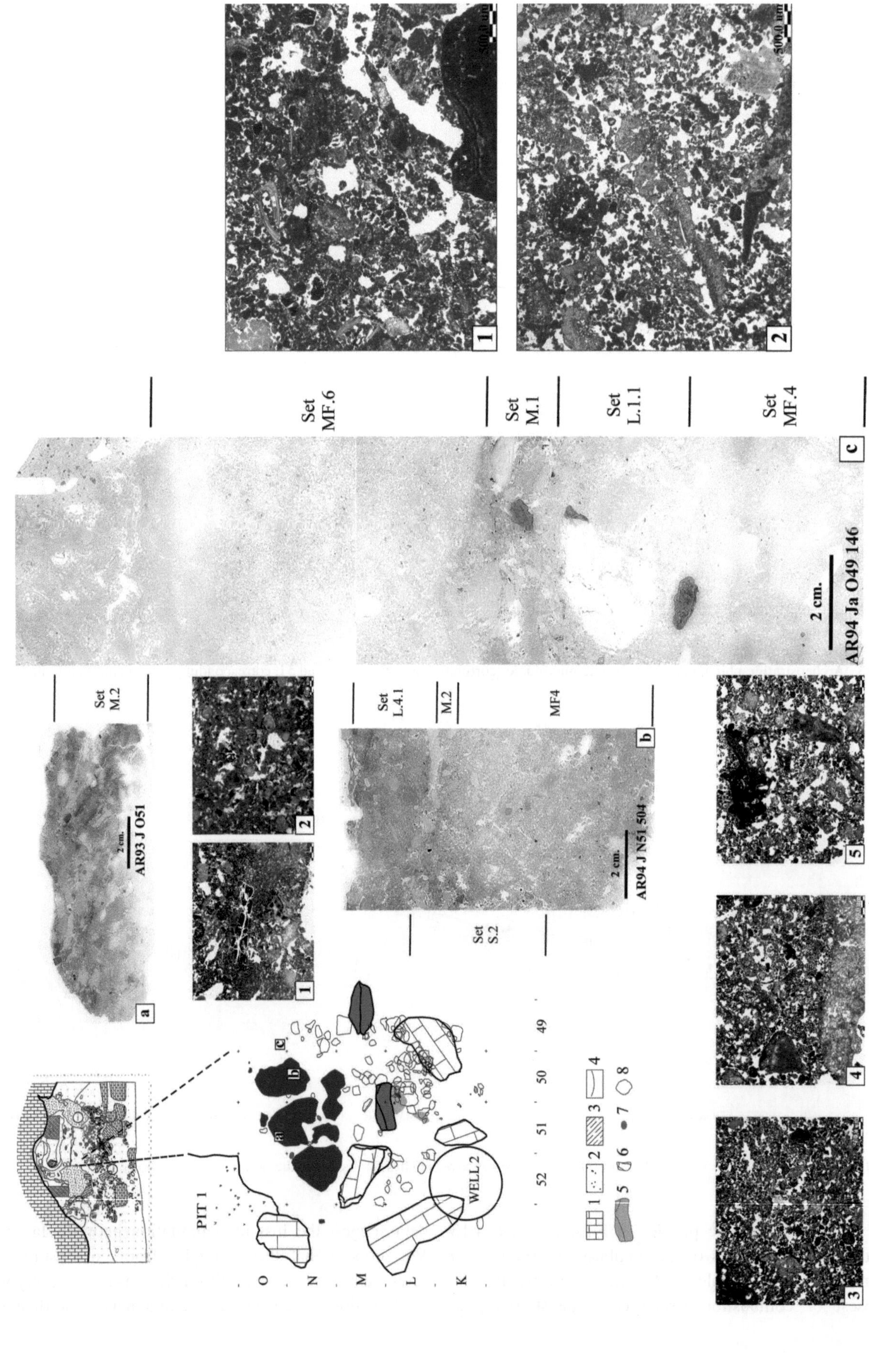
PIT 1
WELL 2
O
N
M
L
K
52
51
50
49
a
b
c
1
2
3
4
5
6
7
8
Set
M.2
2 cm.
AR93 J O51
Set
L.4.1
M.2
MF4
Set
S.2
2 cm.
AR94 J N51 504
Set
MF.6
Set
M.1
Set
L.1.1
Set
MF.4
2 cm.
AR94 Ja O49 146

◀**Fig. 4.15** Clastic and anthropic microfacies combustion structure group 2 of sublevel Ja central zone. Position of the scanned thin sections a, b and c. *1* rockshelter wall. *2* early archeological excavations. *3* stratigraphic outcrops. *4* main topographic discontinuities. *5* J occupation surface boulders. *6* structural blocks and boulders. *7* burned blocks and slabs. *8* slabs. **a** thin section scan made from microlaminated sets of anthropic microfacies M2. *1* low magnification of microlaminated anthropic microfacies with partial lineation and densely packed of spherical sparite with high birefringence. *2* medium magnification of high birefringence spherical sparite set in organo-rich groundmass. **b** thin section scan showing the high variability of anthropic microfacies. Note the sedimentary record of the use of fire in microfacies L.4.1 suggesting a cumulative character based on the identification of anthropic limits in good agreement with the dominance of burned carbonate rock fragments (microphotographs 3, 4, 5). Lateral change between S.2 and M.2 anthropic microfacies, in both sides of the thin section, showing sharp and undulated boundary surface that suggests an episode of digging before repeated episodes of fire use (L.4.1 set). **c** thin section scan showing microstratified anthropic microfacies (M1 and L1) and clastic microfacies 6, which suggests limits between environmental and use of space changes. *1* low magnification of microlaminated microfacies in the *upper half* of the microphotograph. *2* microphotograph 2 corresponds to clastic microfacies 4, which do not show any trace of anthropic microvestiges because Ja and Jb are well delimited in this area of the central zone

which have been associated with these Jb microstratigraphic units.

Another difficulty that came up while examining the archeostratigraphic graph of these zones where sublevels Ja and Jb cannot be well delimited was the characterization of the lateral continuity or discontinuity in the thickness of the archeostratigraphic units:

1. A lateral continuity in the thickness of the archeostratigraphic units can be observed in columns 50 and 51 (Fig. 4.13) in rows P–R, which were assigned in the field to sublevel Ja. The anthropic facies identified in the sample from P51 corresponds with the vertical accretion of S (stratified) anthropic microfacies (Fig. 4.13, column 51). It must be emphasized that in this sample, which was studied from the various thin sections that cover both Ja and Jb, a clear microstratigraphic separation exists between Ja and Jb. We also recorded stratified anthropic microfacies in very thick archeostratigraphic units which exhibit neither lens-like morphological characteristics nor a high density of arranged microartefacts, as in P51 (Vallverdú 2002). These samples were taken along row N in sublevel Jb (Fig. 4.14). These stratified anthropic microfacies correspond to burned microfacies. Their distinct geometries and compositions of stratified anthropic microfacies suggest specific management and a cultural preference for a place to dump sediments.
2. A lateral discontinuity in the thickness of the archeostratigraphic units is observed between columns 46 and 49. In the profile of column 49 (Fig. 4.13), sample Ja-Q49 has a stratified M and S anthropic microfacies. The sedimentary microfacies of this sample belong to sublevel Ja. The lateral discontinuity is present towards the interior zone, which is related to the low density of recorded artifacts from columns 46–49 (Vallverdú 2002), however this discontinuity has not been observed in column 50 nor 51 (Fig. 4.13). Finally, the microlaminar microfacies of sample Q49 allows us to propose as an initial hypothesis that an area with few remains, or even an empty area, existed in front of the stalagmite of the *Coveta Ripoll*.

Microstratigraphic Units with Regular Delimitation

Square N51 of sublevel Ja includes a sample in an archeostratigraphic unit that is not as thick as that of P51, and which contains microstratified anthropic microfacies (see Fig. 4.13). Other samples from archeostratigraphic units more than 10 cm in thickness, and not as lens-like morphologically, are found in row O and column 51 of sublevel Jb, while sample M is located in column 49 of sublevel Ja. These are samples of a laminar anthropic microfacies. These archeostratigraphic units contain a low density of archeological remains (Vallverdú 2002). Their thickness may be attributed to scuff marks over occupation surfaces which have sedimentary facies characterized by a high lithological penetrability.

Microstratigraphic Units with Sharp Delimitation

The sedimentary record of thin archeostratigraphic units corresponds somewhat to the presence of microlaminar anthropic microfacies, as can be observed in N46 (Fig. 4.13). However, microlaminar anthropic microfacies are more common in the thicker archeostratigraphic units, where laminar and microlaminar microfacies are interstratified. These thicker units have been documented in sublevel Ja in N48–49 and O49 and in sublevel Jb in columns 44 and 45.

Anthropic Microfacies Boundaries in Level J Occupation Surfaces

Discontinuities between sedimentary lithofacies were observed during fieldwork and annotated by means of sublevels such as Ja, Jb, and Jsup. Other kinds of discontinuity have been observed in thin section. For example, the thin section sample from the combustion structure in N51 of sublevel Ja contains a considerable number of microstratigraphic units of anthropic microfacies. This sample is one of the few where evidence of an anthropic microfacies change occurs (between stratified set type 2, microlaminar 4 and laminar set 4.1 in Fig. 4.15). In Jsup, where sedimentary microfacies 6 are found, the thin section shows stratification that consists of laminar, stratified and burned laminar microfacies upwards (Fig. 4.14). Another example from sublevel Jb is a sample from O51, which displays a significant thickness of the bedded laminar anthropic microfacies; this was later modified by cryoturbation and therefore has no anthropic limits. The thickness of the anthropic microfacies in the Jb samples of N44–45 suggests diachronic human impact because laminar anthropic microfacies form a bedset (Fig. 4.14). However these bedset of anthropic microfacies require careful interpretation because their boundaries are the product of human sedimentary processes. Likewise, the presence of bedded microlaminar anthropic microfacies with anthropic boundaries in both sublevel Ja and Jb suggests the spatial congruence of covered surfaces in the occupation surfaces of level J at the scale of rockshelter topography.

The occupation surfaces of sublevels Ja and Jb can be characterized by their high degree of anthropic impact on the sedimentary record through their cultural formation processes (Schiffer 1987). This makes it more difficult to recognize the record of natural processes needed to produce microstratigraphic diachrony. However, the cumulative character of some sedimentary mechanisms, like the fill sediments or fuel incorporation during fire use, makes it possible to obtain evidence of spatial homogeneity or heterogeneity in the anthropic sedimentary processes. A sample from M56 is a good example of the interference between natural and cultural sedimentary processes (Fig. 4.11, scanned thin section 5). It shows how this cumulative character, or thickness of the anthropic microfacies, can become the source of important information on the use of space and the diachrony of the occupation surfaces, which analyses of prehistoric occupation surfaces have often neglected (Bordes 1975; Rigaud 1994; Dibble et al. 1997; Brochier 1999).

We should mention that the microstratigraphic distinction of diachronic occupation surfaces in level J based on microstratigraphy has not produced promising results. The lateral and vertical changes in the anthropic microfacies in relation to sedimentary, pedological and anthropic boundaries have hindered making such an approximation. This also emphasizes the limits of the excessive microstratigraphic separation which results from only studying cores and small samples of large surface archeological structures or occupation surfaces. Mapping archeological structures is necessary not only for solving the temporal problems of cultural formation processes, but also for interpreting the anthropic sedimentary processes of occupation surfaces. We think that the cumulative character of occupation surfaces must be analyzed by taking into account interdisciplinary data and discussions (Bordes 1975; Leroi-Gourham 1982; Brochier 1999). Curiously enough, in our opinion, advances in contextual archeology have switched to interdisciplinary work and back to the core disciplines of archeological research, analyzing the time resolution of activity episodes (Butzer 1989).

In summary, our microstratigraphies suggest the interstratification of natural and cultural sedimentary processes inside the archeostratigraphic units Ja and Jb. The temporal transgressive properties of these microstratigraphic units suggest living floor diachrony in the archeostratigraphic unit. What we found in some combustion structures, especially the samples from the excavated areas in which the sublevels could not be distinguished, made us "re-label" sublevels Ja and Jb on the basis of microscopic observations. The result is a new map of the combustion structures which more accurately depicts the living floor.

Anthropic Microfacies Mapping of Sublevel Ja

The distribution map of the anthropic microfacies of sublevel Ja takes into account the diachronous occupation surfaces. Figure 4.16 represents both the zones where archeostratigraphic diachrony is evident since only level J has been documented, and those of sublevel J–Ja where a microstratigraphic diachrony exists as well. This cartography has been drawn up by associating the anthropic microfacies of sublevel J–Ja with the topographical and geomorphological characteristics of the excavated area, and the spatial distribution of dumped and modified travertine blocks and combustion structures presented earlier (Sañudo et al. this volume).

Figure 4.16 also presents the contours of sublevel J–Ja without the pits resulting from the previous excavations. It shows the structural mega-blocks and the blocks which proved useful in deducing anthropic modification by thermal impact or their transfer toward the internal and central zones. This hypothetical scheme for the surface of sublevel Ja shows the spatial distribution of anthropic sedimentation and lateral correlations. In the central zone, we show a band of small gravitational blocks in which large faunal remains, blocks and burned sediments were documented (Martínez Molina 1999; Vallverdú 2002). The accumulation of archeological remains, possibly underestimated due to the interference of recurring gravitational falls, suggests an area of discard, analogue to those described in previous ethnoarcheological and prehistoric archeological studies (O'Connell 1987; Binford 1988, 1996; Fisher and Strickland 1991; Bar-Yosef et al. 1992; Mellars 1995; Meignen et al. 2001). In this area the anthropic microfacies are laminar and stratified. The hypothesis of a discard area is also consistent with the 3 m distance between combustion structures, which suggests an axis of external combustion structures (Gamble 1990). Towards the internal zone, the

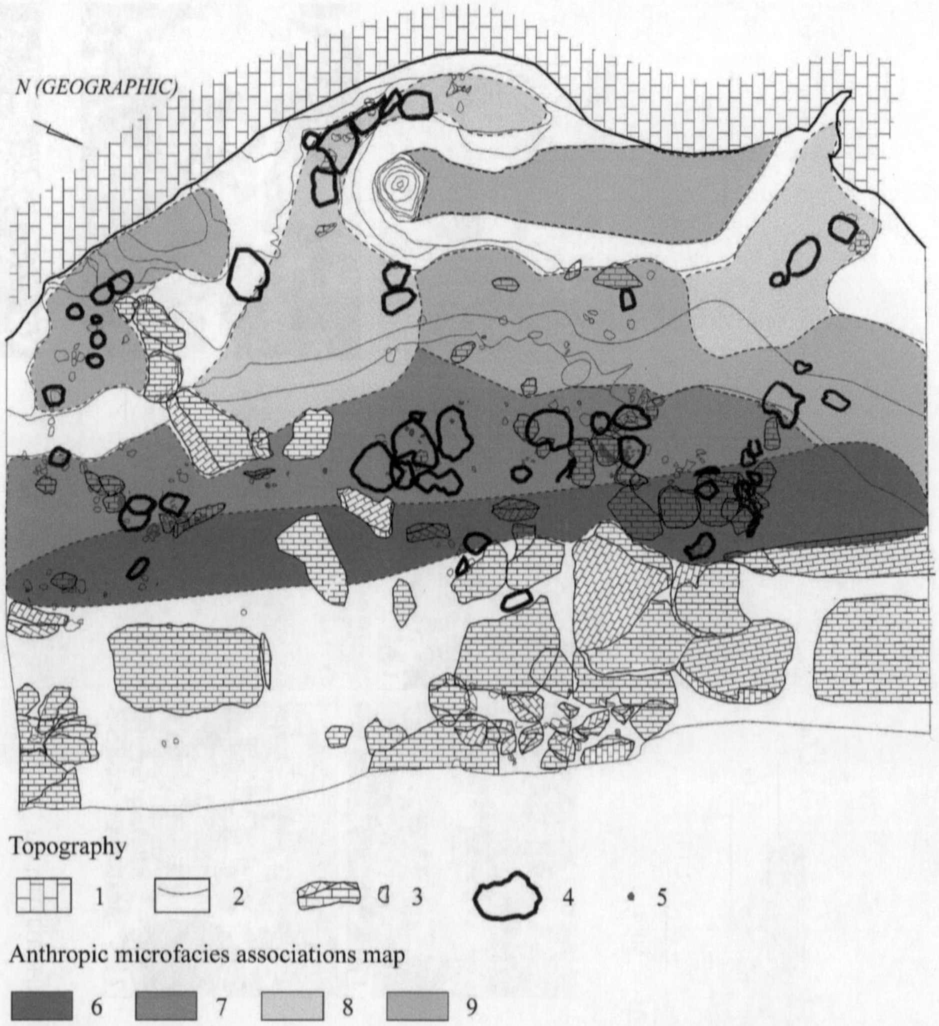

Fig. 4.16 Reconstruction of sublevel Ja topography and shaded map based on the association of anthropic microfacies. Topography: *1* rockshelter wall; *2* main topographical discontinuities of sublevel Jb; *3* structural megablocks and modified blocks; *4* combustion structures; *5* burned blocks and slabs. Anthropic microfacies association map: *6* central-external zones with laminar and stratified microfacies; *7* central-internal zones with laminar and microlaminar microfacies; *8* carbonate ramp and constrained available surfaces of the internal zone: laminar and stratified microfacies; *9* flat surfaces of the internal zone: laminar and microlaminar microfacies

anthropic microfacies of these external combustion structures are microlaminar and laminar. Close to the internal zone boundary, we also sampled stratified anthropic microfacies, which are interpreted as resulting from regular cleaning activities. This explanation of primary waste activity zones fits well with McKellar's principle (Schiffer 1987; Stevenson 1991). Bedsets of stratified microfacies provide evidence of maintenance activities including regular cleaning of the surfaces near combustion structures. This suggests a cultural formation process which can be observed in discard activity areas where dumped archeological remains are found.

Although a more exhaustive knowledge of prehistoric fuels would be useful, the information obtained from the thin-section study suggests that fire was used for a specific purpose in the combustion structure close to where sample N45 was collected (fire pit evidence) (Fig. 4.17). Abundant organic remains of plant or animal origin have also been interpreted as stemming from the presence of mats above humus, carbonized or calcinated sediments that were modified by compaction–dispersion or reused near the combustion structures as construction material for the living floor (microfacies M.2) (Fig. 4.17, scanned thin section b). These processes have also been mentioned in order to explain anthropic burned microfacies of dumped and dispersed ashy sediments in the Mousterian site of Kebara (Bar-Yosef et al. 1992). As mentioned above, we found the microremains of burned hide in one of these trampled deposits with anthropic microfacies L3 (Fig. 4.11, microphotographs 8 and 9 form scanned thin section b in Fig. 4.21; Vallverdú 2002).

The area in front of the *Coveta Ripoll* has a flat surface, and the density of artifacts is very low between spatial units 2 and 3 of the internal zone, limited by a hypothetical wall effect coinciding with a line of placed blocks (Vallverdú 2002). On this flat surface, the anthropic microfacies are microlaminar and laminar. There is a similar flat surface, also with a low density of anthropic remains, in which spatial unit 1 of the internal zone is located (see Sañudo et al. this volume). In this area, the combustion structure contains many small combustion structures, which are close to one another and well delimited. The anthropic

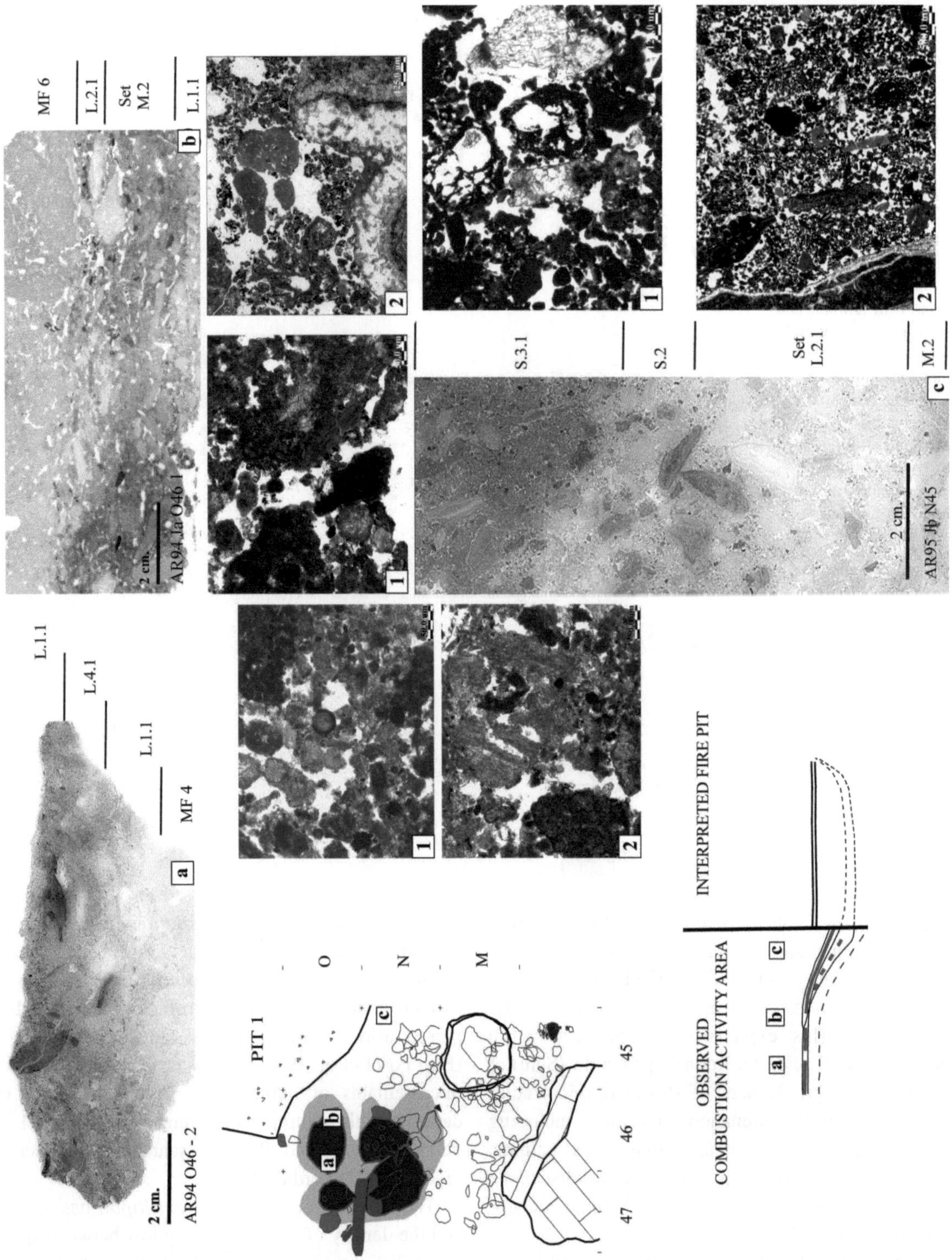

Fig. 4.17 Combustion structure group 3 of the central zone and position of the sedimentary monoliths **a**, **b** and **c**. Anthropic microfacies correlations used to propose sedimentary evidence for defending the fire pit hypothesis in sublevel Ja. **a** Thin section scan and anthropic microfacies succession. Note homogeneous burned microfacies (L.4.1 type). 1 clastic pale spherulite. 2 *ash* pseudomorph. **b** Thin section scan showing anthropic microfacies succession of covered trampled facies (M2). *1* crystal intergrowth of spherulites in dusty cryptocristalline *red* coating on carbonate rock fragment. *2* microcontrasted cryptocristalline nodules by crystal intergrowth of druse microsparite; note a continuous dusty *reddish* coating on carbonate rock fragment. **c** Thin section scan containing dominant and burned organo-rich coarse stratified anthropic microfacies (S). Microartifacts and sedimentary components show intense thermal modifications. *1* vesicular charcoal. *2* coarse carbonate rock fragments have inclined referred distribution pattern to microstratigraphic unit boundary surface. Note the dominant burned organo-mineral components

microfacies were found to be laminar with homogeneous burned components suggesting scarce occupational disturbance. Lastly, in the inner part of the internal zone, we sampled combustion structures located at the entrance of the *Coveta Ripoll*, an area with a limited available surface. These combustion structures display L and S microfacies which suggest an occupational disturbance, including truncated anthropic limits above L3 and L4 microfacies (Fig. 4.18, thin section scan a).

The map of the anthropic microfacies associations of the *Coveta Ripoll* shows that the anthropic microfacies of flat and restricted spaces are clearly juxtaposed. Anthropic microfacies L and M and anthropic microfacies S, associated on the flat surface of this spatial unit 2 in front of the *Coveta*, can be related to the boundary between areas in which a density boundary (between high and low) in archeological material has been documented. This suggests a strong analogy with an intensive outside dumping zone juxtaposed to empty areas, as suggested by the paleoethnographic analysis of huts or dwellings in the Pincevent studies or the exogenous model of small caves proposed by Thomas (Leroi-Gourhan 1976; Thomas 1983).

Conclusion: The Soil Micromorphology of Prehistoric Occupation Surfaces

Archeology and sedimentology propose a contextual approach to archeological sites (Butzer 1989; Schiffer 1995). Evidence derived from anthropogenic (cultural) and natural sediments are an essential source of information for archeological interpretation (Stein and Rapp 1985). It is important to note that in our context the excavation and sampling strategies we used allowing us to obtain a sedimentary record of the Middle Paleolithic was as useful as the fact that we analyzed in large format thin sections.

Advances in our understanding of human evolution require an accumulation of data from many different branches of science, such as models of human behavior, and the cultural ecology of present and past human societies (Isaac 1989; Conard 1994). Anthropic sedimentary processes which form laminar, microlaminar and stratified microfacies have been considered as a comparative tool. To configure this tool it has been necessary to distinguish between natural and cultural sedimentary processes and their associated microfacies, and to become thoroughly familiar with them. Soil micromorphology is technically very effective in identifying this distinction (Dalrymple 1958). This contribution to the archeology of prehistoric occupation surfaces takes a similar route, using the results of soil micromorphology and sedimentary petrography in experimental and historical contexts, to support an archeological interpretation of the Middle Paleolithic occupations of level J.

Though the results derived from the sedimentary record have yet to be subjected to interdisciplinary discussion, we maintain that microarcheological research must include thin section microscopic observations as well as more exhaustive spatial studies and sedimentary analyses of disturbed bulk samples (phytoliths, ashes, etc.). This approach promises a much deeper and rigorous understanding of the different research perspectives applied to the archeological levels at this site. In the future we hope to increase and specify more anthropic sedimentary processes and to expand the number of resulting microfacies types. This will be achieved through experimental and reference studies and the analysis of clear archeological analogues in order to compare features that are not always well preserved in the sediments of archeological sites.

The formation processes of anthropic microfacies have been established mainly through the study of structural transformations caused by thermal impact (burning), trampling and fill sediments. But before these mechanisms were recognized, the natural processes that originally formed these facies had to be identified. Once the formation mechanisms of anthropic facies types are recognized, it is possible to establish a basic record of human activities. The identification of these cultural sedimentary processes illustrates the potential of soil micromorphology to include anthropic sediment research within the strategy of archeological inquiry, and to recognize prehistoric occupation surfaces in a way consistent with other archeological disciplines and analytical techniques.

The ~100 year resolution of the level J sedimentary sequence, with a lithofacies rhythm of approximately 300 years, is indicative of the microstratigraphic context in which the prehistoric occupation surfaces have been documented. This high resolution record runs parallel with the high climatic variability observed in the Marine Isotope Stage 3 (MIS 3) (Dansgaard et al. 1993; Burjachs and Julià 1994). Sedimentary formation processes, characterized by the weathering intensities of the carbonate rock fragments, emphasize the temporal transgressive relationship between sedimentary facies and the human occupation of the Abric Romaní. Sedimentary records of weathering processes by biological mechanisms express the proximity of scales between pedology and the anthropic sedimentary processes in continental sedimentary environments (Holliday et al. 1993).

We propose considering the cultural sedimentary process as empirical evidence to be applied in testing the reality of occupation surfaces once the natural formation processes have been well established. Considering "archeology as sedimentology" (Schiffer 1995) during the evidence collection process was an attempt to integrate both the natural and the cultural sedimentary processes in level J through a conjunction of empirical and systemic analyses. Before interpreting the environmental and cultural inferences of the archeological deposits and remains, microfacies analyses of the sedimentary processes in archeological contexts provide

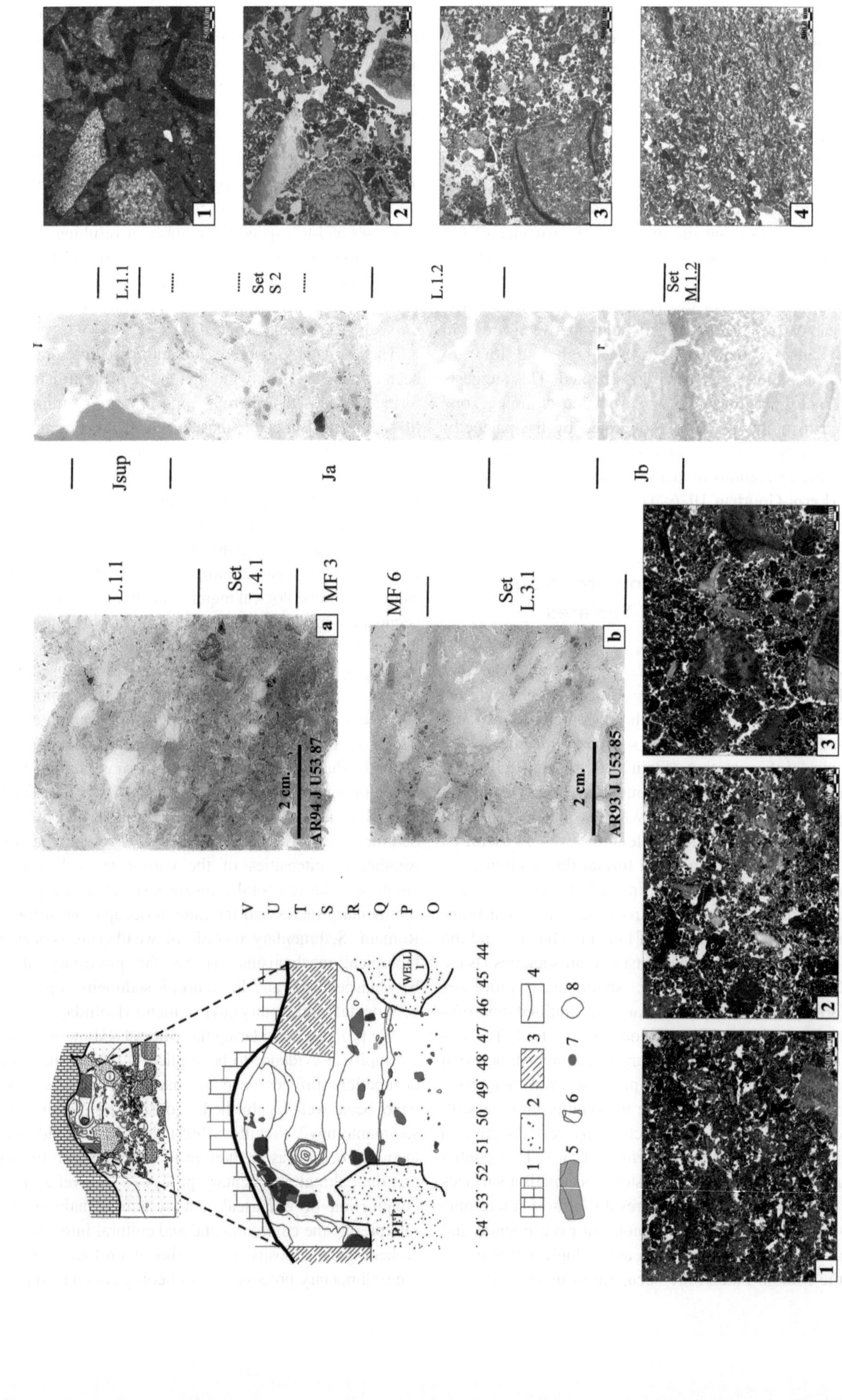
V
U
T
S
R
Q
P
O
PIT 1
WELL 1
54 53 52 51 50 49 48 47 46 45 44
1 2 3 4
5 6 7 8
L.1.1
Set L.4.1
MF 3
2 cm.
AR94 J U53 87
a
MF 6
Set L.3.1
2 cm.
AR93 J U53 85
b
Jsup
Ja
Jb
L.1.1
Set S 2
L.1.2
Set M.1.2
1
2
3
4
1
2
3

◀ **Fig. 4.18** Combustion structure group 2 in the internal zone (*Coveta Ripoll*). *1* rockshelter wall. *2* early archeological excavations. *3* stratigraphic outcrops. *4* main topographic discontinuities. *5* J occupation surface boulders. *6* structural blocks and boulders. *7* burned blocks and slabs. *8* slabs. **a** Thin section scan showing a truncated combustion activity area bedded with laminar and stratified facies. **b** Combustion structure thin section scan showing evident bedding of heterogeneous burned microfacies (L.3.1 type set). 1, homogenous burned microfacies (L.4) below truncation scanned thin section U53 87. 2, L.4.1 homogenous burned microfacies. 3, L.3.1 anthropic microfacies: trampled burned organo-mineral sediment. **c** Set of thin section scans showing the integrity of J deposits in the central area (P52 sample). Ja deposits are composed of stratified microfacies suggesting the palimpsest composition of different environmental and occupational episodes. Bedset of anthropic stratified microfacies (S2 set) show abundant microvestiges and porosity probably originating from the decay of organic matter. *1* and *2* are microphotographs showing very abundant lithic microchips. *3* laminar anthropic microfacies type 1.2 (L.1.2) is cemented and contains ooidal components suggesting water-laid sedimentary process. *4* environmental change imprint in anthropic deposits are mainly related to frost activity in Jb (set M.1.2) and decarbonatation in Jsup on anthropic microfacies L.1.1

and enhance a contextual data set that must be subjected to cross-disciplinary discussion. These contextual data lead to an improved multi-scalar approach, involving feedback concerning the evidence based on field observations, especially for archeological components like combustion structures or less evident empty activity areas at the site. This stimulates critical consideration of the analytical arguments posited with regard to the archeological structures during fieldwork and, by involving all archeological disciplines, incites the reconsideration of issues related to the occupation surface within the long-term processes of cultural sedimentation in archeological layers.

Archeological Implications

With the information derived from the study of cultural sedimentary processes and their related anthropic microfacies, we aim to contribute to the interpretation of the site structure of level J and its activity areas. These geoarcheological observations and interpretations must be coupled with and delimited by other archeological analyses. In the field of archeology various analyses are used to identify archeostratigraphic units made up of artifacts and to reproduce them experimentally. Ethnoarcheological analogues are studied as a reference framework with which to verify and determine inconsistencies in the formation processes of the archeological record (Villa and Courtin 1983; Gould and Yellen 1991; Rigaud 1994). In prehistoric archeology, archeostratigraphic units are commonly established by the early three-dimensional plotting of the artifacts discovered and conducting subsequent quantitative or spatial analyses. The combination of stratigraphy and spatial analysis forms the conceptual core of the discussion that must take place in order to understand what occurred in an archeological layer or on an occupation floor (Bordes 1975). However, excessive stratigraphic control may be a source of confusion, especially when it leads to the identification of false stratigraphies produced by postdepositional processes or anthropic activities (Villa 1982).

In joint stratigraphical and spatial analyses, it is commonly agreed that though it is difficult to establish temporal relationships between vestiges or archeological features (e.g. combustion structures), such relationships cannot be ignored. The conditions that limit the spatial interpretation of the archeological record must also be explained. In other words, archeological methods must define the limits of accumulations and/or groups of anthropic remains (Kroll 1997).

Bordes expressed the difficulties in establishing temporal relationships between vestiges and emphasized stratigraphic control as a link to the validity of spatial reconstructions (Bordes 1975). This approach has gradually been proven worthwhile in the case of large surface excavations and in the discussion of the cultural processes that underlie the archeological record (Schiffer 1972, 1995; Hughes and Lampert 1977). The results of extensive excavations have led to totally qualitative spatial research, based on the presence of favorable "high-resolution" conditions and site structure studies (Leroi-Gourhan 1985; Adouze and Enloe 1997; Pettit 1997). These favorable conditions prevailed at the Abric Romaní: artifact refitting was possible; the archeological level was limited in thickness; combustion structures were well delimited; and specific accumulations of archeological remains were found (Carbonell and Castro Curiel 1992; Allué et al. 1993; Carbonell et al. 1996a, b). However, correct interpretation requires correct orientation and solid theoretical bases (Villa 1977, 1988). Under these favorable conditions various important questions in experimental archeology can be addressed (Villa and Courtin 1983; Schick 1997), and an approach based on the spatial interpretation of the archeological record within a reference framework can be developed (Gifford and Behrensmeyer 1977; Adouze 1987; Brooks and Yellen 1987; Binford 1988, 1996; Kent 1991). Ethnoarcheological research into cultural formation processes has been legitimized by spatial studies of anthropic sediments within complex cultural stratigraphic successions (Stein 1992; Chenorkian 1994). The fact that humans and human groups act as geomorphological agents has been questioned on the basis of the multiple conditions of human activity. Some archeological beds, such as the occupation surfaces of level J, are the result of accumulations which can be understood through paleoenvironmental research and the remarkable record for identifying site structure and occupational episodes that anthropic sedimentary processes and microfacies provide.

Acknowledgments Excavations at the Abric Romaní were carried out with the support of the Departament de Cultura de la Generalitat de Catalunya, Ajuntament de Capellades and Tallers Gràfics Romanyà-Valls. The Generalitat de Catalunya provided financial support to the Research Group in Quaternary Human Autoecology (2005SGR-00702). Fundación Duques de Soria and Fundación Atapuera also provided a grant for this research. We thank the Abric Romaní archeological team for their assistance during fieldwork. Thanks to Dr. Richard Macphail for his thoughtful review and help giving our text a definitive form and correct English. We also thank Dr. Mercè Bergadà Zapata and Dr. Oriol Oms Llobet for their reviews of the manuscript and useful comments.

Appendix 1

Clastic and Anthropic Microfacies Origin in Level J Deposits: Classification and Analysis

We sampled a large number of continuous sections throughout the level J sequence, which afforded us a great degree of control in terms of the spatial and lateral variability across the excavated area. Two major groups of microfacies types were distinguished on the basis of their structural characteristics and basic composition: clastic and anthropic microfacies.

The clastic group comprises all microfacies which were predominantly formed by natural processes and which contain only randomly distributed cultural microvestiges, and in which no human-induced structural influence on the sedimentary components were detected at the microscopic level. Clastic microfacies basically correspond to sterile layers in which no occupational episodes have been determined. Therefore, we will first examine the clastic microfacies, which provide the most reliable data of the record on environmental conditions over the course of the level J deposition processes, as in this group anthropic bias is minimal.

In contrast, the group of anthropic microfacies is associated with the most evident archeostratigraphic units identified during excavation and particularly around combustion structures. In general they display a pronounced anthropic bias that primarily influences the geometry of the basic components, having modified their original properties and governed the incorporation of a variable proportion of cultural microvestiges and organo-mineral components. The diagnostic clues provided by the analysis of the clastic microfacies formed by multiple expressions of natural processes provide a solid data base with which to examine anthropic microfacies in relation to human activities.

Clastic microfacies

Microfacies 0: Gravels and Gray Sands

These microfacies are made of gravels with an open structure and partially filled with sands (Fig. 4.19, thin section scan 1). The CF/FF ratio is less than 8:2. The fine fraction is partially gray, though rarely to occasionally made up of dominating brown carbonate silts. The microstructure is microaggregated in fine and very fine crumbs, but more often consists of loose simple packing of sand without fine fraction. Porosity is considerable (>30%), formed by simple and complex packing with relatively few channels and fissures. The sedimentary structures show stratification and some laminae, emphasized by the orientation of the gravel.

The gravel is angular and subangular in shape, with a parallel plane form and crystalline fabric, while the sands show a variable rounded to subrounded proportion of micrite and microsparite crystals correlated to the increase in corroded fine fraction. The relative CF/FF distribution is monic to enaulic. On rare occasions allochthonous silicate mineral components are present.

The transformations of these microfacies are poorly developed. The textural features primarily consist of rare textural fillings with loose fine sands and silts towards the central zone. Near the wall, we observed textural cappings of non-compacted fine silt and sands above crystalline platy gravels, with a brown carbonate micromass and microdivided anthropic material (charcoal) (Fig. 4.19, scanned thin section 1). The crystalline features are made up of occasional partially corroded micrite and microsparite masses in low concentrations. There are also concentrations around grains of gray micrite, which occasionally encompass fine sands and non-corroded silts. Excremental features are very scarce and biological activity is mostly limited to channels in the samples taken from further away from the internal zone or to those containing higher quantities of organic components. In the central zone, microfacies 0 shows increased corrosion, parallel to an increase in the amount of fine fraction. Anthropic features are occasional and may indicate weak to very intense anthropization (Courty et al. 1989).

Partial Interpretation

The sedimentation of platy crystalline gravel indicates intense fragmentation of the shelter wall and rapid burial, which explains the limited postdepositional evolution linked to the abundance of pedofeatures. Therefore, rapid sedimentation was really a consequence of the fragmentation of the shelter wall. We have been able to distinguish only very low biological activity, notably limited to the presence of the few channels and rare excremental features. Dry conditions are indicated by the poorly developed corrosion in the carbonate rock fragments and only the occasional presence of crystalline features, in the fine fraction as well as in the coarse fraction. This absence of humidity, in concordance with the absence of pedofeatures related to redistribution mechanisms of bicarbonated solutions, indicates sedimentation in a zonal dry environment protected by the shelter. Fragmentation requires numerous cycles of

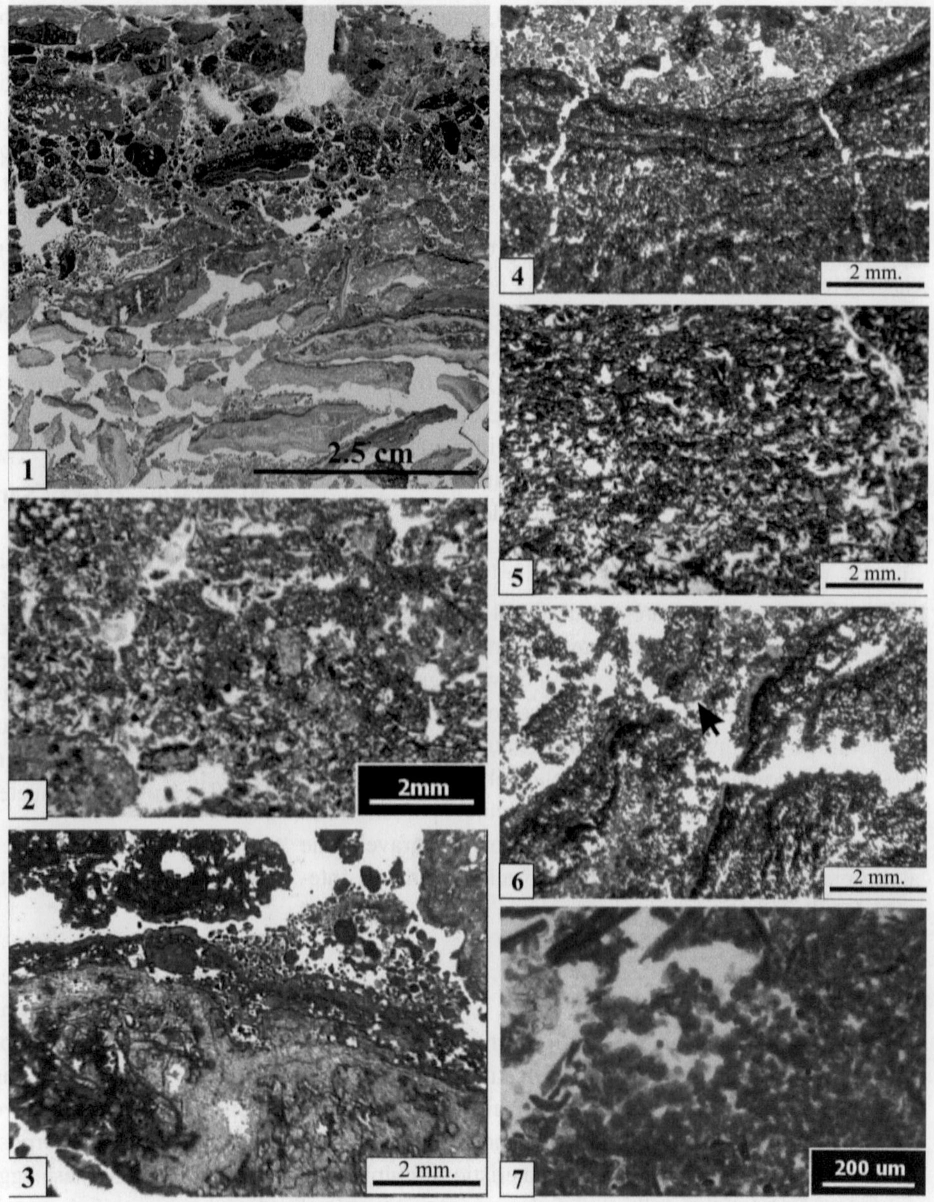

Fig. 4.19 Clastic microfacies 0, 1 and 2 from level J thin sections and microphotographs. *1*, microfacies 0, gray gravels and sands; scanned combustion activity area sample from Q58 in combustion structure group 1 of the internal zone. Note the scarce degree of occupational disturbance. *2* granules and sands included within vesicular microstructure in microfacies 1. *3* exolithic pseudomorphic calcitic features above gravel showing *Rivularia haematites* crystal growths in microfacies 1. *4* cracked algal filaments in microfacies 2 made of *gray* silts and sands. *5 gray* silts and sands—note the abundant fragments of acicular calcite in the coarse fraction and the vesicular microstructure: microfacies 2. *6* microcodium (near *arrow*) and algal filaments; microfacies 2. *7* spherulites in *gray* silts and sands; microfacies 2

wetting and drying, and is evidenced here by the presence of occasional silicate minerals. In fact, the presence of non-compacted textural cappings of brown carbonate silts on gravel with anthropic microdivided material in the samples from near the wall (Fig. 4.19, microphotograph 3) is interpreted as a result of ice-free water percolation during the melting of snow (Vliet-Lanoë 1987). Thus, ice activity was limited to the ground surface. Meanwhile, the fragmentation of the cornice is suggested by the number of dry and mild periods, which favor the appearance of a significant thermal gradient in the rock escarp, where frequent fragmentation has limited the development of weak to strong weathering intensity in Microfacies 0.

The high porosity in this open-structure microfacies gravel favors hydric circulation and aeration. As for postdepositional transformations, and considering the low degree of modifications, the presence of anthropic features stands out. These are mainly related to their microstratigraphic position: these microfacies usually make up the basal layer of the anthropic deposits of level J in the central and internal zones where it has been difficult to distinguish between sublevels Ja and Jb.

Microfacies 1: Gravel and Gray Sandy Silt

These microfacies are composed of medium to coarse angular to occasionally subangular gravels which have an internal crystalline fabric with dissolution cavities and developed corrosion. There is frequently a loss of contour in the gravel in contact with silt (Fig. 4.19, microphotograph 2). Fine fraction is scarce (<8/2) and highly corroded and its relative CF/FF distribution is gefuric. Porosity mainly consists of fissures and vesicles, many of them tubular, beneath the gravel. In the coarse fraction, components of corroded acicular calcite are abundant (*Microcondium*).

The laminar sedimentary structure is underlined by the gravels which show diverse exolithic growths made up of carbonate coatings characteristic of weak weathering intensity (Fig. 4.19, microphotograph 3). Sands are very occasionally present in the form of acicular calcite, fragments of endolithic tufa (pseudomorphic) and fragments of crystalline gravels with irregular outlines. The very abundant silts are subangular to subrounded with rough contours, although there are some with smooth contours. Vertical and sub-horizontal fissures form a well-developed laminar medium to coarse platy microstructure (Fig. 4.19, microphotograph 4).

The transformations are textural and cryptocrystalline. The cryptocrystalline features are important: gray micritic impregnations in dense masses of silt between gravel grains. In a sample close to the wall, we also found reddish cryptocrystalline coatings. The textural features are limited to loose and incomplete fillings of relatively well sorted fine sands. Cemented cappings with vesicular porosity and polyconcave cavities were also found on the gravel. The anthropic features are limited to rare microcharcoals integrated in the impregnated micromass.

Partial Interpretation

These facies contain a coarse fraction with weak alterations and abundant fragmented crystalline features. However, the microstructure and porosity of these facies make the conditions of hydric seasonal contrast on the shelter floor difficult to verify. They are more characteristic of epiphreatic zones marked by the collapse of the microstructure and polyconcave cavities (vughs). The fragmentation of gravel also indicates significant moistening and desiccation. However, the presence of biomineralizations like *Microcondium* is an indication of good drainage and corresponds to seasonal hydric stress (Verrecchia et al. 1995). But they are present in fragmented forms and the presence of fissuration, vesicles under gravel and the detritical character of these biomineralizations indicates the occurrence of seasonal freezing-thawing on the soil surface. Thus, it is possible to generally interpret the micritic impregnation of these microfacies as having occurred through the slow circulation of saturated bicarbonated solutions during periods of thawing, and intense desiccation during freezing. This slow percolation expresses a low and probably lateral infiltration, corresponding to partially confined drainage, enhanced by the micrite impregnations and silty texture of microfacies 1. Finally, the character of these textural features, well-sorted incomplete loose fillings, compact cappings, micrite impregnation, and vesicular porosity, mark the profound effect of freezing on the postdepositional evolution of these microfacies, consistent with the semi-fissural and platy microstructure.

In the internal zone the suggested lateral percolation favors the appearance of mechanical sedimentary processes in accordance with the cryogenic desiccation and fragmentation of tufa. These conditions correspond to cold periods with a maximum degree of humid seasonality, only effective during thawing and accentuated by a low infiltration rate.

Microfacies 2: Gray Sandy Silt with Blocks

The microfacies of gray sandy silt with blocks contain a low percentage of gravel (<10%) and present a significant occurrence of non-corroded calcitic features. The blocks, like the rest of the granulometric classes, show high weathering intensity. Silts and sands have a vesicular and spongy microstructure (tufas). Porosity is characterized by the presence of planar and vertical curved fissures. Some vesicles are interconnected by small fissures.

Vesicular Gray Sandy Silt

Total porosity is low (<30%), revealing a compact appearance upon macroscopic observation. The CF:FF ratio is relatively low, less than 7:3. Its CF/FF distribution is chitonic-gefuric. Gravel of crystalline fabric is not frequent, whereas micritized grains and aggregates are more common. These micritic gravel-sized grains and aggregates are often moss pseudomorphs. In situ sandy disaggregation was also observed in these aggregates, which is expressed by the disappearance of contour between gravels and their cement. Sands also present a micritic fabric and are rounded to subrounded, whereas silts are crystalline. Of the size of the fine sands and silts we saw *Microcondium*, fragments of micritized acicular calcite, along with remains of fragmented moss pseudomorphs (Verrecchia and Verrecchia 1994; Verrecchia et al. 1995; Fig. 4.19, microphotographs 4 and 5).

The transformations we observed are mainly related to cryptocrystalline and textural features. The cryptocrystalline features are manifested in masses of silts impregnated by gray micrite. These form large compact aggregates of well-sorted silts, limited by fissures. They are in the form of well-developed macro-sized laminar or angular blocky aggregate microstructures. The textural features are better developed and consist of infillings and rare micritic-microsparitic coatings. The infillings are formed by well-sorted silts and include incomplete infillings around and under the gravel, some of which has a vertical axis. The vesicular gray silts and sands show very few anthropic features, mostly microcharcoals.

Spongy Gray Sandy Silt (Tufa)

Porosity is high (>30%). It is comprised mainly of sparitic calcite, often peripherally micritized. The sparite is yellowish with pronounced cleavage planes (Courty 1986). This is generally a moss pseudomorph although we also observed abundant *Microcondium*, acicular calcite and other exolithic biomineralizations (*Rivuliaria, Scytonema*) such as biological crusts (Miedema 1997; Fig. 4.19, microphotographs 6 and 7). In some samples there was a pronounced fragmentation of this tufa. Macroporosity is made up of fissures.

Microfacies 2 transformations are mainly related to crystalline pseudomorphic pedofeatures. Tufa and laminated biological crust porosities have loose and incomplete sandy silt infillings. Very fragmented pseudomorphic tufa can be observed with link cappings, textural pedofeatures and vesicular porosity. Fragmented acicular calcite shows a basic distribution pattern. Locally, a fine platy aggregated microstructure is found in the vesicular sandy silt of microfacies 2. Anthropic pedofeatures are rare and limited to a few samples with rubefaction included in the sedimentary record.

Partial Interpretation

The intense gray weathering of these microfacies indicates the slowing down of sedimentation processes. This process is especially active on the cornice of the shelter, which indicates carbonate redistribution. Gravel fabrics are predominantly tufa-clast, which reveals the presence of vegetation on the cornice and, therefore, moisture. Wetting and drying episodes might have been linked if cryosuction occurred more often than seasonal humidity (summer prolongation). The bicarbonate water saturation in these microfacies manifested during thawing. Poor infiltration facilitated the dissolution-precipitation processes that developed the intense gray weathering in the weakly altered fragments that fell from the cornice. The presence of vesicular porosity, occasionally tubular, in the sandy silt facies also indicates water saturation under thawing (Vliet-Lanoë 1985). The water saturation of these facies that are poorly drained during freezing leads to the geliexpansion of the sediments. These geliexpansion phenomena have been observed in the thin section samples in the form of well sorted silty infillings around and underneath the gravel showing evidence of cryoexpulsion (piprakes). These fills are associated with laminar-fissural microstructures which, due to their size, are indicative of deep freezing.

We also observed fabrics of fragmented biocrystals where ice-induced segregation formed link cappings of compact gray silt. In this fragmented tufa a more superficial position of the freezing can be assigned. The presence of slightly fragmented tufa might indicate phases of weaker cryoturbation. These variations in the frost activity record represent a lesser quantity of water relative to the depth of the tufa. Thus, depending on their position relative to the drainage zone, this tufa might have been superficially fragmented or deep frozen in poorly drained samples. Lastly, the abovementioned cryoturbation is clearly a subsurface process. The biological crusts with exolithic organisms identified inside, broad fissures and an undulating appearance have been considered highly elastic forms. These forms were subjected to internal or buried cryogenic geliexpansion, since they are very sensitive to the presence of superficial freezing (piprakes) (Vliet-Lanoë et al. 1995).

Microfacies 3: Gravel and Brownish Gray Sandy Silt

These microfacies contain very abundant gravel (>30%) (Fig. 4.20, microphotograph 1). Occasional to abundant fragments of exolithic crust with high intensity weathering are also present. In the sand fraction rounded to subrounded micritic-microsparitic grains dominate (Fig. 4.20, microphotograph 2). The silt fraction is scarce although there are concentrations of dusty brownish-grey sandy silt. They form very fine crumb microaggregates with a weak crystalline concentration. The aggregation is commonly poorly developed, and when it is well defined by fissural porosity, it is expressed as non-compact fine plate made of scarce gray micromass with a gefuric CF/FF related distribution. There are also occasional silicate silts. The microstructure is not aggregated and is spongy. Porosity is high (>30%), made up of simple packing voids, vughs, tubular vesicles and rare channels (Fig. 4.20, microphotograph 3). The CF/FF ratio is less than 9:1.

The sedimentary structure is poorly stratified with occasional laminar strata suggested by the orientation of the gravel in some samples (Fig. 4.20, microphotograph 1). Gravel is fine to medium-sized, angular to subangular, crystalline and presents a basic distribution pattern on the vertical-orthogonal axis. Pisolithic gravels can be observed at the base of some microstratigraphic units with these microfacies 3.

The transformations are mainly textural and cryptocrystalline. Infillings are abundant in the packing voids of consistently well classified and loose silt and sand. More occasionally textural pedofeatures are microlaminated and fining-upwards brown coarse to fine silt. Other rare textural features are non-compacted and cemented cappings. All

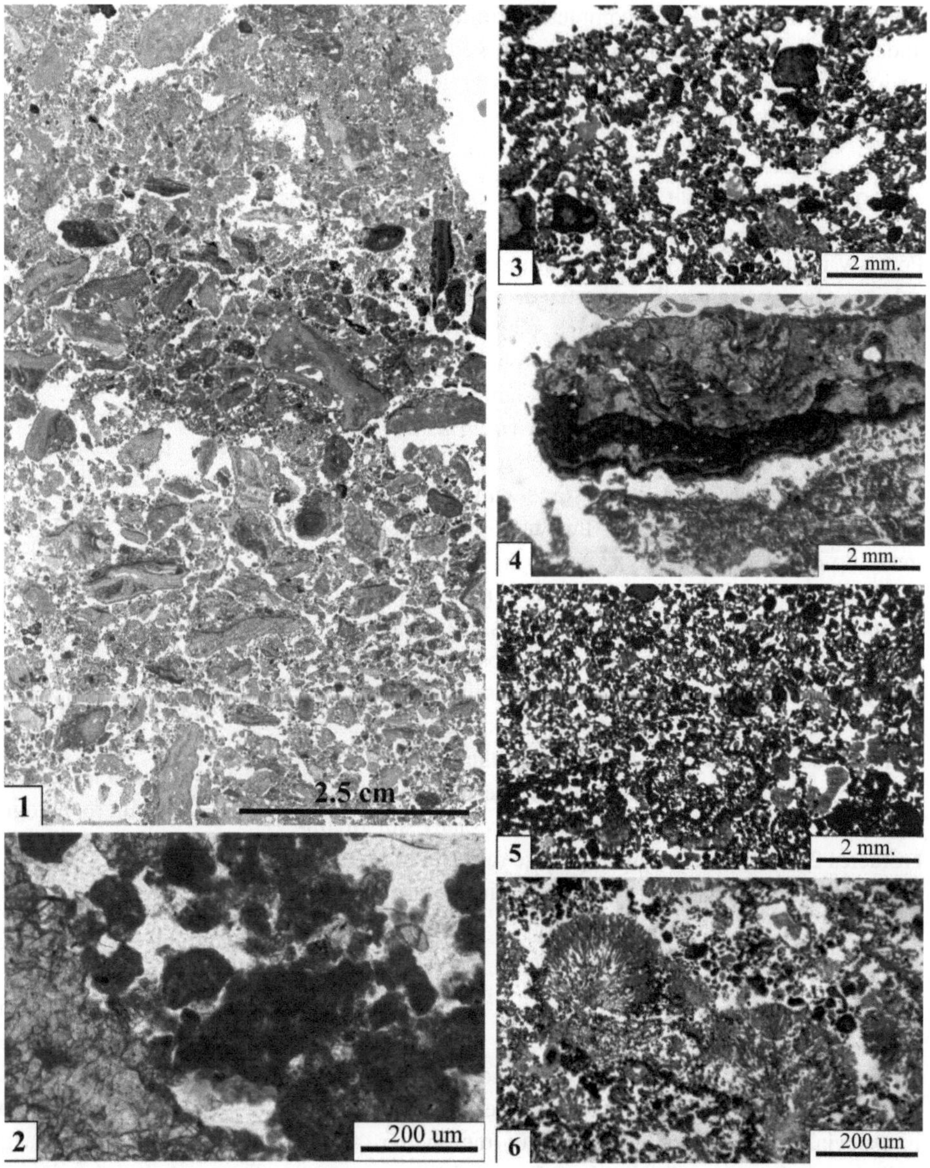

Fig. 4.20 Clastic microfacies 3 and 4: soil micromorphology and sedimentary structures. *1* burned *brownish-gray* gravels and silts combustion structure in M57; scanned thin section. Note the gradual thermal modification of gravels (rubefaction), the sedimentary crust (trampled surface) above erosional truncation surface (human-induced?). *2 grayish-brown* gravels and sands; microfacies 3; ×100 magnification showing high intensity weathering on carbonate rock fragments; *3* vesicular to spongy microstructure in *grayish-brown* gravels and sands in microfacies 3; sublevel Jb; *4* carbonate gravel-sized showing a relict crystalline pedofeature, a pendent, in *grayish-brown* gravels and sands; microfacies 3. *5* complex packing microstructure showing a high weathering intensity in situ crystalline pedofeature (*Rivuliaria*) above carbonate granule-sized rock fragments; microfacies 4

these textural pedofeatures contain very few anthropic microvestiges.

The crystalline pedofeatures are made up of infillings of loose and transparent anhedric microsparite and isolated grains of non-micritized acicular calcite. Rare to occasional anthropic components typically have calcite coating on all sides. Rare to occasional pseudomorphic sparitic calcite coatings are present on the gravel. The cryptocrystalline coatings are gray and are mainly distributed in low-density crystalline masses around grains. They are dominant in all particle sizes with a rough contour.

Anthropic features are very abundant. Burned rock fragments, charcoal, bone fragments, and reddish, gray and yellowish cryptocrystalline aggregates appear in microfacies 3 that make up the sublevel Jb deposit type.

Partial Interpretation

These grayish-brown facies with high intensity weathering indicate that the roof of the rockshelter wall and its carbonate curtain were relatively dry. Exolithic biological activity has occurred in these facies as evidenced by the rare to occasional presence of fragments of exolithic cements in the coarse sand and gravel fractions. These fragments also indicate that this exolithic activity was restricted by the fragmentation of the cornice. The presence of an intense dry-moist thermal gradient at the wall is accompanied by a slight increase in silicate minerals in microfacies 3. However, the observation of abundant crystalline and cryptocrystalline micritic features points to wet conditions in the shelter fill. Consequently this suggests that the carbonate fragments underwent high intensity weathering within the shelter infillings and not on the rockshelter cliff.

At the base of these microfacies of gravel with gray sands we found poorly stratified gravel, which we interpret as stemming from the presence of small sheets of dripping water or floods. There may have been drip ponds which would have caused the enclosed infillings of anhedric calcite. Although the high sedimentation rate favored the burial and preservation of the carbonate rock fragments, the moisture conditions in the shelter ground gave rise to postdepositional transformations indicated by textural, crystalline-microcrystalline and depletion pedofeatures.

Gray cryptocrystalline-crystalline pedofeatures are abundant in the sand fraction. There is a notable spongy microstructure in dusty brown silts with chitonic related CF/FF distribution, which indicates the presence of elluvial washing. This elluvial character is also related to textural features such as sorted coarse infillings, non-compact cappings, and gray pendent coatings on gravels, which point toward the vertical infiltration of water (Fig. 4.20, microphotograph 4).

Other textural features are characterized by the illuviation of brown carbonate silts. In fact, their localization in the lower part of gravel associated with tubular vesicles is very characteristic of surface freezing (piprakes). These illuviated brown silts indicate the textural translocation of corrosion products during cryosuction. This desiccation probably limited the biological activity in the shelter ground.

Gravels and sands in the microfacies indicate a high sedimentation rate, but the postdepositional process might have affected the visibility of the archeological record through, for example, the haloclastism (calcium carbonate) of bone tissues. This can be explained by chemical changes in the infiltrated water or the degree of bicarbonate saturation. In this respect, the development is noticeable in alternating water conditions, slightly loaded with humic substances during cryosuction and containing more carbonates and less acid during ice thawing. During thawing, microfacies 3 was therefore subjected to the intense percolation of water slightly loaded with organic acids. This percolation favored the rise of a temporarily water-saturated hydric regime. Biological activity took place during this wet phase. However, the presence of seasonal dry periods, increased by cryogenic desiccation, induced intense evaporation of a dry to subhumid pluviometric regime which limited endolithic activity in the ground.

Microfacies 4: Bbrownish-Gray Silty Sand with Blocks

These microfacies of sand and gray silt with blocks contain a low percentage of gravel, always less than 10%. In the coarsest fraction fragments of biomineralizations like *Rivuliaria, Scytonema* and microsparitic moss pseudomorphs stand out (Fig. 4.20, microphotographs 5 and 6). The coarse to fine ratio is relatively low (<7:3) and its related distribution is chitonic-gefuric. The shape of the very coarse fraction is rounded to subrounded and it is aggregated biogenic in nature. Its shape inhibited it from following the basically vertical orientation pattern, which some of the large rock fragments possess. The fragments of the very coarse fraction possess a fenestral porosity, are yellowish in color and have low birefringence in analyzed light. The medium-sized sands are also rounded and subrounded and made up of microsparitic-micritic aggregate clasts. The silts have a more angular morphology. Few silt-sized silicate minerals are present.

The sedimentary structure of microfacies 4 is stratified, with some lamination underlined by the presence of horizontally distributed grains. The microstructure consists of open cavities. Total porosity is high (>30%) and macroporosity is characterized by fissures. As for mesoporosity, vesicles associated with rare or occasional tabular vesicles are abundant. Tufa, when present, contains a spongy microstructure and does not show fissural porosity.

The pedofeatures are dominated by crystalline types. Coatings that form a rough gray contour of micrite are abundant in almost all fractions. The textural features consist of well sorted silt coatings, weakly corroded, in the vesicular porosity that is distributed beneath the coarsest fraction. Some infillings, with fine sands and loose silts, are localized in accommodated fissures. Disaggregation is found in the coarse fraction expressed by the gradual grain contour.

Anthropic features are relatively abundant with some fragments of burned rock, microcharcoals and bone fragments. Some pure cryptocrystalline yellow aggregate is also found, isotropic in analyzed light. A basic distribution pattern in bands of similar-sized anthropic components can be observed in some samples.

Partial Interpretation

The brownish-gray silty sand and blocks of microfacies 4 show high intensity weathering, which indicates slow burial. The predominance of tufa-clasts and gravel aggregates in the coarsest fraction points to some humidity in the cornice. This humidity was also detected in the shelter ground through the presence of calcitic crystalline and cryptocrystalline pedofeatures. Exolithic weathering is well represented in the fragments fallen from the cornice. On the floor of the rockshelter, humidity is recorded in the form of tufa (mosses, algal cementations). Likewise, these brownish-gray microfacies contain fragments with intense weathering acquired on the floor of the rockshelter.

The hydric regime in these microfacies is recorded in the spongy microstructure, which indicates an absence of collapse due to moisture. Therefore, these facies correspond to a well drained surface, possibly facilitated by the presence of blocks. Vertical infiltration is also evident in the abundance of textural features mostly consisting of fine silts and, more rarely, loose very fine silts. Their localization around and beneath grains (piprakes), or within vesicular porosity, indicates their formation by superficial freezing and thawing. The effect of freezing-thawing is also expressed in the weakly developed aggregated microstructures. Locally, fissures and cementation show that cryogenic desiccation occurred, probably in the more saturated zones. In this case infillings and micritic cryptocrystalline pedofeatures are more abundant and the constituents coarser. In any case, our interpretation is that a cold setting such as this would result in cryogenic phenomena of a lesser thermal amplitude. The same constituents of the microfacies, with the dominance of the sand fraction, grains and pisoid aggregates, mark a sandy disaggregation sedimentary process from the carbonate curtain. Therefore, cryoturbation would be a mainly postdepositional process on the ground of the rockshelter, above all conditioned by good infiltration that leads to an overestimation of the dry character of the microenvironment during cold periods. Consequently maximal seasoning with high evapo-transpiration could explain the very homogeneous micritization that is observed in the abundant calcitic components and tufa-clast of these microfacies.

Microfacies 5: Gravel and Brown Sandy Silt

The sedimentary structure of these microfacies is laminar and indicated by lines of gravels. Microlaminar structures are present, but less frequent. Porosity is made up of channels, fissures and vughs. Pedality is moderately to well-developed with the presence of mesogranular aggregates to less developed microcrumbs (Fig. 4.21, microphotograph 1). The related CF/FF distribution pattern is enaulic, and more occasionally porphyric. The CF:FF abundance ratio is 6:4.

The gravel is primarily angular to subangular and is crystalline in nature, although there is a considerable proportion of tufa-clast in the sand fraction which shows high intensity weathering (Fig. 4.21, microphotograph 2). Some isolated oolithes and pisolithes are also observed (Leighton and Pendexter 1962). Sands and silts are rounded to subrounded, many are microsparitic, although some are angular. Scarce isotropic subrounded yellow aggregates can also be observed, some with occasional silicate minerals included. There is an increase in the presence of silicate minerals (quartz, feldsparite and mica) in the fine sand and coarse silt fractions. The fine fraction is almost completely corroded, in such a way that locally some slightly weathered calcitic silts may be present.

The pedofeatures described are mainly textural, excremental and crystalline. Textural infillings are rare and made up of loose silt and fine sands in channels and chambers (Fig. 4.21, microphotograph 3). Compact quasi-coatings are also common in well preserved channels with excremental fabric. Finally, crystalline and cryptocrystalline features are observed around grains, as well as sparite coatings, and in abundant masses of low density brown silts. The sparitic coatings are rare and probably inherited from exolithic wall-algae activity. The microcrystalline features of low density masses usually include anthropic microremains like microcharcoals, sedimentary burned fragments, etc. These diffuse masses of brown silts can aggregate into well developed granular aggregates.

Anthropic pedofeatures are rare to abundant. Structural human modification is apparent in the findings of fire use and compaction.

Partial Interpretation

Microfacies 5 contain carbonate rock fragments with weathering of strong intensity. The fine fraction is more abundant below the carbonate curtain of the dripline of the rockshelter, with a correlative increase in biological activity. Biological activity is of intermediate intensity because no coalescence of developed excremental fabrics has been produced (Courty 1990). Postdepositional development is guided by the rare presence of calcitic features, and therefore high sedimentation rates. Only the formation of abundant masses of crystalline weakly concentrated brown carbonate fine silts, which include anthropic microvestiges, is indicative of carbonate dissolution or decarbonatation processes. These processes are implied in the formation of very fine to granular and crumb microaggregation associated with intermediate biological activity as discussed above. This biological activity requires temperature and humidity conditions that favor its development. Therefore, we suggest a medium degree of seasonality prompted by the silicate minerals in microfacies 5. The cornice provides intensely weathered carbonate fragments which indicate permanent humidity at the cornice. The carbonate rock fragments fell during a time of a decarbonatation soil forming process and high sedimentation rate.

Lastly, the hydric regime during the formation of these microfacies is characterized by vertical infiltration in the form of textural infillings, whereas the moderately developed biological activity limited structural collapse when there is no evidence of compaction.

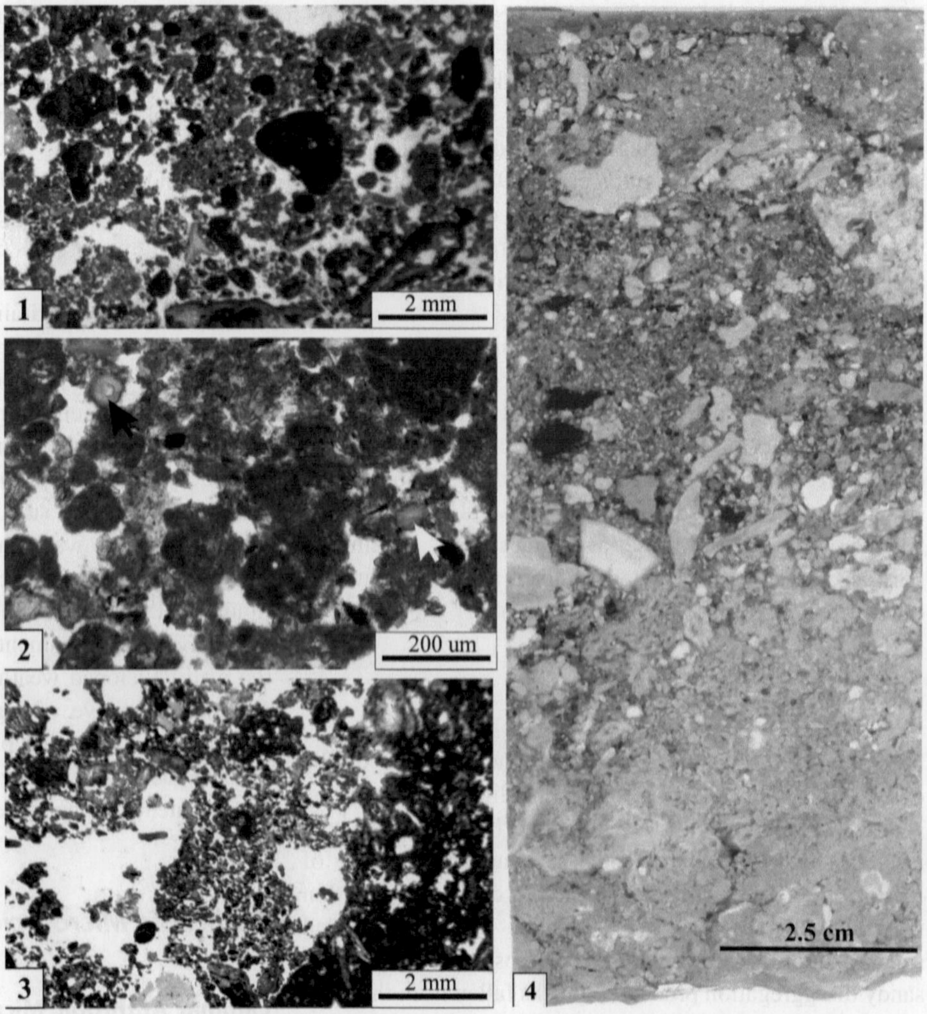

Fig. 4.21 Clastic microfacies 5 and 6: soil micromorphology and sedimentary structures. *1 brown* gravels and sands; microfacies 5. Note the crumby microaggregation with high intensity weathering of carbonate gravel-sized rock fragments. *2* x100 magnification showing diffuse concentrations forming microaggregates (fluffy) of *brown* microspar in microfacies 5. Note the rounded *yellowish-brown* spherical sparite (*black arrow*). *3* loose sand infillings, a textural pedofeature within a chamber in the center of the microphotograph; microfacies 5. *4* impregnated and polished block showing 5 cm of *grayish-brown* sands and silts at the *bottom*. In the *upper half* of the impregnated block there are graded beds of microfacies 5 and 6. Note the discontinuous and fining upward gradation of microfacies 6 showing a sedimentary crust and the partial lineation of gravel and granule carbonate rock fragments below the sedimentary crust (compaction by trampling) with abundant integrated archeological microremains

Microfacies 6: Brown Sands and Silts with Blocks

These microfacies contain a very coarse fraction with few fine gravels (<10%). The coarse fraction (CF) is dominant, rounded and subrounded, with high intensity brown weathering (Fig. 4.21, microphotograph 1). There are a few coarse, yellow, sand-sized grains that are isotropic in analyzed light and subrounded corresponding to microfacies of a late weathering stage. These cryptocrystalline yellow aggregates often contain silicate minerals. They also show black dots in the micromass in normal light and low density masses of microsparitic calcite of low birefringence in analyzed light. The fine fraction (FF) is made of brown carbonate silts. Its sedimentary structure displays poorly expressed fining-upwards graded bedding that is laterally discontinuous (Fig. 4.21, microphotograph 4). Above the graded beds microfacies 6 shows a massive sedimentary structure of corroded silt with some floating clasts and buried sedimentary crusts.

Two microstructures have been documented:

- Microaggregated, with a porosity of channels and fissures, made of large brown carbonate fraction (CF/FF >4/6). Their relative CF/FF distribution pattern is e-naulic-porphiric. Microaggregation gives rise to very fine crumbs that can form medium-sized angular granular aggregates by means of coalescence.
- Loose packing of sands and granules, with less fine fraction (CF/FF >4/6), sets in a porosity of simple to complex packing voids, and a relative CF/FF distribution that is enaulic to chitonic.

The pedofeatures observed in these microfacies are relatively well developed, with common low density masses of brownish carbonate silt and gray crystalline calcitic coatings around grains. Other crystalline features are biocrystalline: they form cryptogamic superficial crusts that cover sedimentary contacts. These cryptogamic crusts cover microlaminated fine sands and brown silts. We were able to identify some of the incrusting species, including *Rivuliaria hematites, Phormidium incrustastum, Scytonema* and mosses (Pentecost and Riding 1986; Freytet 1990).

Depletion pedofeatures stand out as the corrosion of the very coarse fraction. This appears as some small dusty isotropic yellow–red coatings in the porosity of the carbonate rock fragment (Courty et al. 1989). In blocks we observed the dissolution of crystalline cements, which is manifested in the form of zones with brown carbonate silt around coarse gray microsparitic-micritic and aggregate (carbonate pellets) sands.

The anthropic features of these microfacies indicate weak anthropization with the presence of microcharcoals. Some samples also contain significant accumulations of yellow angular to subangular isotropic aggregates. They contain very rare bone fragments including carbonate and sparitic impregnations in tubular pores and scarce to rare silicate components. These aggregates form continuous masses, which led us to interpret them as being related to carnivores (coprolites and regurgitation).

Partial Interpretation

We consider the rounded and subrounded nature of the coarse sands as a product of major endolithic activity at the shelter wall. This activity is supported by the presence of tufa-clast and aggregates, with the abundant presence of *Rivuliaria* and moss fragments. In both cases, in some thin sections we interpreted the formation of these rounded sands of more or less micritized microsparitic fabric as a sandy disaggregation product of blocks fallen from the wall. In other words, some of the facies we examined with high to late weathering came from samples taken at the shelter wall. Furthermore, the presence of subrounded yellow amorphous aggregates that are derived from carbonate rock fragments represent the most extreme stage of dissolution by organic acids and endolithic weathering. Thus, the presence of different stages of weathering, intense and late in this microfacies 6, tells us that the carbonate rock weathering initially took place at the wall and continued on the ground of the rockshelter. This is also indicated by the abundance of calcitic features that show high intensity weathering as well. The wall was probably colonized by heavily incrusted vegetation (mosses and other associated lower plants like lichens) that make up the coarse sand recognized in these microfacies 6. However, it is in the local postdepositional evolution of the facies where we find significant documented variability in the microstructures. We therefore interpret the microstructure of loose packing as suggesting a more rapid burial given the scarce presence of pedofeatures in the sediment. Consequently the most microaggregated facies, with an enaulic and porphyric relation, present a slower sedimentation rate, which is also related to the presence of biological superficial crusts. The presence of cryptogamic crusts attests to a marked degree of stabilization of the sedimentary surface on which an initial pedogenesis took place after the phases of erosion and gradation or simply the slowing down of the sedimentation processes (Boiffin and Bresson 1987). Microaggregation into very fine crumbs and granules suggests that a surface horizon can be identified in microfacies 6. The pedofeatures we discovered suggest soil-forming processes that require a water retention capacity and a well drained hydric regime related to decarbonatation soil-forming processes. The formation of cryptogamic crusts and pedality together with the increase in weathering intensity might also indicate surface sedimentary processes like runoff or simply stagnation of water caused by a well developed vegetation cover.

We can be quite sure that during the formation of microfacies 6 there was significant humidity on the wall and floor of the shelter, especially in the upper surface discontinuity of the deposit, where microfacies of brown silt and sand have been documented. We therefore also expect that the well developed weathering processes in these microfacies are indicative of high biological activity and decarbonatation soil-forming processes.

Level Jsup, containing few anthropic pedofeatures, was documented in microfacies 6.

Anthropic Microfacies

Laminar Anthropic Microfacies (L)

General Characteristics

Laminar microfacies comprise a primary sedimentary structure about 1 cm thick with a continuous and parallel bounding surface. There may be isolated stratum but lamina sets are more common. Microvestiges of meso- and micro-sizes and sedimentary components have a dominant basic parallel orientation and banded or linear distribution. Laminar anthropic microfacies are often made up of fragmented aggregates (clods) of previous occupation surfaces or surface pedosediments (epipedon) mixed with archeological remains. The relevant pedofeatures are textural and consist of loose or dense fillings and sedimentary crusts related to depositional microenvironments of structural crust formation (Boiffin and Bresson 1987). The microstructure is apedal with subhorizontal fissures delineating poorly developed laminar aggregates or joining vesicles in boundary surfaces and lithological discontinuities. Sedimentary and anthropic constituents with thermal modifications range from rare to abundant.

Interpretation

Laminar anthropic microfacies are result of the vertical and horizontal incorporation of archeological remains and sedimentary transformations of the occupation surfaces. Laminar anthropic microfacies showing very dominant incorporation of carbonaceous (charcoal-rich) components and thermal transformation of rock fragments suggest the use of the fire in combustion structures. Mixtures of anthropogenic and sedimentary components in the composition of laminar anthropic microfacies suggest a scattered occupation surface affected by occupational disturbances. Hughes and Lampert (1977) propose that occupational disturbance was caused simply by people moving about on a deposit during the routine activities of daily life. The parallel arrangement of the boundary surfaces, the linear basic distribution of the components and the appearance of cracking may have been caused by trampling (Goldberg and Whitbread 1993).

Laminar 1

Laminar microfacies 1 contain very few to few burned sedimentary constituents (VCFb/VCFnb: 0/10–2/8; and CF + FFb/CF + FFnb: 0/10–2/8) and microaggregated brown silts.

Interpretation

The burned components of laminar microfacies 1 indicate occupational disturbances at the periphery of scattered combustion structures or on the bare occupation surface. Microaggregation of the brown silts is typical of moderate corrosion by organic acids during humification and was probably also enhanced by human-induced organic incorporation.

General Characteristics L.1.1

The porosity consists of channels, chambers and subhorizontal fissuration. The pedal microstructure shows gray platy aggregates with partially accommodated fissures. The related C/F distribution is enaulic and porphyric.

Anthropic features are rare to abundant, mostly consisting of high birefringence spherical sparite (<20%), highly micro-divided charcoals (2–10%), bones (2%), and microcontrasted yellow cryptocrystalline aggregates. These yellow aggregates either display a pure, isotropic matrix or contain silicate inclusions that are rarely carbonate bearing.

There are occasional excremental features and rare to occasional aggregated allochtonous sedimentary components. The textural features are limited to loose and incomplete infillings in chambers and channels. Channels showing textural hypocoatings are common.

Interpretation

The presence of subhorizontal fissuration and platy peds indicate trampling. Channel porosity, aggregates and excremental features point to medium to low intensity biological transformations. The textural hypocoatings along with the channels come from traces of pedofauna (*dwelling* traces or *resting* traces). The relative C/F distribution is interpreted as resulting from the joint effect of compaction and dispersion on an occupied floor surface that was alternatively wet and dry (enaulic related distribution) or only wet (porphyric related distribution).

General Characteristics L.1.2

Porosity is predominantly vesicular, the related distribution varies from chitonic to gefuric and the fissure microstructure consists of gray, poorly developed platy peds.

Anthropic features are rare to abundant, with a few rare burned fine sands, and finely-fragmented microcharcoals. Bones are rare, occasionally fissured, some with punctures. Pure and microcontrasted, yellow and yellowish-red, isotropic and subrounded aggregates are rare to abundant.

Occasional to abundant textural features occur as infillings of fine sands and grayish-brown silts distributed around or beneath the very coarse fraction and as grayish-brown coatings of silt and fine sand in vesicles.

Interpretation

Although these facies do not display obvious structural modifications due to occupational disturbances, the few effects of trampling are shown by the common microaggregated pedality of brown silts. Repeated freeze–thaw cycles, well illustrated by the vesicular porosity and dense particle packing, are most probably responsible for the poorly visible human-linked microstructuration. These anthropic microfacies L.1.2 thus show major transformation of their primary origin, human activities being only recognizable by the thermal transformations of the sedimentary materials and the incorporation of a few anthropic components (subtype L.1.2.1).

Laminar 2

In laminar microfacies 2, burned sedimentary constituents and microaggregation of silts are frequent to common (VCFb/VCFnb: 2/8–4/6 and CF + FFb/CF + FFnb: 2/8–4/

6). The microstructure is vughy with complex packing voids and fissures that outline secondary platy peds and clods. The basic orientation of the very coarse fraction (VCF), and especially of the coarse fraction (CF), can occasionally be dominantly vertical and parallel.

Interpretation

The wide range of thermal modifications shows the mixture of materials from various activity areas, including the combustion structures. This heterogeneity associated with the complex packing voids, the vughy microstructure and the vertical orientation of the very coarse and coarse sedimentary components provide strong evidence for sediments dumped or scattered by intense occupational disturbance. On the other hand, secondary platy peds and clods and fissures may have resulted from trampling. These features, dumped and trampled sediments, suggest an occupied area affected by intense occupational disturbance.

General Characteristics L.2.1

The porosity consists of fissures, cavities and channels. Brown granular aggregates are well developed. The related distribution is enaulic. Excremental pedofeatures are rare.

The abundant anthropic pedofeatures consist of frequent burned sands and silts, occasional bones with coloration resulting from exposure to different temperatures, microcharcoals and occasional to common fragmented macrocharcoals, and rare to occasional isotropic and cryptocrystalline yellow aggregates. Fragments of the coarse and very coarse carbonate rock-fragments show thermal alteration at differing degrees of burning.

Interpretation

The scarcity of pedofeatures is indicative of rapid burial. This is consistent with the abundance of microvestiges pointing to an occupational disturbance affected by low to medium biological activity and suggesting short subaerial exposure.

General Ccharacteristics L.2.2

Porosity consists of simple packing voids and vesicles. The granular and platy microstructures shows poorly developed pedality, with an enaulic to chitonic related distribution.

The rare textural pedofeatures consist of complete and dense infillings around and underneath the very coarse fraction and coatings of fine sandy silt around vesicles. Crystalline pedofeatures are more abundant in the form of incomplete loose infillings of subhedral sparite in vesicles and voids in some samples.

The abundant anthropic pedofeatures consist of micro-contrasted and pure isotropic subangular yellow aggregates, scarce microcharcoals and, occasionally, slightly fragmented macrocharcoals, and abundant bones. The bones are occasionally fissured with black dots, showing different degrees of burning to no burning phenomena. Bones, weathered at their periphery, which show an accentuated isotropy in analyzed light, are only recognizable by the presence of osteon morphologies. Rare coprolites are observed (N45 in the Ja sample). Coarse carbonate rock fragments burned to different degrees are common.

Interpretation

These type 2.2 laminar sedimentary structures, although they present more pedofeatures, also show evidence of a specific-purpose activity area as indicated by the abundance of fragmented small-sized bone remains and the dominant simple packing void porosity. The pedofeatures are significantly attributed to freezing and thawing, and also to biochemical processes such as bone destruction by biological activity.

Laminar 3

Burned coarse and fine fractions are dominant (VCFb/VCFnb: 2/8–4/6 and CF + FFb/CF + FFnb: 4/6–8/2) in laminar 3. Organo-mineral microaggregation is well developed with abundant microcharcoals mixed with high birefringence microsparite.

Interpretation

The high frequency of burned sedimentary materials of the fine and coarse fraction indicates the proximity of combustion structures. However, the presence of non-burned sedimentary constituents, especially of very coarse fraction, also indicates occupational disturbance by trampling. Organo-mineral microaggregation and microcharcoal suggest a considerable accumulation of carbonaceous sediments by human incorporation.

General Characteristics L.3.1

The porosity consists of channels and subhorizontal fissuration. The microstructure shows brown moderately to well developed pedality of silty microaggregates, with enaulic related c/f distribution, and very few pedofeatures.

Plant-origin microvestiges consist of abundant fragmented macrocharcoals; phosphatised ashes; and subrounded microcontrasted aggregates that are yellowish to reddish in analyzed light with crystalline microsparite druse intergrowths. Other rare to common remains are isotropic and detritic subangular cryptocrystalline aggregates that appear yellow and yellowish-red in normal light. Charcoals are found within dense spherical sparite and have a fused appearance. Charred charcoal display high birefringence pseudomorphic microsparite (Fig. 4.11, microphotograph 4). Silicate minerals burned at different intensities are rare to occasional. Carbonate spherulites are rare to occasional, isolated or aggregated, loose or formed by crystalline growth (Fig. 4.11, microphotograph 7).

High birefringence on yellow densely merged spherical sparite L.3.1.1.

High birefringence on yellow pseudomorphic microsparite L.3.1.2

Interpretation

The abundance of organic and mineral components of plant and animal origin with different combustion intensities is consistent with heterogeneous burned sediments. We interpret these anthropic microfacies to be evidence of occupational disturbances during human fire-use in combustion structures. The occurrence of high birefringence charcoal and sparite suggests undetermined episodes of carbonate rock fragmentation and ash incorporation due to burning wood as fuel. The formation of high birefringence pseudomophic microsparitie in charred charcoals and spherulites could indicate consecutive use-abandon-reuse of combustion structures. Lastly, these accumulations with channel porosity and moderate to well developed pedality indicate that transformations were produced by low to medium biological activity and short surface exposure.

General Characteristics L.3.2

These display a vesicular porosity, with poorly developed silt microaggregation and a gefuric related C/F distribution. Anthropic features are numerous, with finely microdivided microcharcoals, abundant isotropic microaggregates that appear yellow and yellowish-red in plane polarized light and occasional druse intergrowth of microsparite in the micromass. Others microaggregates are isotropic and a few are microcontrasted with silicate minerals. Isotropic microaggregates are occasionally subangular and others are often subrounded. Rare to occasional burned bones, both red and yellow, and rare schist fragments are found.

The abundant crystalline calcitic pedofeatures consist of incomplete and loose acicular and anhedric calcite infillings. The textural pedofeatures observed are cappings between grains (link capping) in the sand fraction.

Interpretation

Also characterized by the presence of the non-burned very coarse and coarse fraction, these anthropic microfacies stand out in that they contain isotropic aggregate with crystal intergrowths of calcite in druse arrangement, which can be identified as ashes derived from leaves (Pobeguin 1954; Courty et al. 1989). Similarly, these microfacies lead us to conclude that there were accumulations of burned leaves and burned fine fraction with dispersed microcharcoals, as we found considerable heterogeneity in the evidence of fire use in combustion structures. Lastly, we emphasize the presence of pure and microcontrasted isotropic organo-mineral aggregates, especially subangular ones, with abundant silicate minerals. It is hypothetically possible that these subangular organo-mineral materials have been in contact with silicate allochthonous materials and were subsequently integrated into these burned accumulations. Although their specific identification might be difficult, these subangular elements with silicate components can be associated with materials of animal origin. On an initial examination, we would interpret them as remains originating from hide working and cutting. Finally, these anthropic microfacies show cryoturbation, although some microstratigraphic units which display few transformation features were probably protected by considerable sedimentary accretion as indicated by the abundance of microremains of human origin.

Laminar 4

General Characteristics L.4

Burned very coarse, coarse and fine fraction and organo-mineral microaggregation of charcoals and highly birefringent microsparite are very dominant (VCFb/VCFnb: 4/6–10/0 and CF + FFb/CF + FFnb: 8/2–10/0).

Interpretation

The presence of dominant to very dominant homogeneous thermal modifications and carbonaceous incorporation suggests minimal occupational disturbance in the use of the fire in the combustion structures. The incorporation of carbonaceous components and homogeneous thermal modification is reflected in the microaggregation made up of highly birefringent microsparite and microcharcoals. These highly birefringent microsparites are the byproduct of thermal impact leading to the microfragmentation of sparite into microsparite and the incorporation of ash. In addition to the formation of this fine and coarse fraction with homogeneous thermal modification, rare non-burned rock fragments and remains were seen, especially in the very coarse fraction, suggesting that anthropic microfacies L 4 represent the final episodes of burning fuel in the combustion structure.

General Characteristics L.4.1

They display a porosity of channels and fissures, a porphyric and enaulic related distribution, and a moderately developed pedality of granular and crumby aggregates of brown silts.

Anthropic features are rare to frequent. Carbonate rock fragments with homogeneous thermal modification show at their periphery two burned groundmasses: highly birefringent yellowish-brown microsparite and ash pseudomorphs, and black carbonaceous masses made up of weakly fragmented micro- and meso-charcoals. The carbonaceous groundmass carbonate rock fragments feature heterogeneous thermal modification. Burned bones are absent to abundant, with distinct coloration ranging from yellow, yellowish-red, and red to whitish-gray. Bones are also found with distinct peripheral coloration or double coloring, especially yellow and red. Occasional aggregates are present, which are isotropic, pure and microcontrasted, yellow and yellowish-red, subrounded and subangular. Microcontrasted isotrope aggregates may be detritic (silicate) or with druse crystalline intergrowths of microsparite.

Pedofeatures are made of rare coatings and dense infillings in channels, and locally of incomplete loose calcitic infillings.

Heterogeneously burned and unburned organo-mineral components L.4.1.1.

Homogeneously burned organo-mineral components L.4.1.2.

Interpretation

The distinction between the homogeneity and heterogeneity of thermal modifications and carbonaceous sedimentary materials, combined with the very dominant frequency, is indicative of the temperature intensity. Thus, we distinguish two extremes in the micro-stratigraphic units of sampled combustion structures. The high variability of thermal modification reflects low temperatures and occupational disturbance. Homogeneous thermal modifications and carbonaceous fine mass and ash components suggest high temperatures and scarce occupational disturbance. Lastly, the distinction between carbonaceous ground mass with highly birefringent yellowish-brown microsparite and ground mass formed of yellowish-brown microsparite and ashes allows us to identify three different intensities or temperatures.

The presence of few homogeneous micro-stratigraphic units with a very dominant frequency of burned components is unusual among the sedimentary records of combustion structures. Along with the fact that these facies often present sets of superimposed laminae, this indicates the presence of repeated episodes of fire use in these combustion structures. An important characteristic of laminar anthropic microfacies is their thickness which may indicate a limited surface available for fire use. Thus, the use of fire appears to be organized around small cores. The lateral displacement of these units with a homogeneous burned microfacies process might explain the scattered composition of heterogeneous burned microfacies (L.4.1.1).

Lastly, the rare pedofeatures, which indicate very low biological activity, and the local presence of crystalline infillings of acicular calcite, can be attributed to condensed humidity and rapid evaporation-crystallization.

General Characteristics L.4.2

They display a porosity made up of vesicles and fissures, a gefuric-chitonic related distribution, and poorly to moderately developed gray platy aggregates. Anthropic features are rare to occasional. In the units with rare anthropic features only very microdivided microcharcoals are found. The burned bones vary in color from yellow, red and bicolor to white.

Pedofeatures are rare to abundant, consisting of frequent textural pedofeatures of loose cappings on the very coarse fraction, slightly compact and coarsening upwards, and between grains of coarse sands (link cappings), well sorted infillings, around and under the very coarse fraction, of dense and incomplete silt and fine sand, cryptocrystalline dusty yellowish-brown coatings with high birefringence microsparite. Micro-divided charcoals are finely integrated to the ground mass of the platy aggregates.

Heterogeneous burned and unburned carbonate rock fragments. L.4.2.1.

Homogeneous burned carbonate rock fragments L.4.2.2.

Interpretation

Although the sedimentary components are expressed similarly to the characteristics which define L.4.1, which is based on the frequency and variability of burned carbonate rock fragments and carbonaceous/ashy sedimentation, the microstructure and pedofeatures display significant postdepositional modifications in some of these microfacies types. Therefore, these L.4.2 characteristics are interpreted as resulting from the development of post-depositional process, mainly due to cryogenesis. Some characteristics, such as thickness or lamination (lamina-set bedding), mainly when very dominant frequency and homogeneity of the groundmass are manifested, point to combustion processes analogous to those of type L.4.1. However, cryoturbation in the coarse and fine fraction suggests frost activity previous or subsequent to fire-use episodes.

Microlaminar Anthropic Microfacies (M)

General Characteristics M

Thickness varies between 2 and 0.5 mm. These microlaminae display a non-parallel disposition and can be discontinuous or appear in the shape of meso- to macro-sized clods (fragmented aggregates), with components showing a basic linear distribution pattern. The lithological characteristics are very distinctive. They consist of dense packed rounded to subrounded fine sands and silts, moderately to well sorted. They generally contain occasional to abundant well sorted subrounded, isotropic, yellow and yellowish-brown organo-mineral components. There are occasional to abundant yellowish-brown cryptocrystalline impregnations. The microstructure is massive or of isolated vughs. A prominent and abrupt to clear superior lithological boundary stands out, which is visible at low magnifications (x10). Sets of these microlaminar sedimentary structures can be observed in some samples.

Pedofeatures are rare to abundant. Compact hypocoatings in channels and also crystalline incomplete loose infillings and sparitic crystalline intergrowths stand out.

Anthropic features are rare to abundant. Many microremains present common fractures. In the micro-stratigraphic units with abundant anthropic features, components of plant origin have been noted, such as fragmented mesocharcoals and rare ashes. Fragmented burned bones are rare. Lastly, rare pure subangular cryptocrystalline microaggregates occur, isotropic in analyzed light as well as others that are microcontrasted with silicate components. Within these silicate components, anthropic components of fibrous form are made up of a micromass with a fabric of micro-medium sized anhedric grains. Subangular microvestiges with evidence of thermal impact and with clear black-brown histological structures appear to contain arranged fibrous-planar forms, which we attribute to a fragment of hide (Wattez 1990).

Interpretation

A prominent sharp or clear upper boundary surface and particles formed by densely packed well sorted microaggregates have been described in soil micromorphology hand-books as covered surfaces of houses in *tells* and non-constructed prehistoric occupation surfaces (Gé et al. 1993). Impregnations and well sorted organo-mineral aggregates are also noteworthy within the context of moderately to well sorted fine sands and silts. The rare to occasional presence of incorporated microvestiges, only abundant when burned, may indicate an impervious surface. Thus, we understand the incorporation of these burned microvestiges to have occurred prior to the formation of the microlaminar anthropic microfacies. However, between these rare to occasional incorporated microremains there are subangular cryptocrystalline microaggregates probably pointing to the in situ formation of late weathering carbonate coarse sands and granules. These subangular cryptocrystalline microaggregates with fragments of hide are not easily assimilated due to the loss of the original histological arrangement. However, the abundance of cryptocrystalline components in all the microlaminar anthropic microfacies indicates humidity, probably enhanced by an impervious mat.

Microlaminar 1

These contain few to very few burned sedimentary constituents (CF + FFb/CF + FFnb: 0/10–2/8) and microaggregates of brown silts.

Interpretation

The scarce presence of burned components indicates a buried surface peripheral to the combustion structures.

General Characteristics M.1.1

They display a porosity of channels and chambers and complex packing of 0–30%, moderate to well developed brown aggregates of granular to platy-fissural form, and a relative C/F distribution that is enaulic to porphiric. Anthropic features are rare to occasional, showing pure and microcontrasted isotropic yellow cryptocrystalline aggregates, yellowish-brown cryptocrystalline impregnations, rare to occasional fragmented microcharcoals, and occasional to abundant highly birefringent spherical fine sands.

The pedofeatures that stand out most are excremental and abundant channels with compact hypocoatings. Incomplete loose infillings are present which are associated with channels and chambers.

Porphyric related distribution M.1.1.1.

Enaulic related distribution M.1.1.2.

Interpretation

The relative distribution of the coarse and fine fractions allows us to distinguish the postdepositional evolution of a covered surface by compaction, as expressed by the development of subhorizontal fissural porosity, produced in a more humid (M.1.1.1.) or dryer environment (M.1.1.2.). Microstructural characteristics and pedofeatures indicate a surface deposit with a medium level of biological activity (epipedon).

General Characteristics M.1.2

They display a vughy and vesicular porosity, poorly developed platy-fissural brown aggregates, and a relative C/F distribution that is chitonic and gefuric.

The rare to occasional anthropic features consist of well sorted yellow and yellowish-red cryptocrystalline aggregates and impregnations, and rare microcharcoals.

The rare textural pedofeatures occur as incomplete infillings of sand and silt around the very coarse fraction. Crystalline features occur as infillings and crystal intergrowths of acicular and subhedral hyaline calcite in the vesicular porosity and in the micromass of, for example, yellow isotropic aggregates and cryptocrystalline carbonate coatings in the vesicular porosity.

Gefuric related distribution M.1.2.1.

Chitonic related distribution M.1.2.2.

Interpretation

The M.1.2 microfacies type suggests a decrease in the preservation of the microlaminar sedimentary structure. The formation of textural pedofeatures like infillings of silty sand and the dominance of vesicular porosity indicate the presence of cryogenic modifications. However, in some samples we observed loose and incomplete infillings of acicular calcite, together with calcitic crystalline intergrowths, which are also indicative for rapid crystallization of protected surfaces that retain moisture (Gé et al. 1993). The preservation of these infillings, and especially the crystal intergrowths, indicate a preservation of this microfacies type and microstructure, which are generally associated with a related gefuric distribution.

Microlaminar 2

These display very abundant channels and chambers, a complex packing porosity of 20–0%, brown moderate to well develop aggregates and a related C/F distribution that is porphyric and enaulic. Burned components are frequent to common constituents in the fine fraction (CFb/CFnb: 0/10–2/8 and FFb/FFnb: 2/8–4/6) together with yellow organo-mineral silty microaggregates.

The anthropic features are occasional to abundant, consisting of very abundant isotropic microaggregates that are yellowish and red, pure and rarely microcontrasted, detritic, some well sorted and others meso-sized and unsorted, subangular and subrounded and have characteristics of burned animal fat (Wattez 1990; Fig. 4.22, microphotograph 1). The absence of microbiological attacks on these isotropic aggregates indicates a burned origin for these microaggregates. Also present are abundant yellow and reddish cryptocrystalline impregnations, few to many microfragmented microcharcoals and macrocharcoals. Rare to occasional bone fragments are often burned to yellow and red colors (Fig. 4.22, microphotograph 6). Occasional schist fragments and abundant spherical sparite of high birefringence are present (Fig. 4.22, microphotograph 7). The rare pedofeatures occur in channels which rarely display loose infillings with excremental fabric.

Interpretation

The frequency of thermal modification in the fine fraction and the organo-mineral microaggregation indicate that these anthropic microfacies incorporated significant organo-mineral components in proximity to combustion structures. However, components of the coarse fraction with few thermal modifications indicate occupational disturbance. As a buried surface, they present low structural modification by trampling, which is expressed by the formation of dense packed microlaminar sets, and the frequency of fragmented charcoal remains and fissural microstructural characteristics. We thus interpret this anthropic microlaminar microfacies type 2 as covered organo-mineral accumulations near the combustion structures, probably related to a faunal processing activity area, given the presence of yellow aggregates attributed to burned fat.

Pedofeatures indicate a medium level of biological activity with the presence of abundant excremental fabrics.

Microlaminar 3

These display a dominant to very dominant burned fine fraction (CF b/CF nb: 2/8–10/0 and FF b/FF nb: 4/6–10/0), brown organo-mineral microaggregates with microcharcoals and highly birefringent fine silts.

Interpretation

The dominant to very dominant presence of thermal modification on the constituents of the fine fraction is indicative of the sample provenance near combustion structures.

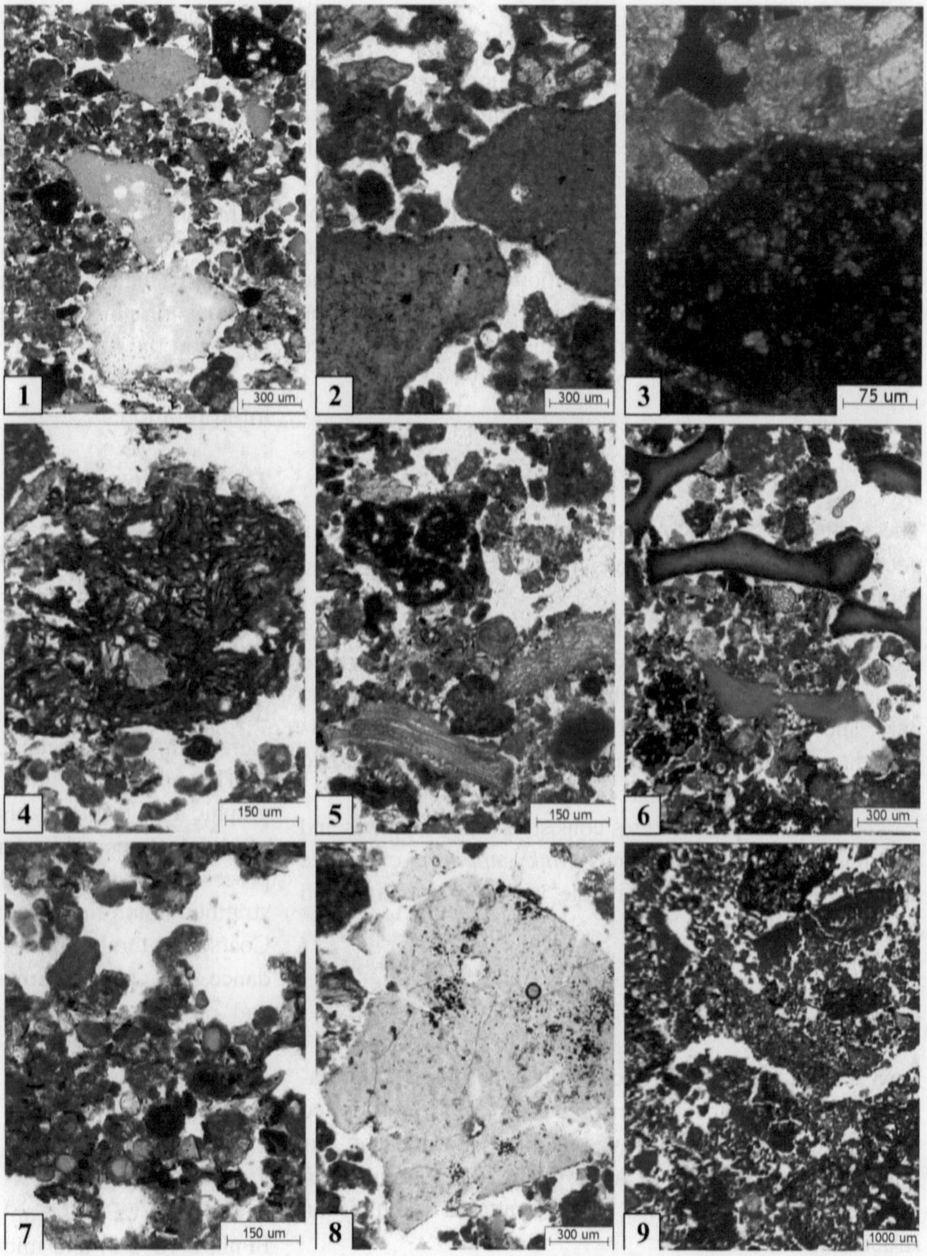

Fig. 4.22 Some microscopic remains of human origin found at the level J. All microphotos in plain polarized light, except for 3: in plain analyzed light. *1* cryptocrystalline pure and microcontrasted *yellow* and *yellowish-red* microaggregates. *2* microcontrasted *reddish* cryptocrystalline aggregate. *3* detail of the micro-contrasted *reddish* aggregate of microphoto 2 showing crystalline intergrowths of druse microsparite. *4* fibrous-shaped cryptocrystalline *reddish-brown* microaggregate with clastic carbonate sand-sized included. *5* pseudomorphic microsparite (*ash*). *6* bones with and without double *yellow* and *red* coloration. *7* highly birefringent spherical sparite in plain analyzed light showing *yellowish-red* color. *8* micro-contrasted and fissured aggregate with silicate material, with black *dots* pointing to microbiological attack. *9* fragment of angular exogenous silicate lithoclast surrounded by *grayish* carbonate rock fragments burned at high intensities (*lime*)

Likewise, the very variable presence of non-burned constituents in the coarse fraction indicates mixing processes by occupational disturbance. Its microlaminar character leads us to interpret it as related to a covered living floor with scattered combustion structures by trampling.

General Characteristics M.3.1

These display a porosity of channels and subhorizontal and subvertical fissuration, poorly to moderately developed brown aggregates of platy and granular form, limited by fissures and with a porphyric and enaulic related distribution.

The anthropic features of plant origin are very abundant as fragments of interconnected fissured charcoal of all sizes. There are rare ash pseudomorphs. Many isotropic microaggregates also stand out, angular and subangular, pure and microcontrasted by silicate or microsparitic crystal intergrowths. There are rare micro- and meso-fragments of red and yellow bones, others with double coloration. There are some micro- and meso-spherulites with high birefringence, either aggregated or isolated, rarely blackened (Fig. 4.11, microphotograph 7).

Pedofeatures are limited to hypocoatings in channels, rare excremental fabrics and incomplete loose infillings in vughs and channels.

Interpretation

Porphyric related distribution and fissures suggest compaction of burned and carbonaceous sediments. We therefore interpret these as covered anthropic microfacies which underwent previous occupational disturbance. In some of the samples a succession is observed which forms microlaminar sets, possibly indicating repeated dispersion and covering episodes. This interpretation coincides with the burned plant and animal components, with heterogeneous burned microfacies. The rare excremental pedofeatures are related to low and medium biological activity.

General Characteristics M.3.2

These display a vesicular porosity with fine fissures, a poorly developed gray platy clods and a porphyric and gefuric related distribution.

The rare to abundant anthropic features consist of microcharcoals finely integrated into the groundmass; rare to occasional subrounded cryptocrystalline microaggregates, pure and microcontrasted by silicate and crystal intergrowths of microsparite druse; and fragments of burned bones with different thermal modifications.

The abundant pedofeatures consist of relict cappings and continuous intergrain cappings (*link capping*), carbonate hypocoatings and coatings in the vesicular porosity, and rare loose infillings.

Interpretation

Few general criteria for microlaminar anthropic microfacies, such as prominent upper boundary surface and closed packing, are clearly distinctive, leading us to interpret this microfacies type M.3.2 as most likely representing cryoturbated covered surfaces over scattered combustion structures.

Stratified Aanthropic Microfacies (S)

The micro-stratigraphic units have a thickness of over 2 cm and their boundaries are rarely concordant. In some samples these micro-stratigraphic units are bedded, forming groups or sets, some interrupted by laminar units and others showing continuity between these anthropic microfacies S, always displaying faint boundaries. The coarse components are randomly distributed and oriented, although there is a common to dominant an inclined referred orientation pattern to boundary surface. The compound microstructure presents simple to complex packing, and the related enaulic and gefuric distribution.

Anthropic features are rare to abundant. In the microstratigraphic units with very few anthropic features, those containing rare to occasional microdivided charcoal stand out. Deposits with a low degree of diversity and others with high diversity of anthropic microvestiges are present.

Pedofeatures are rare. Excremental fabrics, channels with compact hypocoatings, fissuration and loose incomplete infillings of sands and silts stand out. There are crystalline calcitic pedofeatures of incomplete and loose infillings of anhedral calcite and acicular calcite which also appear in the form of crystal intergrowths.

Interpretation

These microstratigraphic units are considerably thick, indicating sedimentary accretion. Their non-concordant nature and commonly sloped surface boundaries, sometimes poorly represented at a laminar scale, indicate a spatial limitation of these S microfacies. Furthermore, the simple and complex packing compound microstructure and the basic vertical orientation of very coarse and coarse fractions, reinforce the interpretation that these microfacies were dumped on occupation surfaces.

The rapid sedimentary accretion which these microstratigraphic units display is also indicated by the scarce incidence of pedofeatures. This scarce incidence of postdepositional processes is reflected in the difficulties of distinguishing the stratified microfacies coming from sublevels Ja and Jb. Coarser textural composition, or gravel particle-size abundance, of S anthropic microfacies through sediment removal and moving on the occupation surfaces accentuates the anthropogenic singularity of these microfacies.

Stratified 1

These display very few to few burned sedimentary constituents (VCFb/VCFnb: 0/10–2/8 and CF + FFb/CF + FFnb: 0/10–2/8) and microaggregation of brown silts.

Interpretation

Dump sediments as well as few or very few burned components indicate surface modifications to the occupation surface by human activities, such as digging, but not activities related to combustion structures due to the lack of thermal impact identified.

General Characteristics S.1.1

These display a porosity of channels and fissures, C/F related distribution that is porphyric and enaulic and well developed yellowish-brown granular aggregates.

The abundant anthropic features consist of very homogeneous anthropic material: microcontrasted and pure subangular isotropic microaggregates with internal fissures and scarce incidence of punctuations in the micromass (Fig. 4.22, microphotograph 8) and rare to occasional silicate minerals inside these yellow isotropic microaggregates, some of them displaying thermal impact, and rare schists.

Pedofeatures occur in channels in the form of dense hypocoatings and occasional to abundant excremental loose and incomplete infillings.

Interpretation

The as of yet unresolved identification of this homogeneous accumulation of organo-mineral or anthropogenic mineral remains hinders interpretation at the thin-section analytical level. These accumulations also present the characteristics of mineralization originating from animal biomass: coprolites (omnivores and carnivores) or fats. These yellow organo-mineral components suggest a provenance other than a combustion structure due to the scarcity of burned and carbonaceous sediments. Therefore, stratified anthropic microfacies type 1 are interpreted as dump sediments derived from an activity area in which the transformation of animal or mineral matter took place.

General Characteristics S.1.2

These display complex packing void porosity, a related c/f distribution that is gefuric to enaulic, poorly developed gray aggregates and rare anthropic features in the form of microcharcoals.

Excremental pedofeatures and loose and incomplete textural infillings of fine sand are present.

Interpretation

The scarcity of burned sedimentary materials along with the general characteristics of stratified anthropic microfacies 1 indicates dump sediments without traces of anthropogenic materials, suggesting that they come from a bare occupation surface.

Stratified 2

These display frequent to dominant burned coarse fraction and fine fraction (yellowish brown spherical sparite and microsparite of high birefringence) (VCFb/VCFnb: 0/10–2/8 and CF + FFb/CF + FFnb: 2/8–4/6) and microaggregation of brown silts.

The anthropic microvestiges are rare to abundant with a high degree of morphometry. In some samples these anthropic microremains are distributed in sets of stratified anthropic microfacies. Fragments of chert, meso- and micro-sized schists, bones burned at different intensities and of different sizes, and isotropic and microcontrasted microaggregates are common. Few fragmented microcharcoals and macrocharcoals occur.

The rare pedofeatures consist of textural pedofeatures made up of incomplete and loose infillings, crystalline pedofeatures of calcite, anhedral and acicular crystal intergrowths in the primary porosity.

Interpretation

Highly birefringent microsparite and spherical sparite and bones burned at different intensities may be related to previously burned microfacies. The general characteristics of microfacies S suggests that these deposits probably result from the maintenance activities (cleaning) of combustion structures. The presence in some samples of interstratification between stratified (S) and laminar (L) anthropic microfacies indicates the spatial congruence of these deposits. We interpret these interstratifications as resulting from repeated activity in order to move sediments, forming dumping areas. These may have been interrupted or modified by occupational disturbances in the form of laminar microfacies. However, as suggested earlier, microlaminar microfacies are more difficult to verify with certainty, given their high porosity, which is enhanced by the complex packing of the microstructure of microfacies S. Nevertheless, anthropic microfacies S type 2 shows rare to occasional calcitic features which can be interpreted as indications of a covered living floor over dump sediments (Gé et al. 1993). In fact, we observed the association of these stratified anthropic microfacies beneath microlaminated microfacies. However, in several stratified anthropic microfacies we were only able to observe these crystal growths and calcitic infillings under units with laminar microfacies or overlying other stratified ones. In these cases, and bearing in mind the difficulties of researching covered occupation surfaces in these very loose microfacies, we interpret the calcitic features as a record of local conditions such as dripping, suggesting interferences between natural and anthropic sedimentary processes.

General Characteristics S.2.1

These display a porosity made up of simple and complex packing voids, fissures and channels, with an enaulic related C/F distribution, and granular and platy brown aggregates with moderate to well developed pedality.

The rare pedofeatures consist of textural features of loose and incomplete infillings, rare channels with carbonate hypocoatings and coatings in the primary packing voids.

Interpretation

Although the general characteristics provide evidence of dump sediment, fundamentally these anthropic microfacies S type 2.1 show occupational disturbance. Evidence of biological activity does not contain significant traces of pedofauna. Therefore pedality is probably inherited and not indicative of aggressive postdepositional conditions or slow burial. Rapid burial in some of these stratified microstratigraphic microfacies is represented by the weak intensity weathering of calcitic pedofeatures, such as carbonate hypocoatings and coatings.

General Characteristics S.2.2

These display a porosity made up of simple packing voids, vesicles, channels and fissures. The relative C/F distribution is gefuric and chitonic. Pedality consists of poorly to moderately developed gray platy aggregates.

The rare to abundant pedofeatures consist of textural features made up of loose and incomplete infillings of sand and silt and carbonate coatings in the packing voids.

Interpretation

The microstructural characteristics and the presence of rare to abundant pedofeatures indicate that these anthropic microfacies S type 2.2 represent different taphonomic situations. Although all the microstructural characteristics and the aggregation point towards cryogenic modifications, some of the microstratigraphic units show poorly developed frost activity. These conditions are related to rapid burial, which is possible when anthropic microfacies are microstratified.

Stratified 3

These display very dominant thermal modification in the fine and coarse carbonate rock fragments (VCFb/VCFnb: 8/2–10/0 and CF + FFb/CF + FFnb: 8/2–10/0). There is a microaggregation of highly birefringent silts, ashes and charcoals in poorly to moderately developed brown and gray granular aggregates. Relative C/F distribution is gefuric to enaulic. Porosity is made up of complex packing voids, vesicles, scarce channels and fissures.

The few pedofeatures consist of incomplete and loose infillings of subhedral calcite, and carbonate coatings in their primary porosity.

Interpretation

Very dominant thermal modifications in anthropic microfacies S type 3 indicate combustion structure origin. The mineral stability caused by burning phenomena in this microfacies suggests cultural sedimentary accretion by a discard activity area on the occupation surface. The rare pedofeatures have signs of thermal impact. Only the dynamics of carbonate dissolution and precipitation has partially affected the particle integrity of the microfacies.

General Characteristics S.3.1

These display dominant reddish burned carbonate rock fragments and abundant to very abundant subrounded to subangular microcontrasted yellowish-red to reddish cryptocrystalline aggregates by crystal intergrowth of drusc microsparite (Fig. 4.22, microphotographs 2 and 3). There are also: rare to occasional isotropic and subangular aggregates; opaque loose spherulites; rare to occasional reddish-black (roasted) charcoals with pseudomorphic microsparite (ash) (Fig. 4.10, microphotograph 5); fibrous black aggregates with occasional to numerous silicate minerals and crystal intergrowths of druse microsparite (Fig. 4.22, microphotograph 4); rare fragments of schists; and rare burned bones.

Interpretation

The abundant yellowish-red to reddish components with druse integrowth of microsparite is most like derived from plant leaves (Courty et al. 1989). The possible presence of aggregates derived from work on hides, hypothetically assigned to the aggregates described as fibrous black aggregates, is especially noteworthy. This combination indicates the presence of microvestiges from plant and animal sources. We have interpreted anthropic microfacies S type 3.1 as dump sediment which corresponds to a specific purpose activity area.

General Characteristics S.3.2

These feature dominant burned carbonate rock fragments with a blackish-gray microcrystalline fabric (lime) and pronounced microfissuration. Rare to occasional pure and microcontrasted cryptocrystalline organo-mineral microaggregates are present, as well as many complete microsparite pseudomorphs of ashes (Fig. 4.22, microphotograph 5), rare microcharcoals, blackened loose spherulites, and rare to occasional burned bones. An interesting large burned silicate subangular allochthonous mesoaggregate was also observed (Fig. 4.22, microphotograph 9).

Interpretation

Anthropic microfacies S type 3.2 consists of very homogeneous coarse components (granules), showing heterogeneity in the thermal modification of the carbonate rock fragments. Frequent calcination of carbonate rock fragments is dominant, suggesting temperatures of about 450–600°C, which is consistent with the ashes found. The scarce to rare spherules manifest an opacification, indicating a reduction phase in the combustion mode, possibly consistent with the thermal modification of non-calcinated sedimentary components.

These microfacies are interpreted as having been subjected to high temperatures. They were exposed to a well aerated oxidizing combustion mode, which had previously undergone a reduction mode. Afterwards it was dumped, with rare to occasional leaf ashes in which a large silicate aggregate stands out, indicating the presence of allochthonous sediments resulting from activities carried out near the combustion structures.

The stratified anthropic microfacies of type 3.2 points to dumped deposits moved from high temperature combustion structures to special purpose activity areas, illustrated as fire pit combustion structure evidence in combustion structure group 3 of the central zone (see Fig. 4.17).

General Characteristics S.3.3

These display reddish and blackish mixed burned carbonate rock fragments and scarce to rare anthropic components with rare to occasional microcharcoals.

Interpretation

A mixture of reddish and blackish carbonate rock fragments is related to the occurrence of oxidation and reduction combustion processes. Reddened carbonate rock fragments point to rubefaction in oxidative combustion processes. Blackened carbonate rock fragments suggest hydrocarbon fixation which occurs at low temperatures in a reductive burning mode. The mixture of burned components we interpret as indicative of the displacement of sediments burned at low to medium temperatures, which is in accordance with the carbonaceous (charcoal) composition.

Stratified anthropic microfacies of type 3.3 point to a dump deposit in a discard activity area moved from low temperature combustion structures.

Appendix 2

Location of the sedimentary monoliths, sampling strategies and number of thin sections (Tables 4.8, 4.9, 4.10, 4.11, 4.12, 4.13)

Table 4.8 Sedimentary monoliths location in the Abric Romaní grid system and the number of manufactured thin sections by monolith in order to characterize the microstratigraphic transect in the central zone of sublevel Ja

Year	Level	Square	Number	X	Y	Zsup	Zinf	Thin section by monolith (n)
AR94	Ja	V49	141	35	65	68	78	1
AR94	Ja	R49	10	10.0	15.0	292.0	310.0	1
AR94	J	Q49	12	10	12	305	325	1
AR94	Ja	P49	87	52.0	0.0	300.0	316.0	1
AR94	Ja	O49	146	5.0	10.0	305.0	325.0	2
AR94	Ja	N49	157	7.0	10.0	309.0	329.0	2
AR94	Ja	M49	67	7	33	301	320	2
AR94	Ja	L54	45	10	90	383	400	1
AR94	Ja	K54	80	10.0	85.0	394.0	410.0	1

Table 4.9 Location of sedimentary monoliths in the Abric Romaní grid system and number of thin sections manufactured by monolith in order to characterize the combustion structures

Year	Level	Square	Number	X	Y	Zsup	Zinf	Thin section by monolith (n)
AR94	Ja	U53	87	85.0	70.0	278.0	284.0	1
AR94	Ja	U53	85	70.0	45.0	270.0	282.0	1
AR93	Ja	S42	–	–	–	–	–	1
AR94	Ja	R52	7	60.0	85.0	324.0	330.0	1
AR94	J	Q50	–	60	50	–	314	1
AR93	J	Q58	–	–	–	–	–	1
AR94	Ja	P50	185	100.0	80.0	310.0	320.0	1
AR95	JA/JB	P50	–	10	60	–	330	1
AR93	J	P58	–	–	–	–	–	1
AR93	J	046-1	Este	–	–	–	269	1
AR93	J	046-2	Oeste	–	–	–	271	1
AR93	J	O50	–	34	19.5	–	325	1
AR94	Ja	N49	149	30.0	95.0	300.0	307.0	1
AR94	Ja	N51	497	52.0	53.0	331.5	341.0	1
AR94	Ja	N51	504	61.0	88.0	331.5	342.5	1
AR94	Ja	M57	–	–	–	–	–	1
AR94	Ja	M56	116	–	–	–	–	1
AR94	Ja	L47	–	–	–	–	–	1

Table 4.10 Supplementary sedimentary monoliths location in the Abric Romaní grid system and the number of manufactured thin sections by monolith

Year	Level	Square	Number	X	Y	Zsup	Zinf	Thin section by monolith (n)
AR94	J	O49	119					1
AR94	J	O51	390	70	80	332	342	1
AR94	Ja	N48	119	55	65	293	305	1
AR94	Ja	N49	35	22, 0	89, 0	0, 0	313, 0	1
AR94	Ja	M55						1
AR95	Jb	M57						1
AR94	Ja	M59						1

Table 4.11 Location of sedimentary monoliths in the Abric Romaní grid system and number of manufactured thin sections by monolith in order to complete J level and sublevels stratigraphy

Year	Level	Square	Number	X	Y	Zsup	Zinf	Thin section by monolith (n)
AR94–95	Jb	M45	30	100.0	90.0	264.0	294.0	2
AR94–95	Ja–Jb	P52		95	20	322	352	3

Table 4.12 Location of sedimentary monoliths in the Abric Romaní grid system and number of thin sections manufactured by monolith in order to characterize microstratigraphic transect in the central zone of sublevel Jb

Year	Level	Square	Number	X	Y	Zsup	Zinf	Thin section by monolith (n)
AR95	Ja–Jb	Q51	–	–	–	–	–	1
AR95	Ja–Jb	O51	–	–	–	–	–	2
AR95	Jb	N51	–	–	–	–	–	1
AR95	Jb	M50	–	–	–	–	–	1

Table 4.13 Location of sedimentary monoliths in the Abric Romaní grid system and number of thin sections manufactured by monolith in order to characterize the combustion structures in sublevel Jb

Year	Level	Square	Number	X	Y	Zsup	Zinf	Thin section by monolith (n)
AR95	Ja–b	N44	165	15	28	285	293	1
AR95	Jb	N45		40	90		292	1
AR95	Jb	N47	321	82	63	312	321	1
AR95	Jb	N48	212	92	40	321	332	1
AR95	Jb	N48	213	35	56	322	332	1
AR95	Jb	N49	226	98	52	324	334	1
AR95	Jb	N49	227	36	43	336	343	1

References

Adams, A. E., MacKenzie, W. S., & Guilford, C. (1997). *Atlas de rocas sedimentarias*. Barcelona: Masson.

Adouze, F. (1987). Des modèles et des faits: les modèles de A Leroi-Gourham et de L. Bindford confrontés aux résultats récents. *Bulletin de la Société Préhistorique Française, 84*, 343–352.

Adouze, F., & Enloe, J. (1997). High resolution archaeology at Verberie; limits and interpretations. *World Archaeology, 29*, 195–207.

Allué, E., Boj, I., Caceres, I., Carbonell, E., Castro, Z., Cebrià, A., et al. (1993). Abric Romaní. *Modelo paleo-etnográfico para el pleistoceno superior. Revista de Arqueología, 15*, 6–15.

Bar-Yosef, O., Vandermeersch, B., Arensburg, B., Belfer-Cohen, A., Goldberg, P., Laville, H., et al. (1992). The excavations in Kebara Cave, Mt. Carmel. *Current Anthropology, 33*, 497–550.

Bertran, P. (2004). Dépôts de pente continentaux. *Dynamique et facies. Quaternaire, Hors-série, 1*, 1–251.

Binford, L. R. (1988). *En busca del pasado. Descifrando el registro arqueológico*. Barcelona: Crítica.

Binford, L. R. (1996). Hearth and home: The spatial analysis of ethnographically documented rock shelter occupations as a template for distinguishing between human and hominid use of sheltered space. In N. J. Conard & F. Wendorf (Eds.), *Middle Palaeolithic and Middle Stone Age settlement systems. Proceedings of the XIII UISPP congress* (pp. 229–239). Forli: ABACO.

Bischoff, J. L., Julià, R., & Mora, R. (1988). Uranium-series dating of the Mousterian occupation at Abric Romani, Spain. *Nature, 332*, 68–70.

Boiffin, J., & Bresson, L. M. (1987). Dynamique de formation des croûtes supperficielles: apport de l'analyse microscopique. In N. Fedoroff, L. M. Bresson, & M. A. Courty (Eds.), *Micromorphologie des sols Actes de la VII Réunion Internationale de Micromorphologie des sols* (pp. 393–399). Paris: AFES.

Bordes, F. (1975). Sur la notion de sol d'habitat en préhistoire paléolithique. *Bulletin de la Société Prehistorique Française, 72*, 139–144.

Brochier, J. É. (1999). Couche archéologique, sol archéologique et distributions spatiales: quelques réflexions (géo)archéologiques sur un vieux problème. In *Geoarqueologia i Quaternari litoral. Memorial Maria Pilar Fumanal* (pp. 91–96). Valencia: Universitat de Valencia.

Brooks, A. S., & Yellen, J. E. (1987). The preservation of activity areas in the archaeological record: Ethnoarchaeological and archaeological work in Northwest Ngamiland, Botswana. In S. Kent (Ed.), *Method and theory for activity area research an ethnoarchaeological approach* (pp. 63–106). New York: Columbia University Press.

Bullock, P., Fedoroff, N., Jonguerius, A., Stoops, G., & Tursina, T. (1985). *Handbook for soil thin section description*. Wolverhampton: Waine Research Publications.

Burjachs, F., & Julià, R. (1994). Abrupt climatic changes during the last glaciation based on pollen analysis of the Abric Romani, Catalonia, Spain. *Quaternary Research, 42*, 308–315.

Butzer, K. W. (1989). *Arqueología*. Una ecología del hombre. Barcelona: Bellaterra.

Callot, G., Mousain, D., & Plassard, C. (1985). Concentration de carbonate de calcium sur les parois des hyphes mycéliens. *Agronomie, 5*, 143–150.

Campbell, C. V. (1967). Lamina, laminaset, bed and bedset. *Sedimentology, 8*, 7–26.

Canti, M., & Linford, N. (2000). The effects of fire on archaeological soils and sediments: Temperature and colour relationships. *Proceedings of the Prehistoric Society, 66*, 385–395.

Carbonell, E., & Castro Curiel, Z. (1992). Palaeolithic wooden artefacts from the Abric Romani (Capellades, Barcelona, Spain). *Journal of Archaeological Science, 19*, 707–719.

Carbonell, E., Cebrià, A., Allué, E., Caceres, I., Castro, Z., Diaz, R., et al. (1996a). Behavioural and organizational complexity in the Middle Palaeolithic from the Abric Romani. In E. Carbonell & M. Vaquero (Eds.), *The last Neandertals, the first anatomically modern humans. A tale about the human diversity* (pp. 385–434). Tarragona: Universitat Rovira i Virgili.

Carbonell, E., Cebrià, A., Rosell, J., Sala, R., & Vaquero, M. (1996b). Els conjunts III i IV de l'Abric Romaní de Capellades (campanyes 1988–1994). *Una seqüència d'ocupacions del paleolític mitjà. Tribuna d'Arqueologia, 1994–1995*, 27–38.

Carozzi, A. V. (1960). *Microscopic sedimentary petrology*. New York: Wiley.

Caudwell, C. (1987). Étude expérimentale de la formation de micrite et sparite dans les stromatolithes d'eau douce à Rivuliaria. *Bulletin de la Société Géologique de France, 8*, 229–306.

Cecil, C. B. (2002). The concept o autocyclic and allocyclic controls on sedimentation and stratigraphy, emphasizing the climatic variable. In C. B. Cecil (Ed.), *Climate controls on stratigraphy* (pp. 13–20). Tulsa: SEPM.

Conard, J. (1994). On the prospects for an ethnography of extinct hominids. *Current Anthropology, 35*, 281–282.

Courty, M. A. (1986). Quelques faciès d'altération de fragments carbonatés en grottes et abris sous roche prehistoriques. *Bulletin de l'Association Française pour l'Étude du Quaternaire, 3–4*, 281–289.

Courty, M. A. (1990). *Environnements géologiques dans le nord-ouest ed l'Inde. Contraintes géodynamiques au peuplement protohistorique* (*Cassin de la Ghaggar—Saraswati —Chautang*). Ph.D. Dissertation, Université de Bourdeaux I.

Courty, M. A., Goldberg, P., & Macphail, R. I. (1989). *Soils and micromorphology in archaeology*. Cambridge, MA: Cambridge University Press.

Courty, M. A., Goldberg, P. & Macphail, R. I. (1994a). Ancient people—lifestyles and cultural patterns. In L. Wilding & K. Oleshko (Eds.), *Micromorphological indicators of anthropogenic effets on soils. ISSS, 15th international conference of soil science* (pp. 250–269), Acapulco.

Courty, M. A., Marlin, C., Dever, L., Tremblay, P., & Vachier, P. (1994b). The properties, genesis and environmental significance of calcitic pendents from the high artic (Spitsbergen). *Geoderma, 61*, 71–102.

Chartres, C. J., Bresson, L. M., Valentin, C. & Norton, L. D. (1994). Micromorphological indicators of anthropogenically induced soil structural degradation. In L. Wilding & K. Oleshko (Eds.), *Micromorphological indicators of anthropogenic effets on soils. ISSS, 15th international conference of soil science* (pp. 206–228), Acapulco.

Chenorkian, R. (1994). Étude spatiale en milieux de rejets. *Préhistoire, Anthropologie Méditerranéennes, 3*, 43–56.

Dalrymple, J. B. (1958). The application of soil micromorphology to fossil soils and other deposits from archaeological sites. *Journal of Soil Science, 9*, 199–209.

Dansgaard, W., Johnsen, S. J., Clausen, H. B., Dahl-Jensen, D., Gundestrup, N. S., Hammer, C. U., et al. (1993). Evidence for general instability of past climate from 250-yr core record. *Nature, 364*, 218–220.

Davidson, D. A., Carter, S. P., & Quine, T. A. (1992). An evaluation of micromorphology as an aid to arcaheological interpretation. *Geoarchaeology, 7*, 55–65.

Dibble, H. L., Chase, P. G., McPherron, S. P., & Truffreau, A. (1997). Testing the reality of a "living floor" with archaeological data. *American Antiquity, 62*, 624–651.

Douglas, L. A., & Thompson, M. L. (1985). *Soil micromorphology and classification*. Madison, WI: SSSA.

Elf-Aquitaine (1975). *Essai de caractérisation sédimentologique des dépôts carbonatés. 1. Éléments d'analyse*. Centrres de Recherches de Boussens et de Pau: Elf-Aquitaine.

Fédoroff, N., & Courty, M. A. (1994). Organisation du sol aux échelles microscopiques. In P. Duchaufour & B. Souchier (Eds.), *Pédologie 2 Constituants et propiétés du sol* (pp. 349–375). Paris: Masson.

Fisher, J. W., & Strickland, H. C. (1991). Dwellings and fire places: keys to Efe Pygmy campsite structure. In C. Gamble & W. A. Boismier (Eds.), *Ethnoarchaeologial approaches to mobile campsites* (pp. 215–236). Ann Arbor, MI: International Monographs in Prehistory.

FitzPatrick, E. A. (1993). *Soil microscopy and micromorphology*. Chichester, England: Wiley.

Freytet, P. (1990). Contribution a l'étude des tufs calcaires (édifices stromatolitiques) du bassin de Paris: les organismes constructueurs, aspects microscopiques. *Bulletin du centre de géomorphologie de Caen, 38*, 35–54.

Freytet, P., & Verrecchia, E. (1989). Les carbonates du pourtour méditerranéen: microfaciès et milieux de formation. *Méditerranée, 2–3*, 5–28.

Gamble, C. (1990). *El poblamiento paleolítico de Europa*. Barcelona: Critica.

Gé, T., Courty, M. A., Mattews, W., & Wattez, J. (1993). Sedimentary formation processes of occupation surfaces. In P. Goldberg, D. T. Nash, & M. D. Petraglia (Eds.), *Formation processes in archaeological context* (pp. 149–163). Madison, WI: Prehistory Press.

Gifford, D. P., & Behrensmeyer, A. K. (1977). Observed formation and burial of a recent human occupation site in Kenya. *Quaternary Research, 8*, 245–266.

Goldberg, P., & Mcphail, R. I. (2006). *Practical and theoretical geoarchaeology*. Malden, MA: Blackwell Science.

Goldberg, P., & Whitbread, I. (1993). Micromorphological studies of Bedouin tent floors. In P. Goldberg, D. T. Nash, & M. D. Petraglia (Eds.), *Formation processes in archaeological context* (pp. 165–188). Madison, WI: Prehistory Press.

Gómez Gras, D. (1997). Estudi de les roques clàstiques amb el microscopi. In J. C. Melgarejo (Ed.), *Atles d'associacions minerals en làmina prima* (pp. 167–182). Barcelona: Edicions de la U.B., Fundació Bosch.

Gould, R. A., & Yellen, J. E. (1991). Misreading the past: A reply to Binford concerning hunter-gatherer site structure. *Journal of Anthropological Archaeology, 10*, 283–298.

Greene, R. S. B., Chartres, C. J., & Hodgkinson, K. C. (1990). The effects of fire on the soil in a degraded semi-arid woodland. I. Criptogam cover and physical and micromorphological properties. *Australian Journal of Soil Research, 28*, 755–777.

Holliday, V. T., Ferring, C. R., & Goldberg, P. (1993). The scale of soil investigations in archaeology. In J. K. Stein & A. R. Linse (Eds.), *Effects of scale on archaeological and geoscientific perspectives* (pp. 29–37). Boulder, CO: Geological Society of America.

Hughes, P. J., & Lampert, R. J. (1977). Occupational disturbance and types of archaeological deposit. *Journal of Archaeological Science, 4*, 135–140.

Humbert, L. (1976). *Élements de pétrologie dynamique des systèmes calcaires*. Tome II. Paris: Technip.

Isaac, G. L. (1989). Chronology and the tempo of cultural change during the Pleistocene. In B. Isaac (Ed.), *The archaeology of human origin* (pp. 33–76). New York: Cambridge University Press.

Jongerius, A. (1970). Some morphological aspects of regrouping phenomena in Dutch soils. *Geoderma, 4*, 311–331.

Kahle, C. F. (1977). Origin of subaerial Holocene calcareous crusts: Role of algae, fungi and sparmicritisation. *Sedimentology, 24*, 413–435.

Kent, S. (1991). The relationship between mobility strategies and site structure. In E. M. Kroll & T. D. Price (Eds.), *The interpretation of archaeological spatial patterning* (pp. 33–59). New York: Plenum Press.

Kroll, E. M. (1997). Lithic and faunal distributions at eigth archaeological excavations. In G. L. Isaac & B. Isaac (Eds.), *Koobi Fora Plio-Pleistocene archaeology* (pp. 459–543). New York: Oxford University Press.

Leighton, M. V., & Pendexter, C. (1962). Carbonate rock types. In W. E. Ham (Ed.), *Classification of carbonate rocks* (pp. 33–61). Tulsa, OK: American Association of Petroleum Geologist.

Leroi-Gourhan, A. (1976). Les structures d'habitat au Paléolithique supérieur. In H. de Lumley (Ed.), *La Préhistoire Française: les civilisations paléolithiques et mésolithiques de la France* (pp. 656–663). Paris: Editions du CNRS.

Leroi-Gourham, A. (1982). Leçon Inaugurale faite le vendredi 5 décembre de 1969. Collège de France. In A. Leroi-Gourham (Ed.), *Les racines du monde* (pp. 262–284). Paris: Pierre Belfond.

Leroi-Gourhan, A. (1985). Las excavaciones y la doctrina de investigación. In A. Leroi-Gourhan (Ed.), *La Prehistoria* (pp. 149–154). Barcelona: Labor.

Limbrey, S. (1975). *Soil science and archaeology*. London: Academic Press.

Macphail, R. I., Courty, M. A., Hather, J., & Wattez, J. (1997). The soil micromorphological evidence of domestic occupation and stabling activities. In R. Maggi (Ed.), *Arene Candide: A functional and environmental assessment of the Holocene secuence (Excavations Bernarbo' Brea-Cardini 1940–1950)* (pp. 53–88). Roma: Memorie dell'Istituto Italiano di Paleontologia Umana.

Macphail, R. I., Cruise, G. M., Allen, M. J., Linderholm, J., & Reynolds, P. (2004). Archaeological soil and pollen analysis of experimental floor deposits; with special reference to Butser Ancient Farm, Hampshire, UK. *Journal of Archaeological Science, 31*, 175–191.

Magaldi, D., & Tallini, M. (2000). A micromorphological index of soil development for the Quaternary geology research. *Catena, 41*, 261–276.

Martínez Molina, K. (1999) *Organización y funcionalidad de la producción lítica*. Ms. Thesis, Universitat Rovira i Virgili.

Martínez Molina, K., & Rando Moral, J. M. (2000). Organización espacial y de la producción lítica en el desarrollo de las actividades durante las ocupaciones del Paleolítico Medio. Nivel Ja del Abric Romaní. In J. V. Oliveira (Ed.), *3° Congresso de Arqueologia Peninsular Paleolitico da Península Ibérica* (pp. 215–234). Vila Real: UTAD.

Matthews, W., French, C. A. I., Lawrence, T., Cutler, D. F., & Jones, M. K. (1997). Microstratigraphic traces of site formation processes and human activities. *World Archaeology, 29*, 281–308.

Meignen, L., Bar-Yosef, O., Goldberg, P., & Weiner, S. (2001). Le feu au Paléolithique moyen: recherches sur les structures de combustion et le statut des foyers. *L'exemple du Proche-Orient. Paléorient, 26*, 9–22.

Mellars, P. (1995). *The Neanderthal legacy: An archaeological perspective from Western Europe*. Princeton, NJ: Princeton University Press.

Miall, A. D. (1996). *The geology of fluvial deposits*. Berlin-Heildelberg: Springer.

Miedema, R. (1997). Applications of micromorphology of revelance to agronomy. *Advances in Agronomy, 59*, 119–169.

Mücher, H. J., Chartres, C. J., Tongway, D. J., & Greene, R. S. B. (1988). Micromorphology and significance of the surface crusts of soils in rangelands near Cobar, Australia. *Geoderma, 42*, 227–244.

Nadel, D., Weiss, E., Simchoni, O., Tsatskin, A., Danin, A., & Lislev, M. (2004). Stone Age hut in Israel yields world's oldest evidence of bedding. *Proceedings of the National Academy of Sciences, 101*, 6821–6826.

O'Connell, J. F. (1987). Alyawara site structure and its archaeological implications. *American Antiquity, 52*, 74–108.

Pentecost, A., & Riding, R. (1986). Calcification in Cyanobacteria. In B. S. C. Leadbeater & R. Riding (Eds.), *Biomineralization in lower plants and animals: Systematics association* (pp. 73–90). Oxford, England: Clarendon Press.

Pettit, P. B. (1997). High resolution Neanderthals? Interpreting Middle Palaeolithic intrasite spatial data. *World Archaeology, 29*, 208–224.

Pobeguin, T. (1954). Contribution à l'étude des carbonates de calcium. Précipitation du calcaire par les végetaux. *Comparaison avec le monde animal. Annales Sciences Naturelles Botanique et Biologie végetale, II*, 29–109.

Rabenhorst, M. C., West, L. T., & Wilding, L. P. (1991). Genesis of calcitic and petrocalcic horizons in soils over carbonate rocks. In W. D. Nettleton (Ed.), *Occurrence, characteristics, and genesis of carbonate, gypsum, and silica accumulations in soils* (pp. 61–74). Madison: SSSA Special Publication.

Rando Moral, J. M. (1999) *Análisis espacial del registro arqueológico del subnivel Ja. Abric Romaní (Capellades, Barcelona). Patrones conductuales de asentamiento en un nivel musteriense de 50.000 BP*. Ms. Thesis, Universidad Rovira y Virgili.

Rigaud, J.-Ph. (1994). L'évaluation contextuelle préalable à l'analyse de la répartition spatiale des vestiges. *Préhistoire, Anthropologie Méditerranéennes, 3*, 39–41.

Schick, K. D. (1997). Experimental studies of site-formation processes. In G. L. Isaac & B. Isaac (Eds.), *Koobi Fora Plio-Pleistocene archaeology* (pp. 459–543). New York: Oxford University Press.

Schiffer, M. B. (1972). Archaeological context and systemic context. *American Antiquity, 37*, 156–165.

Schiffer, M. B. (1987). *Formation processes of the archaeological record*. Alburquerque: University of the New Mexico Press.

Schiffer, M. B. (1995). *Behavioral archaeology*. First principles. Salt Lake City UT: University of Utah Press.

Selby, M. J. (1993). *Hillslope materials and processes*. Oxford, England: Oxford University Press.

Steijn, H. V. (2002). Long-term landform evolution: Evidence from talus studies. *Earth Surface Processes and Landforms, 27*, 1189–1199.

Stein, J. K. (1992). *Deciphering a shell midden*. New York: Academic Press.

Stein, J. K., & Rapp, G. (1985). Archaeological sediments: A largely untapped reservoir of information. In N. C. Wilkie & W. D. E. Coulson (Eds.), *Contributions to Aegean archaeology* (pp. 143–159). Minneapolis, MN: Center for Ancient Studies, University of Minnesota.

Stevenson, M. C. (1991). Beyond the formation hearth-associated artifact assemblages. In E. M. Kroll & T. D. Price (Eds.), *The interpretation of archaeological spatial patterning* (pp. 269–299). New York: Plenum Press.

Stoops, G. (2003). *Guidelines for analysis and description of soil and regolith thin sections*. Madison, WI: Soil Science Society of America.

Thomas, D. H. (1983). *The archaeology of monitor valley: Gatcliff shelter*. New York: American Museum of Natural History.

Tucker, M. (1988). *Techniques in sedimentology*. London: Blackwell.

Tucker, M. E. (2001). *Sedimenary petrology: An introduction to the origin of sedimentary rocks*. Oxford, England: Blackwell.

Vallverdú, J. (2002). *Micromorfología de las facies sedimentarias de la colección de referencia de la Sierra de Atapuerca y del nivel J del Abric Romaní. Implicaciones geoarqueológicas y paleoetnográficas*. Ph.D. Dissertation, Universitat Rovira i Virgili.

Van Vliet-Lanoë, B. (1985). Frost effects in soils. In J. Boardman (Ed.), *Soils and quaternary landscape evolution* (pp. 117–158). Chichester, England: Wiley.

Van Vliet-Lanoë, B. (1987). Dynamique périglaciaire actuelle et passée. Apport de l'étude micromorphologique et de l'expérimentation. *Bulletin de l'Association française pour l'étude du Quaternaire, 3*, 113–132.

Van Vliet-Lanoë, B. (1997). Application of micromorphology in cenozoïc geology and geomorphologie—temperate to cold environments. In *Fourth European intensive course on soil micromorphology II application to soil micromorphology* (pp. 165–175). Granada.

Van Vliet-Lanoë, B., Poch, R. M., Porta, J., & Gilarou, A. M. (1995). Sòls sempre glaçats. In R. Folch (Ed.), *Biosfera, 9 Tundra i insularitat* (pp. 25–33). Barcelona: Fundació Enciclopèdia Catalana.

Vaquero, M. (1997). *Tecnología lítica y comportamiento humano: organización de las actividades técnicas y cambio diacrónico en el Paleolítico medio del Abric Romaní (Capellades, Barcelona)*. Ph.D. Dissertation, Universitat Rovira i Virgili.

Vaquero, M., & Pasto, I. (2001). The definition of spatial units in Middle Palaeolithic sites: The hearth-related assemblages. *Journal of Archaeological Science, 28*, 1209–1220.

Verrecchia, E. P. (1989). Lexique des termes utilisés pour décrire les carbonates continentaux à l'échelle de la lame mince. *Méditerranée, 2–3*, 75–83.

Verrecchia, E. P., Freytet, P., Verrecchia, K. E., & Dumont, J. L. (1995). Spherulites in calcrete laminar crust: Biogenic $CaCO_3$ precipitation as a major contributor to crust formation. *Journal of Sedimentary Research, 165*, 690–700.

Verrecchia, E. P., & Verrecchia, K. A. (1994). Needle-fiber calcite: Acritical review and proposed classification. *Journal of Sedimentary Research, 64*, 650–664.

Villa, P. (1977). Sols et niveaux archéologiques d'habitat du paléolithique inférieur en Europe et au Proche Orient. *Quaternaria, 19*, 107–134.

Villa, P. (1982). Conjoinable pieces and site formation processes. *American Antiquity, 42*, 276–290.

Villa, P. (1988). A propos d'analyse spatiale et de taphonomie. *Les nouvelles de l'archeologie, 31*, 55–59.

Villa, P., & Courtin, J. (1983). The interpretation of stratified sites: A view from underground. *Journal of Archaeological Science, 10*, 267–281.

Wattez, J. (1990). *Dynamique de Formation des Structures de Combustion de la Fin du Paléolithique au Néolitique Moyen. Approche Méthodologique et Implications Culturelles*. Ph.D. Dissertation, Université de Paris I.

Weedon, G. (2003). *Time-series analysis and ciclostratigraphy*. Cambridge, MA: Cambridge University Press.

Chapter 5
Neanderthal Landscapes and Their Home Environment: Flora and Fauna Records from Level J

Ethel Allué, Francesc Burjachs, Ana García, Juan Manuel López-García, Maria Bennàsar, Florent Rivals, Hugues-Alexandre Blain, Isabel Expósito and Jordi Martinell

Abstract The aim of this chapter is to understand, from a multidisciplinary approach, Neanderthal landscapes and environment from layer J records from Abric Romaní rockshelter. The proxy data used are paleontological and paleobotanical records with natural or anthropic orgin. This study includes pollen, charcoal, small vertebrates (micromammals and anphibians), large mammals and malacofauna. The data yielded through these disciplines approached on the basis of different methodologies (anthracology, palynology, paleontology, and dental microwear and mesowear) provide an assemblage of data showing different aspects of the Neanderthal landscapes and environment. These proxies show a diverse landscape with forested and open landscape areas under a cold environment, which locally has yielded taxa reflecting more humidity rates and the presence of Mediterranean taxa reflecting a cooler climate. In this context, Neanderthals exploited a variety of biotopes for the obtaining of resources, pointing out their adaptability capabilities.

Keywords Paleoenvironment • Pollen analysis • Paleontology • Small vertebrates • Malacofauna

E. Allué (✉) · A. García · J. M. López-García · M. Bennàsar · H.-A. Blain · I. Expósito
Institut Català de Paleoecologia Humana i Evolució Social (IPHES), Universitat Rovira i Virgili (URV), C/Escorxador s/n, 43003 Tarragona, Spain
e-mail: eallue@iphes.cat

A. García
e-mail: anagarbo@gmail.com

J. M. López-García
e-mail: jmlopez@iphes.cat

M. Bennàsar
e-mail: mlluc@prehistoria.urv.cat

H.-A. Blain
e-mail: hablain@iphes.cat

I. Expósito
e-mail: iexposito@iphes.cat

F. Burjachs
Institució Catalana de Recerca i Estudis Avançats (ICREA), Institut Català de Paleoecologia Humana i Evolució Social (IPHES), C/Escorxador s/n, 43003, Tarragona, Spain
e-mail: francesc.burjachs@icrea.cat

F. Rivals
Institució Catalana de Recerca i Estudis Avançats (ICREA), Institut Català de Paleoecologia Humana i Evolució Social (IPHES), Universitat Rovira i Virgili (URV), C/Escorxador s/n, 43003 Tarragona, Spain
e-mail: florent.rivals@icrea.es

J. Martinell
Departament d'Estratigrafia, Paleontologia i Geociències Marines, Facultat de Geologia, Universitat de Barcelona, Campus de Pedralbes, Barcelona, Spain
e-mail: jmartinell@ub.edu

Introduction

The Abric Romaní sequence spans a chronological period of ca. 40–70 ka BP with level J dated at ca. 50 ka BP (Bischoff et al. 1988). Different paleoecological studies on the flora and fauna have already provided a vast range of data, with very little variability among their results (Carbonell et al. 1996; Burjachs and Julià 1994; Cáceres et al. 1998; Arteaga et al. 2001; Vaquero et al. 2001; Chacón et al. 2008).

The geological landscape of the Abric Romaní provides different biota for the development of flora and fauna assemblages. The shelter is located at an altitude of 350 m above sea level, faces north-east, and is approximately 200 m from the Anoia River. The geomorphology of the region features different areas, including plains, travertine cliffs and mountain chains (see Vallverdú et al. this volume). According to the geomorphological study, a water source caused dampness in the area above the shelter.

The main objective of this chapter is to provide an overview of the environment surrounding the Abric Romaní in order to contribute further information about the supply

E. Carbonell i Roura (ed.), *High Resolution Archaeology and Neanderthal Behavior: Time and Space in Level J of Abric Romaní (Capellades, Spain)*, Vertebrate Paleobiology and Paleoanthropology,
DOI: 10.1007/978-94-007-3922-2_5, © Springer Science+Business Media B.V. 2012

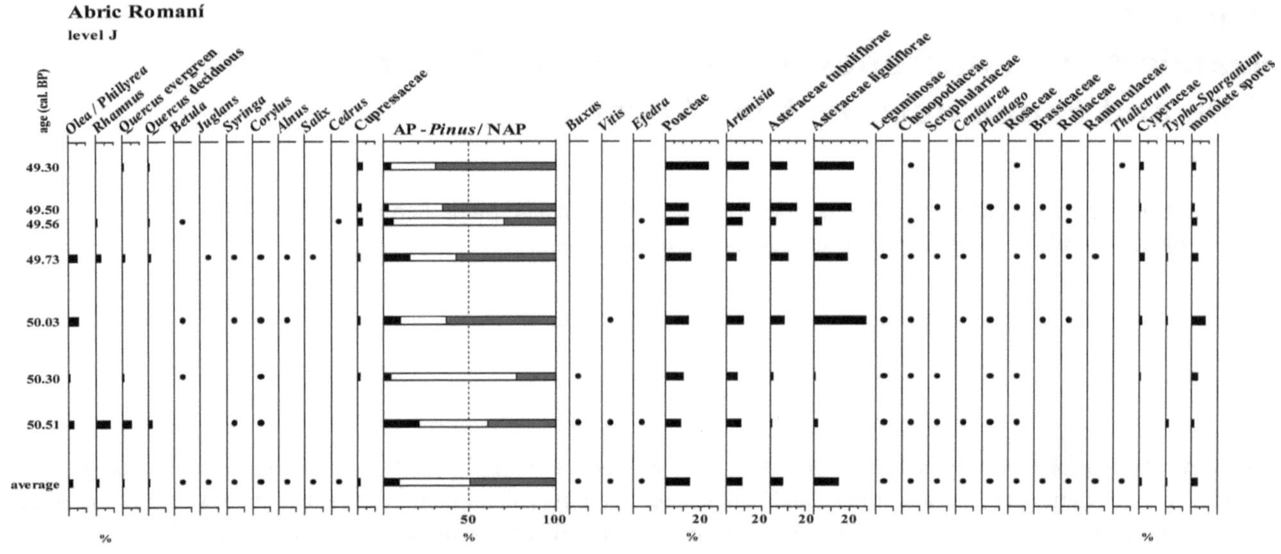

Fig. 5.1 Pollen diagram from level J. (*AP* arboreal pollen; *NAP* non-arboreal pollen)

and distribution of the biotic resources on which the Neanderthals depended. This approach will also define the natural landscape from a regional and local perspective. This information, based on the distribution of fauna and flora, will allow a greater understanding of the environment and climate during MIS 3. It will also define the Neanderthals' framework of adaptation to this environment. Finally, we hope to describe the shelter environment as it relates to habitability during the occupations.

This study incorporates different disciplines for the study of macromammals, microvertebrates and malacofauna and the botanical assemblage including pollen and charcoal.

Paleobotany

The paleobotany record from sublevels Ja and Jb includes pollen and charcoal remains. Pollen analyses are intended to provide information about vegetation, identifying different flora categories, which include herbs, trees and shrubs. The charcoal record, however, is subject to human activities and the data is based considerably on local woody plants (trees and shrubs). Both disciplines complement records on the vegetation that jointly contribute to a clearer interpretation of past landscapes and climates.

Methods

The pollen analysis was conducted using seven samples from level J from the stratigraphical section at square P45. The sediments were treated using the method proposed by Goeury and Beaulieu (1979), slightly modified from Girard and Renault-Miskovsky (1969), and according to the protocol developed by Burjachs (1990) and Burjachs et al. (2003). To the pollen diagram, we have added the average histogram of the seven samples.

The charcoal analysis was conducted using 1205 charcoal fragments that were scattered across the surface of the archeological sublevels or hearths. All the charcoal remains have an anthropic origin related to firewood management (see Allué et al. 2012). Charcoal analyses are based on the taxonomic identification of charcoal fragments through examining the three wood anatomy sections under a metallographic light microscope. A reference collection and a European wood anatomy atlas (Schweingruber 1990) were used to support the identifications.

Results

The results of the pollen analyses show a semi-open landscape with 50% arboreal pollen (AP) (Fig. 5.1). The trees dominating this landscape are pines (*Pinus*) and, to a lesser degree, junipers (cf. *Juniperus*), accompanied in cold protected and sunny areas by mesophilous taxa (deciduous *Quercus*, *Corylus*, *Juglans*, *Rhamnus*) and even Mediterranean mesothermophilous taxa (evergreen *Quercus*, *Olea/Phillyrea*, *Syringa*). We also found evidence of riverside vegetation which would include at least *Alnus*, *Salix*, *Populus*, Cyperaceae and *Typha* or *Sparganium*.

The shrubs we have identified include *Buxus*, *Ephedra fragilis*-type and a liana *Vitis*. The component of the open landscape would mainly have been composed of grasses (Poaceae), Compositae (Asteraceae) and *Artemisia*.

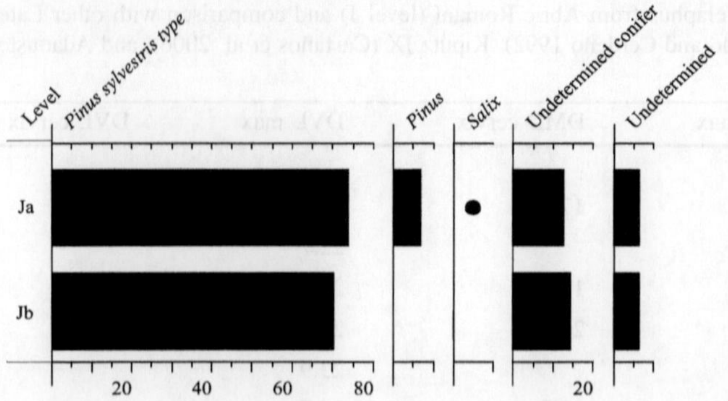

Fig. 5.2 Charcoal diagram from sublevels Ja and Jb

Charcoal analyses reveal only two taxa: *Pinus sylvestris* type (Scots Pine) and *Salix* (willow). *Pinus sylvestris* type is the dominant taxa, whereas *Salix* is represented by a single fragment (Fig. 5.2). There are relatively low percentages of undetermined conifers or undetermined fragments. These categories are due to the size of the fragments or preservation problems which hindered further identification.

Discussion

From these results, we deduced that this region housed semi-open pine groves that would have permitted the proliferation of herbs in their understorey. We suggest that it would have been a rather patchy landscape with forested areas and other open areas dominated by prairies and Mediterranean steppes.

Charcoal analyses suggest that selection occurred concerning the management of resources which were the most abundant in the environment. From this record we can be certain that there was a well developed forested area in the near vicinity of the rockshelter. According to these data, the pine forests were probably mono-specific forests that may have had junipers as the understoried vegetation, which is represented in the pollen record but was not found in our charcoal analyses. Furthermore, the identification of *Pinus sylvestris* (Scots Pine) in the charcoal record points to the fact that during the period under study mountain range pine forests grew 300 m below their present day location.

In fact, the scarce arboreal vegetation cover would have been directly associated with the semi-cold interstadial (IS 14, Dansgaard et al. 1993) as well as with the scarcity of annual rainfall, poorly interspersed throughout the year. Furthermore, detailed analyses of the diagram show that the site is not located in a very cold region, which is suggested by the scarcity of birch (*Betula*) and, more importantly, by the presence of Mediterranean taxa (evergreen *Quercus*, *Olea/Phillyrea* and *Syringa*). In the north-eastern Iberian Peninsula, diagrams of Late Pleistocene vegetation found at similar altitudes to that of the Abric Romaní show an increase in *Betula* in later periods closer to the Last Glacial Maximum (Ros 1987; Burjachs and Renault-Miskovsky 1992), whereas in level J of Abric Romaní there are significant thermo-mesophillous taxa, which point to the existence of refugia.

The MIS 3 climatic conditions that existed during the Dansgaard–Oeschger oscillation corresponding to level J would have been very different from the conditions we find today. The presence of cedar (*Cedrus*) in the vegetation record suggests that dominant meteorological patterns with easterly blowing winds allowed the pollen to arrive from the North African Atlas mountains, since this tree was extinct at the end of the Pliocene and early Pleistocene in Europe (Beaulieu and Reille 1973; Barrón 1999; Carrión et al. 2000; Ravazzi et al. 2005).

Large Mammal Paleontology

Large mammal remains recovered in level J are fragmentary, especially the postcranial bones (Rosell et al. 2012), and so paleontological identification has mainly been performed using teeth (or a few mandibles or maxillae). Ten species were identified: five Carnivora, two Perissodactyla, and three Artiodactyla.

Systematic Study

Carnivora Bowdich, 1821

Carnivores are not very abundant. Among the carnivore remains, however, there was a relatively high degree of diversity, with a sample made up of five different species:

Bear, *Ursus* sp. was identified in sublevel Ja with a first phalanx.

Wolf, *Canis lupus* Linnaeus, 1758 was identified with a cervical vertebra and a deciduous tooth, both in Ja.

Fox, *Vulpes vulpes* Linnaeus, 1758 with one metapodial in sublevel Ja.

Lynx, *Lynx* sp. with a second phalanx in Ja, and an undetermined phalanx in Jb.

Hyena, *Hyaena* sp. was identified with a deciduous tooth in sublevel Jb.

Perissodactyla Owen, 1848

Equidae Gray, 1821

Horse, *Equus ferus* Boddaert, 1785

The horse remains from Abric Romaní were described by Sánchez (1989). On the upper teeth, the fossettes present a maximum of two foldings, in the mesial and distal parts. The

Table 5.1 Measurements (in mm) of the permanent molars of Cervus elaphus from Abric Romaní (level J) and comparison with other Late Pleistocene localities from southern Europe: Cueva Millán, (Pérez-Legido and Cerdeño 1992), Kiputz IX (Castaños et al. 2006), and Adaouste (Defleur et al. 1994)

	Locality	N	DMD max	DMD cervix	DVL max	DVL cervix
Upper teeth						
M1	Abric Romaní	5	21.0	17.2	21.9	20.5
	Cueva Millán	2	20.2		22.7	
	Adaouste	1	19.7	18.5	23.3	
M2	Abric Romaní	7	24.6	21.2	23.9	22.6
	Cueva Millán	1	27.0		23.9	
	Adaouste	2	23.4	21.7	24.7	
M3	Abric Romaní	3	26.1	22.2	25.5	23.7
	Cueva Millán	3	25.7		25.4	
	Adaouste	2	24.8	22.8	22.8	
Lower teeth						
p2	Abric Romaní	3	11.6	9.3	7.1	5.1
	Cueva Millán	2	13.0		9.3	
p3	Abric Romaní	3	16.2	14.8	9.9	8.1
	Cueva Millán	2	17.8		10.7	
p4	Abric Romaní	3	18.7	16.7	11.5	10.3
	Cueva Millán	3	19.0		11.6	
m1	Abric Romaní	4	21.9	18.7	12.9	13.1
	Cueva Millán	2	23.6		15.5	
m2	Abric Romaní	3	26.3	23.1	14.9	15.0
	Cueva Millán	2	25.5		16.2	
m3	Abric Romaní	3	37.0	36.2	15.3	15.2
	Cueva Millán	2	34.2		14.8	
	Kiputz IX	6	34.2		14.3	

Abbreviations: *DMD* Mesiodistal diameter, *DVL* Vestibulolingual diameter

caballine fold is well developed on the premolars and slightly outlined on the molars. The styles are well developed—on the premolars, they present grooves (striations) and on the molars, there is a tendency toward the splitting of the parastyle in two. On the lower teeth, the flexids (pre- and postflexids) are not highly folded. On the premolars, the hypoflexids (vestibular groove) do not penetrate the flexids, and they have a well developed caballine fold. On the molars, the ectoflexid moves toward the double knot but does not reach it; the caballine fold is well outlined only on some specimens. Comparing the metrical data with that from other Spanish sites indicates that the population at Abric Romaní is comparable to the horses of Cueva Horá and Cueva Negra (Sánchez 1989). The Abric Romaní horse would have been an intermediate form between the two Würmian subspecies *germanicus* and *gallicus* (Sánchez, 1989).

Rhinocerotidae Gray, 1821

Rhinoceros, *Stephanorhinus hemitoechus* (Falconer, 1868)

Rhinoceros remains are represented in level J by a deciduous premolar, a permanent third or fourth premolar, a phalanx of an immature individual, as well as a few fragmentary remains. This species was previously identified as *Stephanorhinus hemitoechus* in other levels at Abric Romaní where more significant remains were discovered.

Cetartiodactyla Montgeland et al., 1997

Cervidae Goldfuss, 1920

Red deer, *Cervus elaphus* Linnaeus, 1758

For the Cervidae, morphological features were examined on the upper and lower permanent dentition (Lister 1996). All the remains recovered from level J belong to a single species, *Cervus elaphus*. As described below, no evidence of fallow deer, *Dama dama*, has been found. All upper and lower teeth are represented in the assemblage from level J; however Table 5.1 only shows measurements of large samples.

The upper molars have columns, buccal cones and well pronounced styles. The central lingual column and its associated cingulum are also strongly marked. Finally, the

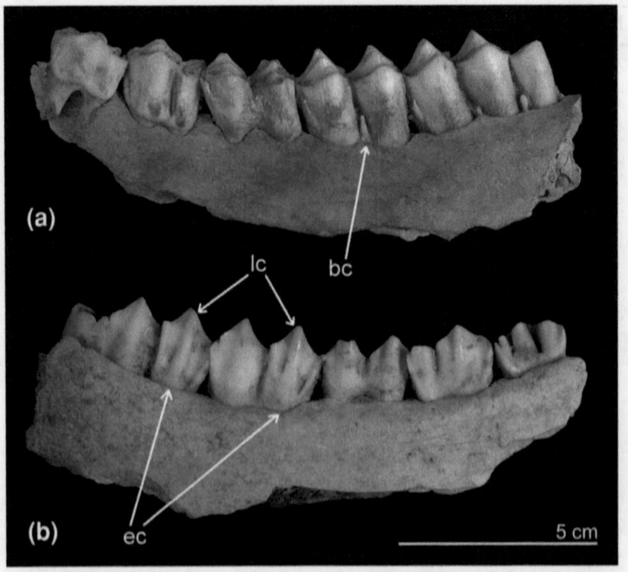

Fig. 5.3 *Left* mandible of *red* deer (*Cervus elaphus*) from Abric Romaní (AR'94 Ja P51). Buccal view (**a**) and lingual view (**b**). Abbreviations: *bc* buccal column; *lc* lingual column; *ec* entoconid column

antero-buccal edge of the teeth is straight toward the base, and not angled as in *Dama*.

On the lower premolars (Fig. 5.3), the entoconid and the entoconid wing are oriented following the transverse axis of the teeth, and the lingual division between the entoconid and the posterior wing of the hypoconid is not as deep as found in *Dama*.

On the lower molars (Fig. 5.3), the lingual columns and furrows in between are stronger than in *Dama*. In addition, unlike in *Dama*, the entoconid column is straight and does not curve anteriorly. More specifically, the central buccal column is stronger in M2 than in M1. Finally, on M3, there is no clear step between the second and third lobes (entoconid and hypoconid).

The measurements were compared to other Late Pleistocene locations in Spain and southern France (Table 5.1): Cueva Millán, Burgos (Pérez-Legido and Cerdeño 1992); Kiputz IX, Gipuzkoa (Castaños et al. 2006); and Adaouste, Bouches-du-Rhône (Defleur et al. 1994). The population from Abric Romaní fits within the limits of variations of Late Pleistocene populations, with some larger and some smaller dimensions. These differences are certainly related to sample size.

Bovidae Gray, 1821

Aurochs, *Bos primigenius* Bojanus, 1827

Identifying the dental remains of aurochs (*Bos primigenius*) and bison (*Bison priscus*) is not easy, but several criteria have been proposed to help in making the distinction. Molars are particularly useful for identifying to the genus level (Slott-Moller 1990). In all the specimens analyzed, we only found morphological criteria of *Bos*. Measurements are presented in Table 5.2.

On both the lower and upper molars, the central buccal column is straight, starts on the cervix (cementoenamel junction) and extends up to the occlusal surface. Other criteria for identification are the shape of the pillars on the protocone and hypocone (upper teeth) and the protoconid and hypoconid (lower teeth). In *Bos*, these pillars are straight, shaped like a column without thickening at the cervix (Moullé 1992).

The upper third and fourth premolars, in mesial view, have sub-parallel lingual and labial edges that do not converge towards the occlusal surface as in *Bison*. On the upper third molar, the lingual connection between the two lobes forms an open angle. The hypocone is V shaped.

The lower third molar is one of the characteristic remains used for distinguishing between the two genera, and particularly the shape of the third lobe (Slott-Moller, 1990). In *Bos*, the third lobe forms a continuity with the lingual side of the tooth (paraconule and metaconule), and forms a largely open angle with the labial side (protoconule and hypoconule) (Fig. 5.4).

Pyrenean chamois, *Rupicapra pyrenaica* Bonaparte, 1845

Remains identified as Caprinae in level J are very rare. Only four specimens were attributed to the genus *Rupicapra*: an upper third molar, fragments of an upper second or third molar, a lower third molar (bud), and a fragment of a lower molar. From these specimens, we chose to focus our paleontological analysis on the upper M3 because it is the only complete, well preserved, and measurable specimen.

The third upper molar (AR93-Ja-U43-3) has a reduced metastylar wing (Fig. 5.5) and the lingual side of the second lobe has a rounded section without the typical pinching of *Hemitragus*. The metastyle forms a small metastylar wing and the distal interstylar surface is smaller than the proximal surface. These features are typical of the species *Rupicapra pyrenaica*.

The metrical analysis (Table 5.3 and Fig. 5.6) shows that the population from Abric Romaní bears a very strong resemblance to the population from Cova de l'Arbreda (Serinyà, Spain), south of the Pyrenees, compared to the reference population from Portel-Ouest (Loubens, France) in the northern Pyrenees. This similarity indicates homogeneity of size and morphology in the populations of chamois in the southern Pyrenees.

Rupicapra pyrenaica is often depicted as a mountain ungulate, but its Pleistocene distribution indicates that it is not only adapted to high altitude habitats. The species preferentially inhabits pastures and rocky areas in diverse mountain regions of Europe. It is found in many Pleistocene locations at low altitudes, such as at Arago Cave (Crégut-

Table 5.2 Measurements (in mm) for the upper and lower teeth of Bos primigenius from Abric Romaní

Specimen nb.	Tooth	DMD max	DMD cervix	DVL max	DVL cervix	Level
Upper teeth						
AR'94.L56.92	P4	22.0	18.5	20.5	19.8	Ja
AR'95.I60.1	M1	33.5	28.8		29.7	Ja
AR'93.I57.17	M1	34.3	23.5	24.7		Ja
AR'93.I57.7	M2	36.4	29.0	25.1		Ja
AR'96.J61.41	M2	39.0				Ja
AR'94.K53.28	M3			26.3		Ja
AR'94.K54.79	M3	37.2		25.8		Ja
Lower teeth						
AR'94.L49.22	p2	13.9		10.66		Ja
AR'94.L49.22	p3	22.7		13.3		Ja
AR'93.N53.42	p4	25.8	20.8	16.2	15.6	Ja
AR'93.N54.78	p4	22.2	18.8	14.5	13.3	Ja
AR'94.L49.22	p4	25.7		13.4		Ja
AR'94.L49.22	m1	29.4		17.8		Ja
AR'94.L49.32	m1	31.8	24.9	18.7	17.7	Ja
AR'95.N52.18	m2	34.58		19.45		Jb
AR'94.L48.27	m3	49.40		19.15		Ja

Abbreviations: *DMD* Mesiodistal diameter, *DVL* Vestibulolingual diameter

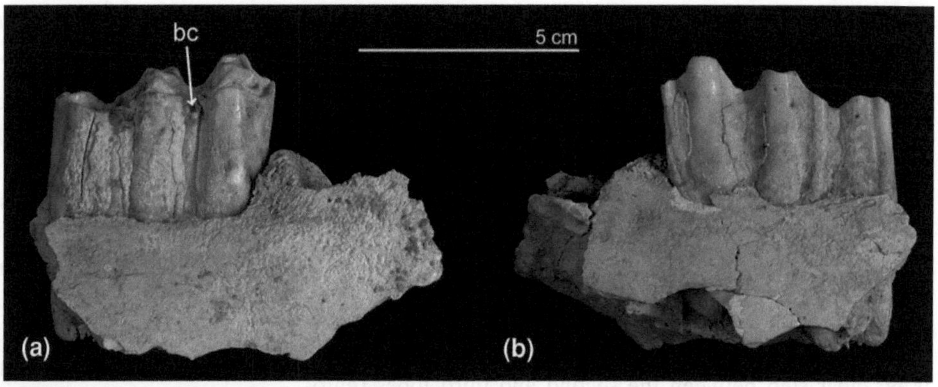

Fig. 5.4 *Right* m3 of aurochs (*Bos primigenius*) from Abric Romaní (AR94 Ja L48 27). Buccal view (**a**) and lingual view (**b**). Abbreviations: *bc* buccal column

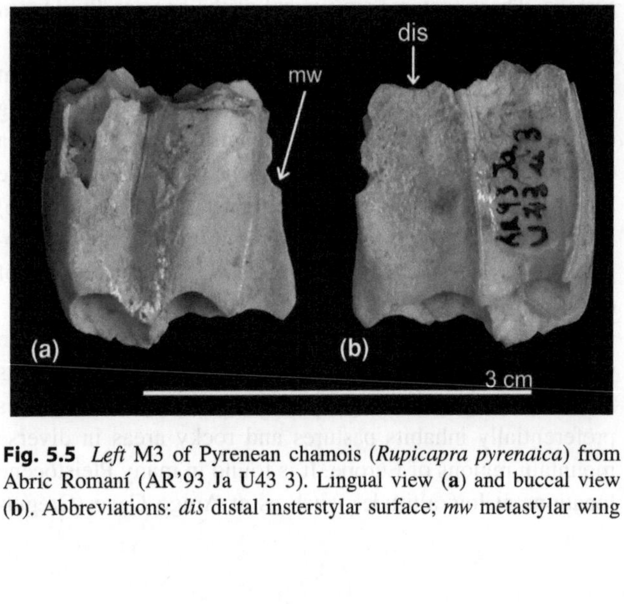

Fig. 5.5 *Left* M3 of Pyrenean chamois (*Rupicapra pyrenaica*) from Abric Romaní (AR'93 Ja U43 3). Lingual view (**a**) and buccal view (**b**). Abbreviations: *dis* distal insterstylar surface; *mw* metastylar wing

Bonnoure 1979) or Les Cèdres (Defleur et al. 1989). The presence of Pyrenean chamois around Abric Romaní is not exceptional for the Mediterranean area. The population may have taken advantage of the travertine cliff.

Dental Microwear and Mesowear Analyses: Reconstruction of Paleodiets and Paleoenvironments

Large mammals, and especially ungulates, are commonly used to create paleoenvironmental reconstructions through various types of proxies (hypsodonty, body mass, biodiversity) and methods of analysis (ecological groups,

Table 5.3 M3 Measurements (in mm) for the M3 of Rupicapra pyrenaica from the Abric Romaní specimen and from other comparative localities

Measurements		N	1	3	4	5	6	7	8	10	11	References
	Abric Romani	1	16.9	9.1	8.7	10.1	8.3	7.5	8.9	8.4	16.2	
R. pyrenaica	Cova de l'Arbreda	2	16.6	8.9	7.9	10.1	8.5	7.7	9.2	9.2	12.9	Rivals 2004
R. pyrenaica	Portel-Ouest	6	13.8	8.3	7.6	11.0	7.5	6.6	9.5	5.1	12.2	Rivals 2004
R. pyrenaica	Bois du Cantet	1	13.0		7.7							Clot 1984
R. pyrenaica	C. del Nacimiento	1	12.0		13.9							Alferez et al. 1980
R. pyrenaica	Cueva Millan	2	14.1		10.2							Pérez-Legido and Cerdeño 1992
R. rupicapra	Combe-Grenal	1	13.2		11.4							Guadelli 1987
R. rupicapra	Canalettes	3	14.6		9.5							Brugal 1993

Measurements from Crégut-Bonnoure (1992): (1) occlusal DMD (mesio-distal diameter), (3) occlusal DVL (vestibulo-lingual diameter) of the mesial lobe, (4) occlusal DVL of the mesial lobe without the styles, (5) basal DVL of the mesial lobe, (6) occlusal DVL of the distal lobe, (7) occlusal DVL of the distal lobe without the styles, (8) basal DVL of the distal lobe, (10) Width of the distal interstylar surface, (11) DMD of the lobes at the occlusal surface

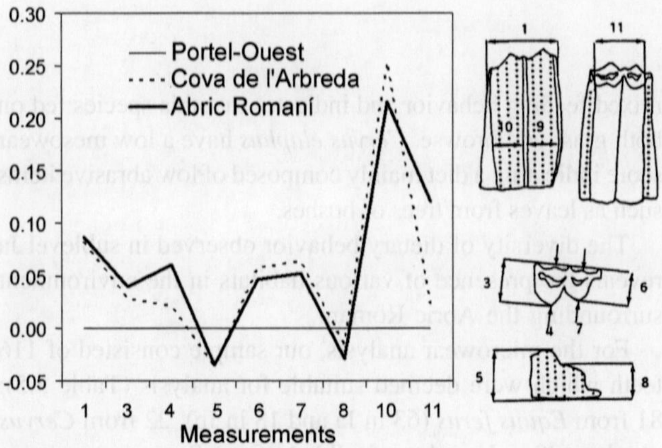

Fig. 5.6 Ratio diagram of M3 average measurements for *Rupicapra pyrenaica* (Standard base: Portel-Ouest cave). See Table 5.3 for description of measurements

climatograms, cenograms, multivariate analysis). However, these methods have some limitations, especially when applied to the Pleistocene. First, they are based on parameters that strongly reflect the phylogenetic history of a group, i.e. they are linked to the adaptations of ancestors. Thus, they provide clues, but do not necessarily indicate the immediate behavior or ecology of a population (Solounias et al. 2004). Also, there are many limitations when these methods are applied to archeological sites, because human behavior implies the intentional selection of the animal hunted and brought back to the site (Valensi and Abbassi 1998). We decided to use dental wear analysis methods (meso- and microwear) because they provide insight into the diet of the individuals, i.e. into the immediate behavior of the population. The results of the mesowear analysis reflect the diet of the last weeks or months (Fortelius and Solounias 2000), whereas the microwear analysis yields information on the last few days or week (Solounias and Semprebon 2002). These methods therefore provide a snapshot of the local environment around the site at the time Neanderthals occupied the shelter.

Methods

Mesowear analysis was performed in keeping with that described by Fortelius and Solounias (2000). Mesowear is based on physical properties of ungulate foods as reflected in the relative amounts of attritive (tooth-on-tooth) wear and abrasive (food-on-tooth) wear on the dental enamel of the occlusal surfaces. Low abrasion diets, such as those of browsers, generate an attritional (sharpening) relationship between the upper and lower cusps. In contrast, abrasion-dominated (food-on-tooth) wear, associated with grazing diets, results in more rounded and blunted wear facets and less precise occlusion. Only adults are found to have consistent signs of mesowear (Rivals et al. 2007a) and were therefore selected for study. Mesowear was scored on the paracone of the M2. In the original formulation of the method, mesowear was recorded by characterizing the buccal apices of molar cusps as sharp, rounded, or blunt, and the valleys between the cusps as high or low. The original data reported by Fortelius and Solounias (2000) on extant ungulates were converted into a more simplified univariate score representing a continuum of mesowear stages from the highest and sharpest cusps (0) to cusps that are completely blunted with little or no relief (3) (Rivals et al. 2007b). Intermediate stages of mesowear consisting of more rounded cusp apices with higher and lower levels of cusp relief were assigned scores of 1 and 2, respectively. Teeth with low relief and sharps cusps were assigned a score of 2.5. Individual scores were then averaged for each sample.

Table 5.4 Summary statistics for the mesowear analysis of the ungulates from Abric Romaní

Species	Level	n	MWS
Bos primigenius	Ja	6	1.17
Cervus elaphus	Ja	9	0.50
Equus ferus	Ja	27	2.50
Equus ferus	Jb	8	2.13

Abbreviations: *n* number of specimens, *MWS* mesowear score

Microwear analysis was performed using the method described by Solounias and Semprebon (2002) and Semprebon et al. (2004). Further discussion of this method and its application in dietary reconstruction can also be found in Godfrey et al. (2004) and Palombo et al. (2005). Molar teeth (lower and upper M2) were selected by excluding young and old individuals. High-resolution epoxy casts were made for all available upper and lower second molars in sublevels Ja and Jb. The casts were screened using a stereomicroscope and any specimens exhibiting signs of weathering were excluded. Microwear features were identified and quantified on high resolution epoxy tooth casts at 35× magnification using a stereomicroscope. Most microwear features can be categorized as pits and scratches of various sizes and textures. Pits are circular or sub-circular microwear scars. Small pits are relatively shallow, refract light easily, and appear bright and shiny. Large pits are deeper, wider, and less refractive. Large pits are recorded qualitatively as being present or absent on the wear surface of the tooth. Scratches are elongated microfeatures with straight, parallel sides and can be subcategorized as fine or coarse. Scratch texture is evaluated on the basis of general appearance and light refractive properties (0 = fine scratches only; 1 = mixture of fine and course, 2 = only coarse scratches). Cross scratches are oriented somewhat perpendicularly to the majority of scratches observed on dental enamel (Solounias and Semprebon 2002). To approximate pit and scratch frequency, they are counted in a standard 0.4 × 0.4 mm square area on the lingual (inner) band of enamel on the paracone of the upper second molar.

Results

Mesowear scores recorded for sublevel Ja, where all three main species are present, indicate a high diversity of dietary resources ranging from low abrasion for *Cervus elaphus* (MWS = 0.5) to high abrasion for *Equus ferus* (MWS = 2.5) (Table 5.4 and Fig. 5.7). High abrasive resources exploited by *Equus ferus* indicate a definite grazing behavior for that species (as well as in sublevel Jb). *Bos primigenius* plots in an area where all dietary categories overlap, which is characteristic of a

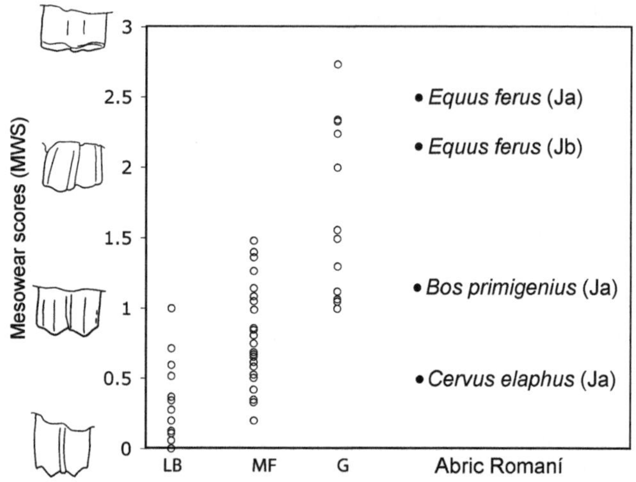

Fig. 5.7 Mesowear scores for the fossil ungulates from Abric Romaní (*solid black dots*) compared to modern wild ungulates (*white dots*). Data on modern species from Fortelius and Solounias (2000). Abbreviations: *LB* leaf browsers, *MF* mixed feeders, *G* Grazers

mixed feeding behavior and indicates that this species fed on both grass and browse. *Cervus elaphus* have a low mesowear score indicating a diet mainly composed of low abrasive items such as leaves from trees or bushes.

The diversity of dietary behavior observed in sublevel Ja reveals the presence of various habitats in the environment surrounding the Abric Romaní.

For the microwear analysis, our sample consisted of 116 teeth which were deemed suitable for analysis (Table 5.5): 81 from *Equus ferus* (63 in Ja and 18 in Jb), 22 from *Cervus elaphus* (20 in Ja and two in Jb), 11 from *Bos primigenius* in Ja, and two from *Rupicapra pyrenaica* in Ja. Samples with fewer than eight specimens are generally considered too small to obtain definite results and are usually excluded from the studies. However, because our study includes a low number of different species, we decided to include the results for the two small samples for the purpose of comparison.

The summary statistics of the microwear features found for the three species are presented in Table 5.5. Two species, *Cervus elaphus* and *Equus ferus*, were found in both Ja and Jb. The results from only one sample for each of these species are presented in Table 5.5 as the samples overlapped. The microwear parameters recorded for these two species are not significantly different in the two sublevels. This indicates that there are no significant differences between Ja and Jb for *Cervus elaphus* and *Equus ferus*: there were no changes in diet for either species during the two occupations (Fig. 5.8).

Microwear data gave definite results for all the species analyzed. *Cervus elaphus*, with a low number of pits and scratches, plots within the browsing morphospace, confirming the low abrasive diet suggested by mesowear.

Table 5.5 Summary statistics for the microwear analysis of the ungulates from Abric Romaní

Species	Level	n	NP	NS	%LP	STS	%CS
Bos primigenius	Ja	11	10.18	24.50	0.18	0.45	0.55
Cervus elaphus	Ja	20	13.63	11.65	0.30	0.90	0.65
Cervus elaphus	Jb	2	15.25	13.25	0.00	1.00	0.50
Equus ferus	Ja	63	10.73	30.91	0.00	0.08	0.40
Equus ferus	Jb	18	11.22	31.58	0.00	0.06	0.39
Rupicapra pyrenaica	Ja	2	26.50	20.75	0.50	1.00	0.50

Abbreviations: *n* number of specimens, *NP* Average number of pits, *NS* Average number of scratches, *%LP* Percentage of specimens with more than four pits, *STS* Scratches texture score (0 = fine; 1 = mixed; 2 = coarse), *%CS* Percentage of specimens with cross scratches

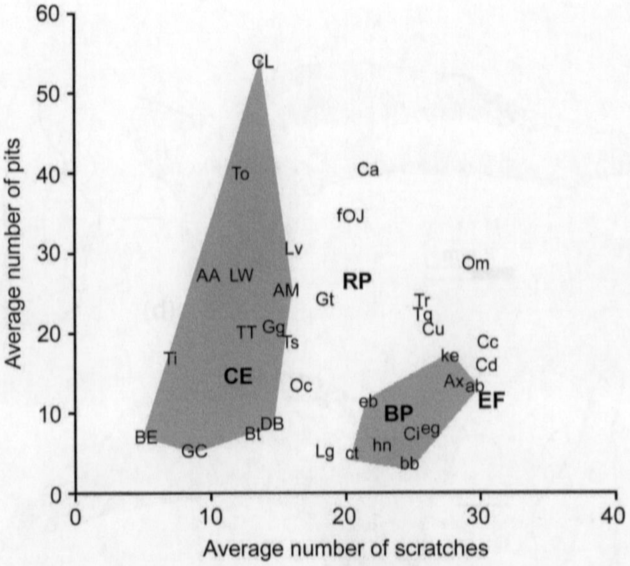

Fig. 5.8 Bivariate plot of the average numbers of pits and scratches in extant ungulates (data from Solounias and Semprebon 2002) and the fossil ungulates from Abric Romaní. Convex hulls are drawn around extant browsing and grazing taxa for ease of comparison. Abbreviations for the fossil species: *BP Bos primigenius CE Cervus elaphus*; *EF Equus ferus*; *RP Rupicapra pyrenaica*. Abbreviations for extant species: see Fortelius and Solounias (2000)

Bos primigenius and *Equus ferus*, with more scratches and fewer pits, plot within or beyond the grazing morphospace. For *Equus ferus*, this confirms a diet with high grass content. *Bos primigenius* falls within the grazing morphospace but the mesowear analysis did not reveal very high abrasive levels. This discrepancy between the two methods is characteristic of mixed feeders (Semprebon and Rivals 2007); it reveals changes in diet on a seasonal basis. *Bos primigenius* was hunted during the season it was preferentially feeding on grasses (Rosell et al. 2012).

Rupicapra pyrenaica falls among the mixed feeders; however, only two teeth were available for analysis. Nevertheless, the pattern observed at Abric Romaní is similar to that found in other samples from other sites in the Mediterranean area (Rivals and Deniaux 2005), indicating a mixed feeding behavior. Compared to other species, the teeth of *Rupicapra pyrenaica* have a relative high number of pits, which is characteristic of exogenous abrasive particles such as grit, dust, or soil (Semprebon and Rivals 2007) found in animals that either feed close to the ground, or that feed on plants covered by grit or dust.

Discussion

The assemblage of ungulates identified at Abric Romaní is typical of the Late Pleistocene fauna of the Mediterranean area. The species have morphologies and dimensions comparable to other populations from late Pleistocene locations in Spain and southern France. The presence of horses, red deer, aurochs and Pyrenean chamois in equal numbers illustrate a diverse environment and a variety of habitats in the vicinity of the site.

Our findings suggest that the ungulates hunted by the occupants of level J had very diverse feeding behaviors and enjoyed a large variety of plant resources. *Bos primigenius* and *Equus ferus* ingested large amounts of highly abrasive plants, such as grasses. This is especially true of *Equus ferus,* which was found to subsist on a diet containing a relatively high abrasive content compared to modern ungulates. Because grass is a larger part of the diet of the fossil species when it is abundantly available in the environment, we propose that at the time of their deaths both *Bos primigenius* and *Equus ferus* were selectively feeding on grasses. However, the mesowear analysis indicated that *Bos primigenius* was a mixed feeder, whose diet changed according to the seasons. These two species were probably living in very similar open habitats where grasses were abundant, such as grasslands. On the other hand, *Cervus elaphus* fed on vegetation with a low abrasive level. It probably inhabited more closed habitats than *Bos primigenius* and *Equus ferus*, and had a diet composed of branches, twigs and leaves of woody plant species.

Despite the small sample size, *Rupicapra pyrenaica* revealed a microwear pattern similar to that found in other Pyrenean chamois populations which inhabited pastures and rocky areas at low altitudes, such as the population from Arago Cave in France (Rivals and Deniaux 2005).

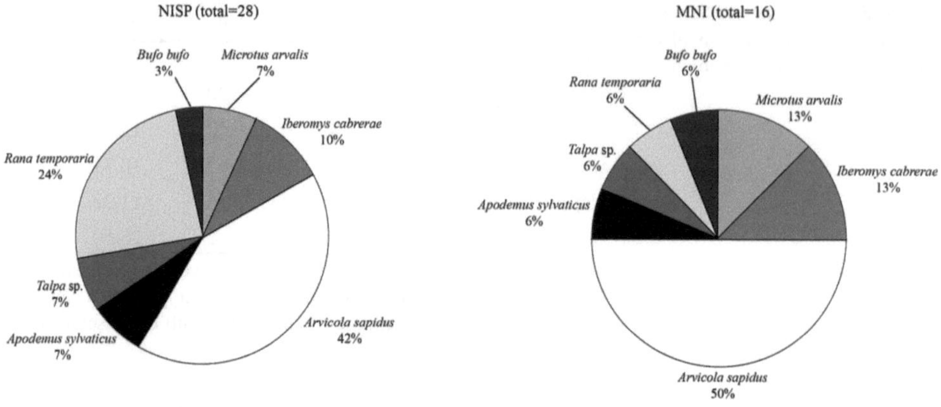

Fig. 5.9 Distribution of small vertebrates in level J of the Abric Romaní in Number of Identified Specimens (NISP) and Minimum Number of Individuals (MNI)

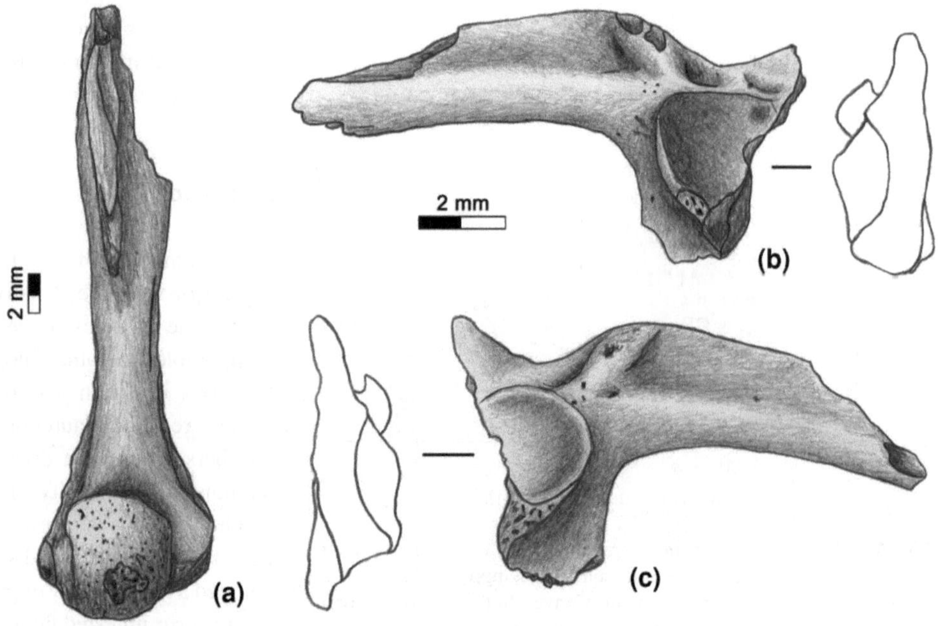

Fig. 5.10 **a** *Bufo bufo*, right humerus of female (AR98/Ja/L42), ventral view. **b–c** *Rana temporaria*, left and right ilia (AR95/Ja/R51), lateral and posterior views

All four species lived in very diverse habitats, thus revealing the diversity of biotopes present around the site during the deposition of level J. This also provides insight into the biotopes exploited by Neanderthals for their subsistence activities at that time. The pattern in level J is quite different from that found in the underlying level M, where *Bos primigenius*, *Equus ferus*, and *Cervus elaphus* were found to live in very similar open habitats (Fernández-Laso et al. 2011).

Small Vertebrates

The small vertebrate assemblage from level J is made up of 201 identified fossils remains. The most representative remains belong to the order of rodents (94.5%), although other taxa such as amphibians (4.5%) and insectivores (1%) are also represented, albeit to a lesser degree (Fig. 5.9).

Systematic Study

We analyzed 29 fossil specimens corresponding to 21 small mammal remains and eight amphibian remains (Fig. 5.9).

Amphibians

Anura Fischer von Waldheim, 1813

Bufonidae Laurenti, 1768

Bufo bufo (Linnaeus, 1758)

One right humerus of a female specimen (Fig. 5.10)

The genus *Bufo* is characterized by a thick humeral shaft without a paraventral crest and a slightly laterally displaced condyle. The poor development of the condyle and epicondyles, in ventral view, as well as the size of the fossil (preserved length = 30 mm) are consistent with *B. bufo*, which is the largest bufonid in western Europe. The absence of a well developed mesial crest and the slightly curved shaft are characteristic of the female.

The common European toad (*B. bufo*) has an extensive Eurasian range. It lives in nearly all environments, even in dry areas. Its only requirement seems to be, during its breeding season, the presence of calm or low energy water, preferably permanent and with vegetation (Lizana 2004). In Catalonia, the distribution of this species is relatively continuous and very common, although it seems to be rarer on the Lleida plain and in the highest mountain areas of the Pyrenees (Llorente et al. 1995).

Ranidae Rafinesque, 1814

Rana temporaria Linnaeus, 1758

Two ilia, one vertebra, one radioulna, two tibiofibulae and one phalanx (Fig. 5.10).

Among the representatives of the family Ranidae, the ilium permits a relatively certain attribution at the species level (Böhme 1977; Esteban and Sanchiz 1985, 1991; Bailón 1999; Gleed-Owen 2000; Blain 2005). In the genera *Rana* (brown frogs) and *Pelophylax* (green frogs), the ilium has a more or less high dorsal crest on the ilial shaft, thee *pars ascendens* is relatively short and the postero-medial side is smooth and without any *sulcus interiliacus*. The two fossil ilia are incomplete, as they lack the anterior part. The juncture ilioischiatica has a fairly straight medial outline, and the ventral acetabular wall is thin, a particular characteristic of brown frogs. The acetabular diameter vs. thickness (d/t ratio, sensu Gleed-Owen 2000) is 3.71 and 3.57, which is well within the range of 2.75–4.00 of brown frogs, and well outside the range of 2.12–2.88 measured for green frogs (Gleed-Owen 2000). The *tuber superior* is unangled and the *pars descendens* is postero-ventrally directed, whereas *R. arvalis* has an angular *tuber superior* and *R. dalmatina* and *R. iberica* a ventrally-directed *pars descendens* (Gleed-Owen 2000 and personal observation). The dorsal crest is rather low although it lacks the typical inflection beyond the *tuber superior*. According to measurements made by Esteban and Sanchiz (1985, 1991), the height of the dorsal crest is consistent with *R. temporaria*. Other bone fragments fit well into the genus *Rana*.

The common brown frog (*R. temporaria*) is a broad-range Eurasian species. In Catalonia, the species only lives in a narrow strip in the north consisting of pre-Pyrenean and Pyrenean areas with high levels of rainfall (average annual precipitation over 800 mm) and cool temperatures (average annual temperature <12°C and average January temperature <5°C) (Llorente et al. 1995, Balcells 1975). At present, its southernmost representatives correspond to a quite isolated population in the Massif of Montseny (Llorente et al. 1995). This population can be found at altitudes ranging from 300 to 2800 m above sea level with larger numbers between 1000 and 1800 m, and lives in various non-Mediterranean environments, such as peat bogs, alpine meadows and wooded river biotopes (Fig. 5.10).

Mammals

Rodentia Bowditch, 1821

Muridae Iliger, 1811

Arvicolinae Gray, 1821

Iberomys cabrerae Thomas, 1906

Three first lower molars (M1) (Fig. 5.11)

The *Iberomys* lineage has *Iberomys cabrerae,* the Iberian vole, as its sole extant representative. It is descended from the *Iberomys brecciensis* species, which appears in the Middle Pleistocene. The Iberian vole is characterized by its relatively long and wide lower molar (M1), the reduction of the triangles of the anteroconid complex (ACC), a long and narrow fifth lingual salient angle (LSA5), which is a measure of its latero-medial asymmetry, also known as labio-lingual asymmetry (Cuenca-Bescós et al. 1995), a fourth buccal salient angle (BSA4) with a quadrangular shape, and enamel completely covering the labial wall of the ACC (Ayarzagüena and López Martínez 1976). These characteristics are present in the specimens from level J (Fig. 5.11). An analysis of the variability in the size of M1 of the Abric Romaní specimens compared to other specimens from the Middle Pleistocene (Galería in Cuenca-Bescós et al. 1999), the Late Pleistocene (Cova del Gegant in López-García et al. 2008) and the Holocene (Cova Foradada; Bronze Age) establishes that the size falls within the range of the extant *Iberomys cabrerae* (Fig. 5.12).

The Iberian vole (*I. cabrerae*) is an endemic species of the Iberian Peninsula, which today has a restricted and discontinuous geographic range (around the Central System, southern Iberian System, sub-bethics Mountains and several points in the Aragon Pyrenees). The Iberian vole is not currently present in Catalonia. This species is characteristic of a Mediterranean climate, with a preferential habitat featuring perennial vegetation and water streams (Blanco 1998a).

Microtus arvalis (Pallas, 1779)

Two first lower molars (M1)

The genus *Microtus* is characterized by the presence of closed triangles (T4–T5) in the lower first molars (M1). Furthermore, the symmetric and parallel disposition of the triangles T4/T5 and especially T6/T7, which display a round morphology of the anteroconid complex (ACC), are specific characteristics of the species *Microtus arvalis.*

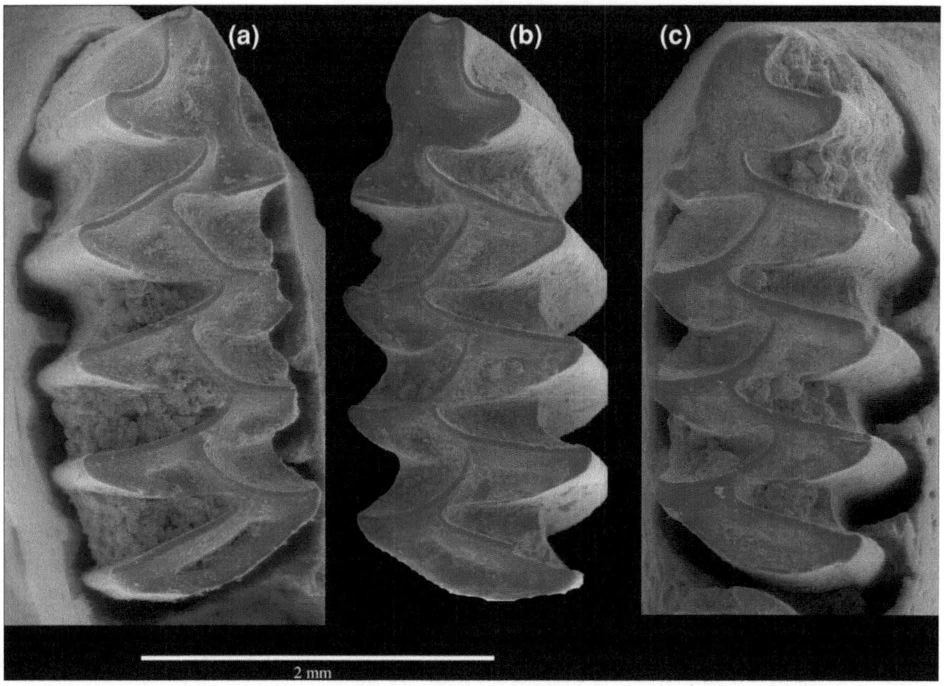

Fig. 5.11 a First lower right molar *Iberomys cabrerae* (AR95/Ja-b/L45), **b–c** first lower left molar *Iberomys cabrerae* (AR95/Ja-b/L45—AR95/Jb/K49). Occlusal view

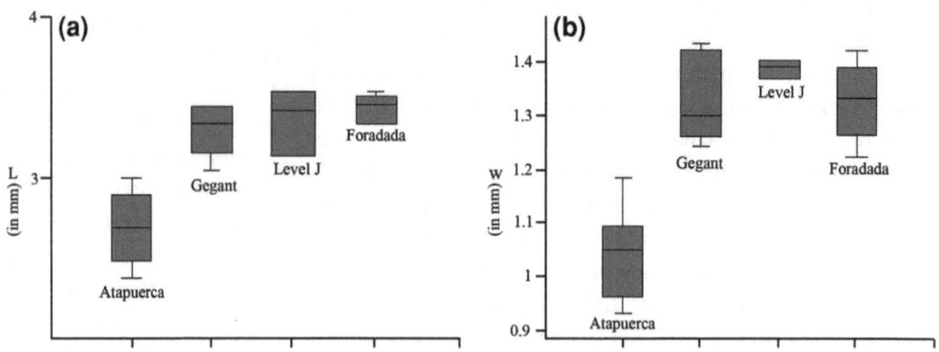

Fig. 5.12 Comparison of the first lower molars (m1) length (A) and width (B) of *Iberomys cabrerae* from level J of the Abric Romaní (n = 3) with *Iberomys cabrerae* from the Cova del Gegant (n = 7) and Cova Foradada (n = 5), and with *Iberomys brecciensis* from the Middle Pleistocene of Atapuerca-Galeria (n = 9; Cuenca-Bescós et al. 1999). Measurements are shown in mm

The common vole (*M. arvalis*) is found in temperate regions of southern Europe. It is a rodent which mainly lives in open and not overly wet meadows. It is normally found at altitudes ranging from 900 to 2000 m above sea level with an average annual precipitation exceeding 800 mm. In Catalonia, this species lives in the upper Pyrenees and at several locations in the northern pre-Pyrenees (Blanco 1998a; Gosàlbez 1987).

Arvicola sapidus Miller, 1908

12 first lower molars (M1)

The genus *Arvicola* is characterized by the presence of a posterior fold in the first lower molar (M1) followed by three closed triangles. The disposition of an acute angle in the lingual side, towards the internal edge of the anterior cusp (AC) in the first lower molar and, the "*Arvicola*" enamel pattern, thicker on the distal side than the proximal side, are morphological characteristics of the species *Arvicola sapidus*.

The southern water vole (*A. sapidus*) has a restricted range in Europe, and can be found throughout most of France and the Iberian Peninsula. In Iberia, it has a homogenous territorial distribution which includes Catalonia. It is a rodent species that prefers permanent water streams, such as rivers or irrigation ditches with slow-moving, deep water at constant levels (Blanco 1998a).

Murinae Illiger, 1811

Apodemus sylvaticus (Linnaeus, 1758)

Mandible (M1–M2) and maxilla (M1–M3)

The genus *Apodemus* is characterized by the presence of a low occlusal surface with six main cusps in the first lower molar (M1). The anterolingual and anterolabial cusps of the M1 are arranged in an X shape. In some specimens, these cusps are separated by a deep, narrow furrow. The posterior cusp (PC) of the M1 is low, rounded and well developed. On the labial side of the M1, there are two or three accessory cusps (C) and one proximal tubercle (TMA) (Cuenca-Bescós et al. 1997). Moreover, the conjunction between the tubercles T4 and T7 on the first upper molar (M1) and the development of the tubercle T9 on the second upper molar (M2) (Pasquier 1974) are the morphologic characteristics that ascribe our material to the *Apodemus sylvaticus* species.

The wood mouse (*A. sylvaticus*) has a large range in eastern and western Europe. It is distributed continuously throughout the Iberian Peninsula, and in Catalonia, it is found from the Pyrenees to the Mediterranean coast. *A. sylvaticus* is a generalist species which lives in a variety of habitat types, although it prefers the marginal zones of deciduous forests with the presence of water streams. It is found at altitudes of up to 2,000 m (Gosàlbez 1987).

Erinaceomorpha Gregory, 1910

Talpidae Fischer, 1817

Talpinae Fischer, 1817

Talpa Linnaeus, 1758

One scapula and one radius

Two species of moles are present on the Iberian Peninsula today: *Talpa europea* (common mole) and *Talpa occidentalis* (Iberian mole). According to Cleef-Roders and Hoek-Ostende (2001) the diagnostic characteristics for distinguishing between the two species are the morphology and measurements of the dentition and mandible and the measurements of the humerus. Due to the scarcity of diagnostic elements, we have considered our specimen to be *Talpa* sp.

The mole is a eurithermic genus, found equally in cold and warm environments, but it prefers deep soils with a high degree of humidity (Blanco 1998b).

Microvertebrate Taphonomy

The accumulation of small vertebrates is usually a consequence of the predation of nocturnal or diurnal birds of prey and small carnivorous mammals (Mellett 1974; Mayhew 1977; Korth 1979). The small vertebrate fossil assemblage cannot therefore be considered an exact reflection of life in the paleocommunity because the specimens found here could have been selected by predator diet criteria. A taphonomic (biostratinomic) study based on a descriptive-systematic method developed by Andrews (1990) allows us to identify the predator which left the accumulation, and these data in turn permit us to make paleoecologic interpretations of the fossil assemblages (Andrews 1983; Andrews and Nesbit-Evans 1983; Andrews 1990; Kowalski 1990; Fernández-Jalvo and Andrews 1992; Denys et al. 1995; Fernández-Jalvo 1995; Cuenca-Bescós 2003).

Table 5.6 A: Rodent skeletal representation by number of identified specimens (NISP) and representation index (IR). B: Pc/C index

A	NISP	IR
Skulls	2	15.4
Maxilla	3	23.1
Jaws	10	38.5
Molar	45	28.8
Incisors	25	48.1
Humerus	11	42.3
Radius	3	11.5
Ulna	3	11.5
Femur	9	34.6
Tibia	11	42.3
Total IR	122	29.6

B	
Index Pc/C	
H + F/S + M+J	1.4

H Humerus, *F* Femur, *S* Skulls, *M* Maxilla, *J* Jaws

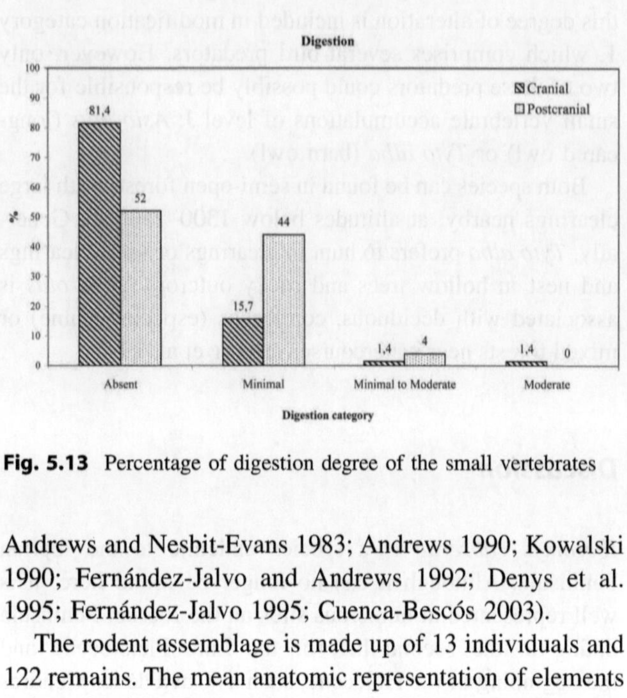

Fig. 5.13 Percentage of digestion degree of the small vertebrates

The rodent assemblage is made up of 13 individuals and 122 remains. The mean anatomic representation of elements is low (29.6%) (Table 5.6A). The Pc/C index (Table 5.6B) indicates that postcranial remains are better represented than cranial elements and we found a very high degree of fracture in the cranial and postcranial elements analyzed. This is related to post-depositional incidents, which break and reduce the number of elements left at the site (Cáceres et al. 2012).

Most of the elements we analyzed did not show signs of digestion (Fig. 5.13). Light digestion is basically only observed on femur and humerus bones. Moderate digestion is only represented on an isolated lower first molar (M1) of

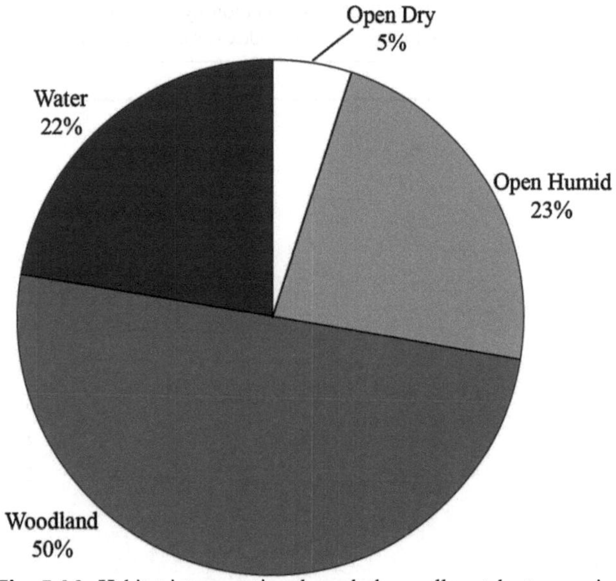

Fig. 5.14 Habitat interpretation through the small vertebrate association. The data represent the percentage of the association taxa by habitat

Microtus arvalis (Fig. 5.14). According to Andrews (1990) this degree of alteration is included in modification category 1, which comprises several bird predators. However, only two of these predators could possibly be responsible for the small vertebrate accumulations of level J: *Asio otus* (long-eared owl) or *Tyto alba* (barn owl).

Both species can be found in semi-open forests with large clearings nearby, at altitudes below 1300–1500 m. Generally, *Tyto alba* prefers to hunt in clearings or semi-clearings and nest in hollow trees and rocky outcrops. *Asio otus* is associated with deciduous, coniferous (especially pine) or mixed forests near watercourses (König et al. 1999).

Discussion

Iberomys cabrerae is a species endemic to the Iberian Peninsula. Although it can no longer be found here, it is well represented at numerous sites on the Iberian Peninsula dating to the second half of the Late Pleistocene and the beginning of the Holocene, such as Cova de les Cendres (Alicante) (Guillem-Calatayud 1999), Cova Bolumini (Alicante) (Guillem-Clatayud 1999), Baños de Mula (Murcia) (Agustí et al. 1990), El Portalón (Atapuerca, Burgos) (López-García et al. 2010a) and Cova Colomera (Lleida) (Oms et al. 2008; López-García et al. 2010b). Their movement from the Iberian Peninsula may be linked to two factors. First, there may have been competition with other species that shared the same ecological habitat, such as *Microtus agrestis* (common vole)—a species that today is very abundant in areas that are not overrun by the Iberian vole (*Iberomys cabrerae*). The second factor may have been environmental changes that occurred after the Bronze Age. The Iberian vole is commonly associated with wet meadows, and its move from the peninsula was quite likely brought about by increased dryness, such as during the Medieval Warm Period between 800 and 1300 AD.

The common frog (*Rana temporaria*) is a Euro-Siberian species with a habitat ranging from eastern Europe to the Urals, reaching as far as Norway in the north. In Spain, it has a continuous distribution along the north, where it finds a high degree of humidity and a cool climate (Esteban and García-París 2004). In the fossil record from the late Lower Pleistocene, Gran Dolina (TD5, Burgos) and Cal Guardiola (Barcelona) have produced the oldest evidence of these frogs on the Iberian Peninsula to date (Blain et al. 2008; Blain et al. 2009). The presence of *R. temporaria* at the Abric Romaní, as well as at Gran Dolina and Cal Guardiola, at the moment represents the only record of this species located slightly south of its present distribution, and is probably related to colder and wetter periods.

The small vertebrate assemblage of level J is dominated by taxa linked to woodland edges (50%) such as *Apodemus sylvaticus* and *Rana temporaria*, and taxa related to open wet meadows (23%) and water streams (22%) such as *Iberomys cabrerae, Talpa* sp., *Arvicola sapidus* and *Rana temporaria*. This association indicates the predominance of wetter conditions than are currently found in the region.

The intersection of the present distributions of all the species occurring in one location may indicate possible climatic conditions (Blain et al. 2008; Martínez and Sanchiz 2005). In order to evaluate paleoclimatic parameters we use the principle of mutual climatic range, which consists of defining the climatic conditions of the area currently inhabited by the extant fauna from the site. This method yields a total of two 10×10 km UTM squares situated in the external ranges of the Pyrenees in Huesca. Such an intersection for level J suggests mean annual temperatures (MAT) (recent data are from Font Tullot 2000) equal to 8 ± 1.4°C, (minimum = 7°C, maximum = 9°C), and mean annual precipitation (MAP) equal to 875 ± 35 mm (min. = 850 mm, max. = 900 mm). The mean temperature of the coldest month (MTC) for level J is 2.75 ± 0.35°C (max. = 3°C, min. = 2.5°C), and for the warmest month (MTW), 20.25 ± 0.35°C (max. = 20.5°C, min. = 20°C) (López-García and Cuenca-Bescós 2010).

Malacofauna

The malacofauna studied comes from sublevel Ja of the Abric Romaní. All the specimens identified belong to continental species, both terrestrial and aquatic, which at

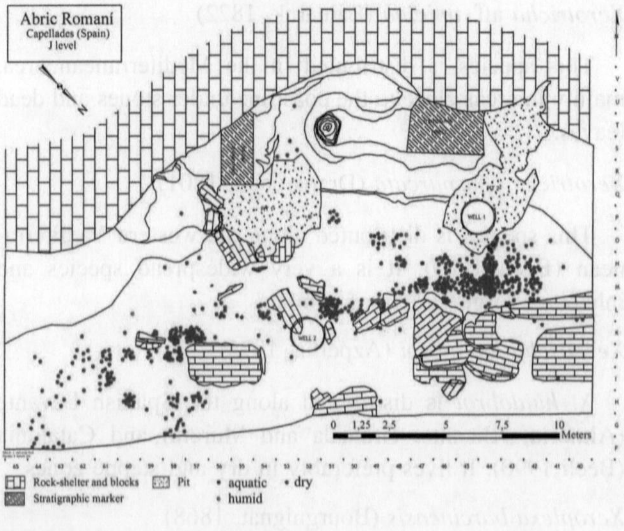

Fig. 5.15 Spatial distribution of malacofauna according to their habitat

present are common in the north-east of the Iberian Peninsula. The information available on their present habitat allows us to understand the environmental conditions in place during their lifetimes and contributes to our understanding of the habitability of the shelter.

Materials and Methods

The studied assemblage includes 746 mollusk remains collected in 90 squares, involving five of the six zones into which the site was divided (Fig. 5.15). The identification of the continental malacofauna is based on Locard (1893, 1894); Haas (1929, 1991), Bech (1990), Martínez Ortí and Robles (2003) and Ruiz et al. (2006).

The conchological study included the following primary aspects: size and description of the general shape; number of whorls in adult specimens; surface sculpture; shape of the aperture; presence and morphology of umbilicus.

To study the distribution of species within the grid established on the site, Diversity (Margalef 1978) (H) and Equitability (Beerbower and Jordan 1969) (E) indices were applied. Diversity provides information about the homogeneity of populations in the assemblage, whereas equitability establishes the relation between the diversity index and its maximum value for that number of species.

Systematic Study

Class Gastropoda

Subclass Prosobranchia

Order Archeogastropoda

Family Hydrobiidae

Mercuria confusa (Frauenfeld, 1863)

Mercuria confusa is found throughout the Mediterranean region and in some northern European countries (England, The Netherlands and Belgium) (Adam 1960). In Catalonia it is a common species, except in the mountains (Bech 1990). It lives in clean, calm, stagnant or little moving waters.

Nehoratia ateni (Boeters, 1969)

Nehoratia ateni inhabits the province of Lleida (Spain) (Bech 1990). This species lives on rocks in warm non-sulfurous moving waters.

Subclass Pulmomata

Order Basommatophora

Family Lymnaidae

Lymnaea palustris (Müller, 1774)

Lymnaea palustris inhabits Europe, Asia, North Africa and probably North America. In Catalonia, it appears along the coast in the provinces of Girona and Barcelona, as well as in the Llobregat river basin (Bech 1990). It lives in marshes and in areas featuring stagnant water.

Lymnaea peregra (Müller, 1774)

Lymnaea peregra is found in Europe, Iraq, the Canary Islands and north-western Africa. In Catalonia it is found in the provinces of Barcelona and Girona (Bech 1990). *L. peregra* inhabits stagnant waters.

Lymnaea (Galba) truncalata (Müller, 1774)

Lymnaea (Galba) truncalata is a holartic species (Hudendock 1951). It is a very common species in Catalonia (Bech 1990) and lives in small masses of stagnant water, and in pits. *L. truncalata* is usually adhered to plants and also lives outside of water.

Order Stylommatophora

Family Succineidae

Oxyloma elegans (Risso, 1826)

Oxyloma elegans is distributed in Europe, North Africa, and western and northern Asia. In Catalonia, it appears in the Llobregat basin, as well as in the delta and in the Montmell range (Baix Penedès) (Bech 1990). Absorbent, it is typical of permanently wet places such as irrigation ditches, pools, spring and creek banks, in herbaceous vegetation, aquatic plants, stones and wet land, as well as in irrigation cultures. Its size seems to diminish with altitude.

Succinea putris (Linné, 1758)

This is a Euro-Siberian species (Kerney and Cameron 1979). In Catalonia it appears in the Llobregat river basin and in the Pyrenean tributaries of the Ebro River (Bech 1990). It lives in very wet locations, stagnant or slow waters, and on aquatic plants.

Family Cochlicopidae

Cochlicopa lubrica (Müller, 1774)

Cochlicopa lubrica is a holartic species (Kerney and Cameron 1979). It is a very common species in Catalonia (Bech 1990), living in wet, shady places. It is found under stones or within dead leaves in wet areas.

Family Zonitidae

Subfamily Gastrodontinae

Zonitoides (Zonitoides) nitidus (Müller, 1774)

This is a Holartic species (Kerney and Cameron 1979). At present, *Z. nitidus* inhabits the entire Iberian Peninsula and is very common in the Catalan region (Bech 1990). It lives near irrigation ditches, in gutters alongside roads, on river banks, in flooded meadows and wooded areas and within riverside vegetation. It is found under stones, within dead leaves or in decomposed vegetation.

Family Helicidae

Subfamily Helicellinae

Cernuella (Cernuella) virgata (Da Costa, 1778)

This species is found in the Mediterranean region and in western Europe (Kerney and Cameron 1979). In general, it lives in two types of habitat, not usually mountainous. It is commonly found in dry locations on the coast and in the steppes of the interior of Catalonia (Bech 1990), generally adhered to plants.

Helicella bolenensis (Locard, 1882)

The distribution and habitat of this species is the western Mediterranean zone (Kerney and Cameron 1979). They live in dry, sandy areas and during dry periods they affix themselves to rocks, whereas in rainy periods they live on grass or in shrubs.

Helicella cistorum (Morelet, 1845)

This species can be found throughout the Iberian Peninsula (Ruiz et al. 2006). It is associated with bush areas (rockroses, heathers, etc.) with or without leafy trees (oaks) and is often found under stones.

Xerotricha aff. *apicina* (Lamarck, 1822)

This species is distributed in the Mediterranean area, mainly in areas close to the coastline under stones and dead leaves.

Xerotricha conspurcata (Draparnaud, 1801)

This species is distributed along the western Mediterranean (Bech 1990). It is a very widespread species and inhabits a large variety of habitats.

Xerotricha huidobroi (Azpeitia, 1925)

X. huidobroi is distributed along the Spanish Levante (Almeria, Alicante, Granada and Murcia) and Catalonia (Bech 1990). It lives preferably in dry and steppe zones.

Xeroplexa barcinensis (Bourguignat, 1868)

This very polymorphic species lives on the Catalan coast, and in Valencia and France (Bech 1990). It inhabits steppic areas near paths. The largest specimens live in low areas, and their size diminishes with altitude. The smallest specimens live at altitudes of 1000–1200 m (Ruiz et al. 2006).

Xeroplexa montserratensis (Hidalgo, 1870)

X. montserratensis is found in the province of Barcelona (Bech 1990) in dry, calcareous areas with poor vegetation.

Xeroplexa murcica (Guirao, 1859)

X. murcica inhabits the Mediterranean area (Bech 1990) in Mediterranean bush regions with pastures and a limestone substratum. It can be found under stones and in vegetation.

Candidula camporroblensis Siro de Fez, 1944

This species lives in the Spanish Levante (Altimira 1959). In Catalonia, it is located in the Baix Camp and Tarragonés areas (Tarragona) (Bech 1990). *C. camporroblensis* inhabits arid and dry zones adhering to plants or on the soil among grasses.

Monacha sp. Fitzinger, 1833

This genus is found throughout the Mediterranean area (Kerney and Cameron, 1979). In Catalonia it has been recorded in the Baix Camp and Alt Camp regions (Bech 1990). The *Monacha* species inhabits submountainous regions, and is also found in wetter zones. It is commonly found in grass.

Fructicum sp. Haas, 1929

The genus *Fructicum* inhabits the Mediterranean area (Kerney and Cameron 1979). It is found in wet areas.

Table 5.7 Terrestrial and aquatic gastropods from sublevel Ja from Abric Romaní

Terrestrial gastropods (dry environment)	Number	Terrestrial gastropods (humid environment)	Number	Aquatic gastropods	Number
Cernuella (Cernuella) virgata	201	*Cochlicopa lubrica*	1	*Mercuria confusa*	2
Helicella bolenensis	69	*Zonitoides (Zonitoides) nitidus*	1	*Nehoratia ateni*	2
Helicella cistorum	2	*Oxyloma elegans*	14	*Lymnaea* (s.s.) *palustris*	4
Helicella sp.	60	*Oxyloma* sp.	3	*Lymnaea* (s.s.) *peregra*	6
Xerotricha aff. apicina	1	*Succinea putris*	1	*Lymnaea (Galba) trunculata*	2
Xerotricha conspurcata	1			*Lymnaea* sp.	1
Xerotricha huidroboi	5				
Xeroplexa barcinensis	15				
Xeroplexa montserratensis	3				
Xeroplexa murcica	1				
Xeroplexa sp.	23				
Candidula camporroblensis	2				
Candidula sp.	1				
Monacha sp.	6				
Fructicum sp.	3				
Helix sp.	103				
Cepaea nemoralis	202				
Theba pisana	11				
Total	709		20		17

Subfamily Helicinae

Cepaea nemoralis (Linné, 1758)

This western European species (Kerney and Cameron, 1979) is very common. It lives in the cold and wet areas of the Pyrenees, as well as in the dry and warm lands of Tarragona (Bech 1990), especially near water courses.

Theba pisana (Müller, 1774)

T. pisana inhabits the entire Mediterranean area (Kerney and Cameron 1979). It lives in warm, dry, arid areas, especially with a marine influence, and adheres itself to plants.

Results

All the identified remains correspond to terrestrial and aquatic continental gastropods. We have identified 28 different taxa: 22 terrestrial and six aquatic species (Table 5.7).

Although the skeletal preservation in the specimens is very good, their fragility has caused many of them to break. In some cases, this taphonomic characteristic impedes identification at the species level. In our case, up to 200 specimens from the total assemblage have been identified only to the genus level (eight genera). The largest number of specimens was concentrated in zone 3 at the site, and included both terrestrial and aquatic gastropods.

The most abundant taxa are *Cepaea nemoralis* and *Cernuella virgata*, both terrestrial and characteristic of dry environments, although *Cepaea nemoralis* can also inhabit cold, wet regions. The next most abundant taxon is *Helicella bolenensis*, which lives in sandy and dry areas. The path distribution of these three taxa in sublevel Ja is very similar, with a concentration in zone 3, zone 4 and zone 6 (Fig. 5.15). The remaining species are dispersed in the same areas of the site. The distribution of the aquatic species is concentrated in zone 3, with some single specimens in zone 4 and one in zone 6. At present, all these taxa inhabit stagnant waters with very little current. The most abundant aquatic species is *Lymnaea peregra* which is distributed in two of the three areas.

The diversity analyses show quite stable environments. The aquatic malacofauna was found to be highly diverse (H = 1.76; E = 0.91) (Table 5.8).

Discussion

Gastropods at the site can be classified into three groups according to habitat: wet terrestrial, dry terrestrial, and aquatic. 97% of the specimens are terrestrial malacofauna, and only the remaining 2.3% are aquatic. The analysis of the bulk of terrestrial gastropods indicates that most of the

Table 5.8 Diversity indexes of malacofauna species from sublevel Ja and level I

	Sublevel Ja	Level I
Total number of species	28	17
Total number of individuals	746	1853
Total number of aquatic species	6	8
Total number of aquatic individuals	17	1316
H	1.76	0.97
E	0.91	0.51
Total number of terrestrial species	22	8
Total number of terrestrial individuals	729	537
Total number of humid environment species	5	4
Total number of humid environments individuals	20	57
Total number of dry environments species	17	4
Total number of dry environment individuals	709	480
H	1.97	1.27
E	0.64	0.46

The index of Taxonomic Diversity (H) used is that of Shannon–Weaver's Information Theory, as given by Margalef (1978). E is defined as the ratio between the observed diversity and the maximun diversity for the same number of species, low values of E shows that the observed diversity is more the result of a large number of species than of a similar number of individuals for each one of them

species are associated with dry environments (97.2%), whereas only 2.75% were found to be associated with wet environments.

The diversity and equitability indices for the aquatic malacofauna suggest a very stable environment, although the low number of specimens in the sample warrants a cautious interpretation. When these results are compared to those previously obtained for level I (Carbonell et al. 2002), various remarkable differences become apparent (Table 5.8). The percentages of aquatic gastropods in level I are higher (71%) than in sublevel Ja, as is the total number of specimens. The relationship between species that prefer wet environments (10.6%) and those that tend to live in dry environments (89.4%) shows an increase in wet-condition species in level I compared to sublevel Ja, although the percentage of dry-environment species is still significantly higher. The diversity analyses applied to the terrestrial mollusks found in level I found a more unstable environment than that found in sublevel Ja. The same occurs with the aquatic gastropod values in level I, which have lower diversity and equitability indices, which would indicate a more unstable environment than that in sublevel Ja.

These data suggest that the malacofauna in sublevel Ja basically corresponds to a dry and stable terrestrial environment, whereas the malacofauna in level I corresponds to a wetter, more unstable environment. Moreover, whereas aquatic malacofauna is dominant in level I, it is clearly less common in sublevel Ja.

Discussion

The results of these multidisciplinary studies paint a picture of a diverse regional landscape in which open woodlands exists side by side with clearings (Table 5.9). Forests were not very dense (48.3% AP in pollen analysis, 50% in small vertebrates), and were probably rather monospecific (99–100% pine in charcoal analyses).

We also know that the open areas had different features depending on their geological subtract, soil type, sun exposure, geomorphology, the dominant direction of wind, and other factors including seasonal variations which affected their appearance. Therefore, we suggest that the area may have functioned as a prairie (13.2% in pollen analysis, 23% in small vertebrates) during the rainy season and as Mediterranean steppe (29.2% in pollen analysis, 5% in small vertebrates) in drier seasons.

In addition, there were wet local areas above the cliff related to the travertine formations of the Capelló and at the Anoia River below the slope of the shelter. Pollen analyses yielded 18.4% hygrophilous taxa and 5.6% riverside trees; small vertebrates show 22% wet prairies; and malacology analyses indicate 28.4% wet-area and 4.7% aquatic taxa.

As for the climate during the period studied, we interpret a cold environment, moderated by the coolness and the proximity of the Mediterranean Sea. This interstadial environment permits the presence of refugia areas of Mediterranean taxa (7.5% Mediterranean taxa in palynology, 2.7% in malacology). The marked seasonality of the Mediterranean climate and the physiographical heterogeneity would have permitted the duality between dry and mesic taxa. Furthermore, the presence of riverside and aquatic taxa points to permanent local water sources throughout the year, whereas the malacological record shows a primarily dry environment inside the shelter, making its conditions for habitability quite favorable.

In short, the environments surrounding the Abric Romaní shelter during the occupation of level J were quite diverse. There were local hygro-hydrophilous components related to water (pools, marshes and rivers), which were responsible for the travertine formations and, on a regional scale, open pine forests, as well as meso-thermophilous taxa forming small forests sharing the territory with scrubland and open areas (plains and cliffs), prairies or Mediterranean steppes.

The study of the environment and landscapes that surrounded the Abric Romaní site during the occupations of sublevels Ja and Jb suggests that biotic resources were available and abundant. Data provided by anthropic and

Table 5.9 Summary data from the fauna and flora assemblage from level J from Abric Romaní

	Record	Landscape		Climate
		Regional scale	Local scale	
Palaeobotany	Charcoal	–	Pine woodland	(cold)
	Pollen	Semi-open (48.3% AP) Pine woodland (39.5%) Steppe (29.2%) Prairie (13.2%)	Riverside (5.6%)	Cool (semi-cold) Mediterranean taxa (7.5%) Hygrophilous taxa (18.4%)
Palaeontology	Large mammals	Open habitats (45%) Woodland habitats (33%)	–	Cool (semi-cold)
	Small vertebrates	Woodland edges (50%) Open humid (23%) Water streams (22%) Open dry (5%)	–	Cool very humid
	Malacofauna	Dry steppe (37.9%) Scrubland (25.1%)	Humid areas (28.4%) Aquatic (4.7%)	Dry (41.9%) Humid (30.3%) Mediterranean taxa (2.7%)

natural records verify that Neanderthals selected their resources from a wide range of possibilities and different biotopes (Finlayson and Carrión 2007; Stringer et al. 2008; Allué et al. 2012; Rosell et al. 2012; Vaquero et al. 2012). Hunting, collecting wood for fuel, and finding wood for manufacturing artifacts depend on specific strategies which include technology, energy waste, organization, and occupation length. Furthermore, biotic resources describe a similar pattern throughout the sequence, which suggests an ongoing behavior. Such behavior is very closely related to the surrounding landscape and the climatic conditions which ensure the supply of essential raw materials.

Plant resources were distributed over a large area, among which the most abundant records are related to wood. According to pollen and charcoal records, woody plants were dominant on nearby slopes and the riverbanks. Pine groves must have covered a vast area near the shelter due to the ubiquitous character of Scots Pine. The abundance of cf. *Juniperus* in the pollen record suggests that it too was important, and we think these were probably small common junipers. Other taxa present in the pollen record might have grown in remoter and more protected areas, suggesting the presence of refugia. The woody vegetation along the river was well established, although the record includes only a single fragment of willow in the charcoal assemblage.

The large mammals hunted by Neanderthals were distributed throughout a large diversity of biotopes. These biotopes were identified through dental wear analyses which indicate the immediate dietary adaptation of the populations and thus the habitat in which the game was preferentially hunted. Horses and aurochs fed mainly on highly abrasive items such as grass, so we can assume that they inhabited and were hunted in open landscapes, probably on the plain below the shelter. On the other hand, the microwear signs found on the dentition of the red deer are characteristic of a diet composed of low abrasive items, such as ligneous plants, indicating that red deer were present in forested habitats. Finally, the chamois probably inhabited and was hunted on or around rocky and abrupt areas which are abundant near the site.

These results indicate that all the biotopes around the shelter were known and exploited by Neanderthal groups during their hunting activities. The woody environment was preferably exploited for wood supply whereas a larger area was explored for hunting or for the exploitation of lithic raw materials.

Acknowledgements We would like to stress our gratitude to the students, researchers and others that have contributed to the field work at the Abric Romaní during the past years. This research has been funded by the Spanish *Ministerio de Ciencia e Innovación* (MICINN) with the projects HAR2008-01984/HIST and HAR2010-19957/HIST and Generalitat de Catalunya SGR813. J.M.L.-G. research has been supported by a postdoctoral grant from Juan de la Cierva Subprogram (JCI-2009-04026), with the financial sponsorship of the Spanish Ministry of Science and Innovation. M.L.B research has been supported by a predoctoral grant from UAU-CSIC. We thank the anonymous reviewers whose comments helped to improve our manuscript.

References

Adam, W. (1960). Faune de Belgique. Mollusques Tome 1.Mollusques terrestres et ducicoles. Bruxelles: Inst. Roy. Sci. Nat. Belg.

Agustí, J., Freudenthal, M., Lacomba, J. I., Martín Suarez, E., & Nägeli, C. (1990). Primeros micromamíferos del Pleistoceno Superior de la Cuenca de Mula (Murcia, España). *Revista Sociedad Geológica España, 3*(3–4), 289–293.

Alférez Delgado, F., Molero Gutiérrez, G., & Bustos Pretel, V. (1980). Los restos fósiles más meridionales de *Rupicapra rupicapra* hallados en Europa. *Col-Pa Publicaciones del departamento de Paleontología, 36*, 53–59.

Allué, E., Cabanes, D., Solé A., & Sala, R. (2012). Hearth functioning and forest resource exploitation based on the archaeobotantical assemblage from level J of Abric Romaní. In E. Carbonell (Ed.), *High Resolution Archaeology and Neanderthal Behavior: Time and Space in Level J of Abric Romaní (Capellades, Spain)* (pp.). Dordretch: Springer.

Altimira, C. (1959). Contribución al conocimiento de la fauna malacologica de la provincia de Tarragona. *Miscel·lània Zoològica, 1*(2), 89–95.

Andrews, P. (1983). Small mammal faunal diversity at Olduvai Gorge, Tanzania. In J. Clutton-Brock & C. Grigson (Eds.), *Animals and archaeology: 1. Hunters and their prey* (pp. 77–85). Oxford, England: BAR International Series 163.

Andrews, P. (1990). *Owls, caves and fossils*. London: The University of Chicago Press.

Andrews, P., & Nesbit-Evans, E. M. (1983). Small mammal bone accumulations produced by mammalian carnivores. *Paleobiology, 9*(3), 289–307.

Arteaga, I., Allué, E., Pastó, I., Vallverdú, J., & Carbonell, E. (2001). Els fogars del Paleolític Mitjà de l'Abric Romaní (Capellades, Anoia). *Cypsela, 13*, 9–19.

Ayarzagüena, J., & López Martínez, N. (1976). Estudio filogenético y comparativo de Microtus cabrerae y Microtus brecciensis. *Doñana Acta Vertebrata, 3*(2), 181–204.

Bailon, S. (1999). Différenciation ostéologique des Anoures (Amphibia, Anura) de France. In J. Desse & N. Desse-Berset (Eds.), *Fiches d'ostéologie animale pour l'archéologie, Série C: varia*. Valbonne: Centre de Recherches Archéologiques-CNRS.

Balcells, E. (1975). Observaciones en el ciclo biológico de anfibios de alta montaña y su interés en el inicio de la estación vegetativa. *Publicaciones del Centro Pirenaico de Biología Experimental, 7*(2), 55–153.

Barrón, E. (1999). Estudio paleobotánico del afloramiento vallesiense (Neógeno) del barranco de Salanca (La Cerdaña, Lérida, España). Aspectos paleoecológicos. *Anales del Jardín Botánico de Madrid, 57*(1), 81–96.

Beaulieu, J.-L. de & Reille, M. (1973). L'interprétation des spectres polliniques de périodes froides: à propos du Cèdre au Quaternaire en Europe. In Actes 9 Congrès INQUA: Le Quaternaire. Géodynamique, Stratigraphie et Environnement. Christchurch: CNRS.

Bech, M. (1990). *Fauna malacológica de Catalunya. Mol.luscs terrestres i d'aigua dolça*. Barcelona: Institució Catalana d'Història Natural.

Beerbower, J.R., & Jordan, D. (1969). Application of information theory to paleontologic problems: Taxonomic diversity. *Journal of Paleontolology, 43*, 1184–1198.

Bischoff, J. L., Julià, R., & Mora, R. (1988). Uranium-series dating of the Mousterian occupation at Abric Romaní, Spain. *Nature, 332*, 68–70.

Blain, H.-A. (2005). *Contribution de la Paléoherpétofaune (Amphibia & Squamata) à la connaissance de l'évolution du climat et du paysage du Pliocène supérieur au Pléistocène moyen d'Espagne*. Ph. D. Dissertation, Muséum national d'Histoire naturelle, Institut de Paléontologie Humaine, Paris, 402 pp.

Blain, H.-A., Bailon, S., & Cuenca-Bescós, G. (2008). Los anfibios y escamosos del Pleistoceno inferior-medio de Gran Dolina (Atapuerca, Burgos, España): una perspectiva paleobiogeográfica. *Studia Geologica Salmanticensia, vol. espec., 8*, 71–84.

Blain, H.-A., Bailon, S., Cuenca-Bescós, G., Arsuaga, J. L., Bermúdez de Castro, J. M., & Carbonell, E. (2009). Long-term climate record inferred from Early-Middle Pleistocene amphibian and squamate reptile assemblages at the Gran Dolina cave, Atapuerca, Spain. *Journal of Human Evolution, 56*(1), 55–65.

Blanco, J. C. (1998a). *Mamíferos de España II. Cetáceos, Artiodáctilos, Roedores y Lagomorfos de la Península Ibérica, Baleares y Canarias*. Madrid: Planeta.

Blanco, J. C. (1998b). *Mamíferos de España I. Insectívoros, Quirópteros, Primates y Carnívoros de la Península Ibérica, Baleares y Canarias*. Madrid: Planeta.

Böhme, G. (1977). Zur Bestimmung quartärer Anuren Europas an Hand von Skelettelementen. *Wissenschaftliche Zeitschrift der Humbold-Universität zu Berlin (Mathematisch-Naturwissenschaftliche Reihe), 36*(3), 283–300.

Brugal, J.-Ph. (1993). La faune des grands mammifères de l'abri des Canalettes—matériel 1980–1986. In L. Meignen (Ed.), *L'abri des Canalettes. Un habitat moustérien sur les grands Causses (Nant, Aveyron). Fouilles 1980–1986* (pp. 89–137). Paris: CNRS éditions.

Burjachs, F. (1990). *Palinologia dels dolmens de l'Alt Empordà i dels dipòsits quaternaris de la cova de L'Arbreda (Serinyà, Pla de l'Estany) i del Pla de l'Estany (Olot, Garrotxa). Evolució del paisatge vegetal i del clima des de fa més de 140.000 anys al NE de la Península Ibèrica*. Ph.D. Dissertation, Universitat Autònoma de Barcelona.

Burjachs, F., & Renault-Miskovsky, J. (1992). Paléoenvironnement et paléoclimatologie de la Catalogne durant près de 30, 000 ans (du Würmien ancien au début de l'Holocène) d'après la Palynologie du site de L'Arbreda (Gérone, Catalogne). *Quaternaire, 3*(2), 75–85.

Burjachs, F., & Julià, R. (1994). Abrupt climatic changes during the last glaciation based on pollen analysis of the Abric Romaní, Catalonia, Spain. *Quaternary Research, 42*, 308–315.

Burjachs, F., López-Sáez, J. A., & Iriarte, M. J. (2003). Metodología arqueopalinológica. In R. Buxó & R. Piqué (Eds.), *La recogida de muestras en arqueobotánica: objetivos y propuestas metodológicas* (pp. 11–18). Barcelona: Museu d'Arqueologia de Catalunya.

Cáceres, I., Rosell, J., & Huguet, R. (1998). Séquence d'utilisation de la biomasse animale dans le gisement de l'Abric Romaní. *Quaternaire, 9*, 379–383.

Cáceres, I., Bennàsar, M., Huguet, R., Saladié, P., Rosell, J., Allué, E., et al. (2012). Level J Taphonomy. In E. Carbonell (Ed.), *High Resolution Archaeology and Neanderthal Behavior: Time and Space in Level J of Abric Romaní (Capellades, Spain)* (pp.). Dordretch: Springer.

Carbonell, E. (2002). *Abric Romaní Nivell I. Models de'ocupació de curta durada de fa 46.000 anys a la Cinglera del Capelló (Capellades Anoia, Barcelona)*. Tarragona: Universitat Rovira i Virgili.

Carbonell, E., Cebrià, A., Allué, E., Cáceres, I., Castro, Z., Díaz, R., et al. (1996). Behavioural and organizational complexity in the Middle Palaeolithic from the Abric Romani. In E. Carbonell & M. Vaquero (Eds.), *The last Neandertals, the first anatomically modern humans: A tale about the human diversity* (pp. 385–434). Tarragona: Universitat Rovira i Virgili.

Carrión, J. S., Munuera, M., Navarro, C., & Sáez, F. (2000). Paleoclimas e historia de la vegetación cuaternaria a través del análisis polínico. Viejas falacias y nuevos paradigmas. *Complutum, 11*, 115–142.

Castaños, J., Castaños, P., & Murelaga, X. (2006). Estudio osteométrico preliminar de los restos de ciervo (*Cervus elaphus*) del yacimiento paleontológico del Pleistoceno Superior de Kiputz IX (Mutriku, Gipuzkoa). *Geogaceta, 40*, 163–166.

Chacón, M. G., Fernández-Laso, M. C., García-Antón, M. D., Allué, E. (2007) Level K and L from Abric Romaní (Barcelona, Spain): procurement resources and territory management in shorts occupations during the Middle Palaeolithic. In M.-H. Moncel, A.-M. Moigne, M. Arzarello, M. & C. Peretto (Eds.), *Proceedings of the XV World UISPP Congress, Lisbon, 2007, Vol.5 Session*

WS23: Raw material supply areas and food supply areas. Integrated approach of the behaviours (pp. 187-197). BAR International Series 1725. Oxford: Archaeopress.

Cleef-Roders, J. T., & Hoek Ostende, L. W. (2001). Dental morphology of *Talpa europaea* and *Talpa occidentalis* (Mammalia: Insectivora) with discussion of fossil *Talpa* in the Pleistocene of Europe. *Zoology Medicine, 75*, 51–67.

Clot, A., Brochet, G., Chaline, J., Desse, G., Evin, J., Granier, J., et al. (1984). Faune de la grotte préhistorique du bois du Cantet (Espèche, Hautes-Pyrénées, France). *Munibe, 36*, 33–50.

Crégut-Bonnoure, E. (1979). *La faune de mammifères du Pléistocène moyen de la Caune de l'Arago à Tautavel, Pyrénées-Orientales*. Ph.D. Dissertation, Université de Provence.

Crégut-Bonnoure, E. (1992). Intérêt biostratigraphique de la morphologie dentaire de *Capra* (Mammalia, Bovidae). *Annales Zoologici Fennici, 28*, 273–290.

Cuenca-Bescós, G. (2003). The micromammal record as proxy of palaeoenvironmental changes in the Pleistocene of the Sierra de Atapuerca (Burgos, Spain). In M. B. Ruiz-Zapata, M. Dorado-Valiño, A. Valdeolmillos-Rodríguez, M. J. Gil García, I. Bardají Azcárate, I. de Bustamante, & I. Martínez Mendizábal (Eds.), *Quaternary climatic changes and environmental crises in the Mediterranean region* (pp. 133–138). Alcalá de Henares: Universidad de Alcalá, Ministerio de Ciencia y Tecnología, INQUA.

Cuenca-Bescós, G., Canudo, J. I., & Laplana, C. (1995). Los arvicólidos (Rodentia, Mammalia) de los niveles inferiores de Gran Dolina (Pleistoceno Inferior, Atapuerca, Burgos, España). *Revista Española de Paleontología, 10*, 202–218.

Cuenca-Bescós, G., Canudo, J. I., & Laplana, C. (1999). Análisis Bioestratigráfico del Pleistoceno medio del yacimiento de Galería (Sierra de Atapuerca, Burgos). In E. Carbonell, A. Rosas, & J. C. Díez (Eds.), *Atapuerca: Ocupaciones humanas y paleoecología del yacimiento de Galería* (pp. 189–210). Burgos: Junta de Castilla y León.

Cuenca-Bescós, G., Laplana, C., Canudo, J. I., & Arsuaga, J. L. (1997). Small mammals from Sima de los Huesos. *Journal of Human Evolution, 33*, 175–190.

Dansgaard, W., Johnsen, S. J., Clausen, H. B., Dahl-Jensen, D., Gundestrup, N. S., Hammer, C. U., et al. (1993). Evidence for general instability of past climate from a 150-kyr ice-core record. *Nature, 364*, 218–220.

Defleur, A., Crégut-Bonnoure, E., & Radulescu, C. (1989). Nouvelles données stratigraphiques et fauniques sur la Grotte des Cèdres (Le Plan d'Aups, Var, France). *Un nouveau gisement d'âge rissien à Hemitragus. Comptes-Rendus de l'Académie des Sciences, 308*, 259–264.

Defleur, A., Bez, J.-F., Crégut-Bonnoure, E., Desclaux, E., Onoratini, G., Radulescu, C., Thinon, M., Vilette, P. (1994). Le niveau Moustérien de la grotte de l'Adaouste (Jouques, Bouches-du-Rhône). Approche culturelle et paléoenvironnements. *Bulletin du Musée d'Anthropologie Préhistorique de Monaco, 37*, 11–48.

Denys, C., Sánchez, V., & Fernandez-Jalvo, Y. (1995). Prédation et fossilisation des micromammifères. Présentation d'un des apports d'une discipline récente: la Taphonomie. *Arvicola, VII*(1), 7–13.

Esteban, M., & García-París, M. (2004). *Rana temporaria*. In J.M. Pleguezuelos, R. Márquez & M. Lizana (Eds.), Atlas y libro rojo de los anfibios y reptiles de España (pp. 131–133). Madrid: Dirección General de Conservación de la Naturaleza-Asociación Herpetológica Española (3a impresión).

Esteban, M., & Sanchiz, B. (1985). Herpetofauna de Erralla. *Munibe, 37*, 81–86.

Esteban, M., & Sanchiz, B. (1991). Sobre la presencia de *Rana iberica* en el Pleistoceno burgales. *Revista Española de Herpetología, 5*, 93–99.

Fernández-Jalvo, Y. (1995). Small mammals taphonomy at La Trinchera de Atapuerca (Burgos, Spain). A remarkable example of taphonomic criteria used for stratigraphic correlations and palaeoenvironment interpretations. *Palaeogeography, Palaeoclimatology, Palaeoecology, 114*, 167–195.

Fernández-Jalvo, Y., & Andrews, P. (1992). Small mammal taphonomy of Gran Dolina, Atapuerca (Burgos), Spain. *Journal of Archaeological Science, 19*, 407–428.

Fernández-Laso, M. C., Chacón, M. G., García-Antón, M. D., & Rivals, F. (2011). Territorial mobility Neanderthals groups: A case study from level M of Abric Romaní (Capellades, Barcelona, Spain). In N. J. Conard & J. Richter (Eds.), *Neanderthal lifeways, subsistence and technology* (pp. 187–202). New York: Springer.

Finlayson, C., & Carrion, J. S. (2007). Rapid ecological turnover and its impact on Neanderthal and other human populations. *Trends in Ecology & Evolution, 22*, 213–222.

Font Tullot, I. (2000). *Climatología de España y Portugal*. Salamanca: Universidad de Salamanca.

Fortelius, M., & Solounias, N. (2000). Functional characterization of ungulate molars using the abrasion-attrition wear gradient: A new method for reconstructing paleodiets. *American Museum Novitates, 3301*, 1–36.

Girard, M., & Renault–Miskovsky, J. (1969). Nouvelles techniques de préparation en Palynologie appliqués à trois sédiments du Quaternaire final de l'Abri Cornille (Istres–Bouches du Rhône). *Bulletin A.F.E.Q., 4*, 275–284.

Gleed-Owen, C. P. (2000). Subfossil records of *Rana* cf. *lessonae*, *Rana arvalis* and *Rana* cf. *dalmatina* from the Middle Saxon (c. 600–950 AD) deposits in Eastern England: Evidence for native status. *Amphibia-Reptilia, 21*(1), 57–65.

Godfrey, L. R., Semprebon, G. M., Jungers, W. L., Sutherland, M. R., Simons, E. L., & Solounias, N. (2004). Dental use wear in extinct lemurs: Evidence of diet and niche differentiation. *Journal of Human Evolution, 47*, 145–169.

Goeury, Cl., & de Beaulieu, J.-L. (1979). À propos de la concentration du pollen à l'aide de la liqueur de Thoulet dans les sédiments minéraux. *Pollen et Spores, XXI*(1–2), 239–251.

Gosàlbez, J. (1987). *Insectivors i Rosegadors de Catalunya. Metodología d'estudi i catàleg faunístic*. Barcelona: Ketres.

Guadelli, J.-L. (1987). *Contribution à l'étude des zoocénoses préhistoriques en Aquitaine (Würm ancien et interstade würmien)*. Ph.D. Dissertation, Université de Bordeaux 1.

Guillem-Calatayud, P. M. (1999). Los micromamíferos (Rodentia, Insectívora y Chiroptera) de la Secuencia Holocena de la Cova de les Cendres y Cova Bolumini. *Saguntum-PLAV, (Extra-2)*, 31–36.

Haas, F. (1929). Fauna malacológica terrestre y de agua dulce de Cataluña. *Treballs Museu de Ciencies Naturals de Barcelona, 13*, 1–491.

Haas, F. (1991). *Fauna malacológica terrestre y de agua dulce de Cataluña. Treballs del Museo de Zoología, 5*. Barcelona: Ajuntament de Barcelona.

Kerney, M. P., & Cameron, R. A. D. (1979). *A field guide to the land snails of Britain and North-West Europe*. London: Collins.

König, C., Weick, F., & Becking, J. H. (1999). *Owls. A guide to the owls of the world*. Sussex: Pica Press.

Korth, W. W. (1979). Taphonomic microvertebrate fossil assemblages. *Annals of Carnegie Museum, 48*, 235–285.

Kowalski, K. (1990). Some problems of the taphonomy of small mammals. In *International symposium on the evolution, phylogeny, and biostratigraphy of Arvicolids* (pp. 285–296). Praha.

Lister, A. M. (1996). The morphological distinction between bones and teeth of fallow deer (*Dama dama*) and red deer (*Cervus elaphus*). *International Journal of Osteoarchaeology, 6*, 119–143.

Lizana, M. (2004). *Bufo bufo*. In J. M. Pleguezuelos, R. Márquez & M. Lizana (Eds.) Atlas y libro rojo de los anfibios y reptiles de España (pp. 103–106). Madrid: Dirección General de Conservación de la Naturaleza-Asociación Herpetológica Española (3a impresión).

Llorente, G. A., Montori, A., Santos, X., & Carretero, M. A. (1995). *Atlas dels Amfibis i Rèptils de Catalunya i Andorra*. Figueres: Edicions El Brau.

Locard, A. (1893). *Les coquilles terrestres. Des eaux douces et saumatres de France*. Paris: Librairie J.-B. Baillière et fils.

Locard, A. (1894). *Les coquilles terrestres de France. Description des familles, genres et espèces*. Paris: Librairie J.-B. Baillière et fils.

López-García, J. M., Blain, H.-A., Cuenca-Bescós, G., & Arsuaga, J. L. (2008). Chronological, environmental and climatic precisions on the Neanderthal site of the Cova del Gegant (Sitges, Barcelona, Spain). *Journal of Human Evolution, 55*(6), 1151–1155.

López–García, J.M., Blain, H-A., Cuenca-Bescós, G., Ruiz-Zapata, M.B., Dorado, M., Gil-García, M.J., Valdeolmillos, A., Ortega, A.I., Carretero, J.M., Arsuaga, J.A., Bermúdez de Castro, J.M., Carbonell, E. (2010a) Palaeonvironmental and palaeoclimatic reconstruction of the latest Pleistocene of El Portalón Cave, Sierra de Atapuerca, northwestern Spain. *Palaeogeography, Palaeoclimatology and Palaeoecology, 292*(3-4), 453–464.

López-García, J.M., Blain, H.-A., Allué, E., Bañuls, S., Bargalló, A., Martín, P., Morales, J.I., Pedro, M., Rodríguez, A., Solé, A., Oms, F.X. (2010b). First fossil evidence of an "interglacial refugium" in the Pyrenean region. *Naturwissenschaften, 97*(8), 753–761.

López-García, J.M. & Cuenca-Bescós, G. (2010). Evolution climatique durant le Pléistocène supérieur en Catalogne (Nord-Est de l'Espagne) d'après de l'étude des micromammifères. *Quaternaire, 21*(3), 249–258.

Margalef, R. (1978). *Ecología*. Barcelona: Omega.

Martínez-Ortí, A., & Robles, F. (2003). *Moluscos continentales de la Comunidad Valenciana*. Valencia: Generalitat Valenciana, Consellería de territori i habitatge.

Martínez Solano, I., & Sanchiz, B. (2005). Anfibios y Reptiles del Pleistoceno medio de Ambrona. In M. Santonja & A. Pérez González (Eds.), *Los yacimientos paleolíticos de Ambrona y Torralba (Soria). Un siglo de investigaciones arqueológicas* (pp. 232–239). Zona Arqueológica, 5, Alcalá de Henares: Museo Arqueológico Regional.

Mayhew, D. F. (1977). Avian predators as accumulators of fossil mammal material. *Boreas, 6*, 25–31.

Mellett, J. S. (1974). Scatological origin of microvertebrate fossil accumulations. *Science, 185*, 349–350.

Moulle, P.-E. (1992). *Les grands mammifères du Pléistocène inférieur de la grotte du Vallonnet (Roquebrune-Cap-Martin, Alpes Maritimes). Etude paléontologique des carnivores, équidés, suidés, et bovidés*. Ph.D. Dissertation, Muséum national d'histoire naturelle, Paris.

Oms, F. X., Bargalló, A., Chaler, M., Fontanals, M., García, M. S., López-García, J. M., et al. (2008). La Cova Colomera (Sant Esteve de la Sarga, Lleida), una cueva-redil en el prepirineo de Lérida. Primeros resultados y perspectivas de futuro (pp. 230–236). In Actas del IV Congreso del Neolítico Peninsular. Alicante: Museo Arqueológico de Alicante.

Palombo, M. R., Filippi, M. L., Iacumin, P., Longinelli, A., Barbieri, M., & Maras, A. (2005). Coupling tooth microwear and stable isotope analyses for palaeodiet reconstruction: the case study of Late Middle Pleistocene *Elephas* (*Palaeoloxodon*) *antiquus* teeth from Central Italy (Rome area). *Quaternary International, 126–128*, 153–170.

Pasquier, L. (1974). *Dynamique évolutive d'un sous-genre de Muridae, Apodemus (Sylvaemus). Etude biométrique des caractères dentaires de populations fossiles et actuelles d'Europe occidentale*. Ph.D. Dissertation, Université de Montpellier.

Pérez Legido, B., & Cerdeño, E. (1992). Los macromamíferos del Pleistoceno superior de Cueva Millán (Burgos). *Estudios Geológicos, 48*, 187–192.

Ravazzi, C., Pini, R., Breda, M., Martinetto, E., Muttoni, G., Chiesa, S., et al. (2005). The lacustrine deposits of Fornaci di Ranica (Late Early Pleistocene, Italian Pre-Alps): Stratigraphy, palaeoenvironment and geological evolution. *Quaternary International, 131*, 35–58.

Rivals, F. (2004). *Les petits bovidés (Caprini et Rupicaprini) pléistocènes dans le bassin méditerranéen et le Caucase. Etude paléontologique, biostratigraphique, archéozoologique et paléoécologique*. Oxford, England: BAR International Series S1327.

Rivals, F., & Deniaux, B. (2005). Investigation of human hunting seasonality through dental microwear analysis of two Caprinae in late Pleistocene localities in Southern France. *Journal of Archaeological Science, 32*, 1603–1612.

Rivals, F., Mihlbachler, M. C., & Solounias, N. (2007a). Effect of ontogenetic-age distribution in fossil samples on the interpretation of ungulate paleodiets using the mesowear method. *Journal of Vertebrate Paleontology, 27*, 763–767.

Rivals, F., Solounias, N., & Mihlbachler, M. C. (2007b). Evidence for geographic variation in the diets of Late Pleistocene and early Holocene *Bison* in North America, and differences from the diets of recent *Bison*. *Quaternary Research, 68*, 338–346.

Ros, M. T. (1987). Anàlisi antracològica de la cova de L'Arbreda. *Cypsela, VI*, 67–7.

Rosell, J, Blasco, R., Huguet, R., Cáceres, I., Saladié, P., Rivals, F., et al. (2012). Occupational patterns and subsistence strategies at Level J of Abric Romaní (Barcelona, Spain). In E. Carbonell (Ed.), *High Resolution Archaeology and Neanderthal Behavior: Time and Space in Level J of Abric Romaní* (*Capellades, Spain*), (pp.). Dordretch: Springer.

Ruiz, A., Cárcaba, A., Porras, A. I. & Arrebola, J. R. (2006). Caracoles terrestres de Andalucía. Guía y manual de identificación. Fundación Gypaetus, Junta de Andalucía.

Sánchez, B. (1989). La fauna de mamíferos del Pleistoceno Superior del Abric Romaní (Capellades, Barcelona). In J. Cins & J. A. Flores (Eds.), *Actas de Paleontología* (pp. 331–347). Salamanca: Universidad de Salamanca.

Schweingruber FH. (1990). *Anatomie europäischer Hölzer ein Atlas zur Bestimmung europäischer Baum-, Strauch- und Zwergstrauchhölzer Anatomy of European woods an atlas for the identification of European trees shrubs and dwarf shrubs*. Stuttgart: Verlag Paul Haupt.

Semprebon, G. M., & Rivals, F. (2007). Was grass more prevalent in the pronghorn past? An assessment of the dietary adaptations of Miocene to recent Antilocapridae (Mammalia: Artiodactyla). *Palaeogeography, Palaeoclimatology, Palaeoecology, 253*, 332–347.

Semprebon, G. M., Godfrey, L. R., Solounias, N., Sutherland, M. R., & Jungers, W. L. (2004). Can low-magnification stereomicroscopy reveal diet? *Journal of Human Evolution, 47*, 115–144.

Slott-Moller, R. (1990). La faune. In J. Jaubert, M. Lorblanchet, H. Laville, R. Slott-Moller, A. Turq, & J.-Ph. Brugal (Eds.), *Les chasseurs d'Aurochs de la Borde : un site du Paléolithique Moyen (Livernon, Lot)* (pp. 33–68). Documents d'Archéologie Française 27. Paris: Maison des Sciences de l'Homme.

Solounias, N., & Semprebon, G. M. (2002). Advances in the reconstruction of ungulate ecomorphology with application to early fossil equids. *American Museum Novitates, 3366*, 1–49.

Solounias, N., Mihlbachler, M. C., Rivals, F., Blondel, C., & Guthrie, D. (2004). Evaluation of diet in mammoths and other Late Pleistocene and Early Holocene ungulates from Alaska. *Journal of Vertebrate Paleontology, 24*, 116A.

Stringer, C. B., Finlayson, J. C., Barton, R. N. E., Fernández-Jalvo, Y., Cáceres, I., Sabin, R. C., et al. (2008). Neanderthal exploitation of marine mammals in Gibraltar. *Proceedings of the National Academy of Sciences, 105*, 14319–14324.

Valensi, P., & Abbassi, M. (1998). Reconstitution des paléoenvironnements quaternaires par l'utilisation de diverses méthodes sur une communauté de Mammifères. Application à la grotte du Lazaret. *Quaternaire, 9*, 291–302.

Vallverdú, J. & Courty, M.-A. (2012). Microstratigraphic analysis of level J deposits. A dual paleoenvironmental-paleoethnographic contribution to the human archaic palaeolithic archaeology at the Abric Romani rock-shelter.

Vaquero, M., Vallverdú, J., Rosell, J., Pastó, I., & Allué, E. (2001). Neandertal behavior at the Middle Palaeolitic site of Abric Romaní Capellades, Spain. *Journal of Field Archaeology, 28*, 93–114.

Vaquero, M, Chacón, G., Cuartero, F., García-Antón, M. D., Gómez De Soler, B., & Martínez, K. (2012). The lithic assemblage of level J. In E. Carbonell (Ed.), *High Resolution Archaeology and Neanderthal Behavior: Time and Space in Level J of Abric Romaní (Capellades, Spain)* (pp.). Dordretch: Springer.

Chapter 6
Taphonomy of Level J of Abric Romaní

Isabel Cáceres, Maria Bennàsar, Rosa Huguet, Jordi Rosell, Palmira Saladié, Ethel Allué, Alex Solé, Ruth Blasco, Gerard Campeny, Montserrat Esteban-Nadal, Cristina Fernández-Laso, Maria Joana Gabucio, Nuria Ibáñez, Patricia Martín, Laura Muñoz, and Antonio Rodríguez-Hidalgo

Abstract The taphonomic study of level J (sublevels Ja and Jb) includes macrovertebrates, microvertebrates, wood and charcoal. The association of level J has, in general, a good state of conservation, although there are several taphonomic modifications. No mechanisms or taphonomic processes which may have distorted any of these associations have been detected. This study shows how Neanderthals were the main collecting agent of animals and wood and the main agent of modification. Moreover, the water activity (water flow, dry periods and reactivitation) were the main postdepositional agent. The burial of the remains was probably relatively fast in both sublevels. There are abundant similarities in the temporary sequences of formation in both sublevels. The fossilization process of level J is consistent with a karst taphosystem in a travertine deposit. The main feature of this taphosystem is a microenvironment characterized by a relative humid environment, marked by wet conditions and reactivations of water flows. Once the remains were buried, the fossildiagenetic phase was tremendously preservative and allowed the exceptional conservation of the remains found at the site.

Keywords Taphonomy • Charcoal • Cutmarks • Toothmarks • Post-depositional processes

Introduction

Taphonomy is the discipline that studies the process of the formation and fossilization of fossil deposits. The term was coined by the Russian paleontologist Efremov in 1940, and is the combination of two Greek terms: *taphos* (burial) and *nomos* (laws). So, etymologically it means the "laws of burial". This discipline was founded with the goal of globally explaining what happens from the time an animal dies until it is buried (Biostratinomy, Weigelt 1927), and then what happens after that burial (Fossildiagenesis, Müller 1963). Efremov (1953) defines taphonomy as the

I. Cáceres (✉) · M. Bennàsar · R. Huguet · J. Rosell · P. Saladié · E. Allué · A. Solé · R. Blasco · G. Campeny · M. Esteban-Nadal · C. Fernández-Laso · M. J. Gabucio · N. Ibáñez · P. Martín · L. Muñoz · A. Rodríguez-Hidalgo
Institut Català de Paleoecologia Humana i Evolució Social (IPHES), Universitat Rovira i Virgili (URV), C/Escorxador s/n, 43003, Tarragona, Spain
e-mail: icaceres@iphes.cat

M. Bennàsar
e-mail: mlluc@prehistoria.urv.cat

R. Huguet
e-mail: rhuguet@iphes.cat

J. Rosell
e-mail: jordi.rosell@urv.cat

P. Saladié
e-mail: psaladie@iphes.cat

E. Allué
e-mail: eallue@iphes.cat

A. Solé
e-mail: asole@prehistoria.urv.es

R. Blasco
e-mail: rblasco@iphes.cat

G. Campeny
e-mail: gcampeny@iphes.cat

M. Esteban-Nadal
e-mail: mesteban@iphes.cat

C. Fernández-Laso
e-mail: cfernan@prehistoria.urv.cat

M. J. Gabucio
e-mail: mjgabucio@iphes.cat

N. Ibáñez
e-mail: nibanez@iphes.cat

P. Martín
e-mail: patricia.martin@estudiants.urv.cat

L. Muñoz
e-mail: lencinar@prehistoria.urv.cat

A. Rodríguez-Hidalgo
e-mail: arodriguez@iphes.cat

E. Carbonell i Roura (ed.), *High Resolution Archaeology and Neanderthal Behavior: Time and Space in Level J of Abric Romaní (Capellades, Spain)*, Vertebrate Paleobiology and Paleoanthropology,
DOI: 10.1007/978-94-007-3922-2_6, © Springer Science+Business Media B.V. 2012

study of the transition of biological entities from the biosphere to the lithosphere.

The general features of Abric Romaní have allowed the conservation not only of fossils belonging to animals, but also of structures, positives and negatives of wood, leaves, hearths, charcoal, etc. The taphonomic study of level J has a markedly multidisciplinary character and has been carried out jointly with zooarchaeological, paleoecological, and archeobotanical analyses, the corresponding chapters of which we will be referring to in the following pages. With this multidisciplinary taphonomic study, we attempt to describe all the alterations observed in the different elements preserved at the site. Most of the taphonomic data come from macromammal remains, although some such modifications and taphonomic processes were also found in micromammals, charcoal, and negatives of wood.

Vertebrate taphonomy has long been applied in archeo-paleontological sites. However, other types of remains have not often been studied from a taphonomic perspective. Nevertheless, taphonomic studies have been undertaken in recent years that aim to examine other preserved elements such as charcoals (Loreau 1994; Chabal et al. 1999; Allué 2002), phytoliths (Albert et al. 2006; Albert et al. 2008) and pollens (Fernández-Jalvo et al. 1999; Carrión 2002).

The fossil associations of level J have a markedly anthropic character. The presence of hearths, the abundance of lithic tools, the evident spatial distribution of the materials, and the abundant modifications directed toward obtaining animal nutrients are testimony to that anthropic character. However, hominids are not the only agents involved in the formation of level J. The action of other agents and taphonomic processes have also been discovered (e.g. carnivores, water activity, vegetation activity, etc.).

This study aims to assess the extent of the damage caused by the identified agents and raises the possibility that the action of carnivores and water may have resulted in taphonomic removals, giving rise to a taphonomic mixture of different associations. We also suggest the possible use of hearths as sanitary tools for the elimination of refuse. Moreover, the existence of multiple hearths linked to the Neanderthals' differential use of space may have been a determining factor in the subsequent intervention of other agents and taphonomic processes.

With this multidisciplinary taphonomic study, we attempt to understand the taphonomic processes (both biostratinomic and fossildiagenetic) that have affected the fossil associations in level J. Obtaining a temporal sequence of the intervention of the agents, mechanisms, and taphonomic processes is one of our main objectives. Another goal is to compare sublevels Ja and Jb, and to identify possible differences or similarities in the formation of these two associations. The paleoecological interpretations reveal the local and regional environment in which the Neanderthals carried out their activities. All of these features will contribute to the characterization of the karstic taphosystem in travertine settings.

Table 6.1 Type and number of remains recovered in sublevels Ja and Jb

Remains	Ja	Jb	Total
Macromammals	6738	1722	8460
Micromammals	114	89	203
Woods	21	1	22
Charcoals	686	321	1007
Total	7559	2133	9692

The following sections describe the results of our study of the modifications to wood, charcoal, and faunal remains. Each section contains a brief introduction to the discipline, the method applied, the results and a short discussion for each type of record (wood, charcoal and faunal remains). Finally, we present the taphonomic sequence formation for level J using all the data provided from all the different types of remains.

The materials included in the taphonomic study consist of 9692 remains (Table 6.1). Most of the remains are faunal, although plant remains are present to a lesser degree. Sublevel Ja provided a much greater number of remains than sublevel Jb, up to double or triple the amount. The taxonomy data for micromammals, charcoal, and wood appear in Allué et al. (2012a, b).

Wood Taphonomy

From a palaeontological perspective, plant taphonomy is especially concerned with the formation processes of the assemblage and is mainly interested in establishing the value of these processes in paleoecological reconstruction (Spicer 1991; Martín-Closas and Gómez 2004). In the case of wood, taphonomic studies are confined to catastrophic phenomena or regional scale studies (e.g. Fritz and Harrison 1985; Gastaldo 2004; Gastaldo and Degges 2007). As a consequence, wood taphonomy studies have usually been considered of low significance on an archeological scale. Wood taphonomy, when applied at archeological sites, is mainly interested in identifying, forming, and modifying agents, trying to advance the knowledge of woodland management. Boyd (1988) considers wood assemblage formation processes in archeological sites as complex systems allowing for variable influence of human and natural processes.

In plant taphonomy, the death of an individual or the disarticulation of its parts generates necrobiotic processes (Martín-Closas and Gómez 2004). It is caused by the concurrence of various biotic and abiotic agents. Disarticulation can be physiological or traumatic. The former may vary, depending on the species and climate, while traumatic disarticulation usually occurs as a result of weather phenomena and especially affects external elements. Prehistoric man, in a taphonomic sense, can be considered a biotic agent responsible for the individual death and traumatic disarticulation of woody plants.

Woody plants might be buried in situ in living position or undergo some sort of transport such as direct gravity, falling, wind transportation, and water transportation (Alexander et al. 1999; Martín-Closas and Gómez 2004). As a consequence of transport, woody material suffers mechanical alterations such as erosion and polishing with differing degradation of tissues depending on the hardness of the material, creating rounded, pointed and angular shapes (Postigo 2003). After a period of transportation a reassociation of woody debris may occur. The character of the profile of the assemblage depends on the transportation media, intensity, and environmental deposition and can be used to infer depositional causes and paleoenvironmental conditions (Alexander et al. 1999).

The potential for fossilization increases if woody parts rapidly enter a water system, where anaerobic conditions and high sedimentation rates favor the preservation of wood (Martín-Closas and Gómez 2004). Transport agents may be responsible for the presence of naturally produced wood debris and/or the disturbance of the anthropic patterns of the abandonment of wood in archeological contexts.

Wood in natural environments is rapidly colonized by organisms. The most significant agents in the biological alteration of wood are fungi, bacteria, and insects. Generally, the degree of colonization depends on environmental conditions. Fungi attacks are based on their capacity to decompose cell walls causing macroscopic cracks, especially in high moisture environments (Unger et al. 2001). Bacteria can go to work in inverse conditions, compared to those needed for other biological agents; they often attack wood in anoxic environments. The presence of bacteria contributes to fungi alteration, promoting the porosity and permeability of wood (Blanchette et al. 1990). Insects are a characteristic agent of wood biodegradation. There are many species of xylophagous insects specializing in different types of wood parts and different phenological and environmental conditions. Their attack is generally based on the construction of galleries, occasionally establishing a symbiotic relationship with the fungi.

Biological damage increases in combination with certain abiotic agents, mainly moisture, which also implies physicochemical variations. The weathering of wood (Feist 1990) is produced by the combination of chemical, mechanical, and light energies. The effects of weathering are grain raises and grain loosening, as well as the production of checks and larger cracks, progressively affecting the wood from the outside in.

Table 6.2 Wood objects recovered in sublevels Ja and Jb

Ja		Jb		Total	
WI	BW	WI	BW	WI	BW
17	4	1	–	18	4
21		1		22	

WI Wood imprint; *BW* Burned Wood

The diagenesis of travertine formations includes dissolution, recrystallization, microbial micritisation, bioturbation, cementation, compaction, oxidation of organic matter, and the formation of authigenic materials (Tucker and Wright 1990) as well as the postdepositional effects on morphology (Pentecost 2005) causing the reworking of the materials encrusted in the form of clasts.

There is not a very significant presence of wood in level J and, in general, the recovered elements are in a state of deficient preservation (Vaquero 1997). In most cases, the objects could not be recovered although 22 elements were documented (Table 6.2 and Fig. 6.1).

These objects were recovered in the exterior area of the shelter, on the spatial boundaries of the Neanderthal occupation. This boundary is in line with the outside blocks and with the area where a curtain of water falls, originating from the cornice of the shelter.

On average, the sediment obtained from Abric Romaní favors the travertine bio-construction of vegetable macroremains as a result of carbonated calcified precipitation on vegetable structures. Examples of such structures are seaweed, moss, stems and leaves of plants, pinecones, and trunks in the "living" position, as well as disarticulated woody elements. The travertine formation process is related to water activation in the shelter; in parallel with the decomposition of the nuclei, the external part is covered and "consolidated" by the precipitation of crystals (Castro-Curel and Carbonell 1995). Then, when the hydric current stops, the wood dries out and eventually disappears leaving behind as negative evidence the space the wood occupied and its impression in the travertine.

The levels in which impressions appear are characterized by sedimentation produced by low energy water flow. They are preserved in specific places within the shelter, which seem to be more affected by these contributions. The travertine platforms are formed from layers of water which often cover the occupied levels, without significantly altering them. This also favors the preservation of large fragments of totally carbonized wood, which can appear without the adaptation of the travertine.

The mechanism which allows for the fossilization of wood in the form of travertine incrustations is the rapid precipitation of calcium carbonate (Fig. 6.2). The vegetation benefits from the fall of CO_2 and the calcium trapped in the water particles of the biochemical precipitation and eventually serves as a substratum for physiochemical

Fig. 6.1 Wood negative of forked branch with the presence of xylophages galleries, in this case formed on the surface of the wood, immediately beneath the bark

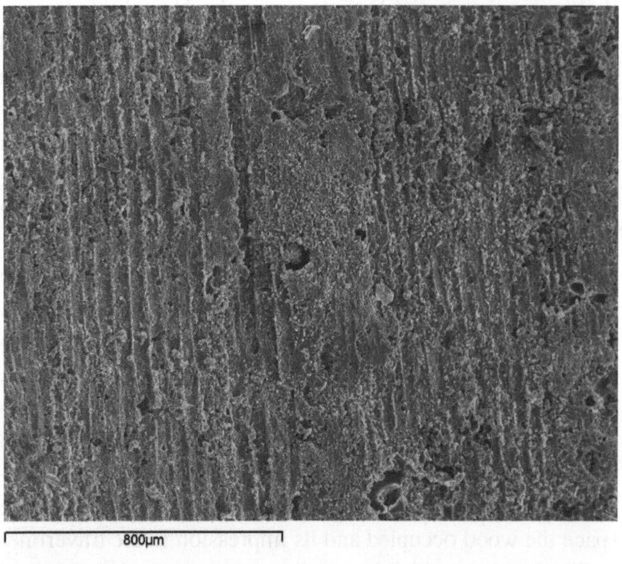

Fig. 6.2 SEM image of silicone cast positivation of wood travertine imprint. Fine-grained calcite precipitation over cellular structure allows histological observation

precipitation. The precipitation of calcium carbonate requires that the water have favorable concentrations of O_2 and Ca^{++} (Pentecost 2005). The representation of these histological structures of wood essentially depends on the size of the crystals. Individual grains of calcium can vary significantly in size—from submicroscopic, (<0.2 μm), to macroscopic, (>10 mm)—while aragonite normally appears in the form of microscopic acicular crystals (Pentecost 2005). In general, the microfacies of cementation in bioconstructions related to pseudomorphs are characterized by the presence of calcitic stromatoliths formed from seaweed associated with mosses and some tubular structures (stems) (Carbonell 2002).

In short, the presence of wood negatives is due to postdepositional processes related to a low energy hydric regime. We were not able to determine the extent to which the wood was altered during the biostratinomic phase or when the fossildiagenetic phase began.

Charcoal Taphonomy

Wood or charcoal taphonomic analyses rely heavily on the anatomical changes of wood cell structure and have been applied from different perspectives. The majority of such perspectives are related to anatomical changes during combustion and aim to deepen taxonomical identification (Rossen and Olson 1985; Prior and Gasson 1993). These analyses may also be used to recognize the properties of wood with industrial objectives (McGinnes et al. 1971; Beall et al. 1974; Juneja 1975; Slocum et al. 1978; Cutter et al. 1980; Baileys and Blankenhorn 1982). Finally, the implementation of taphonomy from an archeological perspective has been treated in relation to fuel management (Fiorentino 1995; Théry 1993; Théry-Parisot 2001; Allué 2002) as well as to methodological objectives related to fragmentation and mass reduction (Loreau 1994; Chabal et al. 1999).

All the analyzed charcoal corresponds to conifers, with *Pinus sylvestris* being the main taxa (Allué et al. 2012a, b) and was found in association with hearths. This has facilitated the task of comparing alterations of the same cell

Fig. 6.3 View of a charcoal transversal section showing vitrification of the cell structure (**a**) *versus* unaltered cell structure (**b**)

structure pattern. The alteration of wood cell structure is always present because combustion processes cause shrinkage, breakage and mass reduction. The latter has not been quantified or analyzed precisely, although there are studies indicating changes related to shrinkage, mass reduction, and fragmentation (Rossen and Olson 1985; Chabal et al. 1999).

The taphonomic study is based on 686 fragments for sublevel Ja and 321 fragments for sublevel Jb. Sublevel Ja contained 45% altered charcoal, whereas sublevel Jb had 78.8% altered charcoal. High percentages of undetermined alterations were also found in both sublevels. In sublevel Ja transversal section cell structure modification was found in 24.3% of the wood examined, and in Jb 43.2%.

Cell modifications that occur during the growth of the plant comprise knots, compressions, and other alterations in the growth rings. The only modifications related to stress (Schweingruber 1983) are compressions on the tracheids, which occur mainly on branches. There are very few of these considering that in general the forest and wood did not undergo any significant stress.

Cell collapse is an alteration in which the cells in late wood collapse for reasons unknown (Fig. 6.3). This alteration has been related to pressure during growth or collapse during combustion that affects only late wood because its cells are more fragile. It has been identified in a few fragments and only quantified in sublevel Jb, at 21.4%.

Most of the alterations are related to combustion and postdepositional processes. Fissures and vitrification are the main alterations related to combustion. Fissures can occur during the growth of the tree but they are particularly marked during combustion. When water and other gases spill out during combustion, cells break up where they are most fragile, producing fissures. These fragile areas can be the anatomical structures of the wood itself (pores, rays, etc.) or the areas affected by microorganisms. Vitrification is a specific alteration of charcoal related to high temperatures and combustion in closed places (Marguerie and Hunot 2007). This alteration was found in 15% of the material in sublevel Jb, and in lesser amounts in sublevel Ja (Table 6.3).

Table 6.3 Main taphonomic modifications observed at charcoal of sublevels Ja and Jb

Processes	Agents	Mechanisms	Alteration	Archeol. levels	
				Ja (%)	Jb (%)
Growth of the tree	Natural/anthropic	Environmental and biogeographic conditions (wind, storm...)	Knots	1.6	2.5
			Alteration on growth rings	1.4	3.0
			Tracheid compression	18.1	10
Decay	Natural	Environmental conditions, decay micro organisms	Micro-organism	1.1	–
Combustion	Anthropic	Carbonization	Fissures	3.8	2.1
			Vitrification	1.9	15.7
			Deformation of cell structure	14.0	2.5
Postdepositional	Natural	Water	Total deformation of cell structure	8.4	18.2

As for postdepositional alterations, there is one in particular related to water activity on charcoal. As mentioned above, a considerable percentage of CaCO3 in water causes its precipitation to act on charcoal as well (Fig. 6.4). This completely deforms the cell structure and prevents identification. In sublevel Ja 8.4% of the studied specimens bear this type of alteration, whereas in sublevel Jb this was only found in 18.2% of the specimens here analyzed.

In general, the charcoal in the level is well preserved as a result of deposit type. The alterations suggest that the wood used for fuel was in good condition due to the low percentage of alterations related to wood decay. Postdepositional processes largely affected sublevel Jb, where water seems to have been the most significant cause of cell alteration.

Vertebrate Taphonomy

The recovered remains of macromammals (8,460) consist of fragments of bones, teeth and antlers (Table 6.4 and Figs. 6.5, 6.6). Sublevel Ja provided the greatest number of remains and in both sublevels bone remains were the most abundant type of remains found. Isolated teeth are more abundant in sublevel Jb (11.09%) and in both levels the fragments of antler are incidental. The taphonomic study was carried out jointly with the zooarcheological study. For this reason, all the data referring to the anatomical and taxonomic identification and the allocation of sex and age as well as other data of paleoeconomic interest can be found in Rosell et al. (2012).

This taphonomic study was carried out on all the recovered faunal remains from the two sublevels. The recovering process of small mammals begins with the collection of sediment, according to the square (X and Y) and depth (Z). The sediments are washed and sieved with water. Later, when the sediment is dry the sorting process begins, which involves separating the archeological and paleontological remains. The small mammal taphonomy interest centers around knowing the collector agent, thereby being able to identify the predator and therefore, carry out the corresponding paleoecological interpretations (Andrews 1983, 1990; Andrews and Nesbit-Evans 1983; Kowalski 1990; Fernández-Jalvo and Andrews 1992; Denys et al. 1995). The remains of micromammals are also subjected to the same mechanisms, processes and taphonomic agents as the macromammal fossils, meaning that they also contribute both biostratinomic and fossildiagenetic information.

Though a preliminary macroscopic examination was conducted, all surfaces were analyzed using stereomicroscope (Olympus EUROPE SZ11 × 110 magnification) and, when necessary, a SEM, (JEOL-6400) and an ESEM (FEI-QUANTA 600). The modifications identified were recorded in a database which includes information on the location of each alteration on the surface, its organization, as well as its arrangement. All the modifications that were possible to measure were measured, noting length, width, and/or depth depending on the alteration. Special attention was paid to the relationship between alterations as well as to the superimposition of modifications.

A study of the spatial distribution of the identified alterations in both sublevels was conducted using Cartesian coordinates (X, Y and Z), orientation and the slope taken during the excavation work.

The fossils of macromammals found in level J are characterized by a considerable degree of fragmentation. This fragmentation is reflected by the fact that only 27 whole skeletal elements were found in sublevel Ja and only 3 in sublevel Jb, which means that 99.57% of the bone remains of sublevel Ja and 99.8% of those in Jb were discovered broken. The high degree of fragmentation of the sample is supported by the circumstances of the elements that appear whole, which are articular bones (carpals, tarsals and sesamoids), phalanxes, and some rabbit bones, all of small dimensions. In both sublevels more than 80% of the remains are less than 5 cm in length.

Fig. 6.4 Macroscopic view of a charcoal fragment affected by CaCO3 deposit

Table 6.4 Fossils recovered in level J grouped by type of remain and sublevel

	Ja		*Jb*	
	NR	%	*NR*	%
Bone	6292	93.38	1530	88.85
Teeth	438	6.50	191	11.09
Antler	8	0.12	1	0.06
Total	6738		1722	

The agents and taphonomic processes that have intervened in the formation of these sublevels led to this high fragmentation rate. As we mentioned earlier, these levels present a high degree of anthropization, which points to the possibility that most of the fragmentation was caused by humans during faunal exploitation (Rosell et al. 2012), mainly while attempting to reach marrow and during fire use in the different stages of the faunal exploitation chain. Nevertheless, we should consider the occurrence of other agents and processes, which have also contributed to the fragmentation of the fossils. The identified taphonomic modifications in the fossils recovered in sublevels Ja and Jb are human damage, carnivore damage, rounding, polishing, trampling, cracking, dissolution, root etching, carbonization, and manganese oxide (Fig. 6.7).

The presence of the modifications varies in each of the sublevels. For example, in sublevel Ja there are modifications related to anthropic activity (41.97%), which is the most common modification type documented, followed by traces left by the action of plants (27.37%). In this sublevel other types of modification are rare, never surpassing 3%. On the other hand, in sublevel Jb the most abundant modification is caused by plants (61.67%), followed by those brought about by anthropic activity, comprising 28.98% of the modifications. Other modification types are more abundant compared to those identified in sublevel Ja. In sublevel Jb, except for toothmarks, polishes, and trampling alterations, which do not even represent 1% of the alterations, the percentages of other modification types range between the 1.22% of manganese oxides and the 16.61% of chemical corrosion.

The anthropic modifications identified consist of cutmarks, evidence of induced fracture and burning (Table 6.5). In both sublevels these modifications are well represented, although the percentage is higher in sublevel Ja than in Jb. It is particularly noteworthy that while the cutmark values and evidence of fracture show similar values in both sublevels, burned remains are much more abundant in Ja than in Jb. It is important to emphasize that remains of burned micromammals were not recovered, which suggests that they were deposited in the shelter subsequent to the human occupancies.

Predator Activity

The intervention of different predators has been identified in sublevels Ja and Jb: birds of prey, which fed on micromammals; and carnivores, which fed on macromammals.

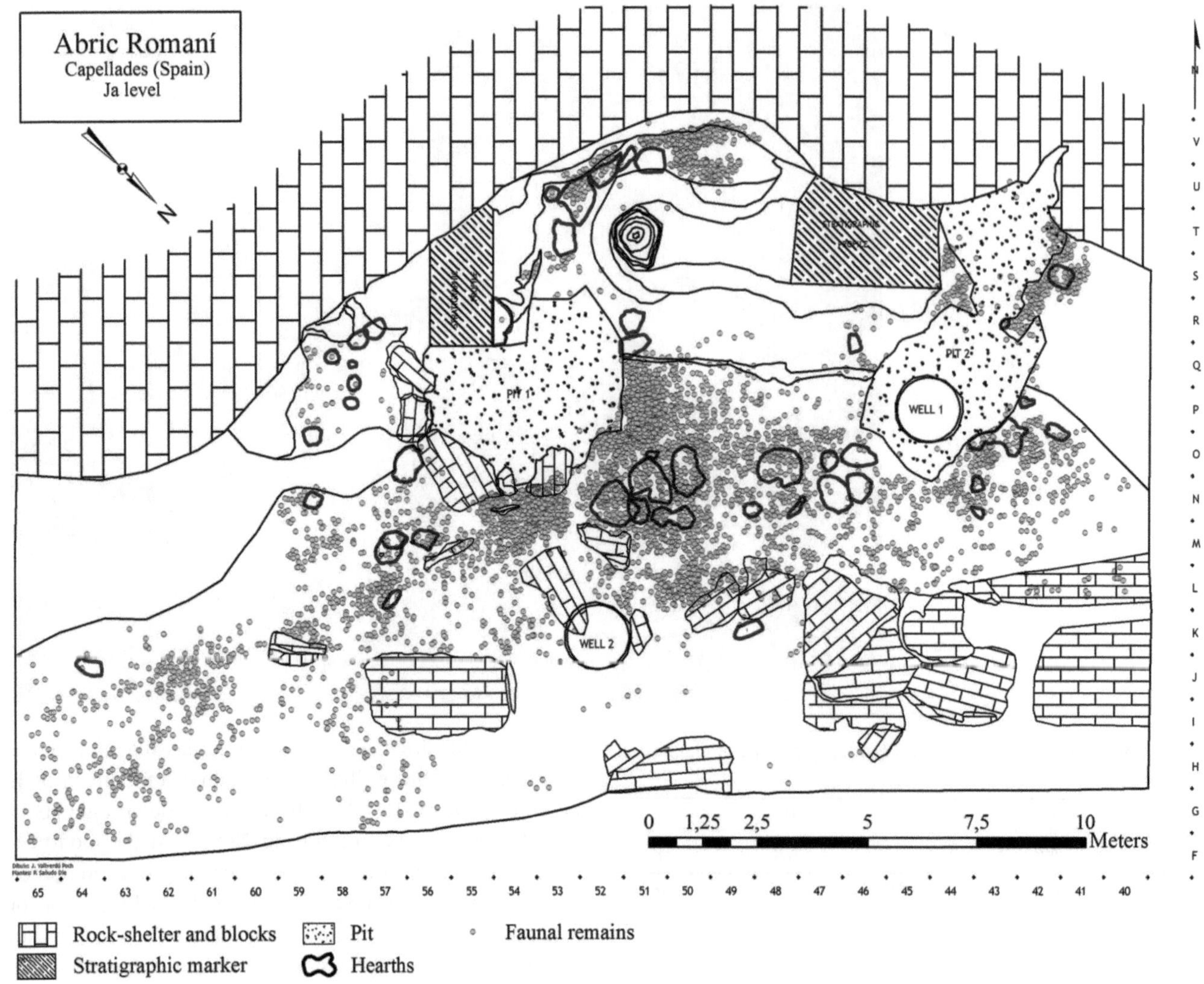

Fig. 6.5 General distribution of faunal remains in sublevel Ja

The modifications identified on the micromammal remains are described in Table 6.6. These data do not permit the identification of the particular species which caused the accumulation, although they do establish that it was a category 1 predator according to the classification described in Andrews (1990).

Carnivore activity is scarce in both sublevels. Its presence has been identified by the presence of toothmarks on the bones, the presence of some digested remains, crenulated edges, and ends affected by the abrasive action of saliva and the tongue during consumption. Coprolites have also been recovered, which is direct evidence of the presence of carnivores in the shelter.

Carnivores produce different types of toothmarks on bones upon trying to extract the soft tissues of the animal carcass and when fracturing bones to access their interior and consume the bone marrow. Various authors have identified different types of toothmarks, (Haynes 1980, 1983; Maguire et al. 1980; Binford 1981; Bunn 1981; Potts and Shipman 1981; Shipman 1981; Blumenschine 1988; Blumenschine and Selvaggio 1988; Blumenschine and Marean 1993; Capaldo and Blumenschine 1994; Stiner 1994). Basically, each type of mark is characteristic and presents its own morphology, however, depending on the number of marks, their location, and their purpose they can be grouped under specific nomenclature. The main identifying types in level J are scores and depressions.

Scores occur when carnivores tear muscle with their teeth. They are scratches of variable length, width, and depth that present a rough, coarse and irregular base, generally in a U-shape. They have a transversal delineation which crosses the longitudinal axis of the bones.

Depressions are caused by the pressure exercised by the molars that crush the surface of the bone. These can be circular or present a more oval morphology. From time to time, the cortical tissue remains sunken in the interior

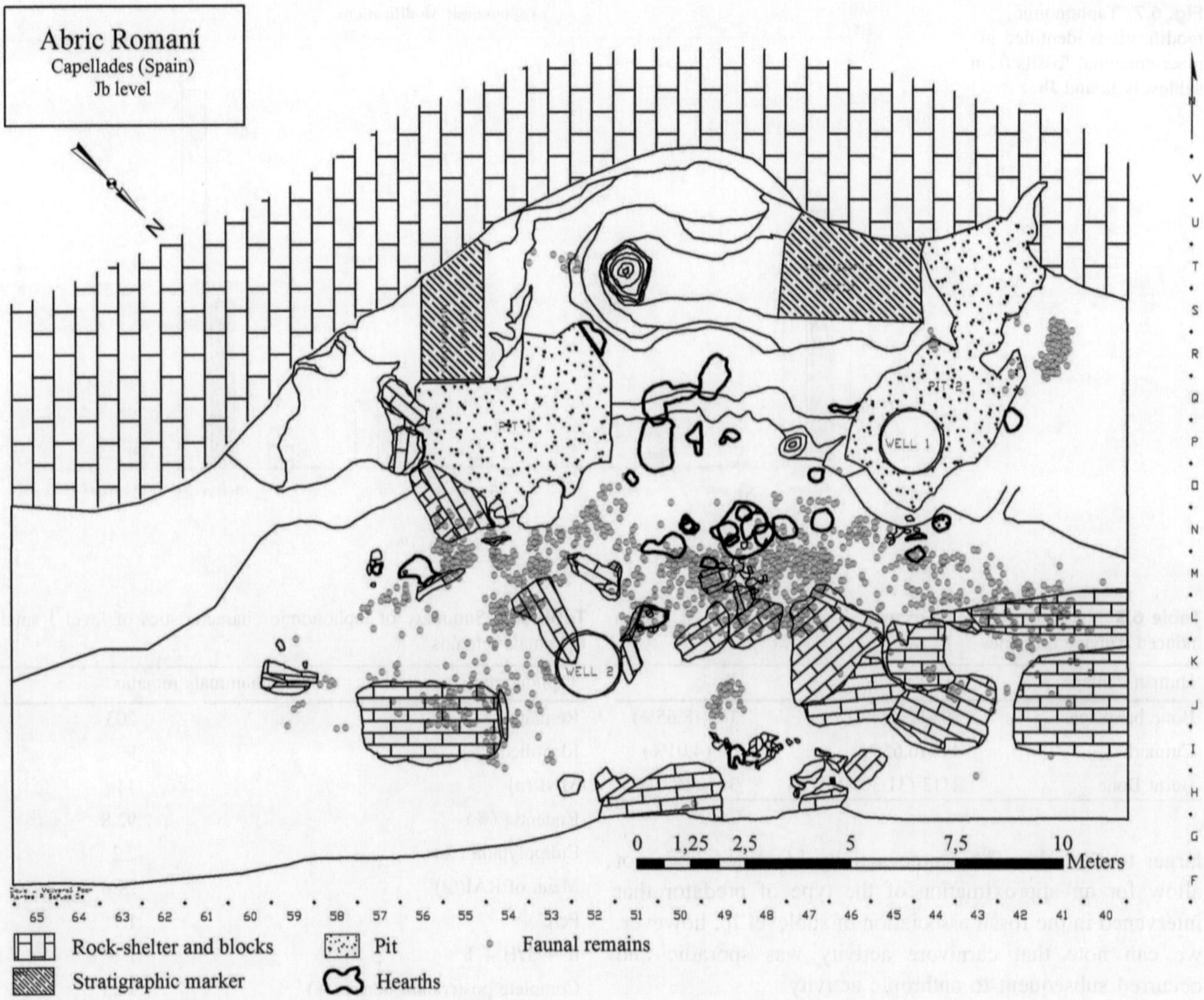

Fig. 6.6 General distribution of faunal remains in sublevel Jb

of the depression, while in other cases it is not preserved and displays an irregular base.

Modifications produced by salivary enzymes are identified by rounding and polishing in the areas that are bitten, and are normally located at the tips of the remains. Pitting is characterized by the presence of abundant small, shallow depressions that appear on the rounded extremes (Binford 1981). As a group, this alteration relates to salvation, biting and erosion caused by the tongue on the same area of the bone.

Most carnivores consume small fragments of bone while eating. Hyenas are the type of carnivore that does this the most frequently, and they are capable of consuming fragments of considerable dimensions. Digested remains present porous surfaces, smooth and small domes, caused by the corrosive action of acids. All the identified toothmarks were measured, except in the cases in which the mark was not well preserved and could therefore contribute to an erroneous measurement. With scores the minimum width is considered, while for depressions the longitudinal and the transversal dimensions are taken as reference. In this study, we apply the method established by Andrews and Fernández-Jalvo (1997) which, in addition to the type of toothmarks and its dimensions, considers the skeletal element and its location, (diaphysis, epiphysis, anatomical edge, fractural edge).

In sublevel Ja, actions of carnivores are present in 0.95% of the remains, while in sublevel Jb their activity does not exceed 0.2%. In sublevel Jb toothmarks have been identified in three fossils (one metacarpus and one medium-sized long bone and one large tibia) and some fragments showing evidence of digestion have also been recovered. The toothmarks are scores and depressions that were not always possible to measure. When measurement was possible, the scores measured 0.36 mm and the depressions around 7 mm. The depressions were found on an epiphysis fragment of a long bone, and it is important to remember that spongy tissues always offer

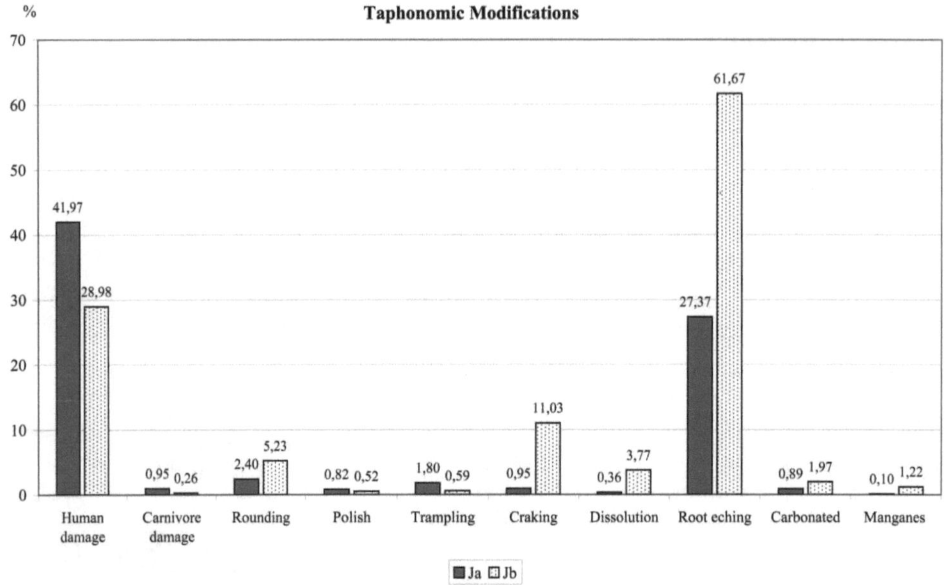

Fig. 6.7 Taphonomic modifications identified in macromammal fossils from sublevels Ja and Jb

Table 6.5 Number of remains recovered and percentage of human induced damage on bones

Human damage	Ja	Jb
Bone breakage	804-(11.93%)	149-(8.65%)
Cutmarks	448-(6.65%)	69-(4.01%)
Burnt Bone	2113-(31.36%)	317-(18.41%)

larger toothmarks. The scarce activity identified does not allow for an approximation of the type of predator that intervened in the fossil association of sublevel Jb, however, we can note that carnivore activity was sporadic and occurred subsequent to anthropic activity.

In sublevel Ja carnivore activity (0.95%) has been observed in remains of all sizes, although activity is most abundant in medium-sized remains, followed by large-sized remains (Table 6.7). Most of the toothmarks appear on appendicular skeletal elements (n = 51), while on the axial segment they appear on seven remains, and on the cranial on only two mandible fragments. One fragment of a carnivorous rib also presents toothmarks.

The toothmarks consist of scores (Fig. 6.8) and depressions. All the types of toothmarks established by Andrews and Fernández-Jalvo (1997) have been identified, except type E. The most abundant types are type B, followed by type A, while other types were present in lesser values (Table 6.8).

The dimensions of the bites are summarized in Table 6.9. These dimensions suggest the intervention of a medium-sized carnivore, although the existence of toothmarks greater than 4 mm indicates the presence of a carnivore of a greater size (hyena), which would be consistent with the recovered coprolites. The fragments that show evidence of carnivorous activity appear dispersed in the shelter without the presence of significant accumulations (Fig. 6.9). The low number of remains showing carnivorous intervention and the absence of accumulations produced by this taphonomic agent suggests that they are the result of occasional visits, probably related to a sporadic prowl strategy. The scarce evidence would indicate that the abovementioned visits would not have been fruitful to the animal involved.

Table 6.6 Summary of taphonomic characteristics of level J small mammals remains

Taphonomic characteristics of micromammals remains	
Remains (n)	203
Identified (%)	99
MNI (n)	14
Rodentia (%)	92.8
Eulipotyphla (%)	7.2
Mean of RAI(%)	29.6
Pc/C	1.7
R + T/H + F	1
Complete postcranial bones (%)	24.3
Complete skulls and maxilars (%)	40
Maxillary teeth loss (%)	50
Complete jaws (%)	10
Mandibular ramus loss (%)	100
Mandibular teeth loss (%)	45.8
Breakage of teeth (%)	65.5
Molar digestion (%)+	6/Light[a]
Incisor digestión (%)+	18.3/Light
Postcranial digestion (%)+	21.5/Moderate

N number, % percentage, *MNI* Minimum number of individuals, *RAI* Relative Abundance Index, *Pc/C* postcranial and cranial index, *R + T/H + F* Index of distal (Radius + Tibia) and proximal bones (Humerus + Femur), + maximum degree of digestion observed, [a] observed isolate remain with moderate digestion

Table 6.7 Skeletal elements from sublevel Ja with carnivore toothmarks grouped by macromammal size classes

Element	Large size	Medium size	Small size	Leporidae	Unidentified	Total
Mandible	1	1	–	–	–	2
Vertebrae	1	–	–	–	–	1
Ribs	–	5	1	–	–	6
Scapulae	–	–	–	1	–	1
Humeri	–	2	–	–	–	2
Radius	–	4	–	–	–	4
Femur	–	2	–	–	–	2
Tibia	1	4	–	1	–	6
Metapodial	1	1	–	–	–	2
Phalanx	–	1	–	–	–	1
Articular bone	–	1	–	–	–	1
Long bone	6	13	1	–	2	22
Flat bone	3	2	1	–	1	7
Unidentified	–	–	–	–	3	3
Total	13	36	3	2	6	60

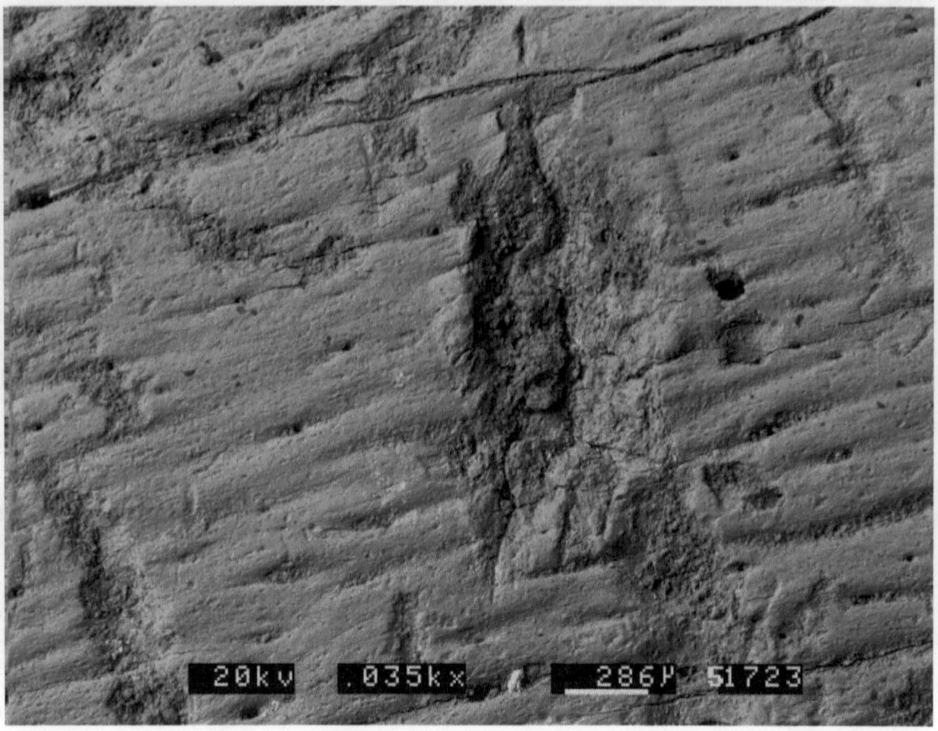

Fig. 6.8 Example of toothmarks (scores type B) in a medium-sized long bone from sublevel Ja

Root Etching and Dissolution

Plant roots are a biological agent whose damage mechanism is chemical. Root etching is characterized by branching scores with variable length, width and depth, more or less rounded and a U-shaped cross section (Behrensmeyer 1978; Morlan 1980; Johnson 1985; Hesse and Wapnish 1985; White 1992; Lyman 1994; Fisher 1995). This type of modification is caused by the dissolution of the bone surface through the acids secreted during the absorption of nutrients or by the action of fungi and bacteria associated with roots. Root etching generally occurs after the burial of the remains (fossildiagenetic stage) whereas mosses, lichens, and algae affect remains when they are on the surface (biostratinomic stage).

Modifications related to plant activity are very common in both sublevels of level J. This modification has been identified mainly as scores and perforations produced by the

Table 6.8 Skeletal elements chewed and type of toothmarks observed in Ja fossils

Element	A	B	C	D	E	F	G	H
Mandible	1	1	–	–	–	–	–	1
Vertebrae	–	–	1	–	–	–	–	–
Ribs	5	2	–	–	–	–	–	–
Scapulae	1	–	–	–	–	–	–	–
Humeri	–	–	2	–	–	–	–	–
Radius	1	2	–	–	–	1	–	1
Femur	–	–	1	1	–	–	–	–
Tibia	2	4	1	–	–	1	1	–
Metapodial	–	2	–	–	–	–	–	–
Phalanx	–	–	1	–	–	–	–	–
Articular Bone	1	–	–	–	–	–	–	–
Long Bone	6	10	2	3	–	–	–	–
Flat bone	3	2	–	–	–	–	–	1
Unidentified	–	–	–	3	–	–	–	–
Total	20	23	8	7	0	2	1	3

Types of toothmarks established by Andrews and Fernández-Jalvo (1997): Type A = Carnivore depressions on bone surface; Type B = Carnivore scores on bone surface; Type C = Carnivore toothmarks on articular surfaces; Type D = Carnivore toothmarks on spiral breaks; Type E = Carnivore toothmarks on transverse breaks; Type F = Carnivore toothmarks on split shafts; Type G = Molar puncture prints made by multicuspid teeth; Type H = Carnivore toothmarks on intact bone edges

Table 6.9 Type of toothmark and size rank identified in sublevel Ja

Type	Size
Type A	0.6–3.4 mm
Type B	0.5–2.5 mm
Type C	Depressions (0.8–5.2 mm); scores (1.4 mm)
Type D	Depressions (0.5–4.7 mm); scores (0.6–2.3 mm)
Type E	No documented
Type F	Depressions (5.1 mm); scores (0.4–1.68 mm)
Type G	No data
Type H	Depressions (0.7 mm); scores (1.13 mm)

See types defined in Table 6.7

biochemical action of plants. Occasionally, the scores are so abundant that they appear over the entire surface of the remains, giving them a rough and uneven appearance. The scores observed in the remains are shallow and there is no presence of reticulated or radial paths. Based on the characteristics of the site and previous works (Cáceres 2002), we might deduce that this modification occurred through the action of bryophyte plants. Along with leaves, stems and branches, mosses are one of the main plant components involved in the formation of travertine (Fig. 6.10).

Root etching is present in 27.37% of the macromammal fossils recovered in sublevel Ja and in 61.67% of those from Jb (Table 6.10). In sublevel Ja the proportions of modified bones (27.67%), teeth (23.06%), and antlers (25%) are similar. Meanwhile, in sublevel Jb, root etching is more commonly found on bone fragments (62.61%) than on dental tissues (19.9%). In contrast, in micromammals, plant activity is more marked in Ja (46.49%) than Jb (14.61%), and mainly affects bone tissues.

Plant activity does not show any selection related to the species, age or the skeletal elements of animals. However, a preference for unburned remains has been noted (Fig. 6.11). In sublevel Ja, 75% of the plant-altered remains do not present any type of thermal impact and 89% of the altered remains in sublevel Jb show no evidence of burning. Combining the number of unburned remains with the number of remains with incipient degrees of cremation (degrees 1 and 2) yields similar results for the two sublevels. The percentage of burned remains (degrees 3–5) altered by roots in sublevel Ja is 3.25% and in sublevel Jb 2.11%. We should emphasize that when the two modifications coincide, the action of the plants is always subsequent to that of fire.

The spatial distribution of plant activity in both sublevels indicates that it is more abundant in the central zone and outside the shelter than in the vicinity of the blocks located near the wall (Figs. 6.12, 6.13). The distribution of this modification coincides with the places with the greatest number of remains and also with the zones where there are a greater number of hearths. The fact that the remains modified by roots are mainly not burned suggests that fire might eliminate the mineral nutrients that the plants seek when they modify the remains, making root-altered burned remains scarce.

Another chemical modification found in the remains is dissolutions. Dissolutions are circular depressions and scores affecting large parts of the bone to convert it into true "bone karst" (Fig. 6.14). This modification appears exclusively in bone remains. The majority of the remains that present dissolution are altered by mosses. This phenomenon occurs in 91.66% of the cases in sublevel Ja and in 73.84% in Jb. This suggests that dissolution is heavily related to moss modification. A phenomenon that supports this hypothesis is the relationship between dissolution and fire. In sublevel Ja, none of the remains that present dissolution are burned, and in sublevel Jb 7.69% show both dissolution and cremation. It is important to stress that in both sublevels, the remains with dissolution that are burned mainly feature degree 1 cremation. There are no bones with this modification that have been burned to higher degrees.

This data lead us to suggest that dissolutions in the remains were most likely produced by moss activity. It is our opinion that these two modifications (root etching and dissolution) occurred during a single taphonomic process: bryophyte colonization in the shelter. These modifications should be understood as biostratinomic and related to water

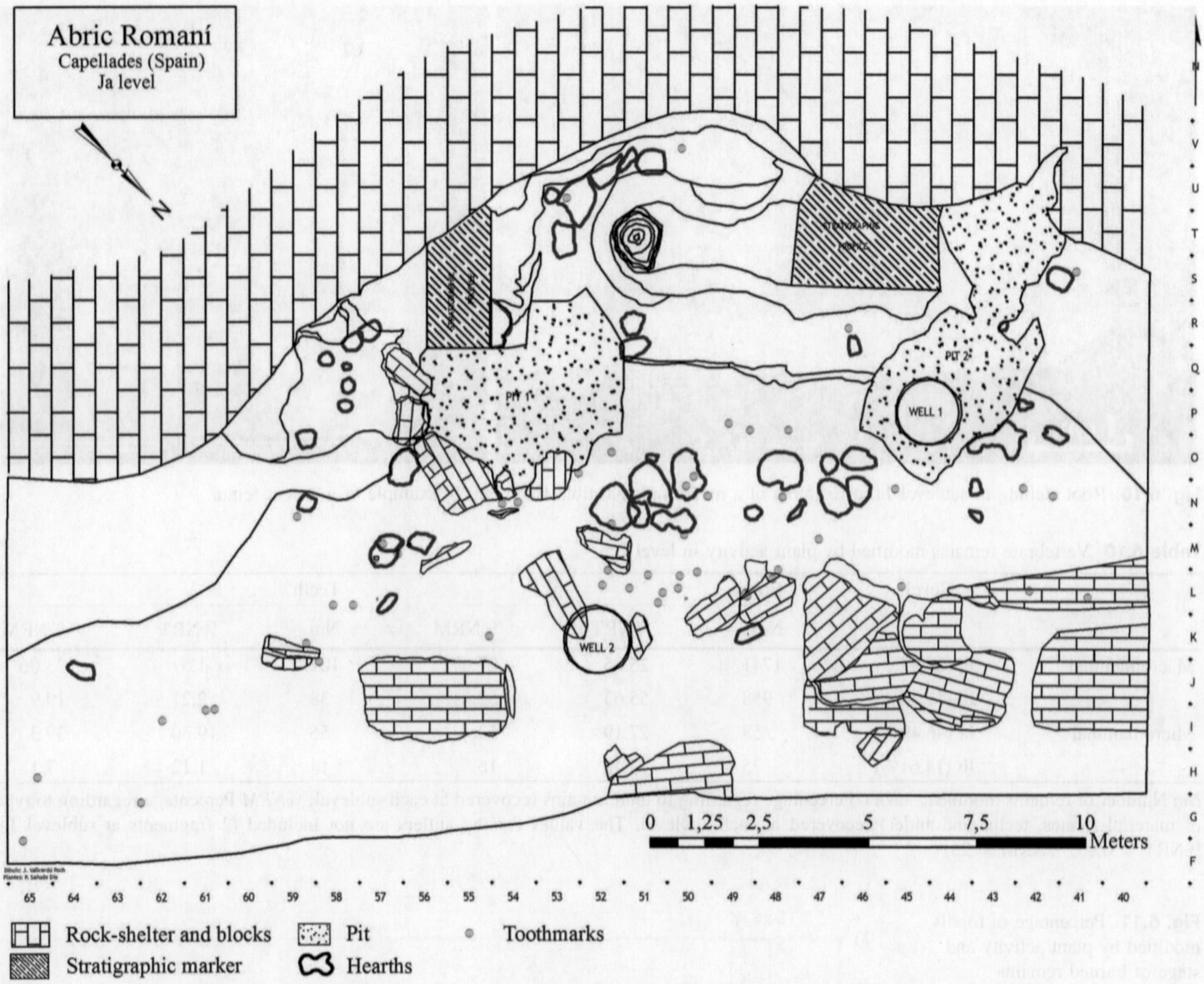

Fig. 6.9 Spatial distribution of bones chewed by carnivores in sublevel Ja

activity, a taphonomic agent that played a critical role in the formation of this deposit.

Water Activity

Water activity has been identified in level J through the presence of rounding and/or polishing on part or over the entire surface of the fossils recovered. These modifications are caused by the friction of sedimentary particle content in water. Therefore, the degree of modification depends on the energy of the current of water, of the type and size of the sediment, and of the state of the bone when modification process began (Fernández-Jalvo and Andrews 2003). The stages of modification indicated in Table 6.11 have been applied in the analysis of this modification (Cáceres 1996, 2002).

Rounding and polishing are not abundantly represented in either sublevel. In sublevel Ja, rounding affects 2.39% of the remains and polishing 0.82%. In sublevel Jb these values are 5.23% for rounding and 0.52% for polishing. Rounding is mainly present in the initial stage (R1) emphasizing that the degree of advanced alteration (R3) is symbolic in Ja and has not been identified in sublevel Jb. In relation to polishing, the most representative stage is P1. The P2 stage only appears in sublevel Ja and the P3 stage is absent in both sublevels. The relationship between these two alterations shows a predominant association of R1/P0, with 43.8% in Ja and 81.1% in Jb (Table 6.12 and Fig. 6.15). In the same way, the effects of water have also been identified in the remains of micromammals in both sublevels: in sublevel Ja, 19.3% of small mammals show R1 stage effects and 2.6% shows P1 stage effects. In sublevel Jb, only the R1 stage is represented (3.4%).

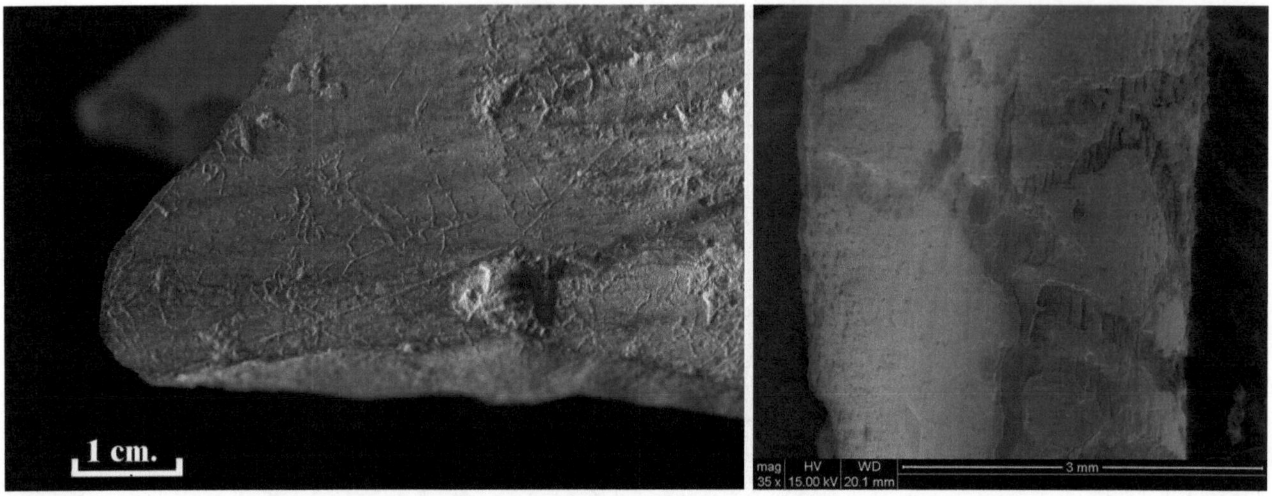

Fig. 6.10 Root etching in sublevel Ja. **a** Example of a macromammal tibia fragment; **b** Example of a rodent femur

Table 6.10 Vertebrate remains modified by plant activity in level J

	Sublevel	Bones			Teeth		
		Nm	%NRT	%NRM	Nm	%NRT	%NRM
Macromammal	Ja (27.37%)	1741	25.83	27.67	101	1.5	23.06
	Jb (57.84%)	958	55.63	62.61	38	2.21	19.9
Micromammal	Ja (46.49%)	58	27.19	53.4	56	19.30	39.3
	Jb (14.61%)	75	13.48	16	14	1.12	7.1

Nm Number of remains modified; *%NRT* Percentage regarding to total remains recovered at each sublevel; *%NRM* Percentage regarding to type of material (bones, teeths and antler) recovered at each sublevel. The values for the antlers are not included (2 fragments at sublevel Ja, %NRT = 0,03; %NRM = 25)

Fig. 6.11 Percentage of fossils modified by plant activity and stage of burned remains

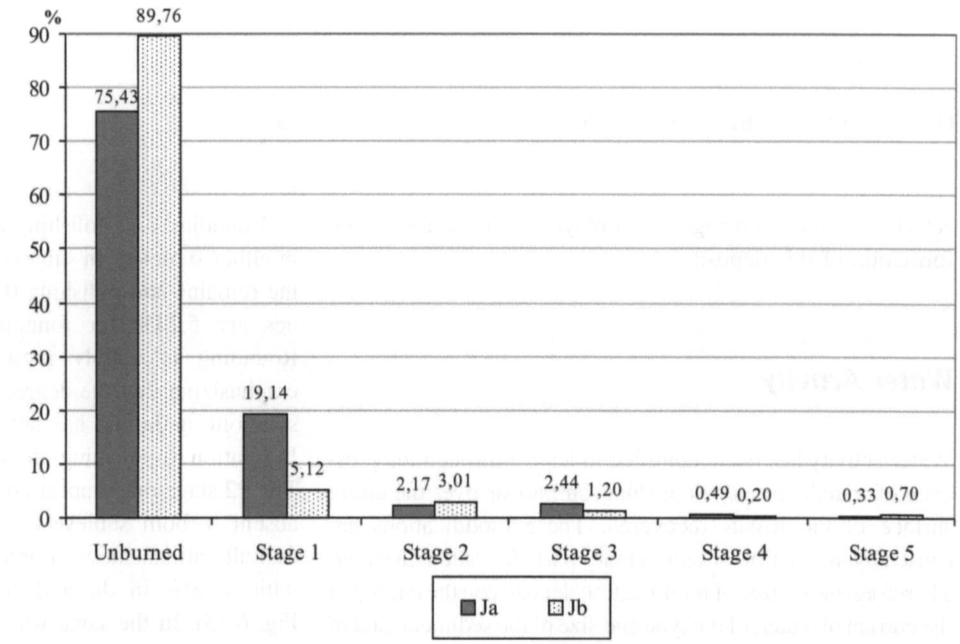

Water abrasion indiscriminately affects all skeletal elements and animals of every size. All the water-altered fossils recovered correspond to bone remains. This taphonomic process has not been documented in the remains of carnivores or rabbits. Remains with evidence of water abrasion present heterogeneous dimensions and in

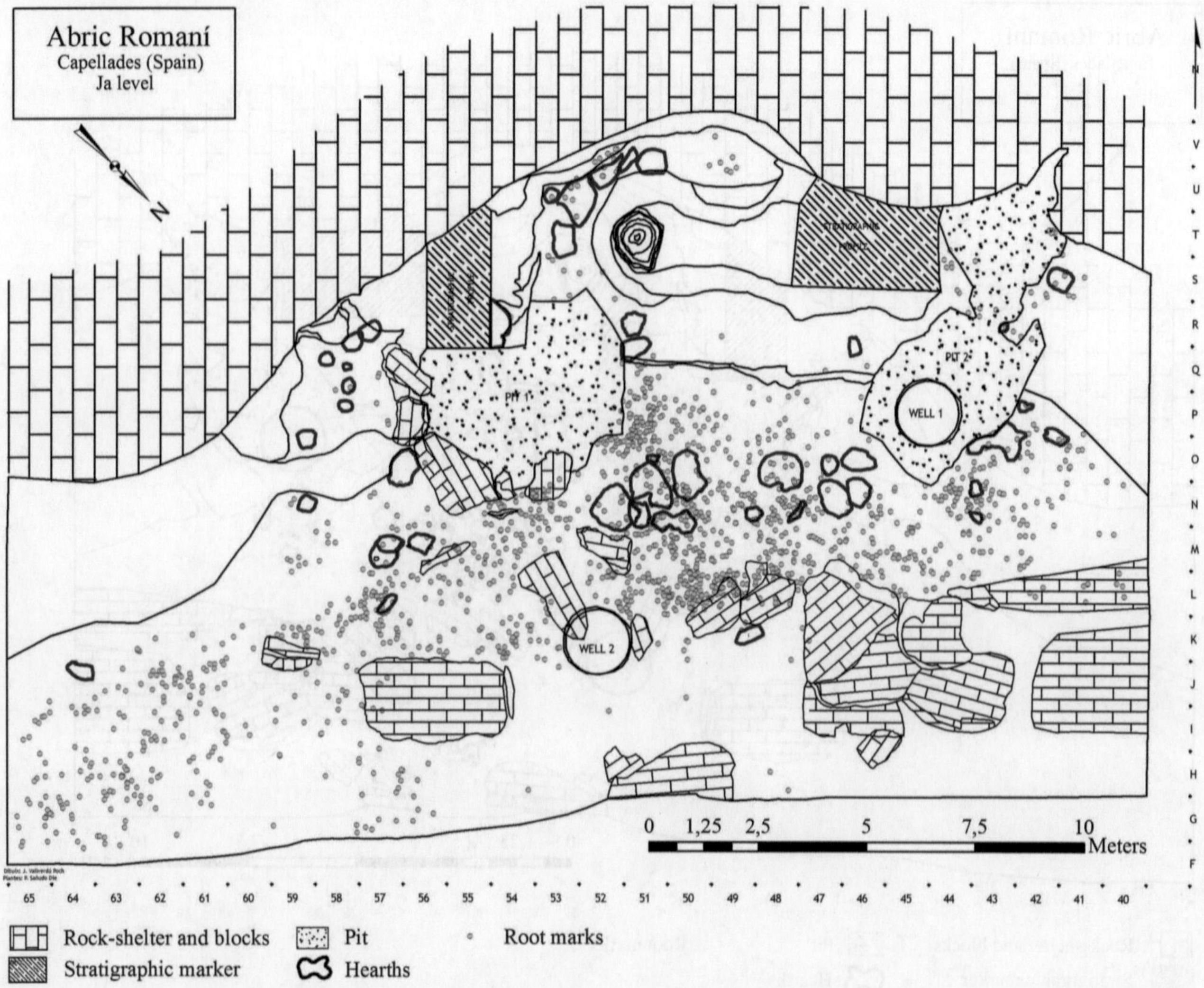

Fig. 6.12 General distribution of faunal remains with evidence of moss activity in sublevel Ja

sublevel Ja (Fig. 6.16) they are found in category B remains (2–5 cm) and category C remains (5–10 cm) with more than 56% of the remains exceeding 4 cm in length. Category B remains with this type of alteration were also found in sublevel Jb, which yielded a greater number of remains, although small-sized remains (A = <2 cms) in sublevel Jb are more abundant than in Ja.

In general, we rule out the possibility of selection by size with regard to the abraded remains recovered, since the dimension categories that contain the most altered remains are also those that dominate the fossil assemblage. This absence of size selection suggests that the water flows that caused these alterations did not have a strong current, which would have probably resulted in the smallest remains being dragged away. It is possible that some of the remains were the object of local resedimentation, since abrasion affects all the sides, which would indicate that the remains were in movement or that they changed position.

The remains of macrovertebrates in both sublevels with evidence of water abrasion are located in the zones where there are a greater number of deposited remains. In sublevel Ja, the remains mainly appear in the central zone and near the line of blocks (line K–O/40–42) and on the exterior of the shelter (G–L/55–65) (Fig. 6.17). In sublevel Jb, the remains are situated in lines M–O–R. No significant groupings have been observed that can be related to dragging by currents of water (Fig. 6.18).

In sublevels Ja and Jb, we found that abrasive processes affect few remains and are somewhat more numerous in sublevel Jb than Ja, and that they basically produce rounding and a light polishing of the anatomical edges and fractures. This modification is consistent with fine lime-type sediments. The abrasion was probably initiated when the state of the bones was semi-dry, because if the size of the sediment was greater or the state of the bone were weathered or fossilized, the degree of abrasion would be

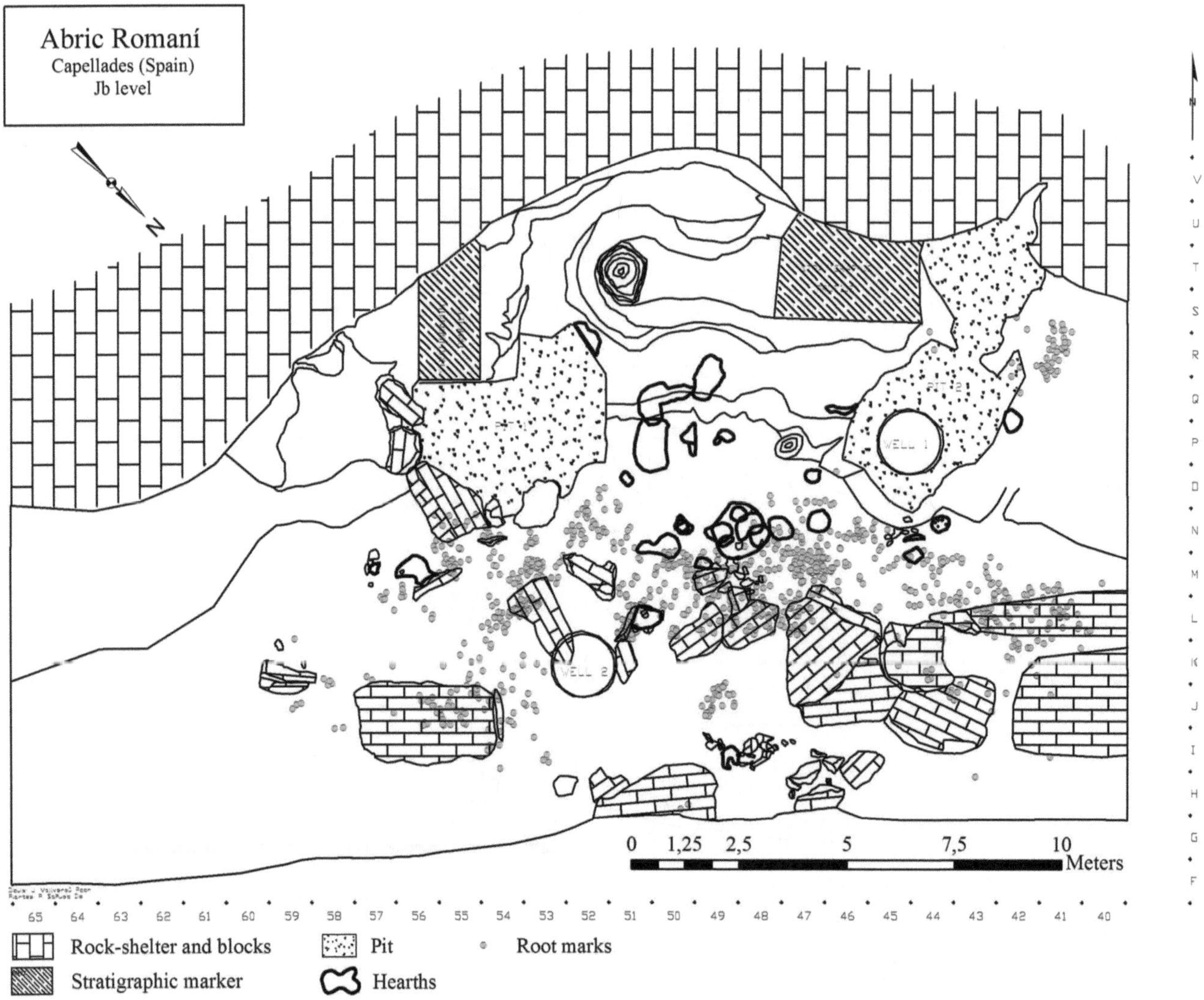

Fig. 6.13 General distribution of faunal remains with evidence of moss activity in sublevel Jb

greater (Fernández-Jalvo and Andrews 2003). The fact that the remains were not completely dry at the time of abrasion is supported by the number of burned remains that appear with abrasions. In sublevel Ja, only one burned remain is abraded to an incipient degree, representing 0.01% of the remains, reaching 0.3% in sublevel Jb. Because fire dries remains, we can assume that at the time at which abrasion occurred the remains were not completely dry, since only remains that had been burned (and were therefore dry) were not abraded.

Water abrasion may have occurred before or simultaneously to moss colonization. 71.6% of the remains showing abrasion in Ja and 77.8% of those in Jb also present alteration by moss. The relationship of these two modifications with chemical corrosion and dissolution is more evident in sublevel Jb, where more than 82% of the abraded remains that show alteration by moss also have corrosion and dissolution. In this sublevel it is possible that the development of the abrasive processes occurred at the same time as the action of the moss, as it has not been possible to establish any type of temporal intervention sequence. On the other hand, in sublevel Ja, the alteration produced by moss is mainly subsequent to the abrasive processes, as the scores caused by plant activity do not appear rounded. However, in 24.7% of the remains the scores also appear rounded and polished, which suggests that water reactivations occurred once the moss had already begun modifying the remains.

The presence of some remains covered by fine layers of travertine suggests that certain remains may have been covered with water that was slowly drying, which permitted the precipitation of the calcium carbonate. This phenomenon is not well represented in either sublevel: comprising only 0.89% of remains in sublevel Ja and 1.97% in sublevel Jb. No grouping was detected in Jb, but in sublevel Ja significant grouping was documented in G–J/57–65, which suggests that this process occurred in this sublevel, mainly locally.

Fig. 6.14 Examples of bone modification produced by mosses and water activity in sublevel Jb. Medular (**a**) and cortical view (**b**) of a large tibia fragment. Fragment of radius of *Equus ferus* with a similar modification (**c**)

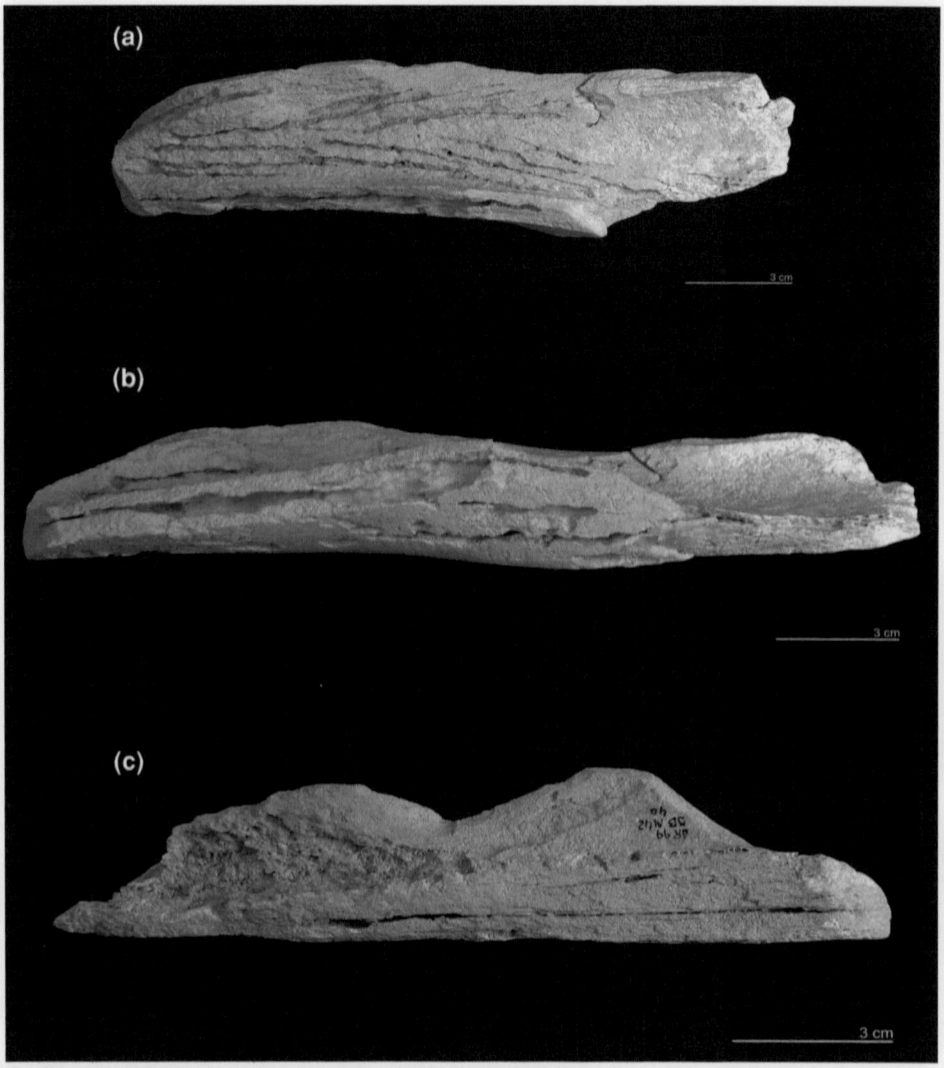

The effects of water have also been identified in the remains of micromammals, basically in the form of rounding and slight polishing (R1/P0). This also suggests that the type of sediment was fine (such as lime) since this sediment has been shown to lightly round microfaunal remains. Sediments of greater granulation mainly tend to produce mechanical fractures in micromammal remains (Fernández-Jalvo and Andrews 2003) and these have not been identified in either sublevel.

A modification that is frequently observed in the macromammal remains as well as in the micromammal remains is oxidation produced by manganese oxides. This oxidation presents as small microscopic arborescence and is dispersed on the surface of the fossils. In sublevel Ja, its presence is incidental among the macromammals (0.1%), and micromammals (0.9%) alike. In sublevel Jb, these values are higher with 1.22% of the remains of macrovertebrates and 4.5% of the microvertebrates affected. Manganese oxides is mainly formed by bacteria that require environmentally humid and aerobic conditions with a pH close to neutral for their development. The bacteria concentrate the metals in the environment, usually from water (Potter and Rossman 1979; Dorn and Oberlander 1981, 1982; Shahack-Gross et al. 1997). It is very likely that the environmental conditions of sublevel Jb were more suitable for the formation of manganese oxide, which is why this modification appears more frequently there than in sublevel Ja.

Trampling

Trampling is the result of a mechanical frictional process or of dragging the bone against the substratum. The effects of trampling are basically more or less linear scratch marks with sectional backgrounds in a U-shape section (and sometimes V-shape section) that may be either isolated or

Table 6.11 Rounding and polish stages used in water abrasion analysis at Abric Romaní (Cáceres 2002)

Rounding		Polish	
Stage	Features	Stage	Features
R0	No modification	P0	No modification
R1	Rounding in anatomical end and edge breakage detectable between 18 and 50 magnifications	P1	Brightness zones in anatomical end and edge breakage detectable between 18 and 50 magnifications
R2	Rounding affects bone surface, detectable macroscopically	P2	Brightness zones affect bone surface, detectable macroscopically
R3	All bone surface is rounded. Loss original morphology	P3	Bone is completely bright.

Table 6.12 Association of rounding and polish in macromammal fossils in sublevels Ja and Jb

	P0		P1		P2		P3		Total	
	NRm	%	NRm	%	NRm	%	NRm	%	NRm	%
Sublevel Ja										
R1	71	43.8	21	13	2	1.23	–	–	94	58.03
R2	34	21	21	13	10	6.17	–	–	65	40.12
R3	2	1.23	–	–	1	0.62	–	–	3	1.85
Sublevel Jb										
R1	73	81.1	8	8.89	–	–	–	–	81	90
R2	8	8.89	1	1.11	–	–	–	–	9	10
R3	–	–	–	–	–	–	–	–	–	–

NRm Number of remains modified

Fig. 6.15 SEM image showing evidence of rounding and polish on an unidentified bone from sublevel Ja. The rounding and polish affect mainly the edge of fragment

concentrated on the surface of the bone (Andrews and Cook 1985; Agenbroad 1989; Fiorillo 1989). This taphonomic process can also result in fractures and displacements (Gifford-Gonzalez 1989a, b; Olsen and Shipman 1988; Yellen 1991) as well as notches (Blasco et al. 2008). In level J, trampling has been identified from scratch marks

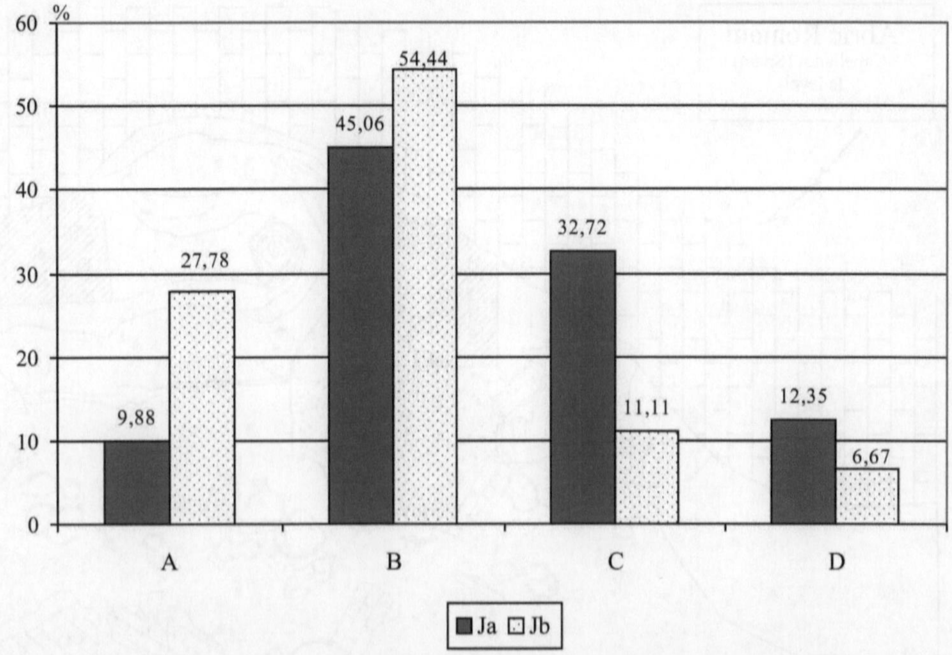

Fig. 6.16 Percentage of water abrasion and size of remains in sublevels Ja and Jb. (A = <2 cms; B = 2–5 cms; C = 5–10 cms; D = >10 cms)

and notches found on remains in sublevel Jb. There is no evidence of fractures or displacements brought about by this taphonomic process.

This alteration is not well represented in either sublevel (Fig. 6.7): 1.80% in Ja and 0.59% in Jb (Fig. 6.19). Trampling indiscriminately affects all skeletal elements and herbivores of all sizes. Trampling has not been documented in the remains of carnivores or rabbits.

The scarcity of remains with this modification in sublevel Jb prevented us from conducting an in-depth analysis of trampling processes. However, the presence of notches in conjunction with the low incidence of carnivore activity in this sublevel suggests that the scarce evidence of trampling that has been detected is due to the movement of the Neanderthals as they went about performing the different activities that they carried out in the course of their occupation of the shelter.

The remains of sublevel Ja with this alteration fit in dimension categories B and C, with more than 53% exceeding 4 cm in length. More than 50% of the remains with trampling have also been burned with degrees of incipient cremation being the most abundant. No changes in color were observed in the scratch marks resulting from trampling and the superficiality of these marks does not permit the sequencing of the dragging against the substratum and the thermal impact of fire. The remains with trampling of Ja are spatially located in the areas with a greater concentration of remains. It must be emphasized that only two remains presented the effects of water abrasion (R1) in which the presence of trampling marks was ruled out due to the impact of the sedimentary particle content in the water. Therefore, we conclude that trampling was caused by the movement of animals and people in the deposition zone of the remains. Because the action of carnivores occurred sporadically, it is possible that most of the scratch marks were caused, just as in sublevel Jb, by Neanderthals.

The fact that trampling marks have not been observed in the remains of microvertebrates in either of the two sublevels reinforces the hypothesis that this modification was induced by the movement of humans, since during the Neanderthal occupation of the shelter, the microvertebrate remains had not yet been deposited.

Cracking and Fissures

Fissures and cracks have been identified in the remains recovered from both of the sublevels studied. These mechanical modifications can result from the action of different taphonomic mechanisms, including weathering, which is frequently the main cause of fissures and cracks. However, the characteristics of the Abric Romaní lead us to conclude that some part of the fissures observed may have been caused by fire (Cáceres et al. 2002) or changes in humidity (Diez et al. 1999).

We distinguished between different types of fissures: *fissures* are considered the initial stage of formation appearing as longitudinal cracks whose edges are not yet separated; and *cracks* are more developed with edges separated from one another. A third type is cracks featuring exfoliations and/or loss of tissues. These stages have been adapted from those established by Behrensmeyer (1978) for

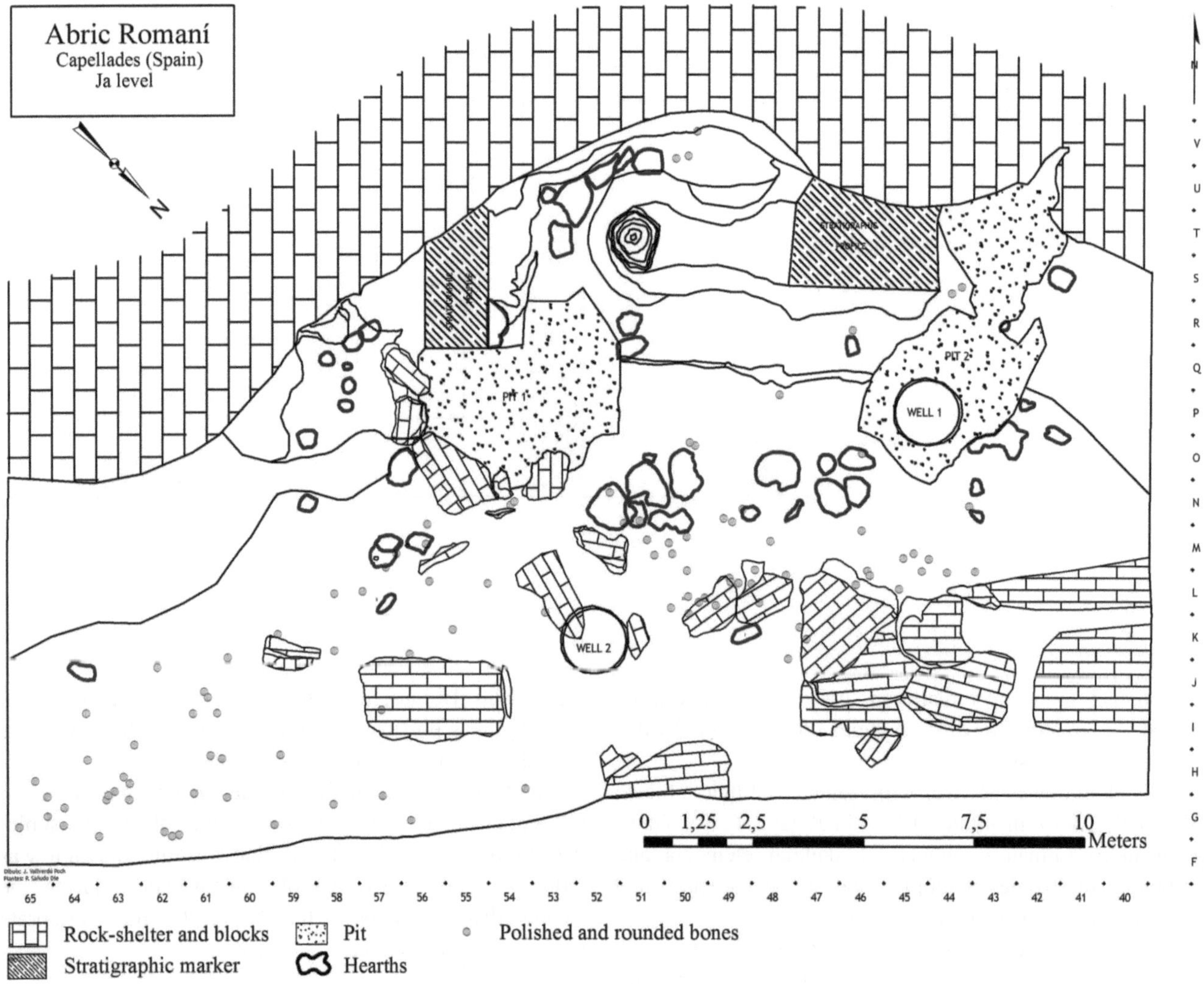

Fig. 6.17 General distribution of faunal remains with evidence of water activity in sublevel Ja

the macrovertebrates and Andrews (1990) for the microvertebrates. These fissures and cracks show a different representation in the sublevels studied. In sublevel Ja, this modification is scarce and has been identified in only 0.95% of the remains, while in sublevel Jb it is more abundant and affects 11.03% of the remains.

In sublevel Ja, the initial stages comprise more than 79.7% of the remains that show fissures whereas the remains with cracks make up 10.9%. Only the remaining 9% show a loss of tissues or exfoliation. Most of the remains with fissures (62.5%) show modifications induced by moss and 14% exhibit the effects of water abrasion. Only four remains are also burned and the degrees of cremation are 1 or 2. This data lead us to suggest that fire cannot have been the cause of the fissures and cracks in this sublevel. At the same time, the low number and incipient stage of alteration does not suggest the action of a mechanism that equally affected the entire assemblage over a long period of time such as the exposure to atmospheric agents. In any case, the presence of vegetation would limit the effects of exposure. Therefore, we might consider that the fissures are owed to the microenvironmental conditions of the shelter and to the changes in humidity that may have occurred. A phenomenon that would corroborate this hypothesis is that in most cases, the fissures and cracks show a particular lifting at the edges that is absent when these are produced by the effects of the sun (Diez et al. 1999; Fernández-Jalvo and Marín 2008).

In sublevel Jb, the representation of the stages of modification is similar to that observed in Ja. Of the remains, 84.2% show incipient fissures, 12.6% cracks, and 3.16% exfoliation or loss of tissues. The latter type of modification stage is where the largest differences of representation occur between the two sublevels, as in Ja this stage was more abundant. The taphonomic study of this modification in sublevel Jb indicates that more than 82% of the remains with fissures are modified by moss and 26% show evidence of chemical corrosion.

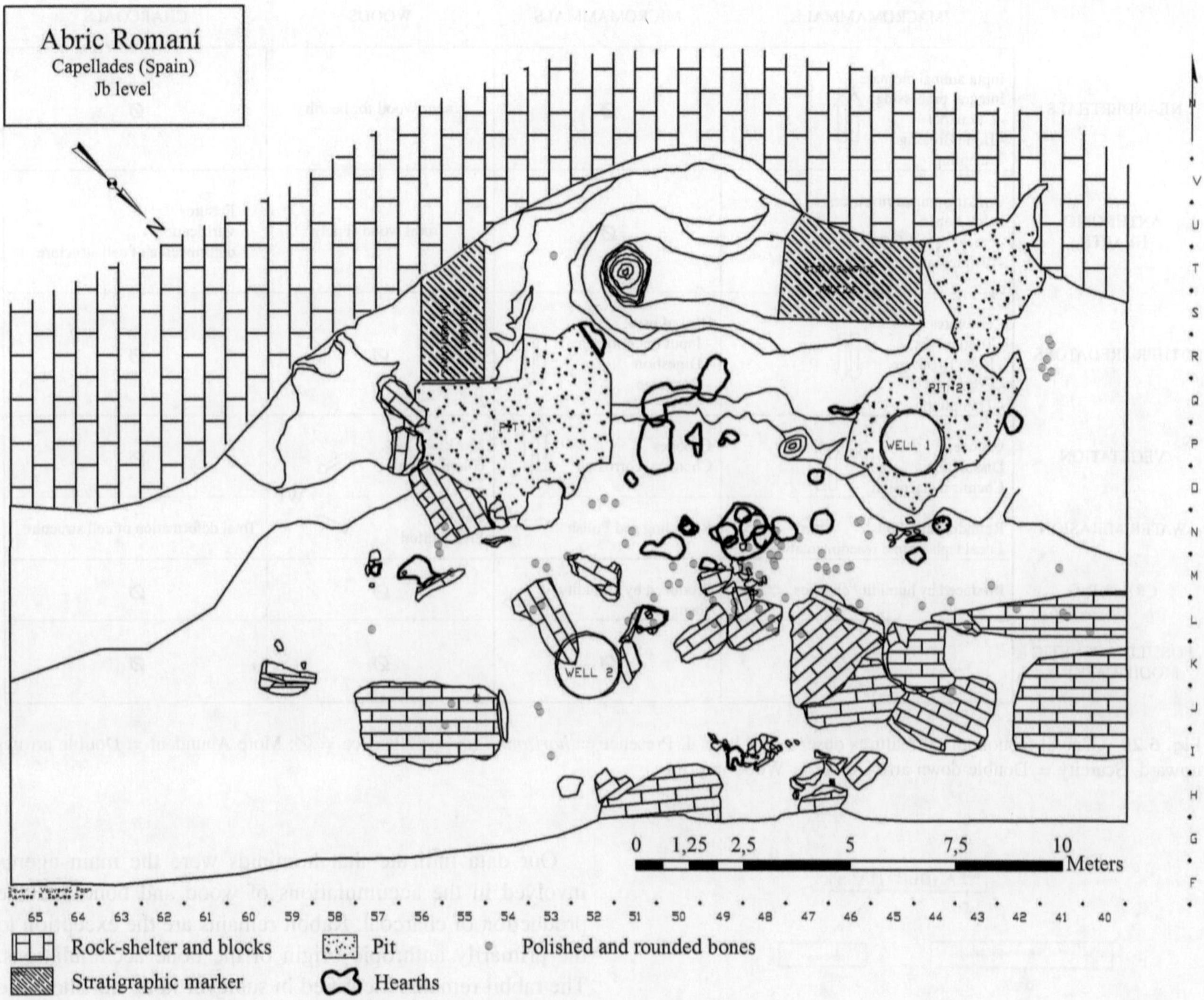

Fig. 6.18 General distribution of faunal remains with evidence of water activity in sublevel Jb

Fig. 6.19 Scratch marks produced by trampling in a medium-sized long bone

Dissolutions are also abundant (7.89%) as is water abrasion (7.37%), which all seem to indicate that the fissures and cracks, as in Ja, are owed to a change in humidity, rather than direct exposure to the sun. As opposed to in sublevel Ja, in Jb there are abundant burned remains (14.21%), which in most cases exhibit an advanced degree of cremation. We therefore cannot rule out that at least some of the fissures resulted from the direct impact of fire on the bones.

Taphonomic Sequence

Throughout this chapter we have shown evidence of the intervention of different agents, mechanisms, and taphonomic processes on the fossil associations of level J (Figs. 6.20, 6.21).

	MACROMAMMALS	MICROMAMMALS	WOODS	CHARCOALS
NEANDERTHALS	Input animal biomass Integral processing: ⇑ - Cutmarks - Bone Breakage Trampling	Ø	Input Wood for hearth	Ø
ANTHROPIC HEARTH	Cooking animal resources Burnt bones: - cracking and fissures ⇑ - disintegration?	Ø	Burnt wood (Fuel)	Fissures vitrification deformacion of cell structure
OTHER PREDATORS	Carnivores: ⇓ - Toothmarks; - Bone Breakage - Coprolites - Trampling	Bird of prey: ⇓ - Input micromammals - Digestion - Breakage	Ø	Ø
VEGETATION	Grooves ⇑ Dissolution Chemical corrosion	Grooves ⇑ Chemical corrosion	Musk colonization ⇘ WI	Ø
WATER ABRASION	Rounding and Polish ⟺ Local taphonomic resedimentation	Rounding and Polish ⟺	$CaCO_3$ Precipatiton ⇗	Total deformation of cell structure
CRACKING	Produced by humidity changes ⟺	Produced by humidity changes ⟺	Ø	Ø
FOSSILDIAGENETIC MODIFICATIONS	Ø	Ø	Ø	Ø

Fig. 6.20 General taphonomical features observed in level J. Presence = *horizontal arrow*; Absence = ∅; More Abundant = Double arrows upward; Scarcity = Double down arrows; WI = Wood imprint

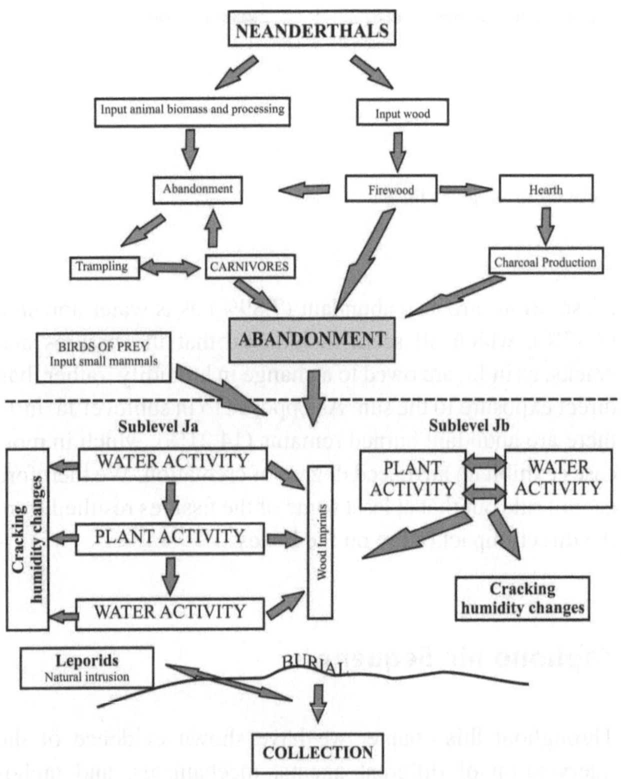

Fig. 6.21 Summary of the temporal taphonomic sequence established for level J

Our data indicate that hominids were the main agents involved in the accumulations of wood and bone and the production of charcoal. Rabbit remains are the exception to the primarily anthropic origin of the bone accumulations. The rabbit remains identified in sublevel Ja do not show the same taphonomic characteristics as the rest of the fauna. They do not present evidence of anthropic activity, and only in two remains carnivore toothmarks have been identified. Moreover, rabbit skeletal elements do not present evidence of water abrasion. This suggests that the presence of rabbits in the shelter is the result of their own natural entry. Such entries would have taken place during the final stages of formation of the level, since the rabbit remains recovered also lack modifications produced by vegetation.

Carnivores can be ruled out as a bone accumulation agent because their activity is low in sublevel Ja and only incidental in Jb. This low evidence of carnivore activity and the absence of groups or dispersions produced by this agent indicate that carnivore intervention should be understood as the result of occasional visits, probably as part of a sporadic prowling strategy at times when Neanderthals were not present. Carnivores did not introduce the micromammal remains either, as these were deposited by birds of prey. The degree of digestion indicates a category 1 predator (Andrews 1990) (Fig. 6.22). The only postdepositional modifications observed in small mammal remains are related to water

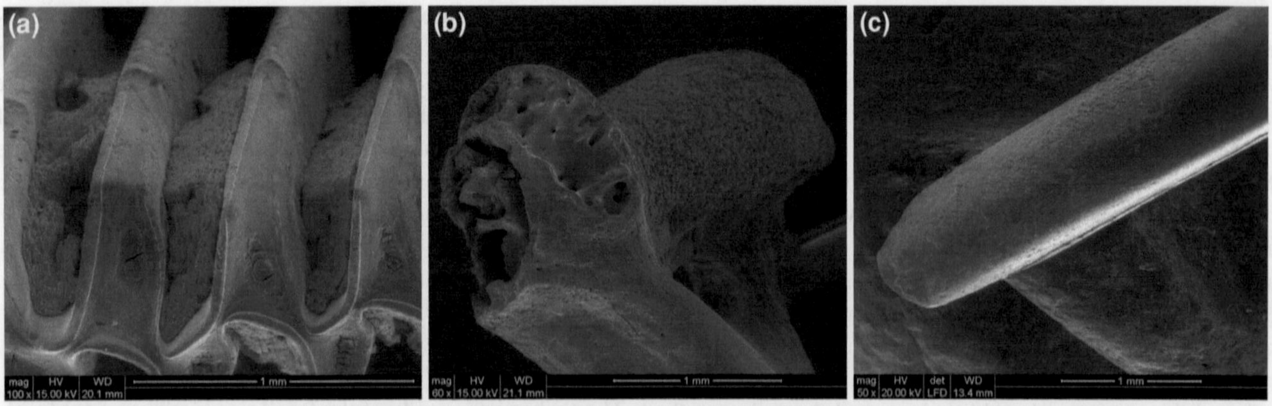

Fig. 6.22 Small mammal digestion in sublevel Ja. **a** Moderate digestion detail in M1 of *M. arvalis*. The enamel is disappearing in the molar angles. **b** Moderate digestion in rodent humerus. **c** Light digestion in rodent incisor

reactivations and moss activity. The absence of cremation and trampling indicate that birds of prey used the shelter subsequent to the Neanderthals' abandonment of the site.

The taphonomic study has provided sufficient criteria to reject the possibility that the fossil associations of sublevels Ja and Jb are the result of the water's current. Although water flows modified the bone remains, producing rounding and polishing, the degree of modification is incipient. This, along with the absence of preferential orientation or selection by size or significant groupings, suggests that the currents flowed at a low level of energy, which would render them incapable of accumulating remains. This incapacity also applies to the ability to induce taphonomic removals in the fossils of the assemblage. The absence of advanced degrees of water abrasion indicates that the remains did not change position during their modification. It is true that some fossils present abrasion on all sides, although in initial or intermediate degrees, which leads us to suggest that if movement did occur it was of a markedly local character.

The initial degrees of water abrasion suggests that this modification was initiated when the state of the bones was semi-dry and the sedimentary particle content in the water was fine, probably consisting of lime (Fernández-Jalvo and Andrews 2003).

The corrosive action of vegetation is, together with modifications produced by the Neanderthals, the most abundant type of modification found in both sublevels. The morphological features of this modification indicate that was produced basically by moss. Moss, leaves, stems and branches are the base of travertine production along with precipitation of $CaCO_3$. As colonizing plants, this type of vegetation does not necessarily need substratum to develop. Therefore, we consider that the modifications to the remains caused by vegetation are basically biostratinomic.

The existence of this vegetation covering the surface limited the incidence of atmospheric agents. Proof of this is the absence of weathering in the assemblage. The fissures in the remains are related to the action of fire or to changes in microenvironmental humidity, based on the characteristic raising of the edges (Diez et al. 1999; Fernández-Jalvo and Marín 2008). The water flows in the shelter were intermittent, occurring only at certain times, thus causing the more or less humid conditions in the microenvironment.

The intermittence of the water flows is identified by the sequence of modification of the water abrasion along with the alterations by moss. In sublevel Jb both modification processes probably occurred more or less simultaneously. In sublevel Ja abrasions generally occurred previous to the moss, although there are some remains in which the scores are also rounded and polished. This suggests that the water flows stopped, the moss alteration occurred, and then water flows began again. Moreover, the existence of fossils in the exterior zone of the shelter, covered by fine layers of travertine, verifies that the currents ceased, since it causes the precipitation of carbonate.

These periods of stopped water flow are verified by the identification of modifications to charcoal. During these periods, vegetation covered the wood deposited in the shelter and the subsequent karst reactivations led to the precipitation of $CaCO_3$ which in turn gave rise to travertine and, consequently, to wood negatives.

In general, the burial process of remains probably occurred faster in sublevel Jb than in Ja. The sequence water-moss-water in Ja indicates a greater exposure time of the remains, characterized by moments of discontinuation and reactivation of water flows.

After burial, the sedimentary conditions of the shelter were optimum for the conservation of the remains. In neither of the two sublevels have fossildiagenetic modifications been identified, indicating that both fossil associations can be considered to be in a good state of preservation. The evidence of the activities carried out by the Neanderthals, the subsequent actions of carnivores, and the intervention of other biostratinomic processes appear without fossildiagenetic disturbances.

These good fossilization conditions reveal some interesting phenomena such as the fact that the Neanderthals were the main agents of bone accumulation and, moreover, the primary modifiers, both of the faunal remains and the physical space of the shelter. The predominance of anthropic modifications (mainly in sublevel Ja) and the use of fire as the organizer of the space verify this. This marked anthropization was an important factor in the subsequent action of other agents or taphonomic processes. The best example is probably the scarce activity of carnivores, which had little interest in the already exhausted nutritional resources the shelter offered, or the action of fire which affected the traces produced by plants and the action of the water flows.

The presence of fire in more or less structured intentional hearths gave rise to intense thermo-alteration, both in the remains and in the substratum. Fire caused fissures, cracks and fragmentation in bones, even leading to the total disappearance of the remains in the most extreme cases. It has also been observed that modifications caused by vegetation activity do not equally affect burned and unburned remains (Fig. 6.11). This suggests that the modification capacity of plants was hampered by the thermo-alteration process, either by modification of substratum properties or by the features of the burned bones. The same phenomenon is observed in the effects of water abrasion, which significantly alters to a lesser extent remains that have been burned to advanced degrees. Advance stage cremation results the total drying of the remains, which when completely dry offer a different resistance to the effects of abrasion.

In general, the taphonomic features of the two sublevel formations are similar (Fig. 6.20). However, some significant differences were found. In sublevel Ja, for example, human activity is more common and more intense (Rosell et al. 2012). On the other hand, in sublevel Jb we have a greater variety of modifications produced by vegetation and by water abrasion. This indicates more humid environmental conditions in Jb than in Ja. The interrelation between taphonomic mechanisms (scores, dissolutions and chemical corrosion—rounding and polishing) suggests that in Jb these mechanisms were more intense and that they occurred practically simultaneously. In sublevel Ja a succession of modifications dominated by the intermittence of water flows has been established. This suggests that in Jb the water flows were more continuous, but at the same low level of energy as in Ja.

The different incidence of water abrasion in the two sublevels has also been observed in wood and charcoal, though in different aspects. The charcoal appears more modified in Jb than in Ja, which would support the idea of continuous water flows. On the other hand, there are a greater number of wood negatives in Ja and more altered micromammal remains.

The relationship between the different taphonomic modifications identified suggests that the burial of the remains in level J occurred relatively quickly. Our data suggest that the remains of sublevel Ja took longer to be buried than those of sublevel Jb. In both sublevels, the state of conservation of the remains is good and there is no differential conservation related to the physicochemical agents.

Conclusions

The taphonomic study of the fossil associations of sublevels Ja and Jb has permitted the identification of the intervention of different agents, mechanisms, and taphonomic processes that have influenced the formation of these associations in different ways:

Neanderthals were the main agent responsible for the accumulation of macromammals remains and wood used as firewood. The presence of carnivores, rabbits and micromammals has a different taphonomic origin. The presence of carnivores and rabbits in the shelter should be understood as a natural intrusion, occurring after human occupancy. The micromammals were introduced by birds of prey while the shelter was unoccupied in the final phases of the formation of the associations

The only bias observed in level J is related to anthropogenic activities (transport, processing, consumption and abandonment of animal resources).

The anthropic dynamic heavily influences the subsequent intervention of other taphonomic agents or processes. For example, the action of fire on the remains conditioned the subsequent action of plant activity and water abrasion.

There are many similarities in the temporal formation sequences in the two sublevels. The main difference is in sublevel Ja, where an intermittence of the water flows has been detected, whereas in sublevel Jb such flows seem to have been more continuous. In both sublevels the water currents were of low energy.

The fossil associations of level J present, in general, a good state of conservation. However, in sublevel Jb the fossils appear more heavily altered.

In general, processes of taphonomic removal have not been observed in either of the two sublevels. Although for sublevel Ja, we propose the possibility of short-distance displacements of a local character due to the action of water.

The burial of the remains was probably relatively fast in both sublevels. The impossibility of determining an intervention sequence of water abrasion and plant activity lead us to deduce that burial may have occurred faster in Jb.

There were no fossildiagenetic modifications in either sublevel, which indicates that once buried, the sedimentary conditions were favorable. No mechanisms or taphonomic processes which may have distorted any of the associations of level J have been detected.

The fossilization process of level J is consistent with a karst taphosystem of a travertine deposit. The main characteristic of this taphosystem is a microenvironment of relative ambient humidity, marked by wet conditions and reactivations of water flows. Once the remains were buried, the fossildiagenetic phase was tremendously preservative and conserved the exceptional remains found at the site.

Acknowledgments We want to thank the excavation team of Abric Romani. This research was supported by Generalitat de Catalunya Grant 2009 SGR 188. The field excavation work was supported by Departament de Cultura i Mitjans de Comunicació de la Generalitat de Catalunya, Diputació de Barcelona, Ajuntament de Capellades and Arts Gràfiques Romanyà-Valls S.A. R. Blasco and M.J. Gabucio are beneficiaries of a FI Grant from Generalitat de Catalunya and financed by European Social Found. A. Rodríguez-Hidalgo is beneficiary of a FPI Grant from Ministry of Science and Innovation of Spanish Government. We thank the reviewers whose comments helped to improve our manuscript.

References

Agenbroad, L. D. (1989). Spiral fractured mammoth bone from nonhuman taphonomic processes at hot spings mammoth site. Bone modification due to burning: Experimental evidence. In R. Bonnichsen & M. Sorg (Eds.), *Bone modification* (pp. 139–148). Orono, ME: Institute for Quaternary Studies, University of Maine.

Albert, R. M., Bamford, M. K., & Cabanes, D. (2006). Taphonomy of phytoliths and macroplants in different soils from Olduvai Gorge (Tanzania) and the application to Plio-Pleistocene palaeoanthropological samples. *Quaternary International, 148*(1), 78–94.

Albert, R. M., Shahack-Gross, R., Cabanes, D., Gilboa, A., Lev-Yadun, S., Portillo, M., et al. (2008). Phytolith-rich layers from the Late Bronze and Iron Ages at Tel Dor (Israel): Mode of formation and archaeological significance. *Journal of Archaeological Science, 35*(1), 57–75.

Alexander, J., Fielding, C. R., & Jenkins, G. (1999). Plant-material deposition in the tropical Burdekin River, Australia: Implications for ancient fluvial sediments. *Palaeogeography, Palaeoclimatology, Palaeoecology, 153*, 105–125.

Allué, E. (2002). Preliminary issues regarding the taphonomic study of archaeological charcoal upon the record from the Abric Romaní (Capellades, Spain). In M. de Renzi, M. Pardo Alonso, M. Belinchón, E. Peñalver, P. Montoya, & A. Márquez-Aliaga (Eds.), *Current topics on taphonomy and fossilization* (pp. 447–452). Valencia: Ayuntamiento de Valencia.

Allué, E., Cabanes, D., Solé A., Sala, R. (2012a). Hearth functioning and forest resource exploitation based on the archaeobotantical assemblage from level J of Abric Romaní. In E. Carbonell (Ed.), High Resolution Archaeology and Neanderthal Behavior: Time and Space in Level J of Abric Romaní (Capellades, Spain). Dordretch: Springer.

Allué E., Burjachs, F., García, A., López-García, J. M., Bennàsar, M., Rivals, Fl., Blain, H.A., Expósito, I., Martinell, J. (2012b). Neanderthal landscapes and their home environment: Flora and fauna records from level J of Abric Romaní. In E. Carbonell (Ed.), High Resolution Archaeology and Neanderthal Behavior: Time and Space in Level J of Abric Romaní (Capellades, Spain). Dordretch: Springer.

Andrews, P. (1983). Small mammal faunal diversity at Olduvai Gorge, Tanzania. Animals and archaeology: 1. *Hunters and their Prey, 163*, 77–85.

Andrews, P. (1990). *Owl, caves and fossils*. London: Natural History Museum Publications.

Andrews, P., & Cook, J. (1985). Natural modifications to bones in a temperate settings. *Man, 20*, 675–691.

Andrews, P., & Fernández-Jalvo, Y. (1997). Surface modifications of the Sima de los Huesos fossil humans. *Journal of Human Evolution, 33*(2–3), 191–217.

Andrews, P., & Nesbit-Evans, E. M. (1983). Small mammal bone accumulations produced by mammalian carnivores. *Paleobiology, 9*(3), 289–307.

Baileys, R. T., & Blankenhorn, P. R. (1982). Calorific and porosity development in carbonized wood. *Wood Science, 15*, 19–28.

Beall, F. C., Blankehorn, P. R., & Moore, G. R. (1974). Carbonized wood-physical properties and use as an SEM preparation. *Wood Science, 6*, 212–219.

Behrensmeyer, A. K. (1978). Taphonomic and ecologic information from bone weathering. *Paleobiology, 4*(2), 150–162.

Binford, L. R. (1981). *Bones: Ancient men and modern myths*. New York: Academic Press.

Blanchette, R. A., Nilsson, T., Daniel, G., & Abad, A. (1990). Biological degradation of wood. In R. M. Rowell & R. J. Barbour (Eds.), *Archaeological wood properties, chemistry and preservation* (pp. 141–176). Washington, DC: The American Chemical Society Publication.

Blasco, R., Rosell, J., Fernández Peris, J., Cáceres, I., & Vergès, J. M. (2008). A new element of trampling: An experimental application on the Level XII faunal record of Bolomor cave (Valencia, Spain). *Journal of Archaeological Science, 35*(6), 1605–1618.

Blumenschine, R. J. (1988). An experimental model of the timing of hominid and carnivore influence on archaeological bone assemblage. *Journal of Archaeological Science, 15*, 483–502.

Blumenschine, R. J. & Marean, C. W. (1993). A carnivore's view of archeological bone assemblage. In J. Hudson (Ed.), *From bones to behavior. Ethnoarchaeological and experimental contributions to the interpretation of faunal remains* (pp. 273–300). Center for Archaeological Investigations, Southern Illinois University, Occasional paper 21.

Blumenschine, R. J., & Selvaggio, M. M. (1988). Percussion marks on bone surface as a new diagnostic of hominid behaviour. *Nature, 333*, 763–765.

Boyd, W. E. (1988). Methodological problems in the analysis of fossil non-artefactual wood assemblages from archaeological sites. *Journal of Archaeological Science, 15*, 603–619.

Bunn, H. T. (1981). Archaeological evidence for meat-eating by Plio-Pleistocene hominids from Koobi-Fora and Olduvai Gorge. *Nature, 291*, 574–580.

Cáceres, I. (1996). Secuencia de los procesos y mecanismos de alteración de la asociación fósil de macromamíferos del nivel I del Abric Romaní (Capellades, Barcelona). In G. Meléndez, M. F. Blasco Sancho, & I. Pérez (Eds.), *II Reunión de Tafonomía y Fosilización* (pp. 73–78). Zaragoza: Institución Fernando el Católico.

Cáceres, I. (2002). Tafonomía de yacimientos antrópicos en Karst. Complejo Galería (Sierra de Atapuerca, Burgos), Vanguard Cave (Gibraltar) y Abric Romaní (Capellades, Barcelona). Ph.D. dissertation, Universitat Rovira i Virgili.

Cáceres, I., Bravo, P., Esteban, M., Expósito, I., & Saladié, P. (2002). Fresh and heated bones breakage. An experimental approach. In M. de Renzi, M. Pardo Alonso, M. Belinchón, E. Peñálver, P. Montoya, & A. Márquez-Aliaga (Eds.), *Current topics on taphonomy and fossilization* (pp. 471–479). Valencia: Ayuntamiento de Valencia.

Capaldo, S. D., & Blumenschine, R. J. (1994). A quantitative diagnosis of notches made by hammerstone percussion and carnivore gnawing on bovid long bones. *American Antiquity*, *59*(4), 724–748.

Carbonell, E. (2002). *Abric Romaní Nivell I. Models d'ocupació de curta durada de fa 46.000 anys a la Cinglera del Capelló (Capellades, Anoia, Barcelona)*. Tarragona: Universitat Rovira i Virgili.

Carrión, J. S. (2002). A taphonomic study of modern pollen assemblages from dung and surface sediments in arid environments of Spain. *Review of Palaeobotany and Palynology*, *120*, 217–232.

Castro-Curel, Z., & Carbonell, E. (1995). Wood pseudomorphs from level I at Abric Romaní, Barcelona, Spain. *Journal of Field Archaeology*, *22*, 376–384.

Chabal, L., Fabre, L., Terral, J. F., & Théry-Parisot, I. (1999). L'anthracologie. In A. Ferdière (Ed.), *La botanique* (pp. 43–104). Paris: Errance.

Cutter, B. E., Cumbie, B. G., & McGinnes, E. A. J. (1980). SEM and shrinkage analyses of southern pine wood following pyrolysis. *Wood Science and Technology*, *14*, 115–130.

Denys, C., Sánchez, V., & Fernández-Jalvo, Y. (1995). Predation et fossilisation des micromammiferes presentation d'un des aspects d'une discipline recente: la Taphonomie. *Arvicola*, *VII*(1), 7–13.

Díez, J. C., Fernández Jalvo, Y., Cáceres, I., & Rosell, J. (1999). Zooarchaeology and taphonomy of Aurora Stratum (Gran Dolina, Sierra de Atapuerca, Spain). *Journal of Human Evolution*, *37*, 623–657.

Dorn, R. I., & Oberlander, T. M. (1981). Microbial origin of desert varnish. *Science*, *213*, 1245–1247.

Dorn, R. I., & Oberlander, T. M. (1982). Rock varnish. *Progress in Physical Geography*, *6*, 317–367.

Efremov, I. A. (1940). Taphonomy: New branch of paleontology. *Pan American Geologist*, *74*, 81–93.

Efremov, I. A. (1953). Taphonomie et annales géologiques (1ère partie). *Annales du Centre d'Études et de Documentation Paléontologique*, *4*, 164.

Feist, W. C. (1990). Outdoor wood weathering and protection. In R. M. Rowell & R. J. Barbour (Eds.), *Archaeological wood properties, chemistry and preservation* (pp. 263–298). Washington, DC: The American Chemical Society Publication.

Fernández-Jalvo, Y., & Andrews, P. (1992). Small mammal taphonomy of Gran Dolina, Atapuerca (Burgos), Spain. *Journal of Archaeological Science*, *19*, 407–428.

Fernández-Jalvo, Y., & Andrews, P. (2003). Experimental effects of water abrasion on bone fragments. *Journal of Taphonomy*, *1*(3), 147–163.

Fernández-Jalvo, Y., & Marín, M. D. (2008). Experimental taphonomy in museums: Preparation protocols for skeletons and fossil vertebrates under the scanning electron microscopy. *Geobios*, *41*(1), 157–181.

Fernández-Jalvo, Y., Scott, L., & Denys, C. (1999). Taphonomy of pollen associated with predation. *Palaeogeography, Palaeoclimatology, Palaeoecology*, *149*, 217–282.

Fiorentino, G. (1995). New perspectives in anthracological analysis. Palaeoecological and technological implications of charcoals found in the Neolithic flintmine at La Defensola (Vieste, Apulia, Italy). *Quaternaria Nova*, *5*, 99–128.

Fiorillo, A. R. (1989). An experimental study of trampling: Implications for the fossil record. In R. Bonnichsen & M. Sorg (Eds.), *Bone modification* (pp. 61–71). Orono, ME: Institute for Quaternary Studies, University of Maine.

Fischer, J. W. (1995). Bone surface modifications in zooarchaeology. *Journal of Archaeological Method and Theory*, *2*(1), 7–68.

Fritz, W. J., & Harrison, S. (1985). Transported trees from the 1982 Mount St. Helens sediment flows: Their use as paleocurrent indicators. *Sedimentary Geology*, *42*, 49–64.

Gastaldo, R. A. (2004). The relationship between bedform and log orientation in a paleogene fluvial channel, Weibelster Basin, Germany: Implications for the use of coarse woody debris for paleocurrent analysis. *Palaios*, *19*, 587–597.

Gastaldo, R. A., & Degges, C. W. (2007). Sedimentology and paleontology of a carboniferous log jam. *International Journal of Coal Geology*, *69*, 103–118.

Gifford-Gonzalez, D. (1989a). Overview-modern analogues: Developing an interpretative framework. In R. Bonnichsen & M. Sorg (Eds.), *Bone modification* (pp. 43–52). Orono, ME: Institute for Quaternary Studies, University of Maine.

Gifford-Gonzalez, D. (1989b). Ethnographic analogues for interpreting modified bones: Some cases from East Africa. In R. Bonnichsen & M. Sorg (Eds.), *Bone modification* (pp. 179–246). Orono, ME: Institute for Quaternary Studies, University of Maine.

Haynes, G. (1980). Prey bones and predators: Potential ecologic information from analysis of bone sites. *Ossa*, *7*, 75–97.

Haynes, G. (1983). A guide for differentiating mammalian carnivore taxa responsible for gnaw damage to herbivore limb bones. *Paleobiology*, *9*, 164–172.

Hesse, B., & Wapnish, P. (1985). *Animal bone archaeology: From objectives to analysis. Manual on archaeology, 5*. Washington, DC: Taraxacum.

Johnson, E. (1985). Current developments in bone technology. *Advances in Archaeological Method and Theory*, *8*, 157–235.

Juneja, S. C. (1975). Combustion of cellulosic materials and its retardance-status and trends. *Wood Science*, *7*, 201–209.

Kowalski, K. (1990). Some problems of the taphonomy of small mammals. In O. Feijfar and W.D. Heinrich (Eds.), International symposium evolution, phylogeny, and biostratigraphy of Arvicolids (pp. 285-296). Prague: Geological Survey.

Loreau, P. (1994). Du bois au charbon de bois: Approche experimental de la combustion. Ph.D. dissertation, Université Montpellier.

Lyman, R. L. (1994). *Vertebrate taphonomy*. New York: Cambridge manuals in archaeology.

Maguire, J. M., Pemberton, D., & Collett, M. H. (1980). The Makapansgat limeworks grey breccia: Hominids, hyaenas, hystricids or hillwash. *Paleontologia Africana*, *23*, 75–98.

Marguerie, D., & Hunot, J. Y. (2007). Charcoal analysis and dendrology: Data from archaeological sites in North-western France. *Journal of Archaeological Science*, *34*(9), 1417–1433.

Martín-Closas, C., & Gomez, B. (2004). Plant taphonomy and palaeoecological interpretations. *A synthesis. Geobios*, *37*, 65–88.

McGinnes, E. A., Kandeel, S. A., & Szopa, P. S. (1971). Some structural changes observed in the transformation of wood into charcoal. *Wood and Fiber*, *3*, 77–83.

Morlan, R. E. (1980). Taphonomy and archaeology in the Upper Pleistocene of the Northern Yukon Territory: A glimpse of the peopling of the New World. Archaeological survey of Canada, Mercury series paper, 94. Otawa, ON: National Museum of Man.

Müller, A. H. (1963). *Lehrbuch der Paläozoologie, 1. Allgemeine Grundlagen, C. Die Fossilisationslehre*. Jena: Gustav Fischer.

Olsen, S. L., & Shipman, P. (1988). Surface modification on bone: Trampling versus butchery. *Journal of Archaeological Science*, *15*, 535–553.

Pentecost, A. (2005). *Travertine*. Berlin: Springer.

Postigo, J. M. (2003). *Contribución al conocimiento de la vegetación pleistocena de la península ibérica. Estudio paleobotánico de macrorrestos vegetales fósiles*. Ph.D. dissertation, Universidad Autónoma de Madrid.

Potter, R. M., & Rossman, G. R. (1979). The manganese and iron-oxide mineralogy of dessert varish. *Chemical Geology*, *25*, 79–94.

Potts, R., & Shipman, P. (1981). Cut-marks made by stone tools on bones from Olduvai Gorge, Tanzania. *Nature*, *291*, 577–580.

Prior, J., & Gasson, P. (1993). Anatomical changes on charring six african hardwoods. *IAWA*, *14*, 77–86.

Rosell, J., Blasco, R., Huguet, R., Cáceres, I., Saladié, P., Rivals, F., Bennàsar, M., Bravo, P., Campeny, G., Esteban-Nadal, M., Fernández-Laso, C., Gabucio, M.J., Ibáñez, N., Martín, P., Muñoz, L., Rodríguez-Hidalgo, A. (2012). Occupational patterns and subsistence strategies in level J of Abric Romaní (Barcelona, Spain). In E. Carbonell (Ed.), High Resolution Archaeology and Neanderthal Behavior: Time and Space in Level J of Abric Romaní (Capellades, Spain). Dordretch: Springer.

Rossen, J., & Olson, J. (1985). The controlled carbonization and archeological analysis of SE U.S. wood charcoal. *Journal of Field Archaeology*, *12*, 445–456.

Schweingruber, F. H. (1983). *Der Jahrring. Standort, Methodik, Zeit un klima in der Dendrochronolge*. Bern und Stuttgart: Verlag Paul Haupt.

Shahack-Gross, R., Bar-Yosef, O., & Weiner, S. (1997). Black-coloured bones in Hayonim cave, Israel: Differentiating between burning and oxide staining. *Journal of Archaeological Science*, *24*(5), 439–446.

Shipman, P. (1981). *Life history of a fossil. An introducction to taphonomy and paleoecology*. Cambridge, MA: Harvard University Press.

Slocum, D. H., McGuinnes, E. A. J., & Beall, F. C. (1978). Charcoal yield, shrinkage, and density changes during carbonization of oak and hickory woods. *Wood Science*, *11*, 42–47.

Spicer, R. A. (1991). Plant taphonomic processes. In P. A. Allison & D. E. G. Briggs (Eds.), *Taphonomy*: *Realising the data locked in the fossil record* (pp. 72–108). New York: Plenum Press.

Stiner, M. C. (1994). *Honor among thieves: A zooarchaeological study of Neandertal ecology*. Princenton, NJ: Princenton University Press.

Théry, I. (1993). *Les deformations anatomiques des charbons de bois: Implications archéologiques (Sites des Canalettes et des Usclades, Aveyron)*. Ph.D. Dissertation, Université de Montpellier.

Théry-Parisot, I. (2001). *Économie des combustibles au Paléolithique. Dossier de Documentation Archéologique 20*. Paris: CNRS.

Tucker, M. E., & Wright, V. P. (1990). *Carbonate sedimentology*. London: Blackwell.

Unger, A., Schniewind, A. P., & Unger, W. (2001). *Conservation of wood artifacts*: *A handbook*. New York: Springer.

Vaquero, M. (1997). *Tecnología lítica y comportamiento humano: organización de las actividades técnicas y cambio diacrónico en el Paleolítico Medio del Abric Romaní (Capellades, Barcelona)*. Ph.D. dissertation. Universitat Rovira i Virgili.

Weigelt, J. (1927). *Rezente wirbeltierleichen und ihre paläobiologische bedeutung*. Chicago: University of Chicago Press.

White, T. D. (1992). *Prehistoric cannibalism at Mancos*. Princeton, NJ: Princeton University Press.

Yellen, J. E. (1991). Small mammals: Post-discard pattering of Kung San faunal remains. *Journal of Anthropological Archaeology*, *10*, 152–192.

Part II
Human Activity

Chapter 7
The Lithic Assemblage of Level J

Manuel Vaquero, María Gema Chacón, Felipe Cuartero, Mª Dolores García-Antón, Bruno Gómez de Soler and Kenneth Martínez

Abstract We present in this chapter the lithic assemblage of level J. Bearing in mind the temporal issues involved in the interpretation of archeological assemblages, we have organized this chapter according to two different analytical levels. On the one hand, we have studied the technological and functional characteristics of the assemblage-as-a-whole, paying special attention to raw material provisioning, knapping strategies and tool manufacture. Raw material provisioning was basically local, although the nearest materials were not the most exploited. Core reduction strategies show a high variability due to the dominance of expedient criteria aimed at the production of small flakes. Finally, tool manufacture was characterized by the clear dominance of denticulates and notches, which were used at different tasks, but especially at hideworking. On the other hand, the second analytical level is focused on the event level and tryes to identify as much single technical episodes as possible. This approach is based on refitting and the identification of Raw Material Units. The spatial distribution of these events allows discussing the temporal dynamics in the formation of the lithic assemblages, highlighting the time-dependent nature of assemblage variability.

Keywords Lithic analysis • Refitting • Raw material unit • Discoidal method • Denticulate • Use-wear • Raw material provisioning

M. Vaquero (✉) · M. G. Chacón · Mª. D. García-Antón · B. Gómez de Soler · K. Martínez
Institut Català de Paleoecologia Humana i Evolució Social (IPHES), Universitat Rovira i Virgili (URV), C/Escorxador s/n, 43003 Tarragona, Spain
e-mail: manuel.vaquero@urv.cat

M. G. Chacón
UMR7194 - Département de Préhistoire, Muséum national d'Histoire naturelle, 1, rue René Panhard, 75013 Paris, France
e-mail: gchacon@iphes.cat

Mª. D. García-Antón
e-mail: madolores.garciaanton@urv.cat

B. Gómez de Soler
e-mail: bruno@prehistoria.urv.cat

K. Martínez
e-mail: kenneth@prehistoria.urv.cat

F. Cuartero
Departamento de Prehistoria y Arqueología, Universidad Autónoma de Madrid Ciudad Universitaria de Cantoblanco, 28049 Madrid, Spain
e-mail: felipecuartero@yahoo.es

Introduction: The Variability of Middle Paleolithic Lithic Assemblages and its Behavioral Implications

The variability of the Middle Paleolithic stone tool assemblages is one of the principal issues in the classic debate about Neanderthal populations, especially because it is one of the main approaches to understanding the behavior of hominids and obtaining information about social organization and evolutionary cultural changes. Since the definition of the Middle Paleolithic, the hypotheses proposed to explain assemblage variability have been numerous, varied and based on different causes. At first, the evolution of Mousterian stone tools was seen as following a unilinear progress from early to more complex types. This idea continued until Peyrony (1930) proposed that Middle Paleolithic industries formed a "complex" characterized by high assemblage variability. François Bordes (1948, 1953, 1961) systematized tool typology and defined the complexity of Middle Paleolithic assemblages. He established the basic terms of the classical Mousterian debate, distinguishing five principal "facies" or assemblage groups in western Europe: Charentian Mousterian, which included the Quina and Ferrasie subfacies, Typical Mousterian, Mousterian of Acheulian Tradition (MTA) and Denticulate Mousterian (Bordes and Bourgon 1951; Bordes 1953). These facies were mainly characterized by the percentages

E. Carbonell i Roura (ed.), *High Resolution Archaeology and Neanderthal Behavior: Time and Space in Level J of Abric Romaní (Capellades, Spain)*, Vertebrate Paleobiology and Paleoanthropology,
DOI: 10.1007/978-94-007-3922-2_7, © Springer Science+Business Media B.V. 2012

of different tool types, although core reduction strategies—especially the Levallois method—also played a role.

From this starting point, the controversy focused on the interpretation of assemblage variability and was marked by the opposition between the cultural and functional paradigms. On the one hand, the cultural hypothesis (Bordes and Bourgon 1951; Bordes 1961, 1992; Bordes and Sonneville-Bordes 1970) proposed that the industrial facies corresponded to different cultures that would have different stylistic conventions. Each type of Mousterian would represent a different group of people characterized by its own traditional way of making tools. On the other hand, the functional interpretation (Binford and Binford 1966) suggested that the Mousterian facies were associated with different toolkits used in different activities.

However, the Bordes/Binford debate came to a dead end, and from the beginning of the 1980s research on Middle Paleolithic variability tended to take on a different perspective. It was becoming clear that other factors beyond style and function influenced the variability of lithic assemblages. The new hypotheses about Middle Paleolithic variability took on a multiform character, converging different theoretical and methodological perspectives (Geneste 1985; Jelinek 1994; Bisson 2000; Moncel 2001). Instead of focusing on any single cause, the new interpretations considered a wide range of possible causes that interacted simultaneously.

Research on core reduction strategies became a key issue in this debate, especially as far as the definition and implications of knapping methods were concerned. For example, the debate about the definition of Levallois and discoidal technologies plays a very important role (Boëda et al. 1990; Boëda 1991, 1993a, b; Dibble 1995b; Jaubert and Farizy 1995; Mourre 2003; Terradas 2003; Vaquero and Carbonell 2003). In addition, retouched artifacts—the primary focus of previous research—were also interpreted in new ways. Both the stylistic and functional hypotheses rested on the assumption that tool types were discrete entities reflecting mental templates. The "reduction model" proposed by Dibble (1984, 1987, 1988, 1991, 1995a) was a significant challenge to this assumption, since he suggested that variability among certain tools—especially sidescrapers—was a function of reduction intensity through the continuous resharpening of the edges.

The most significant contribution of these recent approaches is the central role played by behavioral patterns, particularly those concerning behavioral adaptation to changing economic constraints. The management of lithic resources, conditioned by raw material availability, climatic factors and settlement patterns, explained the differences in tool reduction intensity pointed out by the "reduction model" (Rolland 1981, 1988; Rolland and Dibble 1990; Dibble and Rolland 1992). Provisioning strategies, artifact transport, function of sites and mobility patterns were relevant factors in explaining assemblage variability because they were some of the main constraints in technical behavior (Tavoso 1984; Geneste 1985, 1988, 1989b; Meignen 1988; Féblot-Augustins 1997, 1999; Turq 2000). The transport and method of introduction of lithic resources influenced retouched artifact inventories and knapping methods, since intensity of use was conditioned by the distance to lithic raw material sources.

Raw material economy (Geneste 1985, 1988, 1989, 1991) appeared as a common behavior during the Middle Paleolithic. The distance to raw material procurement areas caused a quantitative and qualitative selection of the lithic items introduced into the site. Some researchers tried to explain variability according to settlement and mobility patterns, very often based on ethnoarcheological studies of modern hunter-gatherers (Binford 1979, 1980). According to these hypotheses, occupation length and mobility ranges were basic factors in lithic assemblage formation. One of the principal applications of these models was carried out by Kuhn for the Middle Paleolithic of Latium (Kuhn 1989, 1992, 1995), which led to the recognition of two kinds of technical contexts:

- High mobility and short-term occupations. Lithic assemblages would be characterized by few remains, high indices of transported artifacts and reduction sequences aimed at obtaining large supports. Scavenging of animal resources was the main subsistence strategy.
- Scarce mobility and long-term occupations. Lithic assemblages would show few exotic objects and reduction sequences aimed at maximizing the number of flakes per core. In these contexts, animal resources were basically obtained through hunting.

However, these behavioral inferences depend on the formation processes of the lithic assemblages. Lithic assemblages are normally defined according to geological criteria. Remains included in an assemblage are those found in the same stratigraphic level. Most assemblages are therefore palimpsests formed during some hundreds or even thousands of years. Technological studies assume that the behavior deduced from such palimpsests is representative of the actual behavior of past populations, regardless their temporal depth. Nevertheless, the influence of temporal resolution in assemblage and inter-assemblage variability should be explored, trying to approach the two temporal levels accessible through archeological research: the "assemblage-as-a-whole" and the single technical event.

The essential tools for distinguishing and reconstructing the different technical events forming the palimpsest are the study of refits and the classification of remains according to the reduction sequence they come from. Spatial analysis and diacritical analysis can also provide valuable information on this matter. If we identify the different synchronic or diachronic single activity events, we may achieve a different perspective on the behavioral issues concerning lithic assemblage formation. Each technical event is an activity carried out by an individual at a specific time in answer to a specific demand (Hill and Gunn 1977; Brooks 1982; Gamble and Porr 2005; Beck et al. 2007). In this sense, descriptions of individual behaviors are indispensable (Cahen and Keeley 1980) and the event-focused analysis implies human behavioral strategies and choices. They are essential to understanding the origin and variability of the archeological record and are also a basic domain for testing hypotheses about the complexity of human populations and interpreting human activities at the site (Vaquero 2008).

An evaluation of technical processes in their spatial and temporal contexts allows for a global approach to lithic variability. The issue must be the interpretation of technical intra-assemblage variability and not the variability of a specific group of artifacts. The development of interdisciplinary studies and a higher temporal resolution are the main procedures involved in understanding the behavior of Middle Paleolithic hominids and their different degrees of variability.

Abric Romaní is a very suitable site for this kind of study due to its sedimentary context—archeological levels embedded between travertine layers—and the fieldwork method employed, which was based on the excavation of the entire occupied surface and the three-dimensional recording of all the archeological remains. In addition to a conventional "assemblage-as-a-whole" approach, we will try to identify specific activity events and to reconstruct their timing during the formation of level J. This method has already been tested in another level of the Abric Romaní sequence (Vaquero et al. 2004; Vaquero 2005, 2008).

Methods and Levels of Analysis

This study has been structured to consider two levels of analysis, which correspond to two different time scales. The first is the assemblage level, in which all the lithic remains are considered as a whole, differentiating only between the two main stratigraphic units (Ja and Jb). This is the highest temporal resolution that can be obtained through exclusively geological criteria, so assemblages constructed using such criteria should be considered as palimpsests formed by an unknown number of technical events. The approach to this level will focus on attribute and use-wear analyses.

The second level of analysis might be called the event level, and focuses on identifying the maximum number of technical episodes. It consists of the highest temporal resolution that can be achieved by means of the archeological record and it is the best approach to "ethnographic time". We will use the refitting and identification of Raw Material Units (RMUs) as the basic empirical procedures with which to tackle this level of analysis. However, these analytical levels should not be viewed as two independent domains. One objective of this study is to test whether an event-focused analysis provides a different picture from that yielded by an assemblage-level analysis. This line of inquiry has important implications concerning the validity of behavioral inferences made from archeological palimpsests. It is important to stress that both levels can provide valuable information. We do not deny the relevance of the assemblage level analysis. However, we think that it is not equally valid as a means to answering certain questions. The assemblage level approach can be especially appropriate for addressing long-term technological changes, while the short-term behavioral responses can be more reliably approached at the event level.

Artifact Attribute Analysis

We conducted an attribute analysis of the main classes of artifacts forming a lithic assemblage: unworked nodules, cores, flakes and retouched artifacts. Some attributes (especially those based on artifact dimensions) were recorded for all these classes, but others were specific to each class. Only lithics larger than 1 cm were included in this analysis. Artifact size is an important characteristic, as it yields information on different areas of research. Some of these topics are related to behavioral patterns (stages of core reduction, goals of production sequences, selection criteria for tool manufacture, etc.), but others are related to natural formation processes and the postdepositional disturbance of archeological sites. Most natural dynamics tend to produce size-sorting phenomena, thus making dimensional analyses a fundamental approach to identifying them (Rick 1976; Schick 1987). In this study, size was analyzed by recording the length, width and thickness of the artifacts. These measurements were use to create five size classes, defined by multiplying length by width: very small (<500 mm^2), small (500–1000 mm^2), medium (1000–1500 mm^2), large (1500–2000 mm^2) and very large (>2000 mm^2). In addition, two indices were calculated for flakes and retouched artifacts: an elongation index and a thickness index. The former is calculated by dividing length by width, the latter by dividing length or width (depending on which of these dimensions is the lowest) by thickness (Laplace 1972). These indices were calculated for unbroken blanks only.

We have distinguished two main domains of activity associated with different artifact classes: the use of unworked nodules in percussion activities, and knapping activities intended for flake and tool production. We have focused special attention on the study of unworked cobbles used as hammerstones. These heavy-duty materials are usually identified or minimally described, but have rarely been analyzed in a functional overview. Our main interest is

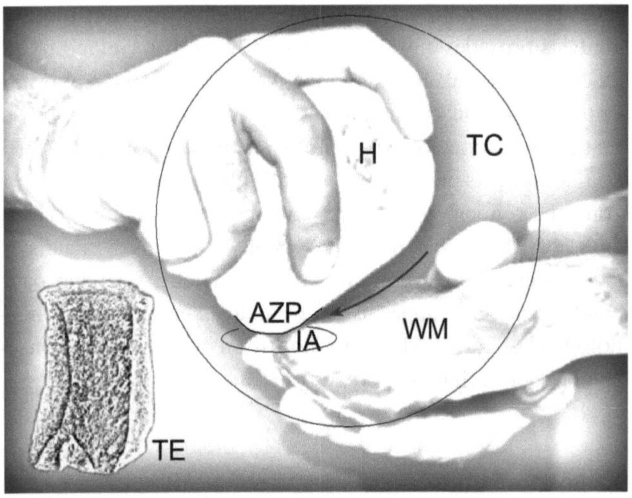

Fig. 7.1 Elements involved in an action of percussion. *H* Hammerstone. *AZP* Active zone of percussion. *IA* Impact area. *WM* Worked material. *TC* Type of contact. *TE* Technical effect

the gesture information that can be deduced from the wear or fracture patterns linked to different percussion activities, but primarily in their interaction with stone tool production. Analyzing breakage and wear traces is one of the best ways to discover the direction and intensity of percussion movements. The appearance and development of stigmata and fractures on cobbles used as percussive tools depends on the properties of the interacting materials as well as the way they are made to interact. The properties of the hammerstone and the worked material, such as elasticity, tenacity, hardness or density, roughness and topography or volume of both surfaces, as well as their structure itself, implies a higher or lower degree of mechanical resistance.

We call the active zone of percussion (AZP) the zone of the cobble that imposes a stress on the impact area (IA) of the worked material (WM) (Fig. 7.1). The combination of both materials in a percussive action with a specific movement is called the type of contact (TC), as the minimal unit of movement in which the angle of incidence on the platform and the direction and velocity of movement are considered. The resulting breakage can be related to the technical effect (TE) of the use of a specific type of contact. These parameters have been analyzed on knapped stone tool production by some authors (Faulkner 1972; Cotterell and Kamminga 1979) and in more specific analyses (Dibble and Pelcin 1995) in order to understand the mechanical principles operating on hertzian fractures in fragile rocks. These studies focused mainly on the products and by-products of knapping, but they also allow us to explain hammerstone breakage. Cotterell and Kamminga (1979) identified three main fracture types: the wedging fracture, the hertzian or conchoidal fracture and the bending fracture, which involves a more or less compressive action of percussion.

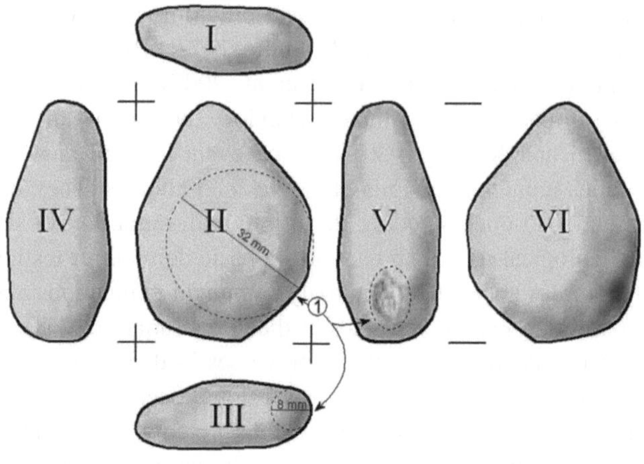

Fig. 7.2 Designation of active zones of percussion and circumference diameter measurement

Hammerstones break when the applied stress surpasses the limit of mechanical resistance supported by the AZP. The scatter resulting from the broken fragment or flake can show the direction and intensity of the blow on the artifact. Other agents such as fissures, chemical alterations or heat shocks can aid, induce or produce similar fractures in rocks with or without the interaction of percussion.

The location of the AZP on the cobble requires a good bi-dimensional projection, which can be accomplished with some graphical conventions. Taking into account the main axis of the cobble, we propose the following orientation criteria that allow us to distinguish six different areas on the cobble surface (Fig. 7.2):

- The most detached surface (the one with the lowest diameter of curvature in the main convexity) will be located in the distal part (I) and the less detached (higher diameter of curvature) in the proximal part (III). They will therefore be the top (I) and bottom (III) of the length axis of the artifact.
- The more convex face will be considered the upper (II) and the flatter will be considered the lower surface (VI).
- The more curved side will be placed to the right (V) and the straighter to the left (IV).

The location of the AZPs has been identified according to these criteria. For example, an AZP might be identified on surface I when it is centered in the distal tip or between I and IV or between I, IV and II when the center of the active area is placed in an oblique position in relation to main axis of the cobble. As the morphology of the cobble may be irregular, with squared or triangular forms, we have also specified whether such morphology conditions this lateralized position. According to the main axis (length), we distinguish between a centered, oblique or lateralized AZP, which can also be described by its position in main views: centered to I, lateralized from III to II, between VI and I, between III, V and II, etc.

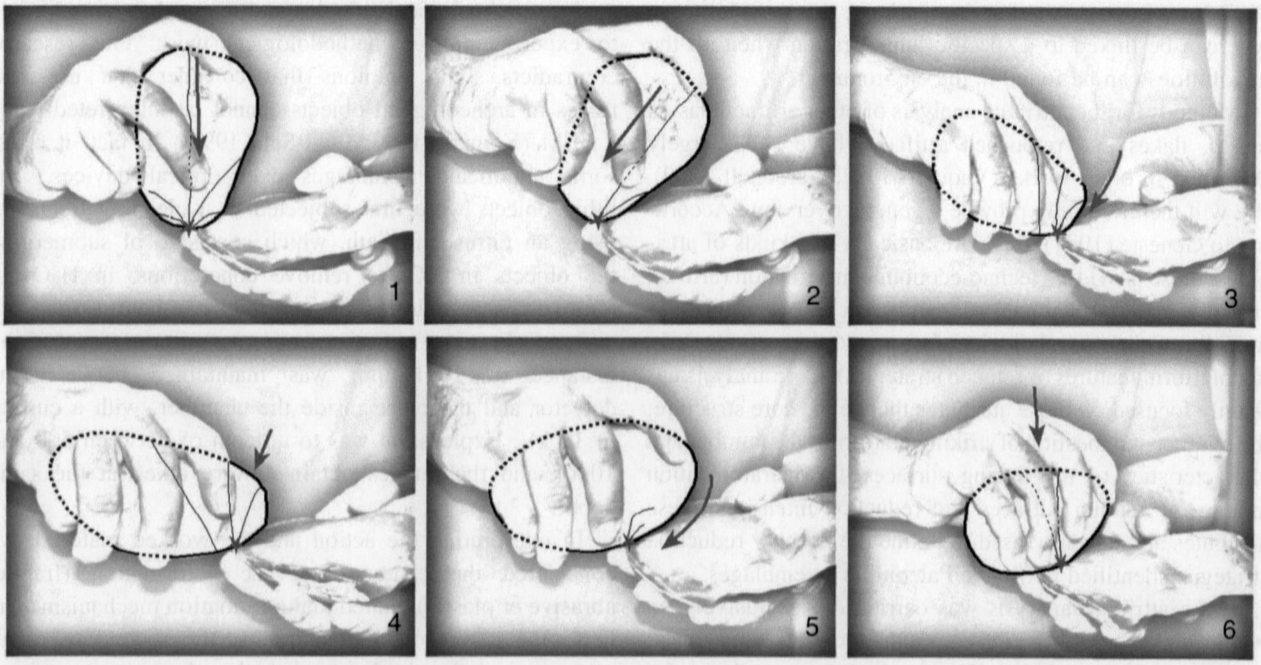

Fig. 7.3 Theoretical movements of a biaxial middle-size cobble and location of the main wear-traces and breakages caused by each one of them. *1* Frontal/longitudinal. *2* Oblique/semi-longitudinal. *3* Tangential/semi-transversal. *4* Oblique/transversal. *5* Frontal/transversal. *6* Tangential/transversal

We can distinguish several types of macroscopic stigmata in hammerstones: percussion points, pits, crushing, hertzian cones, longitudinal and transverse grooves (striations and cuts), and polishes. Their development and intensity depends on the properties of the cobble, the worked material and the stress applied, but especially the rock structure. We will only focus on surface changes and traces, mainly those that indicate a direction of movement. Some of the most explicit stigmata indicating direction are hertzian cones (when partially developed) and grooves, either longitudinal or transversal. Other traces, such as pitted, crushed or polished zones can give more or less information depending on their relative position on the surface of the cobble or their combination-superimposition with other traces.

The different possibilities of percussion use for a cobble in terms of manageability are related to its morphology and size. Prehension has been defined by biomechanical parameters in a distinction between power grip and precision grip (Napier 1962). In lithic tools, a semi-precision grip has also been distinguished, and size ranges for these three options have been proposed (Carbonell 2002). We use these established size ranges with some specificities for rounded forms. Based on their primary axes, we differentiate four basic cobble morphologies: uniaxial, thick biaxial, flat biaxial and triaxial or laminar. We use the terms uniaxial, biaxial or triaxial to define morphological proportions: uniaxial (length = width = thickness); thick biaxial (length > width = thickness); flat biaxial (length = width > thickness); and triaxial (length > width > thickness). These are complemented by all the possible shapes in more or less abstract geometrical forms. Elongated forms (thick biaxial and triaxial) allow for a wider range of prehension and movement and facilitate the reconstruction of their use based on fractures and positions of stigmata.

Considering prehension, movement, and the location of the active zone, we propose six different simple gesture possibilities: (1) frontal/longitudinal, (2) oblique/semi-longitudinal, (3) tangential/semi-transversal, (4) oblique/transversal, (5) frontal/transversal, and (6) tangential/transversal. The possibilities especially relevant to medium-sized and large thick biaxial and triaxial cobbles are simplified to movements one, two and six when produced with uniaxial and flat biaxial artifacts. Within these six options, prehension types for cobbles with elongated morphologies are as follows: the first two imply semi-precise longitudinal prehension with little pressure on the palm; the third and forth imply transversal prehension with the opposing part of the active zone covered by half of the palm and the fourth and fifth fingers; the last two imply whole-palm prehension and lateral support with all the fingers. Fractures and the location and type of stigmata produced by each possible gesture are summarized in Fig. 7.3. The different shapes found in nature provide more options in relation to the active zone chosen. For example, triangular and square shapes offer highly detached AZPs or edges, lateralized or centered on the main axis, and are therefore suitable for lateralized uses.

Selection of one or another morphology, size or volumetric type can be linked to a specific lateralization when all the possibilities can be found in the environment.

Methods for the attribute analysis of other artifact classes (cores, flakes and retouched artifacts) have been largely described in other works (Vaquero 1997; Carbonell 2002). We will therefore only provide a general overview. According to Geneste (1991), there are basically two kinds of attributes: those providing techno-economic information (artifact size, raw material characteristics and cortical cover) and those linked to techno-psychological factors (core structure, striking platform features or retouch strategies). Core analysis has mainly focused on those attributes that define core structure: the number and location of striking platforms, the number and characteristics of the flaking surfaces, the hierarchization among the flaking surfaces and reduction intensity. These attributes are currently used to define the primary reduction strategies identified in Middle Paleolithic assemblages.

Flake attribute analysis was carried out exclusively on whole flakes. Broken flakes, both proximal and distal fragments, were excluded. We recorded three attributes for flake striking platforms: cortical cover (cortical or uncortical), striking platform type (flat, lineal or punctiform), and number of striking platform facets (unfaceted, bifaceted, multifaceted). The amount of dorsal cortex is expressed with the percentage of dorsal cover, and four values have been distinguished: entirely uncortical, entirely cortical, less than 50% cortical cover, and more than 50% cortical cover. Finally, we recorded two attributes for ventral surfaces: bulb of percussion type (marked or diffuse), and ventral curvature (concave, convex, straight or winding).

Retouched artifacts were analyzed using the attributes proposed by Georges Laplace in his *Analytical and Structural Typology* (Laplace 1972). Retouch is characterized by five attributes: retouch angle, depth, direction, delineation, and form. The retouched items have also been classified according to the primary types and typological groups defined by Laplace. We have also used two measures for approaching retouch intensity: the number of sides containing retouch and the retouch perimeter index. The latter is a ratio of the retouched edge relative to the total artifact edge (Hiscock and Attenbrow 2003).

Use-Wear Analysis

We used a scanning electron microscope (SEM) for our traceological analysis. This device, largely for reasons of cost and complexity of use, is not normally used to analyze archeological samples and its use has been limited to experimental or methodological work. Our research contradicts some opinions that consider that use-wear traces in archeological objects cannot be interpreted using a SEM (Yamada 1993; Levi-Sala 1996). In fact, it offers some significant advantages over optical devices. The lithic objects were first subjected to a cleaning treatment using an ultrasound bath, which consisted of submerging the objects in HCl to remove concretions, in H_2O_2 to remove any organic residue and finally in neutral soap and acetone. In terms of observation conditions, a working distance of 10–15 mm was maintained between the detector and the piece inside the chamber, with a current of 15 kV. Exploration was to a depth of between 500 and 1000x and the images obtained were taken at the same depth.

In interpreting the action and the worked material, we considered the character of the deformation (fragile, abrasive or plastic) related to the formation mechanism. The first type corresponds to chipping, the second to rough surfaces and the third to polished and compact surfaces (smooth surfaces contrasting with the natural grainy surface of the chert). The traceological interpretation also took into account the distribution of the deformations along the edge, the relationship between them, and their morphological features (Martínez 2005). Our own experimental work indicated that the differences in use-wear traces in relation to the worked material or the motion used are due to the fact that the use-wear process includes several mechanisms that are activated or alternated depending on the contact conditions between the stone and the worked material. The physical–chemical characteristics of worked materials are fundamental to the development of use-wear traces and the variety of textures of the stones is of vital importance in the degree of development of brittle and plastic deformations produced because of strain on the stone (Astruc 2001). Tribology and structural geology models explain that rubbing between contact points of two bodies causes a deformation, which begins with the loss of material through abrasive brittle and plastic mechanisms and continues with fluid plastic deformation that eliminates inter-crystalline boundaries through the movement of the material at an atomic and molecular level within the crystalline network (Hatcher 1995; Davids and Reynolds 1996). Throughout the deformation process, friction points are reduced and, therefore, so is the edge's resistance to motion. When it reaches this point, the edge dulls and more pressure must be applied in order to cut the worked material again. With increased pressure, the edge fractures, creating new contact points and the described deformation process is repeated (Martínez 2005).

Refitting and Raw Material Units (RMU)

As we stated earlier, the event approach is based on the identification of individual knapping episodes. Refitting and raw material unit identification are the best procedures to employ in order to document these individual events. Refitting methods have been extensively described in the literature (Cziesla et al. 1990; Hofman and Enloe 1992; Morrow 1996a; Schurmans and De Bie 2007). We have used the methods and terminology suggested in Cziesla (1990). Three kinds of connections have been distinguished: production-sequence, breakage and retouch refits. For production and retouch refits, connection lines have been defined according to the chronological order of detachments, from the first refitted flake to the core or tool. In the refits of broken artifacts, connection lines have been established taking into account the contacting surface. We have calculated length and direction for all connection lines.

A raw material unit (RMU) incorporates the artifacts produced during the reduction of a single nodule (Roebroeks 1988). RMUs are defined according to the macroscopic characteristics of the raw materials including color, grain size, texture, inclusions and type of cortex. This procedure, also known as minimum analytical nodule analysis (Hall 2004; Larson 2004; Odell 2004, pp 93–95), is especially useful in assemblages with lithics of variable appearance. In such contexts, it is easier to attribute a series of lithics to the same knapping sequence according to their macroscopic characteristics. This approach is also particularly suitable for assemblages made up of a small number of artifacts, as shown recently in the Broion cave (Peresani and Porraz 2004). Although they come from the same nodule, the artifacts forming a RMU do not necessarily correspond to the same technical event, since different reduction or retouch episodes can be carried out at different times and places on the artifacts detached from a single nodule. Recycling of previously discarded blanks is one of the best examples of how different knapping events can be identified in a single RMU. According to the number and type of remains, RMUs have been characterized taking into account two main features: how they were introduced into the site and what kind of intentional modification was performed on them inside the rockshelter. We have distinguished four ways in which RMUs were introduced:

A. As entire or almost entire nodules.
B. As partially reduced cores.
C. As isolated blanks (retouched artifacts, flakes or flake fragments) produced outside. As we will see, this is a common provisioning method in level J. We will designate these items as transported blanks (TB).
D. As sets of blanks detached outside in the same reduction sequence but introduced together into the site.

For each of these introduction variants, the following possible on-site modifications have been considered:

1. Complete reduction and discard of the core or the retouched item.
2. Partial reduction and transport of the core or the retouched item out of the rockshelter.
3. Discard inside the rockshelter without modification.

The spatial distribution of RMUs and refitting groups is especially important in testing hypotheses on cultural and natural formation processes, postdepositional disturbance, occupation strategies and temporal patterns. The data from this level of analysis will therefore be explained within the framework of this spatial dimension. Refits are especially informative regarding the temporal relationships among different activity areas. Although connections through refits have sometimes been considered as evidence of contemporaneity, it now seems clear that only bidirectional connections can be used to argue that two activity areas or clusters of remains were contemporaneous. Unidirectional connections cannot be used to support contemporaneity. On the contrary, they can provide a good argument in favor of a temporal gap of unknown length between the formations of the two accumulations.

The spatial distribution of RMUs indicates the location of knapping activities. The scattering of RMUs is also important in studying the temporal dynamics in the formation of the lithic assemblage. We have calculated the dispersion radii of all the RMUs made up of more than five artifacts. According to some experimental models, the degree of dispersion of the artifacts resulting from a knapping episode depends on its temporal location in the sequence of technical events forming the lithic assemblage (Stevenson 1985, 1991). Earlier episodes tend to be more widely scattered, since they would have been more affected by intentional and unintentional dispersion factors associated with human occupation. As the knapping events approach the latest phases of occupation, their lithic scatters are less subject to these dispersion processes and they therefore tend to be more clustered.

Raw Material Provisioning

The study of the formations containing lithic raw materials focuses on a 30-km radius around the site. According to Geneste (1989a, b, 1992), and Geneste and Rigaud (1989), between 70 and 90% of the lithic remains from Middle Paleolithic sites of Aquitaine originated from local sources located within a 5 km radius. The same occurs in the archeological sites studied by Meignen (1988) and Turq (1989, 1990) in the same region. Binford (1979, 1982) identified the same pattern in his ethnoarcheological studies among the Nunamiut Eskimo. It seems, therefore, that most Middle Paleolithic assemblages are characterized by low percentages of materials originating in sources at long distances (Féblot-Augustins 1999). We have used these studies as a reference, establishing a 30-km radius around Abric Romaní as our main focus of inquiry. Chert is the dominant raw material in all the lithic assemblages of the Abric Romaní sequence. It is also the raw material exhibiting the most variability as far as the potential provisioning source is concerned. Other materials, like quartz or limestone, are available in the surroundings of the site and tend to be more homogeneous. Therefore, the field research has basically been aimed at identifying chert-bearing formations.

Since the 1970s, the use of the terms 'chert' or 'flint' has been the subject of debate (Luetdke 1992). Some researchers continue using flint (Fernandes et al. 2008), whereas others use chert to refer to the raw material formations and flint for the archeological record (Pettijohn 1975, Verri et al. 2005). Other authors utilize flint only to designate the nodules included in the cretaceous levels of southern England and northern France (Evans et al. 2007). Nevertheless, due to the increasing preference for the term chert in geology and archeology, we have decided to use it in this work.

Geological Formation of Raw Materials

As shown in Table 7.1, the raw materials found in Abric Romaní can be ascribed to 11 geological formations (Fig. 7.4):

- **Paleozoic formations.** These correspond to the basal levels of the Prelittoral Range and are formed by Silurian slates crossed by quartz sills and, occasionally, porphyry dikes. This Paleozoic stratum is scarcely represented in the surroundings of Capellades but further to the south it can be found lying in an east to west direction. In St. Quintí de Mediona, we found the same monotonous series of ash gray slates (García Rodrigo 1957). Slate, quartz, quartzite and porphyry were the Paleozoic materials exploited in level J.
- **Lower Muschelkalk formations**. Bofarull (1997) and Morant (1998) pointed out that chert can be found in the dolomitic limestone of the Lower Muschelkalk. According to these authors, it appears as nodules on Marquès mountain, west of St. Quintí de Mediona.
- **Mediona formation.** This formation is characterized by the presence of *Vidaliella gerundensis* (Bullimus), which suggests a Thanetian age (Upper Paleocene) (Anadón 1978). It is geographically discontinuous and composed of red lutite layers of continental, fluvial and lacustrine origin, inserted in calcrete levels with edaphic-origin chert. The chert-bearing deposits nearest to Abric Romaní were documented at St. Joan de Mediona, near St. Quintí de Mediona creek (Gómez de Soler 2007).
- **Orpí formation.** This is a Lower-Middle Ilerdian formation (Lower Eocene) (Ferrer 1971; Anadón 1978) that appears to the east of Capellades and is distributed throughout the Gaià block and towards the interior of the Ebro basin. We can distinguish three layers in this formation: the basal layer, formed by dolomites and, occasionally, carbonated breccias and cellular dolomites; the intermediate layer, formed by fossil-rich limestone; and the upper dolomitic layer. Limestone is found in the intermediate layer composed of limestone grainstone of foraminifers belonging to the transgressive phase (Morant 1998).
- **Sta. Càndida formation.** This formation is of Upper Ilerdian-Lower Cuisian age (Lower Eocene) and has not yet been prospected. According to bibliographical data, chert is located among dolomites (Anadón 1978). It extends from the northwest border of the Gaià block in Vallespinosa to east of the town of Carme. It is composed of alternating marly levels and lutitic and carbonated levels, mainly limestone and, to a lesser extent, dolomites with cherts. It is an evaporitic formation and the depositional environment corresponds to carbonated lagoons and lacustrine areas of fresh water. The St. Magí-Vallespinosa area, containing a large chert concentration, has lacustrine characteristics.
- **Valldeperes formation.** This stretches 55 km between the towns of Vilaverd and Valldeperes and, according to Ortí et al. (2007), dates to the Bartonian age (Middle Eocene). It is made up of two layers of white calcareous dolomites with chert and limestone that include levels of calcareous marls and dolomites. It is characterized by nodular plasters developed in dolomitic layers, which constitute in some cases the dominant lithofacies (Valldeperes). Due to their vicinity to Abric Romaní, we have centered our attention on the chert-bearing areas of Vallespinosa and Valldeperes (Anadón 1978; Ortí 1990; Bofarull 1997; Morant 1998; Gómez de Soler 2007).
- **Pobla de Claramunt formation.** The chert is located in the area of Carme (Gómez 2007). It is an Upper Lutetian-Lower Bartonian formation (Middle-Upper Eocene) made up of red sandstones and lutites with thick conglomerate levels. The formation has a typically fluvial origin of the channel filling type. The materials were deposited in fluvial environments, corresponding to distal areas of alluvial fans. Chert nodules, although very scarce, were documented in the conglomerates located

Table 7.1 Localisation and description of the lithic raw material formations

Systems and series	Lithostratigraphic units (formals and informals)	Outcrop location (*)	Raw material	Microscopic texture	DRX	Size	Suitability for knapping	Distance
Palaeozoic (Ordovician)	Materials come from quaternary dismantling of the terraces and the colluviums	Capellades Strait (PZC)	Slate, quartz, porphyry and quartzite in primary position	–	–	5–40 cm	Variable knapping suitability	In front of the site ≥300 m
Triassic (Lower Muschelkalk)	Dolomitic limestone formation that contains chert lens nodules	St. Quíntí de Mediona (MED)	Chert in primary position	Cryptoquartz Small mosaic quartz and calcite	Quartz, calcite and dolomite	≤10 cm	Difficult for knapping due to internal fractures	≥8–10 km
Paleocene (Thanetien)	Mediona formation. Red lutite with levels of limeston caliche with chert and conglomerate	St. Joan de Mediona (SJM)	Chert in primary position	(♣)	(♣)	5–20 cm	Difficult for knapping due to internal fractures	≥7–9 km
Eocene (Ilerdian)	Orpí formation. Marine limestone with fossils (*Nummulites perforatis*) and microfossils (severals types of alveolines)	Orpí	Limestone in primary position	Biomicrite	Calcite	≤50 cm	Good for knapping	≥4–10 km
Eocene (◆) (Ilerdian-Cuisian)	St. Càndida formation. Alternance of marls, lutites and carbonate levels, predominately limestones and sometimes dolomitic limestone with cherts	*Non located* St. Magí (SMG)	Chert in primary position	–	–	–	–	≥20–25 km
Eocene (Bartonian)	Valldeperes formation. Palaeogene evaporitic formation with white dolomitic chalks having limestones, gypsum and cherts	Vallespinosa (VLD)	Chert in primary position	Crypto to microquartz and fibrous LFC	Quartz and moganite	≤1 m	Good for knapping	≥20–30 km
Eocene (Lutetian-Bartonian)	Pobla de Claramunt formation. Fluvial deposits formed by lutites, sandstone and conglomerate levels. The last ones with rounded chert nodules	Carme (CME)	Chert in secondary position	Cryptoquartz Small mosaic quartz and larger calcite crystal	Quartz, calcite and moganite	≤5 cm	Good for knapping	≥3 km
Eocene (Priabonian)	St. Martí de Tous formation. Sandy gray lutites with ripples lamination having chert and sandstone with crossed stratification and levels of lenticulates conglomerates	St. Martí de Tous (SMT)	Chert in primary position	(♣)	(♣)	≤1 m	Variable knapping suitability	≥12–15 km
Oligocene (Stampian)	Montmaneu formation. Carbonated lacustrine origin formed by stratigraphic limestones with nodular cherts	Panadella (PAN)	Chert in primary position	(♣)	(♣)	10–30 cm	Good for knapping	≥25–28 km
Pliocene	Guixera series. Stratified conglomerate formation with calcareous and siliceous rounded clasts from the Triassic and the Paleogene of the Ebro basin	St. Quínti de Mediona (SQM)	Chert in secondary position	Crypto to microquartz and fibrous LFC mosaic quartz and euhedral quartz	Quartz	6–18 cm	Variable knapping suitability	≥7–10 km
Quaternary	Fluvial stepped terrace deposits that contains pebbles (**) and cobbles (***)in sandy matrix	Anoia River	Limestone, chert and quartz in secondary position	–	–	≤6 cm (**) ≥25 cm (***)	Variable knapping suitability	≥150 m

(*) location of the samples, (**) pebble size, (***) cobble size, (♣) in process of study. *LFC* Length-Fast Chalcedony. (◆) data (Anadón 1978). Modified from Fernández-Laso et al. 2011, new data from Gómez (2007)

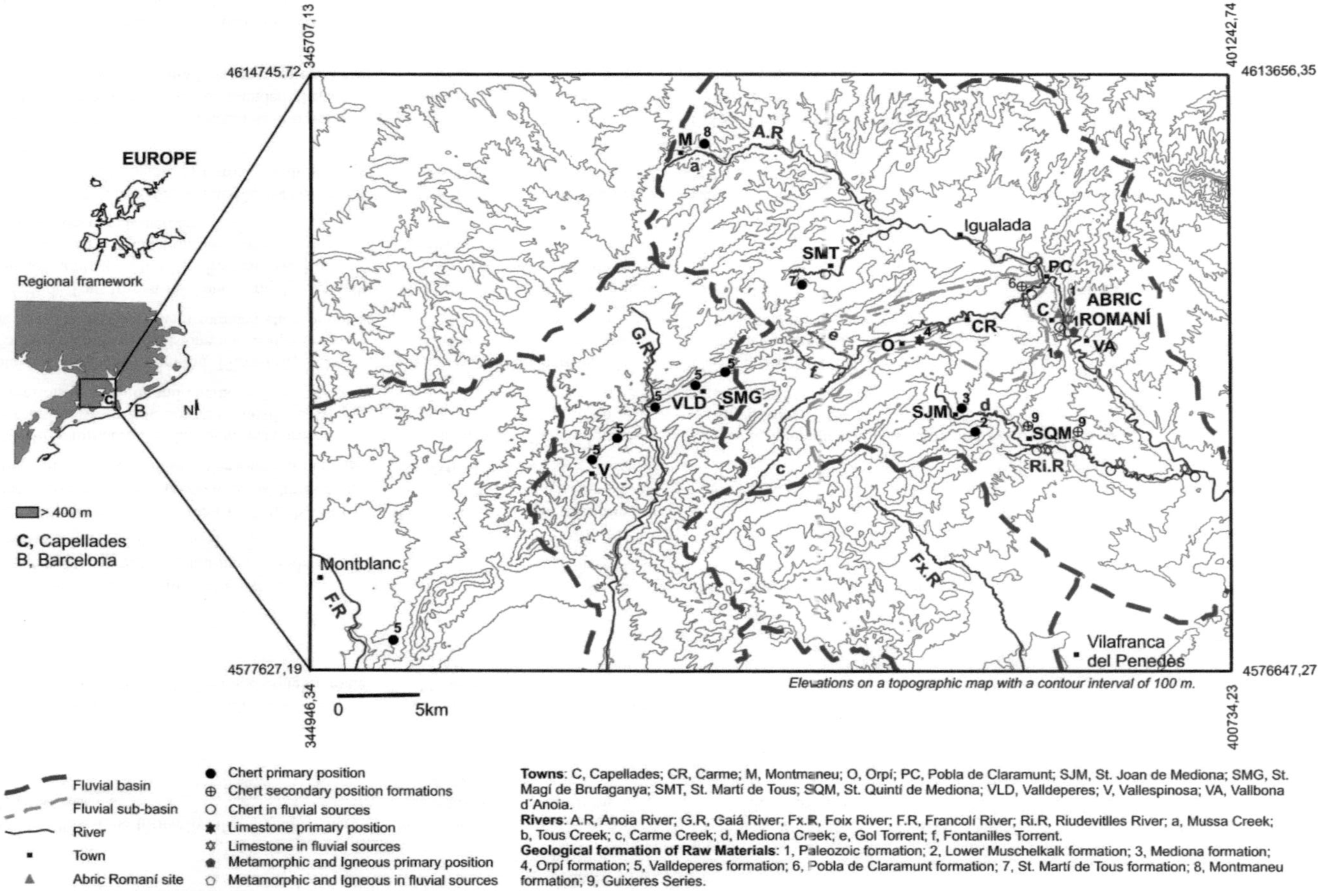

Fig. 7.4 Map showing the distribution of the geological formations with lithic raw materials

between Carme and La Pobla de Claramunt (Morant 1998; Gómez de Soler 2007).

- **St. Martí de Tous formation.** It is a Priabonian (Eocene) formation. Chert appears in the gray lutites of the St. Martí de Tous formation defined by Colldeforns et al. (1994); Gómez de Soler (2007). It is made up of a 50 m-thick succession, formed by gray sandy lutites with ripple lamination that contain metric sandstone levels with crossed stratification and conglomerate lenses. We have located three layers with abundant chert in primary position in the town of St. Martí de Tous.
- **Montmaneu formation.** This chert is located in the Panadella-Montmaneu area (Anadón et al. 1989; Colldeforns et al. 1994; Gómez de Soler 2007). This formation is comprised of deposits from the lacustrine system of La Segarra (Anadón et al. 1989*)*. It is formed by 120 m of stratified light grey limestone with nodular levels of stratified chert of the Lower Oligocene age.
- **Guixera series.** This chert (Gallart 1981; Carbonell et al. 1992; Vaquero 1997; Morant 1998; Gómez de Soler 2007) has been located in the conglomerates of St. Quintí de Mediona and St. Pere de Riudevitlles. It is a Pliocene age series of conglomerates in banks, inserted among red lutite packages with abundant indications of edaphogenesis. It has been suggested (Gómez de Soler 2007) that these sediments originated in the erosion of the Ebro basin sediments during the Miocene. Chert nodules found in these conglomerates would have their source area in the evaporitic deposits of the Valldeperes formation redeposited in the Penedès Depression.
- **Quaternary formations.** Different terrace levels with chert nodules have been documented in the fluvial system of the Anoia River. The Anoia River, the St. Martí de Tous and Carme creeks and the Mediona-Riudevitlles are the fluvial courses containing chert, limestone and quartz (Gómez de Soler 2007).

Characterization of Geological Materials

Siliceous Raw Materials

Lower Muschelkalk Formations (MED)

This Triassic chert can be found primarily in the Lower Muschelkalk levels of the Prelittoral range, west of Sant Quintí de Mediona. It appears in the dolomitic limestone and the tabulated dolomites (Marzo and Calvet 1985; Bofarull 1997; Morant 1998). The average nodule size is 200 mm in length and 80 mm in width. It is not suitable for knapping as inner joints are common and the conchoidal fracture tends to be schistose. This is an opaque grey–brown chert that presents quartz inclusions and microgeodes as well as secondary quartz growths on ancient fissures, possibly of tectonic origin. It has a whitish-yellow alteritic patina with concentrations of ferric oxides in the surface (Fig. 7.5). Its texture is cryptocrystalline, with a high index of carbonatation, dolomite crystals and clay minerals. X-ray diffraction (XRD) shows quartz, calcite and dolomite.

Fig. 7.5 Lower Muschelkalk formation with chert nodule

Mediona Formation (SJM)

This chert appears in primary position in breccias, although in small quantities. Nodules show an angular form and centimetric size, with a maximum thickness of 5 cm. Patina and concretion are absent. The cortex is calcareous, laminated and micritic, generally without transit and with a thickness of

Fig. 7.6 Chert from different geological formations. **a** Mediona formation. **b** Valldeperes formation. **c** Pobla de Claramunt formation. **d** Sant Martí de Tous formation. **e** Montmaneu formation. **f** Guixera series

1.5 mm. It is dark grey–brown with reddish tones due to the presence of oxidation and with opaque translucently. It is saccharoid to the touch and features conchoidal fracture with many fissures, which makes it somewhat difficult to knap. There are many iron oxides and glass growths. It has a laminated sedimentary structure (Fig. 7.6a).

Valldeperes Formation (VLD)

Three geologic levels with chert have been located (VLD1, VLD2 and VLD3), but we will describe them as a whole. Except for VLD3, which appears in a subprimary position with a lens-shaped structure, this chert is found in the primary position as tabulated and metric breccias. Patina and concretion are not common, but a tendency to exhibit white patina has been observed. The 1–2 mm thick cortex is of a calcareous-marly type, irregularly shaped and without transit. Its color goes from white to black in a range of grey–blue and has a translucent appearance. It has a smooth to saccharoid texture with a conchoidal fracture good for knapping. Quartz and gypsum druses have been found, as well as vegetable fibers and gypsum ghosts. Sedimentary structures are not present although some ooids and intraclasts can be observed (Fig. 7.6b).

Pobla de Claramunt Formation (CME)

Chert appears in secondary position and shows a great variability. We have established four groups (CME1, CME2, CME3 and CME4), but we will describe them together. Nodules appear in conglomerates as smooth pebbles generally smaller than 5 cm. In general, the material is not patinated, except CME3 and CME4, which show a white to rosy patina. The neo-cortex is calcareous, in some cases with conglomerate crust, without transit and less than 1 mm thick. It presents a very diverse coloration from brownish-grey to yellowish–brown and grizzly red, which tends to be translucent to opaque. It is smooth to rough to the touch, and in many cases presents a conchoidal fracture with fissures and is generally good for knapping. It presents many oxides and occasional terrigenous remains and ghosts. Its sedimentary structure is very heterogeneous, from a detritic-breccia structure (CME1) to fully laminated (CME2 and CME3) and without structure (CME4). Many alochem elements (pellets, ooids and intraclasts) and in some cases fossils have been found (Fig. 7.6c).

St. Martí de Tous Formation (SMT)

We have located three chert types, one in primary position (SMT3) and two in secondary position (SMT1 and SMT2). Our description will focus on the type of primary origin, which is also the most abundant. Chert appears as subangular and irregular metric-sized blocks. A weak white patina and some clay concretions may be present. The cortex is calcareous, laminated, 1–3 mm thick, without transit, black or brown and has a rough surface. The chert has a translucent appearance and its color varies from light to dark grey in a range of grey–blue. It is smooth to rough to the touch, and conchoidal fracture presents fissures perpendicular to the stratification plane, which gives it a variable aptitude for knapping. Many iron oxides, ghosts, covered quartzes and some geodes have been identified. It has a laminated sedimentary structure. Some pellets, ooids, intraclasts and in some cases fossils (possible spicules) (Fig. 7.6d) have been found.

Montmaneu Formation (PAN)

This chert in primary position appears as lens-shaped layers and pebbles of decimetric measurements in stratified limestone. The material can present some grayish–brown patina. The cortex is calcareous, stratified, grey to black, not gradual and less than 1 mm thick. Its color is blackish-green and has an opaque appearance, but near the cortex it may have brownish tonalities. When oxidations are present, its color varies towards light brown. It is smooth to the touch, and it presents a conchoidal fracture with very good knapping properties. Iron oxides have been found. Its sedimentary structure is laminated and the alochem elements are ooids, pellets and some bioclasts such as gastropods and spicules (Fig. 7.6e).

Guixera Series (SQM)

This chert appears in secondary position and is very heterogeneous due to the geological formation of origin: the Pliocene stratified deposits of conglomerates derived from the Triassic deposits of the Prelittoral range and the marine Paleogene of the Ebro basin border (Gallart 1981). Therefore, this chert would have its origin in the Muschelkalk of the Prelitoral Range and especially in the Valldeperes formation of the Ebro basin. This origin is supported by the macroscopic similarities of these cherts. In general, nodules are centimetric to decimetric. White patina and clay concretions are sometimes present. It presents a calcareous neo-cortex, without transit and less than 1.5 mm thick. Its color is very heterogeneous, and is typically in the blue–grey and brownish–grey ranges and of generally translucent transparency. Of smooth to saccharoid texture, it presents a conchoidal fracture, in some cases with fissures that do not always allow for good knapping. Some oxides are observed and sedimentary structures do not seem to be present. Alochem elements are very scarce (Fig. 7.6f).

Metamorphic and Igneous Materials

A great variety of Paleozoic and plutonic rocks can be found in the area surrounding the site. An intrusive pluton is located opposite Capellades and extends to the southeast to Vallbona d'Anoia. These materials suffered processes of regional metamorphism. They can be located in primary and subprimary positions and nodules exhibit angular forms and medium sizes (10–50 cm). Erosive processes brought great quantities of these rocks to the riverbeds. Therefore, they also appear in secondary position in the river terraces and other colluvial formations.

Granodiorite and Porphyry

Near the site, igneous rocks penetrate the Paleozoic substrate, without sudden temperature changes. In this case, infiltration seems to have been generated with a gradual increase of temperature, and therefore the rock aureoles caused by contact metamorphism are not very well developed (Guigon et al. 1973; Serra 1990). The pluton is composed of a fairly homogeneous granodiorite mass. This medium to coarse-grained rock shows an equigranular texture with a dominance of quartz, plagioclase and biotita. However, it is not good for knapping and is absent from level J. Moreover, the morphologies of the fragments represented in secondary deposits are unsuitable as percussors. The penetration of the granitic pluton favored the infill of fractures of the Paleozoic substrate, with the appearance of microporphydic granodioritic porphyries, which are green due to the presence of chlorite (Serra 1990; Peón et al. 1975). These porphyry dikes are also located throughout the pluton. The texture of this rock is finer than the granodiorite and some artifacts made from this material have been found in

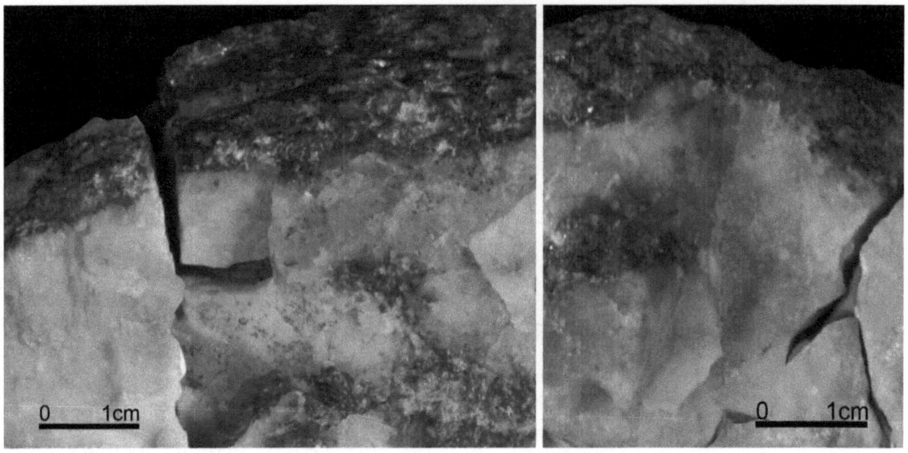

Fig. 7.7 Quartz fragment with slate veins from the Paleozoic levels around the site

level J. It is not common in secondary deposits due to its origin in dikes that traverse the Paleozoic.

Slate

The Paleozoic levels are mainly formed by metamorphic lutites that became slates of green schist facies (Fig. 7.7). These fine-grained, grey–blue slates show oxides on the schistosity planes. Slate was not used for knapping in level J. However, it was intentionally introduced in other archeological levels such as level E (Vaquero 1997), and even used in production sequences in level M (Fernández-Laso et al. 2011).

Metaquartzite

Quartzite is interstratified as thin layers between the slates. Two outcrops of dark quartzite have been documented (Serra 1990), the most significant of which is located north of Capellades, near the village of La Pobla de Claramunt, and the other is southwest of the site. It is dark-grey to black and sometimes red and its texture is cryptocrystalline granoblastic, including some quartz porphyroblasts in the matrix (Peón et al. 1975). It is an opaque rock, with conchoidal fracture, and its macroscopic characteristics are similar to those of chert. It is very good for knapping, although it is scarcely found in level J.

Quartz

There are veins of quartz in the Paleozoic slates, which are also represented among the granodiorites and are formed by white quartz with veins of grey–blue slates with iron oxide inclusions in orange to brown fissures and sometimes incorporated in their crystals. This kind of quartz was selected for knapping in level J and represents more than 90% of the quartz remains found in this level. This material is common near the site and was probably collected in primary or subprimary position.

Limestone

The limestone used in level J varies widely. In general, limestone cobbles were used as percussors, but some reduction sequences have also been identified. Gray to grayish-yellow limestone is especially noteworthy as it exhibits different types of fracture, with some nodules showing a conchoidal fracture suitable for knapping, while others exhibiting a very poor quality. Cortical surfaces clearly indicate that all the limestone nodules found in level J were of a secondary origin.

Orpí Limestone

Among the different types of limestone recovered in level J, only Orpí limestone has been identified (Fig. 7.8). The Orpí formation is located from the east to the southeast of Capellades. It is a Paleogen formation developed during an Ilerdian marine transgression. It contains orange dolomitic limestone of micritic texture. Conchoidal fracture is sometimes coarse-grained, but in general, it is good for knapping. It is easy to identify due to its high content of alveolines derived from its marine origin. Its texture has been described using SEM (scanning electron microscope) (Krinsley et al. 2001). Grain size is homogeneous, and calcitic replacement of the alveoline shells is characterized by a crystalline growth structure. X-ray diffraction shows calcite crystalline phases. Dolomite picks have not been detected, although this mineral has been described (Ortí 1990). Furthermore, microanalysis using scan electron (MASE) shows an elemental composition of aluminum, silica, magnesium, potassium, chlorine and iron in very low proportions.

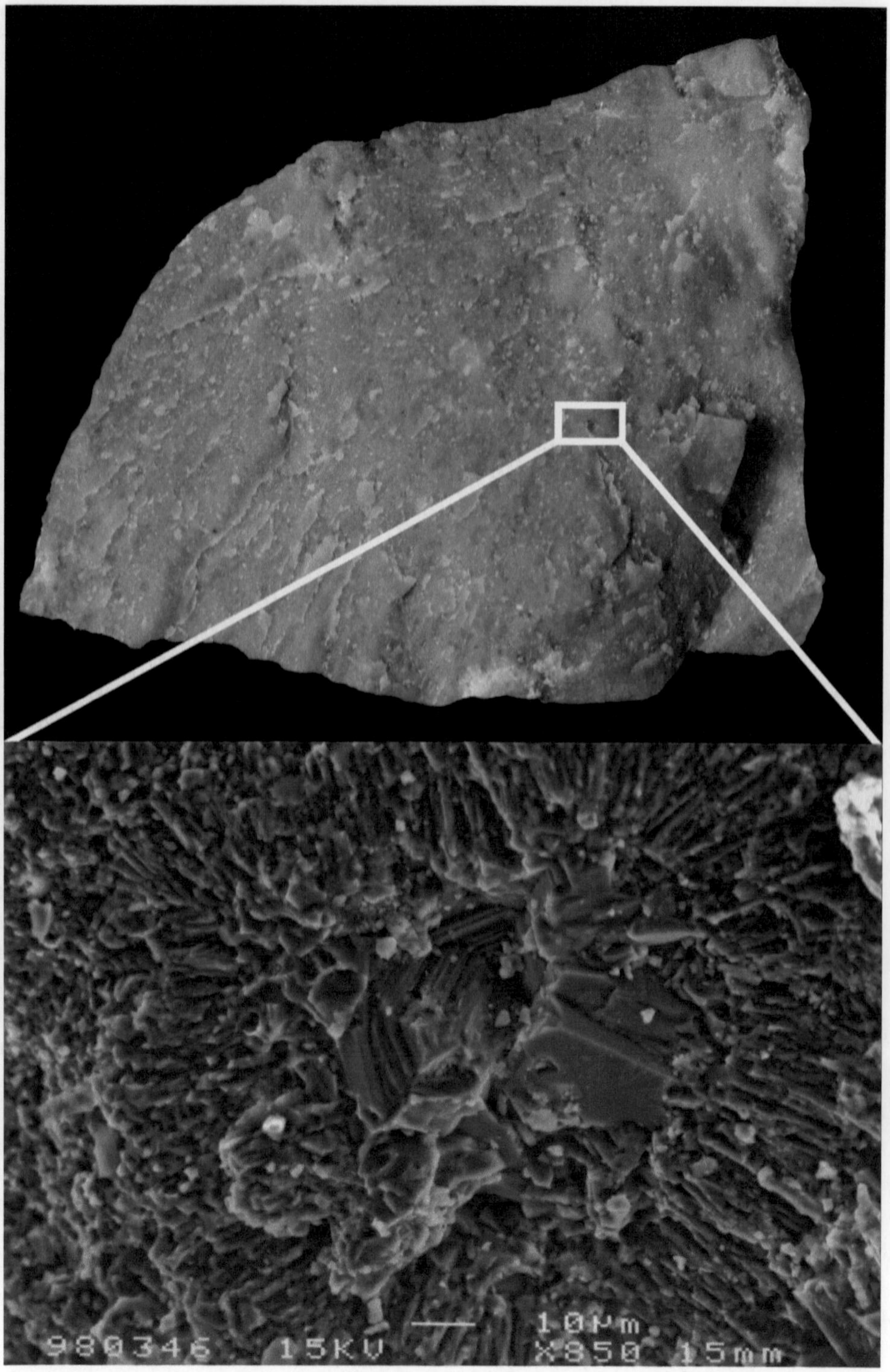

Fig. 7.8 Macroscopic and microscopic details of the Orpí limestone. The SEM image shows a detail of radial calcite replacement growth in an alveoline (SEM photograph by Nuria Morant)

Undetermined Limestone

There is a wide array of limestone gathered in secondary deposits, mainly fluvial terraces. Grayish-green and grayish-yellow limestones suitable for knapping—dominate in the Anoia terraces. Grey–brown and orange limestones with sparitic and microsparitic structures are less common. The texture of some of these good quality grey limestones has been analyzed. Like the Orpí limestones, they have been characterized using SEM and X-ray diffraction. These limestones show iron oxide inclusions in the form of small red points. The microanalysis by scan electron (MASE) indicates the presence of iron, as well as other elements forming part of clayey and saline minerals, such as aluminum, silica, phosphor, chloride, sulfur, potassium and sodium. Calcite is dominant in the X-ray diffraction, whereas

Fig. 7.9 Views of neo-cortex from La Guixera chert samples. **a** Development of a white weathering patina, and sub-rounded ridges. **b** White weathering patina covered by yellow patina, and sub-angular ridges

dolomite picks are less marked. These samples are characterized by a micritic texture, although crystal size presents some heterometry; it is larger than 10 μm in some concentrations. This composition coincides with the description of the basin-border limestone found in the marginal deposits of the Ebro basin (Ortí 1990; Colldefons et al. 1994), although we cannot ascribe them to any of the known facies.

Analysis of Corticality and Neo-Corticality

Cortical cover provides information on provisioning areas. It has sometimes been used intuitively, but there are currently an increasing number of wsorks that analyze these features in a new way, in an attempt to overcome the interpretations so far obtained from the archeological record (Fernandes and Raynal 2006; Fernandes et al. 2008). Two main types of cortex can be distinguished. The first is the original cortical surface, which presents a texture differentiating chert from the surrounding rock. Cortex type also indicates the bedrock where the chert was formed (Fernandes et al. 2006). Resedimentation processes formed the second type of cortex (neo-cortex). During resedimentation in colluvial and fluvial environments, rocks are affected by abrasion and alteration processes. Chert fragments found in secondary deposits tend to show high indices of patina. This characteristic patina and a low index of edge rolling is the neo-cortex. This neo-cortex is different from the patina formed in the archeological site.

The neo-cortex identified in the La Guixera conglomerates shows a strong surface alteration on orange or yellowish subangular chert nodules. It has been also found in the terrace deposits of the Anoia River and Tous creek (Fig. 7.9). These cortical surfaces sometimes exhibit impact damage, which indicates high-energy fluvial transport. This damage has also been documented in the cortical surface of other materials presenting high rolling indices (Fig. 7.10). When this impact evidence or a high rolling index has not been clearly observed, neo-cortex type has not been distinguished. There are therefore two types of neo-cortex: the first is associated with fluvial terraces, whereas the second indicates an indeterminate secondary origin that cannot be attributed to fluvial or colluvial formations.

Provisioning in Secondary Deposits

We conducted a preliminary study on the raw material represented in the Anoia fluvial deposits. It mainly provided information on non-chert materials. Sampling was taken from a 5-km radius of the site, between the villages of Pobla de Claramunt and Vallbona d'Anoia, north and southeast of

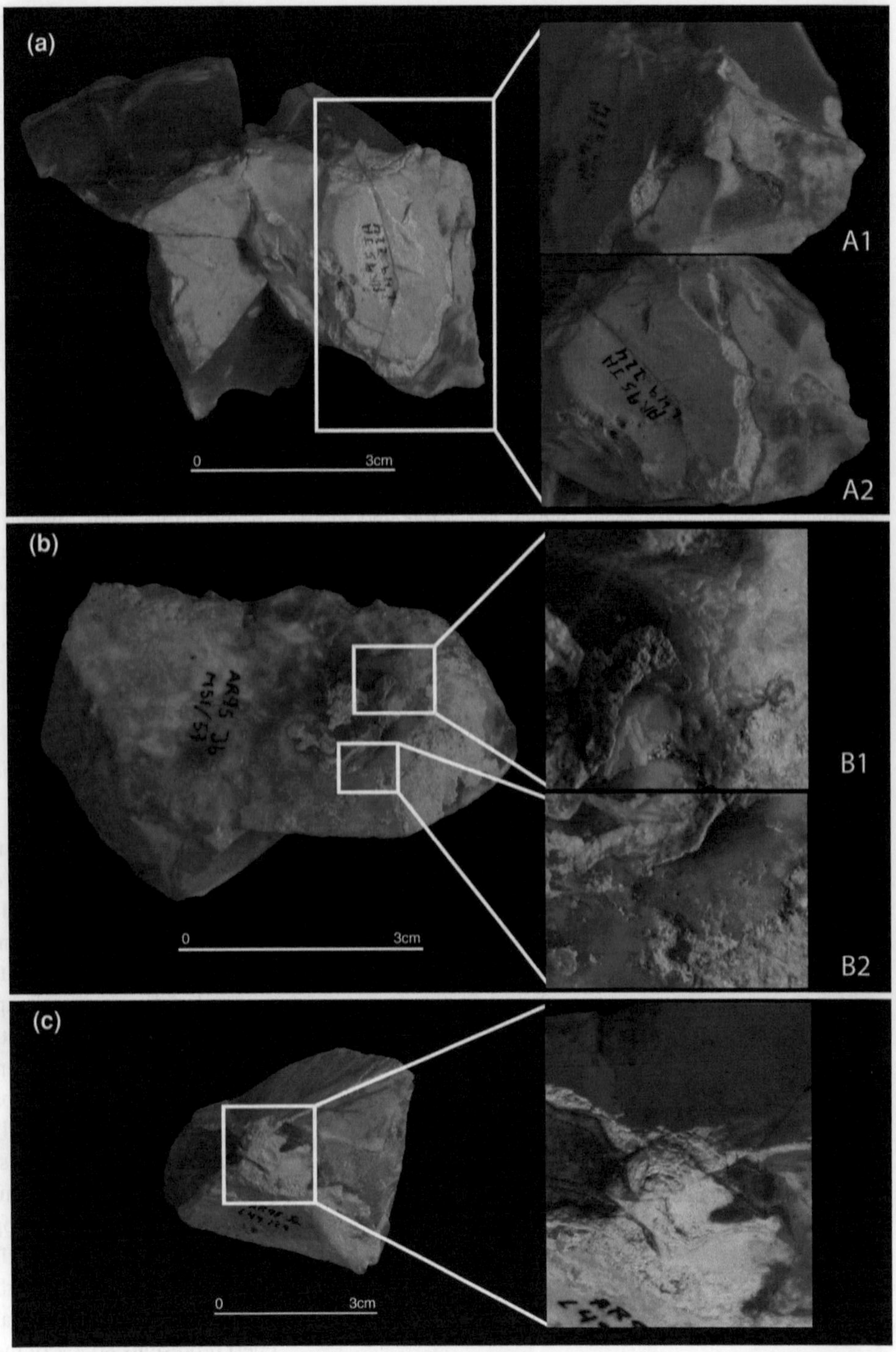

Fig. 7.10 Photographs showing the neo-cortex of some archeological chert samples. **a** AR95 Ja L49–224, well developed white patina; *A1* detail from neo-cortical sub-angular edge; *A2*, double patina, very fine white patina weathering in the site and neo-cortical well developed white patina. **b** AR95 Jb M51-57, rounded neo-cortex by flow steam origin; *B1*, rounded ridge with random arc groove strokes; *B2*, Well rounded neo-cortical surface. **c** AR95 Ja L49–229, very well developed white patina with secondary cover of yellow patina

Capellades respectively. We sampled an area of 1 m^2, adapting the method used in hydraulic engineering (Fripp and Dipplas 1993; Bunte and Abt 2001). The roundness-sphericity index was calculated using Krumbein's visual chart (Krumbein 1941). All the samples included at least 40 cobbles.

The deposits we sampled did not contain chert nodules. These were documented in some fluvial terraces near Sant Martí de Tous and Igualada, and in the fluvial formations near Vilanova del Camí. Other types of sampling should be done in the future, taking into account the scattering of chert nodules on the terrace surface (Shelley 1993; Church 2000). Primary chert formations are not widespread and chert nodules are not dominant in the lithological charge of the rivers.

We took three samplings from the current bed of the Anoia River (Anoia 1, Anoia 3 and Anoia 4), and one sampling from the T2 (+25 m) terrace (Anoia 2), corresponding to the Late Pleistocene (Gallart 1991). The samples recovered upstream

Table 7.2 Distribution of lithic remains by artifact classes and sublevels

	Unworked nodules	Cores	Flakes	Flake fragments	Retouched artifacts	Fragments	Total
Ja	33	81	1950	2724	150	508	5446
	0.6%	1.4%	35.8%	50%	2.7%	9.3%	100%
Jb	5	11	614	746	39	55	1470
	0.3%	0.7%	41.7%	50.7%	2.6%	3.7%	100%
Total	38	92	2564	3470	189	563	6916
	0.5%	1.3%	37%	50.1%	2.7%	8.1%	100%

from Capellades were characterized by the absence of granodiorite; this material appears as angular fragments in the samples located downstream. However, other Paleozoic materials such as quartz and slate are present in all the samples and have angular shapes. This is due to the hardness of these materials and the proximity of the source area. Meanwhile, sedimentary rocks, which are softer and more soluble, tend to exhibit high roundness indices between rounded (0.7) and subrounded (0.5).

Two types of sandstone have been distinguished. The most common is a dark grey to dark reddened brown greywacke that includes slate fragments. The second presents a carbonated matrix and a slightly pink color. This type has been found in sublevel Ja (RMU Lim-011, Lim-072 and Lim-074) and Jb (RMU Lim-20) as well-rounded cobbles. In general, sandstone is characterized by the highest roundness index due to its matrix and cement. Gray micritic limestone is dominant in the Anoia 1 sample; brown and light brown limestone is less common. In Anoia 2 both types are equally represented, but in Anoia 3 and 4 samples of gray limestone are scarce.

The Assemblage Level

The lithic assemblage of level J is made up of 6,916 artifacts larger than 1 cm: 5,446 from sublevel Ja, and 1,470 from sublevel Jb. There are also thousands of micro-remains (less than 1 cm maximum dimension), mostly recovered during the sieving of sediments. These objects have not been included in this study. This is one of the largest lithic assemblages of the Abric Romaní sequence, at present only surpassed by level M. Knapping products form the bulk of the assemblage, especially flakes (37%) and flake fragments (50.1%) (Table 7.2). Cores and retouched artifacts exhibit low percentages, 1.3 and 2.8% respectively. In general, the different artifact classes show similar percentages at both sublevels. The main differences correspond to flakes, more abundant in Jb, and fragments, more abundant in Ja. The blank to core ratio (the number of whole flakes and retouched artifacts divided by the number of cores) is significantly higher in sublevel Jb than in sublevel Ja (58.3 and 25.1 respectively). These differences could be attributed to changes in the uses of raw materials, since quartz and limestone were more commonly knapped in sublevel Ja and these materials exhibit higher indices of broken flakes. Nevertheless, differences in core to blank ratios between sublevels Ja and Jb remain the same when we consider only chert remains (26.2 and 60.1 respectively). In principle, this would indicate that core reduction sequences yielded more blanks in sublevel Jb. On the other hand, the flake to retouched artifact ratio (the number of whole flakes divided by the number of whole retouched artifacts) is the same in both sublevels (21.9).

Raw Material Provisioning

Chert is the dominant raw material, especially if we only take into account the remains directly derived from knapping (Table 7.3). Limestone and quartz are also well represented and were used in reduction sequences. Nevertheless, limestone is dominant among the unworked nodules and fragments that do not show knapping attributes (Table 7.4). This is related to the use of limestone cobbles as hammerstones for other activities. The differences between sublevels Ja and Jb may be partly related to those activities, since fragments are more abundant in sublevel Ja. Consequently, limestone is also more abundant in this sublevel. However, the higher percentage of limestone in sublevel Ja is not exclusively caused by these non-knapping activities; the use of this material for knapping was also more frequent in this sublevel. If we consider only the artifact classes directly associated with knapping, fragments and unworked nodules excluded, chert represents 80.2% of the lithics in sublevel Ja and 90.4% in sublevel Jb. The highest proportion of chert appears among retouched artifacts, which suggests that this material was preferentially selected for retouch.

Chert

Chert from the SMT formation is dominant in sublevel Ja, with a percentage of about 50%, and VLD chert is the

Table 7.3 Distribution of lithic remains by raw materials and sublevels

	Ja	Jb	Total
Chert	3953	1280	5233
	72.5%	87%	75.6%
Limestone	787	63	850
	14.4%	4.3%	12.2%
Quartz	634	107	741
	11.6%	7.3%	10.7%
Slate	54	5	59
	0.9%	0.3%	0.8%
Quartzite	3	11	14
	0.05%	0.6%	0.1%
Calcarenite	7	1	8
	0.1%	0.06%	0.1%
Sandstone	3	1	4
	0.05%	0.06%	0.05%
Porphyry	1	1	2
	0.01%	0.06%	0.02%
Galena	3		3
	0.05%		0.02%
Granite	1		1
	0.01%		0.01%
Speleothem		1	1
		0.06%	0.01%
Total	5446	1470	6916
	100%	100%	100%

second most represented type (30%). The rest of the chert types identified in the assemblage were found in very low proportions: exogenous (6%), PAN (6%), indeterminate (3%), and CME (3%). RMUs assigned to the SMT chert are generally made up of few artifacts. None of these RMUs is composed of more than 40 artifacts, and units with fewer than 15 artifacts are common. The most characteristic RMUs from this formation (including the secondary formations SMT1 and SMT2) can be seen in Table 7.5. These nodules are characterized by highly variable knapping properties. The RMUs from the VLD formation exhibit good properties for knapping. Chert from the PAN formation is very homogeneous and is characterized by good knapping qualities. RMUs attributed to the CME formation are scarce and correspond to small cobbles. Chert from this formation is very heterogeneous and, in general, it is fairly good for knapping.

The RMUs attributed to the exogenous chert present a terrigenous appearance, which suggests that their origin was probably the Triassic formations of the Prelittoral Range or the carbonated and detritic deposits of the La Segarra lacustrine system (Sequence III according to Anadón et al. 1989). Only Chert-031 suggests a primary origin. This RMU is formed by 20 pieces, including a core and a retouched artifact, and two refits have been found. It seems clear that the reduction sequence was carried out inside the shelter and the significant presence of cortical remains indicates that this RMU was introduced as a complete or nearly complete nodule, although small remains are clearly dominant. It is a grayish–blue, opaque, good-quality chert that is smooth to the touch, and shows a medium gray patina and a calcareous cortex.

Chert-010 is formed by four artifacts found in square O49. It is also a good-quality chert, characterized by a brownish-yellow color and an almost opaque appearance. It is smooth to rough to the touch and scarcely patinated, with oxides and quartz geodes. Some artifacts present neo-cortex.

Chert-018 is formed by nine artifacts, six of them conjoined in the same refit (including a core and a retouched artifact). It corresponds to a very small cobble of good knapping quality. It seems that most of the reduction sequence was carried out inside the site and artifacts with neo-cortex are abundant. This opaline chert is grayish–brown, opaque, polished and smooth, without patina. It presents oxides, a laminated sedimentary structure and quite a lot of allochems and fossils.

Chert-047 is the exogenous RMU with the most artifacts (43). Refits and cortical artifacts suggest that most of the reduction sequence was carried out inside the shelter. This chert is highly laminated, brownish-gray with maroon shades, opaque, smooth in the interior and saccharoid on the exterior of the nodule. Oxides are abundant.

Finally, RMU Chert-050 is also a large group (14) with several refits and cortical elements, which indicates that it was introduced as an almost complete cobble and entirely reduced inside the shelter. This is a smooth, brownish-red, opaque chert with very suitable properties for knapping. Its sedimentary structure is laminated and inclusions are scarce.

Of the RMUs identified in sublevel Jb, 34 have been attributed to one of the chert-bearing formations described above, as well as 49 transported artifacts. This sublevel shows less chert variability than sublevel Ja. Only chert from the SMT, VLD and CME formations has been clearly identified. The SMT chert is the most abundant (22 RMUs) (Table 7.5), showing a dominance of nodules gathered from secondary sources ($N = 15$). RMUs indicating a primary origin are less abundant ($N = 7$).

Only three RMUs correspond to the VLD chert, all of them from secondary deposits. The RMUs of fluvial origin are characterized by a high number of artifacts. Finally, a RMU from the CME formation was also identified. It is an orangish–brown chert with subrounded cortical cover and without patina.

As far as the transported blanks are concerned, SMT and VLD cherts have been identified, showing the same

Table 7.4 Distribution of lithic remains by artifact classes and raw materials

	Unworked nodules	Cores	Flakes	Flake fragments	Retouched artifacts	Fragments	Total
Chert		78	2269	2694	177	15	5233
		84.7%	88.4%	77.6%	93.6%	2.6%	75.6%
Limestone	35	6	118	230	6	455	850
	92.1%	6.5%	4.6%	6.6%	3.1%	80.9%	12.2%
Quartz	1	8	164	538	4	26	741
	2.6%	8.6%	6.4%	15.5%	2.1%	4.6%	10.7%
Slate				1		58	59
				0.02%		10.3%	0.8%
Quartzite			9	4	1		14
			0.3%	0.1%	0.5%		0.1%
Calcarenite			1	3		4	8
			0.03%	0.08%		0.7%	0.1%
Sandstone	2		1			1	4
	5.2%		0.03%			0.01%	0.05%
Porphyry			2				2
			0.07%				0.02%
Galena						3	3
						0.03%	0.02%
Granite						1	1
						0.01%	0.01%
Speleothem					1		1
					0.5%		0.01%
Total	38	92	2564	3470	189	563	6916
	100%	100%	100%	100%	100%	100%	100%

percentage (20.4%). However, most of the transported blanks (59.2%) correspond to rare chert types of unknown origin that can be considered as exogenous cherts. There is also a small set of metamorphic and igneous rocks that, as we will discuss later, seem to correspond to the same provisioning event. Most of the transported artifacts present remains of cortex on their dorsal surfaces (53.1%). Twelve artifacts attributed to exogenous chert were uncortical and only one showed a primary cortical cover.

Five transported artifacts from the SMT or VLD formations exhibited a secondary neo-cortex. There were also five artifacts with primary cortex, three from SMT and two from VLD. The primary cortex was already recorded among the SMT units knapped on the spot, but this kind of cortex was not documented among the RMUs from the VLD formation.

We identified several RMUs presenting different degrees of neo-corticality (Table 7.5). Most RMUs from the SMT chert, which represents 75.8% of the remains, were gathered in secondary deposits, while artifacts of primary origin are relatively scarce. Five RMUs from this kind of chert suggest a fluvial origin. Fluvial deposits were the dominant origin for the VLD chert, which makes up 16.9% of the analyzed remains. 6.75% of the remains are of unknown origin, although they probably come from non-prospected tertiary formations. Almost half of these artifacts show a secondary origin. Eight RMUs have been attributed to this exogenous chert. Two of them (Chert-30 and 25) exhibit macroscopic characteristics in between those of the VLD and SMT cherts, but cannot be clearly assigned to either of these origins. The rest of the RMUs cannot be attributed to any of the chert sources documented in the prospected area and the outcrops from which they came may be located outside of the 30 km radius around the site.

Chert-00 is a substitution chert that can be considered as a silicified limestone. Color ranges from light greenish-grey to dark greenish-grey. It is opaque and matt, although a waxy shine has been observed in some artifacts. Neither allochemical components nor other intrusions have been documented. It is very fine-grained and its conchoidal fracture is very regular, making it very apt for knapping. Chert-00 is represented by two refitted *débordant* flakes. One of the artifacts exhibits a well-developed fluvial neo-cortex.

Chert-09 is also a carbonated substitution chert. Color ranges from light grey to light yellowish–brown. It is matt and opaque, and contains many millimetric spicules. It is

Table 7.5 Distribution of chert remains according to the geological formation of origin and the type of cortical surface

Sublevel	Neo-cortex type	Outcrop original	N	%	RMU
Ja	Secondary conglomerate	CME	17	1.3	048 and 079
	Secondary unknown	SMT	189	14.3	009a, 009c, 025a, 035a, 035b, 037, 040 and 041
		VLD	75	5.7	011, 059 and 064
		Unknown	44	3.3	047
	Secondary fluvial	SMT	96	7.3	002b, 012, 013, 032, 060, 077 and 078
		VLD	92	7.01	044, 065, 067b, 071a and 071b
		PAN	21	1.6	022
		Unknown	78	5.9	007, 010, 014 and 050
	Primary	SMT	171	12.9	003, 004a, 005, 020, 026b, 033, 039, 055, 058, 066, 070 and 076
		VLD	138	10.4	024, 029, 043, 046 and 052
		PAN	34	2.6	006, 008 and 068
		Unknown	20	1.5	031
	Non cortical	SMT	185	14.0	001, 002a, 004b, 009b, 017, 019, 042, 045, 057, 061, 067a, 074 and 075
		VLD	124	9.4	016, 021, 023, 026a, 030, 036, 053, 062, 063, 072 and 073
		Unknown	37	2.8	015, 018 and 025b
Jb	Secondary conglomerate	CME	9	1.3	07
	Secondary unknown	SMT	373	52.5	01, 04, 06, 12, 13, 23, 27, 29, 34 and 38
		VLD	7	1.0	33
		Unknown	2	0.3	30
	Secondary fluvial	SMT	83	11.7	03, 14, 15, 17 and 18
		VLD	113	15. 9	02 and 19
		Unknown	14	2.0	00, 16
	Primary	SMT	71	10.0	05, 08 and 21
		Unknown	2	0.3	39
	Non cortical	SMT	12	1.7	11, 20, 24 and 26
		Unknown	25	3.5	09, 25, 31 and 32

Only the artifacts attributed to some RMU have been considered

characterized by a laminated structure and veins of secondary chalcedony. Cortex indicates a probable primary origin, which would correspond to outcrops interstratified between marly limestone layers.

Chert-16 is the exogenous RMU with the highest number of artifacts, since most of the reduction sequence, including the first stage, was carried out inside the shelter. It is mainly saccharoid to the touch, although it tends to be rough. It is opaque, matt and white to light yellow in color, exhibiting a conchoidal fracture that is fairly good for knapping. It shows a silicied and partly rolled cortical surface. This chert contains trabecular-like vacuums sometimes filled with oxides and clays. It also includes millimetric ghosts of long rhombohedral crystals, which could correspond to some kinds of halure or dolomite crystals. This nodule was probably obtained in a fluvial deposit. However, one of the transported artifacts, probably corresponding to this kind of chert, shows a subprimary cortex and ghosts of grouped acicular crystals. These acicular crystals have not been observed in Chert-16, possibly due to the rolling of the cortical surface. The presence of this artifact suggests that the primary location of this chert was also visited by the occupants of Abric Romaní.

Chert-31 corresponds to a bright reddish–brown chert. It is smooth to the touch, translucent and shows a greasy brightness. Its structure is homogeneous and we have not identified inclusions or allochemical elements. It presents a white patina and its origin may be located in the conglomerates of the CME formation.

Chert-32 is only formed by two fragments of a flake affected by a thermal fracture. They are completely patinated and their original color cannot be documented. This chert shows oxides and ooids clustered in a laminated structure. One of the transported artifacts may correspond to this type of material. Fracture is conchoidal and regular and its aptitude for knapping seems to be very good.

Table 7.6 Dimensional analysis of flakes, cores and retouched artifacts

		Length	Width	Thickness	IA	IC
Flakes	Ja	22.37	21.53	7.13	1.08	3.06
		12.20	10.34	4.55	0.41	1.19
	Jb	22.49	21.70	7.05	1.07	3.10
		11.79	9.80	4.37	0.40	1.25
	Total	22.39	21.57	7.12	1.08	3.06
		12.10	10.21	4.51	0.41	1.21
Cores	Ja	46.03	38.58	26.06		
		19.75	16.64	11.96		
	Jb	41.66	31.29	19.87		
		8.38	7.68	4.61		
	Total	45.47	37.64	25.26		
		18.7	15.93	11.46		
Retouched artifacts	Ja	40.48	35.42	12.89	1.2	2.67
		10.85	10.35	4.04	0.36	0.87
	Jb	44.31	42	15.66	1.1	2.59
		11.67	9.03	5.98	0.36	0.82
	Total	41.46	37.11	13.6	1.17	2.65
		11.14	10.40	4.75	0.36	0.86

Mean and standard deviation of the dimensional attributes and indexes

Finally, Chert-39 is also formed by two conjoined artifacts and corresponds to a short reduction sequence carried out on a flake introduced into the site as an isolated item. It presents a subprimary cortex and is rough to sacharoid to the touch. It is matt and orangish in color. There are gray inclusions that can be attributed to carbonates and fissures filled with hematite oxide. It becomes easily patinated and is not very good for knapping in spite of its conchoidal fracture. Although its origin is unknown, it seems to come from tertiary deposits.

Limestone

In general, both the limestone types used as hammerstones and those exploited in reduction sequences correspond to raw materials documented in the fluvial deposits at the foot of the site. The only exception is a retouched artifact from sublevel Jb made on a dark brown, almost black, limestone that does not come from the geological formations surrounding the site. As we will see later, this artifact is part of a set of spatially associated exogenous materials that suggests a specific input event. Gray limestone is dominant, although it exhibits a very heterogeneous knapping quality, as nodules with irregular fracture but conchoidal tendency are abundant. Gray limestone tends to exhibit angular forms, due to its greater degree of hardness. Brown limestone is also fairly heterogeneous, but it is in general more suitable for knapping, although it is weaker and its edges are therefore less resistant. The dominance of gray limestone suggests that limestone provisioning was focused on the fluvial terraces located at the foot of the site.

Quartz

With the exception of quartz, Paleozoic materials are scarcely represented. Quartz remains, always associated with knapping, are more common in sublevel Ja. Low-quality materials were more frequently exploited in this sublevel, whereas the dominance of chert in knapping was more marked in sublevel Jb. Quartz and slate were gathered as angular blocks, which suggests that they were collected in the Paleozoic formations close to the site. Slate appears mainly as small fragments without the typical attributes associated with knapping, but its presence in the rockshelter could be linked to the exploitation of quartz. The quartz nodules found in secondary position in the slope deposits commonly exhibit slate remnants on their surfaces. The slate fragments might be considered as the cortical products of quartz exploitation. The coincidence between the spatial distribution of slate and quartz remains corroborates this interpretation.

Few quartz RMUs (Quartz-004 and 005 from Ja and Quartz-01 from Jb) exhibit fluvial neo-cortex, which indicates that they were not collected in the quartz outcrops close to the site. Quartz-004 is a pinkish, milky quartz, characterized by an irregular fracture and brownish-red oxides.

Although its ultimate origin may be in a quartz vein or dike, the cortical surface suggests that it was collected in a fluvial deposit. Quartz-005 is a gray quartz with irregular fracture that corresponds to a subangular/subrounded nodule collected in alluvial deposits. Finally, Quartz-01 is a rock crystal very suitable for knapping. It is gray and translucent with a vitreous appearance. This RMU was also gathered in secondary position, but its primary formation is not clear.

Other Materials

Apart from chert, limestone and quartz, the remaining raw materials are very scarce. Quartzite and porphyry are present mainly as isolated flakes or retouched artifacts, coming from reduction sequences developed outside of the shelter. Nevertheless, a short knapping event on quartzite, providing only small flakes, has been identified in sublevel Jb. Calcarenite remains have primarily been found in sublevel Ja and they correspond to a single reduction episode. Granite and sandstone are represented by unworked nodules and fragments associated with non-knapping activities. We should also mention the recovery of an artifact made on a speleothem fragment formed in the rockshelter as the use of autochthonous rocks is uncommon in Abric Romaní.

Finally, three galena fragments were recovered in sublevel Ja. This material, the natural mineral form of lead sulfide, has only been found in this archeological level of the Romaní sequence. These fragments were characterized by X-ray microdiffraction (μDRX) and environmental scanning electron microscopy with X-ray microanalysis (ESEM-EDX). The largest galena fragment measures 38.2 × 32 × 26 mm and weighs 130.4 g (Fig. 7.11). The dimensions of the smallest one are 20.3 × 12.3 × 11.4 mm and it weighs 10.3 g. The third fragment measures 38 × 21.5 × 10.7 mm and weighs 31.5 g. We have observed cube-shape exfoliation planes in the three fragments. The largest and the smallest have a dark grey patina covered by another pale grey patina, both of them with a matt tarnish. The medium-sized fragment has a pale grey patina all over the surface. The μDRX and ESEM-EDX indicate that both patinas are cerussite, a lead-carbonate product of galena alteration (Cronyn 1990; Ettler et al. 2003). Some areas of the patina were detached in all the galena fragments, making it possible to see their interior, which was lead-grey and showed metallic brightness. In these cases, the analysis of the metallic areas characterizes the fragments as galena.

Galena has been reported in other Middle Paleolithic, Chatelperronian and Upper Paleolithic assemblages (Cabrera 1984; Poplin 1988; Esparza 1993; Meignen 1993, pp. 251–252), and in Ancient Egypt galena was used as khol, a mixture applied around the eyes as a protection against eye illness. This mineral was available close to the rockshelter. Galena outcrops have been documented in the Paleozoic formations of the Capellades area and other nearby locations (Mata 1990). However, its precise use in level J remains unknown.

Fig. 7.11 Galena fragment found in sublevel Ja

Unknapped Lithics: The Use of Unworked Nodules

Some studies, especially about Upper Paleolithic assemblages, might be considered as taking a holistic approach to the functional role of cobbles in percussive activities (de Beaune 1997, 2000, 2004). Diaphyse bone fragments used as retouchers in the Middle Paleolithic have also been a focus of interest in recent years, and different studies have been conducted on their stigmata and use-wear patterns (Pathou-Matis 2002). The presence of micro-polishes, grooves and micro-residues on the striking platforms of experimental knapped pieces have also been referred to in a few use-wear analyses (Ibañez et al. 1987; Byrne et al. 2006). Various studies have focused their attention on recognizing knapping techniques, inferring them from characteristic stigmata and statistics (Gallet 1998; Wenban-Smith 1999; Roussel 2005). Some approaches (Bourguignon 1997), attempt to draw a correlation between hand movement and the materials employed for Quina scraper retouch through identifying the different properties of "soft stones" such as schist, limestone or sandstone as nearer to that of organic hammers than hard stones such as quartz or quartzite.

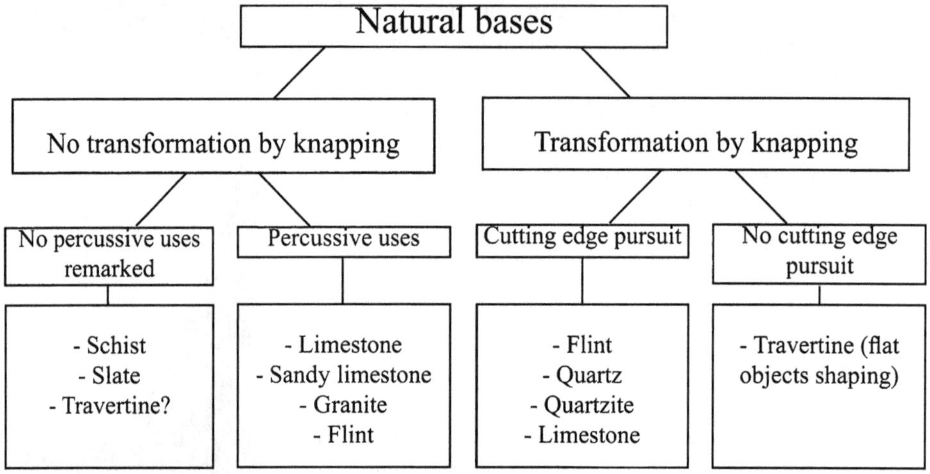

Fig. 7.12 Main goals of knapped and unknapped lithic raw materials represented at Abric Romaní level J

Cobbles in level J: Breakage and Raw Material Units

The study of percussion tools must be undertaken holistically, taking into consideration all of the elements involved. In Middle Paleolithic assemblages linked to habitat occupations (perhaps with the exception of very short occupations or those that occur at a raw material source) it is common to find the intense use and re-use of lithic raw materials, which has sometimes been described as characteristic of this period (Bourguignon et al. 2004). These processes, which can provide rich technological information, usually erase data on initial tool uses or steps in the *chaîne opératoire*. It is therefore essential that pieces that may have been employed both for percussion and for knapping are accurately identified.

The largest part of the lithic assemblage of level J is made on chert. The second most significant raw material, both in number and in weight, is limestone, usually appearing as rounded cobbles. Other raw materials that are present as cobble blanks are calcarenite and granite as well as some chert and quartz nodules. All of them except granite are potential raw materials for knapping. Other than chert, most of these materials, especially the low quality ones, may reflect raw material collection linked to space-conditioning objectives or percussion activities, even if they were also knapped materials. Below we attempt to synthesize some of these possibilities schematically (Fig. 7.12).

Some materials such as slate, which appears highly fragmented and without apparent macroscopic traces, do not permit deeper study. We analyzed 505 remains: 462 items from sublevel Ja and 43 from Jb. These were all the artifacts classified as unworked nodules and most of the fragments. However, after the refit study and the identification of RMUs, these items, most of them in limestone, were grouped into 82 minimal units. The non-knapping use of cobbles was clearly most common in sublevel Ja, which shows higher percentages of unworked nodules and fragments. We therefore focused our analysis on this sublevel.

The properties of limestone, the most frequently selected material for cobbles, allow for easy interpretation of breaks and the longitudinal striations of use-wear traces. Furthermore, limestone can exhibit fire exposure traces. Because of its fragility, softness, and suitability for fire structures, it has a high degree of breakage. Just eight of the 505 pieces are whole cobbles. However, many of them present fractures related to percussion activities, with high potential for reconstructing movements and intensity of use.

The items have been grouped into five categories depending on their degree of fracture and the possibility of recognizing their original volume: whole cobbles (category 1); cobbles with percussion damage affecting a small part of the AZP, but which can still be reconstructed (category 2); cobbles also affected by percussion damage in which the AZP cannot be reconstructed but at least two dimensions of three maximum axes are present (category 3); cobble fragments whose morphology, whole size and weight cannot be reconstructed (category 4); finally, pieces that had been intensely broken or knapped but re-used as percussion tools (category 5). The interpretative potential of each category is different. Categories 1, 2 and 3 allow the recognition of the original blanks, while categories 1, 2 and 5, when percussion marks are present, allow identification of the selected morphologies as well as the direction or intensity of use. Category 4 only allows an approach to the original size and maximum length of the pieces, whether they show signs of percussion use or not. Category 4 comprises two-thirds of the RMUs (N = 55) while the other categories include 27 units. Only the latter will be used for determining morphological or metrical selection. The study of gesture and movement identification can also be carried out with these 27 units.

Raw Material Collection and Properties for Percussion Uses

Our approach does not address the exact identification of the sources, but a general lithological attribution based on macroscopic observation and the evaluation of mechanical properties. We have defined four different groups of raw materials, according to their abundance at the site (Fig. 7.13).

Group I—*Micritic light-gray limestone*. It is a fragile low-density limestone varying in color from light to dark gray, with irregular fissures. Although most remains seem to come from smooth rounded cobbles, some of them have rough surfaces and a low degree of rounding, which suggests different degrees of movement from the primary source. Some pebbles tend to present a calcite crust perhaps related to karstic springs or streams with a saturation of CaCO salts, although their formation at the site cannot be ruled out. Blank size shows a wide range of variation, from a few centimeters up to 20 cm. There were 53 units in this group.

Group II—*Micritic yellow-greenish limestone*. It is a compact, fine-grained micritic limestone, with iron oxides in its composition. Original blanks are well-rounded and smooth cobbles, always at least 8 cm and never over 15 cm. There are 17 units in this group.

Group III—*Other limestones*. This group consists of four different types of limestone: one silicified limestone unit, three units of sandy limestone, one unit of limestone with big calcite crystals, and four sparitic limestone units. These last objects are especially well rounded and smooth, dense and consistent.

Group IV—*Other raw materials*. This group includes a big chert cobble without knapping transformation and a fragment of granite cobble. The chert cobble is slightly rounded and quite large (over 15 cm). It is very dense and its quality for knapping is poor. It has been tested with two removals but not deeply knapped. Some cores of this same low quality chert could be linked to percussion activities, although crushing related to intense use was rarely found.

The first group of raw materials may indicate a nearby collection source, due to the high number of units in which all sizes are represented, and the importance of the heaviest weight categories. This group has the widest range of weights and sizes, with different shapes and degrees of rounding, and there is an abundance of square and triangular forms with biaxial volumetric types. This seems to be the most frequently used raw material for fire and other space conditioning activities, which, along with intense percussion uses, might explain its high rate of breakage. This may be the most abundant material in the environment near the site.

The most rounded cobbles always used for percussion activities are of micritic limestone (group II), sandy limestone and sparitic limestone or granite. These cobbles, which

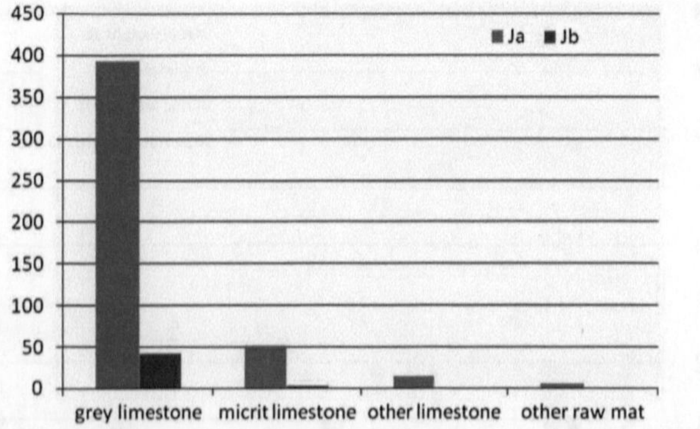

Fig. 7.13 Distribution of cobble remains and percussion tools by raw material groups

can sometimes be nearly triangular or square, exhibit a high degree of rounding. These materials may have been found in river deposits (possibly the Anoia River). Collection can sometimes be linked with knapping activities, as in the case of chert and sometimes greenish limestone, and probably silicified limestone, but this does not imply that these materials have not also been used for percussion activities.

Morphological Selection for Use and Primary Trends in Gesture Options

Of all the RMUs from sublevel Ja, 48 have percussion marks or breakage linked to percussion activities. However, 26 of them are fragments included in category 4; the rare cases of whole cobbles without percussion stigmata are always over 15 cm length, weigh up to 1,200 g and are selected in low quality materials for knapping. Therefore, 22 units from sublevel Ja and four from Jb will be used for gestural recognition.

Morphological, Volumetric and Weight Selection

The weight of hammerstones and percussion tools has long been considered a determining factor in raw material selection. Morphology and the longitudinal proportions of the main axes may also have determined a specific use. Although there are few objects in categories 1, 2 and 3, there is a clear preference for the thick biaxial volumetric types. Morphologies and the degree of rounding are more variable, but in general, groups II and III tend to have more rounded forms. However, the less rounded objects in group I are usually associated with square and triangular shapes.

In terms of weight, some groups are quite noteworthy, although significance was relatively low due to their small number (Fig. 7.14). We did not identify any percussion tools weighing between 200 and 300 g or between 600 and

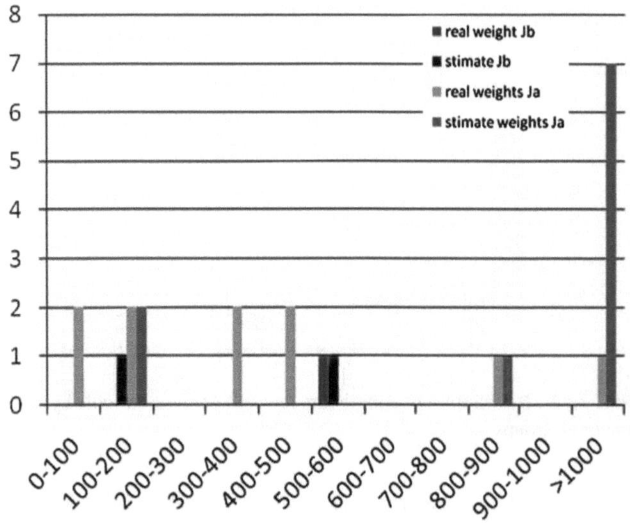

Fig. 7.14 Distribution of cobbles by weight (*in grams*)

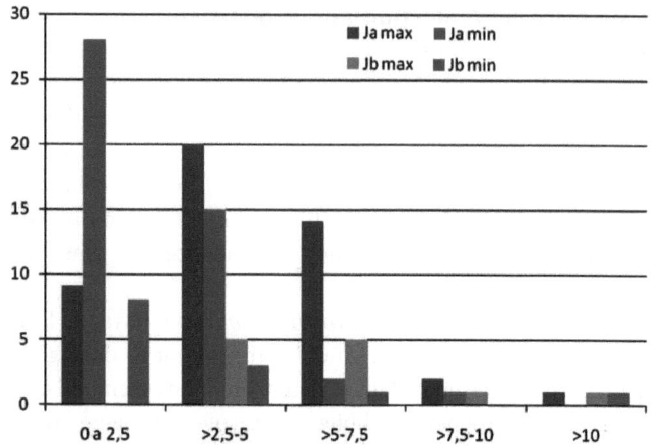

Fig. 7.15 Maximum and minimum AZP's diameter groups (*in centimeters*)

800 g. There is also a significant gap in objects weighing between 900 and 1300 g. This allows us to distinguish four weight groups. The lightest two groups (40–195 and 359–593 g) show intense stigmata and fractures, while in the heaviest groups stigmata are rarely seen but some breakage has been documented.

Selection of Active Zones of Percussion

Specific characteristics of the active zone may have been chosen depending on the percussion activity to be carried out. The position of the AZP on the cobble as well as its type and surface size can have a great influence on the possible range of movements that can be performed. Furthermore, the precision or tenacity of the AZP depends on its size and position on the cobble. Well-rounded cobbles will provide higher mechanical resistance, but a lesser degree of precision. Squared or triangular shapes provide lateralized AZPs with great accuracy but less tenacity. We will attempt to apply objective parameters for locating and measuring the active zones employed.

Another important criterion is the direction indicated by stigmata and fractures. There may be a use pointing towards the center of the main convexity, or alternatively from the center towards the periphery of the cobble. Surfaces of AZPs can be measured in cm of circle diameters. Some cases will require a broad range of expression, mainly for elliptic forms, where two different values or a media will be reflected. Surface type, direction of stigmata and breakage related to the main convexity, and size can be read in every category; location of the surface, and therefore gestural option or direction of use will only be analyzed in categories 1, 2, 3 and 5.

For all AZPs, the surface types were either a natural (smooth) surface (56 AZPs) or a crushed-edge surface (12 AZPs), and in only one case (a chert cobble) a cortical surface. AZPs resulting in removals with polyhedral surfaces may be related to a stress in the use of cobbles or perhaps to a specific, more compressive type of percussion linked to specific activities. We prefer to use the term "natural surface" for cobbles (rounded by mechanical erosion) instead of "cortex", which implies for us an external coat or crust with a different petrological composition—related to the formation of the rock—and therefore different mechanical properties.

The diameters of the AZPs are consistently low (under 5 cm), which may be related to the pursuit of precision (Fig. 7.15). In order to correlate the size of the AZP to its compression capability we have developed a compressive index that can indicate differences in use. This index is obtained through dividing the weight in grams by the diameter in millimeters. The result is expressed in values between 0 and 1, where a value closer to 1 corresponds to low compression capability, and closer to 0 higher compression capability. A similar index is used to calculate the stress on a surface, usually expressed in pounds per square inch. Several high values stand out with results nearing 1, which always belong to small percussion tools, most of them from category 5. This may be related to the search for mechanical resistance in these small objects made from fragile materials. However, most values tend to be low, about 0.2 (Fig. 7.16).

The main stigmata identified on the surface of the AZPs are usually unspecific in terms of direction, although crushing and isolated percussion points can be distinguished on all of them. Striations (longitudinal grooves) related to percussion were only identified on one AZP. The position of stigmata in relation to the main convexity of the AZP is very often centered in square or triangular thick biaxial forms or lateralized in the case of well-rounded morphologies.

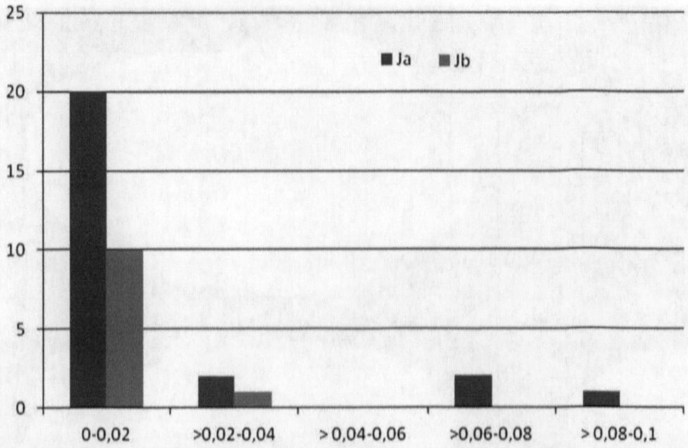

Fig. 7.16 Distribution of the AZPs according to the compressive index (minimum diameter of the AZP/weight)

Prehension and Gestural Choices: A Gestural Reconstruction

The location of stigmata and the direction of breakage indicate a high degree of lateralization in the use of percussion tools, mainly for small cobbles of raw materials from group I. Medium-sized cobbles from groups II and I seem to be used in a more oblique direction, although sometimes lateralized as well. In the case of greenish micritic limestone, whose morphology tends to be rather more rounded and medium sized, stigmata are nearly always found close to the main convexity, pointing toward it. Striae or small breaks show impacts produced on the periphery of the main convexity and movement toward the detached point of this main convexity. Fractures, when present, tend to eliminate distal or proximal tips, usually following a transversal direction.

Some of the gestural possibilities proposed here for thick biaxial and triaxial cobbles seem more plausible. Big cobbles up to 800 g, which do not show percussion marks but in some cases manifest breakage, seem to have been used with a more frontal-longitudinal movement (possibility #1). Weight groups under 600 g, on the other hand, sometimes seem to have been used with positions 1 or 2, but mainly 3 and 4, which implies the lateralization of their use. These two groups (40–195 and 359–593 g) seem to be the more intensely used percussion tools, and always feature percussion stigmata and nearly always significant breakage, which can eliminate up to ¼ of the entire volume of the object. Small percussion tools under 100 g belong to category 5, and may be linked to a precision grip, while those between 100 and 200 g could be associated with a semi-precise grip. Medium-sized cobbles between 300 and 600 g are associated with semi-precise grips used by bearing down with the palm, while big cobbles up to 800 g were probably used with a powerful grip.

Main Uses Inferred on Cobbles: Lifespan Dynamics in Percussion and Non-Percussion Tools

We have proposed only a few possibilities concerning the tribology of percussion activities, which can help in the interpretation of wear traces and breakage. Gestural recognition is one of the objectives of our approach. One of our main goals, however, is to infer a use or action for every element related to percussion. An objective attribution can be accomplished through controlled experiments. The steps taken by each percussion element in its interaction with activities involving breakage must also be explored in order to identify less apparent elements, rather than only whole recognizable hammerstones.

Interpretation of Main Uses of Cobbles

Based on the primary morphological and weight groups and their wear traces or fractures, knapping activities can be attributed to objects weighing less than 600 g. Those weighing between 40 and 195 g would rarely have been used for activities other than knapping (except nut-cracking, maybe). The smallest objects, always on category-5 polyhedral forms can be more precisely linked to retouch or perhaps more commonly to small core knapping.

The mechanical properties of the second weight group (300–600 g) make them suitable for knapping as well as for bone breakage. In any case, now we can only identify intense stigmata and breakage due to knapping activities, so the interaction of both uses remains a mere hypothesis.

The third group (800–900 g), with some stigmata and intense longitudinal breakage may have been associated with some knapping activities, such as the initialization of large cores, which has rarely been documented in the level. It is possible that these were previously used outside and introduced into the shelter with these traces. Intense breakage may perhaps be related to the initialization of limestone cobbles whose morphology provides them with a higher degree of mechanical resistance. Focusing on marrow extraction, some breakage may also have occurred due to an erroneous motion, such as a hammerstone slipping on a bone being crushed on the anvil.

The fourth group seems to be related to marrow extraction and space conditioning activities. Some of the largest cobbles could have been used as anvils, especially the squared ones.

The objects in the first and second weight groups, which are more easily recognized as knapping tools, seem to have been used with tangential or oblique movements, on detached tips. This suggests a preference for precision over mechanical resistance. The type of contact in this percussion is highly compressive, allowing the production of short flakes with sharpened edges, which seems to be a goal of

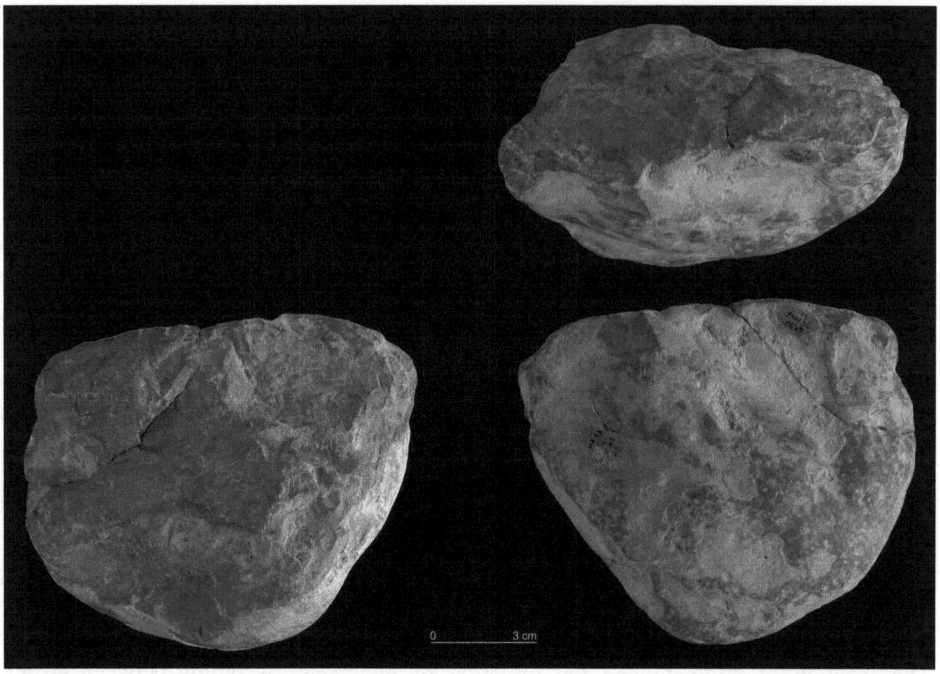

Fig. 7.17 Cobble fragment presenting use traces in the edge

centripetal-discoid technologies. Striking on flat or even concave platforms allows for better preservation of AZPs with this kind of force which, when used on faceted platforms, are easily worn and broken. The selection of square and triangular shapes could be intentional because they offer just these kinds of detached AZPs, with a transversal position in relation to the main axis, easily manipulable with a semi-precise grip and allowing a wide range of oblique and tangential movements.

One group of used cobbles deserves special attention. They are cobbles broken in half that show a plano-convex profile and exhibit a flat breakage surface giving rise to a cutting edge on the boundary with the natural surface of the cobble. The most acute section of this edge shows evidence of intense use in the form of crushing and small removals (Fig. 7.17). We have identified five artifacts corresponding to this category, which were scattered in different areas of the site, although two of them were associated in the southwest corner of the excavated surface. All belong to the heaviest weight group and exhibit a trend to dimensional standardization—length ranges from 120 to 95 cm, width from 82 to 71 cm and thickness from 60 to 35 cm. Although their use cannot be specified, it seems clear that they were employed for tasks that required tools featuring a cutting edge along with a considerable weight. None of these cobbles have been reconstructed through refits, which suggests that they were introduced already broken into the shelter. This means that humans selected cobble fragments presenting such dimensional and morphological characteristics at the provisioning site in anticipation of a specified use.

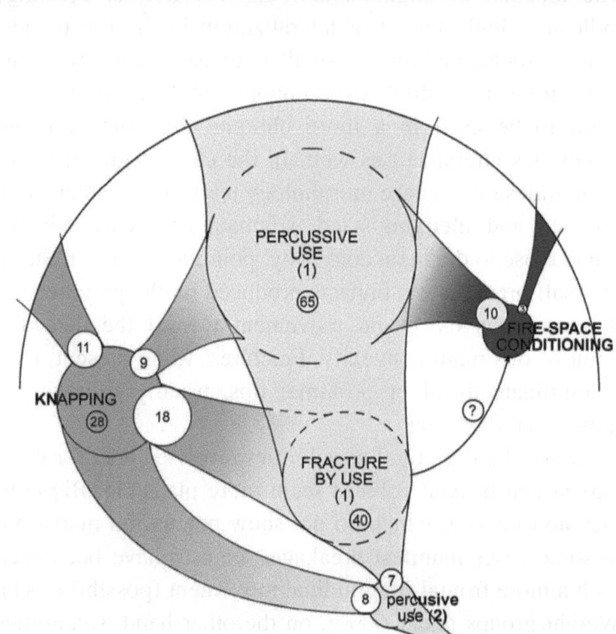

Fig. 7.18 Quantified useful-life steps on cobble units at sub-level Ja

Dynamic Processes on Percussion Tools

We have identified the different steps of cobble use, primarily in sublevel Ja. The greater incidence of breakage and evidence of different uses sometimes seen on a single piece seems to be related to lifelong-use tools, which are usually re-used for different activities. We have attempted to summarize these activities schematically in Fig. 7.18. Sublevel

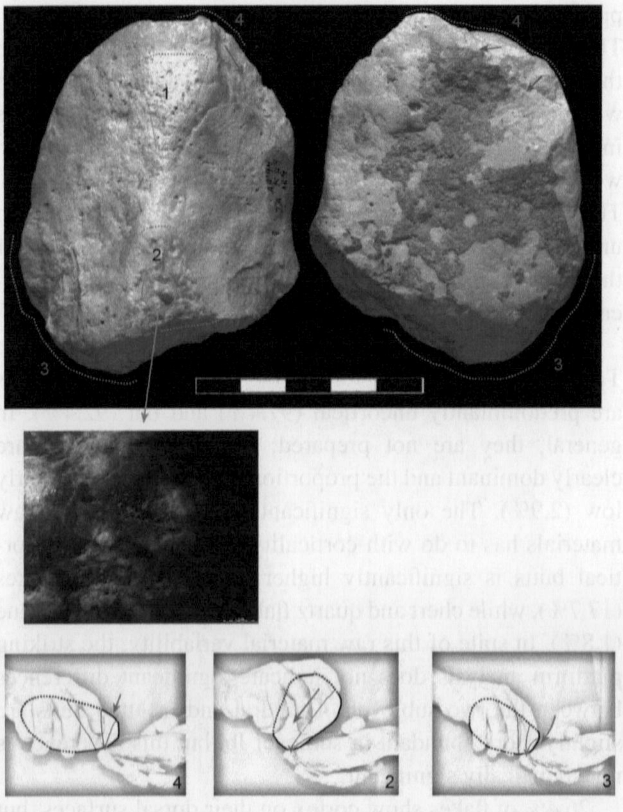

Fig. 7.19 Hammerstone AR'94 Ja K59 # 129 (RMU Lim-065) with detailed views of one AZP and gestural interpretation; numbers on the image indicate different AZPs, limited by *pointed lines*; *arrows* indicate direction of main fractures; numbers of gestural interpretation are referred to gestural options reflected in Fig. 7.3

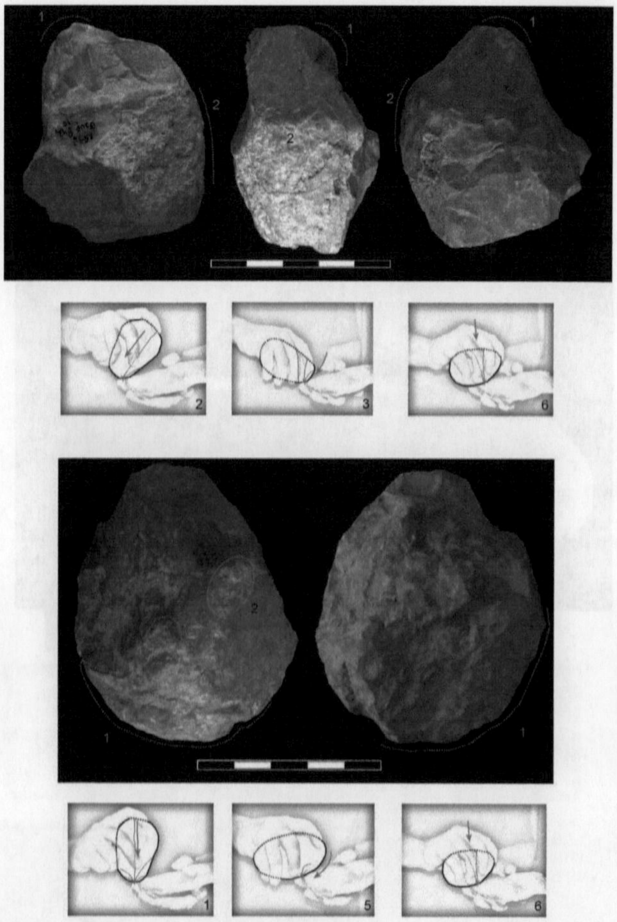

Fig. 7.20 Hammerstone AR'98 Jsup P42 # 10 (*up*) with percussion marks after knapping/breakage and hammerstone AR'94 Ja N49 # 62 (*down*), knapped after its use as hammerstone; numbers on the image indicate different AZPs, limited by *pointed lines*; *arrows* indicate direction of main fractures; numbers of gestural interpretation are referred to gestural options reflected in Fig. 7.3

Ja shows a more intense, dynamic and complex use of cobbles than Jb. The artifacts from sublevel Jb have a lower degree of breakage, which may be related to a relatively shorter period of use. Determining whether some of these dynamic uses of cobbles in Ja occurred outside the site or only during its occupation is extremely complex. However, it is very plausible that most of these heavy-duty tools were collected close to the site and then used and left inside the site. Hunter-gatherers usually discard the heaviest materials they have or those that are no longer useful. Future analyses will provide more information on the maximum distances these tools are transported (Figs. 7.19, 7.20, 7.21, 7.22).

Knapping Activities: Core Reduction and Retouch Strategies

The study of core reduction and retouch strategies is mainly based on the attribute analysis of cores, flakes and retouched artifacts, although other approaches, such as refitting, have also provided some interesting data.

Flake Attribute Analysis

The first characteristic that should be stressed is the marked micro-lithic appearance of the lithic assemblage (Table 7.6). The distribution of flakes by size classes (Table 7.7) shows a clear dominance of very small items (61.5% of all flakes). Large and very large classes comprise less than 10% of the artifacts. The percentages diminish as flake size increases, which is the pattern expected when reduction sequences are carried out in situ. Some differences between raw materials have been observed. Large sizes are more common among limestone flakes, while quartz shows higher proportions of small flakes. The relatively higher proportion of large artifacts in the limestone assemblage can be explained by the character of limestone reduction sequences. Most of them were done on nodules introduced as entire cobbles and sometimes previously used as hammerstones. These cobbles

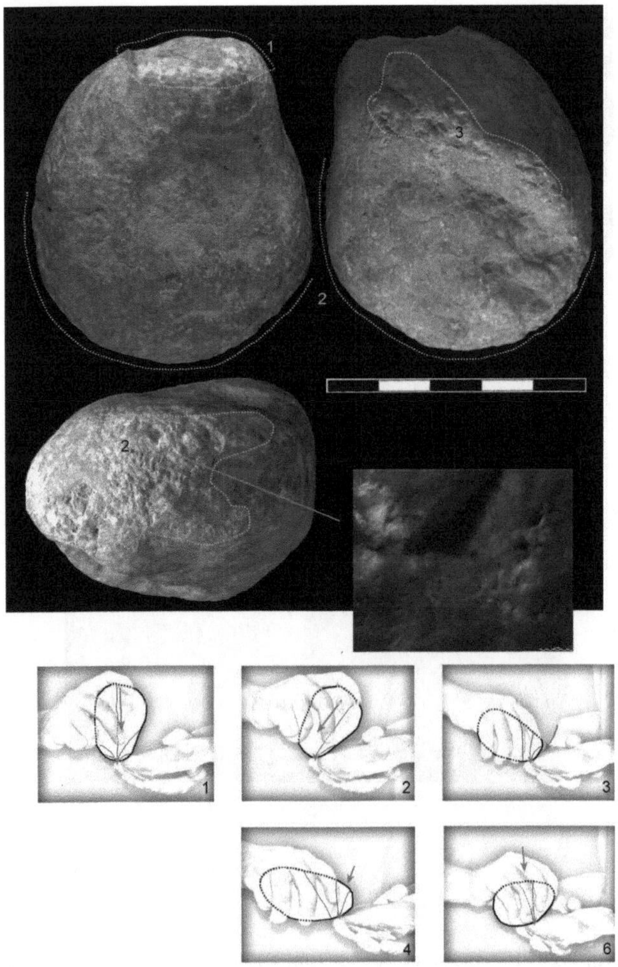

Fig. 7.21 Hammerstone AR'94 Ja N44 # 99 with detailed view of one AZP and gestural interpretation; numbers on the image indicate different AZPs, limited by *pointed lines*; *arrows* indicate direction of main fractures; numbers of gestural interpretation are referred to gestural options reflected in Fig. 7.3

tend to be larger than the chert or quartz nodules used exclusively for flake production and the production of large limestone flakes is therefore more common. In addition, limestone reduction sequences were particularly expedient and limestone cores were not as exhausted as chert cores. However, size distribution does not depend exclusively on the kind of the reduction sequences carried out in the shelter. In spite of these raw material differences, the size distributions of sublevels Ja and Jb are almost identical.

The blade index is very low in both sublevels (Table 7.8). Only 3% of the artifacts are at least twice as long as they are wide. More than the half of the flakes exhibit a length/width ratio of less than 1 (width is higher than length). Reduction strategies tended to produce short flakes. Although the elongation index increases slightly as flakes become larger, these differences are not significant (correlation coefficient of 0.104). This suggests that the length/width ratio remained practically unchanged throughout the reduction sequence. These size-related variations have also been examined for the thickness index. This index shows a negative relationship with flake size, that is, artifacts tend to be thicker as flake size increases. However, the strength of this relationship is even weaker than that of the elongation index ($r = -087$). Therefore, this data indicate that the thickness index also underwent few changes during reduction sequences. Neither the elongation index nor the thickness index exhibit significant differences between the two sublevels (Table 7.9).

The results of the flake attribute analysis can be seen in Tables 7.10, 7.11, 7.12, 7.13, 7.14, 7.15. Striking platforms are predominantly uncortical (97.4%) and flat (92.4%). In general, they are not prepared; unfaceted platforms are clearly dominant and the proportion of facetted butts is fairly low (2.9%). The only significant difference between raw materials has to do with corticality. The percentage of cortical butts is significantly higher among limestone flakes (17.7%), while chert and quartz flakes exhibit the same value (1.8%). In spite of this raw material variability, the striking platform analysis does not indicate significant differences between the two sublevels. Dihedral and facetted butts are slightly more abundant in sublevel Jb, but this difference is not statistically significant.

26.4% of flakes show cortex on their dorsal surfaces, but the percentage of entirely cortical artifacts is considerably lower (1.5%). Cortical flakes are slightly more abundant in sublevel Jb than in sublevel Ja. At first glance, this difference could be attributed to the higher proportion of quartz debitage in sublevel Ja. As described earlier, quartz veins are located in the Paleozoic formations close to the site and most quartz nodules were gathered in the nearby slope deposits. These nodules tend to be only slightly weathered or not at all, making cortical cover difficult to recognize. As a consequence, dorsal cortex shows the lowest value among quartz flakes (11%). However, this is partly compensated by the high index of corticality shown in limestone flakes (40%), which are also more abundant in sublevel Ja. If we consider only chert flakes, differences between sublevels Ja and Jb are still evident (25.5 and 31.3% of cortical flakes respectively).

Ventral surfaces exhibit prominent percussion bulbs (75.7%) and ventral profiles are highly variable, although straight profiles are dominant (36.4%). Both attributes show statistically significant differences according to raw materials. Diffuse bulbs are dominant among quartz and limestone flakes ($X^2 = 247.67$; $df = 2$; $p < 0.001$). These raw materials also present higher proportions of straight and winding ventral profiles ($X^2 = 20.24$; $df = 6$; $p < 0.005$). Both differences can be attributed to the type of fracture exhibited by quartz and limestone, which is less conchoidal than that of chert, especially if we take into account the low-quality characteristic of most quartz and limestone nodules found in level J.

Fig. 7.22 Hammerstone AR'95 Jb M51 # 100 with detailed views of two AZPs with striae and gestural interpretation; numbers on the image indicate different AZPs, limited by *pointed lines*; *arrows* indicate direction of main fractures; numbers of gestural interpretation are referred to gestural options reflected in Fig. 7.3

Table 7.7 Distribution of flakes, cores and retouched artifacts by size classes and sublevels

		Very small	Small	Medium	Large	Very large	Total
Flakes	Ja	1192	438	174	85	61	1950
		61.1%	22.4%	8.9%	4.3%	3.1%	100%
	Jb	385	127	52	28	22	614
		62.7%	20.6%	8.4%	4.5%	3.5%	100%
	Total	1577	565	226	113	83	2564
		61.5%	22%	8.8%	4.4%	3.2%	100%
Cores	Ja	1	13	30	16	18	78
		1%	16.6%	38.4%	20.5%	23%	100%
	Jb		3	2	1	2	8
			37.5%	25%	12.5%	25%	100%
	Total	1	16	32	17	20	86
		1%	18.6%	37.2%	19.7%	23.2%	100%
Retouched artifacts	Ja	2	19	28	26	14	89
		2.2%	21.3%	31.4%	29.2%	15.7%	100%
	Jb		2	6	10	10	28
			7.1%	21.4%	35.7%	35.7%	100%
	Total	2	21	34	36	24	117
		1.7%	17.9%	29%	30.7%	20.5%	100%

Table 7.8 Distribution of flakes by elongation index classes and sublevels

	L1	L2	L3	L4	L5	L6	L7	L8	Total
Ja	56	935	689	210	44	13	2	1	1950
	2.8%	47.9%	35.3%	10.7%	2.2%	0.6%	0.1%	0.05%	100%
Jb	17	299	212	69	15	1	1		614
	2.7%	48.6%	34.5%	11.2%	2.4%	0.1%	0.1%		100%
Total	73	1234	901	279	59	14	3	1	2564
	2.8%	48.1%	35.1%	10.8%	2.3%	0.5%	0.1%	0.03%	100%

Table 7.9 Distribution of flakes by thickness index classes and sublevels

	C1	C2	C3	C4	C5	C6	C7	C8	C9	Total
Ja	8	384	707	546	196	71	26	6	6	1950
	0.4%	19.6%	36.2%	28%	10%	3.6%	1.3%	0.1%	0.1%	100%
Jb	1	114	230	166	63	29	3	4	4	614
	0.1%	18.5%	37.4%	27%	10.2%	4.7%	0.4%	0.6%	0.6%	100%
Total	9	498	937	712	259	100	29	10	10	2564
	0.3%	19.4%	36.5%	27.7%	10.1%	3.9%	1.1%	0.3%	0.3%	100%

Table 7.10 Corticality of flake striking platforms

	Cortical	Uncortical	Total
Ja	51	1899	1950
	2.6%	97.4%	100%
Jb	15	599	614
	2.4%	97.6%	100%
Total	66	2498	2564
	2.6%	97.4%	100%

Table 7.11 Type of striking platform

	Lineal	Flat	Punctiform	Total
Ja	76	1799	75	1950
	3.9%	92.3%	3.8%	100%
Jb	27	570	17	614
	4.4%	92.8%	2.8%	100%
Total	103	2369	92	2564
	4%	92.4%	3.6%	100%

Table 7.12 Preparation of striking platforms

	Cortical	Unfacetted	Dihedral	Facetted	Total
Ja	50	1700	147	53	1950
	2.6%	87.2%	7.5%	2.7%	100%
Jb	15	510	67	22	614
	2.4%	83.1%	10.9%	3.6%	100%
Total	65	2210	214	75	2564
	2.5%	86.2%	8.3%	2.9%	100%

Table 7.13 Corticality of flake dorsal surfaces

	Uncortical	Cortical	Uncortical dominant	Cortical dominant	Total
Ja	1460	29	336	125	1950
	74.9%	1.5%	17.2%	6.4%	100%
Jb	426	10	141	37	614
	69.4%	1.6%	23%	6%	100%
Total	1886	39	477	162	2564
	73.6%	1.5%	18.6%	6.3%	100%

Table 7.14 Bulb of percussion type

	Diffuse	Prominent	Total
Ja	488	1462	1950
	25%	75%	100%
Jb	136	478	614
	22.1%	77.9%	100%
Total	624	1940	2564
	24.3%	75.7%	100%

Table 7.15 Flake ventral curvature

	Concave	Convex	Straight	Sinuous	Total
Ja	594	438	683	235	1950
	30.4%	22.5%	35%	12.1%	100%
Jb	167	125	251	71	614
	27.2%	20.4%	40.9%	11.6%	100%
Total	761	563	934	306	2564
	29.6%	22%	36.4%	11.9%	100%

Débordant flakes are fairly abundant (12.2% of all flakes). Most of them (97.4%) are made on chert, although some *débordant* flakes on limestone, quartz and quartzite have been recovered. This could suggest that the core reduction strategies producing this kind of blanks were more common on chert. However, we will see later that relatively few *débordant* flakes were produced inside the rockshelter. In this respect, it is especially significant that of the eight quartzite flakes recovered in level J, three are *débordant*. Quartzite was a raw material that was mainly introduced into the shelter as individual blanks. Naturally backed flakes are less abundant than *débordant* flakes (only 1%), but are all on chert. *Kombewa* flakes are practically absent (only one has been identified) and this is particularly striking, since, as we will see below, core-on-flakes are common.

Flake attributes do not indicate clear-cut differences between raw materials regarding core reduction strategies, although the higher percentage of *débordant* flakes on chert could suggest that the use of discoid knapping was more common on this material. However, no significant differences among other attributes of raw materials have been documented that would also point to differences in core reduction strategies such as striking platform faceting. Fracture properties and techno-economic factors seem to be a determining factor in flake variability. The first stages of the reduction sequences were more common on limestone that was introduced primarily as entire cobbles. At least some of these cobbles were previously used as hammerstones and knapping of these nodules can be considered an example of recycling. Some limestone flakes show percussion marks on their cortical surfaces and, as discussed below, refitting proves that recycling or secondary use of limestone cobbles was common practice. On the other hand, chert reduction sequences were more fragmented and their first phases were sometimes performed outside of the shelter, especially in sublevel Ja.

However, flake attributes show significant differences depending on flake size, which suggests that the temporal location in the reduction sequence is one of the main factors in flake attribute variability. Cortical dorsal surfaces and striking platforms tend to become more common as flake size increases. For example, only 16.2% of the very small flakes show dorsal cortex, whereas 67.6% of the very large flakes are cortical. This correlation is logical, since both attributes are good indicators of the reduction stage. In spite of the variability caused by differences in nodule size, flake dimensions tend to decrease as the reduction sequence progresses and cortical products are more common in the initial stages of flaking. The type and preparation of the striking platforms also vary depending on flake size. Linear and pointed butts are more common in small flakes, but this is simply the result of the gradual reduction of striking platform area as flake size decreases. Facetted butts tend to increase as flakes become larger (2% for very small flakes, 4.9% for very large flakes), although their percentages always remain low. Diffuse bulbs and straight ventral profiles are more abundant in small and very small flakes, even when only chert flakes are considered. Finally, the percentage of *débordant* products is lower among the very small flakes (7.5%). However, there are not significant differences among the remaining size categories, reaching a maximum of 23.4% in medium flakes.

Core Reduction Strategies

Seven cores were excluded from the dimensional analysis because they presented post-reduction fractures. Distribution by size classes clearly shows the reduced dimensions of most cores (Table 7.7). This is due to the great exhaustion of many cores, but also to the small size of some nodules from the beginning of the reduction sequence. The scatterplot (Fig. 7.23), apart from showing the high positive correlation between length and width, indicates that most cores are concentrated in a well-defined cluster characterized by lengths between 60 and 25 cm, and widths between 50 and 20 cm. This suggests that reduction sequences were guided by a well-defined dimensional threshold that

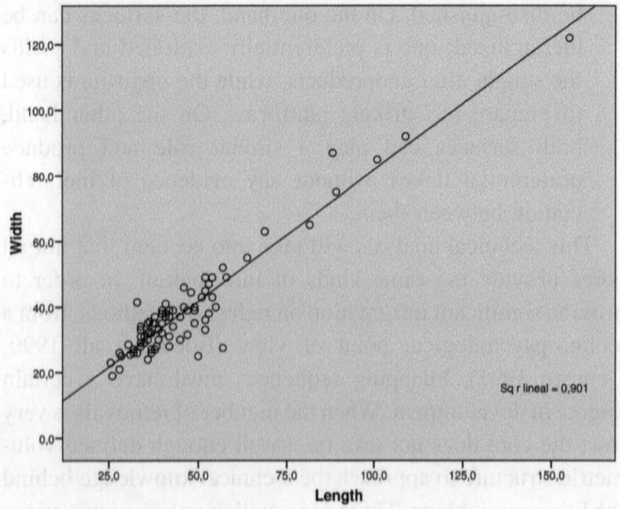

Fig. 7.23 Scatterplot of length and width of cores

determined the point at which core exploitation ended. Only eight unusually large cores fall outside of this cluster: five chert cores, two quartz and one limestone. Four of these large chert cores are spatially located outside of the main accumulations of lithic remains. In addition, two of these large cores can actually be interpreted as pebble tools. The dimensions of the last removals, which tend to correspond to small or very small flakes, also display core exhaustion.

The morphology of most cores corresponds more or less to the discoidal method, as defined by Boëda (1993b), especially regarding two essential criteria. First, cores are structured as two opposing surfaces separated by an intersection plane, and second, the direction of removals is secant to this intersection plane. However, we think that simply characterizing the reduction sequences of level J as discoidal is not sufficient to report the technical variability of this assemblage. The cores of level J exhibit a wide formal variability, although as we will see, the knapping products encountered there do not show significant changes. The attributes defining the variability of discoidal technologies have recently been discussed (Peresani 1998; Mourre 2003; Slimak 2003; Terradas 2003). Among these attributes, two seem to be especially significant:

(a) The angle of removals on each flaking surface. The overall morphology of the core largely depends on this feature. If the intersection plane is located in the middle of the core and the removals have the same angle on both surfaces, the core will have a symmetrical outline. This is the bipyramidal form typical of discoidal cores. However, if the intersection plane is not centered, the flaking surfaces will exhibit different angles and, therefore, the core will show an asymmetrical morphology.
(b) The functional relationships between both surfaces. According to this criterion, two different situations can

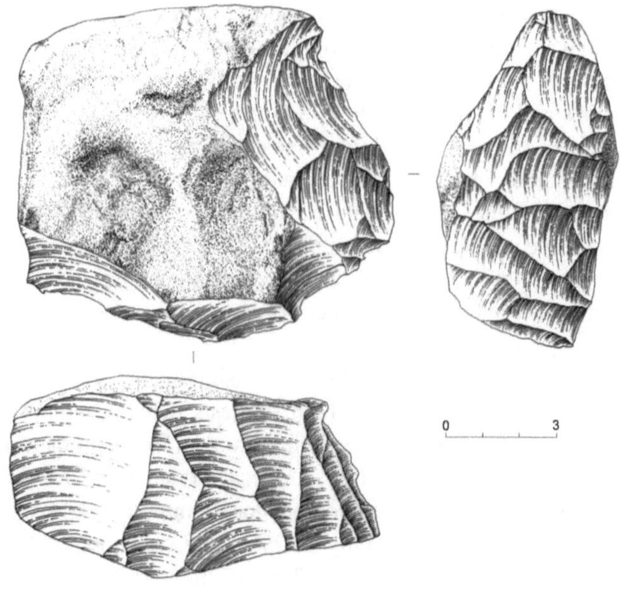

Fig. 7.24 Limestone cobble showing a unifacial reduction that can be interpreted as a pebble tool

Fig. 7.25 Unipolar chert core from which some elongated removals have been detached

be distinguished. On the one hand, the surfaces can be hierarchized; one is preferentially exploited and yields the sought after endproducts, while the opposite is used to prepare the striking platforms. On the other hand, both surfaces can play a similar role and produce preferential flakes, without any evidence of hierarchization between them.

This technical analysis will take into account that not all cores provide the same kinds of information. In order to provide significant information on reduction methods, from a techno-psychological point of view (Boëda et al. 1990; Geneste 1991), knapping sequences must have a certain degree of development. When the number of removals is very low, the core does not take on a well enough defined volumetric structure to approach the technical knowledge behind the lithic assemblage. These cores will therefore play a minor role in identifying reduction methods. However, they commonly provide very interesting techno-economic information. These scarcely reduced cores are well represented (24.7% of all cores), and they may be attributed to two different situations, at both ends of the *chaîne opératoire*:

(a) Very large cores corresponding to tested nodules that were not exploited (n = 3). This was an uncommon situation, especially given the extreme exhaustion of raw materials inferred from other characteristics of the assemblage.
(b) Fragments or flakes from which only some isolated removals have been detached. The behavior explaining these cores is the opposite of that behind the tested nodules explained above. These artifacts exhibited from the start a very limited productive capacity. Their reduction indicates that in some instances practically any object was considered suitable to provide some flakes, which suggests an intense use of raw materials. These artifacts form the bulk of the scarcely reduced cores found in level J (20.4% of all cores). Refitting will provide more data on this subject.

The core structures characterized by two opposing flaking surfaces separated by an intersection plane are dominant. As stated above, this is one of the basic criteria of discoidal strategies and most cores show this structure. However, nine unifacial cores have also been recovered (9.6% of the assemblage). Two of them may in fact be classified as pebble tools as they are a limestone and a quartz nodule showing a series of removals creating a sharp edge with a pointed profile on their distal sides (Fig. 7.24). Some unifacial cores (n = 2) may be characterized as unifacial discoidal cores, since the flaking surfaces exhibit the features typical of discoidal cores: centripetal disposal and secant direction of removals. Others (n = 5) are single striking platform unipolar cores with unidirectional removals that tend to create prismatic or conical shapes. Some of these cores show the detachment of elongated blanks, even of blades (Fig. 7.25). However, these unifacial cores should not be considered the result of specific reduction strategies, but one of the dimensions of variability of a single reduction method.

The symmetrical bipyramidal morphologies typical of discoidal technologies are relatively scarce (n = 12). These cores usually exhibit non-hierarchized patterns; the two faces are used alternatively as flaking surfaces and striking

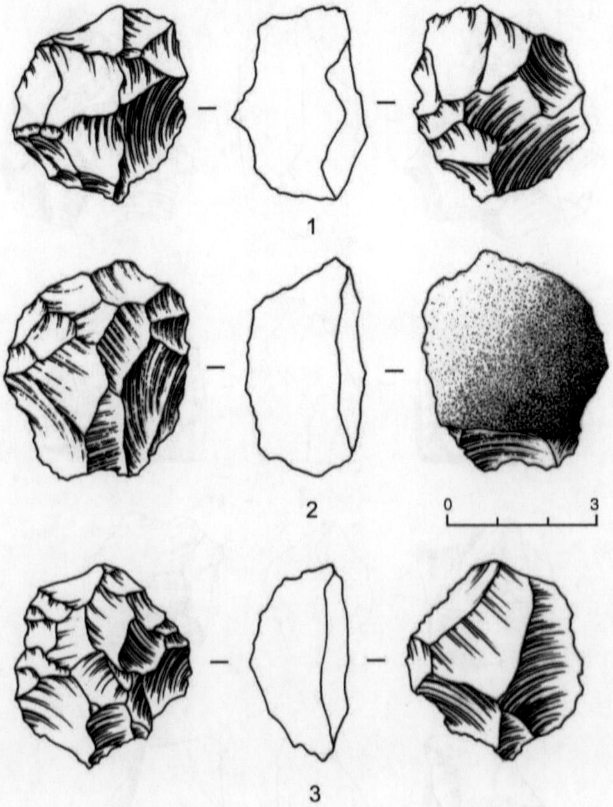

Fig. 7.26 Cores from sublevel Ja

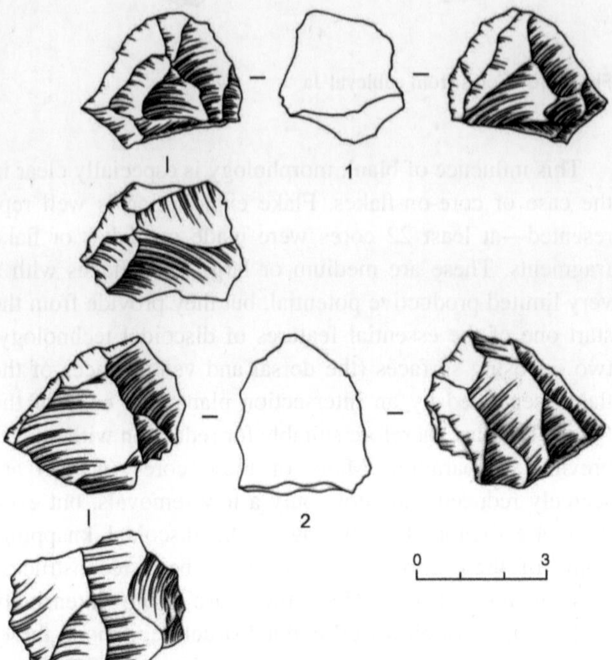

Fig. 7.27 Cores from sublevel Ja

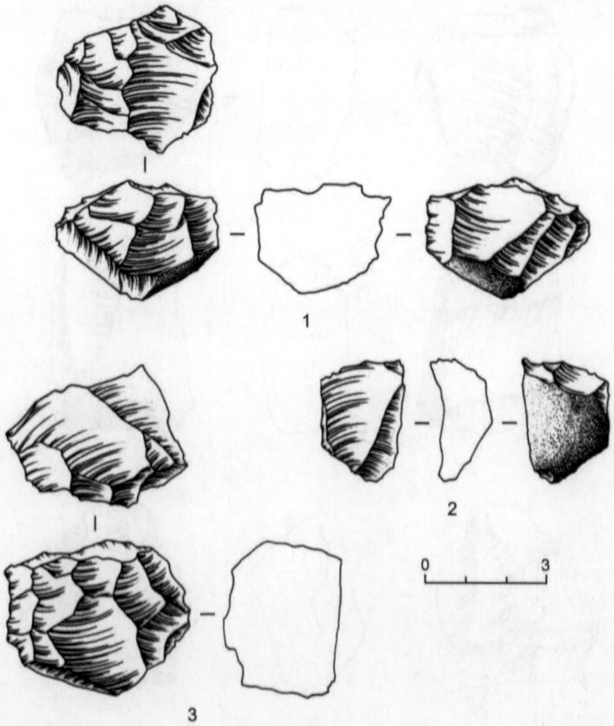

Fig. 7.28 Cores from sublevel Ja

platforms (Fig. 7.30, 1–3). Only one case exhibits a hierarchized structure. Asymmetrical cores are more abundant. In these cores, one surface has a pyramidal shape, while the other has a nearly flat outline. Among these cores, both hierarchized and non-hierarchized patterns have been documented (18 and 16 cores respectively). In the hierarchized pattern, two variants can be distinguished depending on which surface, the flat or the conical, is preferentially exploited. Both of these types of hierarchization are represented and, although in most cases the flaking surface corresponds to the flat face (n = 11), the opposite procedure has also been found (n = 7) (Figs. 7.26, 7.27, 7.28, 7.29, 7.30, 7.31, 7.32).

The cores with a preferentially exploited flat surface are especially interesting, as their morphology is similar to that of typical Levallois cores. In fact, four cores exhibit the scar of a last large flake extending over most of the flaking surface (Figs. 7.28, 2; 7.29, 1). This is reminiscent of the classical definition of the Levallois method (the linear Levallois method according to Boëda 1993a): a strategy designed for producing a single major flake removal. However, in level J small cores sometimes derived from small nodules show this strategy as well, which suggests that it was a procedure aimed at maximizing the size of the flakes in the last phases of the reduction sequences. The rest of the cores with a flat flaking

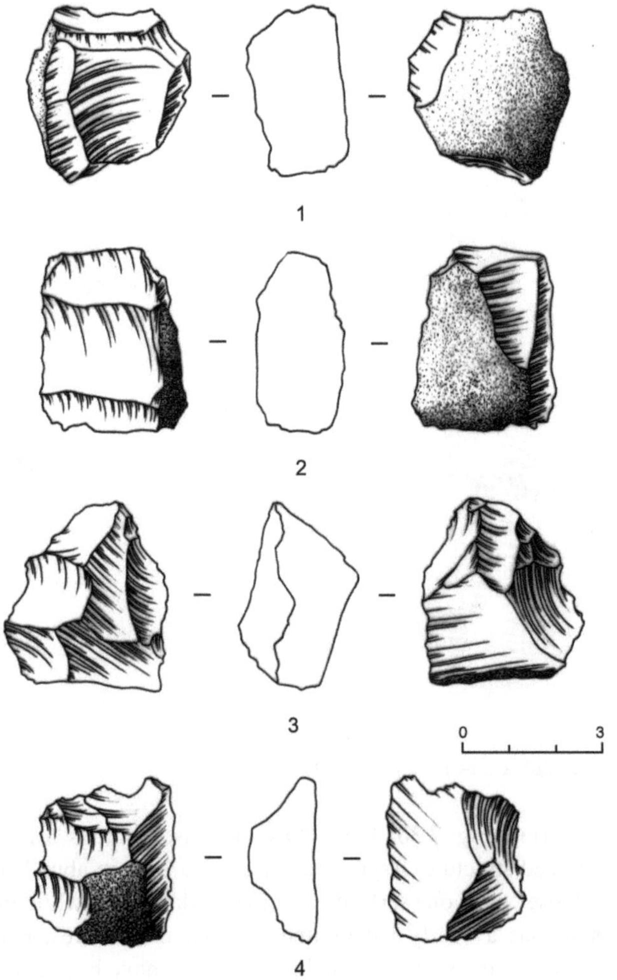

Fig. 7.29 Cores from sublevel Ja

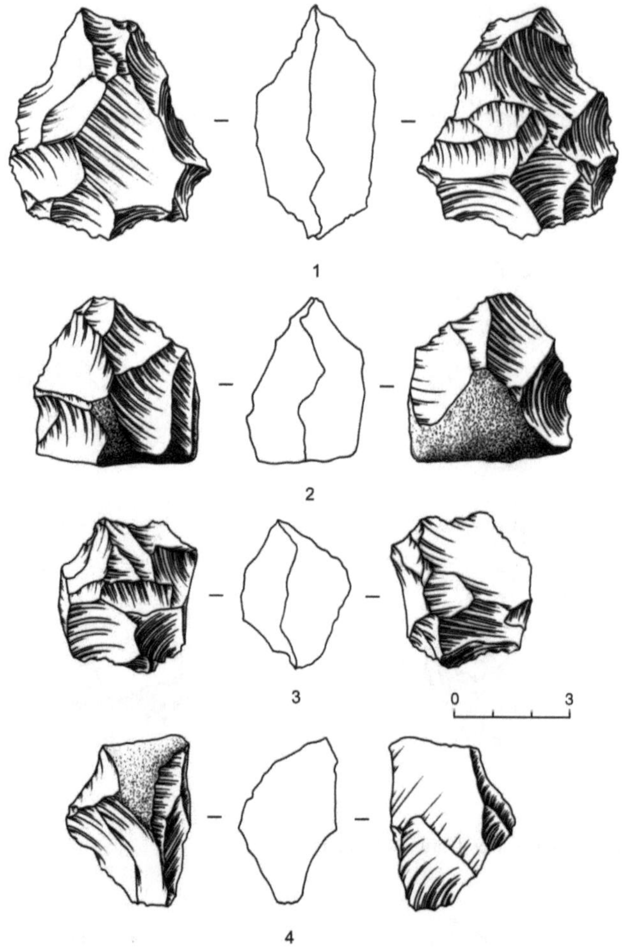

Fig. 7.30 Cores from sublevel Ja

surface tend to show a centripetal scar pattern, like the cores derived from the recurrent centripetal Levallois method. Nevertheless, in spite of their similarity to the Levallois cores, the artifacts from level J do not exhibit the careful preparation of the flaking surfaces that, according to some authors, define the Levallois concept.

It seems that this formal variability was partly related to the initial shape of the core. When the nodules exhibit an asymmetrical morphology, the resulting core also tends to become asymmetrical. For example, the reduction sequence of RMU Lim-012, whose initial stage has been refitted, was conducted on a cobble presenting a plano-convex profile. The natural edge of this cobble was an intersection plane separating the two exploited surfaces and the core therefore reproduced the plano-convex structure of the cobble. Other sequences also show how the knappers took advantage of the natural features of the nodules. The natural edges of some prismatic nodules were used as guides for detaching the first removals, as can be seen in Chert-006 from sublevel Ja and Chert-02 from sublevel Jb (Figs. 7.33, 7.34).

This influence of blank morphology is especially clear in the case of core-on-flakes. Flake exploitation is well represented—at least 22 cores were made on flakes or flake fragments. These are medium or large-sized flakes with a very limited productive potential, but they provide from the start one of the essential features of discoidal technology: two opposing surfaces (the dorsal and ventral faces of the flake) separated by an intersection plane (the edge of the flake). They are therefore suitable for reduction without any previous preparation. Many of these cores ($n = 8$) are scarcely reduced and show only a few removals, but even these are exploited according to the discoidal knapping. Some of these short sequences have been reconstructed through refits (Fig. 7.35); others are more intensively reduced and exhibit a well-defined structure. Among these, asymmetrical shapes are dominant, because flakes are characterized by the opposition of a flat surface (the ventral face) and a convex surface (the dorsal face). The morphology of the core is therefore determined by the initial structure of the flakes. Core-on-flakes produced primarily small and very small flakes, which was one of the main

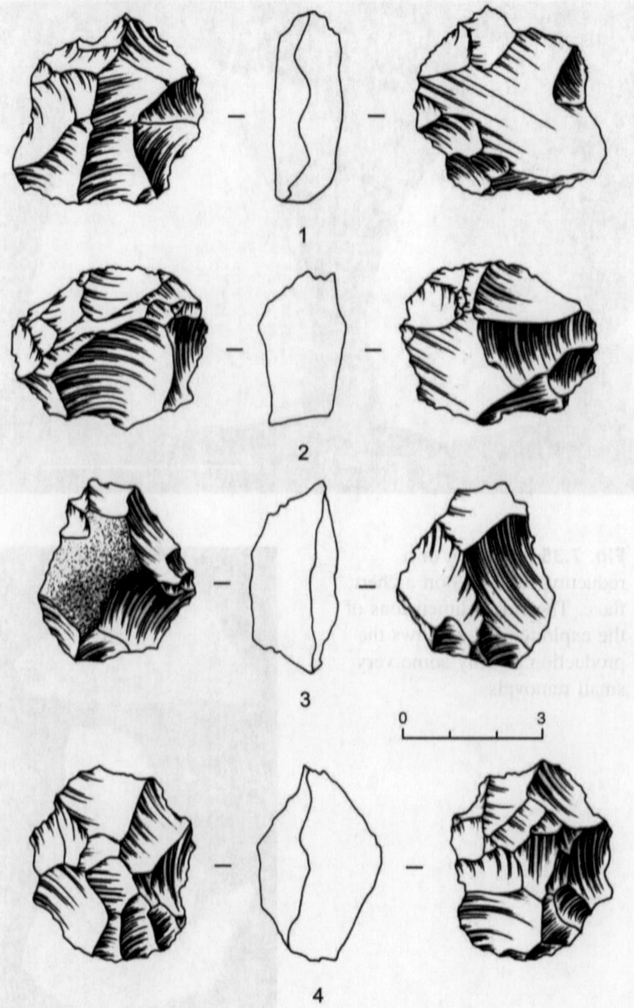

Fig. 7.31 Cores from sublevel Ja

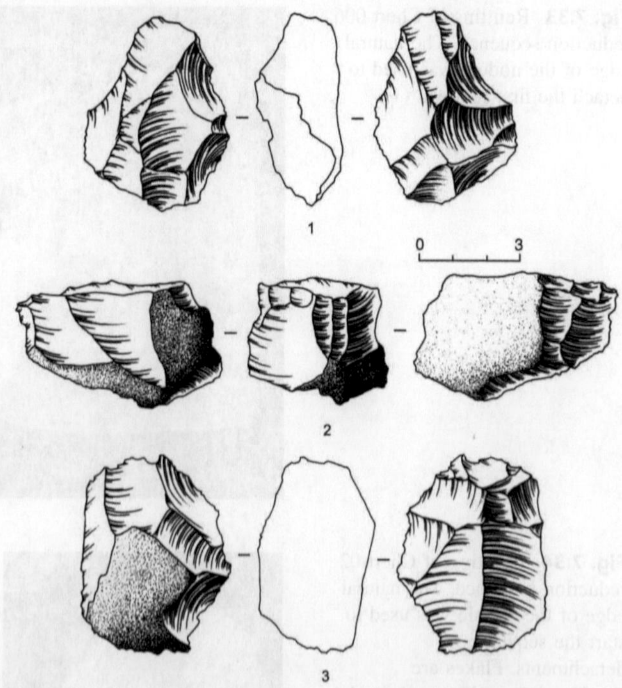

Fig. 7.32 Cores from sublevel Ja

goals of most reduction sequences. In some cases (n = 4), reduction is limited to one side of the flake, as can be seen, for example, in one of the refits of Chert-014 (Fig. 7.36). In this sequence, seven small flakes were detached from the right and distal sides of a *débordant* flake following a discoidal reduction strategy.

Core-on-flakes show both hierarchized and non-hierarchized patterns. In hierarchized cores both the ventral and the dorsal faces of the flake can be used as a flaking surface, although the dorsal faces are slightly dominant. This is especially significant from the point of view of the technical criteria characteristic of level J. Dorsal faces allow secant-direction removals, while ventral faces are more suitable for obtaining flakes parallel to the intersection plane. In a discoidal context, in which the secant direction of removals tends to be the rule, the preferential exploitation of dorsal faces seems the logical choice. This differs from the core-on-flakes documented at other Romaní levels, like level E, where hierarchized Levallois-like strategies are better represented (Vaquero 1999). In these levels, the ventral faces are preferentially selected as flaking surfaces, while dorsal surfaces are used as striking platforms. Consequently, kombewa flakes tend to be more common.

Cores presenting more than one striking platform surface are uncommon. The appearance of a second striking platform surface is associated with both ends of the reduction sequence. On the one hand, this second platform can take advantage of the morphology created following the discoidal strategy. The pyramidal shape common to discoidal cores is characterized by the appearance of an edge at the midpoint of the flaking surface. This edge can be used for removals, although in general these do not significantly modify the discoidal morphology of the core. However, in some cases the removal of flakes from any surface presenting a suitable angle produced polyhedral cores (n = 5). These procedures were characteristic of the last stages of the reduction sequence and they were aimed at maximizing the utility of the raw material before discarding it. On the other hand, more than one striking surface also tends to appear during the first stage of knapping of very large nodules, especially when these are very thick and have a cubic morphology. In these cases, the removal of flakes from different surfaces, taking advantage of the natural planes of the nodule, seems to be an efficient procedure to start the reduction of such large volumes. The three cores found in this initial stage of the reduction sequence show the use of this strategy.

As mentioned in the context of unifacial artifacts, some cores can be characterized as unipolar single-striking-

Fig. 7.33 Refitting of Chert-006 reduction sequence. The natural edge of the nodule was used to detach the first removal (1)

Fig. 7.34 Refitting of Chert-02 reduction sequence. The natural edge of the nodule was used to start the sequence of detachments. Flakes are numbered according to the order of removal

Fig. 7.35 Refitting of a reduction sequence on a chert flake. The small dimensions of the exploited flake allows the production of only some very small removals

platform cores. Ten cores have been included in this category. Five of them are unifacial and the other five present a striking platform prepared by removals, normally a single large one. Detachments were restricted to a small section of the striking-platform perimeter. In general, these cores exhibit relatively short series of removals from blanks characterized by their limited productive potential. In some instances, they correspond to very thin nodules, like that shown in Fig. 7.25. The unipolar character of the reduction seems the best and simplest choice for exploiting such small and prismatic blanks. It would be a quick procedure performed for the purpose of making some use of these scarcely profitable blanks without investing great effort in core structure preparation.

These unipolar cores should be interpreted as another example of the variability characteristic in an expedient technological context, and not as the consequence of the application of a specific reduction method. The laminar character of some removals detached from these cores should be also understood in this context. They were not produced using a laminar method, but are simply an outcome of the wide variability defining this technological system. We think that this could also explain the apparent blade cores found in other late Middle Paleolithic assemblages that are also characterized by the dominance of discoidal technology (Maíllo et al. 2004). In fact, these cores exhibit a morphological continuity with some asymmetrical discoidal cores with a very flat flaking surface.

Therefore, it is clear that the strategy used in the initial stage of the reduction sequence depends on the shape of the nodule or blank. Large prismatic nodules with different suitable surfaces show the removal of flakes in more than one direction and the use of more than one striking platform. Flakes and small nodules, especially those presenting an oval shape, are reduced from the beginning using a discoidal strategy. As far as core structure is concerned, there are no differences between the initial steps of lithic production and the phase of *plein débitage*. This suggests that a separate phase of preparation cannot be distinguished from the rest of the reduction sequence. Decortication should therefore not be considered as a preparation stage, but as part of the production sequence for the endproducts

Fig. 7.36 Refitting of a reduction sequence made on a *débordant* chert flake. Removals were detached following a discoidal strategy

Fig. 7.37 Core-on-flake corresponding to RMU Chert-012

sought. Cortical flakes were often used as blanks for retouched artifacts, and many of the lithics introduced into the shelter as isolated artifacts were also cortical.

These data indicate that recurrence is a fundamental rule in core reduction sequences. The basic goal of debitage is to obtain as many flakes as possible, from the beginning of reduction until core exhaustion. There are no particular requirements for the size or shape of the products. Reduction strategies allow a dynamic development of exploitation, since the continuous reshaping of the convexities needed for flake removal is one of the main advantages of discoidal methods. Moreover, different data indicate a tendency to maximize the profitability of the cores. Although a few almost raw nodules have been found, most cores show a very intense reduction, sometimes reaching the limits of their productive capabilities. The removal scars on these exhausted cores suggest that the last stages of the reduction sequences were aimed at producing a recurrent series of small and very small flakes. Even small flakes and flake fragments were exploited, which means that in some instances any artifact from which removals could be taken was considered suitable for exploitation. It seems, then, that economizing raw material was a main concern.

Many reduction sequences were directed toward the production of small and very small flakes. Core size indicates that in the last stages of reduction many knapping events were able to produce exclusively small blanks. In addition, some reduction episodes were carried out on artifacts that from the beginning had a limited productive capacity, like flakes or fragments that could only provide short series of removals (see for instance Figs. 7.35, 7.36). This intentional production of small blanks is especially clear in some RMUs. For example, Chert-012 shows the exploitation of a cobble that was introduced into the shelter unmodified or slightly reduced. Remains of this RMU were scattered in the central area of the rockshelter (zone 4). Six cores have been identified in this sequence and most of them are core-on-flakes that only yielded small and very small blanks. This nodule was fragmented in an initial stage of the reduction sequence and some products derived from this stage were used to obtain small blanks (Fig. 7.37). This emphasis on small flake production is one of the main characteristics of technical activities carried out in the site, as opposed to the reduction sequences carried out elsewhere and represented in the assemblage as single blanks.

Retouched Artifacts

In order to avoid possible misleading data introduced by post-retouch breakage, dimensional attributes have been recorded only for those unbroken retouched artifacts (n = 117). The mean length, width and thickness of tools

Table 7.16 Distribution of retouched artifacts by elongation index classes and sublevels

	L1	L2	L3	L4	Total
Ja	2	27	39	21	89
	2.2%	30.3%	43.8%	23.6%	100%
Jb	1	11	11	5	28
	3.6%	39.3%	39.3%	17.9%	100%
Total	3	38	50	26	117
	2.6%	32.5%	42.7%	22.2%	100%

Table 7.17 Distribution of retouched artifacts by thickness index classes and sublevels

	C2	C3	C4	C5	C6	Total
Ja	16	49	18	4	2	89
	18%	55.1%	20.2%	4.5%	2.2%	100%
Jb	6	18	2	2		28
	21.4%	64.3%	7.1%	7.1%		100%
Total	22	67	20	6	2	117
	18.8%	57.3%	17.1%	5.1%	1.7%	100%

exceed the mean dimensions for unretouched flakes, suggesting that blanks for retouch were preferentially selected from among the largest flakes (Table 7.6). As the distribution by size class shows, medium and large sizes are dominant, and very large items are well represented (Table 7.7). More than 50% of the retouched artifacts are large or very large, whereas only 7.6% of the flakes belong to these size classes. The retouch index by size classes also demonstrates this preferential selection of large flakes, as large and very large blanks exhibit the highest index (31.8 and 28.9 respectively). This index decreases abruptly in the small and very small size categories (3.7 and 0.1 respectively). The highest retouch indices correspond to large blanks rather than very large ones. This suggests that selection criteria were not simply aimed at maximum size, but they were guided by an optimal dimension above which the increase in blank size did not involve an increase in the retouch index. Retouched artifacts tend to be larger in sublevel Jb. Large and very large tools are clearly dominant in this sublevel, while in sublevel Ja medium-sized artifacts are the most common and small items are well represented. Nevertheless, the distribution by size classes does not show statistically significant differences between the two sublevels ($X^2 = 8.2$; $df = 4$; $p > 0.001$) possibly related to the small sample size of the retouched assemblage in sublevel Jb.

The mean of the elongation index is 1.2. This is a little higher than the mean for all flakes, but is similar to the index of the unretouched large flakes. As mentioned earlier, the elongation index increases slightly with flake size. Therefore, the tendency of retouched artifacts to be more elongated is the result of the selection of large flakes. Furthermore, there are no blade blanks among the tools (Table 7.16), so the selection of blanks for retouch according to this feature is not plausible. Differences between retouched artifacts and unretouched flakes are more significant in the thickness index (Table 7.17). Retouched artifacts tend to be thicker than unretouched flakes (mean = 2.66 and 3.06 respectively), but this difference varies according to size classes. The difference is higher between the small and medium artifacts, it tends to decrease in large artifacts and disappears in very large artifacts (in this last class, unretouched flakes tend to be even thicker than the retouched ones). It therefore seems that the preferential selection of thick flakes was more emphasized as the blanks became smaller.

It can be argued that these differences in relative thickness could be the result of the size reduction produced by retouch. As width and/or length are reduced but thickness remains the same, retouch could produce an increase in the thickness index. The smallest retouched artifacts would correspond to the most reduced blanks, and they would therefore exhibit a higher thickness index. However, different data indicate that the size of retouched artifacts is not related to reduction intensity, that is, the smallest flakes were not more intensively reduced. First, the stepped retouch typical of highly resharpened artifacts is absent. Moreover, if retouch intensity significantly modified the dimensions of the artifact, a significant increase in the elongation index would be expected, since retouch on only one side is clearly dominant (81.2%). As we have seen, this increase does not occur. It seems therefore that the increase in the thickness index as the retouched artifacts become smaller is not the result of retouch intensity. Thicker blanks were intentionally selected for retouch and this criterion was applied more rigorously as blank size decreased.

Other attributes of retouched artifacts can also be related to blank selection according to size. Dorsal cortex is very common (53% of all retouched artifacts), but this percentage is comparable to that of similarly sized unretouched flakes. The percentages of *débordant* and natural backed flakes are also higher among retouched artifacts (15.6 and 8.4% respectively) but, as we mentioned earlier, these values tend to increase with flake size. We cannot propose the intentional selection of these particular blanks for retouch (Table 7.17).

The results of the attribute analysis indicate that most artifacts (81.2%) show retouch on only one edge and lateral retouch is dominant (73.3%). Retouch tends to be located on the longest edge; transversal retouch appears on blanks that are wider than they are long. Dorsal retouch is the most widely represented (70.3%); 15.1% of artifacts show ventral retouch and 14.5% were modified both on the dorsal and ventral surfaces. Finally, 70.3% exhibit deep retouch, whereas 18.1% present only marginal retouch and 9.6%

Table 7.18 Distribution of retouched artifacts according to modes of retouch, typological groups and primary types defined in Laplace's Structural and Analytical Typology

Retouch mode	Typological group	Primary type	Ja	Jb
Simple	Sidescrapers	R11	1	1
		R21	1	1
	Denticulates	D11	1	
		D21	19	5
		D22	3	
		D13	23	5
		D23	74	24
		D24	1	
	Endscrapers	G11	1	
Abrupt	Abrupts	A1	7	
		A2	5	
	Borers	Bc1	1	
Flat		F0	5	1
		F11	4	
Total			146	37

show both deep and marginal retouch. There are no significant differences between sublevels Ja and Jb in any of these attributes. The retouch perimeter index is 0.26 in sublevel Ja and 0.27 in sublevel Jb.

One hundred and sixty four artifacts have been classified according to the types defined in Laplace's *Structural and Analytical Typology* (Laplace 1972); 25 lithics were not classified because they were small fragments of retouched artifacts or presented very slight retouch that was difficult to attribute to any type. Among the classified artifacts, 18 (10.1%) were composite tools and presented more than one primary type; there are therefore 183 primary types. The distribution of retouched artifacts by typological groups and primary types can be seen in Table 7.18. As is common in Abric Romaní, retouch is principally oriented toward creating denticulate edges. In this denticulate-dominated sequence, level J exhibits one of the highest percentages of denticulates (84.6%), especially compared to the upper levels (B–E), where the presence of sidescrapers tends to increase. The rest of the typological groups with simple retouch exhibit very low values, especially sidescrapers (2%).

In the denticulate group, the denticulate sidescraper is the most common primary type, especially the deep variant (D23), which represents 53.5% of all primary types, although the marginal denticulates (D13) are also well represented (15.3%). Notches are less abundant (13.6%) and almost all correspond to the deep variant (D21). Other denticulate primary types such as *épines* (D22) and denticulate points (D24) are proportionally very scarce.

Abrupt retouch is represented by two typological groups: abrupts (6%) and borers, although only one artifact has been assigned to the latter category. Artifacts with abrupt retouch but without formalized shapes form the abrupt typological group. Finally, the flat typological group includes artifacts with flat retouch (5.4%), although they do not present specific morphologies. They are similar to the tools in the denticulate group, but present retouch angles lower than 45°. These low angles are normally produced by the retouch of ventral surfaces. In addition, both abrupt and flat retouches are often associated in the same artifacts with other primary types. Four abrupt and five flat primary types form part of the composite tools, always in association with a denticulate edge.

In spite of the small number of remains recovered from sublevel Jb, some differences between the two sublevels can be distinguished. In general, sublevel Ja exhibits higher variability. In sublevel Jb, retouched artifacts are more concentrated in the denticulate group (91.8 versus 82.8% in sublevel Ja). Except for two sidescrapers and one artifact with flat retouch, all the retouched implements recovered in sublevel Jb are denticulates. This higher proportion of denticulates is basically due to the abundance of the most common primary type, the denticulate sidescraper (66.4% in sublevel Ja and 78.3% in sublevel Jb). Abrupt artifacts are absent in sublevel Jb, and flat retouch is represented by only one artifact. Sublevel Jb is also characterized by the near absence of composite tools. Only one such artifact has been found, whereas in sublevel Ja composite tools represent 11.9% of all retouched artifacts.

Figures 7.38, 7.39, 7.40, 7.41, 7.42, 7.43, 7.44, 7.45, 7.46, 7.47.

Functional Analysis

The initial objective of the functional analysis is to interpret which actions the lithic artifacts were involved in, which materials they worked, and how they worked them. However, we would like to depart from the perception of functional studies as just another complementary analysis in archeological research, which in this specific case is devoted to speculating about the use of objects. The function of objects should invariably be considered within any technological interpretation of lithic production, because such interpretation makes no sense if the artifacts were not used or were not effective. Therefore, the ultimate goal of this functional analysis is to participate in the research on technological development and its behavioral implications for Middle Paleolithic human groups at Abric Romaní.

In determining a given use of a lithic object, we are actually researching the concept of the tool as understood by a certain human group—a conception that is expressed through functional patterns. These patterns emerge when

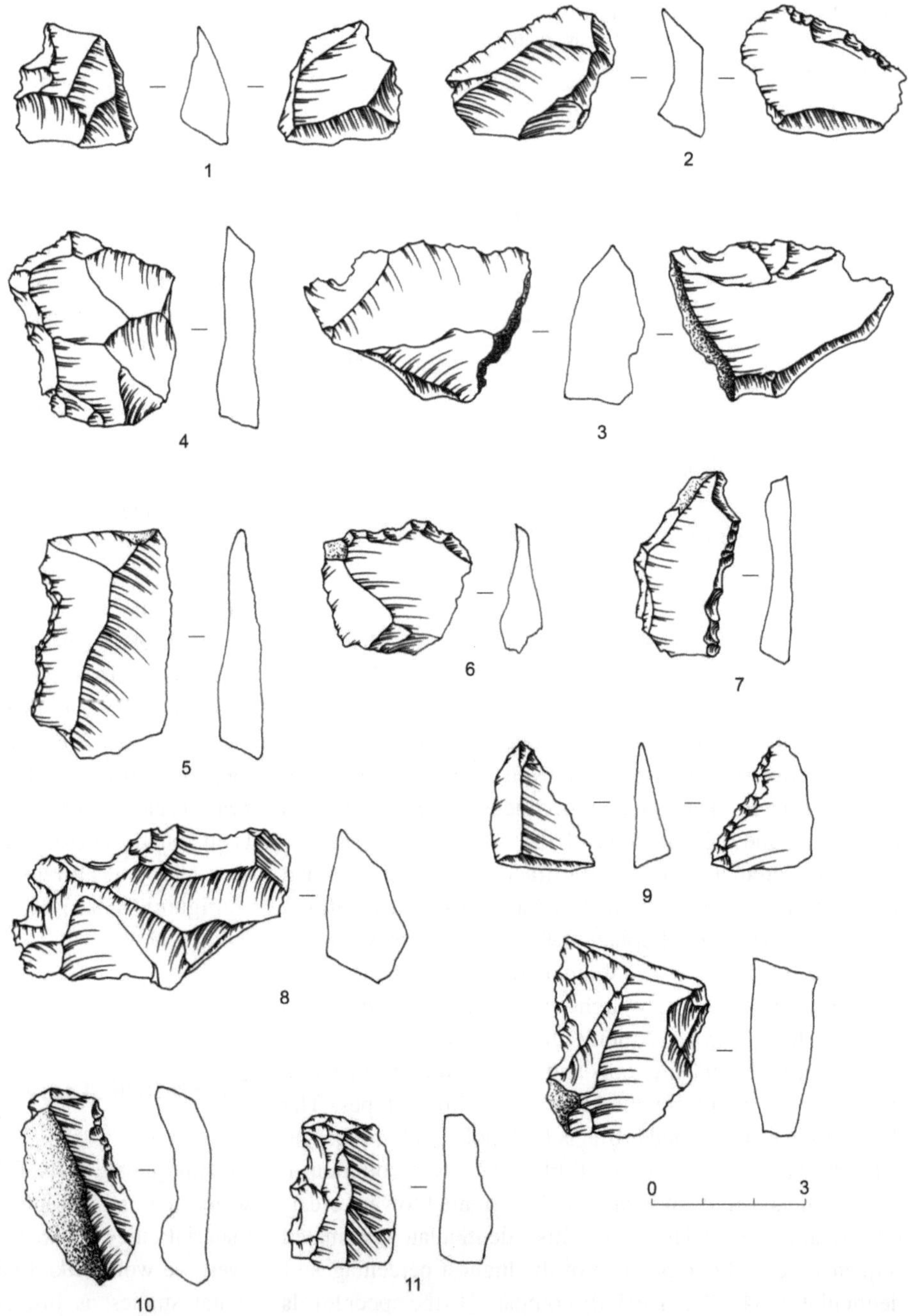

Fig. 7.38 Retouched artifacts from level J

there is a recurrence in the use of certain edge or object morphologies. The existence of functional patterns or the lack thereof, the number of such patterns and the degree of systemization they present reflect the demands imposed by adaptive strategies on the technological abilities of the group. To determine the existence of functional patterns, used objects and edges are described metrically, technically and morphologically (technical use criteria) as is the way they were used—to cut, scratch, scrape, etc. (technical use modes). A close relationship between technical use criteria and technical use modes would imply the existence of specialized tools. In the opposite case, we would be dealing with versatile tools.

Another aspect of functional studies conducted at the site is the interpretation of the spatial organization of activities within the shelter. The spatial location of objects with use-wear traces and their relation to the rest of the archeological record is a primary source of information for evaluating the existence of specialized activity areas or complex spatial structuring.

From sublevel Ja, 101 retouched artifacts were selected and from sublevel Jb, 20 objects were selected. Objects

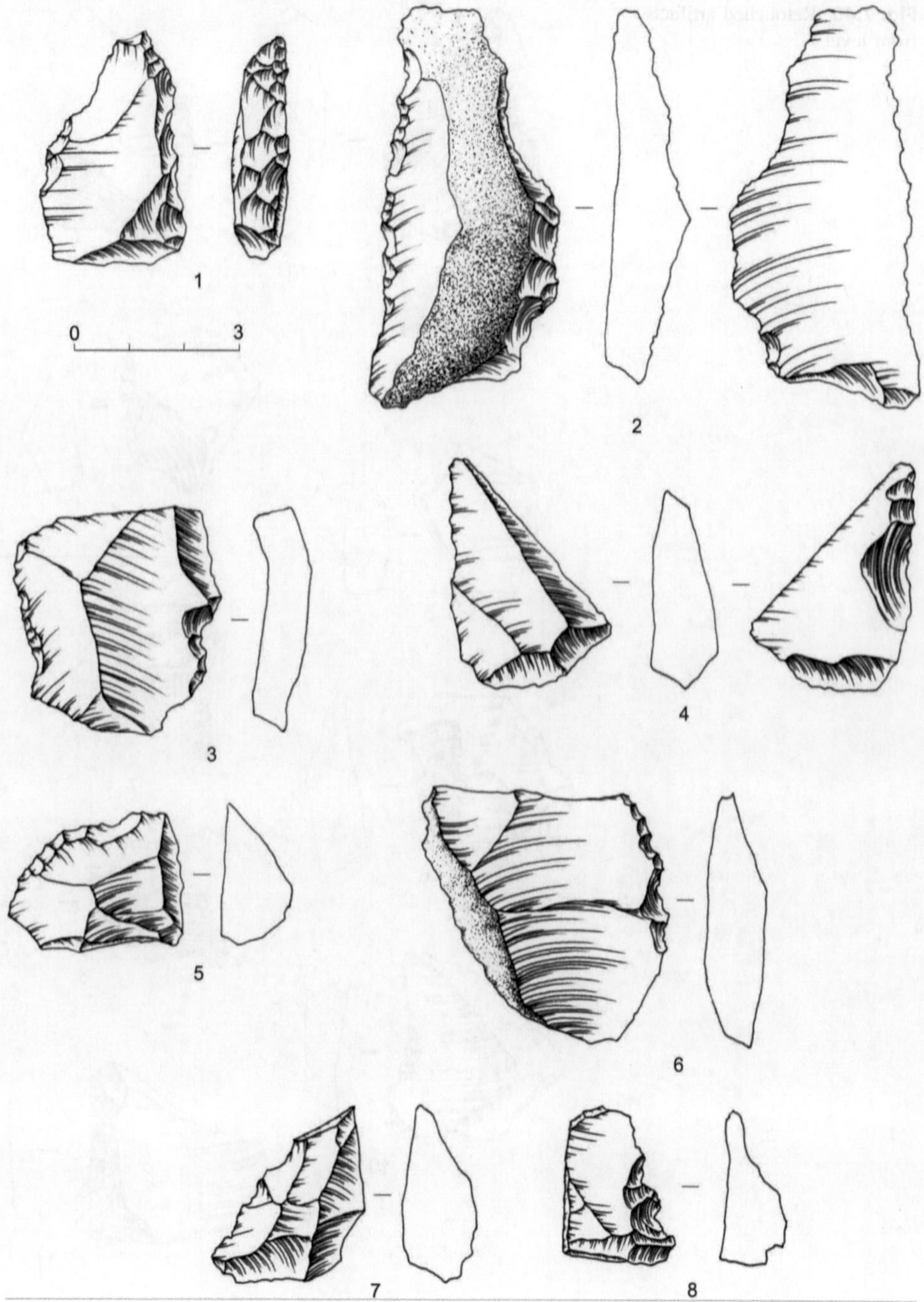

Fig. 7.39 Retouched artifacts from level J

created from raw materials other than chert were excluded because, as more than 90% of the retouched pieces were made from this stone, we decided to limit the effect of petrographic diversity on the appearance of traces (Masson 1982; Lerner et al. 2007). We also rejected objects with thermal damage, with high degrees of erosion and those made on coarse-grained chert. As for the assemblage of flakes, we analyzed 60 pieces of which 47 came from sublevel Ja. They were selected from among the diversity of objects common to the exploitation systems used at the site, with preference given to those of a larger size and excluding those that presented significant alterations. The sample included a significant percentage of large and very large pieces (Table 7.19).

Results of the Functional Analysis

Despite the extensive sample studied from sublevel Ja, we only identified nine retouched objects and four flakes with use-wear traces, and in another nine objects the traces were classified as indeterminate. In sublevel Jb we found two retouched artifacts and three flakes with traces. This reduced

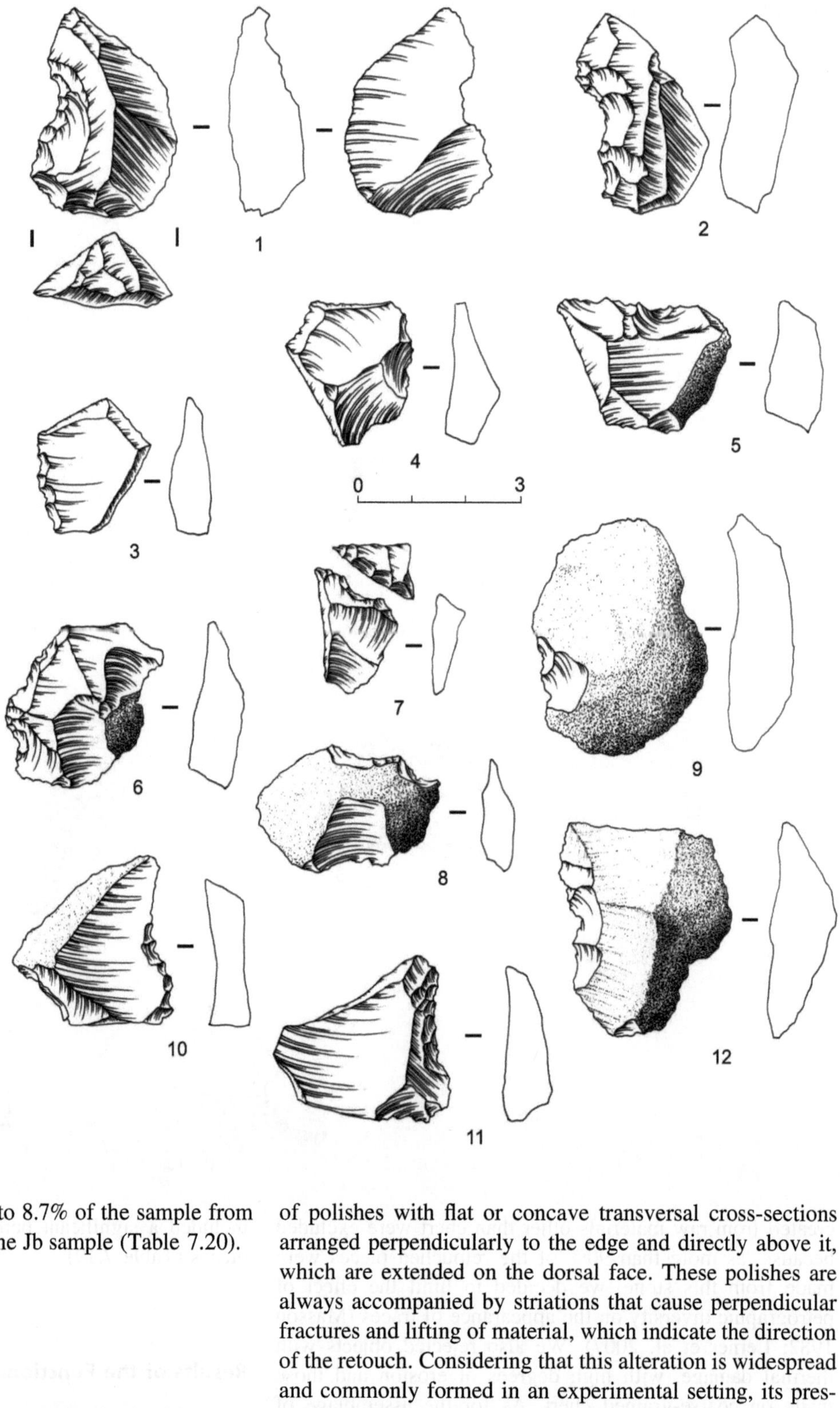

Fig. 7.40 Retouched artifacts from level J

the percentage of identification to 8.7% of the sample from sublevel Ja and to 15.1% from the Jb sample (Table 7.20).

Alterations

a. Technical traces

One of the types of alteration consisted of evidence of retouch on the shaped objects. This alteration, in addition to the fracture of the edges, consists of friction traces in the form of polishes with flat or concave transversal cross-sections arranged perpendicularly to the edge and directly above it, which are extended on the dorsal face. These polishes are always accompanied by striations that cause perpendicular fractures and lifting of material, which indicate the direction of the retouch. Considering that this alteration is widespread and commonly formed in an experimental setting, its presence in the archeological sample is low. This is because, as we have seen in experimental tests, this technical trace includes leftover residues of the hammerstone compressed on

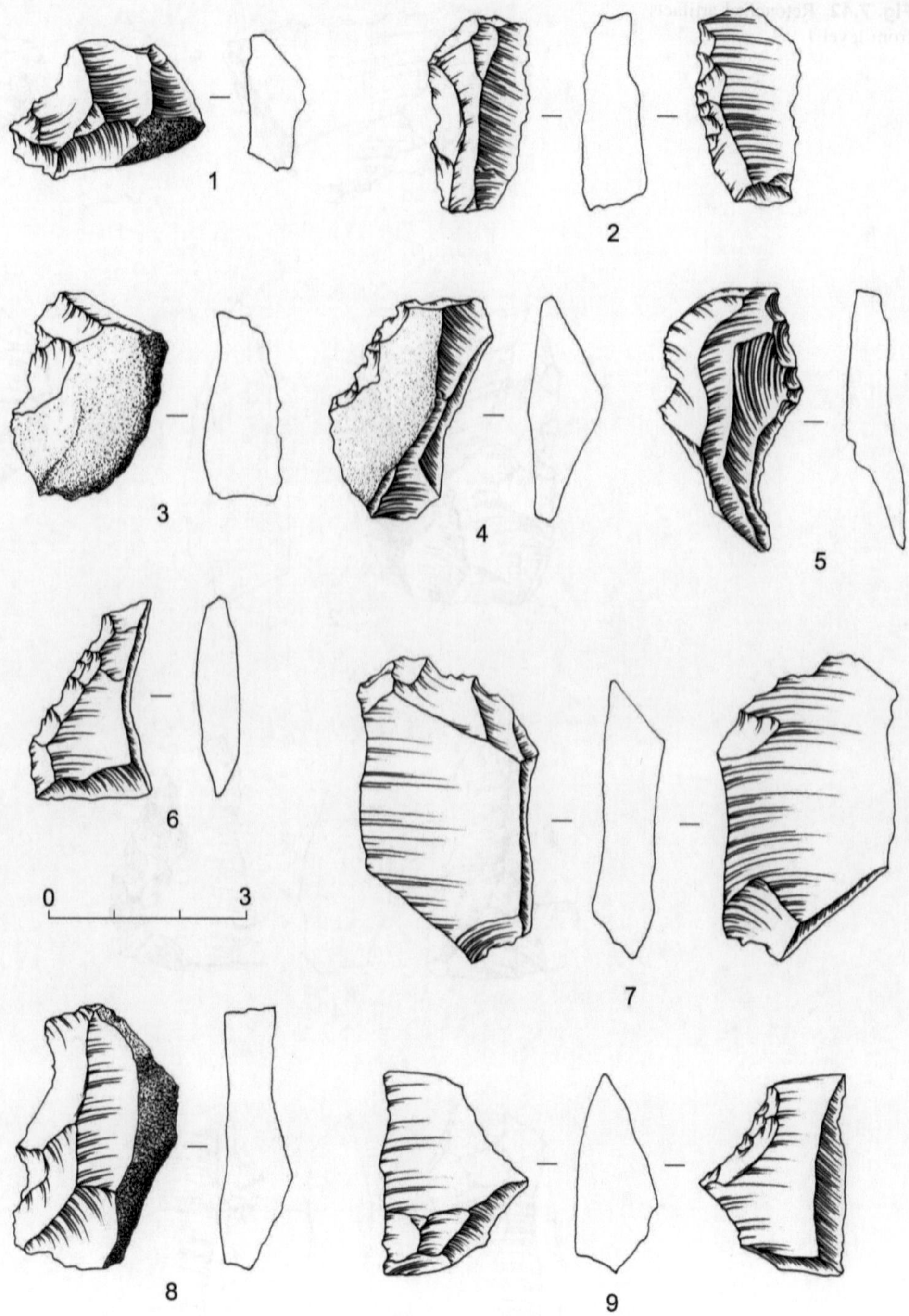

Fig. 7.41 Retouched artifacts from level J

the surface, which is easily eliminated through a cleaning process. Because of this cleaning, the technical trace is reduced to lesser-developed friction traces such as those that are found on these archeological objects.

b. Postdepositional alterations

Nearly half of the sample presented postdepositional alterations. These can be categorized in two groups. The first are friction traces in the form of longitudinal patches of flat polishes with striations that appear discontinuously along the length of the edge and that mark different directions. We attribute these alterations to scratches on the object that occurred while it was buried or as a result of archeological practices. The second group includes pieces with a widespread alteration distributed across the microscopic surface.

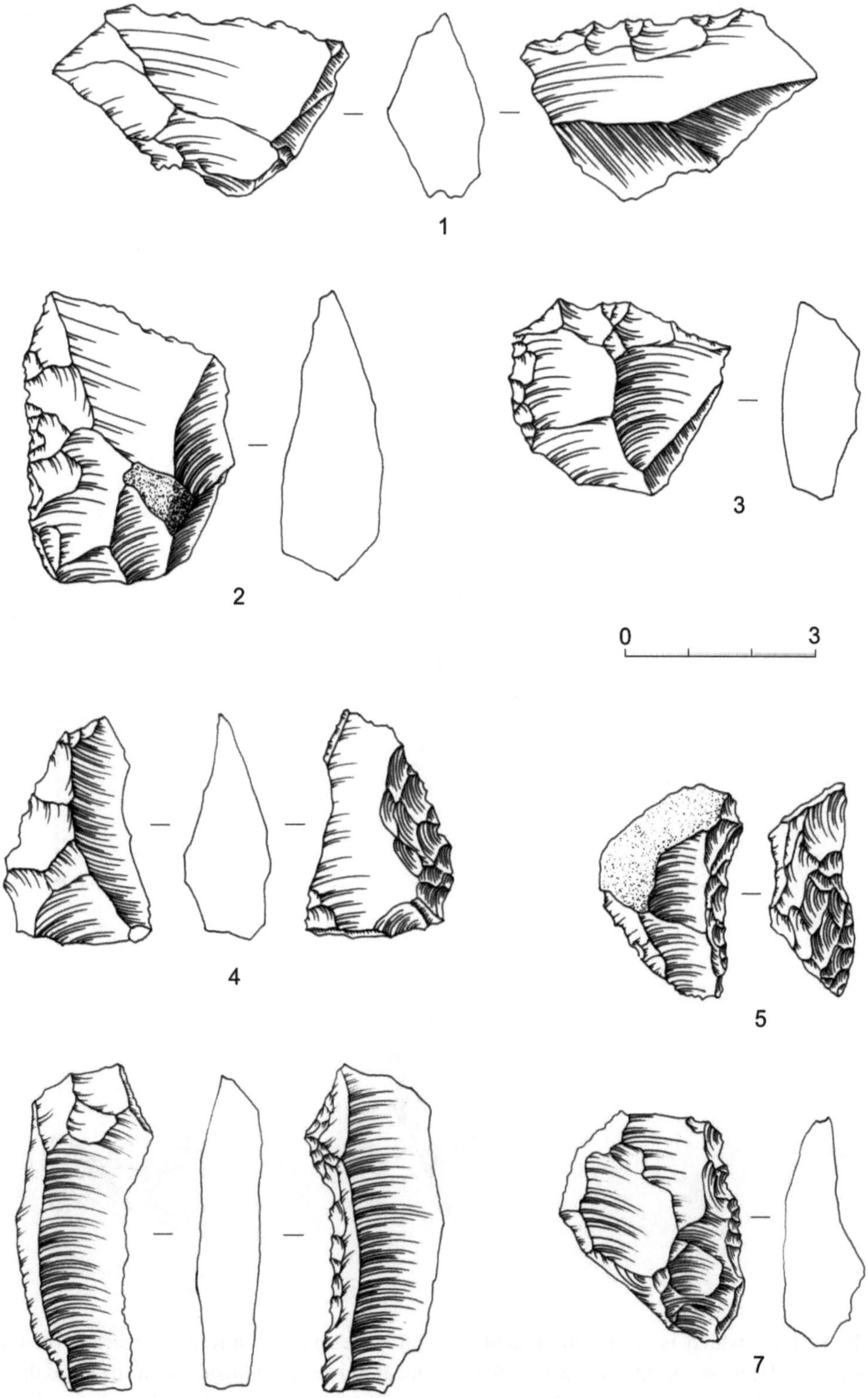

Fig. 7.42 Retouched artifacts from level J

This alteration has two different appearances that reveal the erosive process that provoked them. In one case, the crystalline texture of the objects appears homogenous, smooth and the edge and upper parts are rounded. This alteration would have been produced by the mechanical erosive action of sediment suspended in water as it scratched the surface of the object (Levi-Sala 1986; Ollé 2003; Vergès 2003). In the other case we see a porous, granular crystalline surface with

Fig. 7.43 Retouched artifacts from level J

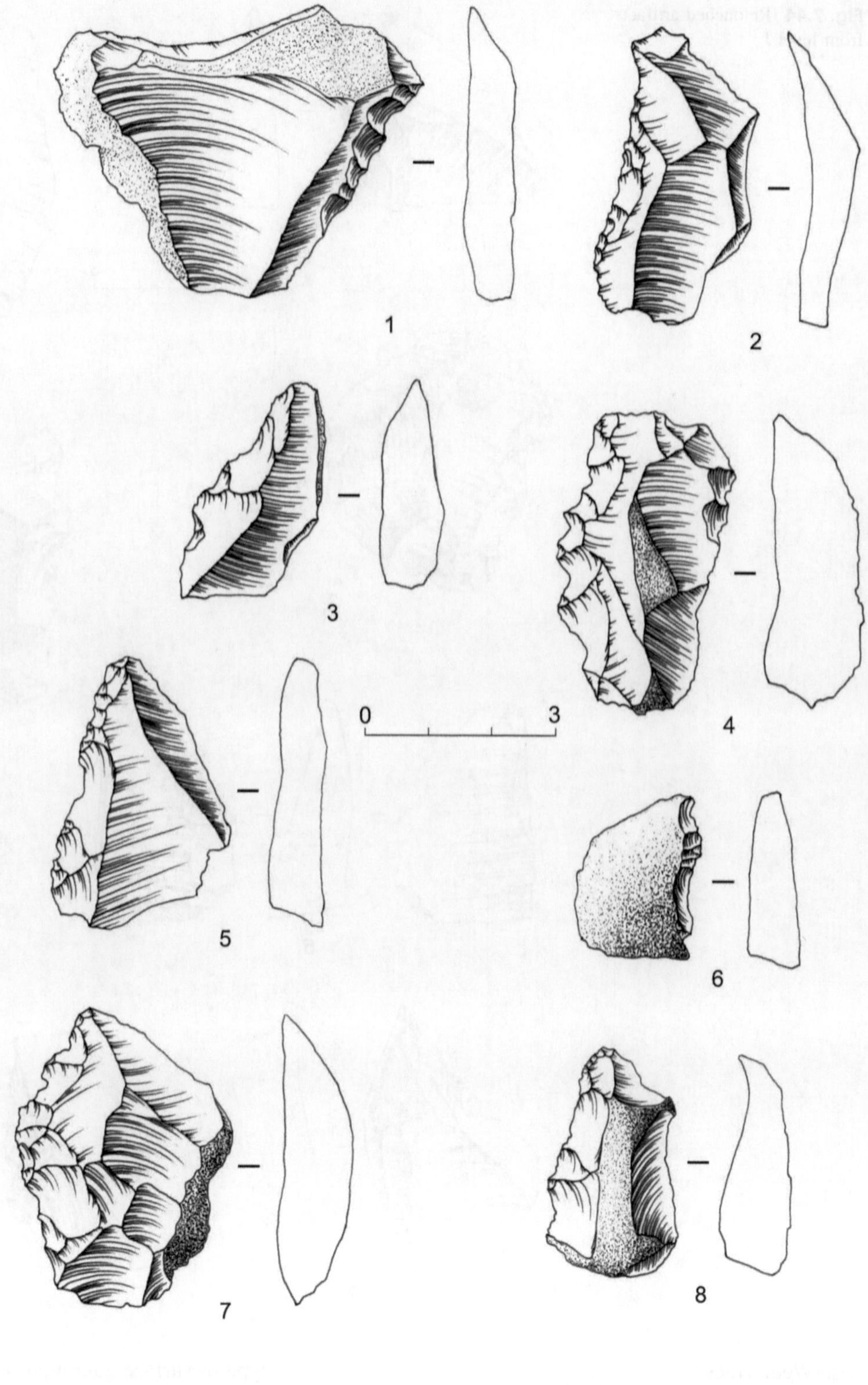

an appearance that has been described as "sugary" (Van Gijn 1989). This texture is caused by the formation of a white patina on the object, caused by the dehydration of the chert (Rottländer 1975; Texier 1981).

White patina affects 75% of the objects to different extents. Despite this patina, a microscopic inspection must be done in order to confirm whether the patina had altered the surface of the object or not, since several tools with use-wear traces were also covered with it. Similarly, we did not find a clear relationship between the extent of the patina on the surface of the object and the degree of alteration to the microcrystalline texture.

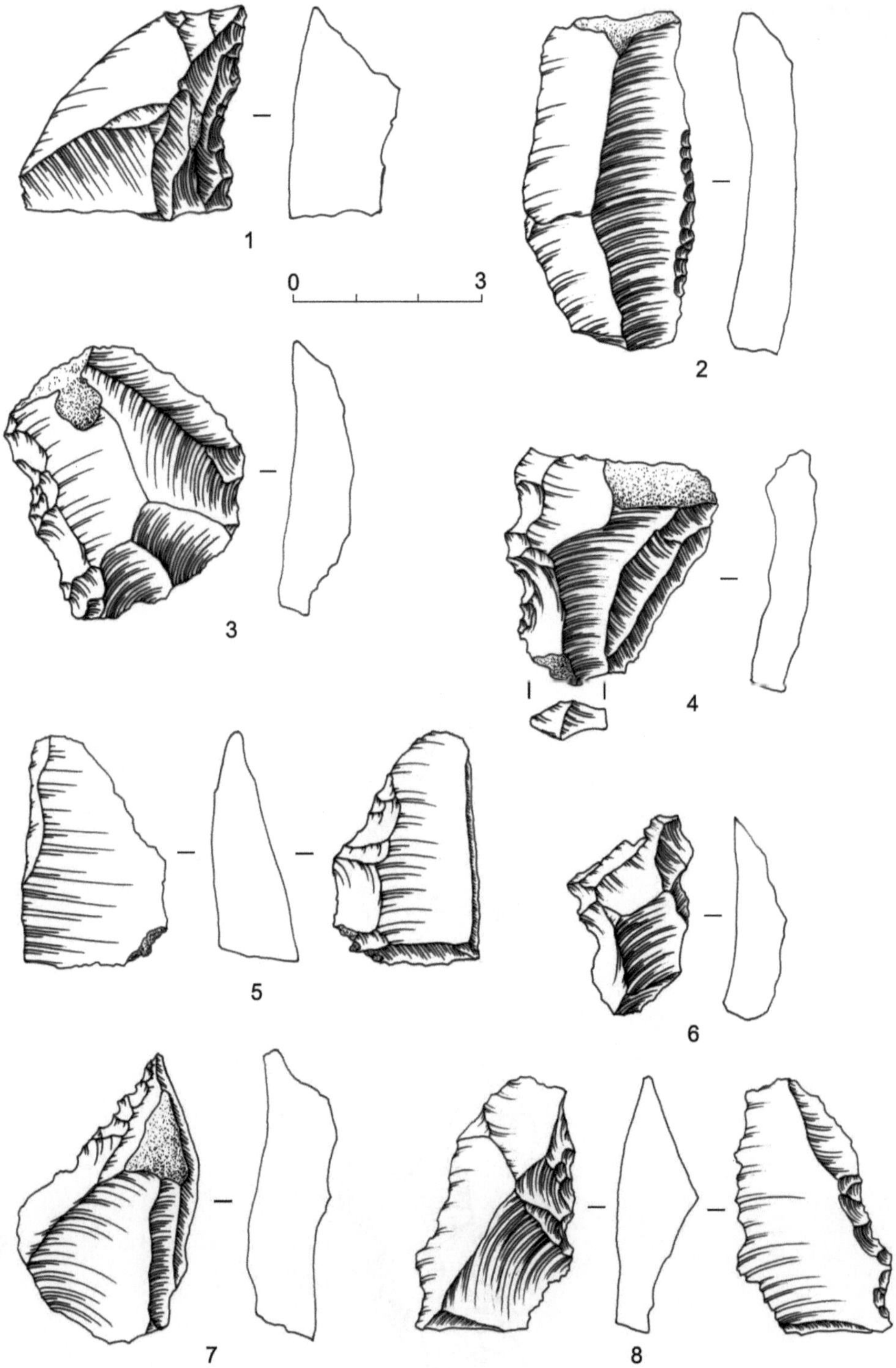

Fig. 7.44 Retouched artifacts from level J

Use-Wear Traces

Use-wear traces were only detected on eleven retouched artifacts and seven flakes, representing respectively 9 and 11.6% of each sample. The remaining objects were divided between those with alterations and unmodified pieces. The latter should not be considered as fresh or unused objects; they are simply items on which we did not recognize any type of surface alteration or modification. We must bear in mind that only 24.8% of the objects are not patinated, meaning that the microscopic surface of these items are altered to different degrees.

Nineteen functional units were identified on the 18 objects whose uses were recognized. A functional unit corresponds to each of the object's active edges on which use-wear traces have been identified (Van Gijn 1989). On these edges,

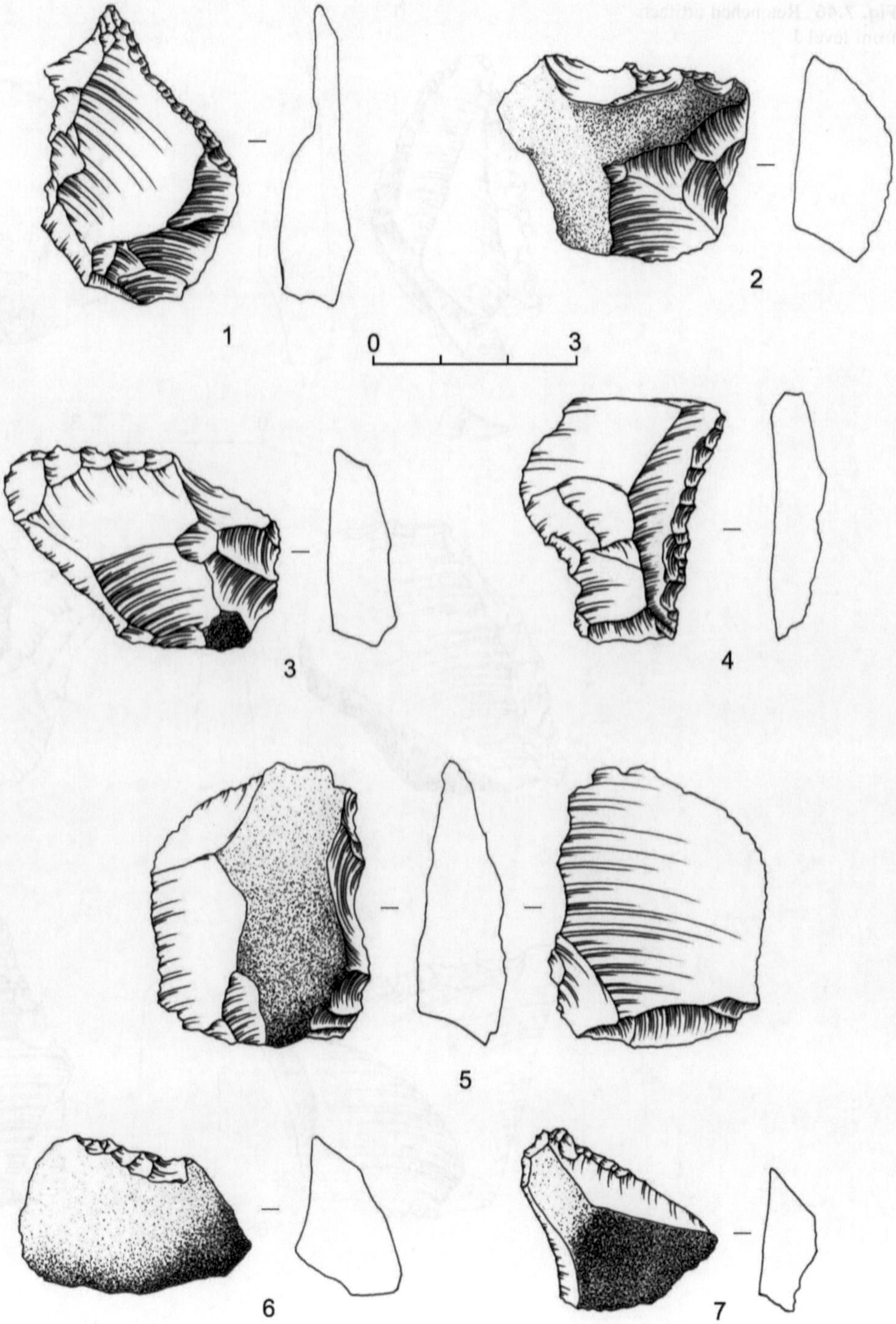

Fig. 7.45 Retouched artifacts from level J

use-wear traces are distributed longitudinally (N = 11), transversally (N = 6) and on two occasions direction of movement was not ascertained, and therefore had to be classified as indeterminate. With the exception of one flake used for planing wood with a positive transversal movement, the remaining flakes were used for cutting with a longitudinal/unidirectional movement. Retouched pieces were used for butchering, with a longitudinal/unidirectional direction (Fig. 7.48); for scraping hides with a negative transversal movement with variations in the working angle; and for defleshing (separating muscle from bone) with a positive transversal movement. In general, these functional units show a reduced degree of trace development when compared with that produced in parallel experiments, although this does not detract from the similarities between the two assemblages, which give us confidence in our interpretations.

In addition to the low use intensity, objects with more than one used side are absent. Furthermore, we did not find traces

Fig. 7.46 Retouched artifacts from level J

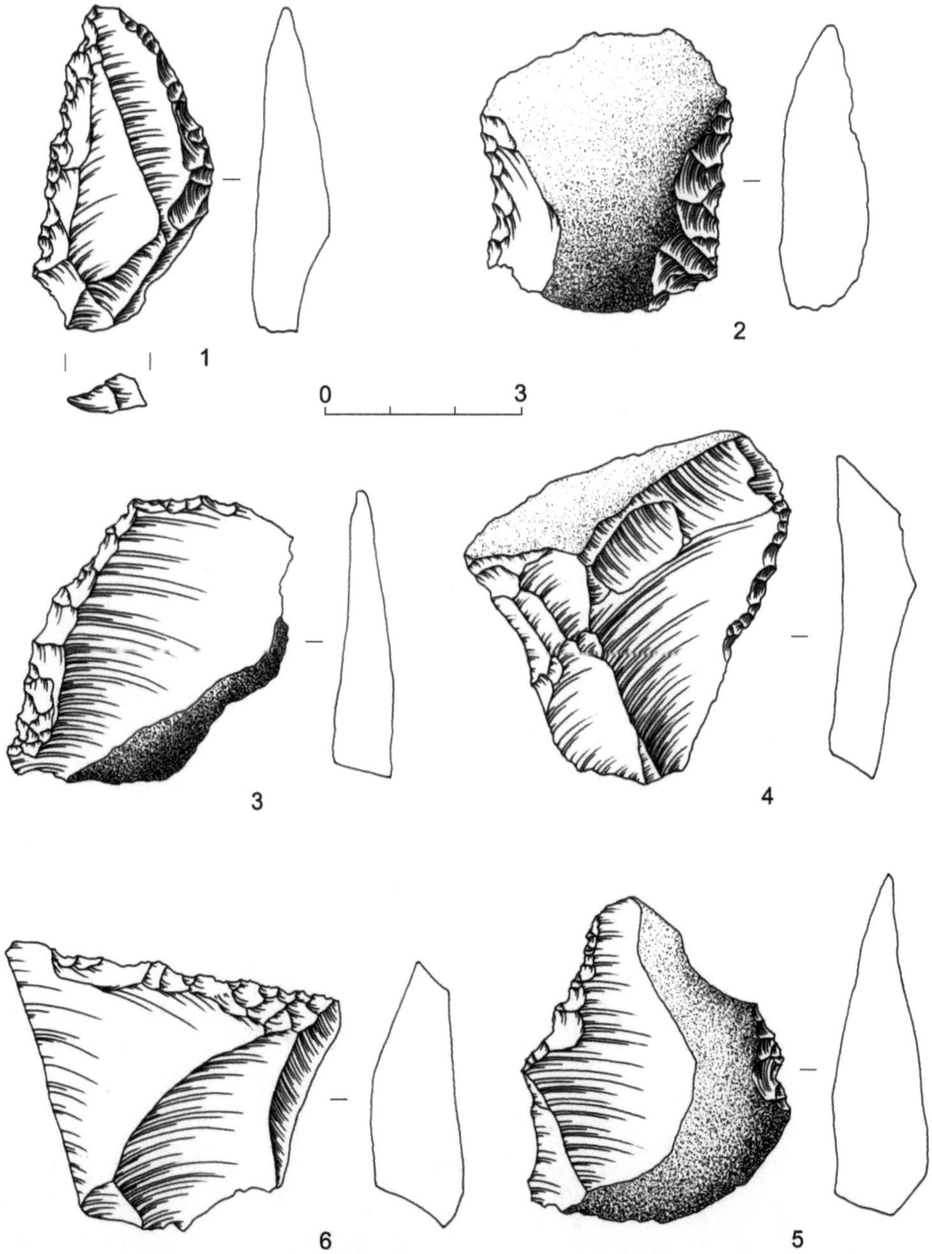

with different characteristics that might be attributed to work on several materials with a single object. In only one object (Ja H60/2) were differences in the orientation of the traces found on the two extremes of the right side, the characteristics of which point to two different actions (Fig. 7.49a, b). On the medio-proximal part, a transversal arrangement can be seen, whereas on the distal part the traces have a longitudinal orientation. For this reason, and because the side presents a uniangular profile, two functional units were distinguished: both are denticulate retouches with straight profiles, but with deep extractions on the distal part and marginal extractions on the medio-proximal portion.

The longitudinal traces are discontinuously distributed along the edge, concentrated at specific points on the side. They correspond to smooth, convex polishes that border the edge and penetrate at points into the interior with an open development. In some cases, plates of smooth or slightly convex polishes and friction traces arranged longitudinally are found on the inside of the edge above high points of relief. Other objects show compacting of the crystalline

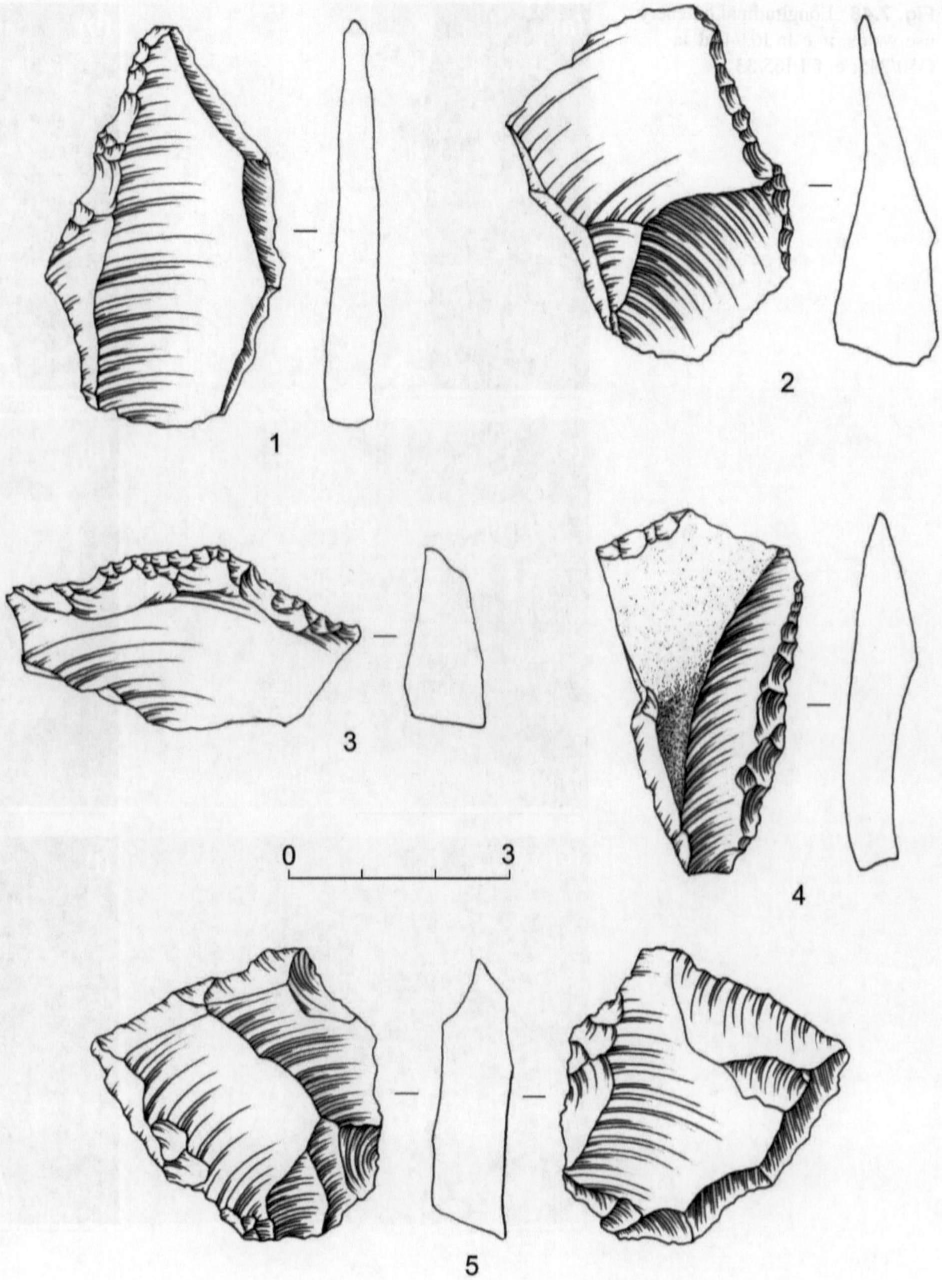

Fig. 7.47 Retouched artifacts from level J

structure or the presence of abrasive traces that cause the loss of surface material and the rounding of prominent parts. These characteristics allow these traces to be related to the actions of cutting soft animal tissue and, in the case of the abrasive traces, they may be related to skinning (Figs. 7.48, 7.50a–c).

The transversal traces show a higher degree of development. The traces correspond to a band of smooth, convex polishes above the edge that, depending on the angle of attack, affect both sides and penetrate the sides to a greater or lesser extent. These polishes show a continual distribution along the length of the edge and tend to have a closed development, a wavy relief perpendicular to the edge and a dominant convex cross-section. In some of the objects, alongside these polishes we found friction traces in the form of exertion bands and striations. The former appear in the most developed polishes and the latter are usually found in objects with initial traces of more open polish development. These characteristics indicate that these objects were used in transversal actions on fresh hides. In all cases, the edges were used with a negative movement, going backward, with the ventral face as the contact surface. This movement was deduced through the identification of a band of polish concentrated on the ventral face of the edge that barely affects the other side of the active edge. This backward movement would prevent the trihedrons of the denticulation

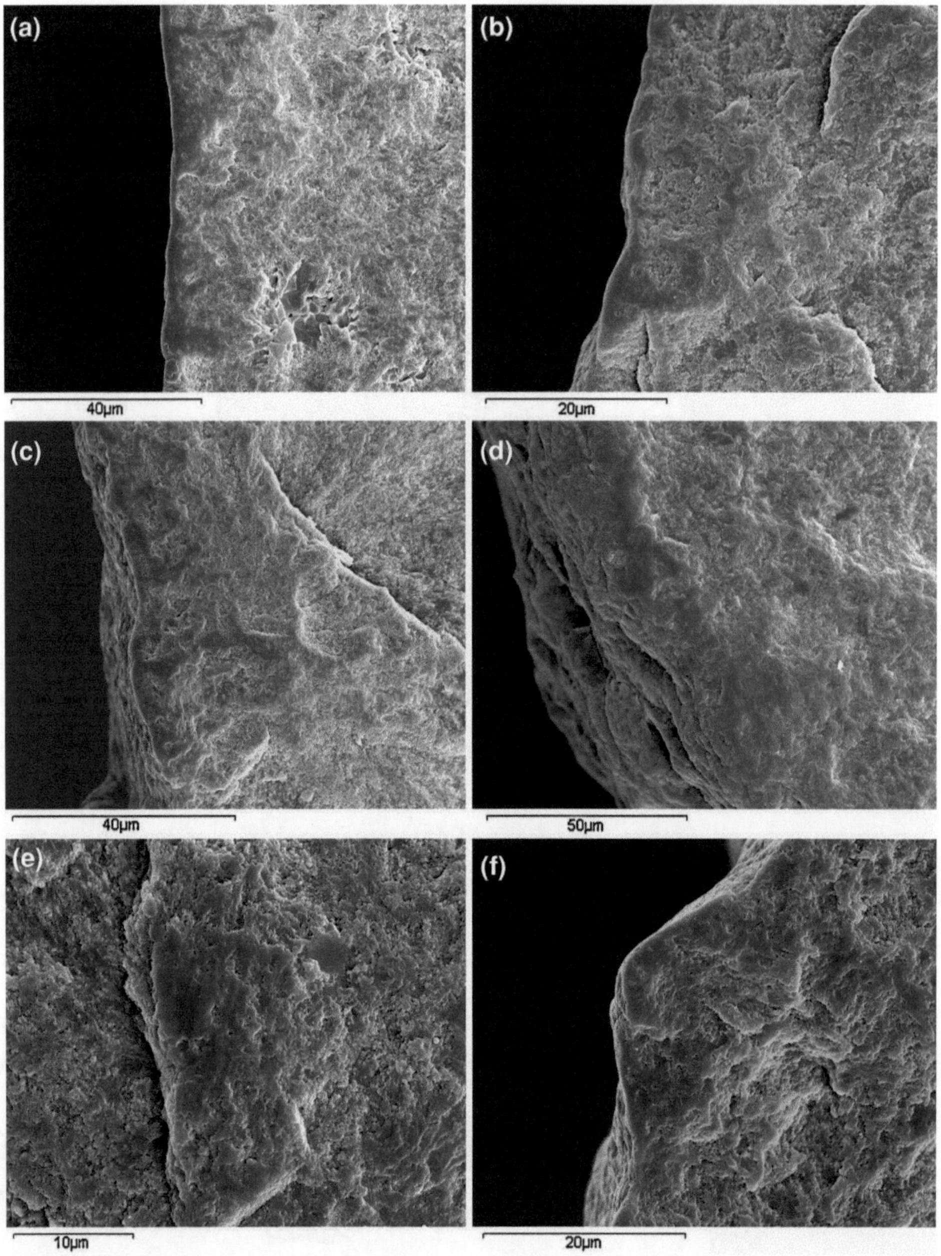

Fig. 7.48 Longitudinal butchery use-wears. **a–c** Ja I63/4, **d** Ja O50/245, **e**, **f** J I65/33

Table 7.19 Sample of flakes and retouched artifacts for use-wear analysis

	Ja	Jb	Total
Retouched			
Fa	101	20	121
Fr	83.4%	16.5%	100%
Total %	138 *(73.1%)*	34 *(58.8%)*	172 *(70.3%)*
Flakes			
Fa	47	13	60
Fr	78.3%	21.6%	100%
Total %	858 (*5.4%*)	126 (*10.3%*)	984 (*6%*)

Table 7.20 Use-wear analysis results

	Ja		Jb	
	Retouched artifacts	Flakes	Retouched artifacts	Flakes
Use-wear	9	4	2	3
Indeterminate	7	2	1	1
PSDM	46	17	5	4
Undiagnostic	39	24	12	5
Total	101	47	20	13
Technical	8		1	

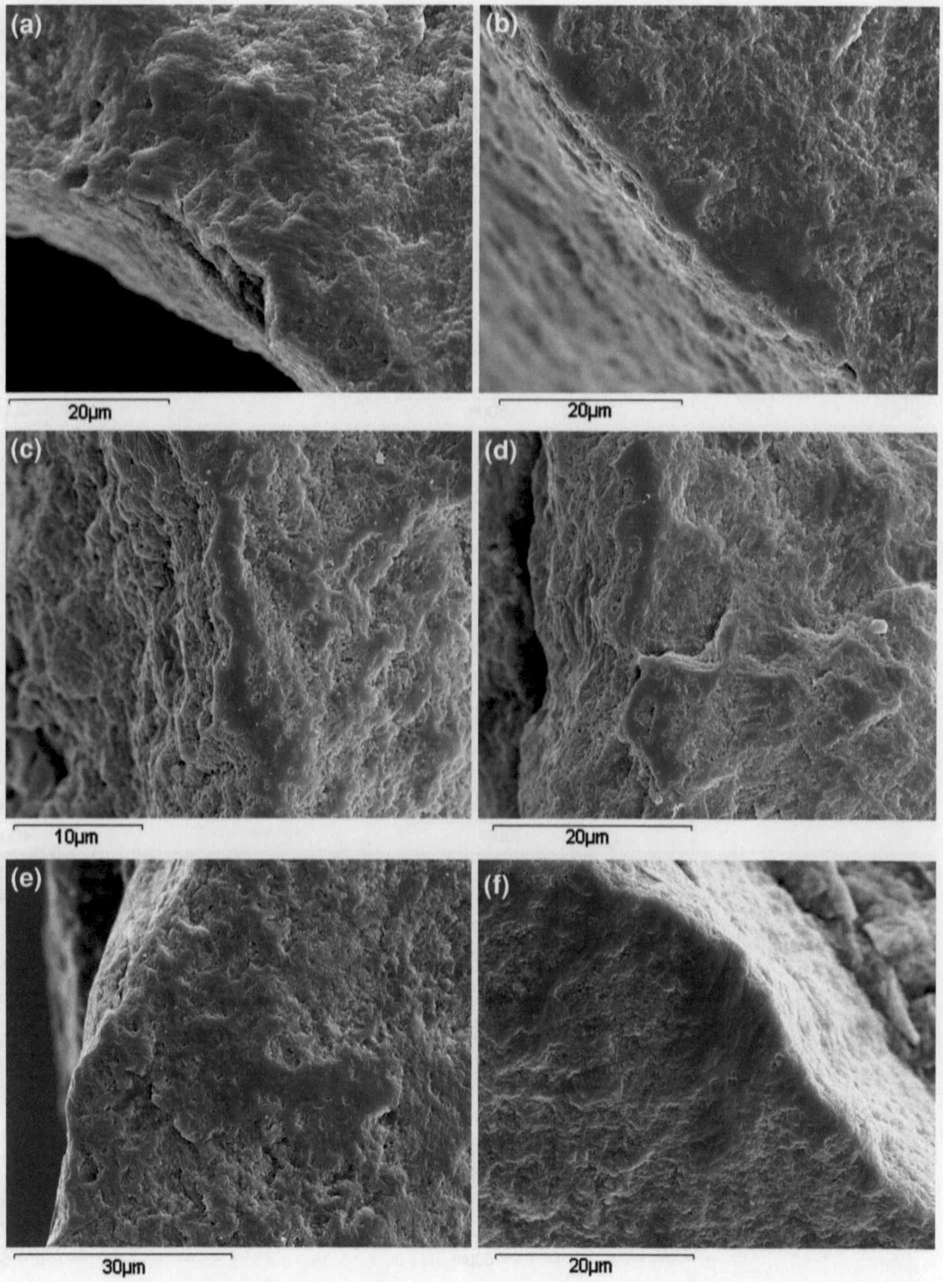

Fig. 7.49 **a** Longitudinal butchery use-wears. **b–f** Scraping fresh hide use-wears. **a, b** Ja H60/2, **c, d** Ja M64/104, **e** Ja Q51/88, **f** Ja V50/378

from getting snagged in the hide as would occur with a positive action (Fig. 7.49c–f).

Use-wear traces perpendicular to the edge were identified on six objects. These are smooth polished plates that penetrate toward the interior, with closed development and a flat cross-section without friction traces. The borders are regular and sharply separated from the non-affected zones. These characteristics correspond to work on a hard material and coincide with the experimental traces generated in defleshing, in which the edge enters into continual contact with the bone to separate the meat. The diagnostic criterion that we used to identify this transversal action on bone consisted of the cross-section of the polishes, which tend to be flat and with an irregular relief. This morphology is produced by the hardness of the material, which does not allow it to adapt to the microsurface of the tool during contact (Yamada 1993).

Only one example of work on plant matter was identified, making wood the least represented material. This object (Jb O45/16) is a *débordant* flake which shows clear traces which we interpret as originating from whittling fresh wood. It is a small object whose used edge measures barely 20 mm in length. The traces are distributed continuously along the edge. The images show the continual front of wide polishes stopping at the edge that penetrate bimodally toward the interior and are sharply separated from the rest

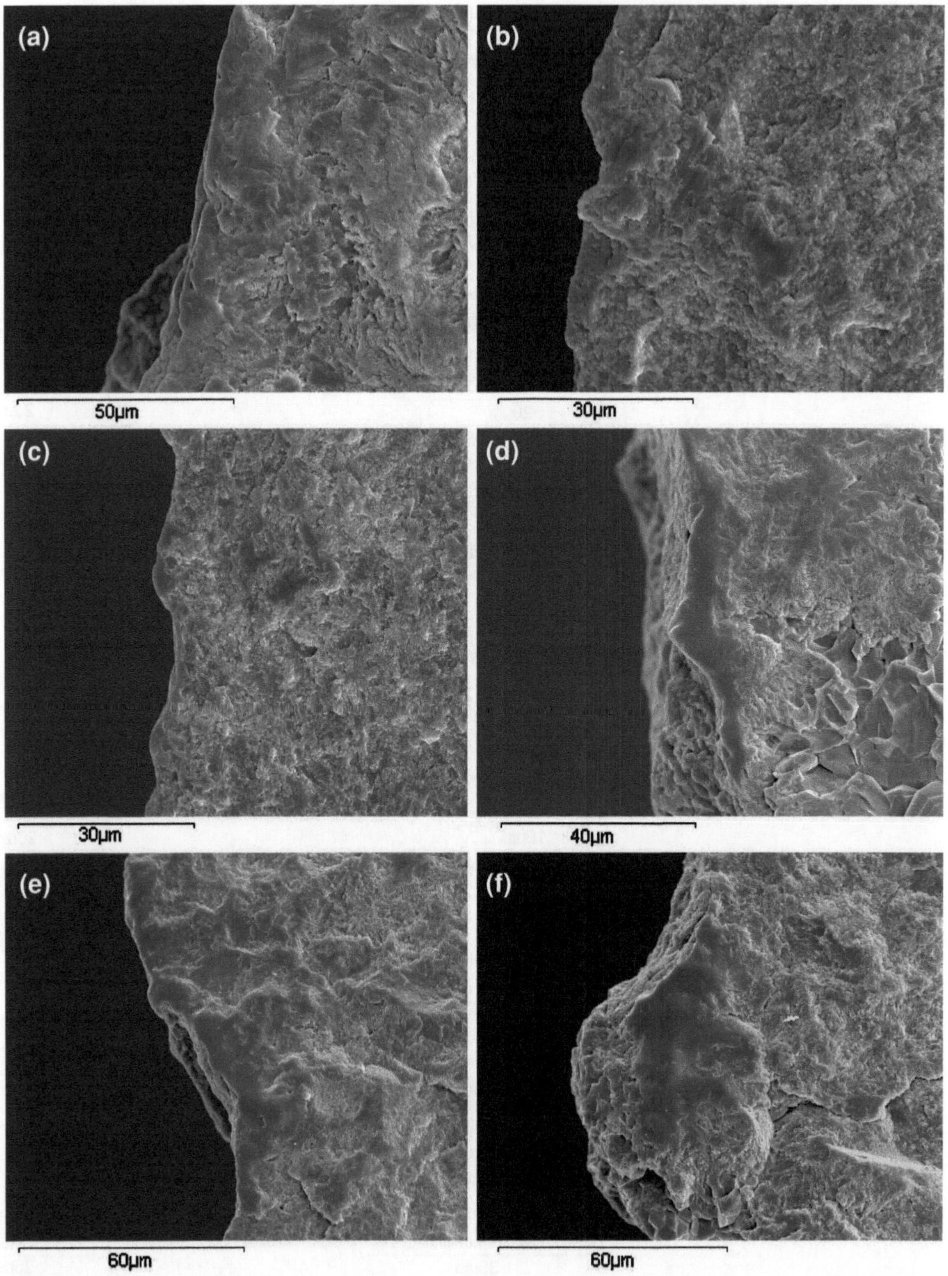

Fig. 7.50 **a–c** Longitudinal butchery use-wears. **d–f** Whittling wood use-wears. **a** Ja L49/8, **b**, **c** Ja O43/165, **d–f** Jb O45/16

of the surface. The polishes are developed and form smooth, voluminous surfaces without relict depressions and have a convex cross-section. In other images, the polishes show a more open, mesh-like development in which there are unaffected depressed areas. At these same points the polishes penetrate toward the interior, which would indicate a lesser angle of attack. These changes in distribution would correspond to variations in the working angle used for whittling actions in order to overcome the resistance of the material (Fig. 7.50d–f).

Both retouched objects and flakes were used to work the same set of materials: soft animal matter (flesh and hides), hard animal matter (bone in butchering actions) and woody plant matter. The movements identified are longitudinal and transversal and, depending on the working angle and the direction of movement, we can distinguish several different actions. Only unidirectional movements were associated with longitudinal movements, leading us to interpret them as cutting actions. Depending on the characteristics of the longitudinal traces, we were able to distinguish between skinning actions, in which we mainly see abrasive modifications and chipping, and dismembering actions, in which plastic deformations are common. On the other hand, in transversal movements over hide, we distinguished objects used with differing angles of attack, depending on the distribution of the traces toward the interior of the ventral face.

We identified four pieces with indications of high angle, unidirectional actions with negative movement. The pieces with polishes related to transversal actions on hard material such as bone were probably used for defleshing actions. This functional evidence allows us to infer that the objects were involved in intense butchering activities, in scraping fresh hides and occasionally in woodworking actions.

Functional Units

In all cases, the used edges correspond to retouched areas and in the case of flakes, to the most potentially suitable edge. The angle of the utilized retouched edges was usually semi-abrupt (SA) with the exception of two simple (S) edges, whereas the opposite was true among the flakes, where we found only one SA edge. Nevertheless, we did not find that the angle differed according to the action performed, as the SA edges of the retouched objects were used both in the longitudinal actions of cutting soft animal material and in the transversal actions of defleshing and hideworking. Similar results were found with flakes in which S angles were used without distinction, even for working wood. The continuity and shape of the edge play an important role in the performance of the object during the task at hand as they determine the quantity of edge that can interact with the material being worked. The shape of the edge on the horizontal plane, as well as the sagittal outline of these edges is extremely variable, with practically all the options having similar percentages. Straight and convex edges, *a priori* the most suitable, were as common as irregular edges. The use of these edges with such different profiles is one of the causes of the discontinuous distribution found in the use-wear traces. In all cases, the retouch outline is denticulated. The concave dihedral of the notches was used in transversal motions on hides, where the polishes border the entire concavity of the dentition. However, traces are primarily found on the trihedral and on its adjacent sides, where they are more developed.

Functional units were found on the left side of the object (N = 8), on the right (N = 5), and on the distal transversal surface (N = 6). These last units correspond to objects that have a greater length perpendicular to the technical axis. Retouched zones were limited on the side on which they were located, as is expressed in the ratio index between side length and retouched length, which is an average of 0.62. The lengths of the used retouched units range between 12 and 47 mm with an average of 25.7 mm. The average of the natural edges of flakes is higher, with a mean of 38.1 mm. These figures give us a clear idea of the small size of the used edges. During our experimentation phase, we discovered how difficult it is to keep hold of these small objects used for butchering actions because the fat and wetness of the animal frequently cause them to slip through the fingers. These problems are less severe with cortical objects and with objects with abrupt sides, which were widely represented in the sample. Such surfaces allow the object to be firmly grasped between the palm of the hand and fingers (Lemorini 2000).

The transversal and sagittal sections of the utilized pieces, in addition to cortical surfaces, have asymmetrical triangular and symmetrical and asymmetrical trapezoidal morphologies. In these objects an abrupt side opposes the used edge, a geometrical model that allows for the comfortable prehension of the objects. Among these objects there are four *débordant* flakes and four naturally backed flakes. The functional units of the *débordant* and backed flakes were used without distinction in longitudinal and transversal motions for butchering, and working hides and wood. These values clearly show the versatility of these geometric models (Beyries and Boëda 1983; Lemorini et al. 2003) (Table 7.21) (Fig. 7.51).

Spatial Distribution

Through the spatial distribution of artifacts showing use-wear traces we hope to reveal whether the space of the shelter was organized into areas with specific functionalities. In addition to postdepositional and sin-depositional processes, which may have altered the original artifact location, it is important to consider that none of the analyzed artifacts formed part of the refit groups, which tend to contribute the most spatial information. This is because this functional analysis was conducted after the refit program, and the refitted artifacts may have suffered microscopic alterations that could be mistaken for use-wear traces.

Bearing in mind the possible distortion in the original spatial distribution, different ethno-archeological works have indicated that in hunter-gatherer camps activities are performed around domestic nuclei, organized around hearths. Only those activities that require more space or are bothersome to others would be done in separate areas, albeit without a specific tool (Yellen 1977; Binford 1980; O'Connell 1987). These types of activities would be those that functionally differentiate the occupations. Each domestic unit is made up of the members of the same family and it is not possible to functionally distinguish between them because they all perform the same activities. In Abric Romaní, exploitation sequences were normally carried out near the hearths in the primary accumulation zones, whereas retouched objects tended to be located in relatively isolated zones outside of the denser areas. This spatial localization of the retouched objects in marginal and peripheral zones may correspond to the spatial organization of the actual and

Table 7.21 Techno-typological features of tools with use-wear and their functional interpretation

Levell	Tool	Edge area	Edge angle	Edge profile	Edge length (mm)	Type	Working angle	Material	Action
Ja	H60-2	Left edge	SA	Straight	47	D23	Open	Fresh hide	Scraping
Ja	H60-2	Left edge	SA	Straight	47	D23	Open	Meat-bone	Cutting
Ja	I63-4	Left edge	SA	Irregular	25	D23	Open	Meat-bone	Cutting
Ja	I65-33	Left edge	S	Straight	19	D23	Open	Meat-bone	Cutting
Ja	M46-104	Right edge	SA	cx	26	D23	Open	Fresh hide	Scraping
Ja	M54-1	Distal end edge	SA	cc	31	D23	Open	Meat-bone	Cutting
Ja	050-245	Distal end edge	SA	Irregular	18	D23	Open	Meat-bone	Cutting
Ja	Q51-88	Left edge	SA	Straight	18	D23	Open	Fresh hide	Scraping
Ja	V50-26	Distal end edge	S	Irregular	33	D23	Medium	Meat-bone	Scraping
Ja	V50-378	Right edge	SA	cc	31	D23	Closed	Fresh hide	Scraping
Jb	M48-83	Distal end edge	SA	Straight	12	D23	Indet	Meat-bone	Indet
Jb	M49-35	Distal end edge	SA	Straight	23	D23	Open	Meat-bone	Cutting
Jsup	049-3	Left/distal end edge	S	Irregular	36/22	Flake	Open	Meat-bone	Cutting
Ja	L49-28	Right edge	S	Irregular	28	*débordant*	Open	Meat-hide	Cutting
Ja	M46-42	Left edge	S	cx	35	N. backed	Open	Meat-bone	Cutting
Ja	043-165	Right edge	SA	Straight	26	Flake	Open	Meat-bone	Cutting
Jb	M57-12	Left edge	S	Straight	30	Flake	Open	Meat-hide	Cutting
Jb	N46-13	Distal end edge	S	Irregular	25	Flake	Indet	Meat-bone	Indet
Jb	045-16	Right edge	S	cx	18	*débordant*	Medium	Fresh wood	Whittling

Table 7.22 Refitting rate by raw materials and sublevels

	Ja		Jb		Total	
	N	%	N	%	N	%
Chert	315	7.9	185	14.4	500	9.5
Limestone	134	17	11	17.7	145	17
Quartz	58	9.1	10	9.3	68	9.1
Calcarenite	4	57.1			4	50
Quartzite			2	18.1	2	15.3
Total	511	9.3	208	14.1	719	10.3

Table 7.23 Refitting rate by sublevels and artifact categories

	Ja		Jb		Total	
	N	%	N	%	N	%
Unworked nodules	6	18.1			6	15.7
Cores	33	40.7	7	63.6	40	43.4
Flakes	186	9.5	95	15.4	281	10.9
Flake fragments	186	6.8	90	12	276	7.9
Retouched artifacts	6	4	2	5.1	8	4.2
Fragments	94	18.5	14	25.4	108	19.1
Total	511	9.3	208	14.1	719	10.3

subactual hunter camps, in which there are activity areas separated from the domestic center.

In sublevel Ja, no spatial differentiation was found in terms of the actions carried out, because in the artifacts with use-wear traces recovered in the domestic areas (around the hearths) all the actions determined in the sample were identified. Hide-scraping was identified next to the hearths even though ethnographic studies have found that tanning activities are carried out in external areas (Yellen 1977). Skinning actions were also identified, which correspond to the first step in carcass processing, and the sublevel Jb artifact used on wood was also found near a hearth. Nevertheless, it is significant that of the nine retouched objects with use-wear traces, four were found in the areas with sparser accumulations, in the exterior area of the shelter. These retouched items found outside of the central areas showed the most developed traces. As there are no functional differences between these objects and those recovered in the central areas, except in the degree of development of the traces, they would have been involved in more intense phases of the processing of carcasses and treating of hides. Three of these objects form part of the accumulations in which the remains of fauna dominate, with evidence of in situ carcass processing (Cáceres 2002). These same accumulations of faunal remains have been related to the central areas through lithic and bone refits (Vaquero 1997). These refits seem to indicate a certain spatial structuring based on the activities performed. Therefore, the only functional difference between the actions performed around the

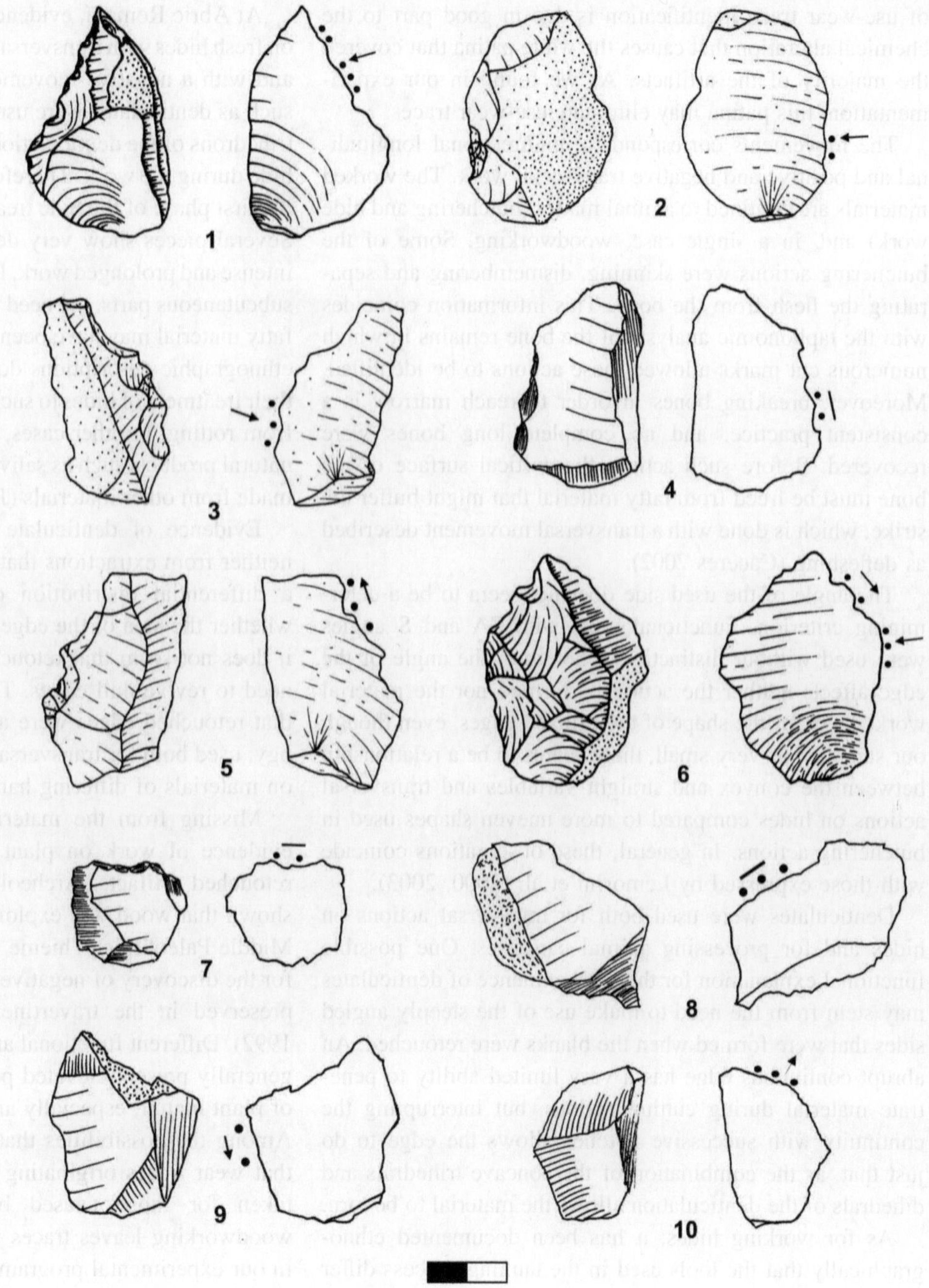

Fig. 7.51 Example of tools with use-wears: (···) use-wear, () direction. *1–7* Retouched tools (V50/378, Q51/88, M46/104, I65/33, I63/4, H60/2, O50/245). *8–10* Flakes (L49/28, O43/165, O45/16)

hearths and those performed in other areas is in the intensity of the work, which we have deduced to be greater in peripheral areas.

Summary

The functional analysis identified 18 objects with use-wear traces, with 19 functional units. Except for one retouched object, which has two units on the same side, the rest of the artifacts present a single functional unit. Furthermore, all the artifacts were used in a single action and on the same material, again except for the piece mentioned above in which different actions were recognized on the same side. In general, the use-wear traces are little developed and appear discontinuously along the edges, especially on objects with butchering traces. This scarce development reflects relatively light work, which may partially explain the low percentage of utilized objects. Above all, traces from butchering actions take the longest to form and may be mistaken for other types of alterations. Furthermore, the low percentage of use-wear trace identification is due in good part to the chemical alteration that causes the white patina that covered the majority of the artifacts. As we found in our experimentation, this patina may eliminate use-wear traces.

The movements correspond to unidirectional longitudinal and positive and negative transversal work. The worked materials are confined to animal matter (butchering and hide work) and, in a single case, woodworking. Some of the butchering actions were skinning, dismembering and separating the flesh from the bone. This information coincides with the taphonomic analysis of the bone remains in which numerous cut marks allowed these actions to be identified. Moreover, breaking bones in order to reach marrow is a consistent practice, and no complete long bones were recovered. Before such action, the cortical surface of the bone must be freed from fatty material that might buffer the strike, which is done with a transversal movement described as defleshing (Cáceres 2002).

The angle of the used side does not seem to be a determining criterion. Functional units with SA and S angles were used without distinction. Therefore, the angle of the edge affects neither the action performed nor the material worked. As for the shape of the utilized edges, even though our sample was very small, there seems to be a relationship between the convex and straight variables and transversal actions on hides compared to more uneven shapes used in butchering actions. In general, these observations coincide with those expressed by Lemorini et al. (2000, 2003).

Denticulates were used both for transversal actions on hides and for processing animal carcasses. One possible functional explanation for the predominance of denticulates may stem from the need to make use of the steeply angled sides that were formed when the blanks were retouched. An abrupt continuous edge has a very limited ability to penetrate material during cutting actions, but interrupting the continuity with successive notches allows the edge to do just that, as the combination of the concave trihedrals and dihedrals of the denticulation allows the material to be torn.

As for working hides, it has been documented ethnographically that the tools used in the tanning process differ depending on the population studied (Jardón 2000). Nevertheless, very few ethnographic studies cite the use denticulates in tanning tasks. The strongest evidence for the tanning of hides during this period comes from the analyses conducted by Lemorini (2000). The objects involved in these tasks correspond to different types of scrapers, which were used in the different phases of the tanning process. This use of a single potential model of object throughout the tanning process differs from ethnographic examples and, in general, does not coincide with archeological observations based on Upper Paleolithic samples.

At Abric Romaní, evidence indicates that work was done on fresh hides with transversal actions, usually at a steep angle and with a negative movement, for which retouched edges such as denticulates were used. This movement prevents the trihedrons of the denticulation from snagging and ripping the hide during the work. Therefore, we can only clearly identify the first phase of the hide treatment—defleshing—at the site. Several pieces show very developed traces, which indicate intense and prolonged work. In defleshing and cleaning off the subcutaneous parts, the need to tear and eliminate tissues and fatty material may have been best met by denticulates. Some ethnographic descriptions document communities that limit their treatment of hides to such cleaning, which prevents them from rotting. In other cases, tanning is done by hand, using natural products such as saliva, urine or with the help of tools made from other materials (Jardón 2000).

Evidence of denticulate resharpening was not found, neither from extractions that interrupt wear traces nor from a differential distribution of wear traces depending on whether the area of the edge is retouched or not. Therefore, it does not seem that retouching occurred in response to a need to revive dull edges. This brings us to the conclusion that retouched edges were a useful and versatile morphology, used both for transversal and longitudinal motions, and on materials of differing hardness.

Missing from the materials we identified was greater evidence of work on plant matter, especially among the retouched artifacts. Archeological research has repeatedly shown that wood was exploited during and even before the Middle Paleolithic (Thieme 1997). Abric Romaní is known for the discovery of negatives and positives of wood objects preserved in the travertine (Carbonell and Castro-Curel 1992). Different functional analyses performed at other sites generally present elevated percentages of the identification of plant matter, especially among denticulates and notches. Among the possibilities that might explain this absence is that wear traces originating from woodworking were mistaken for those caused by other materials. Although woodworking leaves traces with diagnostic characteristics, in our experimental program, and as other researchers have also found, a certain overlap was noticed among the polishes generated by hide, bone and fresh wood (Keeley 1980; Unrath et al. 1986; Van Den Dries and Van Gijn 1997).

Another possible explanation is that a postdepositional chemical alteration especially affected wear traces associated with woodworking. Plisson and Mauger (1988) showed that different chemical compounds are capable of eliminating experimental polishes. They found that, depending on the worked material, polishes are more or less resistant to the same chemical attack. Among the least resistant polishes are those from soft, fresh wood. On submerging the

Table 7.24 Distribution of refitted artifacts by raw materials and size categories

	Chert		Limestone		Quartz		Total	
	N	%	N	%	N	%	N	%
Very small	272	7.6	25	6.2	27	4.6	326	7
Small	138	14.1	33	19.7	19	20.4	190	15.2
Medium	58	14.7	31	31.9	3	12.5	95	18.1
Large	18	10.2	11	20.7	7	36.8	37	14.5
Very large	14	11.3	45	33.5	12	57.1	71	24.6
Total	500	9.5	145	17	68	9.1	719	10.3

objects in alkaline compounds, in NaOH, these objects were left covered in a white patina and the microcrystalline texture of the chert was altered, with all use-wear traces disappearing within a few hours. Nevertheless, our experimental tests indicated that the preservation of traces does not depend on the worked material, but on the type of chert and the resistance of the stone to the chemical attack. Therefore, the white patina equally affects traces originating from different worked materials, only reducing the overall percentage of preserved traces (Martinez 2005).

Therefore, the absence of traces from work on plant matter accurately reflects the importance of each of the worked materials at the site. We must not forget that butchering activities, which are directly related to subsistence, had to be more common than any other. In the first functional studies carried out in the 1980s these activities were rarely detected due to the use of optical devices, and woodworking activities were overrepresented. In recent years, the high percentage of woodworking that appeared in the pioneer studies of Anderson-Gerfaud (1990); Beyries (1987a) ; Shchelinskij (1993) is being looked at with a more critical eye. Lemorini commented that the lithic assemblages studied in these works come from old collections with a very high percentage of altered objects. She points out that the percentage of woodworking interpretation could be distorted by having confused alterations for use-wear traces (Lemorini, 2000, p. 38). An experimental study has recently shown how gelifluction phenomena may have provoked the formation of polishes similar to those obtained from woodworking (Caspar et al. 2003). Lastly, there are many sites at which the percentage of woodworking is low or nonexistent, such as El Salt (Rodríguez et al. 2002) or layers 6 and 3 of Grotta Breuil (Lemorini 2000). Therefore, this evidence leads us to consider that the materials identified and their percentages reflect the activities carried out at Abric Romaní.

Finally, the lack of traceological evidence of hafting and the ergonomic shape of the objects lead us to propose the manual prehension of tools (Lemorini 2000). The objects would have been used by supporting the abrupt side opposite the edge in the palm of the hand and holding them between the fingers, or by placing the index finger on one of the two transversal sides. From this design, we can deduce that both the availability of active edges and the existence of areas to grasp the object were of equal importance in the objects of this period. Nevertheless, the problems in identifying microscopic hafting traces are well known (Jardón 2000), so the hafting of objects should not be completely ruled out (Boëda et al. 1996). More than the absence of clear hafting, we consider that it is the ergonomic design of the objects that indicate they are more suitable for being used manually than as part of a composite instrument.

The Event Level

Refits and Raw Material Units

The refitting program was undertaken in the presence of several negative conditions. First, many artifacts exhibited a highly patinated surface, sometimes as a consequence of burning. This alteration makes it difficult to associate the lithics according to their macroscopic characteristics, which is the initial stage of refitting. The lowest rates of refits correspond to the areas where patinated artifacts are more common (especially the back of the shelter, zones 1 and 2). Another handicap is the small size of most of the artifacts. Small remains are more difficult to refit because the ease with which an artifact can be refitted depends on its surface area and the number of scars it presents (Morrow 1996, p 359). This factor has been especially acute in level J because, as is shown in the core and flake attribute analysis, most reduction sequences were aimed at the production of small flakes and small and very small artifacts are clearly dominant in the assemblage. The very small artifacts show the lowest refit rates (Table 7.24). Among the negative conditions, it is also worth noting the presence of empty zones corresponding to the areas affected by previous

Table 7.25 Distribution of connection lines by kinds of refits, sublevels and raw materials

		Production sequences		Breaks		Retouch sequences		Total	
		N	%	N	%	N	%	N	%
Ja	Chert	148	76.7	43	22.3	2	1	193	100
	Limestone	23	26.7	63	73.3			86	100
	Quartz	22	51.2	21	48.8			43	100
	Others	1	50	1	50			2	100
Jb	Chert	91	77.1	27	22.8			118	100
	Limestone			11	100			11	100
	Quartz	3	42.8	4	57.1			7	100
	Others			1	100			1	100
Total		288	62.4	171	37	2	0.4	461	100

excavations. Most of the remains originally located in these zones are no longer available and this give rise to some uncertainties in the task of refitting, because incomplete groups are more difficult to refit. Some refitting groups include artifacts from previous excavations that have been stored at the Capellades Museum, which confirm that at least some RMUs were scattered throughout these empty areas.

Despite these drawbacks, 10.4% of all artifacts from level J were able to be refitted (Fig. 7.52, 7.53). There were 262 refitting groups, with a total of 719 artifacts. Chert shows a refitting rate of 9.4% (494 conjoinable artifacts and 185 refitting groups) (Table 7.22). Quartz exhibited a similar proportion of refit (9.1%), with 68 artifacts and 20 groups. Limestone showed the highest refitting rate (17.6%); 146 limestone artifacts were refitted, forming part of 54 refitting groups. Finally, two refits of calcarenite artifacts and one quartzite refit were also recorded. The three main raw materials show a higher refitting rate in sublevel Jb, but the difference between Jb and Ja is especially notable for chert. As for artifact categories (Table 7.23), cores had the highest rate of refit (43.4%). This is to be expected because cores tend to show more flake scars and are therefore more likely to be included in a refit. Unworked nodules and fragments also had a relatively high refitting rate; they are basically associated with limestone refits. The low refit rate for retouched artifacts is especially striking, since they tend to be large items with many flake scars. This low conjoinability of retouched artifacts is confirmed by RMU data and can be explained by patterns of lithic provisioning and transport (Table 7.24).

These refits produce 460 connection lines, whose distribution by refitting types can be seen in Table 7.25. Most lines correspond to production sequences (62.8%), while the refitting of breaks represents 36.7%. Very few connections have been attributed to retouch sequences. This can be partly explained by the difficulties in conjoining the very small flakes resulting from retouch, but it may also be related to the scarce manufacture of retouched artifacts carried out inside the rockshelter. As we will see, most

Table 7.26 Mean and standard deviation of connection lines from sublevels Ja and Jb by raw materials and kind of refit

		Mean	S.d.
Ja	Chert	244.54	282.48
	Limestone	258.61	303.5
	Quartz	279.45	336.78
	Others	77.32	18.78
Jb	Chert	11.65	115.98
	Limestone	220.68	228.56
	Quartz	195.63	247.76
	Others	191.63	
Ja	Production	286.14	330.5
	Break	200.49	223.2
	Retouch	187.84	11.64
Jb	Production	126.63	125.4
	Break	133.29	165.39
	Retouch		
Total		215.18	264.04

retouched implements were introduced into the site as isolated artifacts produced outside. Production sequence lines are clearly dominant among chert refits, which is hardly surprising since the bulk of the chert assemblage is derived from core reduction activities carried out in situ. Break refits on chert also correspond to the breakage of knapping products. Limestone shows a different pattern, characterized by the dominance of break connection lines. Most of these connections correspond to the refitting of limestone cobbles used as hammerstones. Quartz exhibits an intermediate pattern; production sequence and break lines present practically identical percentages.

The mean length of the connection lines in sublevel Ja is 251 cm, and 128 cm in sublevel Jb (Table 7.26). This interlevel difference was expected due to the unequal surface size covered by each sublevel. In sublevel Ja, there are no significant differences between the three primary raw materials. However, in sublevel Jb the mean length of the

Table 7.27 Distribution of connection lines by sublevels, type of refit and length intervals

	Ja			Jb			Total
	Production	Break	Retouch	Production	Break	Retouch	
0–100	66	54		49	25		194
	34%	42.5%		52.1%	58.1%		42.1%
100–200	42	30	2	24	11		109
	21.6%	23.6%	100%	25.5%	25.5%		23.6%
200–300	31	15		11	3		60
	15.9%	11.8%		11.7%	6.9%		13%
300–400	14	9		5			28
	7.2%	7%		5.3%			6%
400–500	8	5		4			17
	4.1%	3.9%		4.2%			3.6%
500–600	5	4			2		11
	2.5%	3.1%			4.6%		2.3%
600–700	6	3		1	1		11
	3%	2.3%		1%	2.3%		2.3%
700–800	4	3			1		8
	2%	2.3%			2.3%		1.7%
>800	18	4					22
	9.2%	3.1%					4.7%
Total	194	127	2	94	43		460
	100%	100%	100%	100%	100%		100%

connection lines is higher for limestone (220 cm) and quartz (195 cm), than for chert (115 cm). Nevertheless, it is important to emphasize that limestone and quartz refits are scarce in this sublevel. In addition, the mean length of chert refits is highly influenced by the very short connections found in one of the knapping areas (zone C) formed almost exclusively by a tightly clustered chert scatter. If the refits from this area are excluded, the differences between raw materials tend to disappear. The distribution of connection lines by 100 cm intervals (Table 7.27) shows that 42.1% of the connections are shorter than 1 m. These short connections are more abundant in Jb than in Ja (54 and 37.1% respectively). More than a half of the connections (65.8%) are shorter than 2 m. These short connections coincide with the normal dispersion patterns of lithic remains during knapping (Newcomer and Sieveking 1980; Barton and Bergman 1982; Böeda and Pelegrin 1985). This suggests that knapping areas were not seriously distorted by postdepositional movements, although there are some interzonal differences that must be considered. However, long refits are not uncommon, especially in sublevel Ja, in which 6.8% of the connections are longer than 8 m. These long connections are scarce in sublevel Jb, although again we should mention that the total surface of Jb is smaller than that of sublevel Ja.

Breakage lines (mean length: 200.4 cm) tend to be shorter than production lines (mean length: 286.1 cm). This difference is particularly clear in connections exceeding 8 m, suggesting that intentional movement may explain most of the longest refits. Production movements correspond to the refits of cores or flakes, that is, artifacts with a use value. Broken artifacts lose such use potential and therefore would be less subject to intentional movement, especially in the case of knapping products. In well-preserved assemblages, breakage refits tend to be shorter than production refits (Cziesla 1990). As we will see later, the longest connections of broken artifacts correspond primarily to limestone fragments that could still have been used as hammerstones. The movement of fragments derived from postdepositional breakage is a good indicator of the extent to which spatial distribution has been affected by taphonomic factors. An example is provided by the refit of seven fragments of a flake broken by fire damage. These fragments were scattered in a 166.2 cm radius, in O52, P51–52 and Q49–51, which suggests some postdepositional movement, but not enough to displace fragments to other zones of the rockshelter.

At first glance, the distribution of connection lines by cardinal direction (Table 7.28) shows a clear difference between sublevels Ja and Jb. In sublevel Ja, NE–SW connections are dominant, while in sublevel Jb NW–SE is the more common direction. This suggests that the two sublevels underwent different dispersion dynamics. The dominant direction in sublevel Ja coincides with the natural slope of the stratigraphic deposit. It is therefore worth considering

Table 7.28 Distribution of connection lines by directions and sublevels

	E–W	NE–SW	N–S	NW–SE	Total
Ja	46	149	39	79	313
	14.6%	47.6%	12.4%	25.2%	100%
Jb	16	37	13	53	119
	13.4%	31%	10.9%	44.5%	100%
Total	62	186	52	132	432
	14.3%	43%	12%	30.5%	100%

Only connection lines longer than 10 cm have been included

to what extent the movement of lithic remains in sublevel Ja was influenced by the natural slope of the occupied surface. Furthermore, this pattern characterizes the entire rockshelter, without significant interzonal differences (Table 7.29). Excluding zones 1 and 5, due to the low number of refits in those areas, and zone 2, due to its enclosed nature, zones 3, 4 and 6 are dominated by NE–SW connections. Therefore, it seems that slope affected the movement of artifacts in most of the shelter in sublevel Ja. This should be taken into account when interpreting the spatial significance of these connections, especially the longest ones, since this preferential NE–SW direction is more marked in connection lines longer than 7 m (82.7% have this orientation). As we will see later, when analyzing the spatial scatter of RMUs, a preferential dispersion in the direction of the slope seems clear in some cases. However, a case-by-case approach is needed in order to examine the intentional or unintentional character of these movements. Such slope dynamics had a lesser influence in sublevel Jb, since most connection lines have a counter-slope orientation.

Although the direction of the connection lines does not indicate significant interzonal differences, other data suggest that the intensity of movements was not the same in all the zones of sublevel Ja. Mean length of connection lines is higher in zone 6 than in zones 3 and 4 (232.2 versus 151.2 and 136.2 cm respectively). Connections shorter than 1 m are comparatively scarcer in this zone (Table 7.30). If we only consider the intrazonal connections shorter than 5 m, in order to avoid the misleading data derived from differences in zone size, zone 4 shows the highest percentage of connections shorter than 1 m (53.3%). In zones 3 and 6, these short connections represent 42.8 and 37.2% of all connections, respectively. On the other hand, connections between 2 and 3 m increase in zone 3 and especially in zone 6. This suggests that the dispersion processes were more intense in zone 6 than in the rest of the shelter and caused longer movements. Meanwhile, zone 4 seems to be the least affected by such processes. The scattering of RMUs will provide more data on this subject.

Refits also provide useful archeostratigraphic information. Although most refits were made within each sublevel, some intersublevel connections have been found. On the one hand, refitting shows a strong relationship between artifacts attributed in the field to sublevel Jsup and sublevel Ja. From the 15 connection lines including an artifact previously assigned to Jsup, 13 correspond to refits with Ja. This suggests that lithics from Jsup and Ja are essentially part of the same assemblage, which has been one of the main arguments for incorporating the Jsup items into the Ja sublevel. The two refits linking only Jsup remains could correspond to a different stratigraphic horizon above Ja, but this seems to be a very localized unit including a small number of remains. We will discuss this below in more detail.

On the other hand, refits between sublevels Ja and Jb are less common, although not unheard of. These refits were found along the edges of the principal area in which sublevel Jb was identified, especially along the outer line of blocks. This is the area where sublevels Ja and Jb tended to overlap due to the disappearance of the travertine layer separating the two sublevels in the middle of the shelter. These conjoins correspond to artifacts pertaining to Jb that were assigned in the field to Ja. Two RMUs were principally affected by this mistake: Chert-01 and Chert-16. Some artifacts from these RMUs were found in sublevel Jb, mainly in the area of N50-51 and O50–51, but the artifacts of the same sequences recovered close to the outer blocks (L49–50, K49–51 and J49–51) were erroneously recorded in sublevel Ja. Two more Ja–Jb refits were found in rows O and P. Two flakes from P47 and P50 attributed to sublevel Ja were refitted to another flake from Jb found in O45. Finally, a flake from square O50 assigned to Ja was conjoined with a reduction sequence of sublevel Jb focused in N47.

In level J, 562 RMUs have been identified and 3952 lithics (57.1% the whole assemblage) have been classified according to their RMUs. However, the ability to identify RMUs was not equally as easy for all raw materials (Table 7.31, 7.32). Chert and limestone are very heterogeneous and this approach has been especially suitable for these materials. The RMUs identified for chert and limestone can be associated with confidence to specific technical events. However, the identification of different quartz RMUs is very generic and presents serious difficulties. Most quartz remains have the same geological origin (the veins in the Paleozoic slate formations close to the site) and the degree of macroscopic variability among them is low. These similar characteristics make it complex to distinguish different RMUs, with the exception of some nodules that have a rounded cortex showing a secondary fluvial origin. As a result, the number of quartz RMUs is very low compared to those of chert and limestone. Although a high percentage of quartz artifacts has been classified by RMU, some of these units (especially those having more lithics: Qu-001, Qu-002 and Qu-05) should be considered as large raw material classes, formed by different technical events and individual

Table 7.29 Distribution of connection lines of sublevel Ja according to direction and spatial zones

	E–W	NE–SW	N–S	NW–SE	Total
Zone 1	1		1		2
	50%		50%		100%
Zone 2	3	4	5	4	16
	18.7%	25%	31.2%	25%	100%
Zone 3	4	22	3	12	41
	9.7%	53.6%	7.3%	29.2%	100%
Zone 4	23	63	14	30	136
	16.9%	46.3%	10.2%	22%	100%
Zone 5		2		2	4
		50%		50%	100%
Zone 6	4	28	12	12	56
	7.1%	50%	21.4%	21.4%	100%
Total	35	119	35	60	255
	13.7%	46.6%	13.7%	23.5%	100%

Only intrazone connections longer than 10 cm have been included

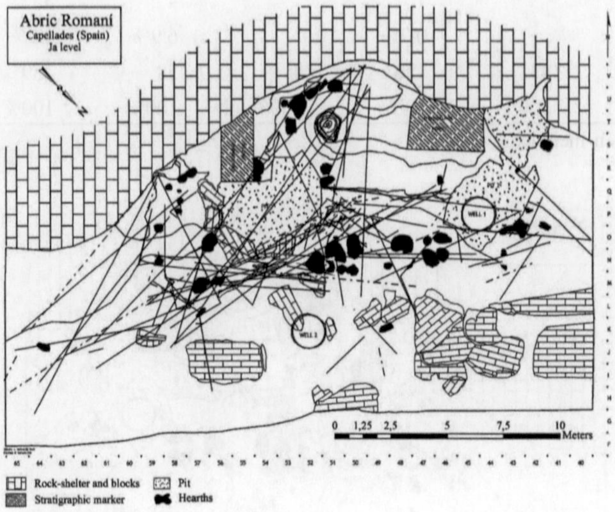

Fig. 7.52 Refitting map of sublevel Ja

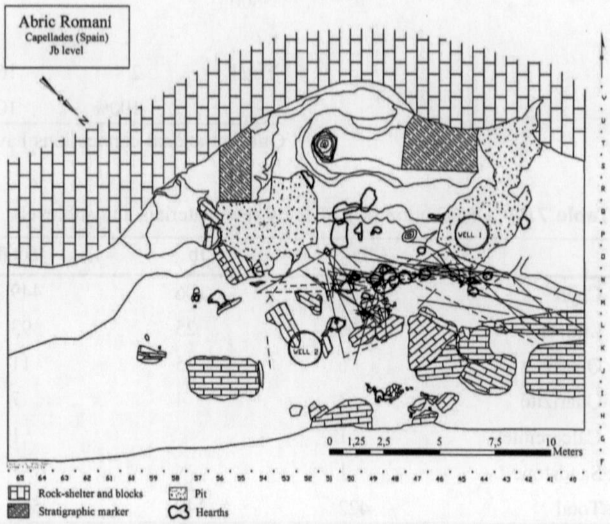

Fig. 7.53 Refitting map of sublevel Jb

nodules. As we will see below, spatial distribution will provide some arguments for identifying groups of higher resolution inside these quartz RMUs. For 460 of these RMUs it was possible to identify how they were introduced into the shelter with relative certainty.

Chert is the raw material with the most RMUs, 343 in sublevel Ja and 106 in sublevel Jb. 238 of these RMUs correspond to blanks detached outside of the shelter and introduced as single artifacts. This is the dominant provisioning strategy, especially in sublevel Ja. Of these single artifacts, 77 are retouched items and 161 flakes or flake fragments. As far as the reduction sequences carried out inside the rockshelter are concerned, partial sequences are clearly dominant (70 RMUs), especially those characterized by the knapping of cores introduced in a more or less advanced reduction stage. Meanwhile, complete sequences, characterized by the introduction of raw nodules that were entirely reduced into the shelter, are relatively scarce (36 RMUs). This last mode of introduction was slightly more common in sublevel Jb than in sublevel Ja (10.3 and 7.2% of the chert RMUs respectively).

Eleven quartz RMUs have been identified, six in sublevel Ja and five in sublevel Jb. Although the number of RMUs is similar in both sublevels, quartz remains are much more abundant in Ja. Although it was not possible to identify the introduction strategy of all the quartz RMUs, three basic modalities have been identified: large angular fragments with quadrangular and rectangular morphologies, unworked cobbles from secondary deposits, and isolated blanks. Although this last modality has been particularly

Table 7.30 Distribution of connection lines of sublevel Ja by length intervals and zones

	Zone 1	Zone 2	Zone 3	Zone 4	Zone 5	Zone 6	Total
0–100	1	9	18	71	1	19	119
	50%	56.3%	41.9%	52.2%	25%	32.8%	45.9%
100–200		2	13	39	1	16	71
		12.5%	30.2%	28.7%	25%	27.6%	27.4%
200–300		2	7	13	2	12	36
		12.5%	16.3%	9.6%	50%	20.7%	13.8%
300–400	1	3	3	6		3	16
	50%	18.8%	7%	4.4%		5.2%	6.1%
400–500			1	4		1	6
			2.3%	2.9%		1.7%	2.3%
500–600				2		1	3
				1.5%		1.7%	1.1%
600–700			1			1	2
			2.3%			1.7%	0.7%
700–800						1	1
						1.7%	0.3%
>800				1		4	5
				0.7%		6.9%	1.9%
Total	2	16	43	136	4	58	259
	100%	100%	100%	100%	100%	100%	100%

Only intrazonal connections have been included

Table 7.31 Distribution of RMUs by raw materials and sublevels

	Ja	Jb	Total
Chert	343	106	449
Limestone	68	25	93
Quartz	6	5	11
Quartzite	3	4	7
Calcarenite	1		1
Sandstone	1		1
Total	422	140	562

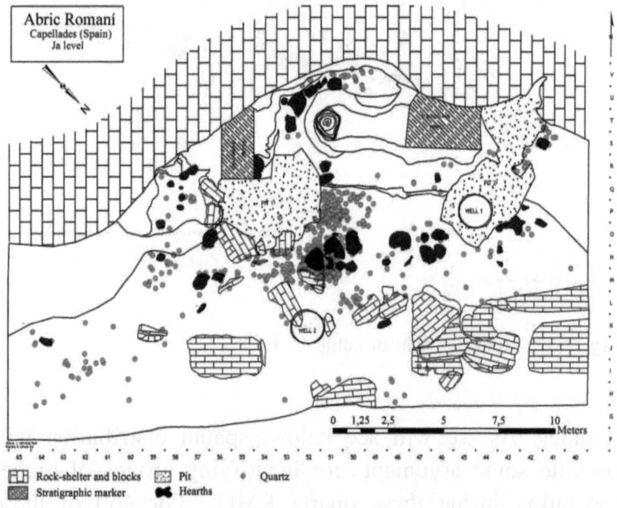

Fig. 7.54 Spatial distribution of quartz remains in sublevel Ja

hard to identify due to the homogeneity of this raw material, one transported artifact has been found. However, it is likely that other transported blanks have been included with the more generic quartz RMUs. Although we have found quartz remains throughout the entire rockshelter, their spatial pattern is very clustered, which makes it easy to identify the main knapping posts of quartz nodules. In the case of sublevel Ja two clear principal accumulations have been identified in zone 4, one situated in M–N/51–52 and another in O–Q/50–51 (Fig. 7.54). In sublevel Jb there are also two clusters, the first focused in N–O/49–50 and the second in L42 (Fig. 7.55).

A total of 93 limestone RMUs have been identified, 68 in sublevel Ja and 25 in Jb. Unlike chert, the transport of limestone to the rockshelter as complete or nearly complete cobbles was clearly dominant: 66 of the limestone RMUs were brought in this way. We have not found any RMUs suggesting the transport of limestone cores in an advanced reduction stage. Both the cobbles used as hammerstones and those used exclusively to produce flakes show this provisioning pattern. The introduction of single blanks has been documented, although this was less common for limestone than for chert (only 12 RMUs) (Fig. 7.56). Unworked cobbles were used above all as hammerstones (in lithic production and other kinds of activities such as processing

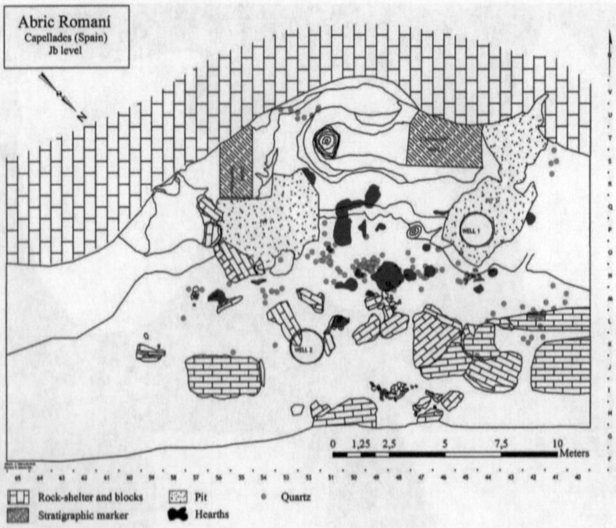

Fig. 7.55 Spatial distribution of quartz remains in sublevel Jb

Table 7.32 Distribution of artifacts included in RMUs by raw materials and sublevels, and the percentage of remains included in RMUs

	Ja		Jb		Total	
	N	%	N	%	N	%
Flint	1852	46.8	877	68.5	2729	52.1
Limestone	471	59.8	58	92	529	62.2
Quartz	566	89.2	107	100	673	90.8
Quartzite	3	100	10	90.9	13	92.8
Calcarenite	7	100			7	87.5
Sandstone	1	33.3			1	25
Total	2900	53.2	1052	71.5	3952	57.1

faunal or plant remains). Some of these were discarded as unbroken cobbles, but many were highly fragmented. Three different behaviors have been identified in relation to the remains derived from these fragmented cobbles: the discard of the fragments, the detachment of a few flakes from some fragments, and the execution of complete and long reduction sequences using the same knapping methods identified for other raw materials. As mentioned earlier, *in situ* reduction sequences on limestone were not carried out in sublevel Jb. The scarce knapping products come from the breakage of unworked nodules during their use as hammerstones. All the flakes identified show percussion marks on the butt.

In sublevel Ja, although limestone remains are dispersed across all zones, the highest numbers of RMUs are located in zones 3, 4 and 6 (Fig. 7.57). Zone 3 shows a well-defined cluster in L–M/44–45, which we will discuss further below. Meanwhile, in sublevel Jb only two RMUs are clustered in a well-defined area (zones A and B); the rest are scattered throughout the entire excavated surface (Fig. 7.58).

Although the number of quartzite remains is very low, the provisioning strategy for this raw material is particularly well defined. Seven RMUs have been identified, three in sublevel Ja and four in sublevel Jb. This is a relatively high number, since only 14 quartzite remains have been found. Six RMUs are single blanks: five flakes and one retouched artifact. The seventh is a small assemblage of six small quartzite flakes and flake fragments from sublevel Jb that were produced by a short knapping event. The remains of this sequence were scattered over an area of six square meters. It is possible that the core was introduced into the site at an advanced stage of reduction. This core has not been found and was perhaps transported to another site, a pattern already described in other levels of the site (Vaquero 1999; Martínez et al. 2005).

Table 7.33 shows the distribution of the RMUs from sublevel Ja by zones and type of technical event. We were not able to confidently establish the kind of modification carried out inside the shelter for 16 RMUs. Zone 4 presents the highest number of RMUs, which indicates that high remain densities correlate to the number of technical events that occurred. This is especially clear in the case of core reduction—45.3% of all the production sequences identified in sublevel Ja are found in zone 4, although core reduction sequences have been documented in all zones. The low-density exterior zone (zone 5) is characterized by the virtual absence of core reduction sequences. However, the number of RMUs is relatively low in the zones near the shelter wall (especially in zone 1) due to alterations affecting lithic remains. The high percentage of patinated artifacts made identifying RMUs in these zones difficult. The use of limestone cobbles is also well documented throughout the rockshelter, except in zones 1 and 2. Finally, the presence of transported blanks produced elsewhere is also common all over the site, particularly in zone 3, where they represent 60% of all the RMUs identified in this zone.

In sublevel Jb (Table 7.34), core reduction sequences are strongly focused in zones A and B. Core exploitation was also carried out in zone C, but all these sequences correspond to RMUs that were also knapped in zone B. The use of cobbles as hammerstones was also concentrated in these zones, as were the highest densities of transported artifacts. These single items are clearly dominant in zones C, D and F.

Looking at the distribution of raw material units, the first characteristic that stands out is the marked variability between the different RMUs in terms of their scatter areas. Some units are widely distributed over the entire excavated area, while others exhibit a much more tightly clustered pattern. As a measurement of this scattering, we have considered the standard deviation of the distance between each artifact and the centroid of the RMU distribution (Table 7.35). The mean deviation for RMUs formed by more than five artifacts is 153 cm, and most RMUs exhibit

Fig. 7.56 Limestone transported artifacts from sublevel Ja. They are large and mostly *débordant* unretouched flakes

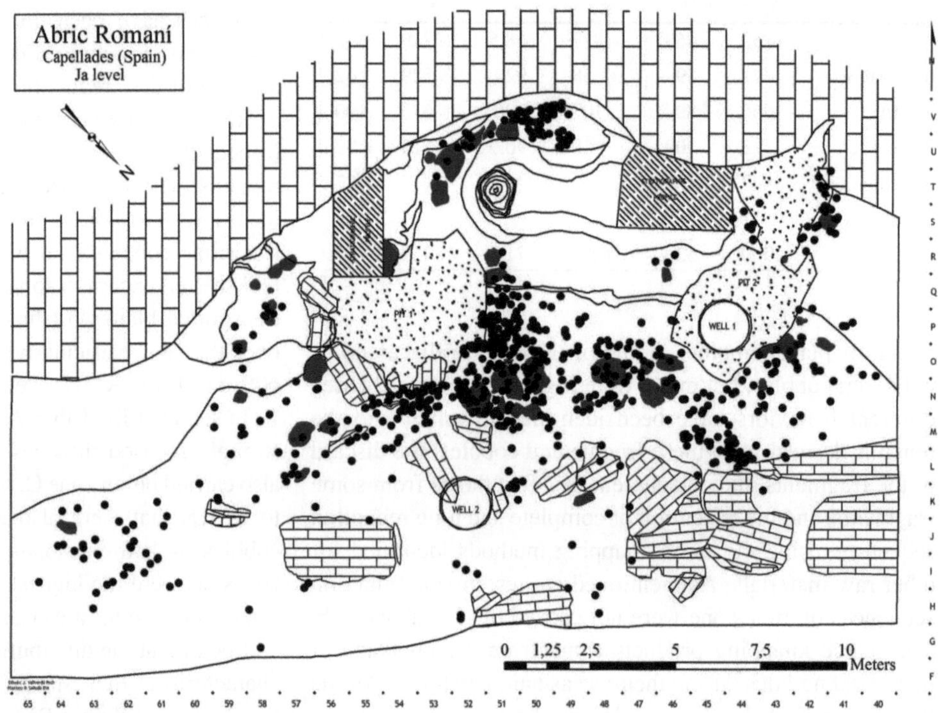

Fig. 7.57 Spatial distribution of limestone remains in sublevel Ja

deviations shorter than 2 m (73.2%). Limestone RMUs tend to be more scattered than chert RMUs (mean deviations are 195 and 131 cm respectively). Both connection lines and RMU scatter areas are larger than those recorded in other Romaní levels, which suggests that artifact movement was more common in this level. The degree of clustering is not related to the number of pieces included in each RMU (r = 0.12).

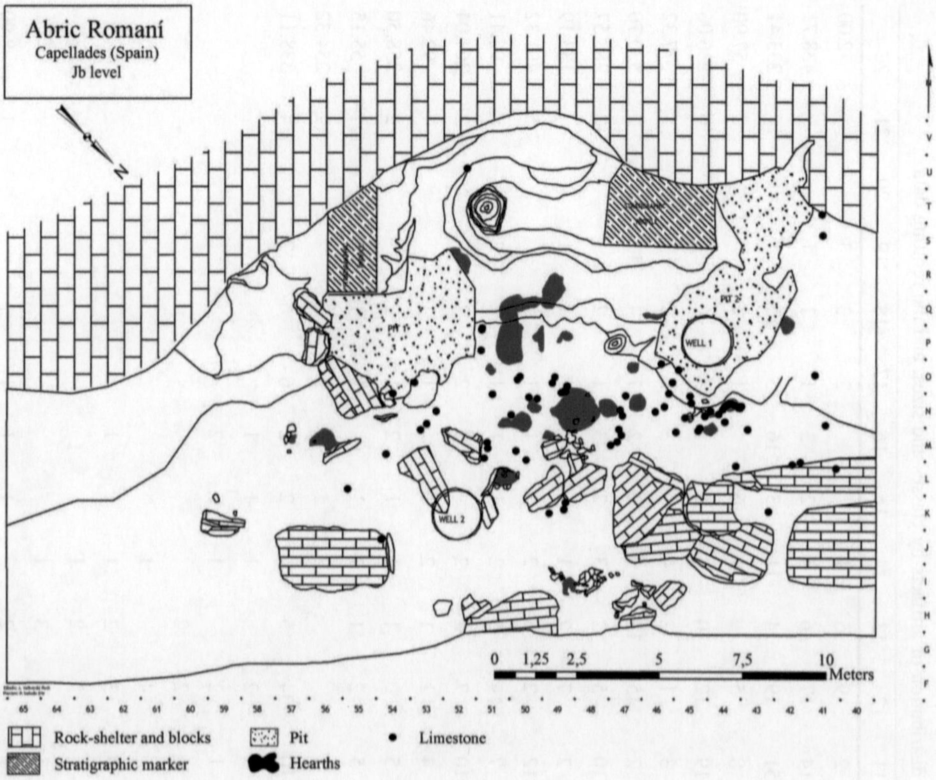

Fig. 7.58 Spatial distribution of limestone remains in sublevel Jb

Table 7.33 Distribution of RMUs from sublevel Ja according to zones and types of technical event carried out inside the shelter

	1	2	3	4	5	6	Total
Core reduction	5	22	37	78	1	29	172
Retouch			1				1
Used cobbles			12	15	3	10	40
Transported blanks	14	15	75	71	9	9	193
Total	19	37	125	164	13	48	406

Table 7.34 Distribution of RMUs from sublevel Jb according to zones and types of technical event carried out inside the shelter

	A	B	C	D	E	F	Total
Core reduction	27	24		1	6	1	59
Retouch							
Used cobbles	5	8		1	2		16
Transported blanks	16	22	10	10	3	4	65
Total	48	54	10	12	11	5	140

Interzonal and intrazonal differences in scatter patterns are especially interesting. RMU scattering data agree with the refit data. The most clustered patterns correspond to the zones near the shelter wall (zones 1 and 2), although this is partly due to their small area and the discontinuity with respect to the rest of the excavated area. Moreover, the physical barrier isolating zone 2 would limit northward and westward movements from this zone, although there are some long distance connections between this zone and the rest of the site. In zones 1 and 5, the scarce number of RMUs limits this approach and again, we believe that the comparison of zones 3, 4 and 6 will prove more informative (Table 7.36). 46.7% of the standard deviations of RMUs in zone 4, 28.6% in zone 3, and 31.8% in zone 6 were less than 1 m. This suggests that remains from the same sequence tend to be more clustered in zone 4, whereas in zones 3 and 6 they tend to be more scattered. This coincides with the conclusions derived from the connection lines, although according to the refitting data, zone 6 stood out as the zone most affected by dispersion processes. RMU scattering does not show such a clear-cut difference between zones 3 and 6, although standard deviations higher than 2 m are more

Table 7.35 List of RMU formed by more than four artifacts, indicating the spatial dispersion data, the distribution of artifacts by classes and sizes, and the refitting data

1	2	3	4	5	6	7	8	9	10	11	12	13	14	15	16	17	18	19	20	21	22
Lim-002	Ja	15	P49 X = 38, Y = 63	356			7	6		2	9	5	1			2	3	3			72.00
Lim-003	Ja	32	O50 X = 86, Y = 76	304.5		3	7	8		14	17	6	4	2	3	1	1	1			438.71
Lim-004	Ja	63	N48 X = 22, Y = 49	196			4	8		51	19	14	11	3	16	5	15	1		14	233.41
Lim-007	Ja	12	L50 X = 33, Y = 11	160.5				4		8	5		1	6		1	2			2	57.90
Lim-008	Ja	29	O52 X = 95, Y = 27	172.5				10		19	12	10	4	1	2	3	6			6	236.06
Lim-011	Ja	8	S50 X = 23, Y = 60	251.5			2	3		3	1	2	4	1		2	2	1		1	77.32
Lim-012	Ja	41	S52 X = 93, Y = 10	346.5			15	19	1	2	15	13	9	2	2	7	14	12		2	530.70
Lim-014	Ja	15	M56 X = 50, Y = 50	209.5			4	1		10	5	2	3	2	3	1	2			2	288.57
Lim-015	Ja	10	O52 X = 31, Y = 65	300		1	4	3		2	7	2	1			1	1	1			133.19
Lim-016	Ja	16	M56 X = 18, Y = 10	266.5		1	2	1		12	2	8	2	3	1	1	5			5	401.32
Lim-017	Ja	23	O51 X = 53, Y = 51	83			9	7		7	14	5	2	1	1	1	2	2			96.41
Lim-018	Ja	13	N51 X = 14, Y = 79	141.5				3		10	7	2	2		2	2	2			2	264.04
Lim-027	Ja	5	V50 X = 69, Y = 15	36				1		4	1	1	2	1		1	1			1	92.46
Lim-028	Ja	5	O49 X = 54, Y = 99	137.5						5		1	1	1	2	1	2			2	258.50
Lim-029	Ja	5	O51 X = 14, Y = 33	55						5	2	1	1	1		1	1			1	55.15
Lim-032	Ja	8	M53 X = 17, Y = 42	612.5			1			7			3	1	4	2	2			2	254.32
Lim-038	Ja	49	O52 X = 39, Y = 80	247.5			13	26		10	34	5	6	2	2	6	6	2		4	338.11
Lim-040	Ja	7	N49 X = 73, Y = 19	253.5	2		1	1		3	2		1	1	1						
Lim-041	Ja	5	S43 X = 92, Y = 51	82.5			2	2		1	2		1		2						
Lim-042	Ja	6	Q42 X = 20, Y = 30	137				5		1	2	2									
Lim-043	Ja	5	N58 X = 53, Y = 43	113			2	3			4		1								
Lim-046	Ja	8	N46 X = 15, Y = 67	91.5	1			4		3	3	2	1	1	1						
Lim-052	Ja	8	P51 X = 75, Y = 36	123.5			3	5			5	2	1								
Lim-053	Ja	6	P50 X = 88, Y = 48	163			1	1		3	2	3	1								
Lim-054	Ja	7	N49 X = 91, Y = 94	94						7	3	2		1	1	1	1			1	8.90
Lim-064	Ja	5	I61 X = 92, Y = 69	205						5			1		4						
Qu-001	Ja	456	N52 X = 77, Y = 60	245.5		5	119	312	6	14	363	52	17	10	14	9	17	7		10	165.90
Qu-002	Ja	91	O50 X = 7, Y = 33	301.5	1		8	66		16	58	17	5	6	5	6	25	15		10	344.12
Qu-004	Ja	6	N45 X = 21, Y = 56	265.5			4	2			5	1				1	1			1	504.00
Qu-005	Ja	9	N49 X = 100, Y = 71	334			5	3		1	7	1			1						
Chert-001	Ja	11	V50 X = 7, Y = 31	72			6	5			11					1	1	1			231.76
Chert-002A	Ja	32	N54 X = 97, Y = 39	48.5		1	9	21	1		17	9	4	2		4	4	2		2	232.7
Chert-003	Ja	12	P51 X = 41, Y = 11	67			8	4			10	2				2	2	2			77.5

(continued)

Table 7.35 (continued)

1	2	3	4	5	6	7	8	9	10	11	12	13	14	15	16	17	18	19	20	21	22
Chert-004	Ja	28	O44 X = 38, Y = 25	118.5			16	12			17	6	4	1		1	1	1			130.64
Chert-007	Ja	38	O51 X = 80, Y = 17	402			17	21			23	6	4	4	1	3	17	15		2	282.29
Chert-008	Ja	14	P51 X = 74, Y = 34	48			11	3			12	2				1	1	1			159.76
Chert-009	Ja	24	L58 X = 45, Y = 64	171		2	11	11			16	4	3	1		2	5	5			419.35
Chert-012	Ja	55	O52 X = 68, Y = 31	186		5	27	23			38	11	6			4	6	6			195.41
Chert-014	Ja	18	L58 X = 47, Y = 47	137.5		1	9	8			14	3	1								
Chert-015	Ja	19	N59 X = 89, Y = 20	237			10	8	1		14	4	1			2	3	3			346.66
Chert-016	Ja	10	O58 X = 14, Y = 67	110.5			8	2			7	1	2			1	1	1			62.09
Chert-017	Ja	6	N53 X = 97, Y = 43	181.5		1	3	1	1		3	1	1	1		1	2	2			188.96
Chert-018	Ja	9	P48 X = 85, Y = 84	246.5		1	3	4	1		7	2				1	4	3	1		314.03
Chert-019	Ja	39	L59 X = 54, Y = 74	269		2	12	25			20	15	2		2	4	4	1		3	482.09
Chert-020	Ja	22	P44 X = 70, Y = 75	160.5		1	6	13	1		19	2	1			1	1	1			125.17
Chert-021	Ja	11	P49 X = 84, Y = 47	105.5			8	3			8	2	1			1	2	2			261.66
Chert-022	Ja	20	O46 X = 24, Y = 91	235.5		2	14	4			12	6	1	1		2	3	3			114.75
Chert-023	Ja	9	U51 X = 36, Y = 99	127.5			6	3			3	3		1	2	1	1	1			28.69
Chert-024	Ja	51	S42 X = 37, Y = 70	81			36	15			37	9	5			1	1	1			314.22
Chert-025	Ja	58	M57 X = 65, Y = 40	342.5		2	25	28	3		33	19	3	3		1	4	4			227.09
Chert-026	Ja	35	N53 X = 68, Y = 79	123		2	16	17			23	9	3			4	5	3		2	153.45
Chert-029	Ja	35	M52 X = 79, Y = 98	75			26	9			31	4				3	5	5			178.81
Chert-030	Ja	5	N58 X = 92, Y = 7	72			4	1			4	1				1	1	1			112.96
Chert-031	Ja	20	P58 X = 98, Y = 25	76		2	10	7	1		11	5	2	1	1	2	2	1		1	91.63
Chert-032	Ja	7	P50 X = 6, Y = 84	80				7			6	1				1	6			6	172.75
Chert-035A	Ja	17	P50 X = 11, Y = 40	62.5			13	4			15	2				2	2	2			175.75
Chert-035B	Ja	18	O48 X = 57, Y = 99	94		1	4	13			17		1			1	2	2			108.93
Chert-036	Ja	5	N53 X = 12, Y = 68	19		1	2	2			4			1		1	1	1			29.73
Chert-037	Ja	34	M59 X = 5, Y = 62	281		1	15	18			20	9	4	1		1	5	5			422.7
Chert-040	Ja	42	P51 X = 94, Y = 36	70.5			14	27	1		27	14	1			5	10	7		3	97.41
Chert-041	Ja	18	P44 X = 68, Y = 45	197.5		1	13	4			8	8	2			2	6	6			322.56
Chert-045	Ja	9	O51 X = 91, Y = 11	200			5	4			6	3				1	1			1	21.37
Chert-046	Ja	32	J62 X = 64, Y = 11	146.5			14	18			16	13			3	2	2			2	159.75
Chert-047	Ja	41	N54 X = 52, Y = 26	330.5		1	15	25			29	6	5	1		4	8	8			372.64
Chert-048	Ja	14	K49 X = 32, Y = 30	187			1	15	1		4	4	5	1		3	3	2		1	191.41
Chert-050	Ja	14	R51 X = 88, Y = 3	329.5		1	7	6			7	7									
Chert-052	Ja	9	N54 X = 89, Y = 87	467			2	7			5	2	2			3	3	2		1	963.45

(continued)

Table 7.35 (continued)

1	2	3	4	5	6	7	8	9	10	11	12	13	14	15	16	17	18	19	20	21	22
Chert-053	Ja	6	M56 X = 2, Y = 23	85		1	3	2			3	1	2			1	1	1			25.31
Chert-055	Ja	9	O43 X = 22, Y = 62	25.5		2	3	4			6	1	2			1	1	1			44.65
Chert-057	Ja	19	N59 X = 59, Y = 26	329.5		1	6	12			15	3	1			1	1	1			902.69
Chert-059	Ja	50	P51 X = 77, Y = 33	89			21	28	1		46	3	1								
Chert-060	Ja	8	O51 X = 20, Y = 17	85		1	2	5			6				1	1	2	2			128.12
Chert-061	Ja	25	P51 X = 71, Y = 52	70			8	17			21	3	1			1	1	1			62.64
Chert-062	Ja	5	N44 X = 29, Y = 63	179		1	1	3			2	2			1	1	2	1		1	183.06
Chert-063	Ja	31	N54 X = 43, Y = 30	79			16	15			30	1									
Chert-064	Ja	21	O51 X = 62, Y = 98	87.5			10	11			19	2									
Chert-065	Ja	41	O51 X = 81, Y = 87	77.5			25	16			29	4	1			2	2	1		1	38.59
Chert-067	Ja	42	P51 X = 80, Y = 32	76.5			24	18			39	3									
Chert-068	Ja	14	P51 X = 66, Y = 22	160			4	10			13	1									
Chert-070	Ja	22	J59 X = 53, Y = 62	218.5		1	9	12			20	1		1		1	1			1	44.01
Chert-071A	Ja	16	O50 X = 55, Y = 87	97.5			9	7			8	4									
Chert-071B	Ja	6	M53 X = 40, Y = 76	193.5		1	2	3			1	3	1	1							
Chert-073	Ja	18	L49 X = 47, Y = 70	14, 5			9	9			16	1	1								
Chert-076	Ja	52	V50 X = 19, Y = 9	158			15	36		1	30	12	5	2	3	1	1	1			10.05
Chert-080	Ja	8	O50 X = 3, Y = 98	95			3	5			6	1	1								
Chert-081	Ja	6	P42 X = 28, Y = 10	2.5				6			6										
Chert-084	Ja	6	V50 X = 21, Y = 29	71.5			3	3			5	1									
Chert-099	Ja	8	V50 X = 71, Y = 30	37.5			3	5			6	2									
Chert-120	Ja	5	I61 X = 57, Y = 83	69			3	2			1	4									
Chert-124	Ja	7	O50 X = 52, Y = 67	105.5		1	3	3			2	4		1							
Chert-126	Ja	8	P51 X = 71, Y = 0	94			3	5			6	2									
Chert-127	Ja	5	P51 X = 53, Y = 89	88.5			3	2			2	2	1			1	1			1	57.94
Chert-130	Ja	6	P48 X = 82, Y = 30	110			2	3	1		6										
Chert-131	Ja	8	O52 X = 38, Y = 58	142			5	3			6	2									
Chert-141	Ja	6	P51 X = 77, Y = 60	127.5			1	5			4	2				2	2			2	16.03
Chert-142	Ja	14	P51 X = 88, Y = 54	94.5			5	9			14										
Chert-145	Ja	8	O51 X = 42, Y = 17	103			7	1			7	1									
Chert-146	Ja	7	P51 X = 42, Y = 7	37.5			7				7										
Chert-147	Ja	5	N53 X = 47, Y = 28	62			1	4			2	3									
Chert-149	Ja	13	N50 X = 9, Y = 57	145.5			10	3			4	7	2								
Chert-150	Ja	6	O52 X = 30, Y = 19	73			2	4			5	1									

(continued)

Table 7.35 (continued)

1	2	3	4	5	6	7	8	9	10	11	12	13	14	15	16	17	18	19	20	21	22
Chert-156	Ja	21	L59 X = 91, Y = 71	180.5			12	9			14	5	1								
Chert-157	Ja	15	Q58 X = 18, Y = 22	48			11	4			11	1	3			1	1			1	131.24
Chert-159	Ja	5	N59 X = 73, Y = 43	82.5		1	1	3			4	1									
Chert-162	Ja	7	V50 X = 63, Y = 40	28			2	5			6		1								
Chert-164	Ja	9	N50 X = 28, Y = 55	93			7	2					2	4	3						
Chert-169	Ja	7	P50 X = 18, Y = 99	47.5			4	3			6	1				1	1	1			20.1
Lim-04	Jb	17	N45 X = 78, Y = 32	116.5						17		3	2		12	2	5			5	128.08
Lim-07	Jb	5	N48 X = 90, Y = 25	239			1			4	2	1	1		1	1	1			1	616.36
Lim-11	Jb	5	O50 X = 24, Y = 71	208.5				2		3	3	1			1						
Lim-12	Jb	5	O47 X = 22, Y = 72	313.5			2	1		2	2	2			1						
Qu-01	Jb	5	M52 X = 16, Y = 86	193.5			4	1			4			1							
Qu-02	Jb	5	N50 X = 2, Y = 37	336			2	3			3			2							
Qu-03	Jb	3	M48 X = 75, Y = 73	483				2		1	1	1			1						
Qu-04	Jb	43	M45 X = 28, Y = 99	320		4	6	20		13	20	13	2	1	7	4	7	3		4	195.60
Qu-05	Jb	106	O49 X = 14, Y = 54	243.5		1	27	73	1	4	84	15	1	2	4						
Quartzite-1	Jb	6	N50 X = 91, Y = 81	107			3	3			6					1	1			1	191.63
Chert-01	Jb	153	M50 X = 48, Y = 67	110.5		2	90	61			114	27	4	3	5						
Chert-02	Jb	75	N47 X = 94, Y = 82	95			27	47	1		42	20	8	2	3	7	7	3		4	163.45
Chert-03	Jb	36	N46 X = 20, Y = 83	117			27	9			27	7	2			4	4	3		1	104.45
Chert-04	Jb	44	N46 X = 4, Y = 94	82.5			23	21			29	12	2	1		3	4	3		1	91.22
Chert-05	Jb	41	N46 X = 59, Y = 45	163.5			15	26			21	16	3	1		4	9	8		1	173.65
Chert-06	Jb	69	N47 X = 75, Y = 85	92			31	40			42	18	6	3	2	4	9	7		2	84.1
Chert-07	Jb	9	N47 X = 90, Y = 62	125			9				9										
Chert-08	Jb	28	N47 X = 36, Y = 71	101			10	18			25	2			1	2	3	2		1	51.12
Chert-09	Jb	6	O52 X = 21, Y = 26	52			1	5			5	1				1	1			1	7.28
Chert-11	Jb	5	N46 X = 34, Y = 45	78			4	1			2	3				1	1	1			105.47
Chert-12	Jb	36	N48 X = 59, Y = 93	79			5	31			29	5	1	1		3	3	2		1	70.13
Chert-13	Jb	47	N51 X = 75, Y = 9	79			23	24			37	10				3	5	4		1	120.76
Chert-16	Jb	20	M50 X = 49, Y = 7	197			10	10			7	6	1	2	4	3	6	4		2	283.73
Chert-17	Jb	5	N50 X = 57, Y = 55	195.5			1	4			2	3				1	1			1	522.01
Chert-18	Jb	19	O49 X = 76, Y = 0	124.5		1	11	7			12	5	2			1	1	1			71.69
Chert-19	Jb	20	N50 X = 34, Y = 85	113.5			13	7			20										
Chert-25	Jb	7	O51 X = 6, Y = 21	166.5			6	1			4	1	2			1	1	1			57.27
Chert-27	Jb	9	O51 X = 74, Y = 1	162		1	4	4			6	1	2								

(continued)

Table 7.35 (continued)

1	2	3	4	5	6	7	8	9	10	11	12	13	14	15	16	17	18	19	20	21	22
Chert-28	Jb	8	N52 X = 44, Y = 56	155.5			4	4			2	5		1		1	1			1	283.34
Chert-29	Jb	12	N52 X = 60, Y = 82	108.5			3	9			11	1				2	2	1		1	91.2
Chert-33	Jb	7	M41 X = 22, Y = 68	43.5			3	4			2	5	1		1	1	1	1			38.01
Chert-34	Jb	13	N48 X = 98, Y = 73	31.5			4	9			12		1								
Chert-36	Jb	10	M53 X = 53, Y = 86	136.5			4	6			9			1							
Chert-37	Jb	18	N47 X = 40, Y = 76	66			12	6			18										
Chert-38	Jb	6	N47 X = 62, Y = 69	84.5			5	1			3	1	2								
Chert-41	Jb	14	N48 X = 67, Y = 63	167.5			5	9			7	5	2								
Chert-42	Jb	12	O51 X = 57, Y = 34	103.5			4	8			12										
Chert-44	Jb	5	M43 X = 86, Y = 21	99.5				5			2	1	2								
Chert-47	Jb	5	N47 X = 26, Y = 60	71				5			4	1									
Chert-52	Jb	7	O47 X = 80, Y = 20	58.5			2	5			7										
Chert-61	Jb	5	O48 X = 49, Y = 0	195			4	1			2	2	1								

1 RMU. 2 Sublevel. *3* Number of artifacts. *4* Centroid location. *5* Standard deviation. *6* Unworked nodules. *7* Cores. *8* Flakes. *9* Flake fragments. *10* Retouched artifacts. *11* Fragments. *12* Very small remains. *13* Small remains. *14* Medium remains. *15* Large remains. *16* Very large remains. *17* Number of refits. *18* Number of connection lines. *19* Production connection lines. *20* Retouch connection lines. *21* Breakage connection lines. 22 Mean length of connection lines

common in zone 6. In order to complete this pattern, it is interesting to look at the intrazonal variability of lithic scatters. For instance, zone 4 contains the most densely clustered RMUs, but we also found the most scattered ones there (15% of the RMUs from this zone show standard deviations higher than 3 m). These intrazonal differences are particularly informative from the point of view of assemblage formation temporality and will be analyzed in detail in the coming paragraphs.

In general, the SMT chert is evenly distributed across the shelter, although a cluster has been documented in zones 3 and 4. RMUs of primary origin tend to be more clustered (Fig. 7.59), while secondary-origin RMUs are more scattered (Fig. 7.60). There are some exceptions however. For example, Chert-033 is an RMU of primary origin formed by three artifacts widely scattered in zones 2 and 3, showing a dispersion pattern more common in RMUs of secondary origin. Nevertheless, one of these lithics is a retouched artifact with use traces, which suggests that this movement was probably intentional. Another chert of primary origin showing a large scatter is Chert-070, which was widely distributed throughout zone 6, although we should bear in mind that this zone is characterized by the greater dispersion of remains. Only one secondary-origin RMU, Chert-055, exhibits a clustered pattern similar to that dominant among the SMT chert of primary origin. This RMU is concentrated in O43.

These differences in spatial pattern according to raw material origin (primary or secondary) can be explained by temporal changes in the provisioning strategies during the formation of the assemblage. Exploitation of secondary deposits would have been more common during the first stages of the assemblage formation. The lithic scatters produced in these earlier phases would have been more subject to the dispersion factors associated with human activity. On the other hand, the more clustered pattern of primary-origin chert suggests that these provisioning areas were more exploited in the last stages. Furthermore, this indicates that more distant lithic sources were more frequently exploited during these last phases.

This difference in the degree of dispersion according to primary or secondary origin was not seen in the case of VLD chert. However, another spatial difference has been documented. Most RMUs of secondary origin are located in the middle of the shelter (zone 4) (Fig. 7.62), while the units of primary origin show a wider distribution and have been found in zones 1, 4 and 6 (Fig. 7.61).

Primary and secondary VLD chert and primary SMT chert are associated with zone 4. They probably correspond to activity events subsequent to those in which primary SMT chert, much more dispersed across the shelter, was exploited. RMUs formed by more than 30 artifacts are more common in primary VLD chert, which corresponds to the furthest

Table 7.36 Distribution by 1 m intervals of the standard deviations of RMUs focused in zones 3, 4 and 6

	0–1 m	1–2 m	2–3 m	3–4 m	>4 m	Total
Zone 3	4	7	3			14
	28.6%	50%	21.4%			100%
Zone 4	28	17	5	6	3	59
	46.7%	28.4%	8.3%	10%	5.1%	100%
Zone 6	7	6	7	2		22
	31.8%	27.3%	31.8%	9.1%		100%
Total	39	30	15	8	3	95
	41%	31.5%	15.7%	8.4%	3.1%	100%

Only RMUs formed by more than five artifacts have been included

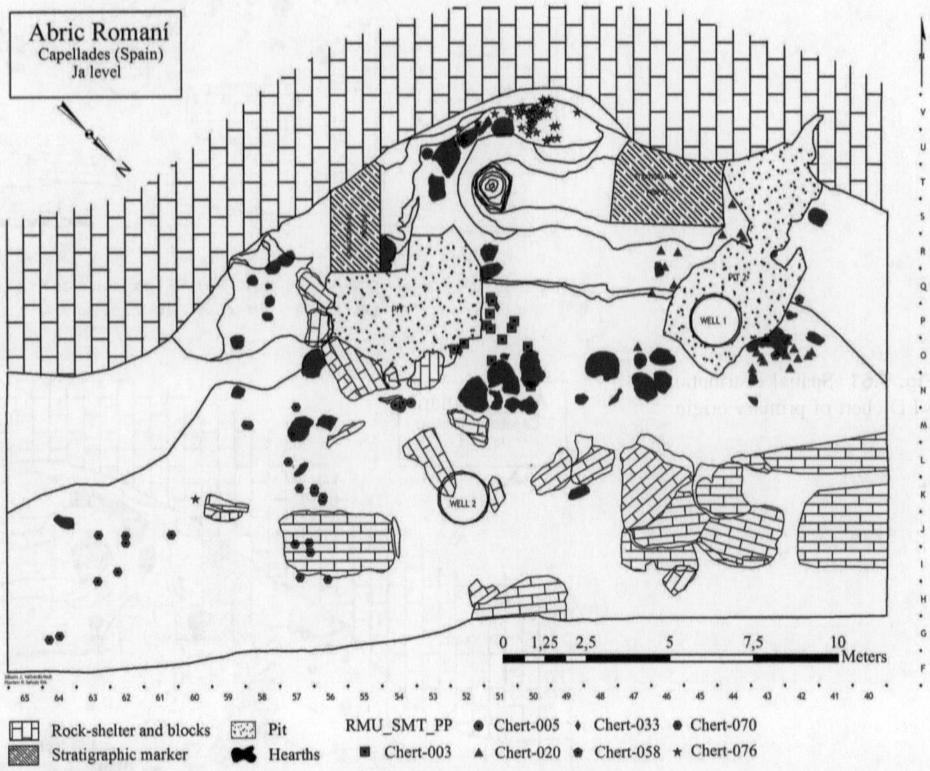

Fig. 7.59 Spatial distribution of SMT chert of primary origin

provisioning area (≥20–30 km). This indicates that the furthest raw material is associated with longer and more productive reduction sequences. SMT chert does not show significant differences in the length of the reduction sequences according to the primary or secondary origin of the nodules. The SMT provisioning area is closer (≥12–15 km) than the VLD area.

The primary RMUs from SMT and VLD tend to be located at the rear of the shelter. Unlike secondary chert, some primary nodules (like Chert-003 and 024) were entirely knapped at the back of the shelter. Therefore, it seems that the shelter was differentially occupied according to the origin of the raw materials and human groups. Those from distant areas especially occupied the rear of the shelter and were present subsequent to the events focused on secondary raw materials.

We do not know the exact origin of the exogenous chert, but more than half was collected in secondary deposits (Chert-010, 047 and 050). Chert-031 is of primary origin, while the primary or secondary origin of Chert-018 cannot be established. Refits indicate that these materials were intensively exploited in the rockshelter, which was predictable due to the high knapping quality of these nodules. These materials tend to show a random spatial distribution (Fig. 7.63). Knapping of exogenous chert was carried out in zones 2 (Chert-050), 3 (Chert-018), 4 (Chert-010, 047 and 050), and 6 (Chert-031). With the exception of Chert-010, all exhibit wide scatters, some of them indicating intentional movement and the succession of different knapping posts (Chert-018 and 050). This suggests an early location in the formation sequence of level J for most of the exogenous materials. Chert-010,

Fig. 7.60 Spatial distribution of SMT chert of secondary origin

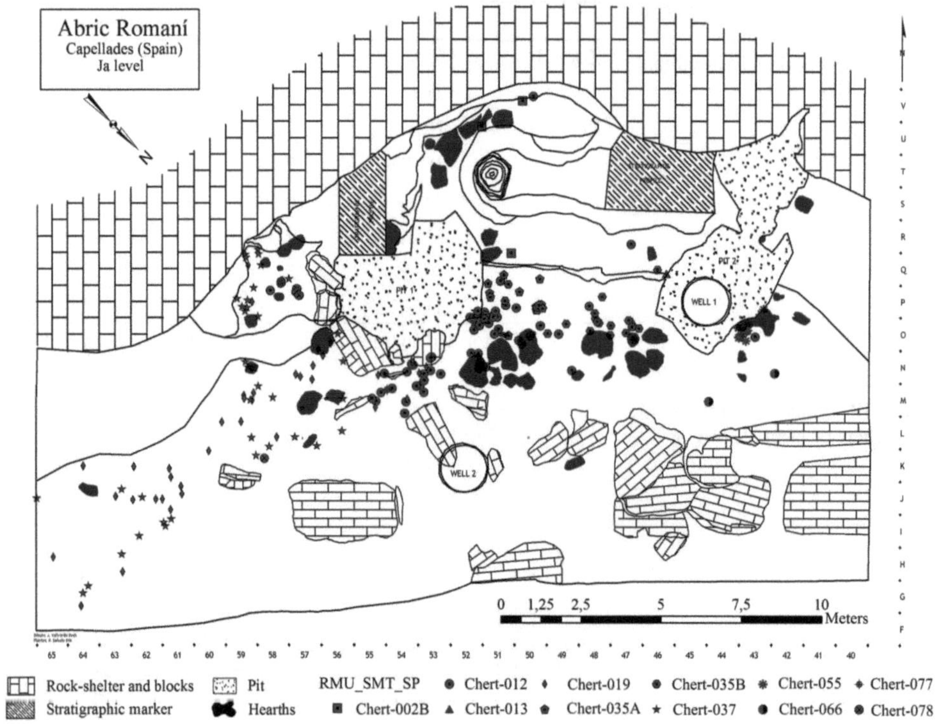

Fig. 7.61 Spatial distribution of VLD chert of primary origin

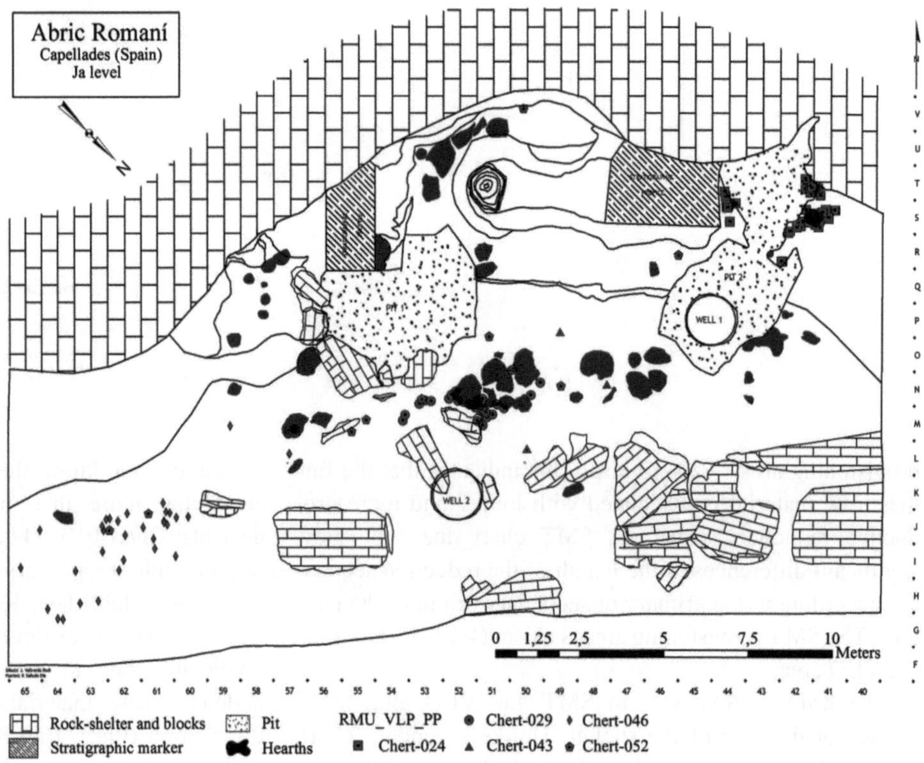

however, corresponds to the last activity events carried out in this level, since it was found at the top of the layer and was highly clustered.

PAN chert does not show any clear spatial pattern (Fig. 7.64). These RMUs are formed by few artifacts (a maximum of 21 remains for Chert-022). Of the four RMUs attributed to this formation, three are of primary origin and one of secondary origin. They are characterized by wide scatters, which suggest relatively early location in the formation sequence of level J.

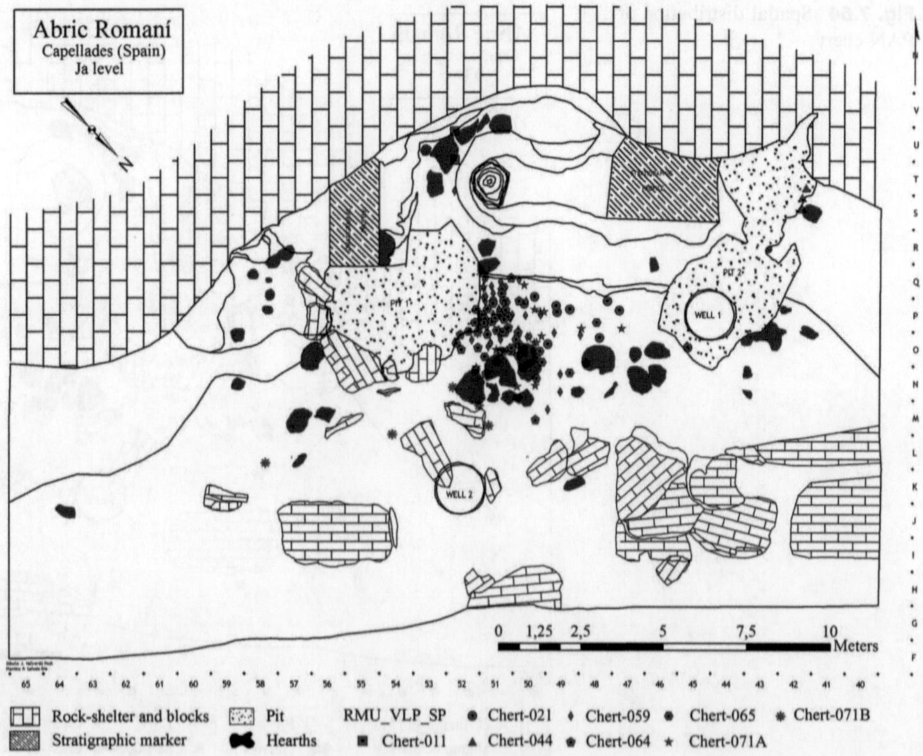

Fig. 7.62 Spatial distribution of VLD chert of secondary origin

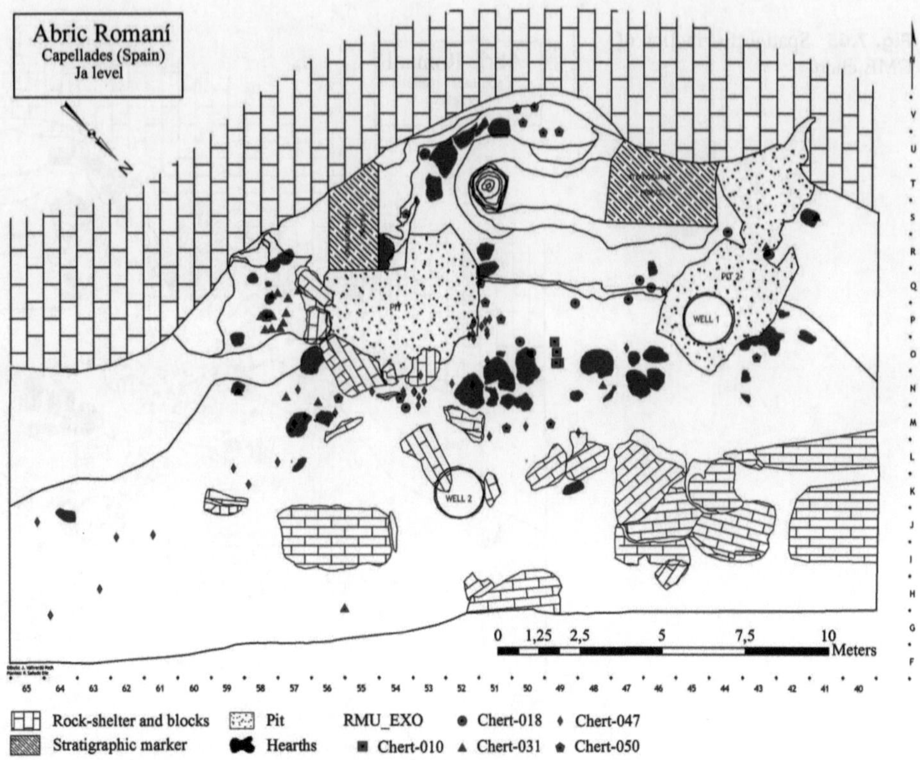

Fig. 7.63 Spatial distribution of exogenous chert

Finally, CME chert corresponds to RMUs with few artifacts (Chert-048 and 079). It can be considered a local chert, and its dispersion shows the same pattern as other nearby materials, like secondary SMT chert (Fig. 7.65). Unlike the other chert types, CME chert tends to be located on the exterior side of the shelter (zones 5 and 6). These are the less protected areas, and we cannot rule out that the wide scattering of these RMUs was partly related to natural postdepositional factors.

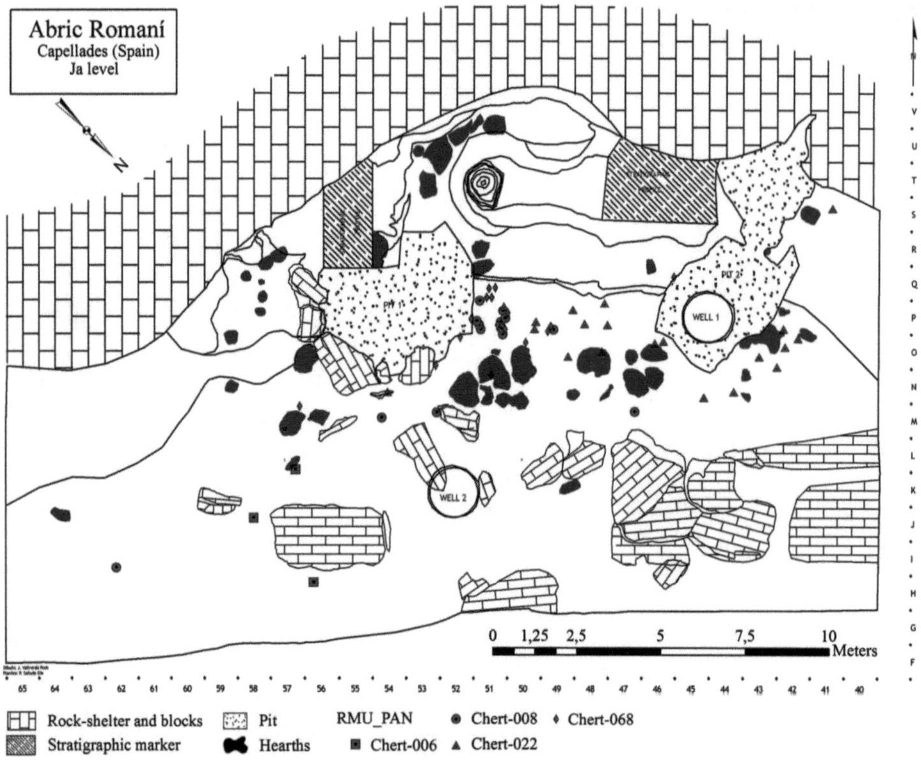

Fig. 7.64 Spatial distribution of PAN chert

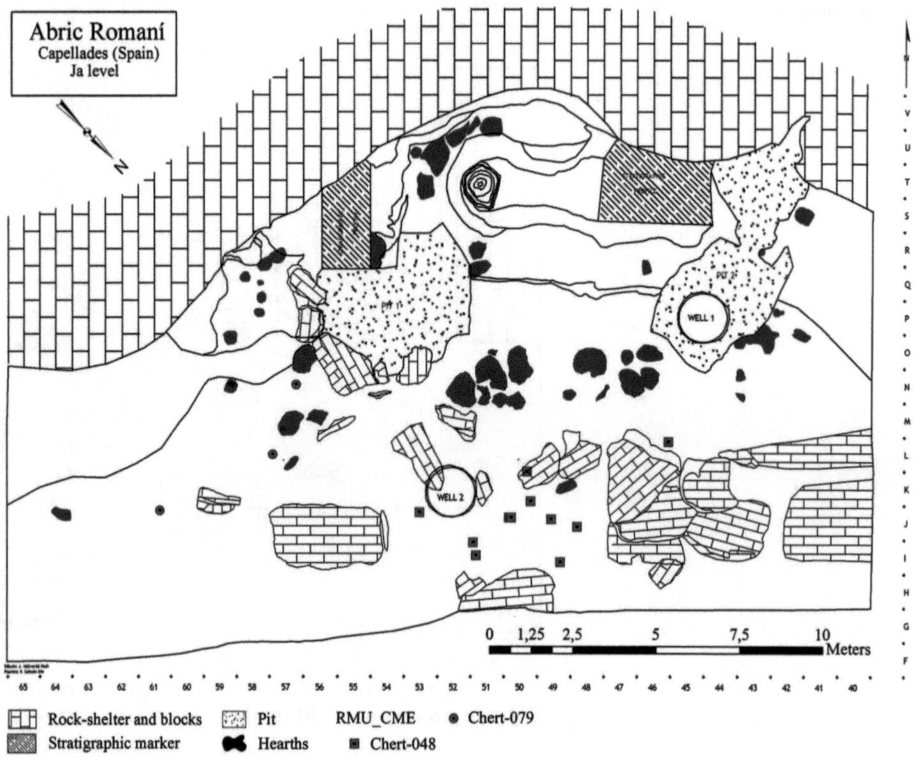

Fig. 7.65 Spatial distribution of CME chert

Spatial and Temporal Patterns

In this section we will focus on the spatial distribution of lithics, paying special attention to the information yielded by refits and RMUs. This spatial layout will be the main basis for the temporal analysis of the lithic assemblage. We will present these spatial data in detail according to several zones that have been defined taking into account site topography and remain distribution. As we have seen, the refitting data and the scattering of knapping events indicate that spatial relationships are quite well preserved at the site, although some postdepositional movements cannot be overlooked, especially in sublevel Ja.

Sublevel Ja

The spatial distribution of level J has previously been tackled in studies focusing on sublevel Ja (Vaquero 1999b; Martínez Molina and Rando 2001). These studies showed that the main factor of spatial variability was related to artifact size. A basic difference was detected between high-density inner areas (from row M to the shelter wall), characterized by a clear dominance of very small remains, and low-density outer areas (beyond the line of blocks), in which large and very large remains were more abundant. This pattern was interpreted as the outcome of a spatial layout guided by functional criteria. The areas close to the wall, inside the dripline, were characterized as hearth-related household areas, where knapping activities producing large amounts of small remains were preferentially carried out. On the other hand, the unprotected zone beyond the dripline would not be an activity locus, but a refuse area where large artifacts would be tossed out from the inner domestic areas. This pattern would be produced by maintenance behaviors similar to those described by Binford (1978) for the distribution of remains around hearths. Inner areas would be the drop zone and outer areas the toss zone. In addition to size sorting, other spatial differences were associated to the segregated location of certain activities requiring specific lithic implements, such as unmodified limestone cobbles. A cluster of limestone fragments found in L44–45 and M44–45 was explained in this way. However, functionally specialized areas were uncommon, since most activities were carried out in the multifunctional hearth-related areas.

We have divided the surface of sublevel Ja into six zones (Fig. 7.66). These zones are simply operational units used to present the spatial data, but they do not have a priori any behavioral meaning. Unlike other Romaní levels (such as levels H, I, K and L), where lithic remains were clustered in various well-defined accumulations (Vaquero and Pastó 2001; Vaquero et al. 2001a, 2004), sublevel Ja exhibits a more or less continuous distribution (Fig. 7.67). The mean density for the entire excavated surface is 26.3 lithic remains/m^2 (only remains longer than 1 cm have been considered) although there were certain density differences between zones. High densities of lithic remains characterize the two isolated zones near the shelter wall (zones 1 and 2). The middle of the site (zone 4) also shows a very high density, especially in the area of N50–54, O50–52, P50–52 and Q50–51. The maximum values are found in V50 (387), P51 (331) and N52 (198). Lithic density is lower in the rest of the site, without significant differences between zones 3 and 6. Finally, the lowest density corresponds to the outer part of the site, beyond the line of blocks (zone 5). Although the cluster pattern is less evident than in other archeological layers, the hearth-focused dynamic seems quite clear, since the highest density areas correspond to the areas where most combustion structures have been documented.

Zone 1

Previous excavations considerably altered this zone, making reading the spatial data particularly difficult. The contribution of this zone to the spatial and temporal interpretation will therefore be very limited. In addition, most chert remains had intensely patinated surfaces, which made identifying the RMUs even more challenging. This zone shows the lowest number of connections, both intrazonal (n = 2) and extrazonal (a connection with zone 3). Compared with the rest of the site, this zone seems to be practically isolated. The number of RMUs concentrated in this zone is also fairly low (n = 19), and most of them correspond to transported artifacts (Table 7.33). However, one RMU (Chert-024) is made up of more than 50 artifacts and can be attributed to a nearly complete core reduction sequence aimed at the production of small and very small flakes. The connection with zone 3 corresponds to a flake fragment detached in a reduction sequence carried out in zone 3 (Chert-041). It belongs to a refit of five flakes conjoined with the core (Fig. 7.68). In general, the artifacts of this sequence are very scattered (average connection length:

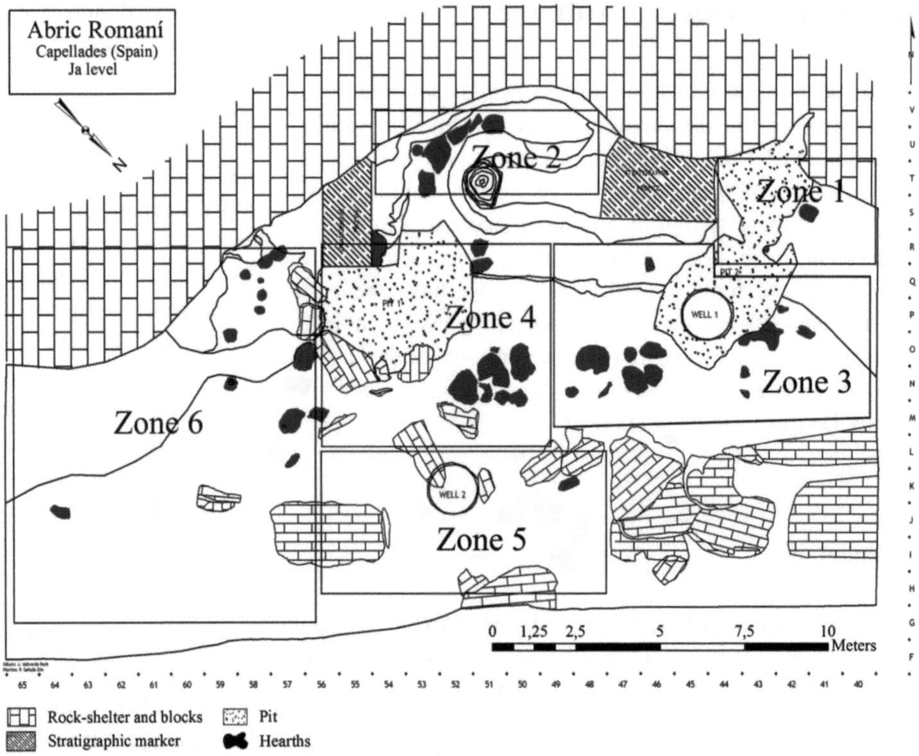

Fig. 7.66 Zones defined for the presentation of spatial data in sublevel Ja

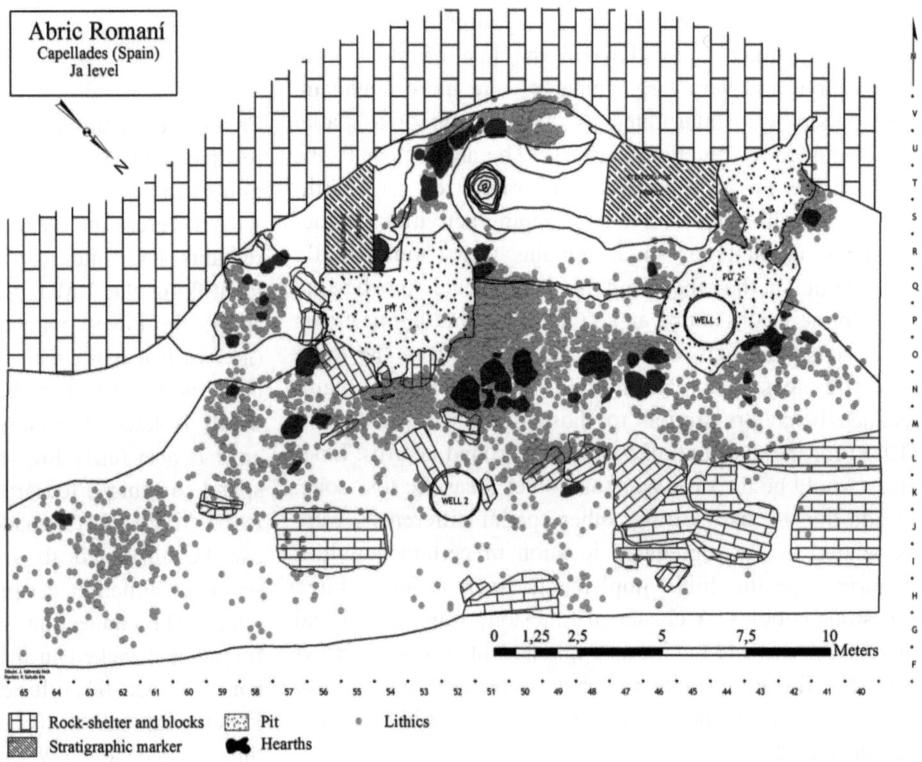

Fig. 7.67 Spatial distribution of lithics in sublevel Ja

322 cm), and the presence of one artifact in zone 1 cannot be attributed to intentional transport.

Limestone and quartz remains are relatively scarce. The only limestone RMU focused in this zone (Lim-041) corresponds to a core reduction sequence made on a cobble of mediocre quality for knapping. It is made up solely of knapping products of different sizes, including some large flakes. The flakes seem to belong to the first phase of the

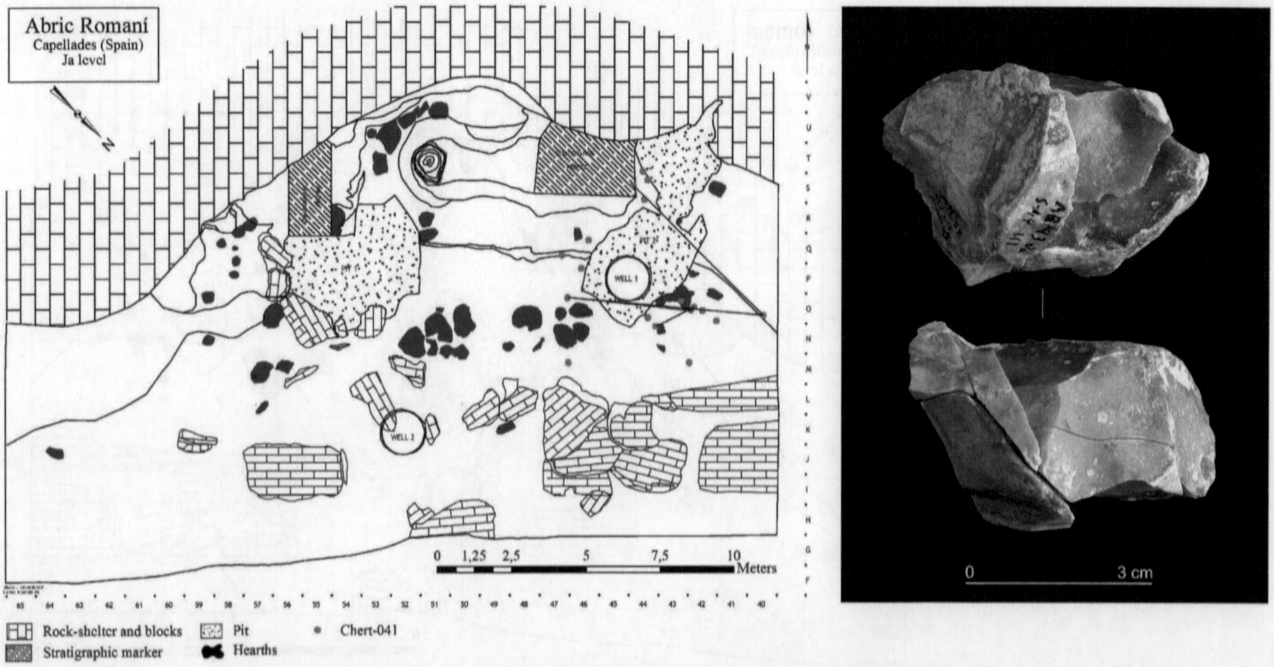

Fig. 7.68 Refitting and spatial distribution of Chert-041. Five flakes were conjoined with the core. All the artifacts were located in zone 3, except one flake recovered in zone 1

reduction sequence because many present cortex. No refitting of these raw materials was found. The few quartz pieces belong to the two general RMUs (Qu-001 and Qu-002), which are dispersed throughout all zones.

Zone 2

Although chert remains also exhibited high percentages of patinated artifacts, the number of connections and RMUs is considerably higher than in zone 1. We have documented 26 connection lines belonging to 23 refits. Sixteen connection lines are intrazonal and ten relate zone 2 to other zones. This is one of the zones with the most interzonal connections and, interestingly, it is the zone with the highest number of long-distance refits. The interzonal connections link zone 2 to zones 3, 4 and 6. There are no connections with zones 1 and 5. In some cases, it was possible to establish the direction of movement. Two denticulates with use-wear traces were located in V50 and were used to work on fresh hide and meat/bone. The denticulate used in hideworking was refitted to a flake found in zone 2. This production refit is particularly intriguing, since no other artifacts have been associated to this reduction sequence (Chert-033). Only three RMUs seem to correspond to complete or nearly complete reduction sequences. One of them (Chert-076) was performed on a nodule of very bad quality and is made up of 56 artifacts, essentially small and very small flakes. All the artifacts associated to this RMU were located in zone 2, with the exception of a very large flake found in zone 6 (square K60).

The number of quartz pieces is very low and there is no quartz refitting. All the elements are small fragments or flake fragments. Limestone provides more information, although the number of RMUs is low and use of limestone cobbles as hammerstones has not been clearly documented. Two limestone RMUs were focused in this zone, providing eight refits and 15 connection lines, most of them from the same reduction sequence (Lim-012). The only calcarenite RMU found in level J was also focused in this zone and it consisted of a short reduction sequence that yielded a few cortical products.

Lim-012 merits special attention. It is a nearly complete sequence that yielded quite a lot of flakes and flake fragments (n = 41) and shows an interesting spatial pattern, suggesting a temporal relationship with other zones. This sequence is distributed among zones 2, 4 and 6 (Fig. 7.69). The high quantity of cortical flakes indicates that the first moment of the production sequence was carried out in the rockshelter in its entirety and that this RMU was introduced as an unworked cobble. From the seven refits included in this RMU, of important note is an assemblage formed by nine artifacts that correspond to the first reduction stage. The first five flakes were widely scattered throughout zone 6 (J65, M58–59 and O59). One of these flakes was retouched; it is one of the few retouched artifacts forming part of refits. The following two flakes were found in zone 2. Finally, the last two lithics (two fragments of a very large flake) were recovered in zone 4

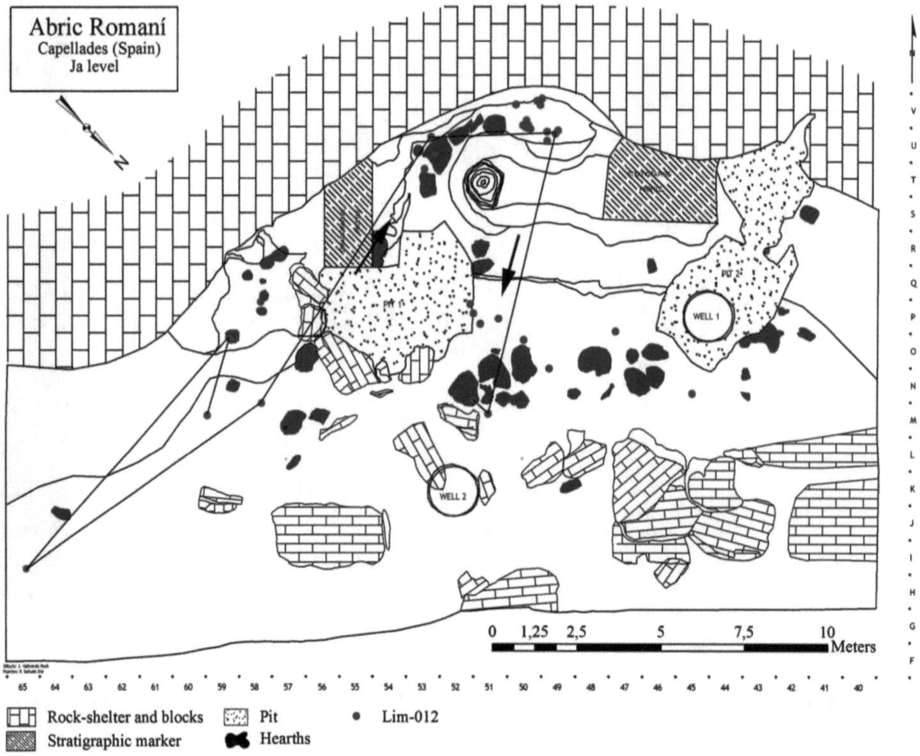

Fig. 7.69 Spatial distribution of Lim-012. *Arrows* indicate the direction of transport

(M51 and N52). The other six refits were located in zone 2, as well as most of the artifacts pertaining to this RMU, which indicates that the main knapping loci were in this zone. The temporal relationship between the artifacts found in zones 2 and 6 seems clear, since remains are spatially segregated according to the chronology of removals. This rules out a postdepositional movement from zone 2 to zone 6. Core reduction was started in zone 6, probably in the area delimited by squares M58–59 and O58–59. The core was later moved to zone 2, where the main part of the reduction sequence was carried out. This refit supports an intentional transport especially because of the counter-slope direction of the movement. The temporal relationship between zones 2 and 4 is more difficult to establish. A new movement of the core to zone 4 may be possible, but refitting data are not conclusive and the movement of flakes previously detached in zone 2 cannot be ruled out. However, the former would be supported by the kind of breakage shown by the flake found in zone 4. This flake was broken by a Siret accident and this breakage is normally produced when the flake is detached from the core.

The following are the other refits connecting zone 2 with the rest of the site:

- Chert-002B. This is a production refitting made up of three small and medium flakes, two of them located in zone 2 (U52 and V50) and the third in zone 4 (Q51). It seems clear that the movement is from zone 2 to zone 4, but the possibility of postdepositional movement cannot be excluded in this case. Other than these flakes, no more artifacts have been attributed to this RMU.
- Chert-009. This is a production refit formed by two very small flakes conjoined with the core (Fig. 7.70). Flakes were found in zone 6 (K58 and L57) and the core was located in zone 2 (V50). This RMU is made up of 25 artifacts, most of them small and very small flakes and flake fragments (n = 20). With the exception of the core discovered in zone 2, all the artifacts were in zone 6, showing a clustering in L57 and M58. These data suggest a movement of the core from zone 6 (probably from the area around L57) to zone 2. Unintentional movement can be ruled out, as the direction of movement is upslope. The second refit forming part of this RMU also shows a movement against the slope: three small and very small flakes located in L58, M58 and O56 conjoined with a core found in P58, in the raised area near the shelter wall. Neither of these two cores was exploited after movement.
- Chert-014. This is also a production refit formed by three small flakes widely scattered across the shelter, in zones 2 (U49), 4 (L50) and 6 (I61). Two of these flakes were located in the northern part of the site and the remaining artifacts belonging to this RMU were clustered in zone 6 (Fig. 7.71). With the exception of the conjoined flake, no more remains of this RMU were found in zone 2. This suggests that zone 2 was again the arrival place of the movement and therefore intentional transport is likely.

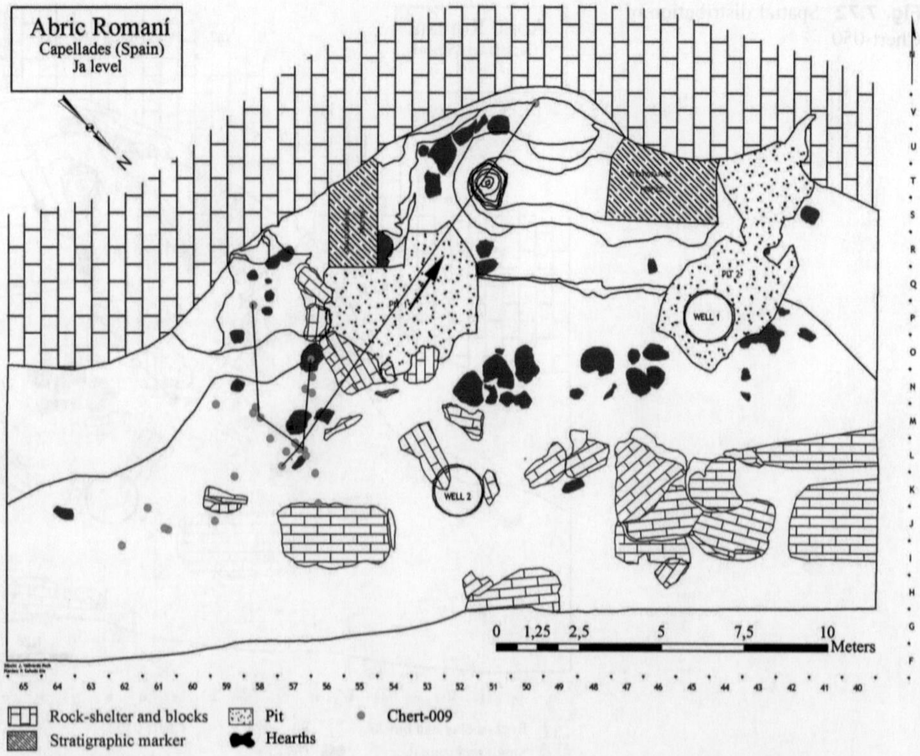

Fig. 7.70 Spatial distribution of Chert-009. *Arrow* indicates the direction of transport

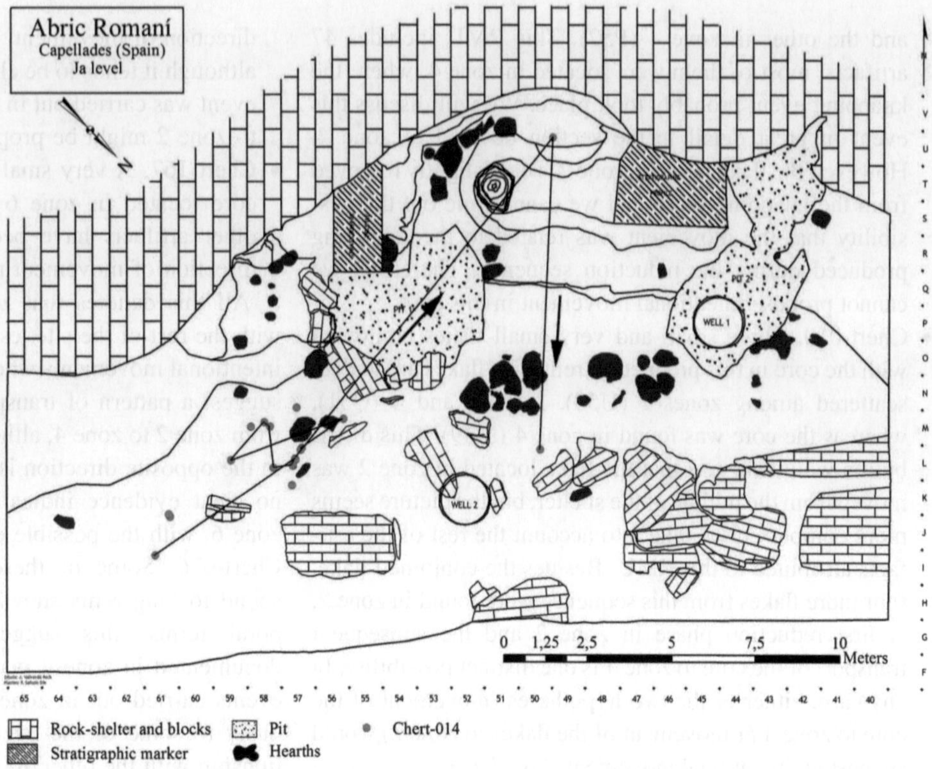

Fig. 7.71 Spatial distribution of Chert-014

- Chert-033. The two artifacts forming this production refit were located in zone 4 (P51) and zone 2 (V50). One of them is the abovementioned denticulate with evidence of hideworking. No more artifacts have been associated to this RMU and therefore direction of movement could not be established.
- Chert-047. This is a production refit formed by two small and very small flakes, one of them located in zone 2 (S53)

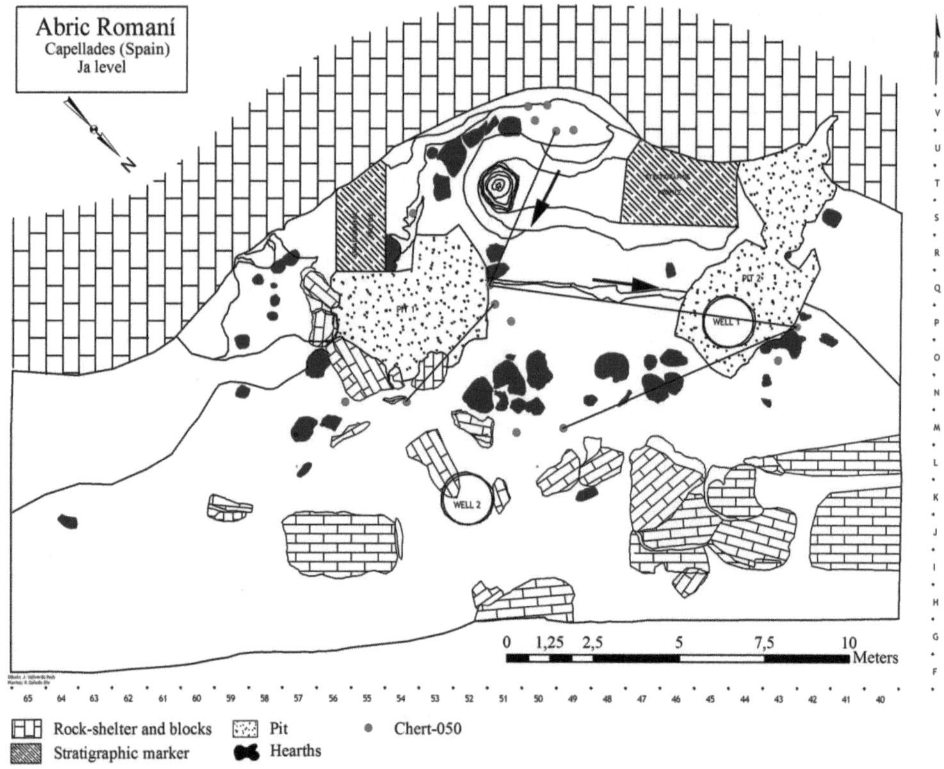

Fig. 7.72 Spatial distribution of Chert-050

and the other in zone 4 (P52). This RMU includes 47 artifacts; most of them were located in zone 4, where the knapping event probably took place. We will discuss this event in great detail in the section devoted to zone 4. However, the flake found in zone 2 is only slightly removed from the knapping focus and we cannot rule out the possibility that the movement was related to the scattering produced during the reduction sequence. Therefore, we cannot propose intentional movement in this case.

- Chert-050. Three small and very small flakes conjoined with the core in this production refit. The flakes are widely scattered among zones 2 (U50), 3 (P43) and 4 (Q51), whereas the core was found in zone 4 (M49). This distribution would suggest that the flake located in zone 2 was moved from the middle of the shelter, but the picture seems more complex if we take into account the rest of the artifacts attributed to this RMU. Besides the conjoined flake, four more flakes from this sequence were found in zone 2. A first reduction phase in zone 2 and the subsequent transport of the core to zone 4 is one distinct possibility. In any case, either of the two hypotheses (movement of the core to zone 4 or movement of the flakes to zone 2) would suggest an intentional movement (Fig. 7.72).
- Chert-052. This is a production refit formed by a small flake located in zone 2 (V50) and a medium flake found in zone 6 (J61). It was not possible to establish the direction of movement. This RMU shows a wide scatter, although it tends to be clustered in zone 4. If the knapping event was carried out in this zone, an intentional transport to zone 2 might be proposed.
- Chert-167. A very small flake recovered in U53 and the core located in zone 6 (K59) form this refit. Since no other artifacts have been attributed to this RMU, the direction of movement remains unknown.

All this denotes that zone 2 had repeated connections with the rest of the site, especially with zones 6 and 4. The intentional movements whose direction has been established suggest a pattern of transport from zone 6 to zone 2, and from zone 2 to zone 4, although some intentional movement in the opposite direction is also possible. However, there is no clear evidence indicating movements from zone 2 to zone 6, with the possible exception of the large flake from Chert-076. Some of these intentional movements correspond to long refits showing a NE–SW direction. In temporal terms, this suggests that the technical events documented in zone 2 occurred after at least some of the events carried out in zone 6 and before or contemporaneously to some of the events of zone 4. A temporal relationship with the other zones cannot be established. Due to the limited scattering of the knapping episodes performed in zone 2, analyzing the intrazonal temporal pattern proved impossible.

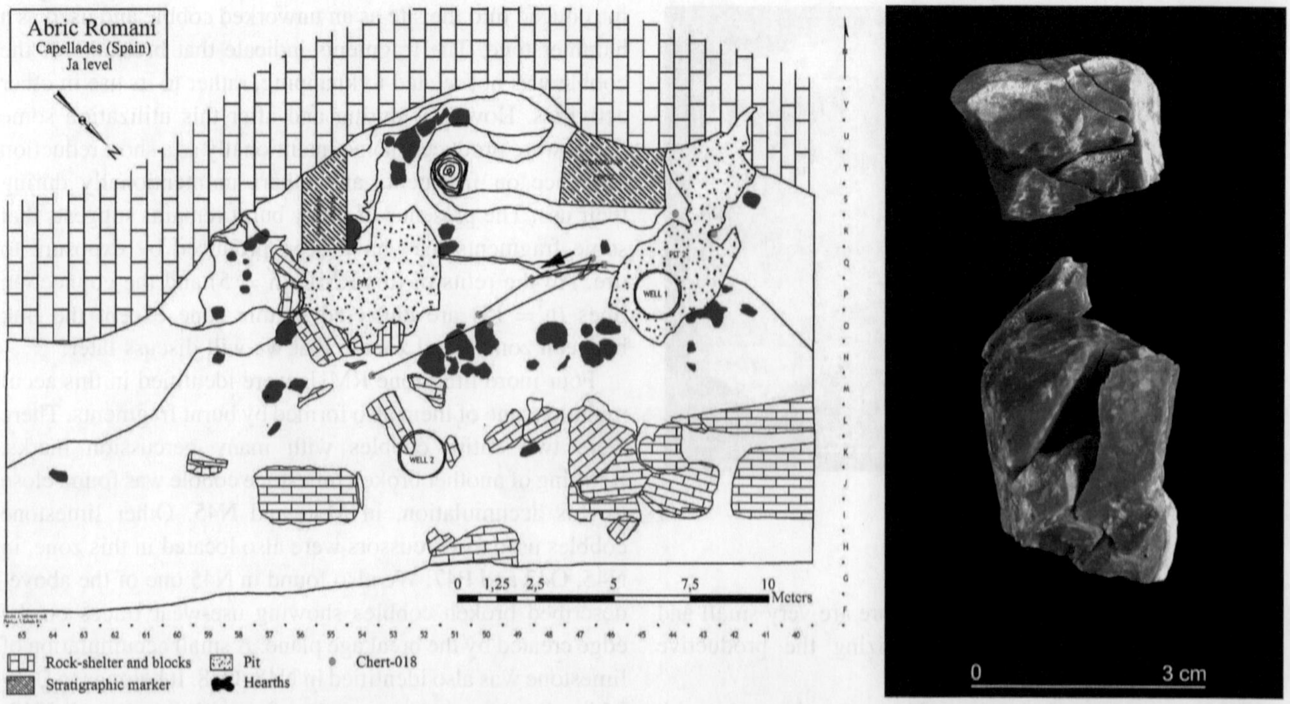

Fig. 7.73 Refitting and spatial distribution of Chert-018

Zone 3

Despite the scattered appearance of the lithic distribution in zone 3, it is possible to recognize some small clusters. However, some of these clusters are probably the result of the discontinuity caused by Pit 1 and they are hardly interpretable in behavioral terms. The clusters focused in O43 and O47 may have originally been the ends of a single accumulation cut off by these former excavations. Both refits and RMU scatters indicate that some sequences are represented on either side of the pit. The cluster focused in L44–45 seems to have more behavioral meaning. This accumulation differs from the other lithic clusters identified in sublevel Ja because of the dominance of limestone remains.

We have found 75 connection lines corresponding to 45 refits and there are 125 RMUs focused in this zone. More than half of them (n = 75) correspond to transported blanks. Use-wear analysis has identified three artifacts with evidence of use: a denticulate used in hideworking and two flakes used for cutting meat/bone. The three used artifacts are transported blanks. Only seven RMUs can be attributed to complete or nearly complete core reduction sequences. Partial reduction sequences are slightly more abundant (n = 9). In this zone we found one of the few refits corresponding to a retouch event: a broken denticulate located in O47 that was linked to a retouch flake found in Q46.

In general, reduction sequences corresponded to small nodules that provided relatively few blanks. The two most productive RMUs are made up of 18 and 20 artifacts. Among the core reduction sequences, two RMUs are especially interesting, because they correspond to good quality chert and show a similar reduction strategy. On the one hand, Chert-041 is a reduction sequence aimed at producing small and very small flakes (n = 16) from a small core that exhibits a hierarchized structure with a flat flaking surface (Fig. 7.68). As stated earlier, these cores show some characteristics of the Levallois strategy. On the other hand, the reduction sequence of Chert-018 has a core with a similar structure, although in this case the nodule was even smaller (Fig. 7.73). This set is formed by nine artifacts whose raw material is an exogenous chert of excellent quality. Both sequences finished with the detachment of a flake that covers most of the flaking surface, which gives the core an appearance typical of the lineal Levallois strategy. Due to the small size of the flaking surface, these last invasive flakes were *débordant*. The last flake of Chert-018 was retouched and transformed into a denticulate. It is one of the few retouch events made from a blank detached in the rockshelter. We will comment later on the spatial significance of this refit. Other sequences found in this zone also exhibit an exhaustive exploitation of the cores, such as Chert-168, formed by two flakes that refit on the core

Fig. 7.74 Refitting of Chert-168

(Fig. 7.74). Both the flakes and the core are very small and demonstrate the attempt at maximizing the productive capabilities of the core.

However, this maximizing trend is apparently at odds with other RMUs documented in the same zone that suggest a waste of lithic resources. For example, Chert-066 is made up of only two artifacts: an almost entire nodule of good quality and the sole flake detached from it. This core was rejected long before its productive capabilities were exhausted.

Five quartz RMUs are represented in this zone. Only two of them (Qu-001 and Qu-002) seem to correspond to nearly complete reduction sequences; the other three are composed of isolated knapping products. There were four refits and four connection lines. There are 19 limestone RMUs concentrated in this zone, but reduction sequences are almost non-existent. Only Lim-042 shows a short and expedient knapping episode on a small cobble. Most limestone RMUs correspond to cobbles used as hammerstones, although five transported blanks have been also found. We have documented 30 connection lines (production n = 2, break n = 28) corresponding to 19 refits. Eight connection lines are interzonal and 22 intrazonal. Seven relate zone 3 to zone 4 and one links zone 3 to zone 6.

One of the most outstanding features of the lithic distribution in zone 3 is an accumulation of limestone remains focused in L44–45 and M44–45. Limestone makes up 59.5% of the lithic remains found in this area, whereas chert is responsible for 37%. Moreover, no chert reduction sequences have been clearly identified in this area and most chert RMUs correspond to transported artifacts. The bulk of the limestone remains has been attributed to the same RMU (Lim-004), formed by 63 artifacts, most of them located in this area or in neighboring squares, although some items have been found in other zones of the shelter (one in zone 4 and another in zone 6) (Fig. 7.75). This RMU was introduced into the site as an unworked cobble and used as a hammerstone. The fragments indicate that breakage of the cobble was not related to knapping, rather to its use in other activities. However, during and after this utilization some flakes were produced, some intentionally as a short reduction sequence on fragments and others unintentionally during their use. The presence of many burnt remains suggests that some fragments may have been produced by exposure to fire. All the refits of this RMU (n = 5) and the connection lines (n = 15) are clustered in this zone except the link between zone 3 and zone 6 that we will discuss later.

Four more limestone RMUs were identified in this accumulation, one of them also formed by burnt fragments. There were two entire cobbles with many percussion marks. Refitting of another broken limestone cobble was found close to this accumulation, in M46 and N45. Other limestone cobbles used as percussors were also located in this zone, in N45, O42 and P47. We also found in N45 one of the above-described broken cobbles showing use-wear traces on the edge created by the breakage plane. A small accumulation of limestone was also identified in N48–O48. It belongs to Lim-054 and was formed by six items (five in O48 and one in N48) in zone 3 and only one in zone 4 (N52). They are all fragments of a broken hammerstone with clear percussion marks. These data indicate that the use of limestone cobbles in percussion tasks was fairly common in this zone.

Among the 75 connection lines documented in zone 3, 43 are intrazonal and 32 interzonal. However, most of the latter (n = 28) are relatively short connections with zone 4 corresponding to RMUs concentrated around the border between zones 3 and 4. They therefore yield little information on the intentional transport of artifacts and the spatial–temporal relationships between the different zones of the site. In addition to zone 4, zone 3 is also connected by refits with zones 1 (n = 1), 5 (n = 1) and 6 (n = 2). The connection with zone 1 was explained in the section devoted to that zone. The following are detailed descriptions of the longest refits (>5 m).

- One of the previously described reduction sequences (Chert-018) shows long distance refits (with a maximum length of 743 cm) with zone 4. The artifacts forming this refitting are widely scattered in zones 3 and 4 (Fig. 7.72). The core and two flakes were found in Q46–47, which probably constituted the main knapping area. Meanwhile, the denticulate and a retouch flake were located in zone 4 (P49–50), which suggests that the retouch event occurred in this zone. The longest connection line links the core (Q47) with a flake recovered in M54. Some artifacts were in zone 1 (R43 and S44), although none of them has been conjoined with the main refit. These data indicate that two technical events can be spatially distinguished: a core reduction event focused around Q46–47 and a retouch event located in P49–50. However, intentional movement can hardly be proposed for the flake in M54.

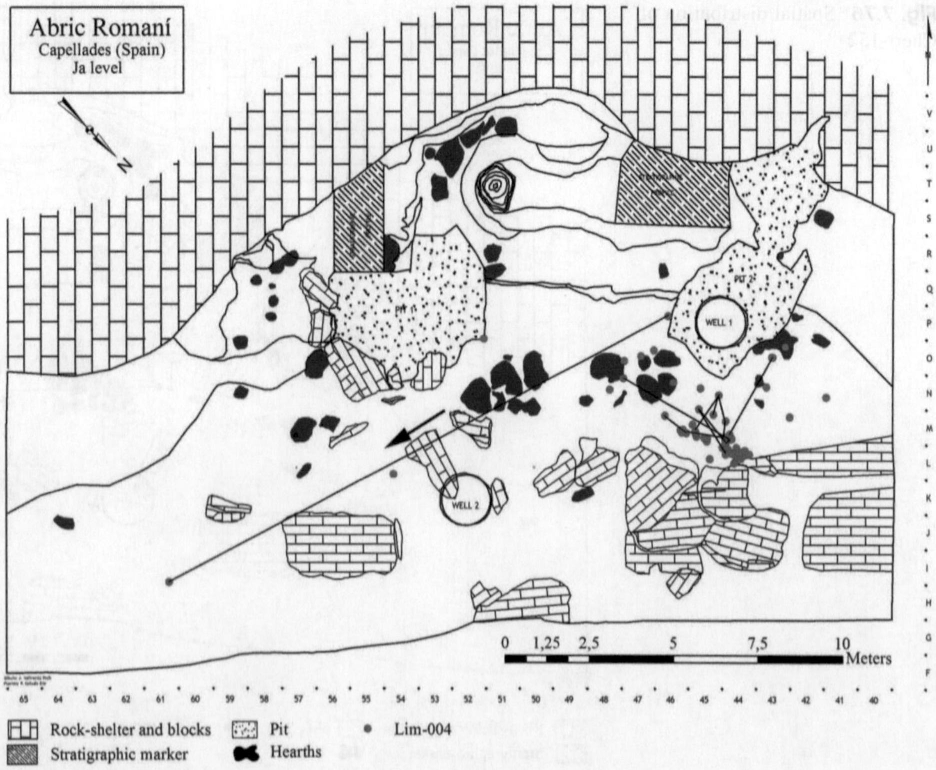

Fig. 7.75 Spatial distribution of Lim-004

- Chert-152 was associated with a production refit connecting a small flake found in N44 and a medium flake in N53 (Fig. 7.76). Other artifacts from this reduction sequence were located in N44–45, which suggests that this was the knapping setting. The direction of the movement was therefore from zone 3 to zone 4.
- A long breakage refit (744 cm) of two quartz fragments links zone 3 (R46) and the northern end of zone 4 (L50). There is a small accumulation of remains corresponding to this RMU in L50, while quartz remains are scarce around R46. This seems to indicate that movement from zone 4 to zone 3 is more likely. It is a counter-slope movement, which suggests intentional transport (Fig. 7.77).
- Another quartz refit connects zone 3 (N45) with zone 4 (N50). It is also a breakage conjunction formed by two very small flake fragments with a 504 cm long connection line. This RMU (Quartz-004) is widely scattered over zone 3 (artifacts in M40, M48, N45 and Q48) and a movement to N50 seems likely. Unintentional movement, however, cannot be ruled out. These two refits correspond to a blank broken by a Siret accident and this kind of breakage is normally produced during flake detachment.
- A third quartz refitting made up of eight artifacts links zones 3 and 4. Seven of these artifacts were found in zone 4, clustered around N51–52 and P51 (Fig. 7.78). As we will see later, this area is the main knapping setting of quartz in sublevel Ja. The eighth artifact (a very large quartz fragment) was located in O42, which represents an 832 cm long connection line. It seems clear that movement occurred from zone 4 to zone 3 and was the result of intentional transport.
- Finally, there is another quartz refit showing a long-distance connection (1,115 cm) between zones 3 (Q46) and 6 (M57). It is a production line formed by a large nodule fragment (M57) with percussion marks and a small flake (Q46) (Fig. 7.79). This line demonstrates intentional mobility not only due to the long distance involved (1,115.8 cm), but especially because the flake was detached before the large fragment was used as a percussor.
- Limestone refits also exhibit interzonal connections. Lim-044 is a cobble broken in two fragments, one of them found in zone 4 (N50) and the other in zone 3 (O41), with a 603 cm connection line. Since it is an RMU formed by only two artifacts, it is not possible to ascertain direction of movement (Fig. 7.80).
- Lim-050 suggests a similar movement. It is also a broken limestone cobble presenting two fragments conjoined by an 1125 cm connection line between P51 and O40 (Fig. 7.81). A third fragment, detached at the time of the breakage, was also recovered in P51, which suggests that this was the breakage locus. The direction of movement was therefore from zone 4 to zone 3, which indicates intentional transport. The two pieces located in zone 4 were burned. The other piece in zone 3 was not burned and showed percussion marks posterior to the fracture. These patterns show that the

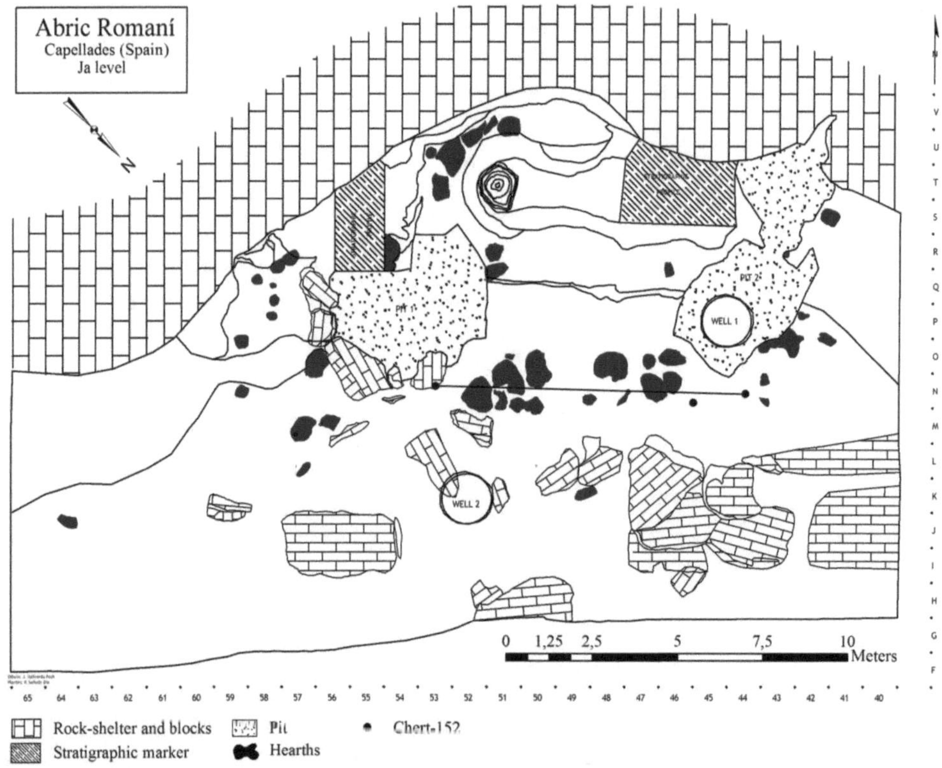

Fig. 7.76 Spatial distribution of Chert-152

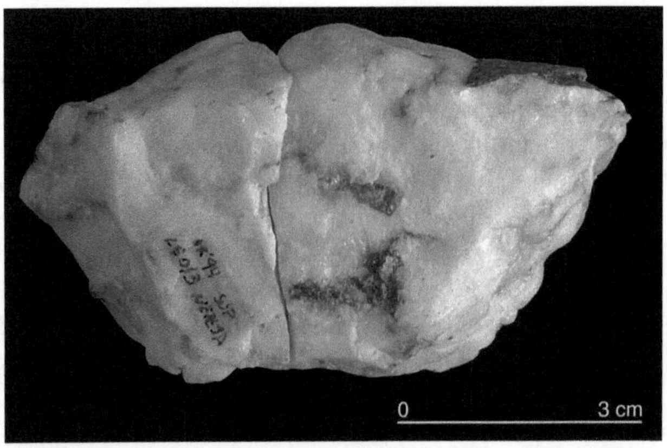

Fig. 7.77 Quartz refitting connecting zone 3 (square R46) and the southern end of zone 4 (square L50)

Fig. 7.78 Refitting of eight quartz artifacts connecting zones 3 and 4. All the remains were located in zone 4, except the largest one (*at the left of the image*) that was recovered in zone 3 (square O42)

fragment was moved to zone 3 before the burning episode and used within this zone as well.

- Lim-008 had one fracture connection line that links zone 3 to zone 4 (Fig. 7.82). The direction was clearly from zone 3 to zone 4. This RMU was focused in zone 4 except for one piece found in zone 2 (V50) and one in zone 3 (N48), which belongs to a refit (n = 5) located in zone 4. Although the distance is in the normal dispersion range (370.13 cm), this connection line between N52 and N48 suggests intentional displacement. This RMU was introduced as an unworked nodule. It was fragmented during knapping use in N52 and a large fragment was moved to N48 and used again as a hammerstone, as its percussion marks occurred subsequent to the breakage.
- Lim-003 was very scattered among zones 3, 4 and 1. It was introduced as an unworked nodule and after breakage some short reduction sequences were carried out. This RMU has a production connection line that links zone 3 to zone 4 (438.71 cm). It is formed by a fragment in O51 and a core in O47. The bulk of this RMU is located in zone 4; direction of movement may have been from zone 4 to zone 3 (Fig. 7.83).
- Finally, another limestone refit connecting zones 3 (P46) and 6 (I61) has the longest connection line found in level

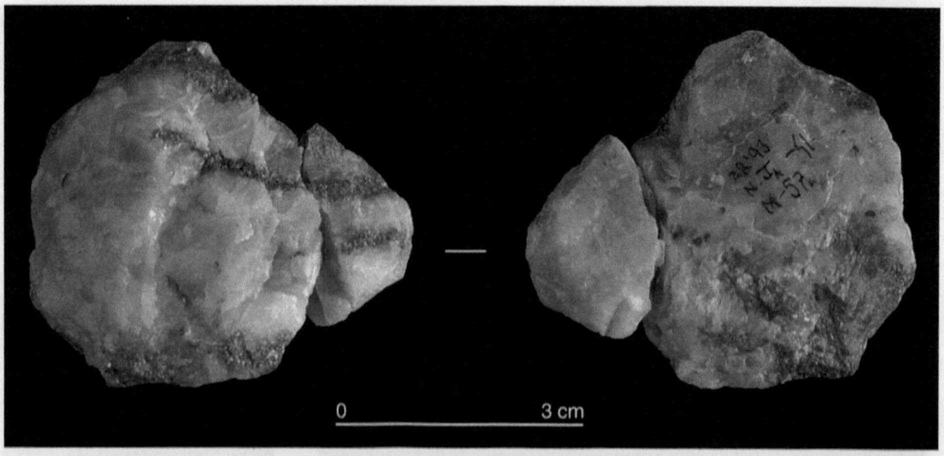

Fig. 7.79 Refitting of RMU Quartz-002 formed by two pieces that show an intentional mobility due to the distance and the location of the percussion marks

Fig. 7.80 Refitting of RMU Lim-044 formed by a cobble broken in two large fragments

J (1,655 cm) and corresponds to the refit of two fragments from Lim-004 (Fig. 7.84). As previously pointed out, most items from this RMU were clustered in the limestone accumulation found in L44-45, but other fragments were widely scattered in zone 3. It therefore seems clear that the movement took place from zone 3 to zone 6. Although this connection follows the direction of the slope, some data suggest that it may have been an intentional movement. On the one hand, we cannot ignore the extraordinary length of this refit, especially if we take into account that the displaced artifact is very large. On the other hand, there is no a generalized scatter of this RMU going in this direction. If the movement was the result of unintentional postdepositional factors, we would expect to find more artifacts from this RMU scattered throughout zones 4 and 6. This, however, is not the case. The artifact in I61 remains an isolated occurrence, while the rest of the RMU is clustered in zone 3.

In this zone, quartz and limestone refits seem to yield more information about interzonal relationships than chert refits. Some of these connections are among the longest refits found in level J. The most robust pattern emerging from these refits is the connection between a sector of zone 3 (O40–42) and the middle of the shelter (N50–52/P50–52). Three refits connect these areas, two of limestone (Lim-044 and Lim-050) and one of quartz (Qu-002). Two of them clearly indicate transport from zone 4 to zone 3. In addition, the kind of artifact transported to zone 3 is the same: large limestone or quartz fragments. It is worth noting that two of these large fragments recovered in zone 3 were attributed in the field to

Fig. 7.81 Refitting of RMU Lim-050 formed by three fragments of a cobble that shows an intentional transport from zone 4 to 3

Fig. 7.82 Refitting of RMU Lim-008. One of its connection lines links zone 3 to zone 4 and shows and intentional displacement due to the percussion marks. They are produce after the breakage

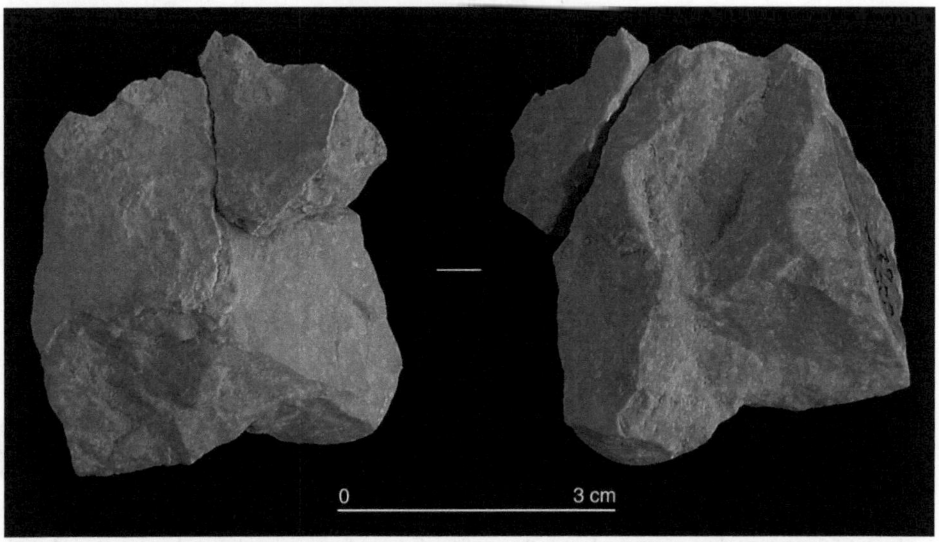

Fig. 7.83 Refitting of RMU Lim-003 with a connection line that links zone 4 to zone 3 formed by a fragment and a core

sublevel Jsup, which indicates that they were found at the top of level J. All these features suggest that these connections correspond to a well-defined event of artifact transport, in which some large fragments initially discarded in zone 4 were moved to zone 3. The stratigraphic location of some of these fragments indicates that this event probably took place during the end of the formation of level J.

Zone 4

This zone shows the highest densities of lithic remains, although the distribution is biased by Pit 1. There is a continuous scatter of remains, but three main accumulations separated by areas of lower density can be identified. They are concentrated in N51–52, N53–54 and P51. One of the goals of the refitting and RMU analysis was to establish the relationships between these accumulations. We have identified 178 RMUs in this zone, more than a half of them (n = 94) attributed to transported artifacts. It is also the zone with the highest number of refits and connection lines. Of the 192 connections, 136 are intrazonal and 56 interzonal. This zone is most linked to other zones, which was predictable given its central position in the rockshelter. In spite of the great number of lithics, the artifacts showing use-wear are not more abundant here than elsewhere. Only

Fig. 7.84 Refitting of RMU Lim-004. It is formed by two fragments and suggests and intentional transport. It exhibits the longest connection-line found in level J. Fragment 1 was located in square P46 and fragment 2 in square I61

four artifacts have been identified: two denticulates and one flake used to cut meat/bone (M54 and O50), and one denticulate used to scrape fresh hide (Q51). Two of these denticulates and the unretouched flake are transported artifacts, and the third has been assigned to Chert-025. This is the zone showing the highest number of complete or nearly complete core reduction sequences. However, there are significant differences between accumulations in terms of reduction sequence type.

Accumulation N51–52. This is formed essentially by quartz remains, most of them from the same RMU (Qu-001 and Qu-002). Both RMUs point to different production sequences realized in this area. There is also one piece from Qu-004, two pieces from Qu-005 and three pieces, including one retouched artifact, from Qu-003. As for other raw materials, knapping on chert is scarcely represented. We have found only one reduction sequence on chert focused in this accumulation (Chert-029). It corresponds to a late stage of reduction aimed at producing very small flakes. From the 39 artifacts associated with this RMU, 35 are very small. The flakes from this sequence display great spatial clustering ($\sigma = 75$ cm) and most of them were located in M51 and N51–52, although there were some remains scattered to the west (M54, M57 and N53) (Fig. 7.85).

It is striking that this accumulation also contains a more widely scattered ($\sigma = 141.5$ cm) limestone RMU (Lim-018). It is a broken limestone cobble whose fragments were distributed among squares N48, N51–54 and O50–52. Other limestone RMUs in this accumulation were also formed by fragments of broken limestone cobbles (Lim-011, Lim-029, Lim-40 and Lim-054). The rest of the fragments in these RMUs are scattered in the rest of zone 4 or in zone 3. Lim-008 shows a cluster in N51–52 (two elements) and another in N53–54 but also has some elements in the P51 accumulation. This RMU was very widely dispersed throughout sublevel Ja but shows two clear clusters in this accumulation (N53–54) and another in P51. The refits are located in each accumulation except one that connects to zone 3 and which is explained above. There is also a worked cobble (Lim-031) in this accumulation. It is a complete cobble with removals all over the distal part/front with clear use traces (see Fig. 7.24). Lastly, we also found some items from Lim-017 clustered mainly in O50–52 but with one element en P51 and two in P47–48 (zone 3). This RMU was introduced as an unworked cobble and a core production sequence was carried out. Some cortical knapping products showed percussion marks indicating its use as a hammerstone at the beginning of the sequence. Some pieces showed different degrees of fire damage subsequent to the knapping event (Fig. 7.86).

Accumulation N53–54. Not including transported artifacts, there were 19 RMUs (most of them chert) concentrated in this accumulation. Three of them seem to correspond to complete or almost complete core reduction sequences and show a similar pattern of spatial distribution (Fig. 7.87):

- Chert-026 is formed by 35 artifacts, essentially small and very small flakes (n = 32). This sequence is fairly dispersed and it is represented in the other two accumulations of zone 4.
- Chert-002A. The 33 artifacts forming this RMU show a clear dominance of small and very small items (n = 28). The raw material is fairly poor for knapping and is characterized by an irregular fracture. Although it is clearly focused in N53–54, there are also quite a lot of remains in the P51 accumulation, as well as some artifacts scattered in zone 6.
- Chert-047. The 43 artifacts of this RMU were also produced from a low quality raw material. It is widely dispersed, although two main clusters can be recognized, one in N53–N54 and the other in P51–52. There also seems to be a preferential dispersion to the west.

Other RMUs focused in this accumulation are characterized by lower degrees of dispersion. For example, Chert-005 is formed by four very small refitted flakes corresponding to a series of unidirectional removals. It indicates a very short knapping event, whose remains were clustered in an area of only 40 cm in diameter. Chert-036 exhibits a similar pattern. It is a production sequence carried out from the ventral surface of a very large flake. Five remains associated with this RMU were clustered in an area 60 cm in diameter. This differs from the dispersion pattern demonstrated by the previously described RMU.

Quartz was very scarce in this accumulation and did not show any clustering. Two limestone RMUs are focused in this zone: Lim-008 (explained in the other accumulation) and Lim-038. The latter was mainly located in zone 4 and showed clusters in the three accumulations differentiated in this zone. It was introduced as an unworked nodule, used as a hammerstone and different production sequences were

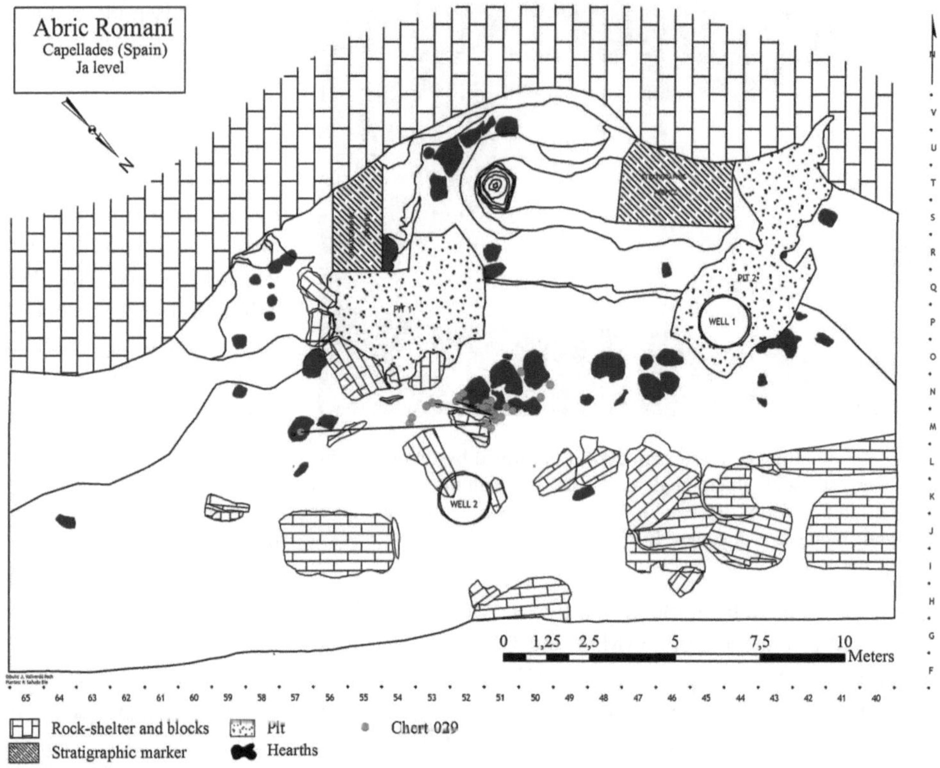

Fig. 7.85 Spatial distribution of Chert-029

Fig. 7.86 Refitting of RMU Lim-017 formed by three elements with different degrees of fire damage

carried out subsequent to fracture. This RMU has a micritic texture making it very suitable for knapping. One refit links zone 4 (O50) to zone 6 (M58) with a break connection line of 832.81 cm. The displacement seems to have been from zone 4 to zone 6. One flake fragment in O50 is burned but another one is not, and this burning occurred after the fracture. The fracture may have occurred in O50 and then either intentional or postdepositional displacement moved the other flake fragment to M58. There are also some cobbles with clear evidence of use: a worked cobble (Lim-030) and a cobble fragment with use-wear traces on the edge created by the fracture plane (Lim-058).

Accumulation P51. This is the level J accumulation containing the highest number of technical events. Excluding the transported artifacts, 57 RMUs are concentrated in this accumulation. Most of them are chert RMUs, but there are also 11 limestone units and one quartz unit. One of the main characteristics of this accumulation is the dominance of reduction sequences whose goal was the production of small and very small blanks. In general, they correspond to very fragmented *chaînes opératoires*, mainly represented by the last stages of the reduction sequences. However, few RMUs include the core corresponding to these sequences. Only two (Chert-040 and 012) can be attributed to complete or nearly complete reduction sequences, although they are also aimed at producing small and very small flakes (97.6% of remains from Chert-040 and 89% from Chert-012). Most of these units show a low dispersion radius, with the exception of Chert-012, which is characterized by a distribution pattern similar to that described previously for the complete sequences of accumulation N53–54. Scatters from most RMUs focused in this accumulation are limited to the area of squares O50–51 and P50–51.

This was the only zone of the rockshelter in which significant development of core reduction sequences on limestone cobbles was found. Five of these sequences were focused in this accumulation, while this kind of event is

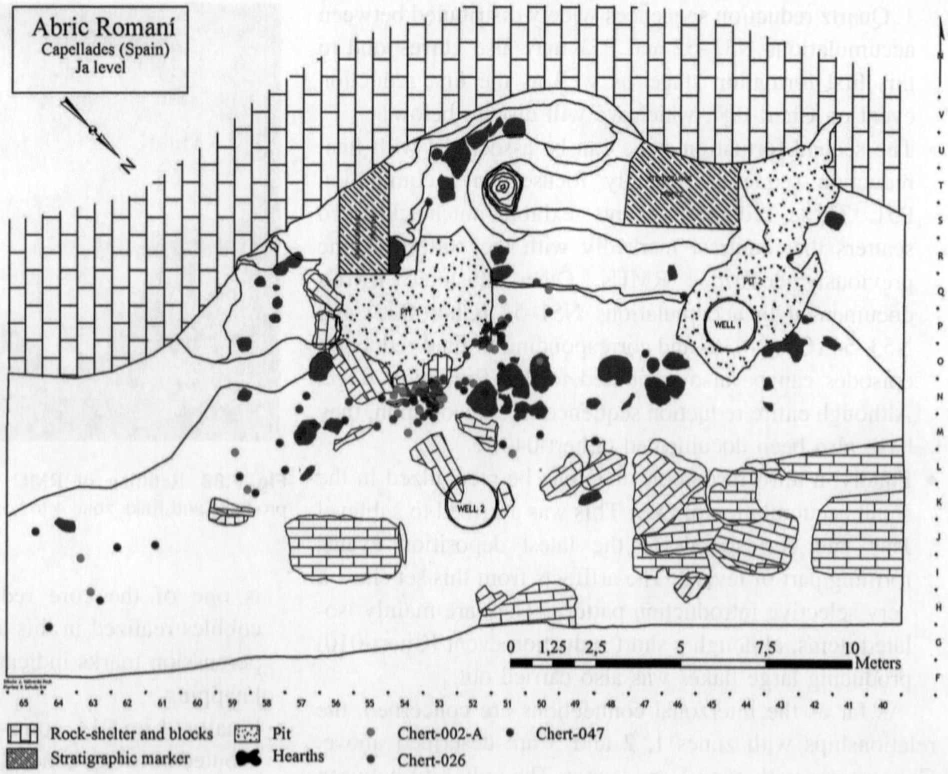

Fig. 7.87 Spatial distribution of Chert-002A, 026 and 047

uncommon or totally absent in the rest of the site, in sublevel Ja as well as in sublevel Jb. These were always very expedient sequences, made on cobble fragments of low quality limestone, which produced an irregular array of products. Unlike most chert RMUs identified in this accumulation, limestone sequences tend to show wider dispersions (Lim-002, $\sigma =$ 356 cm; Lim-003, $\sigma =$ 304 cm; Lim-017, $\sigma =$ 83 cm; Lim-038, $\sigma =$ 247 cm; Lim-052, $\sigma =$ 123.5 cm). The transport of remains towards zone 3 (M42 and P42) zone 6 (L57), zone 2 (U49-50) and zone 1 (S41 and S44) can be proposed for one of these units (Lim-003). As discussed later, other movements of limestone remains exhibit this direction of transport as well, especially to zones 3 and 6. Other limestone RMUs located in this accumulation correspond to whole (Lim-045 and Lim-071) or broken cobbles without evidence of intentional reduction; some of these cobbles show percussion marks (Lim-010, Lim-029, Lim-030, Lim-050) and some knapping products belong to limestone reduction sequences scattered in other zones, especially in zone 3 (ex. Lim-012).

The relationship between these accumulations can be specified a bit further through refits. In terms of intrazonal connections, most are limited to the interior of each accumulation, although there are also several refits between different accumulations. These refits tend to confirm the conclusions derived from RMU dispersion. The quartz remains focused in accumulation N51–52 present several connections with accumulation P51. However, the main chert RMU located in this cluster (Chert-029) shows a refit with square M55, although the rest of connection lines from this unit are much more clustered around the knapping locus. Connections between N53–N54 and P51 are due to the RMU distributed between the two accumulations, which corresponded to complete or nearly complete sequences on low quality chert or limestone. Meanwhile, the RMUs focused in P51 that have generally been attributed to final sequences for producing small and very small blanks show practically no connection with the other accumulations.

It is worth noting an association of remains located in O49 that was attributed in the field to sublevel Jsup. It is made up of eight artifacts, four of them corresponding to Chert-010 and the remaining four to large and very large transported blanks. This association was stratigraphically above the main horizon of sublevel Ja and was one of the few sets of remains found in sublevel Jsup that seems to correspond effectively to an activity event carried out after the formation of sublevel Ja.

Taking these scatter differences as evidence of temporality, three different deposition times can be distinguished in zone 4:

- The first would correspond to the reduction sequences showing more dispersed distributions. These are entire reduction sequences on poor quality materials, mainly distributed between accumulations N53–54 and P51: Chert-002, Chert-012, Chert-026, Chert-047, Lim-002, Lim-003, Lim-008 and Lim-038. Some of these reduction events were probably focused on the area affected by Pit

1. Quartz reduction sequences widely distributed between accumulations N51–52 and P51 may also correspond to this first formation stage, as well as the first reduction event on Chert-007, which we will discuss below.

- The second formation stage can be associated with final reduction sequences, mainly focused in accumulation P51. These reduction events exhibit much clustered scatters that contrast markedly with the pattern of the previously described RMUs. Other clustered RMUs documented in accumulations N51–52 (Chert-029) and N53–54 (Chert-036) and corresponding to final reduction episodes can be also attributed to this formation stage. Although entire reduction sequences are uncommon, they have also been documented (Chert-040).
- Finally, a third formation stage can be recognized in the small accumulation of O49. This was ascribed to sublevel Jsup and corresponds to the latest deposition events forming part of level J. The artifacts from this set show a very selective introduction pattern. They are mainly isolated items, although a short reduction event (Chert-010) producing large flakes was also carried out.

As far as the interzonal connections are concerned, the relationships with zones 1, 2 and 3 are described above. Connections with zone 5 are scarce. The only refit between these zones is formed by two fragments of a limestone cobble. They were located in N49 and K47 and it was not possible to establish the direction of movement. Connections with zone 6 are more common (n = 14); however, most of these are down slope and therefore possibly of an unintentional nature. In general, they correspond to very scattered RMUs. As described earlier, some data suggest that postdepositional movements tend to become more pronounced in the eastern part of the site. However, below we discuss in detail the refits showing the longest connection lines, which may suggest intentional displacement:

- Qu-001 has a refit that proves the movement of a core that was knapped in O51 and P51 where two flakes were detached. The core was subsequently moved and abandoned in N58. This last connection line has a distance of 733.75 cm.
- Qu-002 has a refit of 12 items that connect two of the accumulations of zone 4 (MN51–52 and P51), and zone 4 to zone 6. It is a production sequence realized on a large quartz fragment. The sequence began in P51-Q51, producing fragments and knapping products with a good cutting edge for quartz. There is one fragment that moved to I62, perhaps an intentional displacement but its very small size and the direction of the connection line, which follows the natural slope, make postdepositional displacement a distinct possibility (Fig. 7.88).
- Lim-017 is especially clustered in O50–52, three pieces were found in N51–N52, one in P51, and two pieces in zone 3 (O47–48). It is made up of knapping products and is one of the core reduction sequences on limestone cobbles realized in this accumulation. Some pieces show percussion marks indicating use as hammerstones before knapping.

Fig. 7.88 Refitting of RMU Quartz-002 formed by 12 knapping products and links zone 4 to zone 6 and two accumulations of zone 4

- Finally, Lim-014 corresponds to a broken cobble distributed between zones 4 and 6. The majority of the items are dispersed following the natural slope from MN54–55 to L56–57, apart from three elements in N59, P59, I61 and another one in N50. A short expedient production sequence was later realized on one of the fragments. One refit links these two zones with a connection line of 237.92 cm.

Chert-007 deserves a separate comment, since it is one of the widest scattered of level J. It presents two spatially separated reduction stages, which are characterized by very different dispersion radii. Artifacts from this unit have been recovered in zones 3, 4, 5 and 6 (Fig. 7.89). This RMU was introduced as a complete or nearly complete nodule and the first stages of the reduction sequence show the widest scatter. The artifacts from this stage were clustered around O48, which seems to correspond to the focus of the knapping episode, but some remains were dispersed towards zones 4 and 5. Nevertheless, the end of the reduction sequence, aimed at producing very small flakes, showed a marked cluster in N59. Five of the six very small flakes detached in this terminal stage were in N59 and the sixth in P59. The core was found in the collection from Ripoll's excavation, which indicates that it was moved again to the area of Pit 1.

Zone 5

This zone exhibits the lowest density of lithic remains and RMUs. Only ten RMUs have been identified, six of which correspond to transported artifacts. Of the other four RMUs, one is a reduction sequence on a chert nodule of very low

Fig. 7.89 Refitting of the two reduction stages and spatial distribution of Chert-007

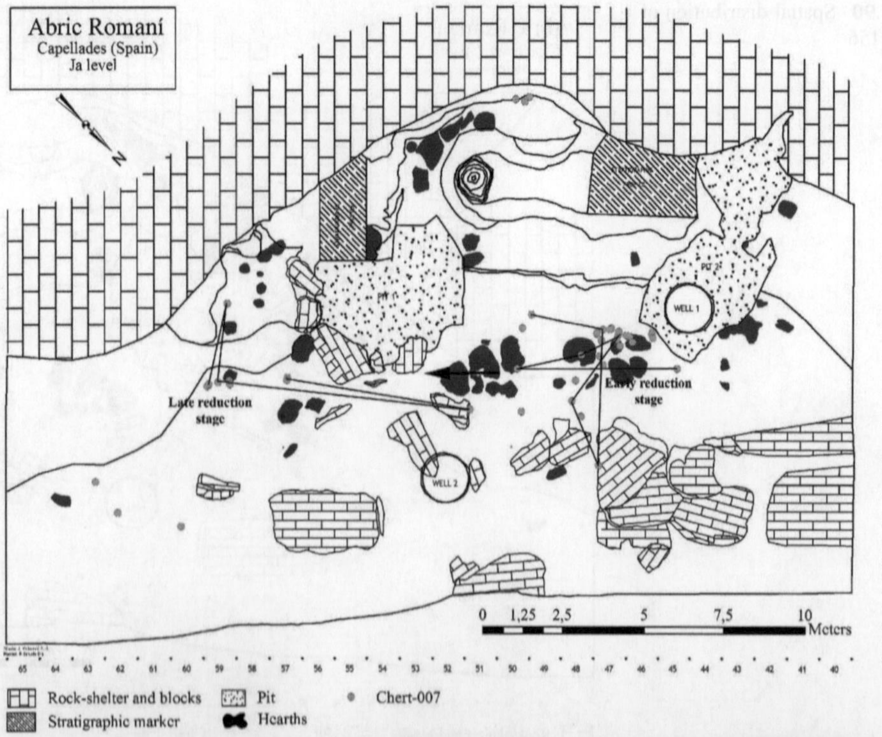

quality and three are limestone cobbles without intentional modifications. The chert knapping event was performed on an entire or almost entire nodule and the remains were scattered in an area delimited by squares I49–51, J48–50 and K50. Although the dispersion radius is relatively high, it is not significantly higher than that documented in other zones. This would support the hypothesis that this zone, although it corresponds to the outer and less protected part of the shelter, was not more affected by postdepositional movements.

Zone 6

Some clusters can be also identified, although they are partly determined by the topography of the surface. For example, the accumulation focused on P58–59 and Q58–59 corresponds to a raised surface separated from the rest of zone 6 by a sudden slope located in row O. However, other small clusters, not immediately associated to topographic factors, have been documented in M59–N59, L57 and I62–

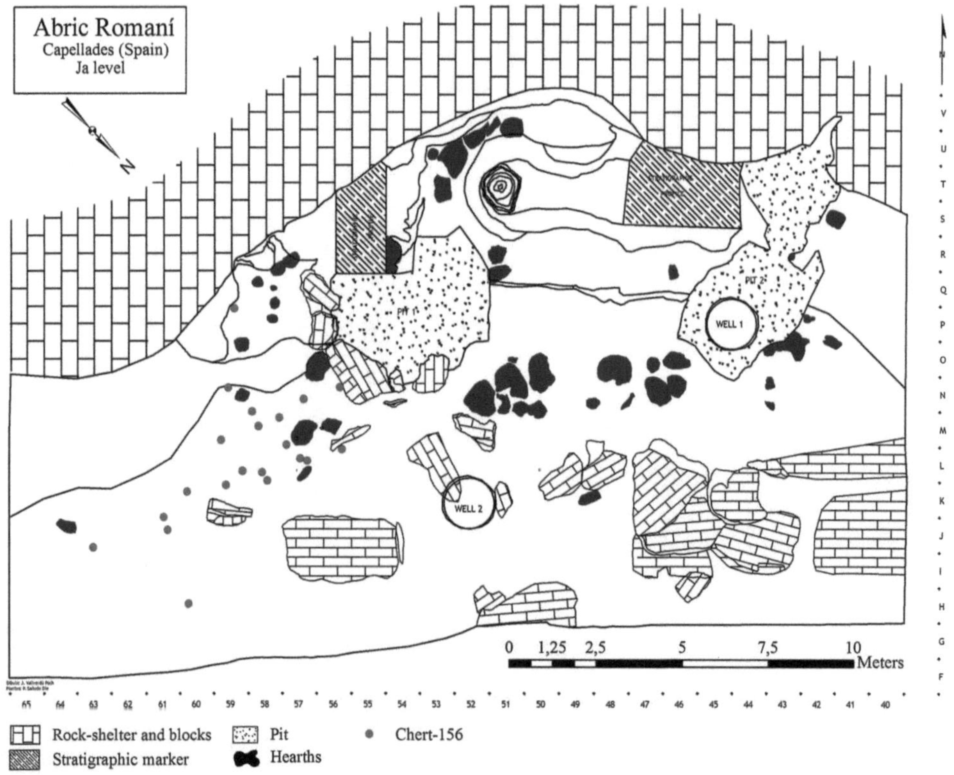

Fig. 7.90 Spatial distribution of Chert-156

J62. We identified 48 RMUs, eight of them attributed to transported artifacts. Besides these transported items, there are 28 chert RMUs and 12 limestone RMUs. In addition, there were also some quartz remains associated to three RMUs focused in other zones. Three denticulates used to cut meat/bone were located in H60, I63 and I65. One of them (H60) also presented another edge used for fresh hide scraping. The RMUs focused in this zone tend to present very wide scatters. With some exceptions, most RMUs had remains dispersed practically throughout the entire zone. In some cases (for example, Chert-156), the dispersion is so marked that it is difficult to even identify the focus of the knapping episode because there are no clusters of remains (Fig. 7.90). The movement of remains from most RMUs towards the NE corner of the excavated surface expresses this dispersion. Nevertheless, we have identified some knapping posts that essentially correspond to the above-cited clusters.

Three RMUs were focused in the I61–J61 accumulation. Two chert RMUs correspond to a nearly complete sequence (Chert-046) and a partial sequence formed by only five artifacts (Chert-120). Both sequences were characterized by a dominant production of small and very small flakes. Qu-002 showed a cluster in I61-63 formed by small and medium knapping products just on the lower part of the natural slope. There is also a limestone RMU represented by a broken cobble with percussion marks. Cobbles and cobble fragments with these marks were relatively common in zone 6, especially in the outer area (J62, K64, J59 and K59). There were two RMUs, Lim-055 and Lim-066, formed by a large cobble fragment with percussion marks that typically indicate that the fracture was produced by percussion on an anvil. This knapping focus suggests that this outer sector was not simply the dispersion area of the sequences carried out up the slope, but that it was sometimes an activity area. Lastly, there is the only piece of Qu-006 (I60), a medium-sized flake that must have been introduced as an isolated piece because no other items of this RMU have been found.

The second knapping center was located around L57. Four chert and two limestone RMUs were concentrated in this accumulation. Among the chert RMUs, there is a nearly complete sequence (Chert-009) and two partial sequences (Chert-070 and Chert-053), all aimed at the production of small and very small flakes. Chert-014 is especially interesting as it shows a differential dispersion according to reduction stage, similar to that of Chert-007. The remains from the first stages are widely scattered and some artifacts have been recovered in zones 2 and 4. The last production event has been refitted and corresponds to the exploitation of a very large flake (Fig. 7.36). Unlike the initial stage, the items from this last event were clearly focused in the L57 accumulation (Fig. 7.71).

The accumulation in M59–N59 includes two knapping sequences, as well as the final stage of Chert-007. There is a complete reduction sequence from low quality chert (Chert-019) and the last phase of a production episode aimed at the

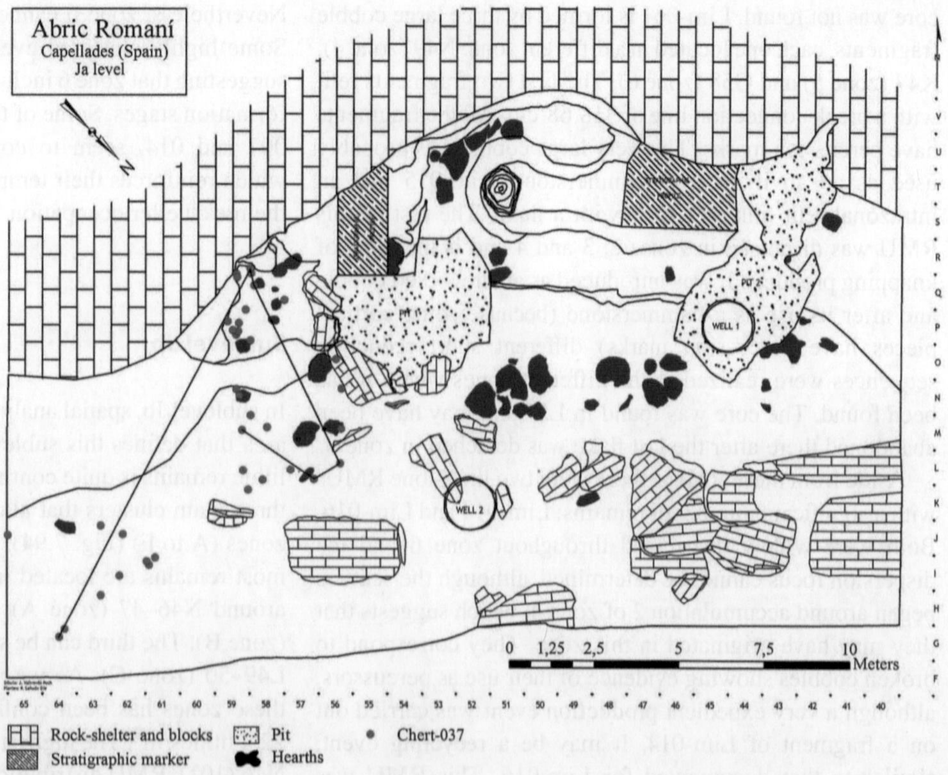

Fig. 7.91 Spatial distribution of Chert-037

detachment of small flakes (Chert-016). The former was widely dispersed ($\sigma = 269$), with artifacts scattered throughout zone 6 and accumulation 2 of zone 4. On the other hand, the latter tended to be more clustered around the main focus ($\sigma = 110$), although there is a preferential dispersion of some remains towards the southern squares (P58–59). This distribution is similar to that of the last knapping episode of Chert-007. Qu-001 showed a cluster in M58–59 and N58–59 that suggests a short reduction sequence. It was formed only by knapping products, one core and a retouched artifact. A refit was carried out that linked this zone to zone 4 (see above).

A quartz cluster (Quartz-002) formed by small and medium-sized fragments and knapping products was located in M56–57. This scatter is linked to one piece in Q46 with a connection line of 1115.79 cm. This line connects a fragment and a flake produced during percussion activity. The movement may indicate intentional displacement from M57 to Q46, because the remaining pieces are located in the cluster and seem to be related to the pieces of this RMU located in MN54–55.

The last knapping locus of zone 6 is focused in P58–Q58, in the raised surface close to the shelter wall. There are five chert reduction sequences, three corresponding to complete or nearly complete sequences and two to partial sequences. On the one hand, three of them show a clear northeastward dispersion, and some items were found in the NE end of the excavated surface ($\sigma = 281$, 237 and 329 cm). A good example of this wide scattering is the refitting of Chert-037, which is formed by artifacts located in G64, L56–57, M59, Q58 and R58 (Fig. 7.91). Chert-015 has a refit that links a flake located in Q57 with one in K64. A similar connection can be seen for Chert-057 between a flake located in P58 and the core found in I64. On the other hand, two sequences were characterized by considerably lower dispersion radii ($\sigma = 76$ and 48 cm), and most of their remains were clustered around the activity focus. Limestone artifacts were scarce in this accumulation (n = 8), although there was one transported artifact, an item from Lim-012 and an artifact from Lim-014, which we will discuss further below. Only Lim-038 showed an association of different remains belonging to the same RMU, although this was ascertained with a breakage refit and there is not enough evidence to suggest a knapping event. This refit exhibits the same long-distance connection as the chert RMU described above: two artifacts located in P58–Q58 conjoined with another fragment found in K63 (720.42 cm connection line). In any case, if we take into account that this RMU was focused in zone 4, the intentional transport of a limestone fragment from zone 4 to zone 6 may be inferred.

In general, the limestone RMUs located in zone 6 were characterized by wide scatters. Among these, it is worth noting the initial knapping episode of Lim-012, already described in the section on zone 2 above, which is dispersed throughout zone 6 without any clustering. Lim-043 was scattered in this zone and was the result of a production sequence. It was formed only by knapping products and the core was not found. Lim-061 is formed by three large cobble fragments, each one located in a different zone: N49 (zone 4), K47 (zone 5) and G58 (zone 6). The first two fragments refit with a break connection line of 318.68 cm. All the fragments have percussion marks. This very large cobble was probably used as an anvil or as a hammerstone. Lim-015 had an intrazonal refit linking a core with a flake. The rest of this RMU was dispersed in zones 2, 3 and 4 and is made up of knapping products. It was introduced as an unworked cobble and after its use as a hammerstone (because some cortical pieces have percussion marks) different short reduction sequences were realized in the different zones where it has been found. The core was found in L57 and may have been abandoned there after the last flake was detached in zone 6.

Aside from these RMUs, there were two limestone RMUs with a significant number of remains: Lim-014 and Lim-016. Both were widely distributed throughout zone 6 and the dispersion focus cannot be determined, although the scatters began around accumulation 2 of zone 4, which suggests that they may have originated in this zone. They correspond to broken cobbles showing evidence of their use as percussors, although a very expedient production event was carried out on a fragment of Lim-014. It may be a recycling event, similar to that documented for Lim-016. This RMU was made up of 16 elements distributed between zones 4 and 6 (Fig. 7.92). They were mainly fragments, three knapping products and a core. It includes a refit that clearly expresses the dispersion pattern of this unit, since it conjoins artifacts located in J62, K58, K61, M59 and P52. One of these fragments presented percussion marks that extended over the fracture plane, which indicates that the use as percussor was subsequent to the breakage event. Another two artifacts of this refit were burnt, while the rest of the set did not show any evidence of fire damage. This also indicates a temporal succession of at least three different events: the cobble breakage, the spatial dispersion of the fragments derived from this breakage, and the exposure to fire of some fragments. A new dispersion event can be also proposed, since the burnt fragments were not located inside a hearth.

As a whole, zone 6 is characterized by the wide dispersion of most lithic scatters. This can be partly explained by the topography of this area, which is characterized by a marked slope to the NE. As stated earlier, dispersion processes seem to have been more intense here than in other zones. However, temporal positioning in the assemblage formation may also be a factor in this spatial pattern. Some of the events represented in this zone show an early location in this formation process. Connections through refits with zone 2 indicate that some reduction events from zone 6 were carried out before those from zone 2. The first reduction stage of Lim-012 is a good example of this temporal schedule. The deep temporal development exhibited by Lim-016 is another example. Nevertheless, zone 6 cannot be interpreted as a single unit. Some highly clustered events have been also documented, suggesting that zone 6 includes events carried out in different formation stages. Some of these clustered events, like Chert-007 and 014, seem to correspond to recycling episodes, which reinforces their temporal setting in the late phases of the rockshelter occupation.

Sublevel Jb

In sublevel Jb, spatial analysis is limited by the small surface area that defines this sublevel. Although the distribution of lithic remains is quite continuous (Fig. 7.93), we can discern three main clusters that allow us to distinguish six different zones (A to F) (Fig. 7.94). The two accumulations with the most remains are located in the middle of the site: the first around N46–47 (zone A) and the second around N49–50 (zone B). The third can be seen close to the line of blocks, in L49–50 (zone C). As we will see below, the definition of these zones has been confirmed by refits. Mean density is 22.4 lithics/m^2. The highest values are seen in N47 (160) and N46 (101). RMU distribution fits this clustered pattern. Most core reduction sequences are focused in zones A and B. Meanwhile, in zone C most remains correspond to the same RMU. Therefore, zones A and B may be characterized as knapping areas where several core reduction episodes were carried out, while the cluster of zone C was primarily formed by a single knapping event. Mean density rapidly decreases eastward of column 52, although this may be due in part to the difficulties in identifying sublevel Jb beyond column 54. This western area has been designated as zone D. The low density area westward of column 45 is zone E. Finally, the isolated area in the SW corner of the excavated area (R41–42, S41–42 and T41) forms zone F, although its correlation to sublevel Jb raises some doubts.

In terms of the content of the three principal accumulations, the main differences correspond to zone C, which is characterized by higher percentages of small and very small remains. Most of the remains of this accumulation stem from the same knapping event. Zones A and B, however, had very similar contents in terms of size of remains and artifact class. A higher proportion of large and very large remains, especially flakes and retouched artifacts, characterized zone D. This pattern was also observed in the low-density areas of sublevel Ja and is related to the scarce development of in situ knapping sequences.

Unlike sublevel Ja, we have not identified production sequences on limestone. All the products come from the fracture of unworked nodules during percussion and/or use as hammerstones. There are two flakes located in zone D

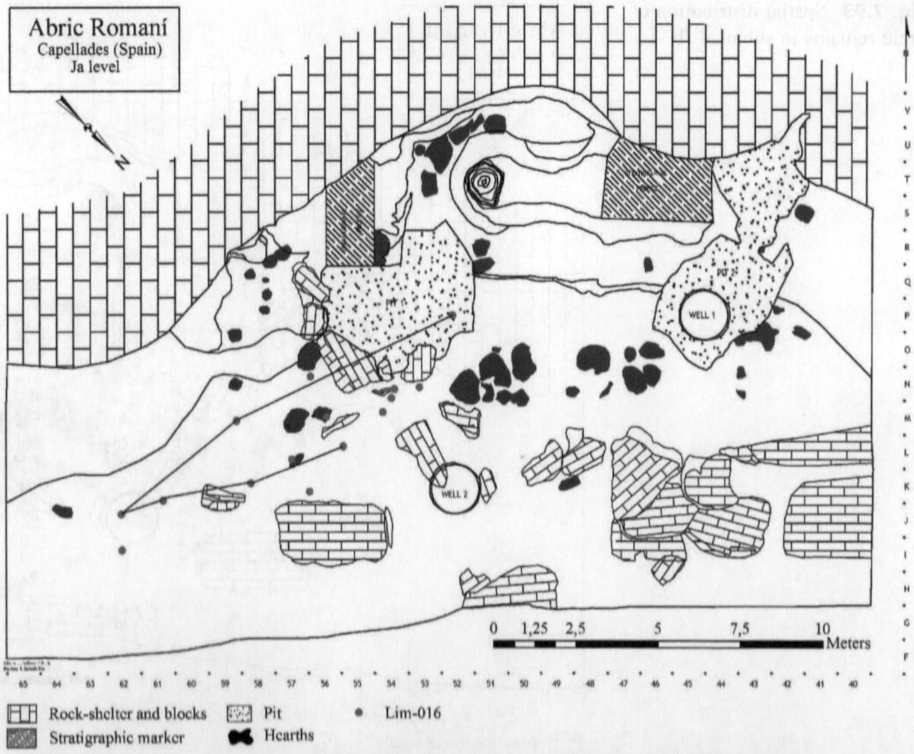

Fig. 7.92 Refitting and spatial distribution of RMU Lim-016 that shows a temporal succession in at least three different events in various zones of the rock-shelter

(O53 and M53) that were introduced as isolated artifacts. The rest are unworked nodules with percussion marks and fragments originating from them as well as some fragments that resemble knapping products but whose percussion marks, especially on the butt, indicate that they stem from fractures produced during percussion.

Zone A

Most knapping episodes identified in sublevel Jb were concentrated in zones A and B. Nevertheless, refits and RMU distribution indicate that they correspond to discrete activity areas. Connections between these zones were

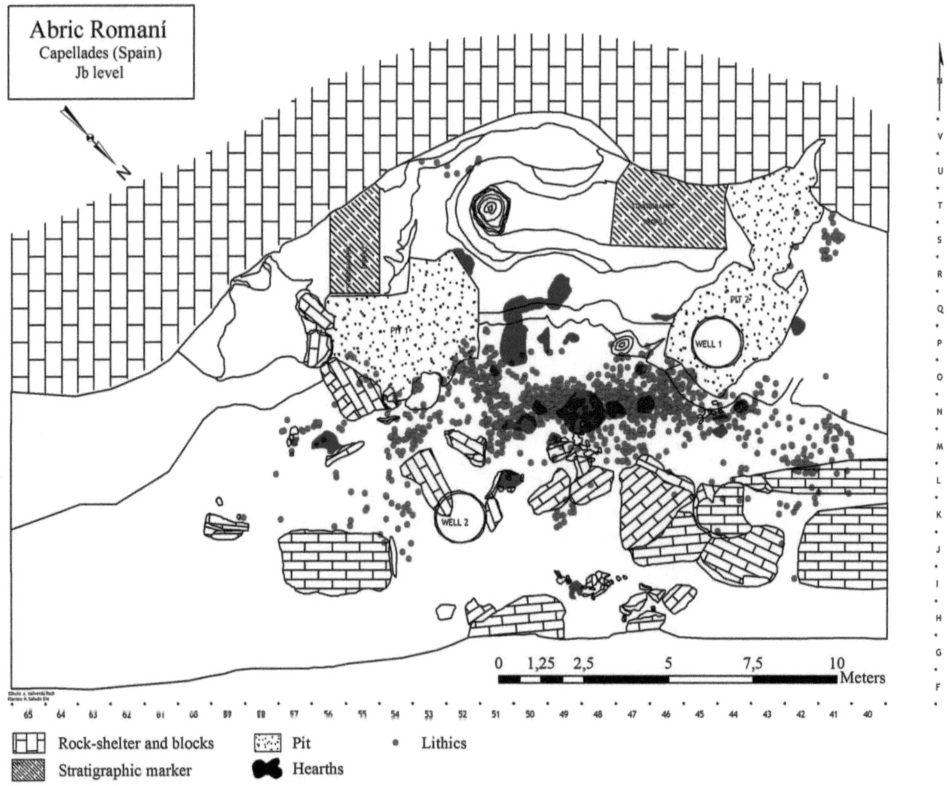

Fig. 7.93 Spatial distribution of lithic remains in sublevel Jb

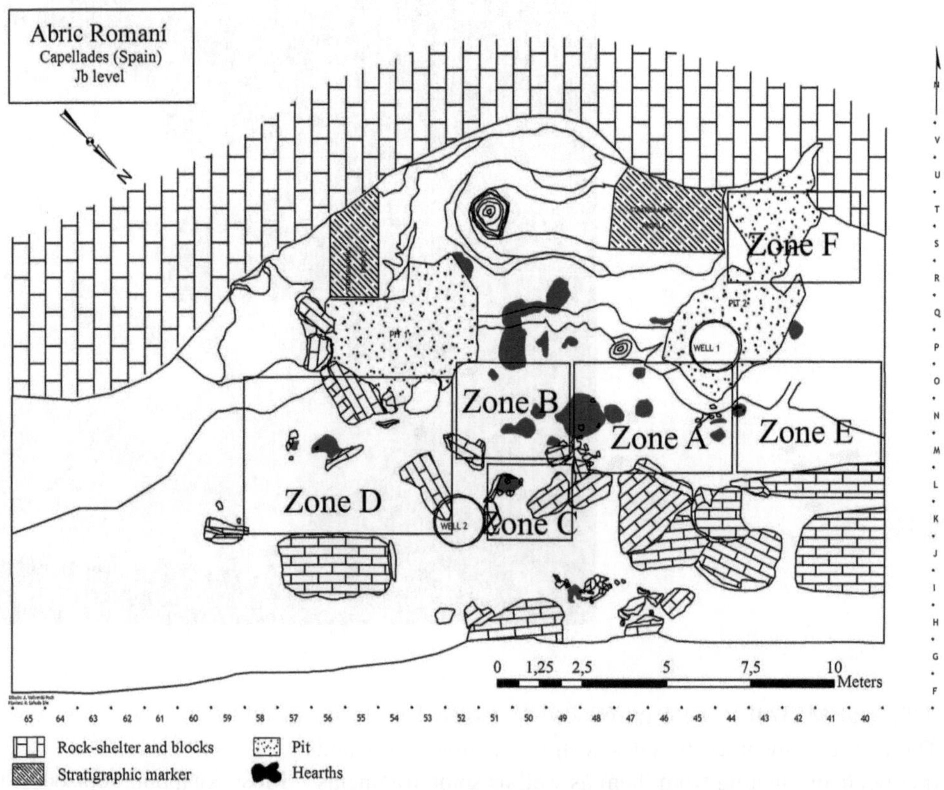

Fig. 7.94 Zones defined for the presentation of spatial data in sublevel Jb

uncommon and those that were found can be attributed to post-depositional factors or the normal dispersion of items during knapping. 38 RMUs were identified in zone A, 19 of them corresponding to transported artifacts. One of the most outstanding features of this accumulation is the relatively high percentage of chert RMUs that can be attributed to complete or almost complete sequences. Six of these sequences have been found, forming the bulk of the lithic assemblage in zone A. They show very similar dispersion radii, between 80 and 117 cm. The sole unit that exhibited a wider scatter (Chert-05) corresponded to the only one affected by intentional movement towards zone E. One denticulate and two flakes with use-wear traces were found. Two of them have been associated with meat-bone cutting and the third one with woodworking (this is the only artifact that showed evidence of this activity). One of these flakes is a transported artifact, whereas the other was assigned to Chert-03 (this is the artifact used on wood).

Of the 54 connection lines found in this zone, 36 were intrazonal and 18 interzonal. The latter connected zone A with zones B and E, except one connection with zone D. There were no links with zones C and F. Most refits with zone B corresponded to isolated artifacts for which intentional movement cannot be implied. Due to the proximity of these accumulations, their connection may be the result of the normal dispersion of items during knapping. Post-depositional movement cannot be ruled out as these connections have a down slope direction. The same is true in the case of the refits of Chert-02, Chert-06, Chert-17, Chert-31 and Lim-07. In general, they are sequences carried out in zone A that have some single items in zone B, although the exact direction of movement could not be always determined due to the scarce number of remains included in some RMUs (Chert-31 and Lim-07). Lim-07 was introduced as a small unworked nodule that was broken by percussion. The only refit identified in this RMU is a breakage that links zone A to D (616.3 cm). It is composed of two small fragments, one located in N47 and another in N53. It was not possible to ascertain its mode of displacement.

Some connections with zone E seem to be the result of intentional transport. Eleven refits have been documented between the two zones and in at least three cases (Chert-05 and Quartz-04) intentional transport is a distinct possibility. They show two different knapping posts, the first in zone A and the second in squares K41–42, L41–42 and M41–42. In the refit of Chert-05 it seems that movement occurred from zone A, where the main stages of the reduction sequence were carried out, to zone E. The refit of Quartz-04 shows the same direction of transport. The direction of movement cannot be determined for the other connections between zones A and E.

Zones B and C

We have identified 47 RMUs in zone B, 22 corresponding to transported items. Only one of the artifacts recovered in this area (a denticulate found in M49) presented evidence of use associated to meat-bone cutting. The main characteristic of this zone is the great amount of remains produced by a single reduction sequence (Chert-01). Of the four RMUs attributed to complete or nearly complete reduction sequences, this was the unit that provided the most artifacts (n = 153), which were distributed between zones B and C. The second most numerous RMU was Chert-13, which yielded 47 artifacts. The other RMUs attributed to entire sequences (Chert-15 and Chert-18) corresponded to small nodules and produced few artifacts. Partial sequences were far more abundant and were aimed at the detachment of small and very small flakes.

Qu-05 is distributed in the same way as Qu-04. There are no refits of this RMU. There is a small scatter area in N–O/48–49. The products of this RMU are mainly knapping products of a very small size, some large flakes and a retouched artifact. It seems that a very big block of quartz was introduced and different production sequences were realized inside the site.

The most outstanding knapping event was the reduction of Chert-01, due not only to the number of products, but to the identification of two spatially separated reduction stages (Fig. 7.95). The first one was widely scattered throughout zone B. The second one formed the bulk of the lithic assemblage found in zone C. The first stage, including the decortication of the nodule (Fig. 7.96), provided a wide array of products. The final stage, however, was exclusively aimed at producing small and very small flakes. This is clearly shown by the refits found in zone C and one of them in particular which is made up of 17 very small and two small flakes that were conjoined on the core (Fig. 7.96). The differences between the two stages were not only related to production goals, but also to the degree of dispersion. On the one hand, the remains from the reduction event carried out in zone B were very scattered. The mean length of the connection lines found in this zone was 154 cm. On the other hand, the second stage was clustered in L49-50, showing a principal accumulation of only 50 cm in diameter. Mean length of connection lines from this last stage was 33 cm. The lithic assemblage of zone C was produced by knapping events carried out from resources transported from zone B. One of the cores found in zone C corresponded to the exploitation of a cortical product detached in zone B, which was transported to zone C and reduced to obtain very small flakes. The other core found in zone C was also transported from zone B. The differences in scatter between the two phases suggest that they were temporally differentiated events. The accumulation of zone C was probably the result of the recycling of artifacts previously discarded in zone B.

There is another RMU represented in zone B that also showed connections with zone C. Chert-16 was distributed between both zones (Fig. 7.97). Some remains were found in J50–51, K49 and L50, while other artifacts were located in zone B (M51–53, N49–52 and O51). Single items from this RMU were also recovered in zones A (O46) and E (M44).

Fig. 7.95 Spatial distribution of Chert-01

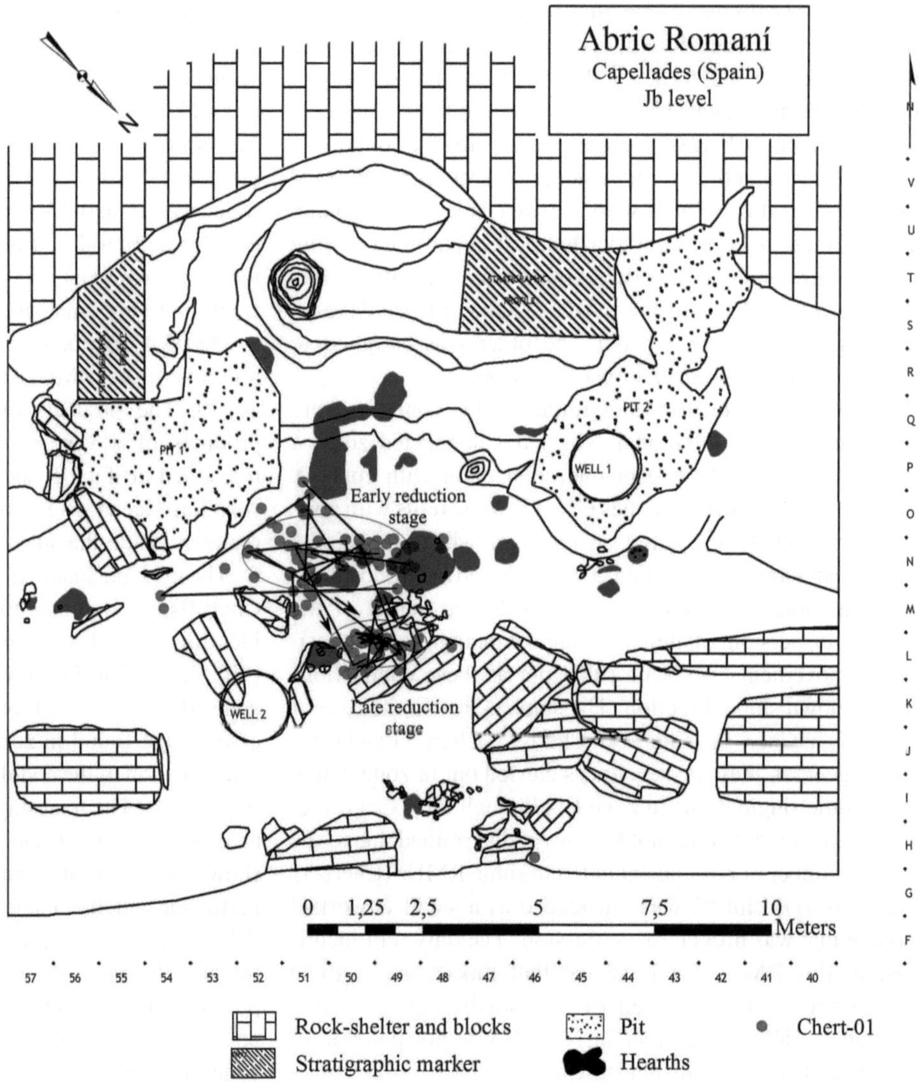

Like the previously described unit, this RMU exhibited a size-sorting pattern, although in the opposite direction of that of Chert-01. Large and very large cortical flakes were preferentially located in zone C, whereas all the remains of this RMU found in zone B were small and very small. The two artifacts located in zones A and E were very large and medium-sized (the first was connected through refits with the outer zone). However, unlike Chert-01, this size sorting does not seem to be related to the succession of two spatially segregated reduction stages, since artifacts from the same stage were distributed between both zones. This indicates that size sorting occurred after the knapping event.

Another outstanding feature of zone B is the clustering of some artifacts made on raw materials uncommon in Abric Romaní. These are different metamorphic rocks (porphyry, quartzite, and schist) and a dark limestone with crystal inclusions very different from the limestone normally found in the vicinity of the site. These raw materials were carried in as large flakes or retouched artifacts (Fig. 7.98), although we have also documented a short reduction sequence on quartzite that only yielded some very small flakes (Fig. 7.99). The transported artifacts were located in M48, M53, N49 and N50, whereas the knapping event was concentrated in N50–O49. The spatial association of these materials, practically absent from the rest of the shelter, and their similar provisioning modalities suggest that these remains were carried into the site in the same introduction event.

Regardless of the connections between zones A and B found through refits, both zones clearly show different patterns in terms of connection-line direction (Fig. 7.53). In zone A, connection lines tend to run in an E–W direction. Meanwhile in zone B they tend to go from N to S. This is partly a consequence of connections with zone C, but the intrazonal refits tend to show this same direction as well. This indicates that these zones were affected by different dispersion dynamics.

Fig. 7.96 Refitting of the two reduction stages of Chert-01

There is a limestone refit connecting zones B and E (Lim-05) with a 709 cm connection line. It is formed by an unworked nodule with percussion marks situated in zone E (K43) and a flake produced by the percussion of the nodule in zone B (O49) (Fig. 7.100). The rest of the artifacts associated with this RMU were found in zones B and D. It seems therefore that movement took place from zone B to zone E. The transport of elements towards zone E was also detected in zone A.

Zone D

Zone D was characterized by the scarce development of knapping sequences. From the 15 RMUs identified in this zone, only two can be attributed to reduction episodes carried out in situ. These two RMUs correspond to short sequences producing very small flakes and were focused in the same area (M54). One of them was done on a quartz cobble that does not correspond to the type of quartz documented in the vicinity of the site. The remaining RMUs were introduced as transported artifacts. An interesting feature of this zone is the provisioning pattern characterized by the introduction of single blanks originating from the same sequence. Three RMUs have been attributed to this kind of transport. In addition, these RMUs were spatially associated, since they were clustered in K54 and L54. They always correspond to sets of two artifacts, flakes or retouched artifacts of large-very large size (with the exception of a medium-sized retouched artifact). Three of the six artifacts forming these RMUs were blanks with an abrupt back opposing a cutting edge (two *débordant* and one naturally backed flake).

Fig. 7.97 Refitting and spatial distribution of Chert-16

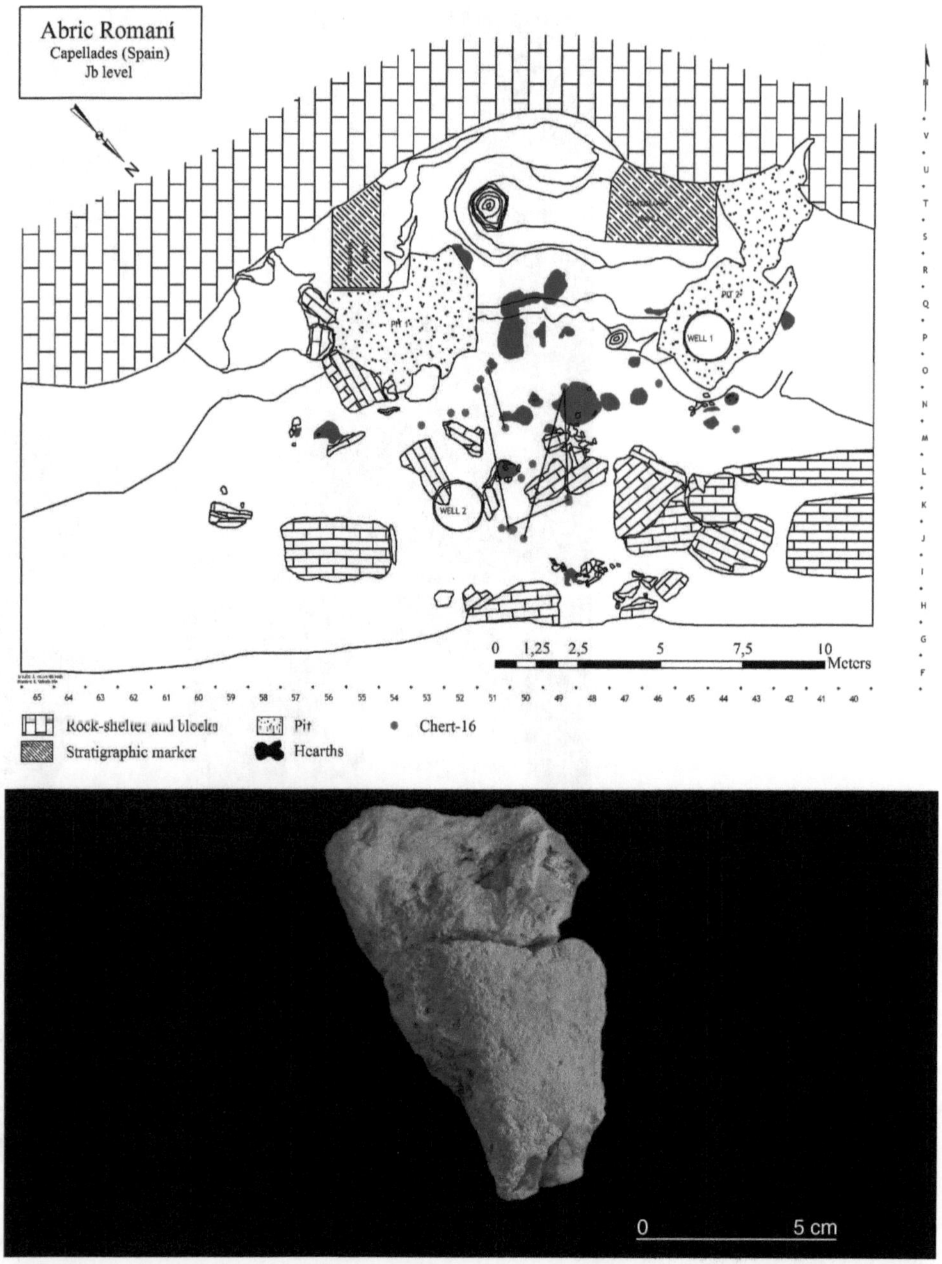

Zone E

Zone E was also a low-density area, although a small cluster was visible in L42. As we have pointed out previously, this cluster has been connected by refits to zones A and B. At least three RMUs show the intentional transport of artifacts from zone A to zone E. There were no complete sequences, with the exception of a reduction event from a small chert nodule that yielded some small flakes. The rest of the RMUs were partial sequences providing few artifacts. We also found five transported artifacts.

Qu-04 was distributed across all zones except zone C, but showed a clear accumulation in K–L–M/41–44. Four refits have been identified, three of two elements and one of five. There are four breakage connections lines and three production sequence connections separated by distances between 1 and 6 m. Only one line belonging to a five-piece refit is interzonal and links zone A to zone E. This line shows the clear intentional movement of pieces belonging to the same production sequence and the temporality of at least this event between these two zones. It begins in O46 where a flake was detached, the core was then transported to N45 and another flake was obtained. Then it was moved to K42 and two flake fragments (broken by a Siret accident) were detached and finally the core was abandoned in M41 (Fig. 7.101).

Lim-04 is the limestone RMU with the most elements in sublevel Jb. This RMU was dispersed in zones A, B and E but showed a small cluster in M–N/43–45. All the elements

Fig. 7.98 Assemblage of exogenous rocks found in sublevel Jb. These artifacts were spatially associated in zone B. They present the characteristics typical of transported items, showing a *débordant* side or a cortical side opposed to a cutting edge

Fig. 7.99 RMU of Quartzite-01 formed by six knapping products of a short reduction sequence carries out inside the site

come from a large unworked nodule fragmented after its use as a hammerstone (Fig. 7.102). The other two refits identified in this sublevel, one intrazonal and the other interzonal, belong to this RMU. The interzonal refit links zone A and zone E and is made up of five fragments, some with percussion marks.

Fig. 7.100 Refitting RMU Lim-05 formed by an unworked nodule and a flake produced by the use for percussion that links zone B to zone D

Fig. 7.101 Refitting of RMU Quartz-04 formed by five knapping products with one connection line that links zone A to zone E and show an intentional displacement of the core

Zone F

Finally, zone F showed only a small assemblage located essentially in S41. The lack of continuity between this zone and the rest of sublevel Jb raises some doubts about their stratigraphic correlation. Furthermore, there are no refits connecting this zone with other areas. We have identified six RMUs, two corresponding to reduction sequences and four to transported artifacts. The first reduction sequence was carried out on a small high-quality chert cobble that yielded a small number of blanks. This accounts for the only refit found in this zone. The second reduction sequence can be attributed to a longer production event done on low-quality chert.

The information detailed above gives us some insights into the formation process of sublevel Jb and the relationships between the different activity areas. This temporal interpretation is made difficult by the absence of a clear temporal relationship between the two main accumulations, those from zones A and B. The weak linkage between these zones and the absence of significant differences in the degree of RMU dispersion inhibit the establishment of any temporal order among these accumulations. However, some data suggests that they were independent activity areas. With the exception of some single artifacts whose intentional movement cannot be ascertained, they do not share RMUs and present different dispersion dynamics, as far as the direction of connection-lines is concerned. On the other hand, they exhibit a strong linkage with other accumulations identified in sublevel Jb. The clearest example is the connection between zones B and C. All of this seems to indicate that the accumulation in zone B was formed prior to the accumulation in zone C, and that the latter was produced by the reduction of artifacts transported from the former. Differences in the dispersion area suggest a temporal difference between both accumulations. The intentional movement of artifacts from zone B to C corresponded therefore to the recycling of previously discarded remains.

The relationship between zones A and the accumulation focused in L42 in zone E shows a similar succession. There was the transport of artifacts from zones A to zone E, which suggests that the accumulation of zone A was formed prior to the L42 cluster. Similar movement was detected between zones B and E (Lim-05). It seems therefore that the accumulation focused in L42 was partly formed by the recycling of remains previously discarded in zones A and B. As a whole, this indicates that the large accumulations of sublevel Jb may be attributed to the first activity events carried out in this sublevel and they were used as provisioning areas during subsequent activity events. As was observed in sublevel Ja, the scarce dispersion of these recycling episodes suggests that they correspond to the last activity events that occurred in sublevel Jb.

Sublevel Jb also has another feature that we have already discussed in relation to sublevel Ja: spatial variability in provisioning and processing patterns. This suggests that the different introduction events that followed one another during the formation of this sublevel had some specific characteristics, and that the variability of the lithic assemblage can be partly interpreted in temporal terms. In some cases, this variability seems to be related to raw material sources and, therefore, to the origin of the human groups. The best example is the association of metamorphic rocks found in zone B, which indicates that the group came from a

Fig. 7.102 Refitting of RMU Lim-04 formed by a large cobble fragments with percussion marks that links zone A to zone E

geographical setting not much frequented by the humans moving towards Abric Romaní. A similar phenomenon was observed in another archeological level of the Romaní sequence, level I, in which a small cluster of transported artifacts made on metamorphic rocks was also identified (Carbonell 2002).

Discussion

Some of the questions raised by the study of the lithic assemblage will be further developed in the general conclusions of this book, especially those related to settlement patterns. In the following paragraphs we will discuss some questions related to the formation and variability of lithic assemblages: provisioning patterns, core reduction and tool manufacture strategies, and the spatial and temporal dimension of lithic assemblage formation. All these topics have been used to explain different aspects of Middle Paleolithic assemblage variability. At first glance, the event-focused approach that we have adopted here has yielded the broadest picture of technical behavior, highlighting some variability phenomena that would go unnoticed in an assemblage-as-a-whole approach. This shows how palimpsests tend to minimize technical variability that becomes evident at the event level.

Provisioning Patterns and Raw Material Economy

Lithic provisioning strategies mostly depend on the availability of raw material sources in the surroundings of the sites. In the last years, several studies have analyzed raw material procurement at Middle Paleolithic sites and it seems clear that raw material availability is one of the main constraints on technical behavior. These studies suggest that these strategies are conditioned by an optimizing behavior, which tends to minimize the weight transported as the distance to raw material sources increases. This argument can be considered as a technological version of the basic principles of the optimal foraging theory, commonly used to explain several domains of hunter-gatherer behavior. According to this minimizing-weight model, the distance from sources to site seems to be a key factor for characterizing these strategies. Raw materials from local sources are dominant, especially during the Lower and Middle Paleolithic (Geneste 1989a, b; Turq 1989, 1990, 1992; Dibble et al. 1995; Féblot-Augustins 1997; Ortega and Maroto 2001; Manzano et al. 2005; Arzarello et al. 2007; Depaepe 2007; Molines 2007; Wisniewski 2007), although there are some Mousterian assemblages in which exogenous materials are the most abundant (Rios 2005). The percentage of remains tends to decrease as their origin becomes more distant, especially when these sources are located more than 10–20 km from the site.

Moreover, the mode in which lithic resources are introduced into the sites also varies according to distance (Tavoso 1984; Geneste 1988). Local raw materials tend to be transported as bulk resources—entire or almost entire nodules—that are processed at the site. As distance to the lithic sources increases, resources tend to be transported in more elaborate forms, like single blanks or prepared cores. In some Middle Paleolithic assemblages, source distance has been shown to influence core reduction technology and tool typology (Kuhn 1991; Dibble et al. 1995; Depaepe 2007). On the one hand, complex reduction methods, like Levallois, are more common among non-local materials, while local sources are exploited using more expedient methods (Wengler 1990; Otte 1991; Bar-Yosef et al. 1992; Moncel et al. 2008). On the other hand, some tool types, like notches and denticulates, are more commonly made on local materials, while other artifacts, like sidescrapers and handaxes, show higher percentages on non-local materials (Geneste 1988; Meignen 1988; Meignen et al. 2007). According to this pattern, sidescrapers would be curated tools, transported across the landscape and more subject to repeated resharpening episodes. Meanwhile, denticulates would show an expedient manufacture and use and their transport to other occupation sites would be uncommon. It is also important to consider that the distance traveled by an artifact partly depends on how long the artifact has been in use. Artifacts with long use histories would therefore be more likely to be resharpened, and would show more reduced morphologies. This relationship between raw material availability and tool morphology is one of the main arguments of the continuous tool-reduction model (Rolland and Dibble 1990; Dibble and Rolland 1992; Dibble 1995a).

The evidence from level J partly agrees with this behavioral model, although there are some discordant aspects that should be taken into account. In the first place, the nearest raw materials were not the most exploited resources. Quartz and limestone are overwhelmingly dominant in the fluvial and colluvial deposits close to the site. They are clearly the more abundant materials within a 5 km radius. As the study of the Anoia fluvial deposits has shown, chert nodules are extremely rare in the fluvial formations of the Capellades strait and they are relatively common in the fluvial deposits located outside the strait, at least 5 km away from Abric Romaní. Nevertheless, chert was preferentially selected for knapping. Although some chert-bearing formations are located in a radius that most authors consider as local, the transport costs of this material was higher than those required for limestone and quartz. This might explain the economizing behavior inferred from core reduction sequences.

Limestone cobbles were commonly brought to the shelter. In fact, if we consider the weight of the lithics found in level J, limestone exceeds both chert and quartz. While there are very large limestone cobbles weighing more than 1 kg, no such sizes in chert and quartz have been identified. However, reduction sequences on limestone were relatively uncommon and most of them were focused in the central area of the shelter (zone 4), where they were associated to the main quartz reduction sequences. Knapping to these strictly local and low quality raw materials is one of the principal differences between sublevels Ja and Jb, and as we will discuss later, seems to be characteristic of a specific occupation period of the level J formation sequence.

Among the chert nodules, we have identified different sources located at different distances from the site, which therefore incur different procurement costs. For instance, VLD, PAN and exogenous cherts come from relatively distant sources, while SMT chert can be considered a more local material. SMT chert would therefore be expected to dominate in both sublevels if the abundance of each type of raw material is based on the distance to the sources. Nodules with primary cortex are especially interesting, since they were gathered near the chert outcrops and are easily interpretable in terms of transport costs. Nodules with secondary fluvial cortex were collected in river deposits that may have been closer to the site than the original outcrops, and they are therefore less reliable in order to infer distance of transport. The longest distance of transport can be attributed therefore to the VLD, PAN and exogenous chert with primary cortex. As expected, these materials are scarce, especially if we take into account the RMUs corresponding to core reduction sequences. Most of the RMUs attributed to these sources exhibit a secondary fluvial cortex. However, three reduction sequences on VLD primary chert and one on exogenous primary chert have been documented in sublevel Ja. Three of them—VLD Chert-024 and 046 and exogenous Chert-031—correspond to complete or nearly complete reduction sequences, which indicates the transport of raw nodules to the site.

In spite of these exceptions, chert provisioning is consistent with the expectations derived from the minimizing-weight model. It seems therefore that environmental constraints operated on the choices previously adopted by humans. Once the preference for chert for reduction sequences was established, the economic factors were taken into account. It is especially significant that exogenous chert is dominant among the RMUs introduced as single artifacts—basically large flakes and retouched artifacts. However, this style of introduction was also used for local materials. Several chert artifacts of local origin, and even some quartz and limestone flakes, were carried into the site as single items. The variability of provisioning methods affected all raw materials, regardless of their origin. Local materials were commonly introduced as bulk resources, but they were also transported in more selective ways, such as prepared cores, flakes or retouched artifacts. This selective mode of transport is dominant for more distant materials, but some bulk procurement events have been also documented. The average behavior fits the predictions of the minimizing-weight model, but a higher variability can be found at the event level, which suggests that all the technical episodes were not affected by the same constraints or economic considerations. As we will see later, the spatial distribution suggests that these events characterized by a specific provisioning strategy could correspond to a particular phase in the formation of level J.

Regardless the origin of raw materials, the distribution of lithics by RMU indicates that two classes of artifacts can be distinguished according to their provisioning strategies. In fact, these classes of artifacts form two different assemblages:

(a) The artifacts produced in the rockshelter during the knapping sequences carried out in situ or derived from the in situ breakage of nodules. Two different provisioning modes can be distinguished according to the

Table 7.37 Distribution by size classes of flakes attributed to some RMUs, comparing transported flakes with flakes from knapping sequences carried out in situ

	Very small	Small	Medium	Large	Very large	Total
Transported flakes	6	28	48	47	40	169
	3.5%	16.5%	28.4%	27.8%	23.6%	100%
Flakes produced in the site	824	295	82	30	28	1259
	65.4%	23.4%	6.5%	2.3%	2.2%	100%
Total	830	323	130	77	68	1428
	58.1%	22.6%	9.1%	5.3%	4.7%	100%

reduction stage in which the core was brought into the site, as entire or almost entire nodules or partially reduced cores.

(b) The artifacts produced outside the shelter and introduced into the site as single items, essentially flakes and retouched artifacts.

This second provisioning strategy is the most common way of introducing lithic resources into the shelter. Of all the introductions of lithic resources identified, 284 (50.2%) corresponded to isolated flakes or retouched artifacts produced outside. Only four of these items were intentionally modified after their transport to Abric Romaní. The RMUs introduced as entire or almost entire nodules are less common, especially if we take into account only those used for flake production. However, these uncommon events form the bulk of the lithic assemblage, which creates a contradiction between the frequency of provisioning strategies and their weight in terms of the number of lithic remains that they represent in the assemblage. In level J, the more common strategy (introducing isolated blanks) produces a relatively small assemblage, while the lesser used strategy (introducing entire or nearly entire nodules) provides most of the lithics, and is therefore the more apparent of the two.

These single blanks make up the toolkit transported by humans during their trips across the land. They correspond to the strategy of provisioning individuals defined by Kuhn (1995) or the "personal gear" described by Binford (1977) among the Nunamiut. This is a useful procedure for making sure that some tools will always be available when needed, even during trips. This toolkit was formed basically by flakes and retouched artifacts and most tools correspond to this provisioning strategy. Some of the cores introduced in a more or less advanced reduction stage would probably have been included in this transported gear. The association between retouched artifacts and this mobile toolkit is abundantly clear in both sublevels. In sublevel Jb, only one retouched artifact can be associated to a reduction sequence carried out inside the shelter. In sublevel Ja, only two retouched artifacts were refitted to core reduction sequences (Chert-018 and Lim-012) and they correspond to small blanks. Retouched artifacts exhibit the lowest refitting rate of all artifact classes. Although more tools from sublevel Ja have been included in RMUs associated with reduction sequences, they represent a very low percentage compared to the retouched artifacts introduced as isolated items. It is worth stressing that there are no typological differences between the retouched artifacts manufactured in the rockshelter and those made outside of the site, since denticulates were clearly dominant in both assemblages.

As far as their morphology is concerned, there are clear size differences between this transported assemblage and the lithics produced in the rockshelter. Among the transported artifacts, medium, large and very large items are dominant, while in situ production was principally aimed at producing very small and small flakes (Table 7.37). For the large and very large categories, the number of transported blanks exceeds that of lithics from in situ knapping events. Although the economic advantages of transporting large blanks have been a topic of debate (Kuhn 1994, 1996; Morrow 1996b), dimensional attributes were a main concern in the selection of Middle Paleolithic mobile toolkits (Close 1996; Morin 2004). Large artifacts are especially suitable for transport, since they allow an extended period of use due to their longer cutting edges and their greater potential for resharpening. Moreover, some functional requirements may lead to the selection of large blanks. However, in spite of the high number of transported blanks, the artifacts typically associated with intense renewal, like sidescrapers, are practically absent from level J. This suggests that there was no need to maximize the durability of transported artifacts.

Furthermore, it seems that the transported toolkit was selected according to some technical attributes. The most outstanding feature is the presence of an abrupt side opposed to the edge; this feature is characteristic of *débordant* flakes, which were preferentially selected for transport (Fig. 7.103). *Débordant* flakes are products typical of discoidal reduction strategies and the preferential selection of *débordant* flakes has long been documented in Middle Paleolithic assemblages (Beyries and Boëda 1983; Meignen 1993). However, such preferential selection not only affected *débordant* flakes, but any blank showing an abrupt side, such as naturally backed flakes or even some flakes with a distal edge opposed to a very large striking platform. In this transported toolkit it is also common to find flakes without an abrupt back, but showing a cortical surface opposite the side presenting a cutting edge along its entire length (Fig. 7.104). These criteria, as well as the preferential selection of large blanks, led to a high

Fig. 7.103 Transported artifacts from level J. *Débordant* flakes

Fig. 7.104 Transported artifacts from level J. Partially cortical flakes

percentage of cortical dorsal surfaces in this transported assemblage. This high index of corticality is significant if we consider that the percentage of cortical products tends to be interpreted as indicating in situ execution of the first stage of core reduction sequences. The evidence from level J indicates just the opposite, since the reduction sequence that resulted in these transported cortical blanks was performed outside of the shelter.

There are not clear-cut technical differences between the reduction sequences from which the transported artifacts were produced and those carried out on the spot. In particular, blanks showing the typical characteristics of Levallois products are rare in the transported assemblage. In some Middle Paleolithic assemblages the Levallois technique is more common among imported blanks. This is not the case in level J. There are no data indicating that the transported artifacts were produced using reduction strategies different to those carried out in the rockshelter. Aside from some quantitative differences derived from the selection criteria, such as the higher percentage of *débordant* flakes in the transported assemblage, most of the imported blanks fit perfectly well within a discoidal strategy.

Nevertheless, the characteristics of the reduction sequences from which the transported blanks were detached differed from those suggested by the knapping sequences carried out in the rockshelter. This difference concerns the size of knapping products. In general, the core reductions performed in Abric Romaní were directed at producing small and very small flakes. Both refits and RMUs provide numerous examples of this kind of sequence. This trend, also inferred from the technical and dimensional analysis of the cores, suggests that small flakes were not simply the lithic refuse generated during the production of larger blanks, but intentional and sought after products. This small flake production was realized using different procedures. In some cases, it appears at the end of long reduction sequences entirely carried out in the rockshelter, which previously produced larger flakes. However, this is the less common process. It is more usual to find sequences whose only purpose was the production of small flakes. The core was sometimes introduced in an advanced reduction stage, and only the last exploitation phase took place in the rockshelter. At other times lithic resources were carried to the site as small nodules only suitable for providing small blanks, although the entire reduction sequence was performed in the site. Finally, recycling of almost exhausted cores or flakes for producing small flakes has been also found. This intentional production of small flakes has recently been documented in other Middle Paleolithic assemblages (Goren-Inbar 1988; Moncel and Neruda 2000; Moncel 2003; Dibble and McPherron 2006). In Abric Romaní, this small-flake-oriented production seems related to the domestic activities carried out around hearths, whereas large flakes tend to be linked to the activities performed during trips.

A peculiar method of provisioning was identified for the transported toolkit: the introduction of sets of artifacts from the same reduction sequence but produced outside of the rockshelter. We have distinguished this kind of provisioning by means of the spatial distribution of these sets. In some areas of the site, we found clusters of lithics (normally two or three items) formed by different artifacts from the same RMU and of a similar size. They are confined to small areas, which suggests that they were introduced during the same provisioning event. They exhibit the characteristics defining the transported toolkit: large size and asymmetrical profile with a cutting edge opposing an abrupt back. The absence of refits and small knapping debris, together with these selected attributes, indicate that the core reduction sequence was not carried out in situ. These items were detached outside, selected for transport and carried to the shelter in the same provisioning episode. We have identified several of these associations, both in sublevel Ja as in sublevel Jb. One example is the set of four medium-large flakes forming the Chert-078 (Fig. 7.105). They have an asymmetrical profile (three *débordant* flakes and one natural backed flake) and three of them were located in the same square (P57); the fourth was found some meters away but in the same zone of the site (K58). Seven flakes of large and very large size, three of which are *débordant*, form Chert-164. All were found in the middle of the shelter (zone 4), in a small radius around N50. In sublevel Jb, we documented three of these sets in K54–L54. The first is formed by a large flake and a very large *débordant* flake; the second by a very large cortical flake and a medium-sized naturally backed flake; and the third by two retouched artifacts, one notch on a *débordant* blank and one denticulate on a very large flake.

Fig. 7.105 Set of flakes corresponding to RMU Chert-078

Limestone was introduced principally as entire or fragmented cobbles, coming from the alluvial formations surrounding the site. However, the introduction of certain limestone items as large flakes or retouched artifacts has been also documented. These correspond to fine-grained varieties of limestone. The use of limestone partly depended on grain size and the properties of the nodules in terms of conchoidal fracture. Knapping and introduction as isolated items has been documented especially for the most silicified materials; on the other hand, the coarse-grained types were preferentially used as hammerstones, although some poor quality cobbles were also reduced for flake production. Finally, the manufacture of pebble tools was uncommon, but at least one limestone artifact seems to correspond to this tool class.

An Expedient Technology

As far as the study of core reduction strategies is concerned, level J is one of the hallmarks of the Abric Romaní sequence. Along with levels E and M, it is one of the stratigraphic units that have provided enough artifacts (especially cores) to allow us to approach core reduction methods. In previous works (Vaquero 1999a; Vaquero et al. 2001b), level J defined one of the technical phases identified in the upper part of the stratigraphy. This phase was characterized by the dominance of discoidal strategies, aimed at producing the maximum number of flakes per core, but with little control over the size and shape of the products. The following phase, represented by level E, showed a significant increase in hierarchical strategies, in which one surface of the core was preferentially exploited and the opposing surface was used to prepare percussion platforms. These differences were associated with other changes in raw material provisioning (e.g. the selection of better quality materials in level E) and tool manufacture (e.g. higher percentages of sidescrapers in level E), leading to the conclusion that level J is a good example of an expedient technology that was not very demanding in terms of the criteria applied to reduction sequences.

We consider expedient strategies as those characterized by a low investment of technical knowledge and low requirements for the metrical and morphological attributes of the products. This expediency can be observed at all the stages of the *chaîne opératoire*, from raw material provisioning to artifact manufacture. There is a common use of raw materials of low or mediocre quality. Limestone and quartz are especially bad for knapping, but also many of the chert nodules exploited on the spot show an irregular fracture, particularly in sublevel Ja. Metrical attributes do not seem to be important criteria either, since reduction sequences on small nodules, flakes and flake fragments has been widely documented. The production of small blanks is one of the main characteristics of the reduction sequences carried out in level J. Both core and flake dimensions, as well as the reconstruction of several reduction sequences indicate this intentional production of small flakes as the main goal of most reduction sequences. The lithic assemblage of level J cannot be interpreted in terms of a unique technology. It is the result of the conjunction of two technical realms: the technology carried out in Abric Romaní and the technology applied outside and represented in the assemblage as single blanks. However, as we have pointed out before, there are not clear-cut differences between these technical assemblages as far as the core reduction strategies are concerned.

Reduction sequences were guided by quantitative criteria. The main goal of most of the core reduction sequences was to produce the highest number of blanks per core. This seems to be the *umbrella plan* (Keller and Keller 1996) of most of the reduction sequences carried out in level J. The technical attributes needed to achieve this goal were simply those allowing the recurrent exploitation of the core: the maintenance of the striking-platform angle and the flaking surface convexities essential to a successful sequence of detachments. Meanwhile, the qualitative requirements for producing particular forms seem to be absent, at least in the reduction sequences that took place inside the rockshelter. The reduction methods should also be understood in this expedient context, in which there was little concern for the size or form attributes of products. The low proportion of flakes with facetted striking platforms is consistent with this low degree of predetermination, since faceting is a procedure aimed at increasing control in flake detachment. It is very difficult to differentiate stages in the reduction sequence according to their preparatory or fully productive character. As the common use of cortical products for transport or retouch indicates, usable blanks were produced from the beginning of the reduction sequence. A distinction between predetermining and predetermined detachments is of little utility.

The concept of expediency is important in interpreting variability in core morphology. In recent years, studies on Middle Paleolithic technology have been aimed at identifying and describing discrete knapping methods. These studies have allowed several methods to be characterized—Levallois, discoidal, quina—that comprise most of the Middle Paleolithic technological variability. These methods are normally understood as mental templates that govern production sequences from their beginnings. These templates consisted of several technical criteria that established the kind of desired products and core structure. Core morphology is therefore one of the basic criteria for identifying these methods in the archeological record. In some assemblages, this assumed correspondence between core morphology and knapping method has produced a proliferation of reduction strategies.

However, we think that these approaches do not place sufficient weight on the different ranges of variability that can be expected in different technical contexts. In a context characterized by a great investment of technical knowledge, the predetermination of certain products leads to the strict control of core structure. In such contexts, a small range of variability and a trend to core standardization can be expected, as is the case in assemblages in which the Levallois method is dominant. On the other hand, seeking out specific products is not the main concern in more expedient contexts aimed at repeatedly producing as many flakes as possible. In such an expedient context, we would expect broad variability in core morphology. This large variability may be interpreted as the result of using different reduction methods, but, in fact, it shows the lack of a method demanding a specific core structure. Level J is a good example of this technical behavior, which consists of applying some simple technical criteria that are continuously adapted to the constraints of each knapping event, especially to the size and form of the exploited nodule.

Discoidal strategies are especially suitable for these quantitative goals. They allow the recurrent exploitation of cores without expensive maintenance of their morphology and provide blanks until productive capabilities are exhausted. Some authors have attributed the discoidal strategy with a high degree of predetermination (Boëda 1993a; Locht and Swinnen 1994; Mourre 2003), based on the production of some blanks with specific morphologies: flakes with asymmetric profiles, especially *débordant* flakes. We are doubtful of the predetermined nature of the discoidal strategy. These alleged predetermined blanks are actually the consequence of maintaining the peripheral convexities of the core and are therefore the logical outcome of applying the recurrence principle. Detachment of these flakes is inevitable if we pursue the dynamic exploitation of the core, but this does not mean that such products were intentionally sought after. The preferential selection of *débordant* flakes for transport or use does not necessarily prove their predetermined nature. They were selected from the wide array of flakes produced throughout the reduction sequence. Flakes with asymmetrical profiles were more often selected, but this selection criterion was not applied solely to *débordant* flakes. Many partially cortical products were also included in the transported assemblage due to their asymmetrical morphology, but no one would argue that cortical flakes were intentionally sought blanks.

However, the discoidal principle does not always give rise to a typical discoidal core. This method is not rigid enough to impose a specific core structure. Therefore, the variability in core structure seen in level J is also a product of this expedient technology. This wide variability has already been recognized in other Mousterian assemblages attributed to the use of discoidal strategies (Locht and Swinnen 1994; Peresani 1998; Mourre 2003). The adaptation to the shape and size of the nodules, taking advantage of their natural morphology, is a perfect expression of such an expedient technical context. The appearance of unipolar, unifacial discoidal or polyhedral cores does not correspond to the use of different reduction strategies, but to the application of the discoidal principle to a wide range of blank forms. Expedient contexts produce a higher variability in core morphology as opposed to contexts characterized by more specialized and elaborate strategies, which tend to produce more standardized cores.

The discoidal and Levallois cannot be therefore interpreted as equivalent methods. Levallois is a true method, since it determines a specific core structure that is applied regardless of the characteristics of the nodule. Discoidal strategy, on the other hand, is not the same kind of method. It is the result of applying the recurrence principle in an expedient technical context. There was not a mental template guiding the core structure. The core morphology was the result of continuously adapting the recurrence principle to changing circumstances. Sometimes this led to the appearance of typical discoidal cores, but other times different core morphologies were produced. This lack of standardization may be mistakenly interpreted as the consequence of applying several independent methods of flake production. Flexibility was therefore a basic feature of the technical behavior dominant throughout the formation of level J. Although this behavior might look opportunistic or unskilled, it reflects the situated character of human technical actions (Keller and Keller 1996). The ultimate goal of production was generic enough to allow multiple acceptable solutions.

Some authors have established a relationship between these different technical contexts and mobility patterns. For example, Kuhn (1995) pointed out that, in the Mousterian assemblages of western central Italy, centripetal or radially worked cores were associated with high-mobility settlement patterns, in which the availability of lithic resources were based on provisioning individuals. In more prolonged occupations, pseudo-prismatic cores were much more abundant. Moreover, Wallace and Shea (2006), taking as a point of departure the study of Parry and Kelly (1987), suggested that low-mobility patterns would be characterized by increasing frequencies of expedient core reduction strategies, which were interpreted as opportunistic methods with little concern for raw material economy. Meanwhile, predetermined or formal strategies would exhibit higher frequencies in high-mobility contexts. Concerning this association, the evidence from level J is not straightforward. It seems clear that reduction sequences were linked to hearth-related domestic areas and therefore correspond to residential occupation events, although the length of these events is difficult to establish. This would support the relationship between expedient core reduction strategies and residential patterns. However, the transported assemblage, which could be associated to high-mobility contexts, does not show clear evidence of having been produced by more predetermined reduction methods.

The main difference between transported and produced-on-the-spot assemblages concerns flake size and selection-for-transport criteria, but both assemblages seem to have employed the same expedient production methods.

Tool Manufacture and Use

Toolmaking was characterized by selection patterns in which size criteria played a central role. The largest flakes were preferentially selected for retouch, especially the thicker ones. As the blanks became smaller, the tendency to maximize thickness increased. This selection of large flakes for retouch has frequently been documented in the Middle Paleolithic (Meignen 1988 and 1993; Dibble and Holdoway 1990). As usual in Abric Romaní, denticulates and notches were the dominant morphotypes. This dominance is particularly marked in level J, since other types—notably side-scrapers—are practically absent. The central role played by denticulates in tool manufacture is one of the most outstanding features of Abric Romaní. Denticulates are dominant at all the archeological levels, although there is a significant increase in the sidescraper index in the uppermost Middle Paleolithic layers (E–B).

Most retouched artifacts belonged to the transported toolkit. Few retouched artifacts have been clearly linked to core reduction sequences carried out in the rockshelter. In some assemblages, denticulates have been characterized as static artifacts on local raw materials that would be manufactured, used and discarded at the same site. They would be an example of expedient behavior as opposed to mobile and curated artifacts, such as sidescrapers and bifaces, which would more likely be transported across the landscape (Geneste 1988; Meignen 1988; Turq 1990; Meignen et al. 2007). Most denticulates from level J do not support this hypothesis, since they were basically mobile artifacts that were manufactured elsewhere and introduced as retouched implements to Abric Romaní. In spite of their mobile character, which may be related to their longer lifespan, retouched artifacts do not show intense reduction. The step-like retouch typical of highly reduced artifacts is entirely absent. In addition, the two measures used to infer retouch intensity do not indicate strong reduction. Retouch limited to one side of the artifact is overwhelmingly dominant and the retouch perimeter index tends to exhibit low values.

Both the retouched objects and the flakes were always used according to the same principle: the use of the longer edge opposed by an abrupt side, on many occasions cortical. In the case of flakes, a simple angle was preferentially used in longitudinal actions. Among the retouched artifacts, semi-abrupt edges were used without distinction in longitudinal and transversal actions. The latter were much better represented among the retouched objects than among the flakes. Defleshing and dismembering actions predominated among the butchering traces on the retouched artifacts. Among the flakes, just the opposite was true, with abrasive traces generated during skinning actions dominating by far. We can suggest a complementarity between the two types of edges (retouched with SA angles and natural with S angles) in carcass processing. The greater resistance of the retouched edges probably adapted better to actions on hard materials that would have fractured the acute edges of the flakes.

Both flakes and retouched items had the same consideration as tools. However, the percentage of retouched artifacts is about 5% of the total number of objects. A similar percentage of use between retouched artifacts and flakes means that the operational capacity of these communities was centered on the former. Although the functional versatility of the objects takes precedence over any other aspect, in general flakes were used in butchering activities, specifically for skinning, while retouched artifacts were used mainly in defleshing and hide treatment actions, indicating dual function.

This seems to indicate that the functionality of the artifacts from level J does not depend on a specialization of uses, rather more likely on other technological or subsistential criteria. The objects demonstrate some homogenous technical use criteria but not technical use modes, as the same edges were used without distinction. The versatility of the objects has priority, eliminating the need to create specialized tools. Objects were not technically adapted to specific activities; rather flexible use prevailed in which the edges that were naturally obtained through knapping were used. The retouching of the objects would not have corresponded to any functional specialization, rather precisely the opposite; it would have increased the versatility of the blanks to strengthen their capacity for use in transversal actions. The operational capacity of these communities would have been based on a single model of a versatile and ergonomic tool selected from among larger or carinated flakes in which an abrupt side opposes the active edge. The usability of this tool would be increased by retouch. The flexibility and versatility of the functional patterns may indicate that social groups were not subjected to pressures to maximize the profitability of lithic resources.

Spatial and Temporal Patterns in Assemblage Formation

Lithics provide fundamental data for addressing the spatial organization of Paleolithic societies. The spatial studies of Middle Paleolithic have been largely based on the distribution of lithic artifacts. Besides the insights provided by the spatial location of the different artifact classes, lithic refitting is especially suited for approaching the relationships between the different activity areas and therefore the formation dynamics of the archeological assemblage. Two issues are particularly relevant in the spatial layout of lithic remains: identifying and characterizing activity areas or secondary refuse areas and establishing the functional linkages between different areas. These issues have been addressed at some Middle Paleolithic sites (Monnier 1990; Meignen 1993; Henry 1998; Brink and Henderson 2001; Henry et al. 2004; Moreno et al. 2004), including Abric Romaní (Vaquero 1999; Vaquero and Pastó 2001; Vaquero et al. 2001 and 2004). In these studies, several features concerning lithic distribution have been outlined, suggesting that different factors were involved in the spatial variability of lithic assemblages:

- When hearths were present, knapping activities were largely focused around them. Hearths were the basic structure defining household areas and knapping was therefore a domestic activity preferentially carried out in this multifunctional space.
- Because of these hearth-focused patterns, lithic scatters show a discrete pattern characterized by well-defined clusters separated by areas of lower density.
- Size-sorting is one of the main phenomena of variability in lithic distribution. According to ethnoarcheological and experimental models, it is interpreted as the result of discarding behavior and the formation of secondary refuse areas. The predominance of small items would indicate the location of primary knapping loci, while large items, which would be more frequently scattered towards the periphery of activity areas, would be accumulated in secondary disposal zones.
- In general, evidence of differential location of activities is not straightforward, but it has sometimes been proposed from the distinctive distribution of certain tool classes (Henry et al. 2004).

Nevertheless, most studies did not sufficiently consider the incidence of the temporal dynamics in the formation of spatial patterns. In general, it is recognized that archeological units are palimpsests that cannot be correlated to single occupations, but this is considered as a blurring factor that tends to obscure the original spatial organization. The spatial patterns remaining in spite of this are interpreted as evidence of a redundant use of space, which indicates that during long periods certain areas were always used for the same purposes. This creates an illusion of contemporaneity and the archeological levels are ultimately interpreted as synchronous entities. However, time is not only a blurring factor, but in some cases it may cause the appearance of spatial patterns. Temporal and spatial analyses should therefore be conducted conjointly, since spatial patterns are strongly linked to temporal dynamics.

The spatial distribution of lithics in level J supports the characterization of most of the rockshelter as a hearth-related domestic space. The main lithic accumulations and most of the knapping events were located around hearths, although the overlapping of these domestic areas blurred the discrete pattern normally generated by these spatial structures. This domestic area exhibits a more or less continuous scatter of lithic remains, in which the main discontinuities are due to ancient excavations and natural features. The preferential focusing of knapping activities around hearths has already been established as the main characteristic of the spatial patterns in Abric Romaní. However, other levels exhibit the discrete spatial distribution typical of this formation dynamic. In level J, hearth-related accumulations tend to be located in the best-protected area inside the dripline, between the shelter wall and the large blocks on the outer part of the site. This indicates that the spatial location of the hearth-related events was conditioned by the natural structure of the rockshelter.

One of the main characteristics of this spatial layout is the size sorting of remains. In sublevel Ja, the interior hearth-related areas are defined by the dominance of small remains, while the outer areas around the large blocks show a higher presence of large remains. Following the drop/toss zone dichotomy, this was interpreted in previous works (Vaquero 1999b; Martínez Molina and Rando 2001) within a framework of the opposition between domestic areas, where the knapping activities were carried out, and refuse areas, where large artifacts where rejected. However, this interpretation has not been fully supported by the data presented here. Refitting data show that connections between the inner and outer areas are scarce in general and practically absent between some zones. A more robust pattern of connections between inner and outer areas would be expected according to the drop/toss zone hypothesis. Most large artifacts located in the exterior areas are unrelated to the knapping activities carried out at the inner domestic areas.

Another explanation can be proposed for the differences in remain sizes. This explanation is related to the differential spatial distribution of the two lithic assemblages defined according to the provisioning method: the artifacts produced on the spot and the artifacts introduced as single items. The first are clearly concentrated in the hearth-related domestic areas and nearly absent in the outer areas. For example, only one core reduction sequence has been clearly documented in zone 5 of sublevel Ja. As we stated earlier, the amount of remains in a zone depends on the number of reduction sequences carried out in that zone. However, the spatial distribution of transported artifacts is not strictly related to the distribution of produced-in-situ artifacts; they are more evenly distributed. This is especially clear if we compare the two zones providing the most RMUs, zones 3 and 4. Core reduction sequences are much more abundant in zone 4 than in zone 3 (78 and 37 respectively), but the number of transported artifacts is practically the same (75 and 71 respectively). This differential distribution is also evident if we take into account the lowest density areas (like zone 5 and sublevel Ja and zone D in sublevel Jb). In these zones, the core reduction sequences are almost nonexistent, but the RMUs corresponding to transported artifacts are well represented. In the statistical analyses used to support the drop/toss hypothesis, the areas where transported artifacts exceed the core reduction units were among those characterized by a high incidence of large remains.

It seems, therefore, that the size sorting observed in level J may be produced by the different spatial signature of the provisioning modalities: the core production sequences are concentrated in the domestic areas, but the transported artifacts tend to be more homogeneously distributed. These differences in the spatial patterns suggest that the formation of the main knapping activities and the discarding of transported blanks were unrelated events, perhaps corresponding to different occupation episodes. Domestic areas were associated with an effective settlement in the shelter, and therefore their spatial location was more carefully selected. On the other hand, if the discard of transported blanks was associated to short visits that did not imply a real dwelling in the shelter, the spatial locations of such events were less constrained by the natural structure and were therefore more evenly distributed. Temporal dynamics would be therefore characterized by the alternation of two different depositional contexts. Size sorting would be produced by the differential spatial pattern of these different depositional contexts that took place during the formation of level J. The short-visit context would tend to create a relatively homogeneous scatter of transported artifacts, without clear accumulations. This scatter would be a sort of intra-site and continuous *veil of stones* (Roebroeks et al. 1992), upon which the discrete accumulations formed during residential events would be superimposed.

This can also explain the spatial distribution of the artifacts showing use-wear traces, which were not concentrated in the main lithic accumulations, but evenly distributed throughout the rockshelter. This spatial pattern has been documented at other Middle Paleolithic sites, like La Combette (Texier et al. 1998), where the highest lithic densities are focused around hearths, but artifacts bearing use-wear traces exhibit a random distribution. The spatial segregation of activities is not clear at level J because the same actions and worked materials were identified in different areas of the shelter. Nevertheless, the most developed traces were identified on objects recovered in peripheral areas. According to this distribution, we might consider the existence of areas of intense activity, although not specialized. However, we should take into account that most of these artifacts can be attributed to the transported toolkit. From the 18 artifacts showing use-wear traces, ten have been attributed to some RMU, eight of them corresponding to transported artifacts and only two to core reduction sequences carried out on the spot. Since longer use lives characterize transported artifacts, a higher presence of use wear can be expected. Three of the four artifacts showing the most intense use-wear traces were spatially associated in the NE corner of the excavated area—in H60, I63 and I65. They may therefore correspond to the same event characterized by the discarding of intensely utilized artifacts. Nevertheless, we should bear in mind that the functional analysis is biased due to the exclusion of refitted artifacts and the dimensional criteria used to select the studied sample, which may have produced an overrepresentation of transported blanks.

Several data indicate the temporal nature of the lithic assemblage. This assemblage is the product of a sequence of technical events that followed one another throughout time and two levels of dyachronicity can be distinguished. The first corresponds to the different depositional contexts described in the previous paragraphs. Domestic residential events and short-visit events alternated during the formation of level J and they constituted the main factor in the spatial variability of the lithic remains. The second axis of the temporal dynamics corresponds to the succession of different residential events. This level of dyachronicity is suggested by the refitting pattern and the differential dispersion of RMUs, especially of those located in the same areas. This indicates that all lithic scatters were not subject to the same degree of post-depositional movement. Whether these post-depositional processes were natural or human, their effects on the lithic scatters vary significantly. Some RMUs show wide scatters, while others located in the same zones are much more clustered. In spite their spatial association, these RMUs were not subjected to the same degree of post-depositional disturbance, which suggests a temporal difference between them.

These criteria, along with archeostratigraphy, have allowed us to identify in the middle of the shelter three phases in the formation of sublevel Ja. Moreover, refitting data also indicates this temporal pattern, since the most robust refitting patterns consist of unidirectional connections that can be interpreted as the result of recycling previously discarded artifacts. Even more importantly, some technical behaviors change according to the location of activity events in this formation process. We present a chronological model for the formation of level J, proposing a minimum number of occupation periods and locating some technical events within these periods. Establishing the temporal location of all the technical events would be illusory, but we think that the temporal relationships identified in this level are enough to draw a general picture of the formation sequence. We do not correlate these periods with single occupations in the ethnographic sense. Maybe the events located in the same period were carried out during the same occupation, or maybe not. This is very difficult to ascertain and perhaps go beyond the scope of the archeological inquiry.

The first periods of the assemblage formation are represented in sublevel Jb. Two main phases can be recognized in this sublevel. The first is associated to the two main lithic accumulations, representing several core reduction sequences from entire or almost entire nodules. Introducing bulk resources was common in this period, although knapping of raw materials available in the immediate surroundings—limestone and quartz—was rare. As we pointed out above, these two accumulations correspond to hearth-related domestic areas and were probably not contemporaneous, although we cannot establish the temporal relationship between them. The second period of activity in sublevel Jb is indicated basically by some recycling events in which some artifacts discarded in the previous period were moved and exploited or used. These recycling events were located at different areas of the site and may correspond to different occupations, since they do not share any RMUs.

At least four different periods can be identified in sublevel Ja. The oldest is associated to RMUs that exhibit the widest scatters, like Chert-007, Chert-009 and Chert-014. These technical events were performed on entire or almost entire nodules. Some artifacts discarded during this period were recycled in subsequent events. The second period would be represented by a series of core reduction sequences carried out in the middle of the shelter that exhibit similar dispersion patterns. These sequences also share other characteristics, since they were performed on entire or almost entire nodules of mediocre quality. Most reduction sequences of limestone and quartz would correspond to this occupation period. This period was characterized by little concern for raw material properties. The third period is mainly represented by the reduction sequences focused in square P51, which show small scatters. Most of them were carried out on partially reduced cores, and the reduction of entire nodules was uncommon. Some recycling events documented at different areas of the shelter may also correspond to this period, as they are also characterized by small scatters. Finally, the last period has been identified by archeostratigraphical data and corresponds to the few events that can be attributed to sublevel Jsup. The clearest was located in O49 and is represented by a small assemblage formed by transported artifacts.

Provisioning strategies change throughout the level J sequence. This temporal variability concerns both the kind of raw materials brought to the shelter and provisioning modalities. Use of limestone and quartz for knapping was not a constant feature throughout the sequence, but it was characteristic of one of the formation periods. Raw material qualities were also different depending on the timing of the technical events. Moreover, according to the intra-site spatial patterns of the different lithic raw materials, it seems that the different temporal events were characterized by different patterns of territorial mobility. Earlier events are associated with chert types originating to the NW (such as PAN chert or secondary SMT chert) and exogenous chert of unknown origin. Their intra-site distribution shows wide scattering due to their longer exposure to cultural and natural dispersion processes.

Later events seem to be associated with materials from the SW and W (primary VLD and SMT chert). The clustered pattern of these RMUs supports the conclusion that the knapping of these nodules occurred more recently. Some recycling of materials introduced in earlier phases also exhibits this grouped pattern. The primary provisioning of VLD and SMT chert suggests the same route from the SW, following the fluvial courses that flow into the middle valley of the Anoia, close to the Capellades gorge.

Secondary VLD chert exhibits a relatively tight clustered spatial distribution focused in zone 4. Although this material also indicates later knapping events, it may have come from a different direction. This chert is especially abundant in the conglomerates of La Guixera, located towards the southeast of the site, in the Sant Quintí de Mediona area. The provisioning of secondary VLD chert at La Guixera formation has been suggested before (Gómez 2007) because there is no secondary material in the fluvial courses crossing the primary VLD formations.

The temporal analysis of archeological levels is not common in Paleolithic research. However, Abric Romaní is not the only site where a time-dependent behavioral variability has been proposed. According to the data provided by archeostratigraphy, refitting and artifact distribution, three main occupation periods where identified in habitation unit Q31 of Étiolles, a Magdalenian site in the Paris basin (Ketterer et al. 2004). Interestingly, each period was characterized by different technical behaviors, especially concerning core reduction sequences. Blade production was dominant in the first period, showing a high level of technical skill. On the other hand, bladelet production was only significant in subsequent periods. Core recycling was also more common in the recent periods of occupation, as well as reduction sequences exhibiting low levels of technical skill.

Recycling is one of the clearest expressions of the temporal dynamics affecting assemblage formation. The importance of recycling in understanding technical behavior has been widely documented in level J. We have documented two kinds of recycling: recycling of cores or blanks for producing short series of small flakes and recycling of cobble fragments for their use as hammerstones. The scatter of remains corresponding to most recycling events, together with the stratigraphical location of some of them, suggest that this behavior was more common in the later phases of occupation. Recycling shows the temporal dimension of behavioral variability. Provisioning choices varied throughout the sequence of events that formed the archeological assemblage. The first occupants of the rockshelter had no other choice than to carry to the site the lithic resources that they needed. As time went by, the succession of introduction events transformed the site into a lithic provisioning area, and the occupants had the opportunity to use the lithics previously discarded by prior visitors. Lithic provisioning may have even driven some visits to the site, as the rockshelter was transformed into a supply location. The recycling of lithic resources during the last occupation phases was also proposed for other archeological levels of the Abric Romaní sequence (Vaquero 2008).

Conclusions

The study of the lithic assemblage of level J has yielded interesting insights on several aspects of Neanderthal technical behavior. This behavior combines some structural features characterizing level J as a whole and other aspects showing strong variability. The latter shows that flexibility was an important characteristic of Neanderthal technical behavior. This flexibility is evident in several aspects, from raw material provisioning to artifact use, and is largely the result of the expediency prevailing in the entire technical system. This expediency allows for permanent adaptation to changing circumstances. Variability is the rule in most behaviors, although certain aspects exhibit a marked stability, like the manufacture of retouched artifacts, which is strongly focused on denticulates and notches. Another structural trend is the preference for chert as a raw material for knapping, although it was less available than other materials in the immediate surroundings of the site. However, this preference was less pronounced during some occupation periods, which underlines the temporal variability of provisioning strategies.

Lithic variability is partly related to the temporal dynamics. The assemblage is a palimpsest formed by an unknown number of events. Some of them were associated to knapping episodes, but others to the introduction and discarding of single artifacts. In fact, two different lithic assemblages can be distinguished in level J, each of them formed in a different depositional context. These different depositional contexts were largely correlated to different types of occupation events: residential campsites and short non-residential visits. These occupation types are two basic components of Middle Paleolithic settlement patterns. In fact, residential campsites may be one of the features defining the Middle Paleolithic as a developmental stage (Rolland 1999), as indicated by the apparent correlation between the beginning of the Middle Paleolithic and the first well-defined hearth-related assemblages. As far as technical activities were concerned, the residential character of the human presence in the shelter represents a threshold in the way that lithic resources were brought and processed. Although residential occupations were not necessarily the most common, they produced the bulk of the lithic remains found in the rockshelter. They were associated with hearth construction and exhibit the repeated use of the best-sheltered areas.

The second depositional context—short non-residential visits—has been also well documented throughout the Middle Paleolithic, although it is one of the most characteristic archeological contexts since Early Pleistocene times. It has been especially well defined at sites in which it was not mixed with a residential component and the single-artifact-transport pattern was therefore easier to identify. There are many examples of these kinds of assemblages at Middle and Late Pleistocene sites. They are mainly formed by large blanks, including high percentages of retouched artifacts. Remains derived from in situ knapping sequences tend to be scarce (Brugal and Jaubert 1991; Defleur and Crégut-Bonnoure, 1995; Ollé et al. 2005; Costamagno et al. 2006). This depositional context corresponds to a common use of cave and rockshelter sites, characteristic of all prehistoric and even historic times. This component would be present in practically all the lithic assemblages. The residential use of caves would determine the differences between sites, since this depositional context would be less generalized, both in diachronic and synchronic terms.

The factors conditioning technical behavior worked at the event level and some technical features were event-specific. The assemblage variability was therefore conditioned by the kind and number of events that contributed to its formation. Since the number of events is partly a function of formation length, differences in the amount of time represented in the assemblages are a relevant factor in explaining inter-assemblage variability. For example, introducing metamorphic rocks or using quartz and limestone for knapping were specific of certain events or temporally restricted occupation phases. They were not characteristic of the whole level J sequence. Characterizing level J from a behavioral point of view seems illusory, since level J is not a behavioral unit. The "technical behavior of level J" is the average of several different and maybe contradictory actual behaviors. The behavioral significance of technical strategies should therefore be sought on an event time scale.

The number of events represented in each assemblage can explain some of the differences between sublevels Ja and Jb. These differences have been especially observed in raw material provisioning and retouched artifact types. At both cases, sublevel Ja is characterized by a stronger variability. Raw materials other than chert are more abundantly represented, especially in core reduction sequences, and typological variability is also increased due to the higher percentage of artifacts with flat or abrupt retouch. Composite tools were also more abundant in sublevel Ja. Since the number of technical events was higher in sublevel Ja, the occurrence of specific and uncommon behaviors was also more likely. Inter-assemblage variability therefore seems to be time dependent, since it is partly related to the number of events represented in the assemblage.

Some studies have tried to find coherence in the lithic assemblage, explaining both core and retouched artifact variability according to the same kind of behavior. However, since different events were conditioned by different constraints, we would expect to find contradictory behaviors in the assemblage-as-a-whole. Level J provides a good example in the raw material economy inferred from core reduction sequences and transported artifact use and discard. On the one hand, cores exhibit an economizing behavior. They were exploited until exhausted and reduction strategies were aimed at maximizing their profitability. On the other hand, transported artifacts were not intensively used, as indicated by the absence of strongly reduced artifacts and the low percentage of tools showing use wear traces. Economizing concerns were not a determining factor in the management of these artifacts. If we take into account that most retouched artifacts correspond to the transported assemblage, this suggests that core reduction and artifact retouch were unrelated phenomena subject to different constraints.

Residential occupations were more conditioned by the availability of raw materials in the immediate surroundings of the rockshelter. For the groups living in Abric Romaní, it would have been less likely to find the preferred raw material (chert), especially if foraging activities were focused on the areas closest to the site. In this context, it seems logical to maximize raw material exploitation. Short non-residential visits, however, in which most retouched artifacts were discarded, were less conditioned by the distribution and availability of lithic raw materials in the vicinity of the site. In these high mobility situations, it would have been more likely to go to the chert provisioning areas located outside of the Capellades gorge. If this movement was anticipated, there would be no pressure for maximizing the lifespan of the transported toolkit. This uneconomical behavior would be more emphasized if the residential base from which these visitors came was located close to chert outcrops. Once again, the coexistence of two different depositional contexts, associated with different occupation types, would explain these apparently contradictory behaviors.

Temporal dynamics are relevant to the interpretation of the spatial patterns of the lithic remains. Spatial studies should not be exclusively spatial. They must also include a temporal analysis, since some spatial patterns can be conditioned by the succession of different depositional contexts showing different spatial distributions. In particular, size-sorting phenomena should be scrutinized from this temporal perspective.

Level J also provides insight into Middle Paleolithic technological variability. We have identified different occupational contexts throughout the formation of level J, but there is not a clear correlation between these contexts and changes in knapping methods. Both transported artifacts and reduction sequences carried out on the spot can be attributed to the same core reduction strategies, whose expedient character is outlined above. Variability in reduction strategies is more evident at the inter-assemblage rather than the intra-assemblage level. These changes are clearer when comparing different Romaní levels—for instance, levels E and J—but can hardly be shown when analyzing only one lithic assemblage. This suggests that technological trends had a temporal pattern and characterized long time spans. Expedient discoidal methods were dominant during the formation of level J, regardless of changes in provisioning strategies or occupation types. This technological conception was applied in the different activity contexts performed by human groups in their annual cycles. Changes in the criteria used in reduction sequences would therefore correspond to long duration dynamics, acting at the longest wavelength of time (Bintliff 1991). These technical criteria defined the initial conditions upon which the variability associated with the adaptation to the specific circumstances took place. Use of complex or expedient reduction methods would not have depended on occupation length or mobility patterns. The technical paradigms were in place previous to the variations caused by such settlement factors.

Although this can hardly be considered as a scientific explanation, it seems feasible than the technological concepts dominant over a given time span were the outcome of historical circumstances, which would determine the general strategies for exploiting the environment. These strategies were the product of specific social, temporal and spatial contexts and they would remain unchanged for long times due to the conservatism characteristic of human societies in the transmission of cultural practices. In this sense, Neanderthal groups should be considered as any other subsequent historical society. The low temporal resolution of most archeological deposits may blur these temporal patterns, since artifacts corresponding to different technical phases are mixed in the same assemblages. Temporal resolution becomes therefore a key factor in evaluating technological variability and, particularly, the coexistence of expedient and predetermined reduction methods in the same assemblages.

Acknowledgments We are very grateful to Pablo Sañudo for the distribution maps included in this chapter, Susana Alonso for the drawings of lithic artifacts and Gerard Campeny for the photographs. We would also like to thank Gala G. Merino for providing us the analytical data of the galena fragments. We also thank Javier Baena, Steven L. Kuhn and Marco Peresani for their helpful comments on an earlier version of this chapter.

References

Anadón, P. (1978). El Paleógeno inferior anterior a la trasgresión "biarritziense" (Eoceno medio) entre los ríos Gaià y Ripoll (Provincias de Tarragona y Barcelona). *Estudios Geológicos, 34*, 431–440.

Anadón, P., Cabrera, L., Colldeforns, B., & Sáez, A. (1989). Los sistemas lacustres del Eoceno superior y Oligoceno del sector oriental de la Cuenca del Ebro. *Acta Geológica Hispánica, 24*, 205–230.

Anderson-Gerfaud, P. C. (1990). Aspects of behaviour in the Middle Palaeolithic: Functional analysis of stone tools from Southwest France. In P. Mellars (Ed.), *The emergence of modern humans: An archaeological perspective* (pp. 389–418). Edinburgh: Edinburgh University Press.

Arzarello, M., Bertola, S., Fontana, F., Guerreschi, A., Liagre, J., Peretto, C., et al. (2007). Aires d'approvisionnement en matières lithiques et en ressources alimentaires dans les niveaux moustériens et épigravettiens de l'Abri Tagliente (Vérone, Italie): une dimension « locale ». In M.-H. Moncel, A.-M. Moigne, M. Arzarello, & C. Peretto (Eds.), *Raw material supply areas and food supply areas. Integrated approach of the behaviours* (Vol. 1725, pp. 161–170). Oxford, England: BAR International Series.

Astruc, L. (2001). La texture des matières siliceuses et son influence sur la nature et le développement des traces d'usure: apport de méthodes expérimentales. L'exemple des cherts de la formation de Lefkara (Chypre). In L. Bourguignon, I. Ortega, & M. Ch. Frère Saulot (Eds.), *Préhistoire et approche expérimentale* (pp. 213–232). Montagnac: Monique Mergoil.

Bar-Yosef, O., Vandermeersch, B., Arensburg, B., Belfer-Cohen, A., Goldberg, P., Laville, H., et al. (1992). The excavations in Kebara Cave, Mt. Carmel. *Current Anthropology, 33*(5), 497–550.

Barton, R. N. E., & Bergman, C. A. (1982). Hunters at Hengistbury: Some evidence from experimental archaeology. *World Archaeology, 14*(2), 237–248.

Beck, R. A., Jr., Bolender, D. J., Brown, J. A., & Earle, T. K. (2007). Eventful archaeology. The place of space in structural transformation. *Current Anthropology, 48*, 833–860.

Beyries, S. (1987). *Variabilité de l'industrie lithique au moustérien. Approche fonctionnelle sur quelques gisements français*. Oxford, England: British Archaeological Report International Series 328.

Beyries, S., & Boëda, E. (1983). Etude technologique et traces d'utilisation des "eclats débordants" de Corbehem (Pas-de-Calais). *Bulletin de la Société Préhistorique Française, 80*, 275–279.

Binford, L. R. (1977). Forty-seven trips: A case study in the character of archaeological formation processes. In R. V. S. Wright (Ed.), *Stone tools as cultural markers* (pp. 24–36). Canberra, ACT: Australian Institute of Aboriginal Studies.

Binford, L. R. (1978). Dimensional analysis of behavior and site structure: Learning from an Eskimo hunting stand. *American Antiquity, 43*, 330–361.

Binford, L. R. (1979). Organization and formation processes: Looking at curated technologies. *Journal of Anthropological Research, 35*(3), 255–273.

Binford, L. R. (1980). Willow smoke and dogs' tails: Hunter-gatherer settlement systems and archaeological site formation. *American Antiquity, 45*(1), 4–20.

Binford, L. R. (1982). The archaeology of place. *Journal of Anthropological Archaeology, 1*(1), 5–31.

Binford, L. R., & Binford, S. R. (1966). A preliminary analysis of functional variability in the mousterian of Levallois facies. *American Anthropologist, 68*(2), 238–295.

Bintliff, J (1991) The contribution of an Annaliste/structural history approach to archaeology. In J. Bintliff (Ed.), *The Annales school and archaeology* (pp. 1–33). New York: Leicester University Press.

Bisson, M. (2000). Nineteenth century tools for twenty-first century archaeology? Why the Middle Paleolithic typology of François Bordes must be replaced. *Journal of Archaeological Method and Theory, 7*, 1–48.

Boëda, E. (1991). Approche de la variabilité des systèmes de production lithique des industries du Paléolithique inférieur et moyen: chronique d'une variabilité attendue. *Techniques et culture, 17–18*, 37–79.

Boëda, E. (1993a). *Le concept levallois: variabilité des méthodes*. Paris: Editions du CNRS.

Boëda, E. (1993b). Le débitage discoïde et le débitage levallois récurrent centripète. *Bulletin de la Société Préhistorique Française, 90*(6), 392–404.

Boëda, E., & Pelegrin, J. (1985). Approche expérimentale des amas de Marsagny. *Archéologie Expérimentale, 1*, 19–36.

Böeda, E., Geneste, J.-M., & Meignen, L. (1990). Identification de chaînes opératoires lithiques du Paleolithique Ancien et Moyen. *Paléo, 2*, 43–80.

Boëda, E., Connan, J., Dessort, D., Muhesen, S., Mercier, N., Valladas, H., et al. (1996). Bitumen as hafting material on Middle Palaeolithic artefacts. *Nature, 380*, 336–338.

Bofarull, J (1997) *Estudi d'un tecnocomplex del Paleolític Mitjà. La industria del nivell E de l'Abric Romaní (Capellades, Anoia)*. Ms. Thesis, Universitat Rovira i Virgili.

Bordes, F. (1948). Les couches moustériennes du gisement du Moustier (Dordogne)—typologie et technique de taille. *Bulletin de la Société Préhistorique Française, 45*, 113–126.

Bordes, F. (1953). Essai de classification des industries "moustériennes". *Bulletin de l'Association Française, 50*, 457–466.

Bordes, F. (1961). *Typologie du Paléolithique Ancien et Moyen*. Bordeaux: Delmas.

Bordes, F. (1992). *Leçon sur le Paléolithique. Paléolithique en Europe*. Vol. II. Paris: Presses du CNRS.

Bordes, F., & Bourgon, M. (1951). Le complexe moustérien: Moustériens, Levalloisien et Tayacien. *L'Anthropologie, 55*, 1–23.

Bordes, F., & de Sonneville-Bordes, D. (1970). The significance of variability in Palaeolithic assemblages. *World Archaeology, 11*, 61–73.

Bourguignon, L. (1997). *Le Moustérien de type Quina: nouvelle définition d'une entité technique*. Ph.D. Dissertation, Université de Paris X.

Bourguignon, L., Faivre, J. P., & Turq, A. (2004). Ramification des chaînes opératoires: une spécificité du moustérien? *Paléo, 16*, 37–48.

Brink, J. S., & Henderson, Z. L. (2001). A high-resolution last interglacial MSA horizon at Florisbad in the context of other open-air occurrences in the central interior of Southern Africa: An interim statement. In N. J. Conard (Ed.), *Settlement dynamics of the Middle Paleolithic and Middle Stone Age* (pp. 1–20). Tübingen: Kerns Verlag.

Brooks, R. L. (1982). Events in the archaeological context and archaeological explanation. *Current Anthropology, 23*, 67–75.

Brugal, J.-Ph., & Jaubert, J. (1991). Les gisements paléontologiques pléistocènes à indices de frequentation humaine: un nouveau type de comportement de prédation? *Paléo, 3*, 15–41.

Bunte, K., & Abt, S. R. (2001). *Sampling surface and subsurface particle-size distributions in wadable gravel-and cobble-bed streams for analyses in sediment transport, hydraulics and streambed monitoring*. Fort Collins, CO: Department of Agriculture, Forest Service, Rocky Mountain Research Station.

Byrne, L., Ollé, A., & Vergès, J. M. (2006). Under the hammer: Residues resulting from production and microwear on experimental stone tools. *Archaeometry, 48*(4), 549–564.

Cabrera, V. (1984). *El yacimiento de la Cueva de "El Castillo" (Puente Viesgo*. Santander). Bibliotheca Praehistorica Hispana, XXII. Madrid: Consejo Superior de Investigaciones Científicas.

Cáceres, I. (2002). *Tafonomía de yacimientos antrópicos en Karst. Complejo Galería (Sierra de Atapuerca, Burgos), Vanguard Cave (Gibraltar) y Abric Romaní (Capellades, Barcelona)*. Ph.D. Dissertation, Universitat Rovira i Virgili.

Cahen, D., & Keeley, L. H. (1980). Not less than two, not more than three. *World Archaeology, 12*, 166–180.

Carbonell, E. (1992). Abric Romaní Nivell H. Un model d'estratègia ocupacional al Pleistocè Superior mediterrani. *Estrat, 5*, 157–308.

Carbonell, E. (2002). *Abric Romaní nivell I. Models d'ocupació de curta durada de fa 46.000 anys a la Cinglera del Capelló (Capellades, Anoia, Barcelona)*. Tarragona: Universitat Rovira i Virgili.

Carbonell, E., & Castro-Curel, Z. (1992). Palaeolithic wooden artefacts from the Abric Romaní (Capellades, Barcelona, Spain). *Journal of Archaeological Science, 19*, 707–719.

Caspar, J.-P., Masson, B., & Vallin, L. (2003). Poli de bois ou poli de glace au Paléolithique inférieur et moyen? Problèmes de convergence taphonomique et fonctionnelle. *Bulletin de la Société Préhistorique Française, 100*(3), 453–462.

Church, T. (2000). Distribution and sources of obsidian in the Rio Grande Gravels of New Mexico. *Geoarchaeology, 15*(7), 649–678.

Close, A. E. (1996). Carry that weight: The use and transportation of stone tools. *Current Anthropology, 37*, 545–553.

Colldefons, B., Anadón, P., & Cabrera, L. (1994). Litoestratigrafía del Eoceno superior-Oligoceno inferior de la zona oriental de la Cuenca del Ebro. Sector Igualada-Santa Coloma de Queralt. *Geogaceta, 15*, 55–58.

Costamagno, S., Meignen, L., Beauval, C., Vandermeersch, B., & Maureille, B. (2006). Les Pradelles (Marillac-le-Franc, France): A mousterian reindeer hunting camp. *Journal of Anthropological Archaeology, 25*, 466–484.

Cotterell, B., & Kamminga, J. (1979). The mechanics of flaking. In B. Hayden (Ed.), *Lithic use-wear analysis* (pp. 97–112). New York: Academic Press.

Cronyn, J. M. (1990). *The elements of archaeological conservation*. London: Routledge.

Cziesla, E. (1990). On refitting of stone artefacts. In E. Cziesla, S. Eickhoff, N. Arts, & D. Winter (Eds.), *The big puzzle. International symposium on refitting stone artefacts. Monrepos 1987* (pp. 9–44). Studies in modern archaeology 1. Bonn, Germany: Holos.

Cziesla, E., Eickhoff, S., Arts, N., & Winter, D. (1990). *The big puzzle. International symposium on refitting stone artefacts. Monrepos 1987*. Studies in modern archaeology 1. Bonn: Holos.

Davids, G. H., & Reynolds, S. J. (1996). *Structural geology of rocks and regions*. New York: Wiley.

De Beaune, S. A. (1997). *Les Galets utilisés au paléolithique supérieur: approche archéologique et expérimentale*. Supplément Gallia Préhistoire. Paris: CNRS.

De Beaune, S. A. (2000). *Pour un Archéologie du geste*. Paris: Éditions du CNRS.

De Beaune, S. A. (2004). The invention of technology. *Current Anthropology, 45*(2), 139–161.

Defleur, A., & Crégut-Bonnoure, E. (1995). *Le gisement paléolithique moyen de la grotte des Cèdres (Var)*. Paris: Editions de la Maison des Sciences de l'Homme.

Depaepe, P (2007) L'approvisionnement en matières premières lithiques en France septentrionale au Paléolithique moyen récent. In M.-H. Moncel, A.-M. Moigne, M. Arzarello, & C. Peretto (Eds.), *Raw material supply areas and food supply areas. Integrated approach of the behaviours* (pp. 87–92). Oxford, England: BAR International Series, 1725.

Dibble, H. L. (1984). Interpreting typological variation of Middle Paleolithic scrapers: Function, style or sequence of reduction? *Journal of Field Archaeology, 11*, 431–436.

Dibble, H. L. (1987). The interpretation of Middle Paleolithic scraper morphology. *American Antiquity, 52*, 189–196.

Dibble, H. L. (1988). The interpretation of Middle Paleolithic scraper-reduction patterns. In M. Otte (Ed.), *L'Homme de Neandertal, 4. La technique* (pp. 49–58). Liège: ERAUL.

Dibble, H. L. (1991). Mousterian assemblage variability on an interregional scale. *Journal of Anthropological Research, 47*, 239–257.

Dibble, H. L. (1995a). Middle Paleolithic scraper reduction: Background, clarification, and review of the evidence to date. *Journal of Archaeological Method and Theory, 2*(4), 299–368.

Dibble, H. L. (1995b). Biache Saint-Vaast, Level IIA: A comparison of analytical approaches. In H. L. Dibble & O. Bar-Yosef (Eds.), *The definition and interpretation of Levallois technology* (pp. 93–116). Monographs in World Archaeology, Vol. 23. Madison, WI: Prehistory Press.

Dibble, H. L., & Holdaway, S. J. (1990). Le Paléolithique moyen de l'abri sous roche de Warwasi et ses relations avec le Moustérien du Levant. *L'Anthropologie, 94*, 619–642.

Dibble, H. L., & McPherron, S. P. (2006). The missing Mousterian. *Current Anthropology, 47*(5), 777–803.

Dibble, H. L., & Pelcin, A. (1995). The effect of hammer mass and velocity on flake mass. *Journal of Archaeological Science, 22*, 429–439.

Dibble, H. L. & Rolland, N. (1992). On assemblage variability in the Middle Paleolithic of Western Europe. History, perspectives and a new synthesis. In H. L. Dibble & P. Mellars (Eds.), *The Middle Paleolithic: Adaptation, behavior and variability* (pp. 1–28). Philadelphia, PA: The University Museum, University of Pennsylvania.

Dibble, H. L., Roth, B., & Lenoir, M. (1995). The use of raw materials at Combe-Capelle Bas. In H. L. Dibble & M. Lenoir (Eds.), *The Middle Paleolithic site of Combe-Capelle Bas* (France) (pp. 259–287). Philadelphia, PA: The University Museum, University of Pennsylvania.

Esparza, X. (1993). Introducción al Paleolítico Superior de la cueva de Lezetxiki (Mondragón, Guipúzcoa). *Espacio, Tiempo y Forma, Serie I, Prehistoria y Arqueología, 6*, 31–60.

Ettler, V., Johanb, Z., & Hradilc, D. (2003). Natural alteration products of sulphide mattes from primary lead smelting. *CR Geoscience, 335*, 1013–1020.

Evans, A., Wolframm, Y. B., Donahue, R. E., & Lovis, W. A. (2007). A pilot study of ''black chert'' sourcing and implications for assessing hunter-gatherer mobility strategies in Northern England. *Journal of Archaeological Science, 34*, 2161–2169.

Faulkner, A. (1972). *Mechanical principles of flintworking*. Ph.D. Dissertation, Washington State University.

Féblot-Augustins, J. (1997). *La circulation des matières premières au Paléolithique. ERAUL, 75*. Liège: Université de Liège.

Féblot-Augustins, J. (1999). Raw materials transport patterns and settlement systems in the European lower and Middle Paleolithic: Continuity, change and variability. In W. Roebroeks & C. Gamble (Eds.), *The Middle Paleolithic occupation of Europe* (pp. 193–214). Leiden: University of Leiden.

Fernandes, P., & Raynal, J.-P. (2006). Pétroarchéologie du silex: un retour au sources. *Comptes Rendus Palevol, 5*(6), 829–837.

Fernandes, P., Raynal, J.-P., & Moncel, M.-H. (2008). Middle Palaeolithic raw material gathering territories and human mobility in the southern Massif Central, France: First results from a petro-archaeological study on flint. *Journal of Archaeological Science, 35*(8), 2357–2370.

Fernández-Laso, M. C., Chacón, M. G., García-Antón, M. D., & Rivals, F. (2011): Territorial mobility of Abric Romaní level M Neanderthals groups (Capellades, Barcelona, Spain). In N. J. Conard & J. Richter (Eds.), *Neanderthal lifeways, subsistence and technology*, pp. 187-202. New York: Springer.

Ferrer, J. (1971). El Paleoceno y Eoceno del borde sur-oriental de la depresión del Ebro (Cataluña). *Mémoires Suisses de Paléontologie, 90*, 1–70.

Fripp, J. B., & Diplas, P. (1993). Surface sampling in gravel streams. *Journal of Hydraulic Engineering, 119*(4), 473–490.

Gallart, F. (1981). Neógeno superior y cuaternario del Penedès (Catalunya, España). *Acta Geológica Hispánica, 16*, 151–156.

Gallart, F. (1991). Estudi geomorfològic de la conca d'Òdena. *Estrat, 3*, 7–45.

Gallet, M. (1998). *Pour une technologie des débitages laminaires préhistoriques*. Paris: Dossier de documentation préhistorique 19, CNRS.

Gamble, C., & Porr, M. (2005). From empty spaces to lived lives. Exploring the individual in the Palaeolithic. In C. Gamble & M. Porr (Eds.), *The hominid indivual in context. Archaeological investigations of Lower and Middle Palaeolithic landscapes, locales and artefacts* (pp. 1–12). London: Routledge.

García Rodrigo, B. (1957). El Valle del Anoia. *Memorias y comunicaciones, XVI*, 45–80.

Geneste, J.-M. (1985). Analyse lithique d'industries moustériennes du Périgord: une approche technologique de comportement des groupes humains au Paléolithique moyen. Ph.D. Dissertation, Université de Bordeaux.

Geneste, J.-M. (1988). Les industries de la Grotte Vaufrey: technologie du debitage, economie et circulation de la matière première lithique. In J.-Ph. Rigaud (Ed.), *La Grotte Vaufrey à Cenac et Saint-Julien (Dordogne). Paléoenvironnements, chronologie et activités humaines* (pp. 441–517). Paris: Mémoires de la Société Préhistorique Française XIX.

Geneste, J.-M. (1989a). Systemes d'approvisionnement en matières premièrs au Paléolithique moyen et au Paléolithique supérieur en Aquitaine. In L'Homme de Neandertal, 8. La mutation (pp. 61–70). Liège: ERAUL.

Geneste, J.-M. (1989b). Economie des resources lithiques dans le mousterien du Sud-Ouest de la France. In *L'Homme de Neandertal, 6. La subsistance* (pp. 75–97). Liège: ERAUL.

Geneste, J.-M. (1991). Systèmes techniques de production lithique: variations techno-économiques dans les processus de réalisation des outillages paléolithiques. *Techniques et culture, 17–18*, 1–35.

Geneste, J.-M. (1992). L'approvisionnement en matières premières dans les systemes de production lithique: la dimension spatiale de la technologie. In R. Mora, X. Terradas, A. Parpal, & C. Plana (Eds.), *Tecnología y Cadenas Operativas Líticas* (pp. 1–31). Treballs d'Arqueología, 1. Barcelona: Universitat Autònoma de Barcelona, Spain.

Geneste, J.-M., & Rigaud, J.-Ph. (1989). Matières premières lithiques et occupation de l'espace. *Cahiers du Quaternaire, 13*, 205–218.

Gómez de Soler, B. (2007). *Áreas de captación y estrategias de aprovisionamiento de rocas silíceas en el nivel L del Abric Romaní (Capellades, Barcelona)*. Ms. Thesis, Universitat Rovira i Virgili.

Goren-Inbar, N. (1988). Too small to be true? Reevaluation of cores on flakes in Levantine Mousterian assemblages. *Lithic Technology, 17*(1), 37–44.

Guigon, I., Benzanquen, M., Martínez, W., Núñez, A., & Cabañas, I. (1973). *Mapa geológico de España. E. 1:50.000. Hoja n° 419 (Villafranca del Penedés)*. Madrid: IGME.

Hatcher, J. R. D. (1995). *Structural geology*. Upper Saddle River, NJ: Prentice Hall.

Hall, C. T. (2004). Evaluating prehistoric hunter-gatherer mobility, land use, and technological organization strategies using minimum analytical nodule analysis. In C. T. Hall & M. L. Larson (Eds.), *Aggregate analysis in chipped stone* (pp. 139–155). Salt Lake City, UT: University of Utah Press.

Henry, D. O. (1998). Intrasite spatial patterns and behavioral modernity. Indications from the Late Levantine Mousterian Rockshelter of Tor Faraj, Southern Jordan. In T. Akazawa, K. Aoki, & O. Bar-Yosef (Eds.), *Neanderthals and modern humans in Western Asia* (pp. 127–142). New York: Plenum Press.

Henry, D. O., Hietala, H. J., Rosen, A. M., Demidenko, Y. E., Usik, V. I., & Armagan, T. L. (2004). Human behavioral organization in the Middle Paleolithic: Were Neanderthals different? *American Anthropologist, 106*(1), 17–31.

Hill, J., & Gunn, J. (1977). *The individual in prehistory*. New York: Academic Press.

Hiscock, P., & Attenbrow, V. (2003). Early Australian implement variation: A reduction model. *Journal of Archaeological Science, 30*, 239–249.

Hofman, J. L., & Enloe, J. G. (1992). *Piecing together the past: Applications of refitting studies in archaeology*. Oxford, England: BAR International Series 578.

Jardón, P. (2000). *Los raspadores en el Paleolítico Superior. Tipología, tecnología y función en la Cova del Parpalló (Gandía, España) y en la Grotte Gazel (Sallèles-Cabardès, Francia)*. Valencia: Diputación Provincial de Valencia.

Jaubert, J. & Farizy, C. (1995). Levallois debitage: Exclusivity, absence or coexistence with other operative schemes (Garonne Basin, SW France). In H. L. Dibble & O. Bar-Yosef (Eds.), *The definition and interpretation of Levallois technology* (Vol. 23, pp. 27–248). Monographs in World Archaeology, Madison, WI: Prehistory Press.

Jelinek, A. J. (1994). Hominids, energy, environnement and behavior in the Late Pleistocene. In M. H. Nitecki & D. V. Nitecki (Eds.), *Origins of anatomically modern humans* (pp. 67–92). New York: Plenum Press.

Keeley, L. H. (1980). *Experimental determination of stone tool uses: A microwear analysis*. Chicago: Chicago University Press.

Keller, C. M., & Keller, J. D. (1996). *Cognition and tool use. The blacksmith at work*. Cambridge: Cambridge University Press.

Ketterer, I., Pigeot, N., & Serra, S. (2004). Le temps de l'occupation. Une histoire des activités et des comportements. In N. Pigeot (Ed.), *Les derniers magdaléniens d'Étiolles. Perspectives culturelles et paléohistoriques* (pp. 235–254). Paris: Éditions du CNRS.

Krinsley, D. H., Pye, K., Boggs, S., & Tove, N. K. (1998). *Backscattered scanning electron microscopy and image analysis of sediments and sedimentary rocks*. Cambridge: Cambridge University Press.

Krumbein, W. C. (1941). Measurement and geological significance of shape and roundness of sedimentary particles. *Journal of Sedimentary Petrology, 11*(2), 64–72.

Kuhn, S. L. (1989). Hunther-gatherer foraging organization and strategies of artifact replacement and discard. In D. Amick & R. Mauldin (Eds.), *Experiments in lithic technology* (pp. 33–46). Oxford, England: BAR International Series 528.

Kuhn, S. L. (1991). "Unpacking" reduction: Lithic raw material economy in the Mousterian of West-Central Italy. *Journal of Anthropological Archaeology, 10*, 76–106.

Kuhn, S. L. (1992). On planning and curated technologies in the Middle Paleolithic. *Journal of Anthropological Research, 43*, 185–214.

Kuhn, S. L. (1994). A formal approach to the design and assembly of mobile toolkits. *American Antiquity, 59*(3), 426–442.

Kuhn, S. L. (1995). *Mousterian lithic technology. An ecological perspective*. Princeton, NJ: Princeton University Press.

Kuhn, S. L. (1996). The trouble with ham steaks: A reply to Morrow. *American Antiquity, 61*(3), 591–595.

Laplace, G. (1972). La typologie analytique et structurale: Base rationnelle d'étude des industries lithiques et osseuses. Banques de données archéologiques. *Colloques nationaux du CNRS, 932*, 91–143.

Larson, M. L. (2004). Chipped stone aggregate analysis in archaeology. In C. T. Hall & M. L. Larson (Eds.), *Aggregate analysis in chipped stone* (pp. 3–17). Salt Lake City, UT: University of Utah Press.

Lemorini, C. (2000). *Reconnaître des tactiques d'exploitation du milieu au Paléolithique Moyen. La contribution de l'analyse fonctionnelle. Etude fonctionnelle des industries lithiques de Grotta Breuil (Latium, Italie) et de La Combette (Bonnieux, Vaucluse, France)*. Oxford, England: BAR International Series 858.

Lemorini, C., Peresani, M., Rossetti, P., Malerba, G., & Giacobini, G. (2003). Techno-morphological and use-wear functional analysis: An integrated approach to the study of a discoid industry. In M. Peresani (Ed.), *Discoid lithic technology. Advances and implications* (pp. 257–276). Oxford, England: BAR International Series 1120.

Lerner, H., Du, X., Costopoulos, A., & Ostaja-Starzewski, M. (2007). Lithic raw material physical properties and use-wear accrual. *Journal of Archaeological Science, 34*, 711–722.

Levi-Sala, I. (1986). Experimental replication of post-depositional surface modifications on flint. *Early Man News*, 9(10/11), 103–110.

Levi-Sala, I. (1996). *A study of microscopy polish on flint implements*. Oxford, England: BAR International Series 629.

Luedtke, B. E. (1992). *An archaeologist's guide to chert and flint*. Los Angeles, CA: University of California.

Locht, J.-L., & Swinnen, C. (1994). Le debitage discoïde du gisement de Beauvais (Oise): aspects de la chaîne opératoire au travers de quelques remontages. *Paléo, 6*, 89–104.

Maíllo, J. M., Cabrera, V., & Bernaldo de Quirós, F. (2004). Le débitage lamellaire dans le Moustérien final de Cantabrie (Espagne): le cas de El Castillo et Cueva Morin. *L'Anthropologie, 108*, 367–393.

Manzano, I., Baena, J., Lázaro, A., Martín, D., Dapena, L., Roca, M. et al. (2005). Análisis de los recursos líticos en la Cueva del Esquilleu: Gestión y comportamiento durante el Musteriense (Comarca de la Liébana, Occidente de Cantabria). In R. Montes & J. A. Lasheras (Eds.), *Neandertales cantábricos, estado de la cuestión* (pp. 285–300). Museo de Altamira, Monografías, 20. Santander: Museo de Altamira.

Martínez, K. (2005). *Análisis funcional de industrias líticas del Pleistoceno Superior. El Paleolítico Medio del Abric Romaní (Capellades, Barcelona) y el Paleolítico Superior de Üçağizli (Hatay, Turquía) y del Molí del Salt (Vimbodí, Tarragona). Cambios en los patrones funcionales entre el Paleolítico Medio y el Superior*. Ph.D. Dissertation, Universitat Rovira i Virgili.

Martínez Molina, K., & Rando, J. M. (2001). Organización y funcionalidad de la producción lítica en un nivel del Paleolítico Medio del Abric Romaní. Nivel Ja (Capellades, Barcelona). *Trabajos de Prehistoria, 58*(1), 51–70.

Martínez Moreno, J., Mora, R., & de la Torre, I. (2004). Methodological approach for understanding Middle Palaeolithic settlement dynamics at La Roca dels Bous (Noguera, Catalunya, Northeast Spain). In N. J. Conard (Ed.), *Settlement dynamics of the Middle Paleolithic and Middle Stone Age II* (pp. 393–414). Tübingen: Kerns Verlag.

Marzo, M. & Calvet, F. (1985). *Guia de la excursión al triásico de las Catalánides*. II Coloquio de estratigrafia y paleogeografia del Pérmico y Triásico de España. La Seu d'Urgell.

Masson, M. (1982). Silice et traces d'usage. *La part du silex. Nouvelles Archéologiques du Musée d'Histoire Naturelle de Lyon, 22*, 31–49.

Mata, J. (1990). *Els minerals de Catalunya*. Barcelona: Institut d'Estudis Catalans.

Meignen, L. (1988). Un exemple de comportement technologique différent selon les matières premières: Marillac, couches 9 et 19. In *L'homme de Neandertal, La technique* (Vol. 4, pp. 71–79). Liège: ERAUL.

Meignen, L. (1993). *L'abri des Canalettes. Un habitat moustérien sur les grands Causses (Nant, Aveyron). Fouilles 1980–1986. Monographie du CRA, 10*. Paris: Editions du CNRS.

Meignen, L., Costamagno, S., Beauval, C., Bourguignon, L., Vandermeersch, B., & Maureille, B (2007) Gestion des ressources lithiques au Paléolithique moyen dans une halte de chasse spécialisée sur le Renne : Les Pradelles (Marillac-Le-Franc, Charente). In M.-H. Moncel, A.-M. Moigne, M. Arzarello, & C. Peretto (Eds.), *Raw material supply areas and food supply areas. Integrated approach of the behaviours* (pp. 127–140). Oxford, England: BAR International Series 1725.

Molines, N. (2007). Paléogéographie, matières premières et territoires exploitables dans l'Ouest de la France au Pléistocène moyen. In M.-H. Moncel, A.-M. Moigne, M. Arzarello, & C. Peretto (Eds.), *Raw material supply areas and food supply areas. Integrated approach of the behaviours* (pp. 61–72). Oxford, England: BAR International Series 1725.

Moncel, M.-H. (2001). Middle Paleolithic blade assemblages in Southeastern France: The question of technical variability during the Middle Paleolithic and its meaning. *Archaeology, Ethnology & Anthropology of Eurasia, 2*, 37–47.

Moncel, M.-H. (2003). *L'exploitation de l'espace et la mobilité des groupes humaines au travers des assemblages lithiques à la fin du Pléistocène moyen et au début du Pléistocène supérieur*. La moyenne vallée du Rhône entre Drôme et Ardèche. Oxford, England: BAR International Series 1184.

Moncel, M.-H., & Neruda, P. (2000). The Külna level 11: Some observation on the debitage rules and aims. The originality of a Middle Palaeolithic microlithic assemblage (Külna Cave, Czech Republic). *Anthropologie, 38*(2), 219–247.

Moncel, M.-H., Brugal, J.-Ph, Prucca, A., & Lhomme, G. (2008). Mixed occupation during the Middle Palaeolithic: Case study of a small pit-cave site of Les Pêcheurs (Ardèche, south-eastern France). *Journal of Anthropological Archaeology, 27*, 382–398.

Monnier, J.-L. (1990). L'organisation spatiale de l'habitat au Paléolithique inférieur et moyen d'après les premiers résultats de la fouille de trois sites armoricains. *Bulletin de la Société Préhistorique Française, 87*(10–12), 321–325.

Morant, N. (1998). *Estudi de les matèries primeres lítiques del nivell I del jaciment de l'Abric Romaní (Capellades, Barcelona)*. Ms. Thesis, Universitat Rovira i Virgili.

Morin, E. (2004). An experimental model for interpreting Levallois patterning and toolkit transport. In N. J. Conard (Ed.), *Settlement dynamics of the Middle Paleolithic and Middle Stone Age II* (pp. 317–341). Tübingen: Kerns Verlag.

Morrow, T. M. (1996a). Lithic refitting and archaeological site formation processes: A case study from the Twin Ditch Site, Greene County, Illinois. In G. H. Odell (Ed.), *Stone tools. Theoretical insights into human prehistory* (pp. 345–373). New York: Plenum Press.

Morrow, T. M. (1996b). Bigger is better: Comments on Kuhn's formal approach to mobile toolkits. *American Antiquity, 61*, 581–590.

Mourre, V. (2003). Discoïde ou pas discoïde? Réflexions sur la pertinence des critères techniques définissant le débitage discoïde. In M. Peresani (Ed.), *Discoid lithic technology. Advances and implications* (pp. 1–18). Oxford, England: BAR International Series 1120.

Napier, J. R. (1962). The evolution of hand. *Scientific American, 207,* 56–62.

Newcomer, M. H., & de Sieveking, G. (1980). Experimental flake scatter-patterns: A new interpretative technique. *Journal of Field Archaeology, 7,* 345–352.

O'Connell, J. F. (1987). Alyawara site structure and its archaeological implications. *American Antiquity, 52*(1), 74–108.

Odell, G. H. (2004). *Lithic analysis.* New York: Kluwer Academic/ Plenum Publishers.

Ollé, A. (2003). *Variabilitat i patrons funcionals en els sistemes tècnics de mode 2. Anàlisi de les deformacions d'ús en els conjunts lítics de Riparo Esterno de Grotta Paglicci (Rignano Garganico, Foggia), Àridos (Arganda, Madrid) i Galeria (Sierra de Atapuerca, Burgos).* Ph.D. Dissertation, Universitat Rovira i Virgili.

Ollé, A., Cáceres, I., & Vergès, J. M. (2005). Human occupations at Galería site (Sierra de Atapuerca, Burgos, Spain) after the technological and taphonomical data. In N. Molines, M.-H. Moncel, & J.-L. Monnier (Eds.), *Les premiers peuplements en Europe. Colloque international: Données récentes sur les modalités de peuplement et sur le cadre chronostratigraphique, géologique et paléogéographique des industries du Paléolithique ancien et moyen en Europe (Rennes, 22–25 septembre 2003)* (pp. 269–280). Oxford, England: BAR Internacional Series, S1364.

Ortega, D. & Maroto, J. (2001). Matières premières et technologie lithique du Moustérien final de la grotte des Ermitons (Pyrénées Méditerranéennes). In J. Zilhao, Th. Aubry, & A. F. Carvalho (Eds.), *Les premiers hommes modernes de la Péninsule Ibérique. Actes du Colloque de la Commission VIII de l'UISPP. Vila Nova de Foz Côa, 22–24 Octubre 1998* (pp. 69–76). Lisboa: Instituto Portugues de Arqueología.

Ortí, F. (1990). Las formaciones evaporíticas del Terciario continental de la zona de contacto entre la Cuenca del Ebro y los Catalánides. In F. Ortí & J. M. Salvany (Eds.), *Formaciones evaporíticas de la Cuenca del Ebro y cadenas periféricas, y de la zona de Levante: Nuevas aportaciones y guía de superficie* (pp. 70–75). Barcelona: Enresa-Universidad de Barcelona.

Ortí, F., Rosell, L., Inglès, M., & Playà, E. (2007). Depositional models of lacustrine evaporites in the SE margin of the Ebro Basin (Paleogene, NE Spain). *Geologica Acta, 5*(1), 19–34.

Otte, M. (1991). Evolution in the relationships between raw materials and cultural tradition in the European Paleolithic. In A. Montet-White & S. Holen (Eds.), *Raw material economies among prehistoric hunter-gatherers* (pp. 161–167). Lawrence, KS: University of Kansas.

Parry, W. A., & Kelly, R. L. (1987). Expedient core technology and sedentism. In J. K. Johnson & C. A. Morrow (Eds.), *The organization of core technology* (pp. 285–304). Boulder, CO: Westview Press.

Patou-Mathis, M. (2002). *Retouchoirs, compresseurs, percuteurs ... : Os à impressions et éraillures.* Fiches typologiques, Cahier X. Paris: Société préhistorique française.

Peón, A., Alonso, F., & Ramírez del Pozo, J. (1975). *Mapa Geológico de España. E.1:50.000. Hoja n° 391 (Igualada).* Madrid: IGME.

Peresani, M. (1998). La variabilité du débitage discoïde dans la Grotte de Fumane (Italie du Nord). *Paléo, 10,* 123–146.

Peresani, M. & Porraz, G. (2004). Ré-interprétation et mise en valeur des niveaux moustériens de la Grotte du Broion (Monti Berici, Vénetie). Etude techno-économique des industries lithiques. *Rivista di Scienze Preistoriche,* LIV, 181–247.

Pettijohn, F. J. (1975). *Sedimentary rocks.* New York: Harper and Row.

Peyrony, D. (1930). Le Moustier, ses gisements, ses industries et ses couches géologiques. *Revue Anthropologique, 40*(3–76), 155–176.

Plisson, H., & Mauger, M. (1988). Chemical and mechanical alteration of microwear polishes: An experimental approach. *Helinium,* XXVIII(1), 3–16.

Poplin, F. (1988). Aux origines néandertaliennes de l'art. Matière, forme, symetries. Contribution d'une galène et d'un oursin fossile taillé de Merry-sur-Yonne (France). In L'Homme de Néandertal, 5. La Pensée (pp. 109–116). Liège: ERAUL.

Rick, J. W. (1976). Downslope movement and archaeological intrasite spatial analysis. *American Antiquity, 41*(2), 133–144.

Rios, J. (2005). Características de la producción lítica al final del Paleolítico Medio en el País Vasco. El caso del Nivel B de Axlor (Dima, Bizkaia). In R. Montes & J. A. Lasheras (Eds.), *Neandertales cantábricos, estado de la cuestión* (pp. 333–348). Museo de Altamira, Monografías, 20. Santander: Museo de Altamira.

Rodríguez Rodríguez, A. C., Galván Santos, B., & Hernández Gómez, C. (2002). Contribución del análisis funcional en la caracterización de El Salt como un centro de intervención referencial de las poblaciones neandertalianas en los valles de Alcoi (Alicante). In I. Clemente, R. Risch, & J. F. Gibaja (Eds.), *Análisis Funcional. Su aplicación al estudio de sociedades prehistóricas* (pp. 121–132). Oxford, England: BAR International Series 1073.

Roebroeks, W. (1988). *From find scatters to early hominid behaviour: A study of Middle Palaeolithic riverside settlements at Maastricht-Belvédère (The Netherlands). Analecta Praehistorica Leidensia, 21.* Leiden: University of Leiden.

Roebroeks, W., de Loecker, D., Hennekens, P., & van Ieperen, M. (1992). "A veil of stones": On the interpretation of an early Middle Palaeolithic low density scatter at Maastricht-Belvédère (The Netherlands). *Analecta Praehistorica Leidensia, 25,* 1–16.

Rolland, N. (1981). The interpretation of Middle Paleolithic variability. *Man, 16,* 15–42.

Rolland, N. (1988). Variabilité et classification: nouvelles donnees sur "Le complexe mousterien". In M. Otte (Ed.), *L'Homme de Neandertal, 4. La technique* (pp. 169–183). Liège: ERAUL.

Rolland, N. (1999). The Middle Palaeolithic as development stage: Evidence from technology, subsistence, settlement systems, and hominid socio-ecology. In H. Ullrich (Ed.), *Hominid evolution. Lifestyles and survival* (pp. 315–334). Gelsenkirchen/Schwelm: Archaea.

Rolland, N., & Dibble, H. L. (1990). A new synthesis of Middle Paleolithic variability. *American Antiquity, 55*(3), 480–499.

Roussel, M. (2005). L'usage de la percussion à la pierre tendre au Paléolithique moyen: approche expérimentale et reconnaissance des stigmates de taille appliquées au Mousterien de type Quina de « Chez-Pinaud » à Jonzac (Charente-Maritime). Ms. Thesis, Université de Bordeaux I.

Rottländer, R. (1975). The formation of patina on flint. *Archaeometry, 17*(1), 106–110.

Sala, M. (1994). Sistema Costero Catalán. In M. Gutiérrez (Ed.), *Geomorfología de España* (pp. 287–304). Madrid: Editorial Rueda.

Schick, K. D. (1987). Experimentally-derived criteria for assessing hydrologic disturbance of archeological sites. In D. T. Nash & M. D. Petraglia (Eds.), *Natural formation processes and the archaeological record* (pp. 86–107). Oxford, England: BAR International Series 352.

Schurmans, U., & De Bie, M. (2007). *Fitting rocks: Lithic refitting examined.* Oxford, England: BAR International Series 1596.

Serra, M. (1990). Geología de 1os materiales Paleozoicos del área de Capellades (Prov. de Barcelona). *Acta geológica hispánica, 25*(2), 123–132.

Shchelinskij, V. E. (1993). Outils pour travailler les bois et l'os au Paléolithique inférieur et moyen de la Plane russe et du Caucase. In P. Anderson, S. Beyries, M. Otte, & H. Plisson (Eds.), *Traces et fonction: les gestes retrouvés* (pp. 309–316). Liège: ERAUL.

Shelley, Ph. (1993). A geoarchaeological approach to the analysis of secondary lithic deposits. *Geoarchaeology, 8*(1), 59–72.

Slimak, L. (2003). Les débitages discoïdes moustériens: évaluation d'un concept technologique. In M. Peresani (Ed.), *Discoid lithic technology. Advances and implications* (pp. 33–67). Oxford, England: BAR International Series 1120.

Solé Sabarís, L. (1958–1964). *Geografia de Catalunya*. Barcelona: Aedos.

Stevenson, M. G. (1985). The formation of artifact assemblages at workshop/habitation sites: Models from peace point in northern Alberta. *American Antiquity, 50*(1), 63–81.

Stevenson, M. G. (1991). Beyond the formation of hearth-associated artifact assemblages. In E. M. Kroll & T. D. Price (Eds.), *The interpretation of archaeological spatial patterning* (pp. 269–299). New York: Plenum Press.

Tavoso, A. (1984). Réflexion sur l'économie des matières premières au Moustérien. *Bulletin de la Societé Préhistorique Française, 81*(3), 79–82.

Terradas, X. (2003). *Discoid flaking method: Conception and technological variability.* In M. Peresani (Ed.), *Discoid lithic technology. Advances and implications (pp. 19–32).* Oxford, England: BAR International Series 1120.

Texier, J.-P. (1981). Désilification des silex taillés. *Quaternaria, 23*, 159–169.

Texier, P.-J., Brugal, J.-Ph., Lemorini, C., & Wilson, L. (1998). Fonction d'un site du Paléolithique moyen en marge d'un territoire: l'abri de La Combette (Bonnieux, Vaucluse). In J.-Ph. Brugal, L. Meignen, & M. Patou-Mathis (Eds.), *Économie préhistorique: les comportements de subsistance au Paléolithique* (pp. 325–348). Sophia Antipolis: Éditions APDCA.

Thieme, H. (1997). Lower Palaeolithic hunting spears from Germany. *Nature, 385*, 807–810.

Turq, A. (1989). Exploitation des matières premières lithiques et exploitation du sol: l'exemple du Moustérien entre Dordogne et Lot. *Cahiers du Quaternaire, 13*, 179–204.

Turq, A. (1990). Explotation des matières premières lithiques dans le moustérien entre Dordogne et Lot. *Cahiers du Quaternaire, 17*, 415–427.

Turq, A. (1992). Raw material and technological studies of the Quina Mousterian in Périgord. In H. L. Dibble & P. A. Mellars (Eds.), *The Middle Paleolithic: Adaptation, behavior, and variability* (pp. 75–85). Philadelphia PA: University of Pennsylvania.

Turq, A. (2000) *Paléolithique inférieur et moyen entre Dordogne et Lot. Paléo*, Supplément 2.

Unrath, G., Owen, L. R., Van Gijn, A., Moss, E. H., Plisson, H., & Vaughan, P. (1986). An evaluation of microwear studies: A multi-analyst approach. *Early Man News*, 9/10/11, 117–176.

Vallverdú, J. (2001). *Micromorfología de las facies sedimentarias de la Sierra de Atapuerca y del nivel J del Abric Romaní. Implicaciones geoarqueológicas y paleoetnográficas.* Ph.D. Dissertation, Universitat Rovira i Virgili.

Van Den Dries, M., & Van Gijn, A. (1997). The representativity of experimental usewear traces. In A. Ramos-Millán & M. A. Bustillo (Eds.), *Siliceous rocks and culture* (pp. 499–513). Granada: Universidad de Granada.

Van Gijn, A. L. (1989). *The wear and tear of flint. Principles of functional analysis applied to Dutch Neolithic assemblages. Analecta Praehistorica Leidensia 22.* Leiden: University of Leiden.

Vaquero, M. (1997). *Tecnología lítica y comportamiento humano: organización de las actividades técnicas y cambio diacrónico en el Paleolítico Medio del Abric Romaní (Capellades, Barcelona).* Ph.D. Dissertation, Universitat Rovira i Virgili.

Vaquero, M. (1999a). Variabilidad de las estrategias de talla y cambio tecnológico en el Paleolítico Medio del Abric Romaní (Capellades, Barcelona). *Trabajos de Prehistoria, 56*(2), 37–58.

Vaquero, M. (1999b). Intrasite spatial organization of lithic production in the Middle Palaeolithic: The evidence of the Abric Romaní (Capellades, Spain). *Antiquity, 73*, 493–504.

Vaquero, M. (2005). Les stratégies de transport d'outils dans un contexte résidentiel: un exemple du Paléolithique moyen. In D. Vialou, J. Renault-Miskovsky, & M. Patou-Mathis (Eds.), *Comportements des hommes du Paléolithique moyen et supérieur en Europe: territoires et milieux* (pp. 121–132). Liège: ERAUL.

Vaquero, M. (2008). The history of stones: Behavioural inferences and temporal resolution of an archaeological assemblage from the Middle Palaeolithic. *Journal of Archaeological Science, 35*, 3178–3185.

Vaquero, M., & Pastó, I. (2001). The definition of spatial units in Middle Palaeolithic sites: The hearth-related assemblages. *Journal of Archaeological Science, 28*(11), 1209–1220.

Vaquero, M., Chacón, G., Fernández, C., Martínez, K., & Rando, J. M. (2001a). Intrasite spatial patterning and transport in the Abric Romaní Middle Paleolithic site (Capellades, Barcelona, Spain). In N. J. Conard (Ed.), *Settlement dynamics of the Middle Paleolithic and Middle Stone Age* (pp. 573–595). Tübingen: Kerns Verlag.

Vaquero, M., Vallverdú, J., Rosell, J., Pastó, I., & Allué, E. (2001b). Neanderthal behavior at the Middle Palaeolithic site of Abric Romaní, Capellades, Spain. *Journal of Field Archaeology, 28*(1–2), 93–114.

Vaquero, M., Rando, J. M., & Chacón, G. (2004). Neanderthal spatial behavior and social structure: Hearth-related assemblages from the Abric Romaní Middle Palaeolithic site. In N. J. Conard (Ed.), *Settlement dynamics of the Middle Paleolithic and Middle Stone Age 2* (pp. 367–392). Tübingen: Kerns Verlag.

Vergès, J. M. (2003). *Caracterització dels models d'instrumental lític del mode 1 a partir de les dades de l'anàlisi funcional dels conjunts d'Aïn-Hanech i El-Kherba (Argèlia), Monte Poggiolo i Isernia la Pineta* (Itàlia). Ph.D. Dissertation, Universitat Rovira i Virgili.

Verri, G., Barkai, R., Gopher, A., Hass, M., Kubik, P. W., Paul, M., et al. (2005). Flint procurement strategies in the Late Lower Palaeolithic recorded by in situ produced cosmogenic 10Be in Tabun and Qesem Caves (Israel). *Journal of Archaeological Science, 32*, 207–213.

Wallace, I. J., & Shea, J. J. (2006). Mobility patterns and core technologies in the Middle Paleolithic of the Levant. *Journal of Archaeological Science, 33*, 1293–1309.

Wenban-Smith, F. F. (1999). Knapping technology. In M. B. Roberts & S. A. Parfitt (Eds.), *Boxgrove: A Middle Pleistocene hominid site* (pp. 384–395). London: English Heritage.

Wengler, L. (1990). Économie des matières premières et territoire dans le moustérien et l'atérien maghrébin: Exemples du Maroc oriental. *L'Anthropologie, 94*(2), 321–334.

Wisniewski, A. (2007). Raw material and technology in Central European early Middle Palaeolithic. In M.-H. Moncel, A.-M. Moigne, M. Arzarello, & C. Peretto (Eds.), *Raw material supply areas and food supply areas.* Integrated approach of the behaviours (pp. 213–226). Oxford, England: BAR International Series 1725.

Yamada, S. (1993). The formation process of 'use-wear polishes'. In P. Anderson, S. Beyries, M. Otte, & H. Plisson (Eds.), *Traces et Fonction. Les gestes retrouvés* (pp. 447–458). Liège: ERAUL.

Yellen, J. (1977). *Archaeological approaches to the present.* New York: Academic Press.

Chapter 8
Occupational Patterns and Subsistence Strategies in Level J of Abric Romaní

Jordi Rosell, Ruth Blasco, Rosa Huguet, Isabel Cáceres, Palmira Saladié, Florent Rivals, Maria Bennàsar, Pilar Bravo, Gerard Campeny, Montserrat Esteban-Nadal, Cristina Fernández-Laso, Maria Joana Gabucio, Nuria Ibáñez, Patricia Martín, Laura Muñoz and Antonio Rodríguez-Hidalgo

Abstract The faunal analysis of level J has been carried out using a zooarcheological method with some contributions from the field of Taphonomy. The data were obtained from the anatomical and taxonomical analysis and from the structural modifications of the bones. Red deer (*Cervus elaphus*) and horses (*Equus ferus*) are the most abundant animals, although they are accompanied by other herbivores like *Bos primigenius, Stephanorhinus hemitoechus* and *Rupicapra pyrenaica*. The data indicate that level J is an anthropic accumulation with a minimal incidence of carnivores. Carcasses were selectively transported into the site, where they were processed and consumed. The temporal dimension and the possibility that the faunal assemblage may be the result of different occupational events have been also considered.

J. Rosell (✉) · R. Blasco · R. Huguet · I. Cáceres · P. Saladié ·
M. Bennàsar · P. Bravo · G. Campeny · M. Esteban-Nadal ·
C. Fernández-Laso · M. J. Gabucio · N. Ibáñez · P. Martín ·
L. Muñoz · A. Rodríguez-Hidalgo
Institut Català de Paleoecologia Humana i Evolució Social
(IPHES), Universitat Rovira i Virgili (URV), C/Escorxador s/n,
43003 Tarragona, Spain
e-mail: jordi.rosell@urv.cat

R. Blasco
e-mail: rblasco@iphes.cat

R. Huguet
e-mail: rhuguet@iphes.cat

I. Cáceres
e-mail: icaceres@iphes.cat

P. Saladié
e-mail: psaladie@iphes.cat

M. Bennàsar
e-mail: mlluc@prehistoria.urv.cat

P. Bravo
e-mail: pbravo@codex.cat

G. Campeny
e-mail: gcampeny@iphes.cat

Keywords Zooarcheology • Seasonality • Carnivores • Carcass transport • Consumption patterns

Introduction

The study of subsistence strategies is fundamental to understanding the ways of life of the human communities during the Pleistocene and their evolution over time. Subsistence strategies are the set of systems that human groups use in a territory to organize themselves socially and economically with the aim of guaranteeing the survival of

M. Esteban-Nadal
e-mail: mesteban@iphes.cat

C. Fernández-Laso
e-mail: mcfernandezlaso@gmail.com

M. J. Gabucio
e-mail: mjgabucio@iphes.cat

N. Ibáñez
e-mail: nibanez@iphes.cat

P. Martín
e-mail: patricia.martin@estudiants.urv.cat

L. Muñoz
e-mail: lencinar@prehistoria.urv.cat

A. Rodríguez-Hidalgo
e-mail: arodriguez@iphes.cat

F. Rivals
Institució Catalana de Recerca i Estudis Avançats (ICREA),
Institut Català de Paleoecologia Humana i Evolució Social
(IPHES), Universitat Rovira i Virgili (URV), C/Escorxador s/n,
43003 Tarragona, Spain
e-mail: florent.rivals@icrea.es

E. Carbonell i Roura (ed.), *High Resolution Archaeology and Neanderthal Behavior: Time and Space in Level J of Abric Romaní (Capellades, Spain),* Vertebrate Paleobiology and Paleoanthropology,
DOI: 10.1007/978-94-007-3922-2_8, © Springer Science+Business Media B.V. 2012

their community (Díez and Rosell 1998). One of the scientific disciplines that contributes the most to this subject is Zooarcheology. The main objective of this discipline is to understand the relationships established between human groups of the past and animals. From this perspective, zooarcheology makes a significant contribution toward uncovering the strategies used by Pleistocene hunter-gatherers for the procurement and management of faunal resources—strategies that reflect the abilities of different types of hominids. Therefore, we consider the study of the acquisition of new techniques from a diachronic perspective to be a good indicator of the evolution of ways of life (Adler et al. 2006). It is obvious, for example, that the behavior of groups with incipient technology and some simple techniques for procuring animal resources cannot be the same as the behavior of groups with a high level of technological sophistication and with the capacity to develop varied and therefore complex hunting techniques.

For most of the past century, the European archeological tradition has fundamentally been concerned with obtaining criteria to classify archeological sites within a chrono-cultural framework and developing an evolutionary framework of prehistory. Research on lithic typology was the primary tool employed for this purpose and the resulting data were crossed- referenced with the corresponding faunal lists to place them within their natural context. During this period in archeology, studies aiming to contribute to the understanding of past human groups through paleoethnographic data about the relationships of human groups within their environments were very limited and lacked a structural base and solid methodological foundations.

In the second half of the century, however, prehistoric archeology took a qualitative step forward with the reopening of the debate about the origin of modern human behavior (Isaac 1983; Binford 1984a). The redefinition of this concept placed the hunting of large mammals toward the top of long list of attributes that hominids had to fulfill to be considered entirely human. This new paradigm gave impetus to zooarcheological studies dedicated to the procurement and processing of faunal resources in different periods and led the discipline to take on a more important role in archeological inquiry.

This debate was well received in Europe. At that time, paleoanthropologists were beginning to deduce that Neanderthals and *Homo sapiens* belonged to different human lineages. Therefore, it was easy for archeologists to assume that their behavior also had to be different in many aspects. Soon, two clearly opposite tendencies arose for explaining the average way of life of human communities in the Paleolithic. The first, derived from the prevailing paradigms in place until the seventies, defended the Neanderthals as efficient hunters entirely adapted to their environment with a high capacity for foresight (Chase 1986, 1988, 1989). This perspective defended an ancient origin of modern human behavior. The second tendency arose out of the qualitative advances in methodologies made in the early eighties (Binford 1981; Brain 1981; Shipman 1981) and painted a more opportunistic picture of the Neanderthal way of life, with systems of animal acquisition fundamentally based on scavenging (Binford 1985, 1989). From these ideas it was inferred that modern human behavior was the exclusive heritage of anatomically modern hominids and that its development was intrinsic to our species alone.

Today, with the discovery of new sites and the evolution of the zooarchaeological discipline, these two views have become less mutually exclusive. More and more new data paint a picture of Neanderthals as skillful hunters with a high capacity for adaptation and an adaptability to numerous and varied ecosystems (Fiore et al. 2004; Krönneck et al. 2004). However, the debate remains alive in many aspects related to the ways of life of these hominids, such as the opportunist or planned nature of their hunting activities, the complexity of their faunal resource processing and their relationships with large carnivores (Martínez-Moreno 2005; Adler et al. 2006).

Level J of the Abric Romaní is a crucial site for addressing these questions. This rockshelter, located in the travertine cliff known as the *Cinglera del Capelló*, contains a 16-m stratigraphic sequence. Approximately 20 clearly separated archeological levels have been recognized, separated by sterile sediments including bioconstructive layers of travertine. All the levels belong to the Middle Paleolithic, except for the uppermost level, which has been attributed to the Early Upper Paleolithic.

Level J was mainly excavated in 1993, 1994 and 1995. Using an excavation method based on Cartesian coordinates, the excavation affected a surface of about 250 m^2. This kind of excavation makes it possible to recover all the preserved elements in a well defined context, in a way that allows the site to be subsequently reconstructed. Abundant combustion structures were recovered in association with a large number of lithic artifacts and bone remains from different mammals. Another noteworthy discovery consisted of plant pseudo-morphs contained inside travertine that are often clearly associated with human activities (Allué et al. 2012).

From a zooarcheological point of view, the complete recovery of all preserved elements gives rise to a wide range of questions related to the ways of life and subsistence strategies of these early hominids. However, some implicit questions should be considered from a more general perspective. The most important of these questions is the matter of the temporality of the assemblage. It is obvious that level J shows a certain diachrony, which became clear during excavation, and resulted in the distinction between three different sublevels: Ja and Jb.

Ja and Jb are distinct enough to be treated as independent assemblages. However, the distinction between sublevels is not clear across the entire excavated surface. There are some places where Ja and Jb are superimposed and it is impossible to distinguish the archeological elements belonging to each of them.

But in addition to this clear diachrony, another less obvious one may also exist, which is related to the large accumulations of archeological objects. Initially, it is easy to fall into the temptation of treating the remains as the result of two separate, high-intensity occupational events, one in Ja and another in Jb. However, the temporal dimension and the possibility that the conserved assemblage may be the result of different occupational events should be considered. This phenomenon has been recorded in other levels at the same site (Carbonell et al. 1996; Vallverdú et al. 2005). Therefore, from a zooarcheological perspective, our goal is to seek out elements of the faunal remains that will allow us to discover the synchronic or diachronic nature of each set, as well as to obtain data to establish the different occupational models that were in place.

An occupational characterization such as this will allow us to look at the animal resource procurement and processing behaviors of the human groups. Only in this way can we define what strategies were used by the hominids to manage faunal resources, and determine whether those strategies changed according to occupation types. All these elements provide important empirical information for considering questions related to the organization and ways of life of the groups.

Methodology

The faunal analysis of level J was carried out using a zooarcheological method with some contributions from the field of taphonomy, and includes the study of all the recovered fossil remains. The data were obtained from the anatomical and taxonomical analysis and from the structural modifications of the bones.

Sometimes, the degree of fragmentation of the bones made specific anatomical attribution difficult. However, these unidentified fragments were classified *a priori* at the skeletal level according to their morphological characteristics: long bones, flat bones, and irregular bones. The bone and tooth fragments were classified in terms of anatomical portions to which they belonged. At a taxonomical level, they were grouped according to approximate weight categories, which depend on the size and age of the animal. Taking the species documented in level J as a reference, three weight sizes were established: (1) large (taxa greater than 300 kg); (2) medium-size (between 100 and 300 kg); and (3) small (less than 100 kg) (Table 8.1).

Table 8.1 Species documented in level J according to weight

Body weight	Species and ages
Large size >300 kg	Juvenile and adult *Stephanorhinus hemitoechus*
	Juvenile and adult *Bos primigenius*
	Adult *Equus ferus*
Medium size 100–300 kg	Infantile *Bos primigenius*
	Infantile and juvenile *Equus ferus*
	Juvenile and adult *Cervus elaphus*
	Juvenile and adult *Ursus* sp.
Small size <100 kg	Infantile *Cervus elaphus*
	Infantile *Ursus* sp.
	Infantile, juvenile and adult *Rupicapra pyrenaica*
	Infantile, juvenile and adult *Crocuta* sp.
	Infantile, juvenile and adult *Lynx* sp.
	Infantile, juvenile and adult *Canis lupus*
	Infantile, juvenile and adult *Vulpes vulpes*

Following anatomical and taxonomical identification, the remains were then inventoried to assess the completeness of the sample. To do this, the following characteristics were calculated: NR (Number of Remains); NISP (Number of Identified Specimens); MNE (Minimum Number of Elements); MNI (Minimum Number of Individuals) and Skeletal Survival Rate (Brain 1981; Lyman 1994), which estimates the proportion between the elements recovered and that which would be expected in a similar context. The MNE was calculated based on age, portion and size. To calculate the Skeletal Survival Rate (Brain 1969), the following formula was used: $\%\ \text{survival}_i = \text{MNE} \times 100/$ number of elements in the animal skeleton × MNI.

In determining the animals' ages at death, the best indicator was the tooth replacement sequence and dental wear (e.g., Silver 1969; Bökonyi 1972; Riglet 1977; Mariezkurrena and Altuna 1983a, b; Azorit et al. 2002): infantile (with deciduous teeth), juvenile (without third molar), adult (with third molar) and senile (with an advanced dental wear). However, the degree of epiphysis fusions of the bones was also used in some cases (Silver 1969; Barone 1976), as well as the type of cortical tissue (compact in adults or more porous in young animals).

The seasonality for the ungulates was estimated by studying deciduous dentition and/or teeth with reduced use-wear (wear stage corresponding to the eruption of the tooth out of the gum). Mandibles and maxillae, as well as isolated teeth, were used for detecting seasonality for *Equus ferus*, *Cervus elaphus*, and *Bos primigenius*. Observations on the fossil assemblages were compared to published reference data on related modern species. The development stages of teeth and their wear were identified and assigned to age categories according to a scheme developed for the modern horse (Smuts 1974; Levine 1982; Guadelli 1998; Fernandez et al. 2006), modern red deer (Mariezkurrena and Altuna 1983a, b; Carter 1998, 2005), and modern cattle and bison (Grigson 1982; Brugal 1999; Johnson et al. 2005).

The surface alterations generated by the biological agents (mainly hominids and carnivores) are treated at both a macroscopic and microscopic level (Olympus SZ stereomicroscope from 18 to 110 magnifications) and ESEM (environmental scanning electron microscope). In terms of hominid modifications, special attention was placed on cutmarks (Binford 1981; Potts and Shipman 1981; Shipman 1983; Shipman and Rose 1983; Shipman et al. 1984), which were grouped into incisions (slicing, scraping and sawing marks) and chopmarks. Incisions are striations with a linear outline of variable length, width, and depth. Incisions have a V-shaped section and display internal microstriation (Potts and Shipman 1981). In some cases Hertzian cones (Bromage and Boyde 1984), shoulder effects and barbs (Shipman and Rose 1983) were found. Scrape marks are shallow sub-parallel cutmarks (Noe-Nygaard 1989) caused when a stone tool is dragged transversally along the length of the bone. Chopmarks were the result of blows using a bladed cutting instrument that caused short deep cuts, which were generally isolated, and in which no internal microstriations were observed. The analysis of the cutmarks took into account the number of striations, the location on the anatomical element, the distribution over the surface (isolated, clustered, crossed) and the orientation regarding the longitudinal axis of the bone (oblique, longitudinal, transverse). Maximum and minimum lengths were taken in millimeters. The action performed (viscera removal, disarticulation and defleshing) was recorded according to morphology, emplacement and distribution of anthropogenic incisions.

Three types of carnivore toothmarks have been identified: pits, punctures, and scores (Maguire et al. 1980; Binford 1981). Pits consist of superficial marks lacking a longitudinal axis, and punctures are deep pits that penetrate through cortical bone or along the edges of intact or broken bone. Scores are surface marks with a longitudinal axis of more than four times the perpendicular axis (Cáceres et al. 2012).

The breakage pattern was analyzed by measuring all the remains using digital calipers to quantify their maximum length, width, and thickness (Morales Pérez 2008). We used a modified method from Bonnichsen (1979) and Bunn (1983) and completed by Villa and Mahieu (1991) to determine breakage pattern. This method considers fracture outline (transverse, curved/V-shaped, longitudinal), fracture angle (oblique, right, mixed) in relation to the cortex of the bone, and fracture edge (smooth, jagged).

Surface damage caused during the breakage of the bones was also analyzed and the diagnostic elements of anthropogenic fracturing were documented: percussion pits or percussion marks (Blumenschine and Selvaggio 1988; Pickering and Egeland 2006), percussion notches, impact flakes (Capaldo and Blumenschine 1994; Pickering and Egeland 2006), adhering flakes (Díez et al. 1999; Fernández-Jalvo et al. 1999) and peeling (White 1992). Percussion marks refer to pits and striations, which are sometimes but not always associated with percussion notches. The diagnostic morphology and configuration of percussion pits and striations are described and illustrated by Blumenschine and Selvaggio (1988). Pits are often closely associated with and/or emanate from the patches of striations that result from the slippage of stone against bone during impact events. Percussion notches are semicircular shaped indentations on fracture edges with corresponding negative flake scars on medullar surfaces of limb bones (Pickering and Egeland 2006). Impact flakes refer to shaft fragments produced by hammerstone percussion that display the same basic technical percussion attributes as found on stone flakes (Texier et al. 1998). Flakes have a platform at the impact point and a percussion bulb below the platform and they occasionally show 'ripple' marks and/or hackle (stress) marks originating at or near the platform or bulb (Fisher 1995). Adhering flakes are bone flakes that adhere to the fracture surface of a skeletal element. Peeling is a term coined by White (1992) and defines a roughened surface with parallel grooves or a fibrous texture produced when fresh bone is fractured and peeled apart, similar to bending a small fresh twig with two hands (Díez et al. 1999).

Burning has also been identified. Various different studies have described the alterations that fire produces on bones, which allow us to identify their state at the time of thermo-alteration (fresh bones, dry bones, bones with flesh, without flesh, etc.) (Shipman et al. 1984; Nicholson 1993; Buikstra and Swegle 1989; Gifford-Gonzalez 1989; Costamagno et al. 2005, 2009; Théry-Parisot et al. 2005). The most obvious change on bones exposed to fire is the effect on their natural coloration. Depending on temperature and exposure time, bones acquire brown tonalities that progressively become blacker. From there, black gives way to a bluish-grey until it becomes white (Correia 1997). Stiner et al. (1995) distinguish between six degrees of coloration based on these criteria: (1) degree 0 represents fresh bones; (2) degree 1 concerns bones with a light brown tonality but that still maintain part of their collagen; (3) degree 2 bones are partially black; (4) degree 3 includes completely black bones without collagen; (5) degree 4 consists of bones with grey tonalities and; (6) degree 5 includes bones that have very noticeable grey or white tonalities.

Table 8.2 NISP, MNE, MNI by sublevels of the different taxonomic groups recorded in level J

	Ja			Jb		
	NISP	MNE	MNI	NISP	MNE	MNI
Equus ferus	351	88	15	139	27	4
Cervus elaphus	497	202	12	96	37	4
Bos primigenius	88	47	5	15	9	2
Stephanorhinus hemitoechus	33	17	3	27	4	2
Rupicapra pyrenaica	6	5	2	2	2	1
Ursus sp.	1	1	1	–	–	–
Lynx sp.	1	1	1	1	1	1
Canis lupus	2	2	2	–	–	–
Vulpes vulpes	1	1	1	–	–	–
Crocuta sp.	–	–	–	1	1	1
Carnivora indet.	4	2	–	–	–	–
TOTAL	984	366	42	281	81	15

Fire also affects the mineralogical structure of bone. Shipman et al. (1984) observe a considerable reduction in mass due to alterations of the microscopic morphology and crystal clear structure of the burned bones in parallel with these changes in color. 60–70% of the weight of fresh bones is attributable to apatite crystals (calcium fluophosphate). At room temperature, these crystals gradually decompose due to diagenesis. However, their disappearance can be accelerated by exposure to the sun or weathering, or occur instantaneously when bones are exposed to the high temperatures of fire. The result is an increase in the fragility of the bone. At archeological sites, burned bones are usually small (Schiffer 1983; O'Connell 1987). The fragmentation is not the result of an anthropic cultural pattern, but rather it is caused by the accelerated drying processes induced by fire in contact with bone and by human trampling (Stiner et al. 1995). The bones exposed to fire undergo similar modifications to those observed as a result of weathering: first longitudinal fissures appear due to contractions brought about by the sudden change of temperature, which become mosaic-like fissures when the bone is dry (Thurman and Willmore 1980; Gilchrist and Mytum 1986; Buikstra and Swegle 1989; Spenneman and Colley 1989).

Faunal data from Level J

A total of 8,460 large mammal remains have been recovered from level J. Of these, 6,738 belong to sublevel Ja and 1,722 to sublevel Jb. In sublevel Ja 984 remains (14.6%) have been anatomically identified, using taxonomic groups, whereas in sublevel Jb 281 (16.3%) have been identified. The high fragmentation of the bone remains, especially in sublevel Ja, has meant that in most cases they do not preserve sufficient morphological features to be attributed to an anatomical element or taxonomic group.

Taxonomic and Anatomic Representation

Red deer (*Cervus elaphus*) and horses (*Equus ferus*) are the most abundant animals found in the two assemblages, in terms of the number of skeletal remains (NISP), the number of identified elements (MNE), and the minimum number of individuals (MNI) (Table 8.2). These two taxonomic groups are accompanied by herbivores (*Bos primigenius, Stephanorhinus hemitoechus*) related to plain and forest ecosystems, which indicates a directionality of the accumulator agent towards the spaces at the ends of valleys. To a lesser extent, some elements from animals that today are found in rocky areas, such as the chamois (*Rupicapra pyrenaica*), have also been recovered.

In spite of the scarce skeletal representation of carnivores in sublevel Ja, a wide specific variability has been documented: an ursid (*Ursus* sp.) represented by a first phalanx, two large canids (*Canis lupus*) (one infantile identified from a deciduous tooth and one adult documented from a cervical vertebra), one small canid (*Vulpes vulpes*) documented from a metapodium and a lynx (*Lynx* sp.) represented by a second phalanx. Two rib fragments of a large carnivore and two rib fragments of a medium-sized carnivore have been recorded, which could not be attributed to any taxa.

In sublevel Jb only two carnivore remains, belonging to two individuals of different species, have been documented: an infantile hyaenid (*Crocuta* sp.) represented by a deciduous tooth and an adult lynx (*Lynx* sp.) represented by a phalanx.

There is a marked quantitative difference between the skeletal representation of cervids and equids and other taxonomic groups. The remains of red deer exhibit a greater anatomical integrity, whereas among the equids, some skeletal parts are generally absent (Table 8.3).

In sublevel Ja, the least abundant animals (rhinoceroses and large bovids) are mainly represented by elements of the appendicular skeleton. In Jb this trend can only be seen in the large bovids.

The survival rates of the remains recovered reinforce the hypothesis that as a taxonomic group red deer has the highest degree of skeletal integrity. Rhinoceros and caprid elements have very low survival rates, especially in sublevel Jb (Table 8.4).

Except for equids, the most abundant anatomical segments in the two faunal assemblages are the long bones, mainly from the proximal zone of the appendicular skeleton (stylopodial and basipodial bones).

In all taxa a low number of axial skeleton and pelvic elements have been documented, especially in the larger taxa such as equids, rhinoceroses and large bovids. Red deer shows the highest survival rates for these segments. One of the most relevant data in relation to the skeletal representation of the individuals is the scarce presence of basipodial and acropodial bones, which is in stark contrast to the high number of long bones of the extremities (stylopodial, zygopodial and metapodial bones) recovered.

Ages at Death and Seasonality

Adult individuals are most common in level J (21 for Ja and 8 for Jb) although the sum of immature individuals (infantile and juvenile) is also high (17 for Ja and 6 for Jb). On the other hand, the number of senile animals is low, (two equids and two cervids in sublevel Ja and one equid in Jb) (Table 8.5).

The deciduous and permanent dentition of nine specimens of *Cervus elaphus* was used to assess seasonality in sublevel Ja (Fig. 8.1 and Table 8.6). Four individuals compare favorably with modern animals which died between 4 and 8 months, one individual between 16 and 17 months, and one between 27 and 30 months. Taking into account that the birth of the modern red deer occurs in late spring (Nowak 1999), we can suggest that *Cervus elaphus* was hunted from autumn to winter. For *Equus ferus*, the dentition of three individuals with unworn deciduous dentition is compared with very young animals (two lower dp3 and one upper dp2) and indicates that death either occurred during the prenatal period (a few weeks before birth), or in the first 2 weeks after birth. Considering that modern equids are born in spring (Bennett and Hoffmann 1999), we suggest that these individuals were all obtained in spring. The Neanderthals either killed a pregnant female horse (close to parturition) or foals. For *Bos primigenius*, only one specimen, an upper third molar (AR'94 K54/79), corresponds to an individual aged between 30 and 33 months which may have died in the autumn or spring (Table 8.6).

In sublevel Jb only *Cervus elaphus* and *Equus ferus* provided material for the analysis of seasonality. For *Cervus elaphus*, two teeth have the same wear pattern observed in sublevel Ja. A deciduous premolar Dp3 (AR'95 K54/49) is compared to modern animals aged between 5 and 8 months and an unworn lower molar (m1 or m2) is compared to animals between 4 and 8 months. As in sublevel Ja, these two individuals probably died in autumn or winter. The teeth sampled for *Equus ferus* were three unworn molars and an unworn incisor. Ageing is not definite for adult individuals and age could only be estimated at between 2.5 and 3 years with the possible season of death in autumn or winter. Results correlate with data gathered on *Cervus elaphus*.

In short, in sublevel Ja two death periods were identified, late autumn or early winter for *Cervus elaphus*, and spring for *Equus ferus* and *Bos primigenius*. Unlike sublevel Ja, both *Cervus elaphus* and *Equus ferus* from sublevel Jb probably died during the same period, i.e. in late autumn or early winter. Our results are consistent for all individuals from the same species, thus suggesting a clear seasonality in the Neanderthal occupations at Abric Romaní.

Unidentified Remains

In both of the assemblages analyzed the percentage of faunal remains not assigned to any taxonomic group is high (85.4% for Ja and 83.7% for Jb). In most cases these remains have been anatomically identified and ascribed to one of the predetermined weight sizes. The remains which have been impossible to assign to a specific anatomic element have been grouped into three anatomical categories.

From these non-taxonomically identified remains, the most abundant elements in sublevels Ja and Jb are medium-sized animal remains (mainly Cervidae). Small animals are the least represented in these sublevels.

The anatomical representation is very similar at both sublevels, although in sublevel Ja the results are more significant. The cranial elements (dental remains) and the long bones from the extremities are the most abundant, whereas remains from the axial skeleton and the pelvis present the lowest values. The remains from the axial skeleton mainly belong to medium-sized and, to a lesser extent, small

Table 8.3 NISP (and MNE) by taxonomic groups of herbivores for sublevels Ja and Jb

	Ja						Jb					
	Equidae	Cervidae	Bovinae	Rhinocerotidae	Caprinae	Total	Equidae	Cervidae	Bovinae	Rhinocerotidae	Caprinae	Total
Horn/Antler	–	9(3)	–	–	–	9(3)	–	1(1)	–	–	–	1(1)
Cranium	1(1)	4(2)	2(1)	–	–	7(4)	–	1(1)	–	–	–	1(1)
Maxilla	6(19)	5(9)	–(2)	–(2)	–(1)	11(33)	0(6)	0(2)	–	0(1)	–	0(9)
Mandible	33(19)	18(16)	10(4)	–(2)	–(1)	61(42)	11(8)	2(4)	0(2)	0(3)	0(1)	13(18)
Isolated teeth	232	54	17	16	3	322	99(0)	23(0)	4(0)	27(0)	1(0)	154(0)
Hyoids	–	3(2)	–	–	–	3(2)	–	–	–	–	–	–
Vertebra	1(1)	7(6)	–	–	–	8(7)	–	–	–	–	–	–
Ribs	–	48(15)	1(1)	–	–	49(16)	–	–	–	–	–	–
Pelvis	3(1)	1(1)	–	–	–	4(2)	–	1(1)	–	–	–	1(1)
Scapula	–	12(5)	4(2)	–	–	16(7)	–	1(1)	–	–	–	1(1)
Humerus	18(8)	47(22)	13(9)	5(4)	–	83(43)	2(1)	14(4)	5(2)	–	–	21(7)
Radius–ulna	16(9)	40(12)	12(8)	5(4)	–	73(33)	2(1)	5(4)	1(1)	–	–	8(6)
Metacarpus	3(2)	71(24)	4(3)	–	–	78(29)	2(1)	10(3)	–	–	–	12(4)
Carpus	–	2(2)	–	1(1)	–	3(3)	–	–	–	–	–	–
Femur	5(5)	42(20)	14(10)	1(1)	–	62(36)	4(3)	7(3)	2(1)	–	–	13(7)
Patella	–	1(1)	–	–	–	1(1)	–	–	–	–	–	–
Tibia	15(11)	51(22)	6(5)	1(1)	–	73(39)	14(5)	3(3)	2(2)	–	–	19(10)
Metatarsus	6(4)	58(22)	4(2)	–	–	68(28)	1(1)	21(4)	1(1)	–	–	23(6)
Tarsal	–	1(1)	–	1(1)	–	2(2)	–	2(2)	–	–	–	2(2)
Lateral metapodium	5(4)	–	–	–	–	5(4)	–	–	–	–	–	–
Metapodials	7(4)	6(2)	1(0)	–	1(1)	15(7)	4(1)	2(1)	–	–	–	6(2)
Phalanx	–	12(10)	–	1(1)	2(2)	12(10)	–	–	–	–	1(1)	1(1)
Sesamoids	–	5(5)	–	–	–	5(5)	–	3(3)	–	–	–	3(3)
Long bones	–	–	–	1(0)	–	1(0)	–	–	–	–	–	–
Flat bones	–	–	–	1(0)	–	1(0)	–	–	–	–	–	–
Total	351(88)	497(202)	88(47)	33(17)	6(5)	975(359)	139(27)	96(37)	15(9)	27(4)	2(2)	279(79)

Table 8.4 Skeletal survival rate of the species identified in sublevels Ja and Jb of Abric Romaní. MNE (% Survival)

% Survival	Ja					Jb				
Taxa	Equidae	Cervidae	Bovinae	Rhinocerotidae	Caprinae	Equidae	Cervidae	Bovinae	Rhinocerotidae	Caprinae
MNI	15	12	5	3	2	4	4	2	2	1
Cranial skeleton	39(52)	32(50)	7(28)	4(26.7)	2(20)	14(70)	8(40)	2(20)	4(40)	1(20)
Axial skeleton	1(0.1)	21(4.4)	1(0.3)	–	–	–	–	–	–	–
Girdles	1(1.7)	6(12.5)	2(10)	–	–	–	2(12.5)	–	–	–
Stylopodials	13(21.7)	48(87.5)	19(95)	5(41.7)	–	4(25)	7(43.7)	3(75)	–	–
Zygopodials	20(33.3)	55(71)	13(65)	5(41.7)	–	6(37.5)	7(43.7)	3(37.5)	–	–
Basipodials	–	9(4.2)	–	2(2.8)	–	–	5(6.9)	–	–	–
Metapodials	14(23.3)	48(100)	5(25)	–	1(12.5)	3(18.7)	8(50)	1(6.25)	–	–
Acropodials	–	10(3.5)	–	1(0.7)	2(4.2)	–	–	–	–	1(4.2)

Table 8.5 MNI by age groups in sublevels Ja and Jb

	Ja					Jb				
	Infantile	Juvenile	Adult	Senile	Total	Infantile	Juvenile	Adult	Senile	Total
Horse	3	2	8	2	15	–	1	2	1	4
Red deer	3	2	5	2	12	1	–	3	–	4
Auroch	–	2	3	–	5	–	1	1	–	2
Rhinoceros	1	1	1	–	3	1	–	1	–	2
Chamois	–	1	1	–	2	–	1	–	–	1
Bear	–	–	1	–	1	–	–	–	–	–
Wolf	1	–	1		2					
Fox	–	–	1	–	1	–	–	–	–	–
Lynx	–	–	1	–	1	–	–	1	–	1
Hyaena	–	–	–	–		1	–	–	–	1
Total	9	8	21	4	42	3	3	8	1	15

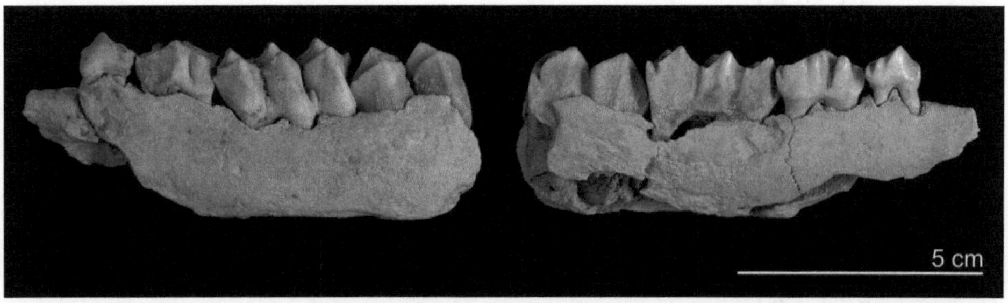

Fig. 8.1 Red deer mandible with deciduous teeth. This animal died around springtime

animals. Very few axial skeleton remains from large animals were recovered (Table 8.7).

Classifying the recovered remains into weight categories allowed us to compare data on taxonomically classified remains with those that could not be identified at a taxonomic level. Therefore, the faunal remain data described below are presented by weight category.

A similar trend is observed in the two sublevels, although it is more marked in Ja. Adult individuals predominate in both assemblages. Most of them are large and medium-sized animals. Immature individuals (infantile and juvenile) correspond primarily to small animals (Table 8.8).

The most abundant elements correspond to medium-sized animals (Table 8.9). This is related to the fact that the individuals from this weight category present the greatest anatomic integrity. Large animals are represented mainly by elements from the appendicular and cranial skeleton, especially in Ja. Small animals present the smallest number of skeletal elements.

Table 8.6 Age estimations for *Cervus elaphus* from sublevel Ja

Number	Teeth	Age (months)
AR'95 I63 17	Mandible with dp2, dp3, dp4, and unworn m1	3–5
AR'98 L40 3 AR'98 L40 3	dp4 m1 unworn	4
AR'95 K48 42	m1	4
AR'98 K40 1	M1	4
AR'94 M49 12	M2 unworn	4
AR'94 S53 36	Mandible with dp2, dp3, dp4, and m1 slightly worn	5–8
AR'93 I59 3	Mandible with dp4 and M1 slightly worn	5–8
AR'96 I63 27	M2 slightly worn	16–17
AR'94 L48 13	Maxillary with M1, M2, M3	27–30

Table 8.7 Number of remains (minimal number of elements)—NR(MNE)—by weight sizes from sublevels Ja and Jb of Abric Romaní

	Ja					Jb				
	Large size	Medium size	Small size	Unident.	Total	Large size	Medium size	Small size	Unident.	Total
Cranium	57(5)	37(4)	3(1)	1	98(10)	3(1)	9(1)	3(1)	–	15(3)
Maxilla	24(4)	2(1)	–	–	26(5)	–	–	–	–	
Mandible	52(4)	3(2)	–	–	55(6)	7(1)	3(2)	–	–	10(3)
Isolated teeth	66(–)	7(–)	–	43	116(–)	10(0)	3(0)	–	24	37(0)
Scapula	12(5)	6(2)	–	–	18(7)	–	–	–	–	
Humerus	14(2)	12(4)	3(1)	–	29(7)	5(1)	1(1)	–	–	6(2)
Radius-ulna	11(6)	1(1)	1(1)	–	13(8)	–	–	–	–	
Vertebra	1(1)	55(23)	13(9)	–	69(33)	2(2)	6(5)	2(2)	–	10(9)
Rib	8(2)	125(9)	12(5)	–	145(16)	6(2)	19(3)	8(3)	–	33(8)
Metacarpus	–	–	–	–	–	–	–	–	–	
Carpus	–	–	–	–	–	–	1(1)	–	–	1(1)
Pelvis	1(1)	–	–	–	1(1)	1(1)	–	–	–	1(1)
Femur	5(4)	3(3)	3(2)	–	11(9)	–	1(1)	–	–	1(1)
Tibia	12(7)	5(4)	2(2)	–	19(13)	4(2)	–	–	–	4(2)
Metatarsus	–	–	–	–	–	–	–	–	–	
Metapodium	–	–	–	–	–	1(1)	–	–	–	1(1)
Tarsus	–	–	–	–	–	–	1(1)	–	–	1(1)
Phalanx	1(1)	–	–	–	1(1)	–	1(1)	1(1)	–	2(2)
Flat bones	145(4)	374(6)	53(3)	19	591(13)	27(1)	154(4)	48(3)	5(0)	234(8)
Flat (cranium/girdle)	454(10)	215(10)	10(2)	3	682(22)	17(1)	30(1)	1(1)	27(0)	75(3)
Flat (axial)	12(2)	444(15)	60(3)	0	516(20)	1(1)	14(1)	12(1)		27(3)
Long bones	897(23)	655(18)	101(4)	7	1660(45)	204(6)	365(10)	144(6)		713(22)
Articular	32(6)	15(7)	3(2)	4	54(15)	0	0	0	0	0
Unidentified	346	206	88	986	1626	9(0)	9(0)	2(0)	256	272
Total	2150(87)	2165(109)	352(35)	1063	5730(231)	297(20)	617(32)	221(18)	308	1443(70)

The anatomical segments with the highest survival rate are the long bones, specifically the stylopodial and the zygopodial bones from medium-sized and large animals. In the case of small animals, the elements of the cranial skeleton are the most abundant followed by the proximal bones from the appendicular skeleton. Interestingly, large animals present very low values regarding the axial skeleton. In sublevel Ja small animals also present low values for this anatomical segment (Table 8.10 and Figs. 8.2, 8.3).

Table 8.8 MNI by weight sizes and age from sublevel Ja and Jb

	Ja					Jb				
	Infantile	Juvenile	Adult	Senile	Total	Infantile	Juvenile	Adult	Senile	Total
Large size	–	3	12	2	17	1	1	4	1	7
Medium size	–	4	5	2	11	–	1	3	–	4
Small size	7	1	1	–	9	1	1		–	2
Total	7	8	18	4	37	2	3	7	1	13

Table 8.9 MNE by weight classes from sublevels Ja and Jb faunal assemblages

	Ja				Jb			
	Large size	Medium size	Small size	Total	Large size	Medium size	Small size	Total
Horn	–	3	–	3	–	1	–	1
Cranium	7	6	1	14	2	3	1	6
Maxilla	14	14	3	31	6	2	1	9
Mandible	20	13	4	37	12	7	2	21
Scapula	6	7	0	13	2	3	2	7
Humerus	23	20	2	45	6	8	1	15
Radius-ulna	19	11	3	33	3	7	1	11
Metacarpus	5	19	1	25	2	5	1	8
Carpus	1	2	–	3	–	2	–	2
Vertebra	1	23	15	39	5	10	6	21
Rib	2	17	10	29	4	8	6	18
Pelvis	2	–	–	2	2	2	1	5
Femur	16	17	3	36	5	6	1	12
Patella	–	1	–	1	–	–	–	–
Tibia	20	22	3	45	12	4	2	18
Tarsus	2	6	–	8	–	1		1
Metatarsus	6	15	–	21	3	5	1	9
Metapodium	8	2	1	11	3	1	0	4
Phalanx	2	9	3	14	0	2	4	6
Flat bones	1	7	5	13	1	4	3	8
Flat (cranium/girdle)	6	10	5	21	1	1	1	3
Flat bones (axial)	1	9	7	17	–	–	–	
Long bones	15	21	9	45	1	1	1	3
Unidentified	2	8	4	14	6	10	6	22
Total	179	268	79	527	76	93	41	210

This table combine all the faunal remains (identified and unidentified remains)

Differential Conservation and Utility rate

The low representation of skeletal elements with low structural density (for example, axial elements) raises the question of the differential post-burial destruction of the remains. To check the validity of this hypothesis, the density measurements of some elements of different animals were statistically contrasted with the survival rate of the bones. For the densities of large animals, we used horse bone density (Lam et al. 1999), for the densities of medium-sized animal remains, we used deer bone density (Lyman 1985) and for small animals, the densities of sheep were used (Lyman 1985; Tables 8.11, 8.12).

The results obtained both in Pearson's r and in Spearman's *Rho* show that there is no correlation between the skeletal representations (% Survival) and the density of the

Table 8.10 NISP (% Skeletal Survival Rate.) by anatomic segments of the different weight categories from sublevels Ja and Jb faunal assemblages

	Ja			Jb		
Sizes	Large size	Medium size	Small size	Large size	Medium size	Small size
MNI	17	11	9	7	4	2
Cranial skeleton	41(48.2)	36(65)	8(17.8)	20(57.1)	13(65)	4(40)
Axial skeleton	18(1.6)	61(10.7)	35(7.5)	10(3.6)	19(11.9)	13(13.5)
Girdles	8(11.8)	7(15.9)	–	4(14.3)	5(31.2)	3(37.5)
Stylopodials	39(57.3)	37(84.1)	5(13.9)	11(39.3)	14(87.5)	2(25)
Zygopodials	39(57.3)	33(75)	6(16.7)	15(53.6)	11(68.7)	3(37.5)
Basipodials	3(0.9)	9(3.1)	–	–	3(4.2)	–
Metapodials	19(27.9)	36(81.8)	2(5.5)	8(28.6)	11(68.7)	2(25)
Acropodials	2(0.6)	9(3.4)	3(1.9)	–	2(8.3)	4(8.3)

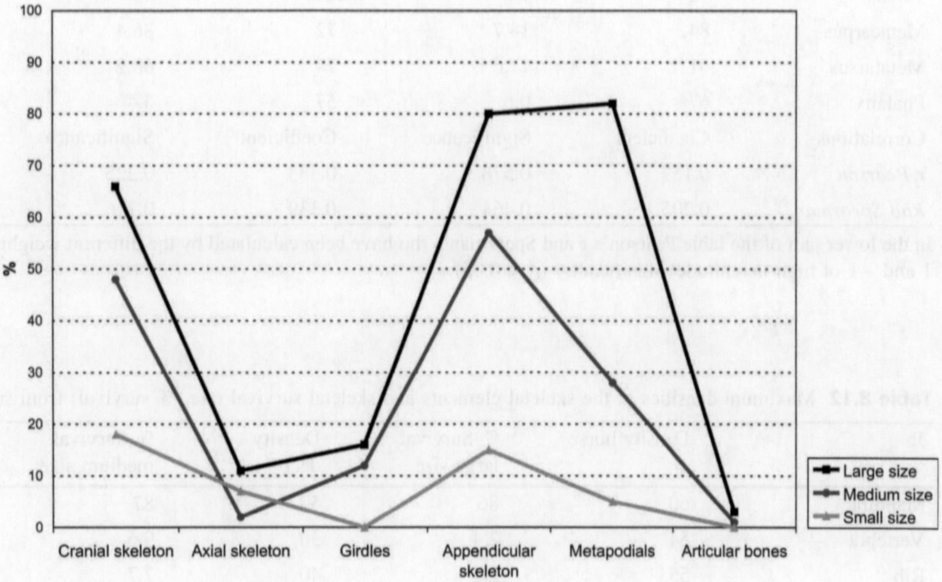

Fig. 8.2 Representation of % Skeletal Survival rate by anatomical segments from level Ja assemblage

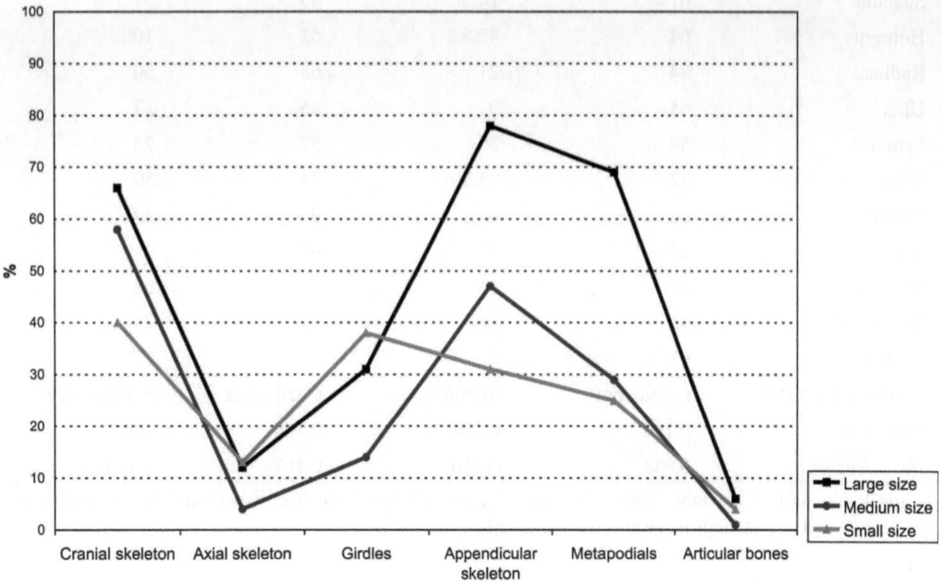

Fig. 8.3 Representation of % Skeletal Survival rate by anatomical segments from level Jb assemblage

Table 8.11 Maximum densities of the skeletal elements and skeletal survival rate (% survival) from sublevel Ja

Ja	Density horse	% Survival large size	Density deer	% Survival medium size	Density sheep	% Survival small size
Mandible	60	59	57	59	55	22.2
Vertebra	54	0.2	30	8	26	6.4
Rib	55	0.4	40	5.9	37	4.3
Pelvis	65	5.9	49	0	49	0
Scapula	67	17.6	49	31.8	33	0
Humerus	64	67.6	63	90.9	42	11.1
Radius	84	41.2	68	40.9	52	16.7
Ulna	65	14.7	45	9.1	26	0
Femur	59	47	57	77.3	36	16.7
Tibia	82	58.8	74	100	59	16.7
Carpus	62	0.4	98	1.3	68	0
Tarsus	72	1	64	4.5	63	0
Metacarpus	84	14.7	72	86.4	67	5.5
Metatarsus	91	17.6	74	68.2	68	0
Phalanx	67	0.6	57	3.4	55	1.9
Correlations	Coefficient	Significance	Coefficient	Significance	Coefficient	Significance
r Pearson	0.157	0.578	0.333	0.225	−0.026	0.927
Rho Spearman	0.205	0.464	0.339	0.216	0.112	0.690

In the lower part of the table Pearson's r and Spearman's rho have been calculated by the different weight sizes. Correlation (coefficient between 1 and −1 of high magnitude) significantly ($p < 0.05$)

Table 8.12 Maximum densities of the skeletal elements and skeletal survival rate (% survival) from sublevel Jb

Jb	Density horse	% Survival large size	Density deer	% Survival medium size	Density sheep	% Survival small size
Mandible	60	86	57	87	55	50
Vertebra	54	2.4	30	9.6	26	11.5
Rib	55	2.2	40	7.7	37	11.5
Pelvis	65	14.3	49	25	49	25
Scapula	67	14.3	49	37	33	50
Humerus	64	42.8	63	100	42	25
Radius	84	21	68	50	52	25
Ulna	65	0	45	37	26	0
Femur	59	36	57	75	36	25
Tibia	82	85.7	74	50	59	50
Carpal	62	0	98	3.6	68	0
Tarsal	72	0	64	2	63	0
Metacarpal	84	14	72	62	67	25
Metatarsal	91	21	74	75	68	25
Phalanx	67	0	57	2.2	55	8.3
CORRELATIONS	Coefficient	Significance	Coefficient	Significance	Coefficient	Significance
r Pearson	0.150	0.594	0.149	0.597	0.034	0.903
rho Spearman	0.068	0.810	0.167	0.553	−0.009	0.974

In the lower part of the table Pearson's r and Spearman's rho have been calculated by the different weight classes. Correlation (coefficient between 1 and −1 of high magnitude) significantly ($p < 0.05$)

Table 8.13 General Utility Rate by weight sizes in sublevel Ja according to Emerson (1993)

Ja	General utility	Alimentary utility	Fat	Medullar volume	% Survival		
					Large size	Medium size	Small size
Cranium	25.5	10.4	–	–	41	54.5	11.1
Vertebra	47.4	61.7	100	0	0.2	8	6.4
Rib	62.3	71.3	93	0	0.4	5.9	4.3
Pelvis	34.7	39.8	54	3.9	5.9	0	0
Scapula	25.5	28.4	30.4	40.6	17	31.8	0
Humerus	27.5	28.4	30.4	79.8	67.6	90.9	11.1
Radius-Ulna	19.1	19.7	22	69.4	56	50	16.7
Femur	100	100	76.7	93.5	47	77.3	16.7
Tibia	57.7	58.1	48.6	100	58.8	100	16.7
Carpal	8.6	10.6	13	43.4	0.4	1.3	0
Tarsal	19.4	30	26.9	60.6	1	4.5	0
Metacarpal	5.4	6	8.4	30.3	14.7	86.4	5.5
Metatarsal	10.8	15.9	16.1	40.8	17.6	68.2	0
Phalanx	3.9	8.4	9.2	22.2	0.5	3.4	1.4
Correlations	*r Pearson*	General utility		Coefficient	0.289	0.196	*0.537*
				Significance	0.316	0.502	*0.047*
		Alimentary utility		Coefficient	0.139	0.068	0.407
				Significance	0.635	0.817	0.149
		Fat		Coefficient	−0.61	−0.168	0.264
				Significance	0.843	0.584	0.383
		Medullar volume		Coefficient	0.822	0.703	*0.683*
				Significance	*0.001*	*0.007*	*0.010*
	Rho Spearman	General utility		Coefficient	0.170	0.218	0.418
				Significance	0.562	0.454	0.137
		Alimentary utility		Coefficient	−0.018	0.020	0.232
				Significance	0.952	0.946	0.425
		Fat		Coefficient	−0.083	0.019	0.287
				Significance	0.788	0.950	0.341
		Medullar volume		Coefficient	0.784	0.591	0.455
				Significance	*0.002*	*0.033*	0.118

Correlation (coefficient between 1 and −1 of high magnitude) significantly ($p < 0.05$)

elements. Therefore, there is no differential preservation of elements. Another explanation must be found for the anatomical bias observed in both assemblages.

The Survival Rates of the different weight categories were also compared with the Utilities Rates calculated by Emerson (1993). The results show that in sublevel Ja there is correlation between the medullar volume and the Skeletal Survival Rate (% Survival) of the three weight sizes according to Pearson's *r*, nor with the large and medium-sized animals according to Spearman's *Rho*. For the small animals, the calculations resulting from Pearson's *r* also indicate a correlation between the Skeletal Survival Rate and the General Utility Rate (Table 8.13). In sublevel Jb a correlation was found between medullar volume and the Skeletal Survival Rate in large and medium-sized animals according to the correlation methods (Table 8.14).

These results show directionality towards the selection of skeletal elements with high energetic contents like fat or bone marrow in both assemblages and for all taxa/weight categories. These elements are mainly the long bones of the extremities, mainly stylopodial and basipodial bones with high representation values. On the other hand, elements of the axial skeleton with low fat content and medullar volume are uncommon.

Table 8.14 General Utility Rate by weight classes in sublevel Jb according to Emerson (1993)

Jb	General utility	Alimentary utility	Fat	Medullar volume	% Survival		
					Large size	Medium size	Small size
Cranium	25.5	10.4	–	–	28.6	75	50
Vertebra	47.4	61.7	100	0	2.4	9.6	11.5
Rib	62.3	71.3	93	0	2.2	7.7	11.5
Pelvis	34.7	39.8	54	3.9	14.3	25	25
Scapula	25.5	28.4	30.4	40.6	14.3	37	50
Humerus	27.5	28.4	30.4	79.8	42.8	100	25
Radius-Ulna	19.1	19.7	22	69.4	21	50	12.5
Femur	100	100	76.7	93.5	36	75	25
Tibia	57.7	58.1	48.6	100	85.7	50	50
Carpal	8.6	10.6	13	43.4	0	3.6	0
Tarsal	19.4	30	26.9	60.6	0	2	0
Metacarpal	5.4	6	8.4	30.3	14	62	25
Metatarsal	10.8	15.9	16.1	40.8	21	75	25
Phalanx	3.9	8.4	9.2	22.2	0	2.2	8.3
Correlations	*r Pearson*	General utility		Coefficient	0.397	0.139	0.205
				Significance	0.160	0.636	0.483
		Alimentary utility		Coefficient	0.275	−0.013	0.050
				Significance	0.341	0.964	0.865
		Fat		Coefficient	0.052	0.158	0.038
				Significance	0.865	0.606	0.902
		Medullar volume		Coefficient	*0.740*	0.584	0.338
				Significance	*0.004*	*0.036*	0.259
	Rho Spearman	General utility		Coefficient	0.460	0.201	0.313
				Significance	*0.098*	0.490	0.276
		Alimentary utility		Coefficient	0.216	−0.064	0.016
				Significance	0.458	0.828	0.957
		Fat		Coefficient	0.270	0.036	0.186
				Significance	0.371	0.907	0.543
		Medullar volume		Coefficient	0.613	0.466	0.304
				Significance	*0.026*	0.108	0.312

Correlation (coefficient between 1 and −1 of high magnitude) significantly ($p < 0.05$)

Fragmentation

One of the main characteristics of the bone assemblages from level J is the small size of the remains. 38% of the bone remains in sublevel Ja and 26% in sublevel Jb are smaller than 2 cm. This is a phenomenon typical of the archeofaunal record and is markedly different from what would be expected in carnivore dens (Villa et al. 2004). Only 3.4% of the remains in Ja and 5% in Jb exceed 10 cm. Grouping the bone remains by size, a correlation emerges between size and number of remains, the small fragments being the most abundant. This is statistically supported by a highly significant regression line (Figs. 8.4 and 8.5).

Only 0.2% of the assemblage in Ja and 0.1% in Jb are made up of complete bones (Table 8.15). These are mainly represented by the acropodial and basipodial bones of red deer, with the exception of one caudal vertebra from the same species found in sublevel Ja. There are no complete bones of large animals (*Bos* or *Equus*).

Few fragments of adult epiphyses have been recovered; therefore anatomical identifications have been conducted using diaphysis remains. The graphs below show many similarities in the results from sublevels Ja and Jb. Proportionally, epiphysis fragments are more abundant in Ja (3.1%), whereas in Jb metaphyses (11.7%) are the most frequently represented after diaphyses (Figs. 8.6, 8.7).

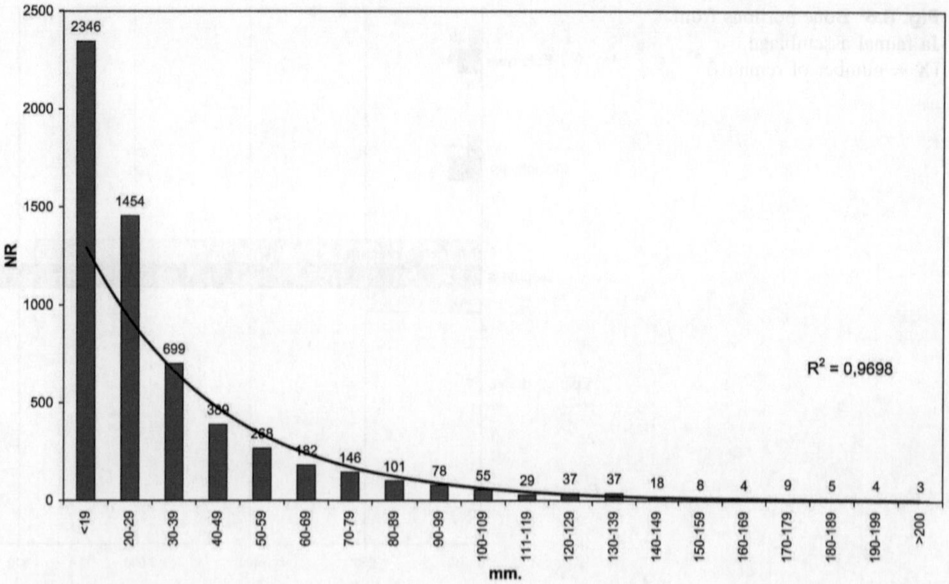

Fig. 8.4 Histogram with bone lengths and exponential curve from sublevel Ja

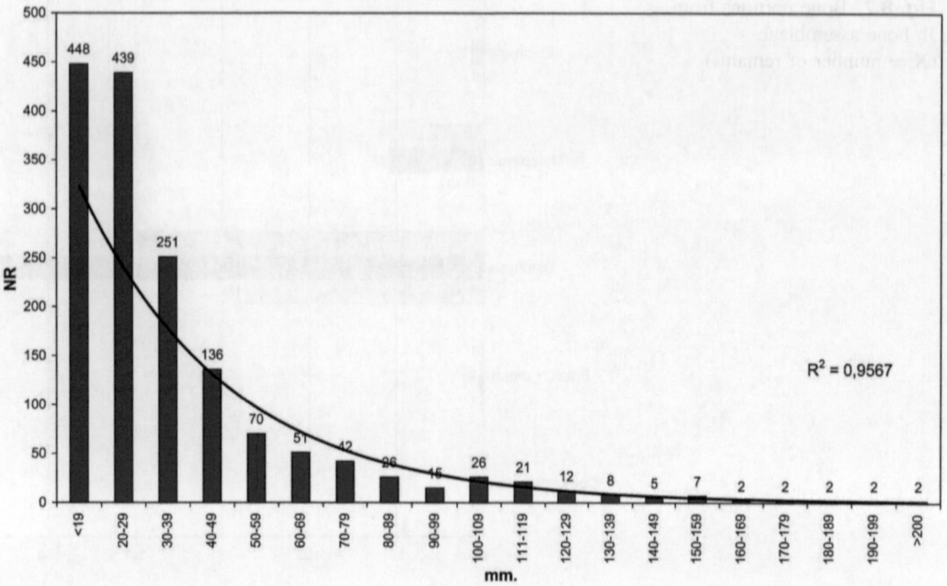

Fig. 8.5 Histogram with bone lengths and exponential curve from sublevel Jb

Table 8.15 NISP of entire bones by species from sublevels Ja and Jb of Abric Romaní

	Ja				Jb		
	Vertebrae	Basipodials	Acropodials	Total	Basipodials	Acropodials	Total
C. elaphus	1	4	6	11	–	–	–
R. pyrenaica	–	–	1	1	1	1	2
S. hemitoechus	–	1	1	2	–	–	–
Total	1	5	8	14	1	1	2

The proportions of the different categories are similar in the two sublevels for the three weight classes. This would indicate that the agent responsible for the absence of epiphyses followed the same pattern in both sublevels.

Due to the high degree of bone fragmentation, the breakage planes were studied in order to establish their origin, which may be attributable either to fresh breakage or postdepositional causes. The breakage planes on all bones

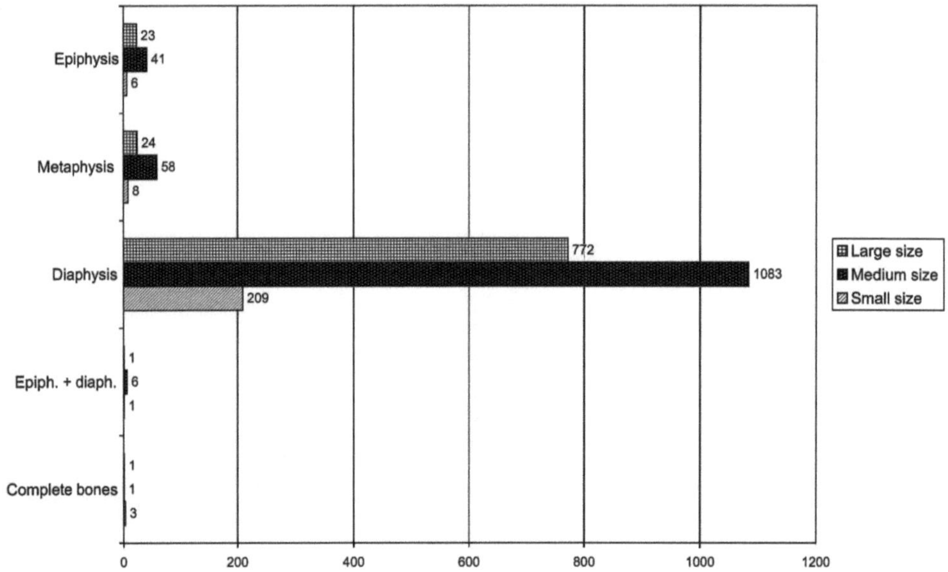

Fig. 8.6 Bone portions from Ja faunal assemblage (X = number of remains)

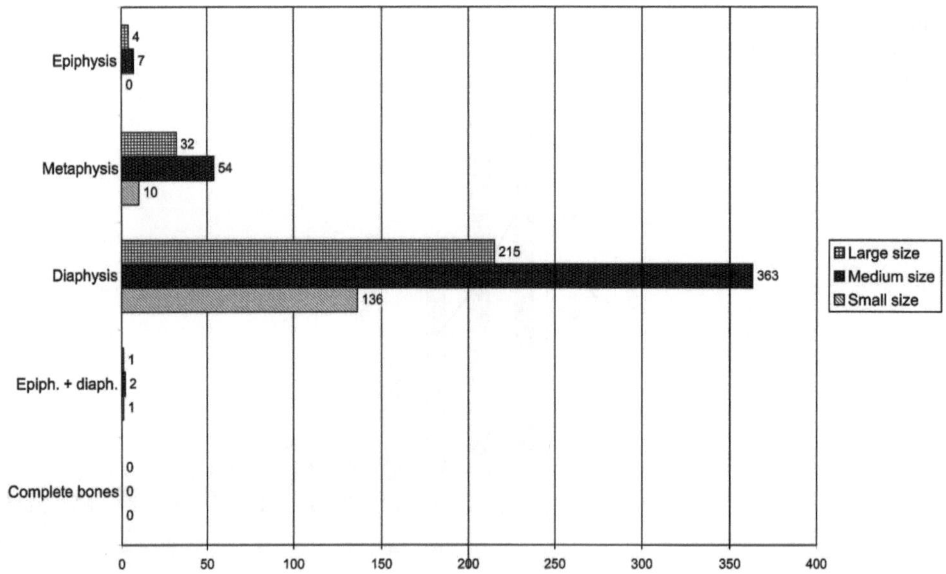

Fig. 8.7 Bone portions from Jb bone assemblage (X = number of remains)

larger than 5 cm were analyzed: 3,702 breakage planes on 1231 bones in Ja and 1509 on 400 remains in Jb. Only 30 bones present more than a third of their total original circumference.

The sample from sublevel Ja contains 894 breakage planes with longitudinal delineation, 1,162 transversal, 1,630 curved, and 16 that fall into other categories (Tables 8.16, 8.17). In the sample from sublevel Jb, 506 breakage planes with a longitudinal delineation have been identified, 289 transversal and 714 curved (Tables 8.16, 8.17). The presence of a large number of longitudinal fractures with right or oblique angles and soft surfaces, and transversal fractures with oblique angles and soft surfaces indicates the predominance of fresh breakage (Rosell 2001). Fragmentation of dry bones is practically nonexistent (Tables 8.16, 8.17).

Evidence of Anthropogenic Activity

The hominid activities in level J are mainly documented by cutmarks, intentional bone breakage and burning. The analysis of bone remains allowed us to distinguish between the use of the external animal resources (primarily meat, tendons, viscera, skin) mainly through the presence of cutmarks, and the use of internal resources (marrow and fat) through the information provided by breakage evidence. Fire exposure

Table 8.16 Fracture angle according to delineation. Long.: longitudinal, Trans.: transverse, Curv.: curved/V-shaped

	Ja								Jb					
	Long.	%	Trans.	%	Curv.	%	Others	%	Long.	%	Trans.	%	Curv.	%
Right	386	43.2	342	30	539	33.3	6	37.5	229	45.2	41	14.2	121	16.9
Oblique	415	46.4	490	41	771	47.7	3	18.8	198	39.1	72	24.9	372	52.1
Mixed	93	10.4	330	28.9	320	19	7	43.8	79	15.6	176	60.9	221	30.9
Total	894	100	1162	100	1630	100	16	100	506	100	289	100	714	100

Table 8.17 Fracture edge surface according to delineation. Long.: longitudinal, Trans.: transverse, Curv.: curved/V-shaped

	Ja								Jb					
	Long.	%	Trans.	%	Curv.	%	Others	%	Long.	%	Trans.	%	Curv.	%
Soft	749	83.8	590	50.1	1234	75.7	12	75	429	84.8	173	59.8	581	81.4
Jagged	145	16.2	572	49.2	396	24.3	4	25	77	15.2	116	40.1	133	18.6
Total	894	100	1162	100	1630	100	16	100	506	100	289	100	714	100

has also been documented. Thermal alterations are the most abundant modifications of anthropic origin in level J, whereas cutmarks are the least abundant (Table 8.18).

Table 8.18 Number of remains (NR) with anthropic origin alterations and percentage of total recovered remains by sublevels

	Ja		Jb	
Total remains	6738		1722	
Anthropic breakage	804	11.9%	149	8.6%
Cutmarks	448	6.6%	69	4%
Burning damage	2113	31.3%	317	18.40%

Cutmarks: obtaining external resources

The use of the external nutrients of animals is established from the cutmarks present on the bone remains. The characteristics of these cutmarks are generally conditioned by the animal's physiology (morphology of the bones by species), and change according to size and/or taxon. Nevertheless, the technology used by the hominids and their specific cultural patterns, which can differ widely from group to group, give rise to a certain margin of variability in the techniques employed in the use of carcasses. Therefore, cutmarks types, their position and disposition on the anatomical elements make it possible to establish cultural processing patterns.

Few remains with cutmarks have been documented in level J and the percentage of remains on which they are present does not exceed 4% in any case. Incisions are the most abundant type of cutmark, although scrape marks, saw marks and chopmarks have also been identified. Incisions are mainly found in large and medium-sized animals and affect all skeleton segments. Scrape marks, more abundant in sublevel Ja (2.2%) than in Jb (1.4%), have mostly been documented on the long bones of large and medium-sized animals and to a lesser extent on the flat bones. Both saw marks and chopmarks are marginally represented. Chopmarks have been found on equid humeri and metatarsals and on one flat bone from a large animal. Saw marks in sublevel Ja have been documented on a rib and a long bone from a medium-sized animal (Table 8.19). In sublevel Jb they have been recorded on one long bone from a large animal and one red deer metatarsal (Table 8.20).

Generally, cutmarks types are related to the meat of animals (incisions) and to the use of the nutrients closest to the cortical portion of the bones, i.e. fat and periosteum (scrape marks). However, chopmarks and saw marks have also been identified, which are normally related to disarticulation and dismembering processes or to the recovery of particular parts, such as tendons.

The analysis of the typology and location of the cutmarks indicates a series of activities that were performed on the bones and contributes to reconstructing the use sequence of external resources. However, several factors must be considered. For example, the absence of certain anatomical parts and the juxtaposition of postdepositional modifications, such as the high root mark rate detected in Jb, may have disguised or destroyed many of the modifications of anthropic origin located on the cortical surface of the bones (Cáceres et al. 2012).

The sequence of use of external resources is described according to weight classes. We have identified more activities related to faunal processing in sublevel Ja than in Jb. This difference between the two faunal assemblages is mainly due to the smaller number of remains recovered from sublevel Jb combined with a higher rate of bones with postdepositional modifications.

Skinning

Normally, the first activity in carcass processing is the removal of the skin. It is not uncommon to find cutmarks on

Table 8.19 Cutmarks groups according to skeletal element and taxa from sublevel Ja

		Groups	N° striations by group	Type of cutmarks	Location	Orient.	Action performed
Skull	Bovinae	3	2–3	i (slicing marks)	Supraorbital area	obl	Skinning
	Large size	1	1	i (slicing marks)	Frontal process	tr	Skinning
	Medium size	2	1–3	i (slicing marks)	–	obl-tr	Skinning
Maxilla	Equidae	2	1–2	i (slicing marks)	Incisor (I2) M1 (vestibular face)	tr	Defleshing
Mandible	Bovinae	4	2–4	i (slicing marks)	Mandibular condyle horizontal ramus	tr obl-tr	Disarticulation, defleshing
	Cervidae	2	1–2	i (slicing marks)	Horizontal ramus (M2–M3)	obl	Defleshing
	Equidae	11	1–4	i (slicing marks)	Horizontal ramus (M2) horizontal ramus (M2–M3) horizontal ramus (P3–P4)	obl	Defleshing
	Large size	14	1–7	i (slicing marks)	Mandibular border horizontal ramus	tr-obl	Skinning, defleshing
	Medium size	1	1	i (slicing marks)	Horizontal ramus	tr	Defleshing
Teeth	Equidae	9	1–2	i (slicing marks)	D3 sup: vestibular face P2,P4,M1, M3 sup: vestibular face P3, M2 inf: vestibular face	obl	Defleshing
Rib	Cervidae	6	2–5	i (slicing marks)	Diaphysis: dorsal face neck: dorsal fase epiphysis	obl	Defleshing, defleshing disarticulation
	Large size	1	4	i (slicing marks)	Diaphysis: vertral face	obl	Viscera removal
	Medium size	32	1–10	i (sawing marks) i (slicing marks)	Diaphysis: dorsal face neck: dorsal fase diaphysis: ventral face	obl-tr	Defleshing, defleshing, viscera removal
	No identified	3	1–3	i (slicing marks)	Diaphysis: neck: dorsal face	obl	Deflesing
Vertebra	Cervidae	2	2–3	i (slicing marks)	Spinous process	obl	Defleshing
	Medium size	5	1	i (slicing marks)	Superior articular facet	obl	Disarticulation
	No identified	1	1	i (slicing marks)	Transverse process	obl	Defleshing
Pelvis	Equidae	3	2–4	i (slicing marks)	Ischial tuberosity	obl	Defleshing
Scapula	Bovinae	2	1–2	i (slicing marks)	Scapular neck	obl	Defleshing
	Large size	5	1–2	i (slicing marks)	Infraspinous fossa	long-obl	Defleshing
	Medium size	1	5	i (slicing marks)	Subscapular fossa	obl	Defleshing
Humerus	Bovinae	13	1–30	i (slicing marks) scrapes	Diaphysis-metaphysis metaphysis-epicondyle prox. metaphysis-diaph.	obl-tr tr	Defleshing, periosteum removal
	Cervidae	8	1–2	i (slicing marks) scrapes	Diaphysis-metaphysis Diaphysis	obl-tr	Defleshing, periosteum removal
	Equidae	14	1–10	i (slicing marks) scrapes chopmarks	Diaphysis-epicondyle diaphysis distal metahysis	obl-long long-obl	Defleshing, disarticulation Defleshing
	Rhinocerotidae	2	1–7	i (slicing marks) scrapes	Diaphysis	obl long	Defleshing
	Large size	4	1–2	i (slicing marks)	Diaphysis prox.metahysis	long-obl	Defleshing

(continued)

Table 8.19 (continued)

		Groups	N° striations by group	Type of cutmarks	Location	Orient.	Action performed
Radius	Bovinae	1	1–2	i (slicing marks)	Diaphysis	obl	Defleshing
	Cervidae	2	1–15	i (slicing marks)	Radial tuberosity –diaph.	obl	Defleshing
	Equidae	6	1–22	i (slicing marks) scrapes	Diaphysis-prox.metahysis diaphysis	obl-tr long-obl	Defleshing
	Large size	3	2–4	i (slicing marks)	Diaphysis	obl	Defleshing
Ulna	Cervidae	4	1–3	i (slicing marks)	Diaphysis	obl	Defleshing
Femur	Bovinae	2	1	i (slicing marks)	Diaphysis	obl-tr	Defleshing
	Cervidae	14	1–3	i (slicing marks) scrapes	Diaphysis	obl-long long	Defleshing
	Equidae	2	1–3	i (slicing marks)	Diaphysis	tr-obl	Defleshing
	Large size	3	2–3	i (slicing marks)	Diaphysis	obl	Defleshing
Tibia	Cervidae	3	1–3	i (slicing marks) scrapes	Soleal line-diaphysis diaphysis	obl-long	Deflesing
	Equidae	8	1–5	i (slicing marks)	Diaphysis-prox.metahysis	obl-long	Defleshing
	Rhinocerotidae	1	1	i (slicing marks)	Diaphysis	obl	Defleshing
	Large size	3	1–2	i (slicing marks)	Diaphysis prox. metahysis	long-obl	Defleshing
	Medium size	2	4-7	i (slicing marks)	Diaphysis: cranial face	obl	Defleshing
Metacarpal	Cervidae	6	1–5	i (slicing marks) scrapes	Diaphysis	obl-tr	Skinning, periosteum removal
	Equidae	2	1–3	i (slicing marks) scrapes	Diaphysis	obl-long	Skinning, periosteum removal
Metatarsal	Cervidae	2	1–2	i (slicing marks)	Diaphysis	obl-tr	Skinning
	Equidae	8	1–3	i (slicing marks) scrapes, chopmarks	Diaph.-prox.metaphysis diaph.-prox.metaphysis Diaphysis	obl-long obl	Skinning Tendon removal
Metapodial	Equidae	2	1–2	i (slicing marks)	Diaphysis	obl	Skinning
Long bones	Large size	95	1–6	i (slicing marks) scrapes	Diaphysiss metaphysis	obl-tr	Defleshing, periosteum removal
	Medium size	91	1–10	i (sawing marks) i (slicing marks)	Diaphysis epiphysis	obl-tr	Defleshing, disarticulation
	Small size	3	1	i (slicing marks)	Diaphysis	obl	Defleshing
	No identified	7	1–2	i (slicing marks)	Diaphysis-metaphysis	obl-tr	Defleshing
Flat bones	Large size	36	1–9	i (slicing marks) scrapes chopmarks	–	ob-long-tr	Defleshing
	Medium size	80	1	i (slicing marks)	–	obl-tr	Defleshing
	Small size	3	1	i (slicing marks)	–	obl-tr	Defleshing
	No identified	14	1–2	i (slicing marks)	–	obl-tr	Defleshing
Unident.	Large size	1	3	i (slicing marks)	Diaphysis	obl	–
	Medium size	8	1- > 8	i (slicing marks)	Diaphysis	obl-tr	–
	No identified	22	1–2	i (slicing marks)	–	obl-tr	–

Table 8.20 Cutmarks groups according to skeletal element and taxa from sublevel Jb

		Groups	N° striations by group	Type of cutmarks	Location	Orient.	Action performed
Mandible	Large size	1	1	i (slicing marks)	Horizontal ramus	obl	Defleshing
Ribs	Large size	2	3–7	i (slicing marks)	Diaphysis: ventral face diaphysis: dorsal face	tr	Viscera, removal defleshing
	Medium size	2	2–3	i (slicing marks)	Diaphysis: dorsal face	obl	Defleshing
	Small size	1	1	i (slicing marks)	Diaphysis: dorsal face	obl	Defleshing
Vertebrae	Medium size	1	2	i (slicing marks)	Spinous process	obl	Defleshing
Pelvis	Cervidae	1	1	i (slicing marks)	Ilium	obl	Defleshing
	Large size	1	1	i (slicing marks)	Acetabular fossa	obl	Dismembering
Humerus	Cervidae	4	2–5	i (slicing marks)	Diaphysis-metaphysis	obl	Defleshing
	Large size	2	1–2	i (slicing marks), scrape	Diaphysis	ob-long	Defleshing, periosteum removal
Femur	Bovinae	1	2	i (slicing marks)	Diaphysis	obl	Defleshing
Tibia	Equidae	3	3–5	i (slicing marks)	Diaphysis	obl	Defleshing
	Large size	3	1–10	i (slicing marks)	Diaphysis	obl	Defleshing
Metatarsals	Cervidae	4	1–2	i (slicing marks) i (sawing marks)	Diaphysis	obl long	Skinning Tendon Removal
Metapodials	Large size	1	1	i (slicing marks)	Diaphysis	obl	–
Long bones	Large size	12	1–6	i (slicing marks) i (sawing marks)	Diaphysis	obl-tr	Defleshing
	Medium size	26	1–6	i (slicing marks)	Diaphysis	obl-tr	Defleshing
	Small size	5	1–3	i (slicing marks)	Diaphysis	obl	Defleshing
Flat bones	Large size	1	2	i (slicing marks)	Diaphysis	obl	Defleshing
	Medium size	8	1–3	i (slicing marks)	Diaphysis	obl	Defleshing
	Small size	1	1	i (slicing marks)	Diaphysis	obl	Defleshing
Unident.	No identified	1	1	i (slicing marks)	–	obl	–

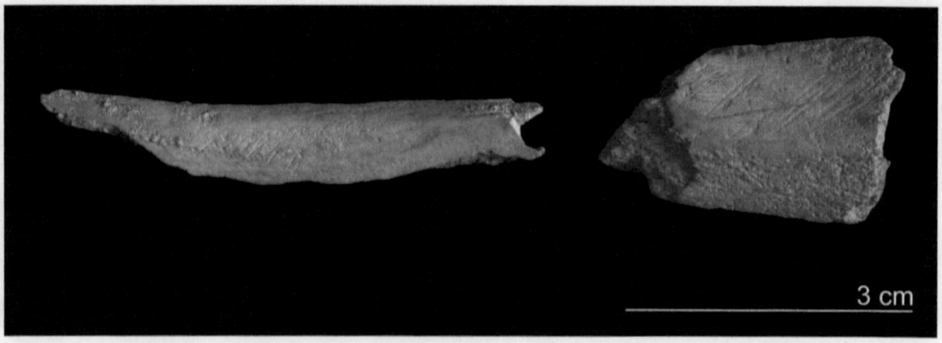

Fig. 8.8 Cutmarks related to viscera removal on the ventral side of ribs from Ja

the cranial skeleton and the distal part of the extremities as a result of this activity. In sublevel Ja we identified this type of mark on cranial elements of large and medium-sized animals. On three frontal bones of unidentified animals, incisions of different lengths (3–12 mm) were found close to the eye sockets. Some of the marks found on the maxillae were also related to skinning activities. These marks are incisions close to the incisors and canine teeth used to separate the lips from the bone. The scarce presence of phalanxes, articular bones such as the calcaneus and certain epiphyses of long bones (proximal part of the ulna, distal part of the tibia, or those from the metapodium) has prevented us from identifying skinning activities on the appendicular skeleton.

Viscera removal

The action of removing the viscera does not generally leave behind a high number of marks, but when they are present they tend to be focused on the elements of the axial skeleton, mainly ribs. The scarcity of these remains in the two faunal assemblages has hindered the identification of this activity. While six ribs from large and medium-sized animals with incisions on the ventral side have been related to evisceration in sublevel Ja (Fig. 8.8), the only evidence of evisceration in sublevel Jb is an incision on the inner part of a rib of a large animal.

Dismembering

Dismembering separates the animal's trunk from its limbs. The marks left behind by this activity are often located on the scapulae and the acetabular fossa of the pelvis. Only one incision group was observed on one pelvis from a large animal in sublevel Jb.

Disarticulation

Disarticulation marks are not very abundant because such activity leaves little evidence on bones. The absence of epiphyses contributes to the scarce representation of this type of mark.This activity has been identified only in sublevel Ja.

Defleshing

Defleshing is one of the activities that leaves the highest quantity of cutmarks, and in level J (both Ja and Jb) this activity affected the highest number of remains. In sublevel Ja marks related to defleshing have been documented on 60 bones and on 15 in Jb. In both faunal assemblages these defleshing marks have been found on all anatomical segments and in all weight classes. It is very difficult to clearly determine whether defleshing was performed before or after the disarticulation of the bones, although these activities are normally carried out simultaneously.

Defleshing marks on the cranial skeleton are mainly found in large animals (mainly represented by *Equus ferus*), since these animals have a large quantity of muscle mass attached to their mandibles. This mass decreases significantly in medium-sized and small animals. Incisions on the mandibles are mainly found on the gums, on both sides of the bone. The defleshing of mandibles in sublevel Jb is only documented from the incisions identified on a single mandible remain from a large animal. On the other hand, in sublevel Ja this activity on the cranial skeleton has been documented on 15 mandibles from different large animals.

Defleshing marks on the axial skeleton occur when the meat of back is removed from the carcass. These incisions, which on some occasions are sawed, are found on the dorsal and lateral sides of the ribs, mainly on the metaphysis and the medial area of the bones. Most of these marks are found on the axial skeletons of medium-sized animals in sublevel Ja. In sublevel Jb defleshing marks have been identified on a rib as well as on the spinous process of a dorsal vertebra from a medium-sized animal.

Incisions related to defleshing have also been identified on some pelvic bones. In sublevel Ja a scapula was recovered with cutmarks on its caudal side that have been related to muscle mass removal. However, in sublevel Jb marks associated with procuring these resources have only been identified on one cervid pelvis.

The elements of the appendicular skeleton, specifically the long bones, present a higher number of marks associated with defleshing (Fig. 8.9). These marks have been found on large animals as well as on medium-sized and small animals. Some of the long bones with defleshing marks show similar patterns, especially in sublevel Ja where more remains with defleshing marks have been documented.

In sublevel Ja, 17 humeri remains with defleshing marks have been identified. These humeri belong to animals of different taxa and weight classes. Most of the cutmarks are found grouped in the zones of nutrient foramens, although

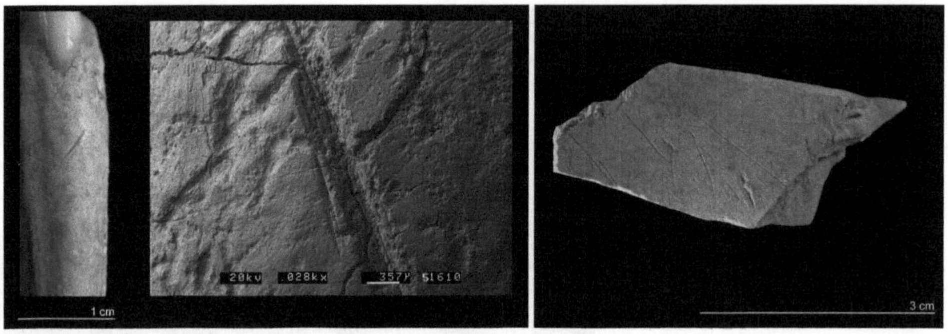

Fig. 8.9 Defleshing marks on equid molar (*left*) and on the long bone of a medium-sized animal (*right*) from Ja

in some cases there are isolated incisions on the metaphysis of the inner distal condyle and on the proximal part of the diaphysis. Most of the defleshing marks located on these bones exceed 10 mm.

Defleshing marks on the remains of radii and ulnas are scarce. In sublevel Ja these incisions are grouped on the medial side of the bone and are longer than 10 mm.

In the case of the femur, defleshing marks were mainly found on cervids. Incisions longer than 20 mm are located on all four sides of the bone, although they are most frequently found on the cranial side.

Among the tibias, defleshing patterns differ depending on the size of the animal. Whereas the tibias of medium-sized animals present marks of about 10 mm on the proximal area of the diaphysis, on the lateral side, the tibias of large animals present incisions between 20 and 40 mm long on the medial side as well as on the two lateral sides. Deep, isolated incisions have also been documented on the caudal side of large animals.

Tendons and fat removal

In addition to the nutrient-seeking activities described above, other animal resources were exploited such as fat, the periosteum, and tendons. Evidence of such use is mainly exhibited in the form of scrape marks, which are normally located on appendicular skeleton elements. In sublevel Ja, scrape marks, which typically appear as a result of procuring fat and periosteum, have been documented on different elements of the extremities, mainly humeri and metapodials.

In the case of sublevel Jb, the presence of one scraped humerus from a large animal can be related to the extraction of fat and periosteum. However, not all the scrape marks identified are necessarily related only to obtaining the nutrients closest to the bone such as fat and periosteum. They may also have resulted from making maximum use of the small portions of meat still stuck to the bone after defleshing.

Some authors relate this activity to preparing the bone for subsequent fracturing, as the periosteum and fat absorb the effects of direct percussion (Brain 1981). This activity is often associated with a thermal treatment that dries the bone and removes its elastic properties. As a consequence the bone becomes much more fragile and more sensitive to the effects of percussion impact (Bonfield and Li 1966; Brain 1981).

Evidence derived from obtaining tendons has been observed mainly on metapodials. In sublevel Ja it was found on two equid metatarsals (posterior side). In sublevel Jb, a saw mark with two oblique incisions on the dorsal side of the central diaphysis of a red deer metatarsus is clearly associated with the action of cutting the tendon located on the anterior channel of the skeletal element.

There are few identified cutmarks in level J, especially in sublevel Jb. Among the activities documented, only in sublevel Ja does the repetition of marks associated with defleshing activities allow us to establish defined processing patterns of external animal resources. These patterns can be seen on specific pieces, such as the humerus, and they allow us to infer the existence of similar patterns on other bones. The use of muscle mass is intensive, as can be deduced from the marks related to defleshing, which are located in areas of muscular insertion or that are difficult to access.

Anthropogenic breakage: obtaining internal resources

Level J has yielded more data related to the use of interior resources than to the acquisition of external resources. In sublevel Ja, diagnostic elements of anthropogenic fracturing were identified on 804 bones and in sublevel Jb such evidence was found on 149 osseous remains (Table 8.21). In both sublevels, the most abundant diagnostic elements are percussion cones and cortical flakes in large and medium-sized animals, while percussion marks and percussion impacts are less frequent. In sublevel Ja, these traces have allowed us to identify specific breakage patterns.

Intentional anthropogenic breakage is more frequent in large animal remains (equids, bovids, and rhinoceroses) than among cervids and medium-sized and small animals. This difference is more strongly marked in Ja than in Jb. In both assemblages diagnostic elements in small animal bones are scarce. The bones of large animals are denser and thicker (Lyman 1984), and breaking them would have therefore proved more difficult. Perhaps for this reason, more repetitive evidence of breakage appears among the large animals of the two assemblages. Peeling fractures are scarce in either assemblages, but especially in Jb. This type of modification mainly appears on the bones of medium-sized and small animals, mostly on flat bones from the axial skeleton.

Table 8.21 Number of remains (NR) with diagnostic signs of anthropic breakage at sublevels Ja and Jb

	Percussion marks		Percussion notches		Impact flakes		Adhering flakes		Cortical flakes		Medular flakes		Peeling		Countercoup	
	Ja	Jb	Ja	Jb	Ja	Jb	Ja	Jb	Ja	Jb	Ja	Jb	Ja	Jb	Ja	Jb
Equids	7	–	46	6	3	–	1	–	1	–	–	–	–	–	3	–
Cervids	4	–	76	8	1	–	2	1	4	–	4	–	7	–	4	1
Large bovids	1	–	33	2	3	–	1	–	2	–	1	–	–	–	2	–
Large size	24	1	49	12	112	24	1	–	19	5	1	1	4	1	3	–
Medium size	6	1	21	6	103	34	4	1	38	10	–	7	18	1	4	–
Small size	2	–	–	1	–	1	–	–	–	3	–	2	3	–	–	–
Unident.	5	–	8	–	102	10	–	–	26	9	–	1	7	–	–	–

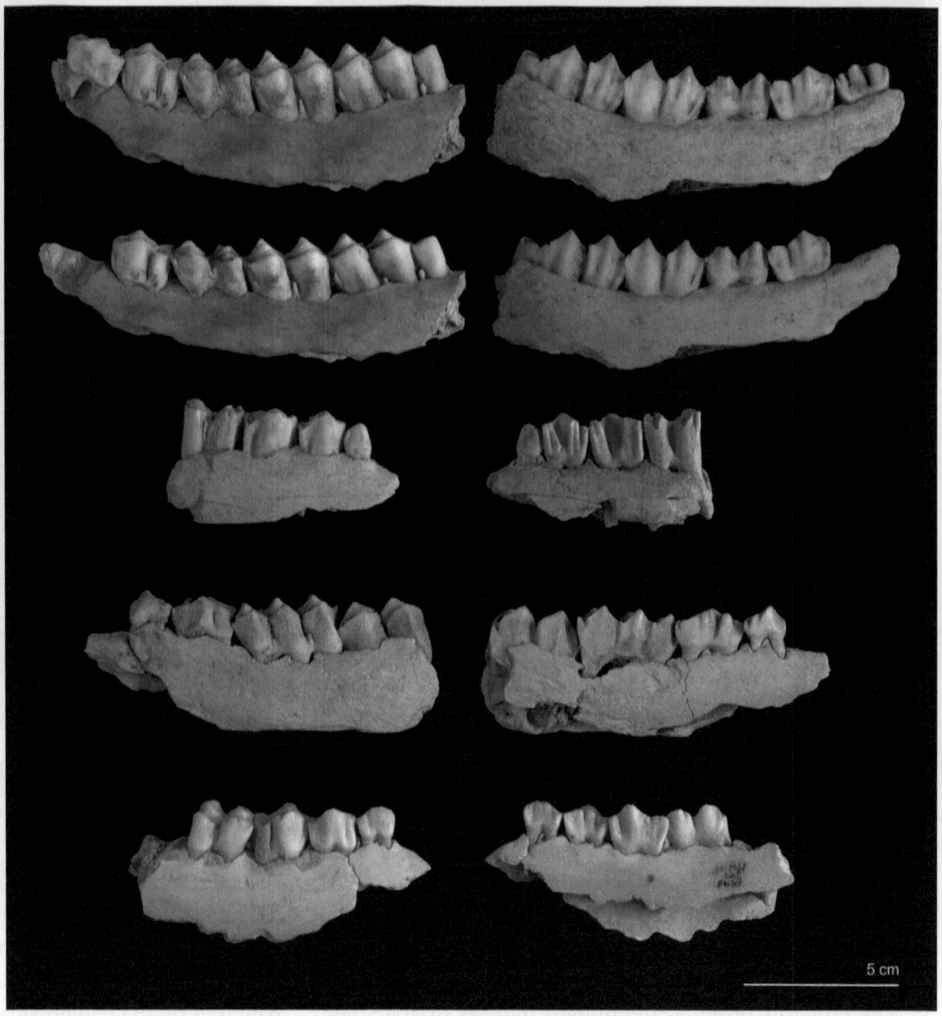

Fig. 8.10 Anthropic breakage on cervid mandibles from Ja: vestibular (*left*) and lingual (*right*) side

A series of breakage patterns have been established through the study of the faunal assemblage in Ja. The low number of anatomically identified fractured bones in sublevel Jb prevents us from establishing any type of breakage pattern in this sublevel.

The breakage patterns established in Ja consider both the location (side and portion) of the diagnostic elements on the bone and the resulting morphotypes. In the cranial skeleton it was possible to specify breakage patterns on the maxillae and mandibles. In both cases, repeated morphotypes have been documented, not only in Ja but also in other levels of the site (Vallverdú et al. 2005). The most common morphotype is documented on the mandibles, which are fractured on the horizontal branch in order to reach the marrow located in this area (Fig. 8.10). Many dental fractures (mainly on the roots) are the result of this activity. In the case of the equids, the

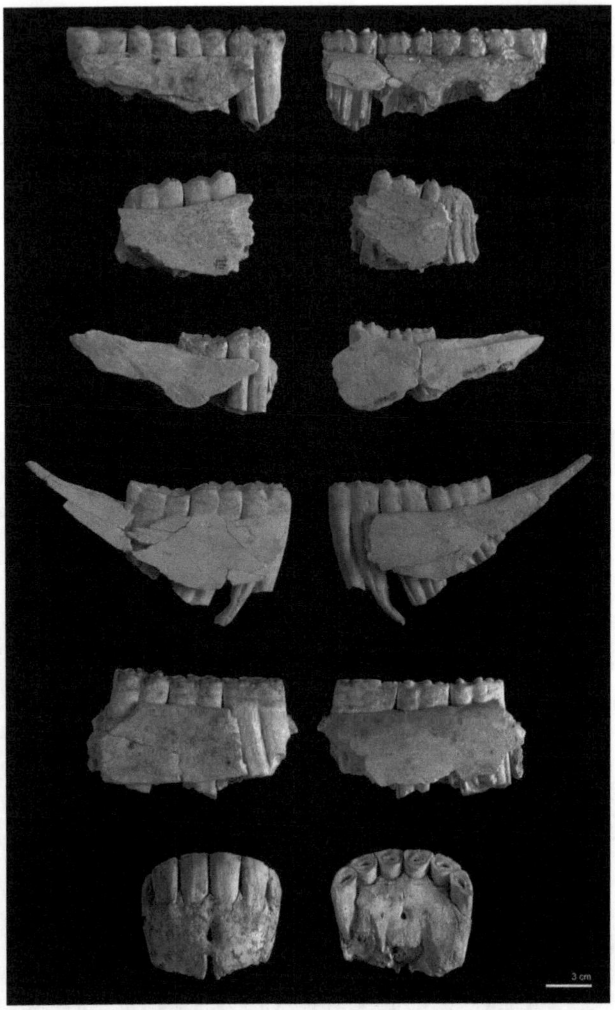

Fig. 8.11 Anthropic breakage on Ja equid mandibles: vestibular (*left*) and lingual (*right*) side

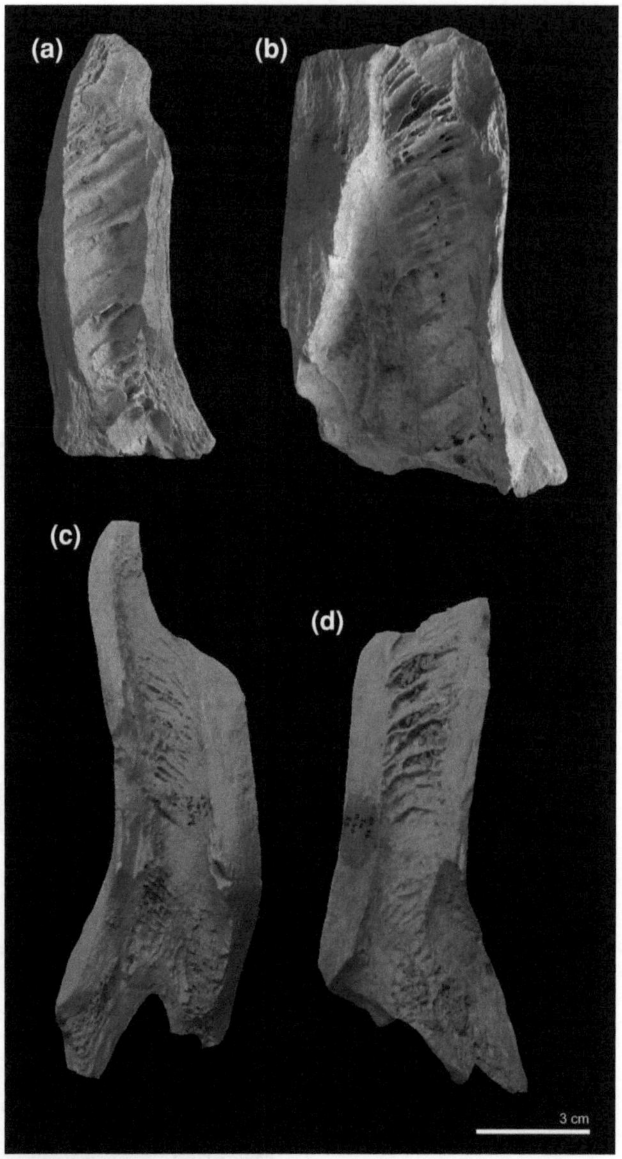

Fig. 8.12 Anthropogenic breakage on humeri from Ja: red deer (a, c and d) and auroch (b)

structure of their hemimaxillae and their hemimandibles fused by the symphysis gives rise to a different morphotype. These skeletal elements are fractured for the purpose of separating the incisors from the rest of the cranial skeleton (Fig. 8.11).

Peeling of the ribs has been documented in the axial skeleton. The peeling fracture is located in the first quarter of the total length of the bone.

The humerus provides the highest number of specimens of the appendicular skeleton with which to establish a pattern. These are mainly fractured on their distal metaphysis, which generates a very specific morphotype in the form of diaphysis fragments that practically cover the whole cranial side of the bone, from the proximal part to the area where the olecranon fossa begins (Fig. 8.12).

In the case of the radius and ulna, the scarcity of diagnostic elements prevented us from establishing a pattern. We can only note that the percussion impacts are located in two preferential zones: (1) the proximal part of the caudal side and (2) the central part of caudal and medial side.

The femurs present impacts in different zones according to taxa. Among the perissodactyls, percussion impacts are centered on the caudal side of the proximal metaphysis. For artiodactyls, mainly cervidae, percussion notches have been documented on the proximal metaphysis and the caudal side of the central areas of the diaphysis. Sometimes these diagnostic elements move to the distal area of the medial side of the diaphysis.

Tibiae also show high variability in the location of impacts. Nevertheless, it was not possible to distinguish a specific pattern for each taxon. The impacts identified are centered in four main zones: (1) the center of the diaphysis

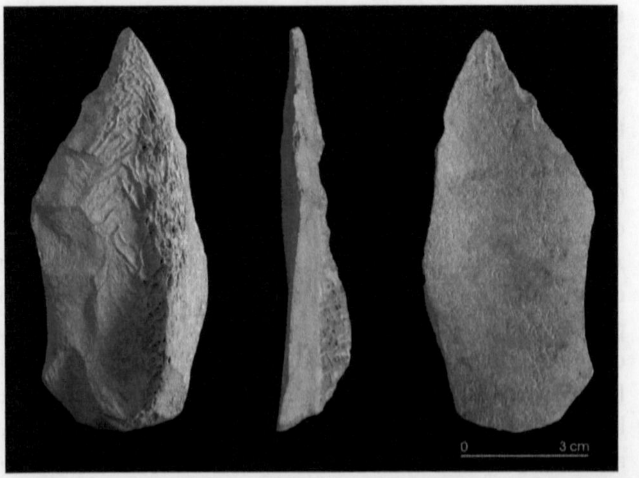

Fig. 8.13 Pointed tool on aurochs mandible from sublevel Ja

on the caudal side; (2) the distal part of the medial side; (3) the medial proximal zone of the lateral side; and (4) the proximal zone of the medial side. The percussion impacts on metapodials are concentrated on the medial area of the bone, on all sides.

Occasionally, some of the elements are shaped for a subsequent use. In sublevel Ja only one bovid mandible was found, which indicates that it was used as a raw material for tool production. It has been made pointed by means of a sequence of continuous removals on the lateral edges (Fig. 8.13).

Burning damage

A total of 2,430 burned remains have been recovered in level J: 2,113 in Ja and 317 in Jb, representing 31.3 and 18.4% of all remains, respectively (Table 8.22). Most of the burned bones are faunal remains that were unidentified at the anatomical and, consequently, taxonomical level. As a general rule, the categories with the highest number of remains also hold the highest number of burned remains. Nevertheless, in level J the burning rate decreases as animal size increases, except in the case of the equids. Gifford-Gonzalez (1993) and Oliver (1993) have proposed from the ethnographic studies that there is a pronounced trend to cook small animals for consumption while the marrow of large animals is most commonly consumed raw.

In sublevel Ja, the same trend emerges when analyzing the burned remains by anatomical segments. Except for among the small animals, whose percentages are distorted because of their scarce representation, the burning rate increases as size decreases. In large anatomical units, few significant differences between segments have been found (except for the zygopodial bones), especially between anterior and posterior extremities.

Stiner et al. (1995) defined six degrees of burning using color criteria, from simple rubefaction to total calcination.

Table 8.22 Percentage of burned bones from Ja according to anatomical portions and weight categories (N° of burned bones = 2113)

Ja	Large size	Medium size	Small size
Cranial skeleton	16.7	19.5	33.3
Axial skeleton	5.5	12.5	16.7
Girdles	21.9	3.5	0
Stylopodials	14.9	17.4	33.3
Zygopodials	13.8	6.9	25.0
Metapodials	13.2	24.8	66.6
Acropodials	0	45.5	50.0

Table 8.23 Number of burned bones from sublevels Ja and Jb according to the classification system adapted from Stiner et al. (1995)

	Stiner et al. (1995)	Coloration	Ja	Jb
Degree 0	Degree 0	Absence of thermo-alterations	4625	1405
Degree 1	Degrees 1–2	Brown	1111	89
Degree 2	Degrees 3–4	Black	698	138
Degree 3	Degree 5	White (calcined)	304	90

For increasing the number of specimens in each category, these six degrees have been grouped into four categories for the remains from level J (Table 8.23).

Brown tonalities (degree 1) stand out in sublevel Ja. The number of specimens decreases as the degrees increase (degrees 2 and 3). In sublevel Jb, burned bones present very few significant values at any degree.

There are many remains that have double colorations. This is due to the fact that the entire surface has not been exposed to fire at the same intensity. Some of these double colorations are located on the same side. This may be related to the habit of roasting pieces of meat previous to defleshing activities. At an experimental level, Rosell (2001) found that double colorations can be produced during roasting. This happens because when the bone is placed on the hearth, the part of the bone without meat changes color faster, taking on tonalities ranging from black to grey and white. Meanwhile, the parts of the bone covered by meat maintain their original color, and on some occasions acquire brown tonalities in places where there is less muscle mass.

Burned remains with cutmarks have also been recovered. The bottoms of these cutmarks exhibit lesser degrees of burning than the total surface of the bone. The areas of the bone exposed directly to fire receive the calorific effects

Fig. 8.14 Double colorations on long bone fragments from Ja faunal assemblage. Note cutmark with a lesser degree of coloration than the entire bone surface

Table 8.24 Number of remains (NR) and percentage of burned bones according to double colouring

	Ja		Jb	
	NR	% of total burned NR	NR	% of total b urned NR
Degree 1/0	24	1.14	7	2.2
Degree 2/1	196	9.32	6	1.9
Degree 3/2	109	5.2	28	8.8

more intensely than areas covered by meat. This indicates that meat was roasted before being separated from the bone (Rosell 2001; Fig. 8.14).

In sublevel Ja, 296 remains with double coloration due to thermal modifications have been documented: black/brown (2/1) and white/black (3/2). The most abundant burned remains belong to the degrees of greater intensity—degrees 2/1 and 3/2. There are very few remains showing evidence of lower intensity burning. In sublevel Jb, double colorations have been identified on 41 remains, making this the most common coloration (degree 3/2) in the sublevel. In most cases, these double colorations are located on the same side (Table 8.24).

In general, burned bones are small. Remains shorter than 20 mm clearly stand out. The burned bone values decrease as the length of the bones increases. The group of bones with the lowest degrees of burning contains most of the large elements; whereas the remains with the highest degrees of burning are mostly smaller than 60 mm (Table 8.25).

Breakage of some remains (82 in sublevel Ja and 8 in Jb) has been related to their exposure to fire. These remains show cracking and mosaic fractures, which are present on all bones with high degrees of calcination.

Carnivore Evidence

Some carnivore remains have been recovered and traces of their activity have been documented. The number of specimens is very low: nine remains (four species) in Ja and two (two species) in Jb. They are isolated remains dispersed over the entire surface of both sublevels without preferential zones of accumulation. Traces caused by humans or other carnivores are very scarce on these remains: only one bear phalanx with degree 1 of rubefaction and two fragments of a carnivore rib with carnivore toothmarks have been identified. The absence of cutmarks or intentional breakage on the bones of these animals indicates that their presence in the assemblage is not related to human intervention. Furthermore, the burning damage on the bear remain provides insufficient data, as its presence at the site may have occurred previous to the formation of the human hearths.

Herbivore remains with toothmarks are not abundant either: there are only 61 remains in sublevel Ja (0.9% of the total) and 4 remains in sublevel Jb (0.2%) (Fig. 8.15). In the latter, depressions have been documented on one equid tibia, one cervid metapodium and two long bones of a medium-sized animal. In sublevel Ja toothmarks have been identified mostly on long bones, mainly on the diaphyses of large and medium-sized animals (Table 8.26). These remains are dispersed over the entire surface of the shelter. Something similar happens with the coprolites. The distribution of coprolites (5 in Ja and 12 in Jb) is similar, scattered over the surface of both sublevels, although they tend to be located inside hearths.

The dimensions of the carnivore toothmarks would indicate that the predator that acted on the assemblage was a medium-sized carnivore such as a canid. However occasional toothmarks from a larger carnivore, probably a hyaenid, have also been documented (Figs. 8.16, 8.17).

Cutmarks as well as carnivore toothmarks have been identified on 18 remains from sublevel Ja. The marks do not provide enough information to determine which of the two agents carried out the primary consumption and which the second (Blasco and Rosell 2009). Cutmarks produced by hominids interrupted by carnivore marks have only been documented in one case (Fig. 8.18).

Faunal accumulation Origin

At a skeletal level, the J level shows a bias in representation. Animals weighing over 100 kg are not fully represented. In general, they are characterized by the predominance of elements from the proximal appendicular skeleton (stylopodial and zygopodial bones) and the cranial skeleton (mainly

Table 8.25 Burned bone lengths in millimeters according to coloration degree

	Ja			Jb		
	Degree 1	Degree2	Degree 3	Degree 1	Degree 2	Degree 3
≤20	510	356	176	40	94	70
21–50	411	206	150	31	38	20
51–100	51	75	18	15	4	0
≥101	11	2	3	5	0	0

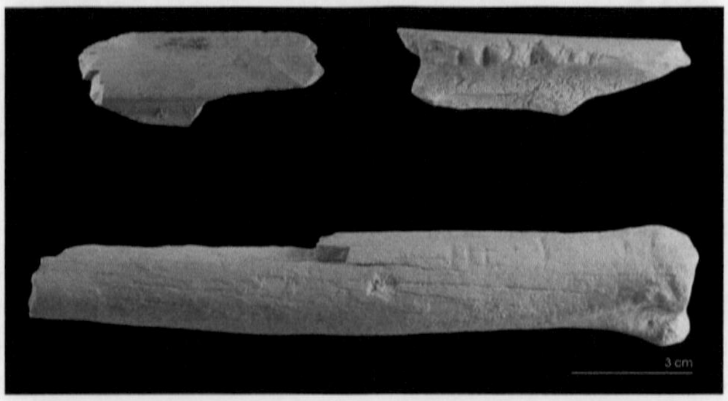

Fig. 8.15 Examples of carnivore toothmarks on metapodials of red deer from Jb faunal set

Table 8.26 Number of remains (NR) with carnivore toothmarks by anatomical elements and taxa from Ja assemblage

Ja		NR	Types of toothmarks	Location
Bovinae	Tibia	1	Score	Diaphysis
	Mandible	1	Score	–
Equidae	Vertebrae	1	Pit	Proximal apophysis
	Metacarpal	1	Score	Diaphysis
	Rib	3	Puncture-pit	Diaphysis
	Humerus	2	Score-pit	Metaphysis
	Radius	1	Score-pit	Diaphysis
Cervidae	Femur	2	Score-pit	Diaphysis-epiphysis
	Patella	1	Pit	–
	Tibia	3	Score-pit	Diaphysis
	Metatarsal	1	Score	Diaphysis
	Phalange	1	Pit	Epiphysis
Large size	Long bones	3	Puncture-score	Diaphysis
	Flat bones	3	Score-pit	Diaphysis
	Ribs	2	Score-pit	Diaphysis
Medium size	Long bones	11	Pits-score-puncture	Broken edge-metaphysis
	Flat bones	7	Pit	Diaphysis
Small size	Long bones	1	Score-punctures	Diaphysis
	Flat bones	1	Pit	Diaphysis
Unident.	Long bones	1	Score	Diaphysis
	Flat bones	1	Score	Diaphysis
	Indet	3	Pit-score	Diaphysis-broken edge

mandibles and maxillae) with elements of the axial skeleton (vertebrae and ribs) being almost absent. On the other hand, the anatomical parts of small animals tend to be more equally represented. This bias in the representation of different animals rules out several possibilities a priori. One such supposition is that herbivores related to the plains or the slopes around the valley entered the shelter under their own volition. These types of herbivores make up the majority of the animals in the assemblage, and none of them exhibits behaviors that would point to such an action (see exceptions e.g. Moncel et al. 2008). On the other hand, the sporadic presence of carnivores with specific cave-use habits, may be due to natural intrusions. The anatomical bias is the consequence of some physical, chemical or biological action that has impeded the entire representation of the skeletons.

This anatomical bias rules out the possibility that the rockshelter may have acted as a natural trap. The morphology of the enclave itself, the ages of the individuals (mainly healthy adult animals), the anatomical representation and the absence of entire elements and anatomical connections or semi-connections exclude this possibility (Brugal and Jaubert 1991; Díez 1992; Wang and Martin 1993; Rosell 1993, 1998; Huguet 1997, 1998; Huguet et al. 1999; Costamagno 1999; Brugal et al. 2006). Therefore, the causes that gave rise to the accumulation must be found in phenomena capable of moving remains from the exterior of the shelter to the interior using some selection criteria. This transport from the outside to the inside of the shelter is based on a priority given to elements that have higher medullar content and which are nutritionally richer. So, any anatomical selection as a consequence of postdepositional processes prior to burial (for example, the action of hydric currents on the carcasses, etc.) can be ruled out. On the other hand, fossil-diagenetic processes do not explain the absence of certain types of remains.

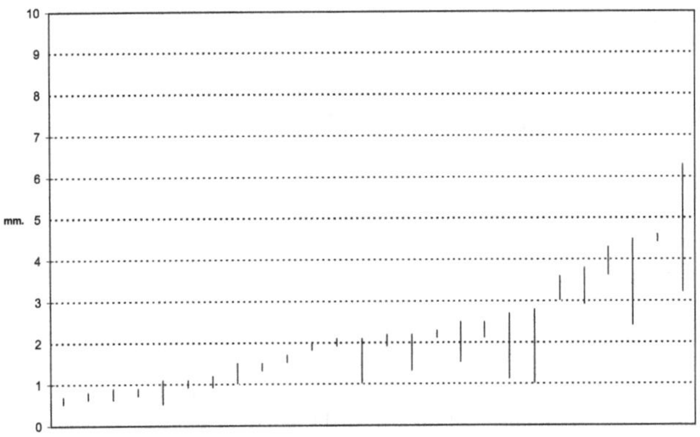

Fig. 8.16 Graphic representation of puncture dimensions from sublevel Ja (X = toothmark; Y = (Max. * min.) in millimeters)

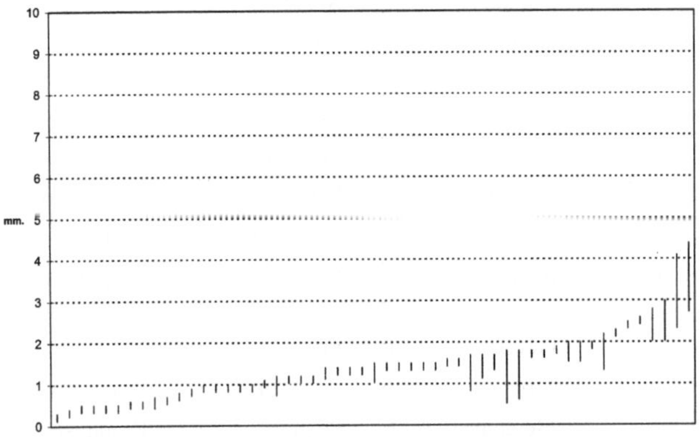

Fig. 8.17 Graphic representation of score dimensions (L * l) from sublevel Ja (X = toothmark; Y = (L * l) in millimeters)

Bones, once buried, do not seem to undergo any process that would cause differential conservation based on density. Only the medium-sized animals show apparently significant correlation values, which contrast with the low rates obtained among the large and small animals. From this perspective, it is not possible to surmise that the absence of elements from medium-sized animals might be due to a problem of differential conservation without it affecting the large and small animals as well.

Therefore, the phenomenon that caused the accumulation is restricted to the action of specific biological agents with the capacity of transporting and accumulating remains according to a preconceived selection. From this perspective, hominids as well as carnivores may have accessed the carcasses with similar objectives and occupied the space alternatively. Both predators have left traces of their activity in the two assemblages. Their incidences, however, are very different. The proportion of modifications produced by one agent or the other shows predominant anthropic intervention. 41.9% of the remains analyzed in Ja and 28.9% in Jb show evidence of the intervention of hominids (cutmarks, intentional breakage

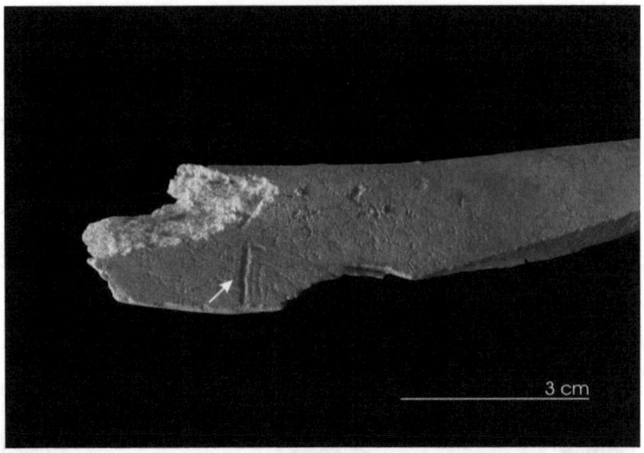

Fig. 8.18 Overlapping of carnivore toothmarks over cutmarks from Ja

and burning). Only 0.9% in Ja and 0.2% in Jb show modifications by carnivores. These low percentages, along with the absence of cylinders and digested bones, mainly immature or senile animals, and a larger representation of the axial skeleton of herbivores show that carnivores were not the main agent behind the accumulation (Cruz Uribe 1991). On the contrary, the data indicate that level J is an anthropogenic accumulation with a minimal incidence of carnivore activity.

Reconstruction of the Faunal Chaîne Opératoire

The Concept of the Chaîne Opératoire

The concept of the *chaîne opératoire* (Leroi-Gourhan 1964) arose from the need to determine the technical activities involved in making lithic tools. Artifacts are made in different stages, the succession of which form a sequence or *chaîne opératoire*. The interaction between different *chaîne opératoires* forms a Technical System and at the same time, the interactions between different Technical Systems constitute Social Organization (Lemonnier 1982).

Plant remains and bones of consumed animals belong to the Eco-social Dynamic of human groups and they require the development of their own *chaîne opératoire*. The dynamic conception that Leroi-Gourhan and the Logical Analytic System (Carbonell et al. 1983) designed for lithic remains can be applied to other materials recovered at archeological sites. The different *chaîne opératoires* documented at a site interact to establish a temporal order that makes it possible to distinguish between production *chaîne opératoires* and final *chaîne opératoire* or Consumption *chaîne opératoires* (Rosell 2001).

Table 8.27 Results of the combinations of the different elements distinguished in the acquisition of animals showing the different existing modalities (Rosell 2001)

Access type	Access time	Access modality	Implications
Primary	Immediate	Active	Hunting
			Scavenging, before consumption, of prey snatched from other hunters or scavengers
Primary	Immediate	Passive	Scavenging of recently dead animals who died a natural death
Primary	Late	Active	Scavenging of the remains of carcasses, dead for some time, through confrontation with other scavengers
Primary	Late	Passive	Scavenging of the remains of carcasses, dead for some time
Secondary	Immediate	Active	Scavenging, with confrontation, of prey partially consumed by other predators or scavengers
Secondary	Immediate	Passive	Scavenging of carcasses abandoned by other predators or scavengers
Secondary	Late	Active	Scavenging, with confrontation with other scavengers, of carcasses dead for some time
Secondary	Late	Passive	Scavenging of the remains of a carcass previously consumed and abandoned by other predators or scavengers

The *chaînes opératoires* of production are dedicated to obtain functional elements with preconceived morphologies, as in the case of lithic industry. On the other hand, the purpose of final *chaînes opératoires* or consumption *chaînes opératoires* is to process consumable materials. One of the clearest examples is the processing of animals for purely nutritional purposes (Rosell, op.cit.).

According to the temporal sequences which are organized hierarchically in the faunal *chaîne opératoire*, four main processes can be distinguished: the acquisition of the animal, the mode of transporting the animal, processing techniques, and the consumption of nutrients and the subsequent disposal of the remains (Brugal 1995). The different segments of the sequence are highly interrelated. During the analysis of faunal remains, all the observed parameters are required in order to prove the modality adopted at each step of the sequence.

Identifying the method of obtaining animals, as well as animal consumption and abandonment patterns, is fundamental in determining the behaviors and, consequently, the degree of complexity of the human groups that occupied sublevels Ja and Jb, and plays a key role in addressing the question of site function(s).

Acquisition Methods and Differential Transport

The acquisition of animals is understood as the set of strategies or techniques that human groups use for the provision of faunal resources. Hunting and scavenging are the main two modalities. Distinguishing between these two techniques is not easy at the archeological level. Bunn and Ezzo (1993) established the following criteria:

1. Access modality: describes the possible confrontations with other predators or with the prey itself at the time it is obtained. This can be active if there is a confrontation and passive if there is none.
2. Access type: shows the predators' intervention sequence on the carcass. It can be primary if the hominids arrive first or secondary if they access the prey once it has been obtained by other agents.
3. The time that passes between the animal's death and its acquisition. This can be immediate or late access.

On the basis of the criteria established by Bunn and Ezzo (op. cit.), hunting is defined only by active, primary and immediate access. All the other possibilities are related to different scavenging modalities (Table 8.27).

In an accumulation with a large anthropic component as is the case of level J, there is ample evidence that contributes information about the modalities of animal acquisition. Many researchers suggest that the main component to consider is the skeletal representation of the different taxa (e.g. Binford 1978, 1981; Brain 1981; Klein 1989; Lyman 1994; Stiner 1994) at a given site. The age of the animal at death must also be taken into consideration (Gaudzinski and Roebroeks 2000) as well as the signs of carcass processing found on the recovered bones: (1) the location and purpose of the cutmarks; (2) the presence of carnivore toothmarks; and (3) fractures of both anthropic and animal origin (Binford 1981; Bunn 1981; Shipman and Rose 1983; Marean and Kim 1998; Domínguez-Rodrigo and Pickering 2003; Domínguez-Rodrigo and Barba 2006).

On rare occasions the place where the animal resource was acquired coincides with the place where the group lived. Strategies based on the organization of a group with strong social cohesion involve the development of systems that enable the sharing of resources with the rest of the members in the group (Díez and Rosell 1998). This is why human groups move their prey. Sometimes the size of the animal represents a problem for transport. When an animal of considerable weight is acquired, human groups selectively transport only the nutritionally richest anatomic parts. This phenomenon is defined as the "schlepp effect" (Perkins and Daly 1968). Some ethnoarcheological works attempt to establish transport patterns and provide an explanation for the skeletal representation at sites (O'Connell et al. 1988a, b, 1990; Bunn 1993; Oliver 1993; Bartram 1997; Hawkes et al. 2001; Lupo 2001). When parts of the skeleton are being selected for transport, certain determining factors emerge in the decisions made by human groups (O'Connell et al. 1988b, 1990): (1) the size of the prey, (2) the time of day at which the animal is obtained, (3) the number of available bearers, and (4) the distance to the camp. To these, Oliver (1993) adds a new determining factor: the technology available to process the animal, within which he distinguishes between pre-fire, post-fire, pre-boiled bone, and post-boiled bone technologies.

Starting from the anthropic origin of the accumulations in sublevels Ja and Jb, there are several different pieces of evidence that provide information on the modalities of obtaining animals. These modalities have been identified from the skeletal representation, the types and location of the cutmarks, animal ages and the presence of carnivore toothmarks and incisions from anthropic processing.

Both sublevel Ja and Jb contain abundant cranial (mainly mandibles and maxillae) and appendicular elements (stylopodial and zygopodial bones) from large and medium-sized adult animals (mainly cervids and equids). Elements of the axial skeleton, the distal parts of the extremities and small animals are scarce. The skeletal bias in both sets can be interpreted as the consequence of a differential transport of the animals. According to this model, the hominids of level J developed an acquisition strategy that was characterized by the systematic transport of the appendicular and cranial parts of the animals to the shelter. The axial skeletons (vertebrae and ribs) and pelvis of the prey were abandoned at the kill-site.

The differential transport of anatomic segments has been related by different authors to the type of access that hominids have to the animals. So, an assemblage with an abundance of stylopodial and zygopodial bones and an absence of distal elements from the extremities and of elements from the axial skeleton may be related to primary access to prey (Binford 1978, 1981, 1989). This pattern is related to the fact that the long bones of the extremities contain more nutrients (especially marrow) compared to the distal elements. This contrasts with the models developed by Klein (1976, 1989) and Klein et al. (1999a), which indicate that the systematic transport of cranial elements and the distal parts of the extremities is indicative of secondary access (the head and foot pattern). However, in level J the representation of cranial elements is fairly high, although it is combined with abundant stylopodial and zygopodial bones. The systematic reiteration of anatomic parts with high nutritional contents can only be explained by the primary and immediate access model. The modality employed by hominids to secure prey was, therefore, based on cynegetic strategies.

Most of the cutmarks identified correspond mainly to subsistence strategies based on hunting. They are usually oblique and longitudinal incisions and tend to be situated mainly on the diaphysis of the long bones (Capaldo 1995, 1997; Blumenschine 1988, 1989; Blumenschime and Marean 1993; Marean and Kim 1998; Domínguez-Rodrigo and Pickering 2003). On the other hand, hardly any saw marks or deep incisions are found in the form of large groups, especially on the less nutritionally rich anatomic parts.

Another element which reinforces cynegetic practice as a main survival strategy among the hominids of level J is the dominance of adult individuals. For Gaudzinski and Roebroeks (2000) this contributes to the characterization of the hominids' strategies of faunal acquisition. Adult animals are the healthiest and strongest animals and are the most frequently hunted.

There is very limited taxonomic variability in level J. Five species of herbivores have been identified in both sublevels. From these, *Equus ferus* and *Cervus elaphus* are the predominant species, followed by *Bos primigenius* and *Stephanorhinus hemitoechus.* The chamois is not abundant and does not exceed two individuals in either sublevel. This bias of species gives rise to two possibilities: (1) the predominance of certain species in the territory or, (2) the directionality of the human groups towards certain taxa.

From extant parameters, the species present in level J describe different ecosystems. On the one hand, there are taxa linked to the plains or slopes along the valley, such as *Cervus elaphus*, *Equus ferus*, *Bos primigenius* and *Stephanorhinus hemitoechus.* On the other hand, *Rupicapra pyrenaica* is associated with steep and rocky habitats (Allué et al. 2012). So, the faunal composition of level J indicates a certain variability of herbivores in the territory. Nevertheless, the abundance of cervids and equids suggests a certain specialization of the hominids towards these taxa and directionality towards the plains and slopes around the site.

Consumption Patterns

Skeletal and specific representation, cutmarks and the reiteration of their positions on the bones, fractures, and burning damage are used to determine the type of processing activities performed by the hominids. The combination of these characteristics allows us to consider the existence of possible patterns at the time the animals were processed. Some of the activities are represented by scarce cutmarks (viscera removal and skinning) and this makes the observation of clear reiterations, which prove systematization, difficult. However, activities such as defleshing or bone fracturing can indicate the existence of systematic techniques related to the use of animals nutrients.

A model based on the differential transport of elements and the activities related to their use allows us to suggest the existence of different stages in the processing and consumption of animal resources by the hominids at the site:

1. Primary processing consists of the preparation of the animal at the kill-butchery site in order to transport it to the camp. This process involves skinning and dismembering, and the abandonment of a large part of the axial skeleton.
2. Secondary processing is performed at the camp and is comprised of all the activities related to preparation for consumption: defleshing, disarticulation, fracturing of the bones and cooking.

Primary processing and immediate consumption

The systematic abandonment of the axial skeletons at the site at which the animal is obtained does not involve leaving behind the nutrients in the axial skeleton and the abdomen. Viscera have a high nutritional value, but tracking the use of these anatomic parts at an archeological level is impossible. In level J we have located some marks on the scarce elements of the axial skeleton of large and medium-sized animals which have been related to viscera removal and the extraction of the back meat.

Some ethnographic works claim that in a selective transport system of anatomic segments similar to that seen in level J, two consumption modalities of the viscera can be distinguished. The first suggests the processing and consumption at the animal acquisition site and the second involves its transport to the camp (Binford 1978, 1981; Bunn et al. 1988; O'Connell et al. 1988a, b).

In level J, sufficient data have not been obtained to deduce whether this type of consumption occurred at the acquisition site or at the camp subsequent to transport, although it is possible that both modalities may have occurred. This would explain the presence of some axial elements as the result of occasional transports. Evidence of viscera consumption in level J, however scarce, allows us to suggest that the hominids may have immediately consumed the viscera at the kill-site.

The variability of strategies used by hunter-gatherers in the transport and processing of animals appears to be bimodal in level J. On one hand, there are heavy animals on which primary processing was carried out at the acquisition site. On the other hand, there are small and medium-sized animals that, if a similar model was followed to that applied to the large animals, may have also been transported to the shelter whole, where they were then processed in their entirety.

Secondary processing and final consumption

Once the crania and extremities of animals larger than 100 kg were selected and brought to the shelter, the preparation for consumption began. This activity falls in the category of secondary processing which includes defleshing, disarticulation, roasting, fracturing and rejection or use for other purposes, such as burning, etc. The first steps in this sequence are difficult to follow in level J, and especially in sublevel Ja. This is due to the superposition of marks in which the evidence of some activities has been disguised or destroyed by traces resulting from subsequent activities or by modifying agents intervening posterior to human activities. Furthermore, the absence of the distal extremities (epiphysis) of the bones prevents us from documenting marks related to disarticulation.

Most of the cutmarks are associated with defleshing. This activity leaves the most signs during the processing of the animal's external nutrients. Before the animals were defleshed, their skeletal portions were roasted. Double coloration on the same side of the bones and the presence of cutmarks with a lower degree of burning at the bottom verify this sequence (Rosell 2001).

A bimodality in the defleshing of bones is observed in level J, which clearly indicates two different times at which this action was performed. The first defleshing activity was intended to extract a large part of the meat and is identified by means of long longitudinal cuts in the diaphysis of the long bones (incisions longer than 20 mm). The second stage is related to the use of the remaining meat stuck to the bone or wedged in the muscular insertions after defleshing and is identified by marks located on the metaphysis and on the muscular insertions. It is possible that this phase may have ended with scrape marks on the bones to use the periosteum of the superficial fat from the diaphysis and perhaps to prepare for fracturing. Differences between perissodactyls and artiodactyls in relation to the location of the cutmarks on the bones can be observed in this second defleshing. These are determined by the condition of the muscular insertions and the animals' physiology.

After roasting the anatomical portions and defleshing, the hominids proceeded to fracture the bones in order to obtain their internal nutrients. In sublevel Ja fracturing was found to have been produced by direct percussion in specific patterns (Rosell 2001). First, the epiphyses and diaphyses were systematically separated through blows on the metaphysis which produced transverse fractures. The diaphyses were then broken on their longitudinal axis with a

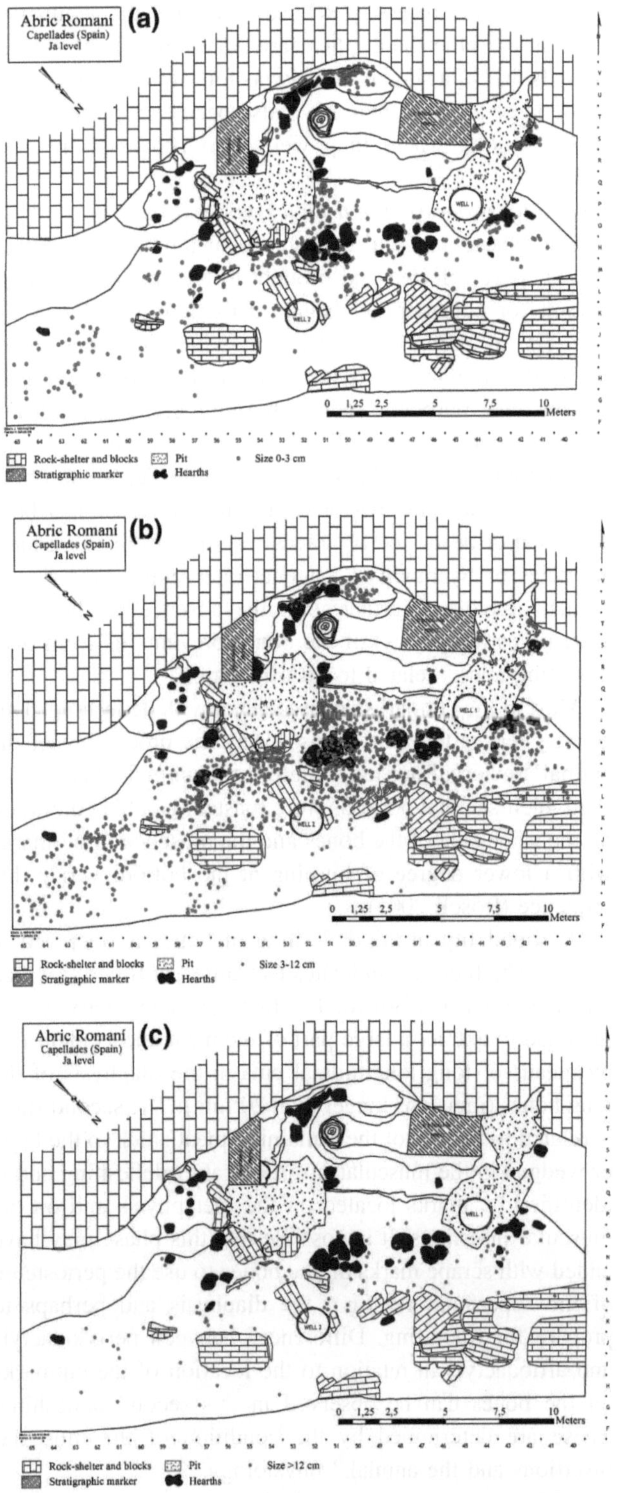

Fig. 8.19 Spatial distribution by bone lengths in sublevel Ja: **a** 0–3 cm; **b** 3–12 cm; **c** >12 cm. The empty zones near the wall of the shelter correspond to the sinters. These travertinic formations are characterized by a pronounced slope and are inconsistent with human activities

variable number of blows so that in most cases the marrow was obtained in excellent condition. Therefore, fracturing has a double intentionality consisting of, on the one hand, obtaining the epiphyses and on the other hand, getting to the marrow content from the diaphysis.

Intensive breakage generates a large number of small remains (smaller than 20 mm), which stand out in the studied assemblages. As the dimensions of the remains increase, the number of specimens decreases. Therefore, we suggest that good use was made of the existing nutrients in level J. No entire long bones have been recovered, which would indicate the nutrients of animals had been exploited in all cases. Occasionally, some of the elements resulting from fracturing were configured for use. In sublevel Ja only one object—a bovid mandible—has been found with indications that the bone was used as another material with which to produce artifacts.

Rejected Elements

Once the nutrients had been removed, the hominids discarded the leftover bone fragments. This discarding of bone waste led to an accumulation of remains, which has made it possible to reconstruct occupational patterns (Vaquero 1997; Vaquero and Pastó 2001; Rosell 2001; Carbonell and Rosell 2004).

Bones are found across the entire surface of both sublevels without significant accumulations in specific zones. However, the analysis of bone measurements shows considerable variations, mostly in sublevel Ja (Fig. 8.19). The small elements are distributed over the entire excavated area, whereas the number of large items (longer than 100 mm) tends to be higher in the zones situated in the center and exterior of the shelter. The absence of large bones close to the walls cannot be due to agents such as water or gravity acting subsequent to human activity, as the preferential slope in both sublevels is not steep enough to bring about significant displacements. Furthermore, water which dripped from the roof or leaked from the wall would have flowed along the original incline of the level without ever picking up a high level of energy (Cáceres et al. 2012). Therefore, the large bones have remained in their original places and only the small remains could have been incorporated into these water courses. If a high-energy current had occurred, we would find few or no small remains in the level and large remains reoriented according to the directionality of the water currents.

We should also not rule out the possibility that some dispersions correspond to the action of non-human predators. Generally, the carnivores that act on an abandoned anthropic accumulation tend to disperse the remains rather than cluster them. However, in level J carnivore activity is too sporadic to have caused selective movements of materials or to have distorted the original position of the elements in any significant way.

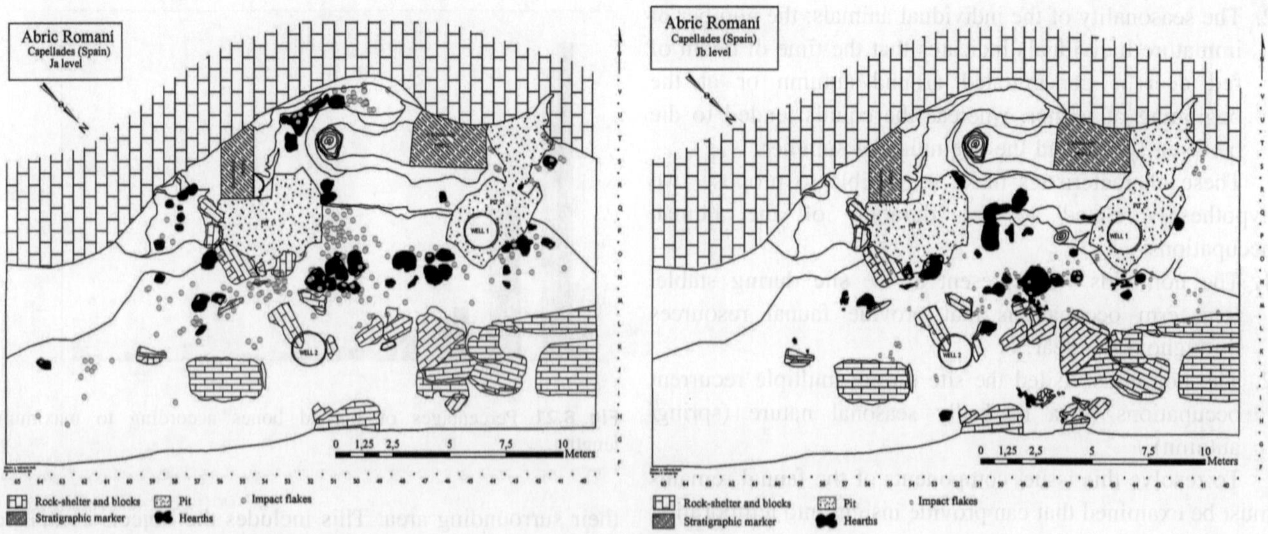

Fig. 8.20 Distribution of impact flakes in sublevels Ja and Jb of Abric Romaní

The differential location of the remains according to their maximum lengths is a result of the action of hominids. This is part of a pattern based on the accumulation of large remains in the peripheral zones of the occupied area. We must take into account that the whole exterior zone of the site is characterized by the presence of large travertine blocks. The scarce presence of combustion structures in this sector indicates that this was a marginal space in terms of its use for human activities.

The spatial distribution of the osseous remains by sizes may be related to the division of the space into two areas with different functions. These may be identified as what Binford (1978) calls *toss* areas and *drop* areas:

1. The external area coincides with the *toss* area (external zone with the large travertine blocks). In this zone large and small remains appear without distinction situated randomly between the travertine blocks and without apparent association in any combustion area.
2. This area might be identified as a waste accumulation zone. The central area and the zones close to the wall belong to the *drop* area. In these zones small elements dominate the assemblage; many of them are impact flakes, associated with the numerous hearths (Fig. 8.20). Several large bones are present but in low density.

At an experimental level we were able to verify that the fracturing process generates many small pieces, which are concealed by sediment or during excavation (Carbonell et al. 1992). Therefore, many of these remains escape notice when larger pieces are moved to the waste areas, and remain in the original place of production. This makes it possible to recognize areas where bones were fractured (accumulations of impact flakes). However, the detection of the loci of prior activities, such as defleshing, is subject to the movement of the remains. These activities are developed on whole bones and, therefore, they do not leave obvious accumulations at their place of production.

Synchrony and Diachrony in the Assemblage

A common tendency in zooarcheological studies is to treat faunal assemblages as synchronic units. But archeological sites normally represent palimpsests in which the preserved material corresponds to different events of various natures. This problem is often exacerbated by the characteristics of the site itself and the absence of an extensive method of excavation, which would cover the entire space used by humans and, therefore, make it possible to clearly identify the preserved activity areas and the possible connections between them.

In level J, the excavated area exceeds 250 m^2. Its surface is marked by the presence of considerable structural discontinuities, such as paleochannels, gours, stalagmites and sinter formations (Sañudo et al. 2012). The faunal remains of sublevel Ja are located practically throughout this entire paleosurface, whereas those from Jb, because of archeostratigraphic problems, are assembled mainly in the central area. This confirms that two archeostratigraphic units clearly separated in time can be distinguished in level J: sublevel Ja and Jb. Nevertheless, in each of the subassemblages there are elements in the faunal remains that suggest a certain diachrony on a shorter time scale (Isaac 1983; Binford 1988):

1. The quantity of counted individuals: the MNI of herbivores in sublevel Ja is 37, whereas in Jb it is 13. The transport and consumption of this quantity of animal biomass by human groups requires some time.

2. The seasonality of the individual animals: the number of immature individuals indicates that the time of death of red deer is concentrated around autumn or at the beginning of winter, whereas the equids tended to die between spring and the beginning of summer.

These characteristics make it possible to propose two hypotheses related to the duration of the human occupations:

1. The hominids were present at the site during stable, long-term occupations that provide faunal resources throughout the year, or
2. The hominids visited the site during multiple recurrent occupations of a markedly seasonal nature (spring/autumn).

To resolve this issue, components of the faunal remains must be examined that can provide insight into temporality. Diachrony on a geological time scale can be evaluated through the archeostratigraphic position of the burned bones post-burial. The synchrony or diachrony of the assemblage on a biological time scale can be determined through: (1) the presence of refits between the different items in the assemblage and (2) the presence of carnivores and their activity in the assemblage.

Burned Bones and Their Location in Relation to the Hearths

Burned bones and their connections with the diachrony of the assemblage

Most researchers agree that the use of fire represented significant advantages for human groups of the past. Fire provides light at night, heat, and protection against animals, but it also means an expansion of the hominids' ecological niche, the socialization of groups and, in short, a considerable variation in their way of life. Fire also represents a considerable advantage in terms of diet. Since they began controlling fire, hominids had become aware of its culinary properties, and a great part of their food was thermally treated before consumption. At archeological sites, the treatment of animal resources with fire is mainly documented from burned bones.

Hominids can generate intentionally burned bones through two different activities (Binford 1981; Stiner 1994):

1. The processing of animals for consumption: roasting the meat and warming the bones to break them; and
2. The deposition of osseous remains in hearths.

However, the burning of bones at archeological sites is not always intentionally caused by human groups. The radius of influence around the hearths can affect elements in

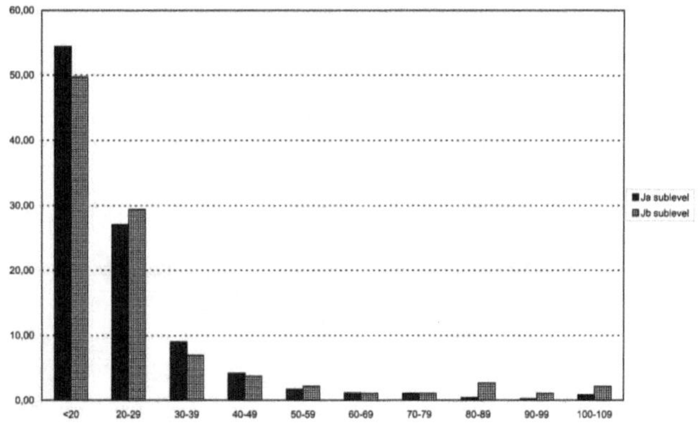

Fig. 8.21 Percentages of burned bones according to maximum lengths

their surrounding area. This includes the objects contained in the sediments where the hearths were located.

From an experimental hearth developed over a period of 7 weeks, De Graff (1961) proved that bones buried approximately 10–20 cm deep are completely carbonized (degrees 3-4-5 of Stiner et al. (1995). Bones located 30 cm beneath the surface also show signs of thermal alteration, although in lesser degrees. Stiner et al. (1995) and Bennet (1999) also conducted similar experimental reproductions in which they introduced variables, such as the composition of the soil, the duration of the hearth, and the temperature of the fire. The researchers agree that the temperature reached is more important than the duration of the hearth. But, despite the ongoing discussion between these authors, their works show that the unintentional post-burial cremation of bones is possible. An archeological example can be found at the Australian Puntutjarpa Rockshelter site (Archer 1977). From this perspective, the hearths and their associated remains can be considered on a biological time scale as *minimum units of synchrony*. However, the presence of bones burned post-burial is a good indicator of the existence of diachrony on a geological time scale.

Bones burned post-burial

In level J, burned bones are not very abundant. In sublevel Ja, they represent 31.3% of the bones recovered, whereas in sublevel Jb, they only represent 18.4%. Brown tonalities (degrees 1 and 2) predominate in sublevel Ja. The number of specimens decreases as the degree of increases (degrees 3–4 and degree 5). In sublevel Jb, burned bones present very few relevant values at any degree.

The most striking characteristic of the burned bones from level J is their small size (Fig. 8.21), most of them do not exceed 3 cm in length. These bones tend to be grouped in specific areas, which in most cases coincide with the location of the hearths (Figs. 8.22, 8.23).

The figures show a concentration of bones with a high degree of burning in the center of the combustion structures.

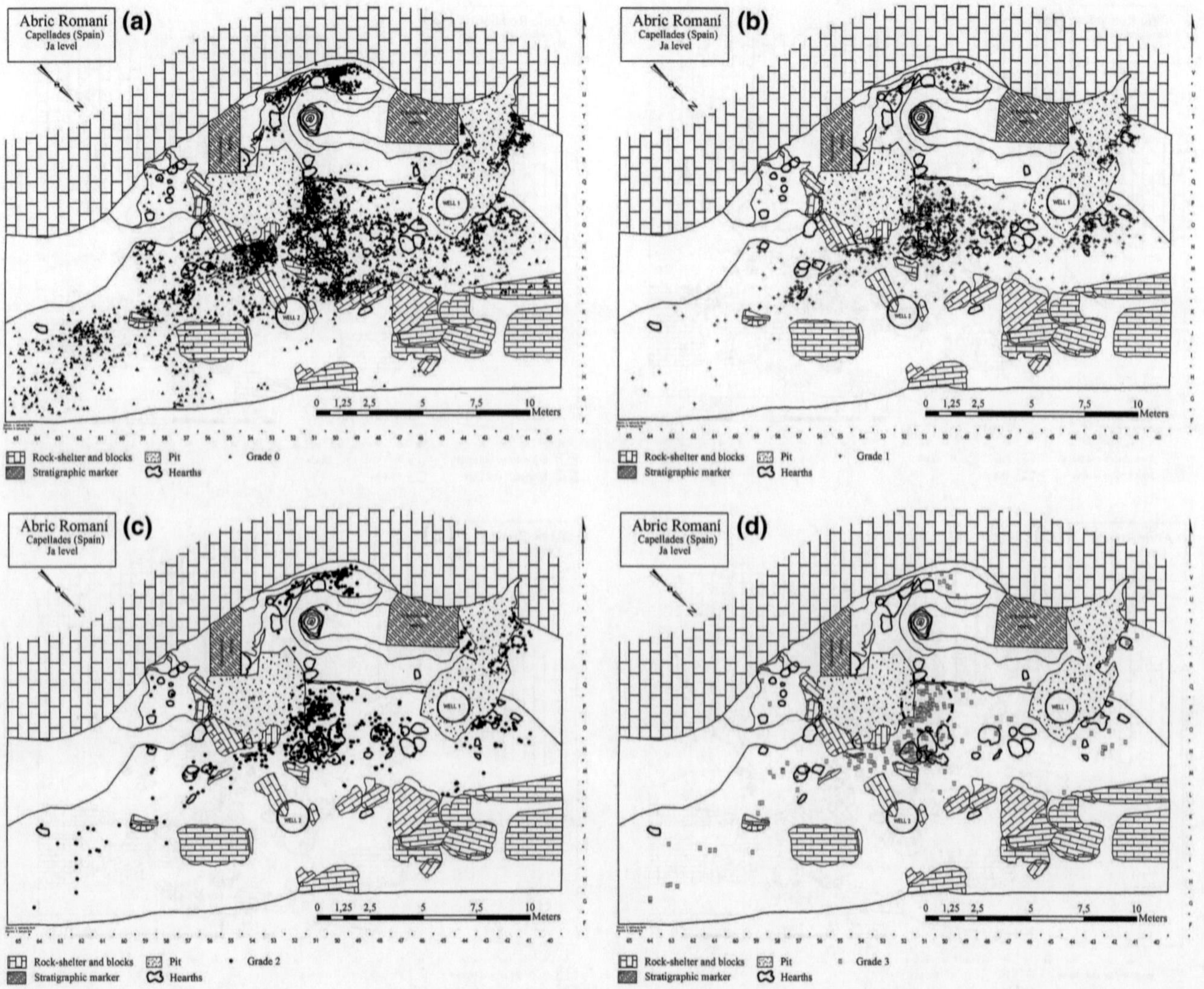

Fig. 8.22 Dispersion of burned bones by degrees from Ja faunal remains: **a** degree 0; **b** degree 1; **c** degree 2; **d** degree 3

However, bones with less burning tend to be located on the periphery of these areas, especially in sublevel Ja. This phenomenon suggests that the position of the bones recovered during excavation is very close to their original position and that in general, there were not significant space disturbances in the assemblage (Sergant et al. 2006).

The vertical projections do not show evidence of superimposed accumulations. The objects appear grouped at similar elevations and burned bones are not found under the baselines of the hearths. However, in square S41 an interesting phenomenon was observed that appears to be related to this matter (Fig. 8.24). A hearth that affects a large part of the surface of sublevel Ja was described in this sector. The number of remains, burned as well as unburned, is high and they are randomly distributed over the entire surface of this sublevel in a maximum thickness of about 10–15 cm. Another accumulation appears underneath, clearly separated from the upper accumulation by a thin travertine layer, which belongs to sublevel Jb. No hearth has been identified in this sublevel, but the accumulation appears much more organized than the upper one (Ja) and its objects are distributed in a narrow strip approximately 5 cm deep. The double coloration documented on these items is located on the same side of the bone. Different coloration has not been observed on the cortical or the medullar side. This absence of differential coloration indicates that the burning of these elements did not take place after their burial. Under this accumulation of sublevel Jb, there is another accumulation made up of elements dispersed in a strip approximately 15 cm thick. This accumulation was detected during excavation and it was called Jb_2. This lower assemblage does not have spatial continuity and is only detected in very specific places where a particular sedimentation phenomenon occurs, such as gours (Vallverdú and Courty 2012). Therefore, in the general zooarcheological analysis we do not distinguish between Jb_1 and Jb_2, but for purposes of

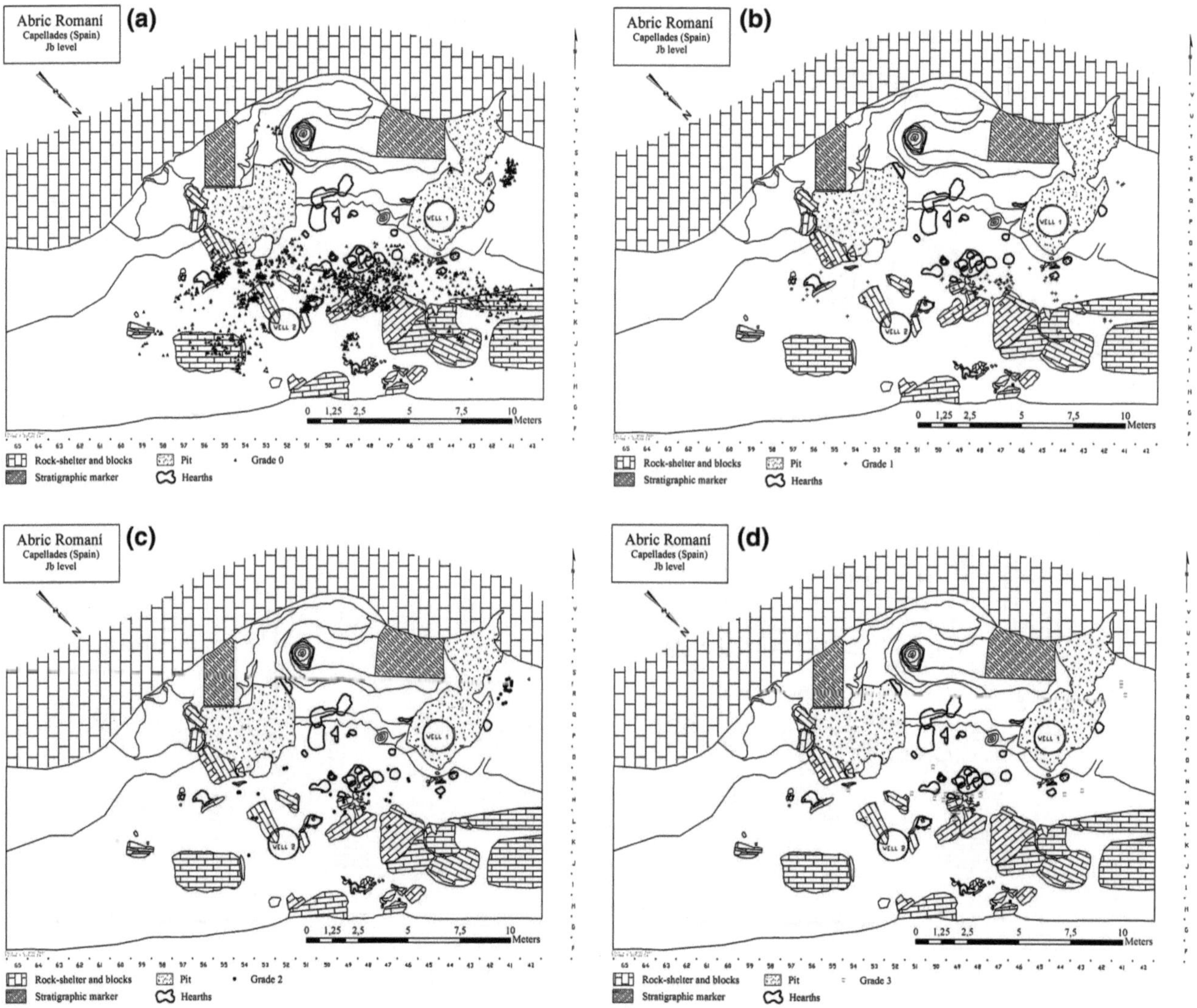

Fig. 8.23 Dispersion of burned bones by degrees from Jb faunal remains: **a** degree 0; **b** degree 1; **c** degree 2; **d** degree 3

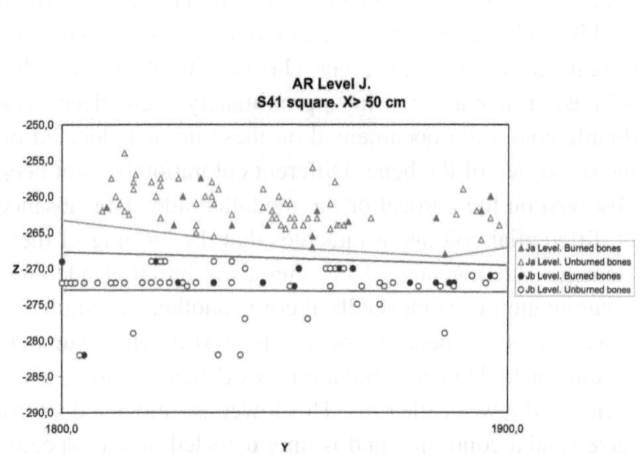

Fig. 8.24 Y-Z projection of burned and unburned bones from square S41. Note the three main accumulations. The *upper* one belongs to sublevel Ja, whereas the two *lower* ones belong to Jb. The surface to depth scale has been changed to obtain higher resolution

temporality, their individualization is important. In Jb_2 thermo-altered remains are very scarce. There is only one fragment of an 8 cm long flat bone exhibiting degree 4 cremation. It is uniformly burned on all sides. The archeostratigraphic position of these bones shows a certain diachrony on a geological scale in the formation of the Jb assemblage. The upper accumulation (Jb_1) is comprised of numerous items arranged linearly in a strip approximately 5 cm thick. However, the lower accumulation (Jb_2) is more dispersed and seems to have resulted from prior events.

The postdepositional modifications identified in this square are not marked (Cáceres et al. 2012). Only root traces are present on a large part of the bones, which is common across the entire assemblage. These are more marked in sublevel Jb than in sublevel Ja. However, the degree of erosion on the bones is very low throughout level

J. Scratch marks or bearing marks from dragging are observed on the bone surface which indicate reworking or a significant transport of materials to this zone.

This archeostratigraphic distinction in sublevel Jb seems to be a local phenomenon that only affects square S41. In the rest of the shelter this diachrony is not so clear. The vertical projections of the faunal remains regarded as the *minimum units of synchrony* (hearths) indicate that post-burial cremation is not a common phenomenon.

From this point of view, the two sublevels of level J appear as two synchronic occupations on a geoarcheological scale that occupy a large part of the shelter: one in Jb and another later one in Ja. However, it is possible that some previous low intensity impacts also occurred, such as that detected in sublevel Jb_2 in square S41, and that they have been disguised by these larger occupations.

The Presence of Faunal Refits

Refit techniques are used with the aim of reconstructing and examining the cultural and natural processes that have played a role in the formation of the archeological remains. Refit techniques yield more precise data on the nature of all the processes that have influenced the deposits so that they are taken into account in the analyses of the assemblages (Hoffmann 1992).

Faunal refits

A faunal refit consists of the effective connection of two or more fragments belonging to the same skeletal element or portion and involves physical contact between their surfaces. There are two types of refits:

1. Articular connections between different bones of the skeleton; and
2. The refits themselves, which affect different fragments of the same bone.

The study of bone refits is an important tool in the attempt to determine the anthropic processes that took place in the formation of an archeological site. Such processes include the intentional transport of remains, which correspond to different rules of behavior such as the cleaning and arrangement of space. However, other types of information are also important, such as the type of bone fracture or the existence at the site of other taphonomic processes, which could alter the structure of the remains and their original position: trampling, the activity of carnivores or other animals, the action of water and ice, bioturbations, etc.

Faunal refits start from an initial analysis of the material based on macroscopic criteria, with the aim of facilitating the association of elements belonging to the same weight category and skeletal element. Along with these premises, spatial criteria are used to make associations between the elements. These criteria are based on the location of the material at the site according to the Cartesian coordinate grid, with the aim of later shifting to a spatial criterion based on the identification of accumulation zones.

Each one of the refits is expressed through one or several connection lines, depending on the number of connected elements. The separation distance of the different connection lines is calculated in centimeters, as is the dispersion radius in cases with refits of more than two elements. These distances contribute interpretive data to the study of the intentional or unintentional movement of the affected elements.

Only the refits of the material from sublevel Ja have been studied. Sublevel Jb presents excessive taphonomic modifications which affect the original morphology of the bones, thereby preventing their successful physical refitting.

Faunal refits in sublevel Ja

During the formation of level J, Abric Romaní represented a unique and ideal space for the humans to carry out their daily activities. It was a wide space clearly separated from the outside by a line of large travertine blocks. The inside tended to be flat with a slight slope, which became more pronounced as one moved away from the center of the shelter. The lobes of the walls provided protected spaces separated from one another by natural pillars.

The extended excavation of this level was only interrupted by earlier interventions (pits and wells). These mainly affected two areas: (1) Pit 2 excavated by Amador Romaní at the beginning of the twentieth-century (squares O–Q/43–45) and, (2) the base of Pit 1 excavated in the 1950s in the center of the shelter by Eduard Ripoll (squares O–R/52–56).

Bearing in mind the structural units of the shelter, the general characteristics of the remains and the existing gaps caused by former excavations, the shelter has been divided into six main zones for study (Vaquero et al. 2012):

- Zone 1: Lobe 2 (R–U/41–44). The area is highly altered by former excavations. It shows some sequences with hearths and abundant faunal remains.
- Zone 2: Lobe 3 or the *Coveta Ripoll* (T–V/48–53). This is a protected space with its own cornice and it is clearly separated from the rest of the shelter by a very large stalagmitic formation. There is a high density of osseous remains associated with hearths.
- Zone 3: Central-western zones (L–Q/40–48). This consists of a flat space delimited from the exterior of the shelter by a line of large travertine blocks fallen from the ceiling of the cavity. Its connection with zone 1 is bordered by the Romaní well and Pit 2. It has abundant

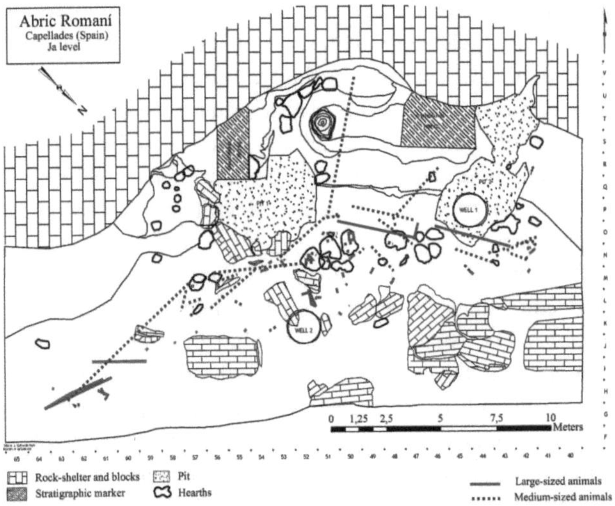

Fig. 8.25 Distribution of faunal refits from the Ja faunal assemblage

faunal remains and some hearths. Its separation from the central zone is diffuse.

- Zone 4: Central zone of the shelter (L–Q/49–55). This is a flat area partially affected by Pit 1. The line of blocks that delimits the exterior space in this zone creates an arch until it reaches the shelter wall. This area revealed the greatest density of osseous remains, lithic industry and hearths.
- Zone 5: Outer north-eastern zone (G–K/40–55). This zone is marked by the presence of large travertine blocks fallen from the ceiling of the shelter. Its incline towards the south is quite pronounced; even forming a considerable slope in the outermost area. Faunal remains are not as abundant as in other zones, but they are characterized by their large dimensions and appear in a disorganized way among the large blocks. There are small isolated hearths.
- Zone 6: Exterior eastern zone (G–R/56–65). This is the drainage zone of the hydrological system of the site. The slope towards the east becomes more and more pronounced as it moves away from the center of the shelter. Osseous remains appear dispersed over the entire surface. The hearths are clustered in the eastern area closest to the blocks.

In sublevel Ja, 92 refits affecting 207 remains have been registered, which represent 3.3% of the faunal remains recovered in this sublevel. These refits are translated into 115 connection lines: 76 refits of 2 elements, 12 refits of 3 elements, 2 refits formed by 4 elements, 1 refit of 5 and another of 6 elements (Bravo 2001; Fig. 8.25).

By taxonomic groups, red deer (*Cervus elaphus*) has the largest number of refits (21), affecting 49 elements. The horse (*Equus ferus*) takes second place, with 14 refits affecting 31 elements. The bovids (*Bos primigenius*) provided two refits, involving four elements. Finally, the least represented species are the rhinoceroses (*Stephanorhinus hemitoechus*) with one refit that affects two elements. In the remaining connections (54 affecting 121 elements), the taxonomic identification of the elements involved was not possible. Among these non-taxonomically identified elements, those belonging to medium-sized animals had the largest number of refitted elements, a total of 104, which translated into 46 refits (50%). The large animals had 39 connections (42.39%) and a total of 89 elements. Small animals only provided one refit (1.08%) formed by two elements. There were also 12 elements that could not be classified by weight classes, which corresponded to six refits (6.52%) (Fig. 8.26).

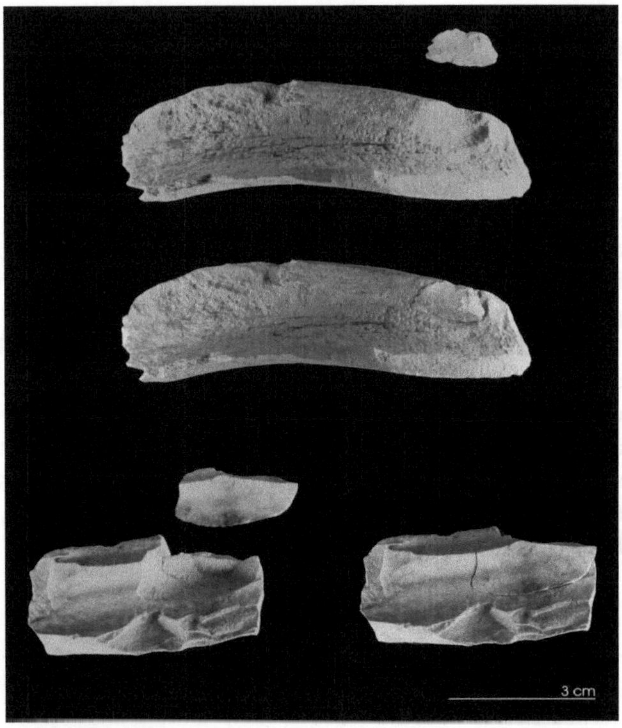

Fig. 8.26 Examples of Ja faunal refits (type 2)

The refits mainly affect the long bones, which are the most abundant elements in the assemblage. Among the identified bones, the humeri and the metapodials are the skeletal elements that had the highest number of connections. The refits of cranial elements from large animals mainly correspond to dental fragments (refits type 1). 71.74% of the type 2 refitted elements show clear signs of anthropic intervention: 7.72% show signs of intentional breakage, 34.78% have differing degrees of thermal impacts and 11.59% show cutmarks (Table 8.28).

Although the refits identified in Ja are spread out over the entire excavated surface, they are most numerous in areas 3, 4 and 6. The highest degree of movement was found among elements from large animals. The average distance of the connection lines between the elements of this size is 78 cm.

Table 8.28 Absolute (aFr) and relative frequency (rFr) of faunal refits by skeletal elements at sublevel Ja, according to established weight categories

		Small size		Medium size		Large size		Unident.		Total	
	Type	Abs.Fr.	Rel..Fr.	Abs.Fr.	Rel..Fr.	Abs.Fr.	Rel..Fr.	Abs.Fr.	Rel..Fr.	Abs.Fr.	Rel..Fr.
Teeth	1	–	–	1	1.09	6	6.52	–	–	7	7.61
Mandible	1	–	–	–	–	3	3.26	–	–	3	3.26
Humerus	2	–	–	3	3.26	4	4.35	–	–	7	7.61
Radius-ulna	2	–	–	2	217	1	1.09	–	–	3	3.26
Ribs	2	–	–	2	2.17	–	–	–	–	2	2.17
Metacarpus	2	–	–	5	5.43	1	1.09	–	–	6	6.52
Femur	2	–	–	2	2.17	1	1.09	–	–	3	3.26
Tibia	2	–	–	5	5.43	2	2.17	–	–	7	7.61
Metatarsus	2	–	–	3	3.26	1	1.09	–	–	4	4.35
Metapodium	2	–	–	1	1.09	1	1.09	–	–	2	2.17
Long bone	2	1	1.09	20	21.74	17	18.48	2	2.17	40	43.48
Flat bone	2	–	–	2	2.17	1	1.09	–	–	3	3.26
Epiphysis	2	–	–	–	–	1	1.09	–	–	1	1.09
Unident.	2	–	–	–	–	1	1.09	3	3.26	4	4.35
Total		1	1.09	46	50.00	40	43.48	5	5.43	92	100

On the other hand, the average distance of the connection lines on the elements from medium-sized animals is shorter, around 60 cm.

The most numerous refits have two osseous fragments and are connected by short distances (see annex tables). They are often connections of small elements with high degrees of burning. All these remains are located in the areas surrounding the hearths and at a similar depth.

Connections between areas

The refits with long distances relate the main central areas to one another. Only zones 1 and 5 remain isolated from the rest. It is important to emphasize the direction of the refits located in zones 3 and 4. Generally, the longest connection lines in these areas follow the direction S–N. This phenomenon goes against the direction of the general drainage of the shelter surface, which lies in a NW–SE direction in this zone. The refits in zone 6 clearly mark the incline of the shelter. This phenomenon is also observed in the westernmost area of zone 4.

There are very few non-anthropic modifications that affect the refitted elements. Root activity had the largest incidence and affected practically the entire assemblage to different degrees. No carnivore toothmarks have been identified in any case. Only six elements present scratch marks from trampling and two show a slight degree of rounding on their breakage planes due to hydric erosion. The taphonomic modifications observed on the surface of the bones do not suggest the existence of agents capable of displacing materials in any significant way. Roots, like trampling, can be responsible for vertical movement. The trampling generated by the circulation of hominids can produce horizontal movements over long distances. Furthermore, evidence of carnivore action in the assemblage is scarce, and it does not seem to be responsible for any notable remobilizations of material. Only the action of water can have affected the spatial dispersion of some remains in a significant way. This does not seem to have happened in the central zones, where the remains are mostly small and appear clustered near the hearths. Only in zone 6, where there is a larger general incline, might the dispersion of some fragments be the result of small remobilizations caused by water.

In the other zones, however, a large part of the long distance connections owe their origin to displacements of anthropic origin. Many of these show evidence of intentional breakage or fire action. The best example is the one that connects the *Coveta Ripoll* or lobe 3 (zone 2) with the central area (zone 4). The morphology of zone 2 is singular; it is a sheltered space with its own cornice and is clearly separated from the rest of the shelter by a large stalagmite formation. The high density of faunal remains found in this space is associated with hearths. In this zone, two connection lines have been identified, and one of them connects this area with the central zone (zone 4).

In short, in sublevel Ja numerous refits have been documented which affect the most represented taxa. The most numerous refits were in local ranges, affecting the remains found in the same square or in adjoining squares. However, there were some refitted elements with significant connection distances. The origin of displacement of these fragments is mostly anthropic. These long-distance connections suggest the existence of two different large areas with a certain temporal integrity.

1. The first is concentrated in the eastern area of the site, and is formed by the two groups of hearths located in the center and in the east of zone 3 (squares L–N/43–45 and O46-47). The presence of Pit 2 and the Romaní well prevented us from drawing possible connections with zone 1.
2. The second area is the largest, covering the entire central area of the site (zones 2 and 4) as well as the hearth located in N–O/48 and the group of hearths in the easternmost section of zone 6. It is possible that a large part of the refit material in the south-westernmost section of the site belongs to this area.

Therefore, two different occupational periods in sublevel Ja can be established: one concentrated in the eastern area and another larger occupation covering the entire central area. In both occupational units, human activities related to fauna were conducted exclusively inside the shelter. The line of large blocks outside represents a natural barrier, which clearly separates the space outside from the inside space.

Evaluating the amount of time between the formations of the two areas is not easy. It is possible to make an approximation from the vertical projections of the materials located in the intersection zones without physical discontinuities. These conditions are clustered in the hearths situated in N-O48 and its surroundings (Fig. 8.27).

The projections situated in these squares show a continuous line of materials between the two areas. There is no clear superposition that makes it possible to distinguish the eastern area from the central area. This may indicate that the geological time between one occupational unit and the other was not long enough for sedimentary processes to act in the shelter, which would allow us to make a clear distinction.

Generally, the study of the refits suggests the existence of two preferential zones in sublevel Ja used at different times. The two zones exhibit high occupational intensity, with activity areas related to fauna and situated around the hearths. The biological time and number of occupational events in each of these zones are difficult to estimate. Nevertheless, the high number of animals in the assemblage and the extension covered by each of the zones can be related to a certain stability of the occupations.

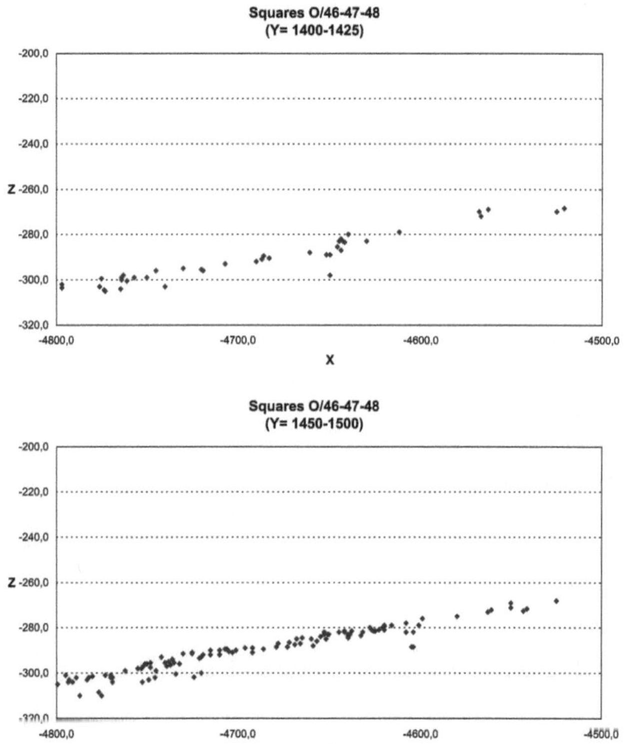

Fig. 8.27 Projection of faunal remains from squares O46-48 where the intersection between the two large zones of sublevel Ja is found. Material superposition is not observed between the two zones

The Presence of Carnivores

Carnivores in archeological sites

The anatomical features of carnivores or evidence of their activity are common at European Middle Paleolithic sites. Such evidence usually comes in the form of skeletal remains of the carnivores themselves, the bones of other animals with carnivore toothmarks, or the presence of coprolites. However, during this period commensalism or domestication and coexistence of hominids and carnivores in the same space was, by definition, impossible. Therefore, the mix of human and carnivore activities must be understood as the result of a diachronic component in the formation of the sites. In this case, humans and carnivores took turns carrying out their activities over short periods of time (biological scale), and the sedimentation rates were not fast enough to produce a clear archeostratigraphic separation between them.

In the places where the remains left by human activities dominate the assemblage, the presence of carnivores can be explained by three different processes (Rosell and Blasco 2009):

1. Direct confrontations between hominids and carnivores, which result in the death of the non-human predator. This option is part of the natural dynamic of human activities and can be related to the use of resources offered by carnivores (skin, meat, bones, claws for ornaments, etc.) or to the elimination of competitors. The presence of anthropic signs on the bones of these animals (cutmarks, fractures or intentional burning of the bones) is the best argument for proposing this option.

2. Installation of dens or refuges at times of human absence. The alternation of humans and carnivores is a common phenomenon in many sites of the European Pleistocene. However, the application of certain criteria can distinguish between the remains left by each of these agents (Cruz-Uribe 1991; Capaldo 1998; Pickering 2002). According to these criteria, the cranial or appendicular skeletal remains of adult herbivores are dominant in human occupations. Cutmarks on the bones of these animals are frequent as well as the diagnostic elements of intentional breakage (Pickering and Egeland 2006). Burned bones are common at sites where there is a controlled use of fire. On the other hand, evidence of immature or senile animals is frequently found in carnivore dens. There are no preferential skeletal representations. Middle shaft fragments and toothmarks on the bones are abundant, mainly in the form of pitting. These elements tend to appear along with coprolites and infantile carnivore remains due to the cannibalistic behavior of some of these species, especially hyaenids (Cruz-Uribe 1991).
3. Carnivore scavenging on the remains left by humans after their occupation. Carnivores often intervene in abandoned human camps. These animals find themselves attracted to the smells from the abandoned faunal remains and hearths, and they try to make the most of the scarce nutrients left. This phenomenon has been observed repeatedly in current and sub-current hunter-gatherer camps (Binford 1978, 1981; Bartram et al. 1991; Yellen 1991; O'Connell et al. 1992). Similar phenomena have also been reproduced at an experimental level (Binford et al. 1988; Blumenschine 1988; Marean and Spencer 1991; Marean et al. 1992; Marean and Bertino 1994). These places are characterized by evidence of intense human activity and by slight carnivore modifications. In these cases, all researchers agree that carnivores show a predilection for elements that have more soft tissue: the epiphysis of the long bones and the axial skeletal bones (vertebrae and ribs). Diaphyses fractured by hominids for the consumption of the marrow hold little attraction for them. From a spatial point of view, the remains left by hominids tend to be consumed in situ. Element remobilizations of some meters only occur when episodes of competition with other scavengers take place. The presence of coprolites is also common in these cases, with a clear tendency of being clustered in the extinguished hearths or their surroundings (Bartram and Marean 1999; Klein et al. 1999b).

To identify which of these three scenarios has occurred in an archeological assemblage, it is necessary to analyze the skeletal representation and the type of carnivore marks on the bones. This data must be accompanied by a spatial evaluation of their position at the site, both at the vertical and horizontal levels. The alternation of the presence of hominids and

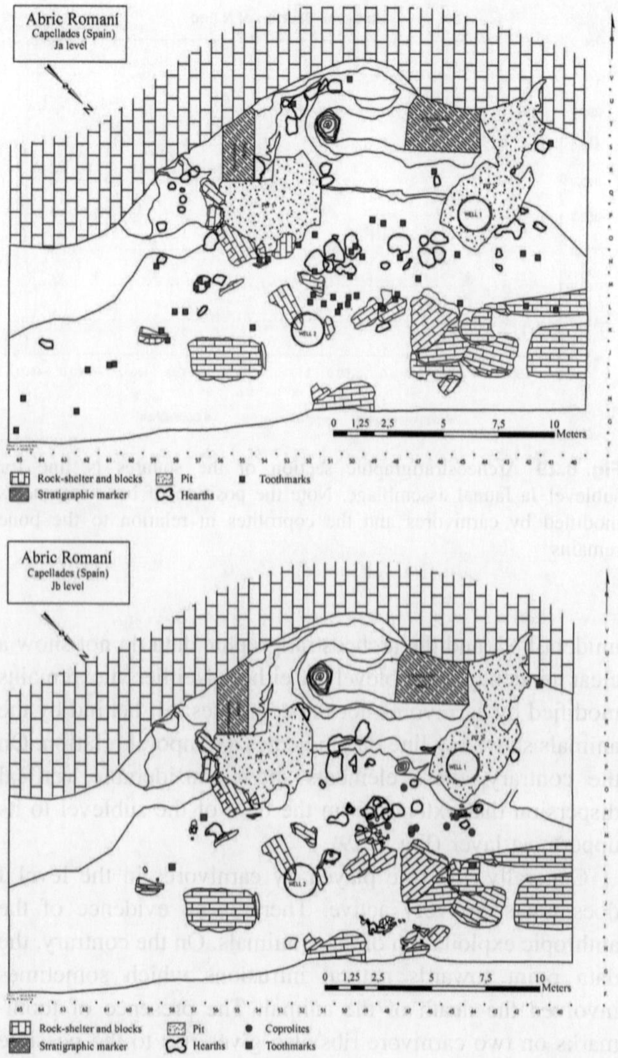

Fig. 8.28 Distribution of coprolites and bone fragments with carnivore toothmarks in sublevels Ja and Jb

carnivores can be recognized in the context of palimpsest from the apparent mix of modified elements. In the opposite case, from an archeostratigraphic point of view, the elements generated by each of them are grouped together.

Carnivores in level J

Sublevel Ja shows more carnivore modifications than Jb. However, neither spatial grouping nor preferential zones of action for these animals has been observed. The same applies to coprolites (only five have been identified, probably from hyenids or large canids), which are dispersed over the entire area of sublevel Ja with a certain preference for areas close to the hearths. Something similar occurs in sublevel Jb where 12 coprolites were recovered (Fig. 8.28). In this case, there are not many remains with toothmarks, although the coprolites tend to be situated in the area surrounding the hearths in the theoretical middle west of the shelter. The animals responsible for these coprolites are

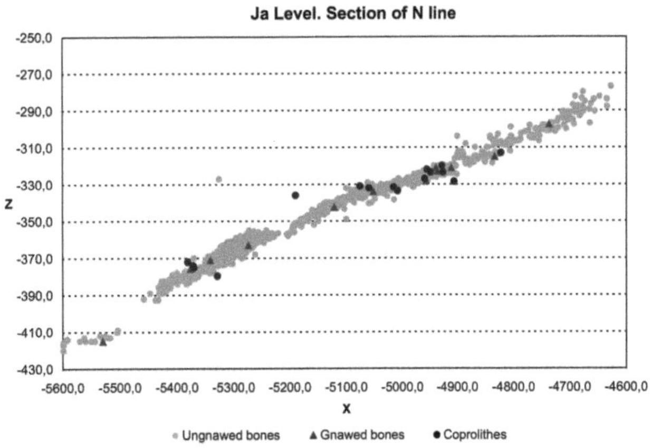

Fig. 8.29 Archeostratigraphic section of the squares N line for sublevel Ja faunal assemblage. Note the position of bone fragments modified by carnivores and the coprolites in relation to the bone remains

unidentified. and the archeostratigraphic data do not show a clear tendency in sublevel Ja either. Neither the remains modified by carnivores nor the coprolites left behind by the animals show any line that indicates a temporal relation. On the contrary, these elements show considerable vertical dispersion that extends from the base of the sublevel to its uppermost layer (Fig. 8.29).

Generally, the role played by carnivores in the level J does not seem very active. There is no evidence of the anthropic exploitation of these animals. On the contrary, the data point towards natural intrusions which sometimes involved the death of the animal. The presence of toothmarks on two carnivore ribs also gives rise to the possible hypothesis that the shelter was used as a refuge for a predator at times of human absence. This intrusion, which involves the transport of isolated elements, seems to correspond to a sporadic event within the assemblage of level J. The frequency of toothmarks identified on the bones and immature herbivore remains is too low to make frequent use of the shelter as a refuge or den feasible (Cruz Uribe 1991).

The distributions of the remains modified by carnivores and of the coprolites indicate that these animals were active in various areas of the shelter, coinciding with the same zones occupied by the human groups. The number of bones altered by carnivores is higher in the strip located toward the outside of the shelter, which is where the main waste areas were located and where the largest elements were recovered. This phenomenon seems to be related to the secondary processing by carnivores of remains abandoned by hominids. It is possible that after the human occupations, scavengers were attracted to the smells of the extinguished hearths and of the remains of previously consumed herbivores. Their activities seem to be limited to the use of the few nutrients left in the bones and to occasional remobilizations.

The vertical dispersion of the elements generated by these animals indicates that they visited Abric Romaní several times during the formation of level J. The reiteration of prowler visits to the camps abandoned by the hominids suggests a diachronic component on a biological scale in the formation of the preserved archeological assemblage.

Synchrony or diachrony in the assemblage?

The geological characteristics of Abric Romaní are unique in that they raise questions related to occupational models and their durations. The rapidity of the formation of the travertine bioconstructions and method of excavation has permitted the paleoethnographical reconstruction of the different occupational events.

Two long occupational periods have been detected in level J, which present two archeostratigraphically separated sublevels (Ja and Jb). The two sublevels show abundant herbivore remains distributed over their entire surfaces. The skeletal remains of these animals are a product of human hunting and of the differential transport of anatomic segments. The crania and the extremities were systematically carried to Abric Romaní, whereas the axial elements were abandoned at the kill site. A strong relationship between the transported elements and the individuals examined was found, suggesting the high integrity of the transported segments (Skeletal Survival Rate). This accumulation of potentially consumable biomass in both assemblages implies a relatively prolonged period of time, which is not always perceptible from an archeological point of view.

To address the question of occupancy duration, the faunal remains have been treated from different perspectives. Each of them allows us to evaluate temporal criteria on a different scale. It is obvious that archeological remains do not allow us to pinpoint the real time of an area's use by human groups (Binford 1988; Blasco 2006; Huguet 2007). Nevertheless, the data obtained from faunal remains can help us to understand the formation time of the assemblage on different scales (Table 8.29):

1. On a geological time scale, the archeostratigraphy of the faunal remains indicates the existence of considerable diachrony. The presence in S41 of three different sublevels without signs of remobilizations or erosions of the material (Ja, Jb_1 and Jb_2) suggests that the assemblage was formed over a long period of time in which human occupations and absences both occurred (formation of travertine platforms).
2. On a biological time scale, there are different elements which provide synchrony/diachrony criteria in the interpretation of the assemblage:
   2.1. The large quantity of animal biomass in the two sublevels and their full use involve either the existence of occupations by small groups for a long

Table 8.29 Summary table of the characteristics of the faunal assemblage with contribute criteria on the temporality of the occupations

Elements indicating synchrony and diachrony	Method	Number of occupations	Time of occupation	Time of human absence
Biomass quantity	MNI	One prolonged or several short	–	–
Seasonality of the animals	Dental eruption of immature animals	Spring and autumn	Seasonal	–
Bones burned post-burial	Projection of remains related to hearths	Ja: >one Jb: >two	–	Seasonal
Connection between areas	Refits	Ja: two Jb:?	Long	Short
Presence of prowler scavengers	Projection of toothmarks and coprolites	Various	–	Short

time or the presence of large groups for a short time;

2.2. The presence of two preferential occupation zones in sublevel Ja, as deduced from the study of the refits and the intrusion of carnivore prowlers at times of human absences, indicate the alternation of occupations of longer or shorter duration.

The various data obtained regarding the temporality of the occupations do not contradict one another. On different temporal scales, we observe a dynamic constant of the human occupations and absences of the shelter, regardless of group size. This dynamic was repeated over a long time (formation of Jb_2, Jb_1 and Ja) and is consistent with the data on seasonality and animal time of death which indicate at least two periods of the year in which human groups visited the shelter: spring and autumn. However, during the formation of level J the human occupations repeatedly used the same methods of animal processing as well as spatial distribution.

Discussion

Hunting Specialization and Seasonality of the Human Occupations

Abric Romaní is located on the right bank of the Anoia River. At this point, the river becomes a strait as it crosses the Prelittoral Range. The Anoia constitutes, therefore, one of the few natural channels of communication between the Mediterranean coastal plains and the inland plains. Consequently, the immediate surroundings of the site have acquired some very particular ecological conditions, mostly determined by the geomorphologic characteristics of the strait. In this area a broad ecotone has been formed consisting of the Anoia River itself, the surrounding plains, and the mountains which delimit the strait. This phenomenon significantly influences the diversity of mammal species present in the immediate surroundings (Carbonell et al. 1996; Cáceres et al. 1998). Therefore, during the formation of level J it would have been easy to locate horses, large bovids, and perhaps even rhinoceroses on the open plains of the valley. The valley and the open forests of the mountain slopes would have been the places preferred by red deer, whereas the chamois would have restricted themselves to steep environments and rocky cliffs in the highest parts of the surrounding mountains (Allué et al. 2012).

The human groups made good use of this variety of ecosystems. The anthropic origin of the accumulation of herbivores at the site indicates that the human cynegetic activities were directed towards all these taxa. However, a preference was shown for two main species: horses and red deer. The quantitative data indicate that these taxa together represent 70% of level J herbivores. This difference is more marked in Ja (73%) than in Jb (61.5%), but in both sublevels a certain directionality of the human groups towards these species is suggested.

The existence of an orientation in hunting in a Neanderthal (or earlier) context has always been the cause of controversy as well as the source of various contradictions (Adler et al. 2006). These phenomena have been interpreted differentially by several researchers depending on the chrono-cultural period of the site and the human species (see a historiographic revision in d'Errico 2003). In Upper Paleolithic contexts clearly associated to *Homo sapiens*, the predominance of a taxon is usually associated with specialization and the capacity of these hominids to develop complex hunting techniques whereas in European Middle Paleolithic contexts, such directionality among Neanderthals is usually attributed to the abundance of those taxa in the territory and to the lack of possibilities for these hominids to exploit different resources.

At French Middle Paleolithic sites, the high representation of a species in the archeological assemblages has been interpreted as a result of its abundance in the immediate surroundings. This is, for instance, the case of the mouflons of level F at the Caune de l'Arago (Monchot 1999; Rivals et al. 2006) or the red deer in the Grotte du Lazaret (Valensí 1996, 2000; Boyle 2000). However, at other sites with similar chronologies, the same phenomenon seems to be related to the specialized cynegetic skills of the human groups, for example, in the case of the horses of Schöningen in Germany (Thieme 1997, 1998). Several Middle Paleolithic sites have been interpreted as accumulations resulting from specialized hunting. This is the case of the aurochs and the bears from Biache-Saint-Vaast in Belgium (Auguste 1995), the bison from Mauran in France (Farizy et al. 1994), the reindeer from Salzgitter Lebenstedt in Germany (Gaudzinski and Roebroeks 2000), and the saiga from Chokurcha, Buran Kaya, Starocelié and Kabazi II in Crimea (Patou-Mathis 2006).

But this apparent contradiction between the two opinions gives rise to several questions:

1. The archeofauna present a biased view of the animals existing in the environment. This bias is influenced by the selection that the accumulator agent (human groups) makes of existing potential prey in the near surroundings. From an archeological point of view, there are several important determining factors in the formation of this bias: (1) the frequency with which the hunting episodes occurred, (2) the cynegetic strategies used (individual predation versus multiple predation or mass predation) and, (3) the existence of archeological palimpsests formed by multiple human occupancies that occurred over time following the same guidelines and traditions.
2. The specialization in a resource is usually directed, in terms of energetic profitability, towards resources with high return rates. According to the predictions of the *Optimal Diet* or *Diet Breath Model*, the equation to calculate these values is simple: energy provided by the resource minus the energy invested in its capture and processing (Madsen and Schmitt 1989). Positive values in the return rates only take place when the selected resource is relatively abundant in the territory. With some animals, mainly herbivores, their numbers increase or decrease in a territory depending on their seasonal diet and according to their eco-ethological features (groupings of herds or the transition of migrations). These population fluctuations have considerable repercussions on the behavior of their predators, who must adjust to these seasonal changes.

Human groups are not unconnected with these phenomena. Their presence or absence in a territory depends on the availability of the resources necessary to guarantee their way of life and, consequently, their success as a group (Díez and Rosell 1998). Due to their complexity as a biological species, hominids possess a wide range of possibilities in terms of ways to adapt to the external situation. The strategies developed to guarantee their survival in a territory match this complexity and are therefore considerably varied (seasonal movements, occupational stability, dispersion and/or grouping of their members by family units, etc.). However, these strategies always coincide with the quantity of potentially available resources in the surrounding territory (Butzer 1989). Therefore, opportunist strategies, such as those defended by some authors, cannot be archeologically substantiated at the sites from the overrepresentation of mainly adult animals belonging only to one or two taxa. Just the opposite in fact, as this type of modality would be reflected in significant specific variations of prey, age groups, sex and periods of death. Therefore, the two opinions to explain the abundance of one or two species in an assemblage can be considered complementary, resulting both from the knowledge of the hunters in the territory and from the behaviors of the animals.

In the case of level J, this knowledge of the ethology of the animals in the area is reflected in the seasonality of the hunted prey. Very few remains were available for use in estimating the period of the animals' deaths in level J (we have to remember that most of the prey were adult specimens). However, these remains show two clearly different periods: spring for the equids and autumn for the red deer.

It is important to recall that determining seasonality using faunal remains depends on the occurrence of a single and defined birth season, and the presumption that fossil species are similar to their modern relatives. The paleoenvironment in southern Europe during MIS 3 was significantly affected by climatic phases and seasonal changes (Allué et al. 2012). Similar observations were made in other parts of western Eurasia (e.g. Henry 1995) where altitudinal and seasonal variations in flora and fauna are especially pronounced. Such information indicates that seasonality was marked and thus affected animal populations due to the need for seasonal migration.

Therefore, Abric Romaní represented a strategic place for hunters. The site is situated on an elevated spot at the north-western entrance to the Capellades strait and it provides a good view of the Anoia river valley. The visual control of this natural passage between the coastal and interior plains would allow the seasonal displacements of herds of large herbivores with migratory habits to be monitored. Such movement would occur at times of transition between the harshest periods of the year—in spring and autumn. The animals would move inland toward the green grasses in spring and they would seek a more temperate climate towards the coast in autumn.

Considering the data on seasonality obtained in level J, this phenomenon could explain the predominance of one of the two most represented taxa, the horse, but it would not be useful to explain the abundance of red deer. Today, this animal displays a much more territorial and ubiquitous behavior and its seasonal displacements are not usually marked. Therefore, it would not have made use of the Anoia valley for the same purpose as the horses. However, the dominant males of this species adhere to a very particular mating strategy consisting of attempting to gather, by means of acoustic signals, the maximum number possible of females in their territory to form harems. Today, this occurs in autumn. For the herds this is a period of significant stress, but to some efficient hunters, like human groups, it is an ideal time to take them by surprise relatively easily.

It is important to bear in mind, however, that very little direct data on animal seasonality was collected at the site. It is therefore possible to propose that the fact that the slaughtering of immature individuals occurred in spring and autumn does not mean that the shelter was unoccupied the rest of the year (winter and summer). But it is also important to remember that the hunting techniques used by the human groups in level J were mainly directed towards adult animals, although occasionally they provided immature specimens. Just as immature specimens are present in spring and autumn, if the human occupancies were stable during the whole year, we would also expect to find some immature animals during the other seasons of the year (winter and summer).

From this point of view, the degree of diachrony observed in the occupational superpositions of level J could coincide, in terms of biological time, with a seasonal diet. So, the human groups would use the area during the spring in order to take advantage of the displacements of the horses, and in autumn to hunt red deer. The exploitation of other taxa, such as rhinoceroses, or other mountain resources, like the chamois, can be considered isolated phenomena. Such strategies of the seasonal occupancy of sites are not uncommon among Neanderthal communities (Patou-Mathis 2006). The sites of Biache-Saint-Vaast and Salzgitter-Lebenstedt mentioned above seem to illustrate this strategy, as do the sites of Ehringsdorf in Germany, Cracovie-Zwierzyniec and Dziezyslaw in Poland, Kulna, Beçov and Sipka in the Czech Republic, and Subalyuk in Hungary, among others (Patou-Mathis 2000).

The absence of young or senile animals is another argument in favor of the orientation or directionality of the hunting practiced by the human groups of level J. The cynegetic techniques developed by the human groups were clearly aimed at obtaining adult specimens or juveniles close to their sexual prime. This is not a new phenomenon at Abric Romaní. The same demographical pattern has also been observed in more recent levels (levels E, H and I) (Carbonell 1992, 2002; Cáceres et al. 1998; Vallverdú et al. 2005) and at other sites of the European Middle Paleolithic, Salzgitter Lebenstedt in Germany being the clearest example (Gaudzinski and Roebroeks 2000).

This selection of prey by age suggests the development of procurement techniques based on the acquisition of one or several individuals during the same hunting event (Steele and Baker 1993). No evidence has been found that would lead us to propose the use of complex techniques, such as multiple predation or mass predation, which are so frequently seen in late Upper Paleolithic and Holocene sites. However, such complex hunting techniques are not unheard of during this period. Oriented hunting episodes have been suggested by Mauran, Biache-Saint-Vaast (Patou-Mathis 2006), Coudoulous I Level 4 (Jaubert et al. 2005) and La Borde (Jaubert and Brugal 1990) in France, Wallertheim (Gaudzinski 1995) in Germany and Il'skaya (Hoffecker and Cleghorn 2000) in Russia. Recently, the practice of cooperative hunting has been suggested in relation to Qesem Cave in Israel (Stiner et al. 2009). But the clearest example is from level XII of Bolomor Cave in Valencia, Spain (MIS 6). At this site, the presence of horse remains of different ages combined with a lineal archeostratigraphic distribution of these fossils, presence of refits among the bones located at this archaeostratigraphical line and clear diagnostic elements of primary access has been interpreted as a result of a multiple predation event (Blasco et al. 2010). Nevertheless, this strategy does not seem to be practiced in a generalized or systematic way during the Middle Paleolithic, or at least not to the same degree, as in more advanced periods. The examples of such techniques are scarce and should be examined more thoroughly.

Processing and Consumption Patterns

Research into carcass processing methods was based on the skeletal parts and the bone portions present in the assemblage, cutmarks and the repetition of cutmark positions, fractures, and burning damage. All these elements point to the development of a complete *Chaîne Opératoire* based on the repetition of some very specific guidelines in level J. The *Chaîne Opératoire* was carried out in several phases and in different places. The animals were prepared for transport at the acquisition site. This involved primary processing steps related to skinning, viscera removal and dismembering. The remaining stages of the *Chaîne Opératoire* were carried out inside Abric Romaní (defleshing and disarticulation, roasting and bone breaking).

Some very well established guidelines emerge, then, in the processing of prey by the human groups of Abric Romaní. These guidelines are reflected in the faunal remains as systematic repetitions of the modifications on the bones (mainly cutmarks and breakage). From this perspective, the evidence of clear patterns in carcass processing represents another element in support of the existence of directionality in the hunting and processing of animals. These patterns are mainly documented in the initial and final phases of the operational sequence:

1. Initial phases. The transport of selected anatomical parts is the main activity that determines the faunal accumulation at Abric Romaní. This consists of carrying only the limb bones and cranial remains. Meanwhile, the axial skeleton (vertebrae and ribs) is systematically abandoned at the prey acquisition site. The almost exclusive presence of long bones in the assemblage indicates the existence of significant planning of animal acquisition and transport methods.
2. Final phases. Clear patterns also emerge among the final stages. On the one hand, defleshing is usually done after roasting for the purpose of extracting large amounts of muscle mass. On the other hand, breakage techniques are always applied in two phases: first the epiphyses are separated from the diaphyses and then the diaphyses are opened lengthwise. Subsequently, the activity areas are cleaned: the large bone fragments are moved to the toss zones and the small bone fragments stay around the hearths or are burned.

The processing techniques employed at Abric Romaní result in an assemblage fundamentally formed by diaphysis fragments of long bones. As mentioned earlier, the absence of bones from the axial skeleton is due to the transport modality. However, this phenomenon does not explain the absence of some osseous portions, which, in theory, should have entered the site with the transported anatomical parts. We are referring to the basipodial bones (carpals and tarsals) and the epiphysis of the long bones. The number of these elements recovered is very low in both sublevels (1.5% in Ja and 1% in Jb). In most cases these are very small soft tissue fragments basically impossible to identify at a skeletal and/or taxonomic level. We must keep in mind that this low percentage does not result from any differential destruction phenomenon based on the density of the bones.

Considering the breakage techniques identified in level J, how the epiphyses were used after this process is difficult to ascertain. Its disappearance might be attributable to the activity of the carnivores that visited the shelter and/or to the anthropic processing sequence.

The possibility of the disappearance of these osseous segments as a consequence of the visits of carnivores subsequent to the human occupancies has been considered in different experimental works. All the authors that have conducted studies on remains formed in these conditions have made reference to the carnivores' preference for the epiphyses abandoned by hominids (Richardson 1980; Blumenschine 1986, 1988; Marean and Spencer 1991; Marean et al. 1992; Brantingham 1998; Selvaggio 1998). The high fat content in these anatomic parts makes them very attractive to non-human predators.

Many researchers agree on the low degree of preservation of nutrients in bones in open-air contexts, where the atmospheric agents dry them out and degrade them relatively quickly (Bunn et al. 1988; O'Connell et al. 1992).

Table 8.30 Total number of recovered epiphyses in relation to the total number of epiphyses with carnivore toothmarks and the corresponding percentage

	Total number of epiphyses	Epiphyses with toothmarks	%
AR-Ja	78	2	2.5
AR-Jb	18	0	0
Firts hominid—Experimental series	73	16	21.9
First hominid—Natural model	3	2	66.7
Kobeh Cave	270	64	23.7
First carnivore—Experimental series	32	13	40.6
First carnivore—Natural model	2	2	100.0

Data extracted from Marean and Kim (1998), Blumenschine (1988), Marean and Spencer (1991) and Marean et al. (1992). In the case of Abric Romaní carpal and tarsal bones are also included

However, Marean and Bertino (1994) observe that in closed or semi-closed environments, such as Abric Romaní, the bones can preserve fat and other components that would be attractive to carnivores for up to 1 year after being abandoned. Brain (1981) also documents a similar phenomenon in the desert of Namibia and he suggests that, in dark and damp conditions, such preservation can even last for over 7 years. Therefore, after the human occupancies of level J the abandoned bones remained exposed to the action of carnivores for a certain period of time. In the experiments carried out by Marean and Spencer (1991) and by Marean et al. (1992) on hyenas in assemblages of anthropic origin it was found that around 20% of the preserved epiphyses had toothmarks. This percentage is maintained in the study carried out by Marean and Kim (1998) at the site of Kobeh Cave in Iraq and it increases significantly in the natural models documented by Blumenschine (1988).

However, the remains in level J are not consistent with these proportions. The percentage of epiphyses with carnivore toothmarks is very low, indicating that there is a more noticeable destruction of these bone segments at Abric Romaní than at other sites or in different experiments (Table 8.30). Furthermore, very few epiphyses were recovered at Abric Romaní (78 fragments in Ja and 18 in Jb) and more than 90% present signs of exposure to fire.

The disappearance of the epiphysis, therefore, does not seem to be related to posterior carnivore activity. However, the objective of the breakage techniques employed by the hominids was the full recovery of these osseous segments. It is possible that the hominids used a technique related to these segments that involved their thermal treatment and subsequent disappearance (bone as a combustible). The absence of these elements prevents us from determining whether this disappearance occurred during the consumption process or because of its exposure to fire. Therefore, the lack of epiphyses at level J must be attributed to human activity.

Another question related to consumption is the systematic transport of a considerable volume of potentially consumable animal products (meat, viscera, fat, and marrow). The hunting systems were basically directed towards obtaining large and medium-sized animals. Therefore, the human groups obtained a large quantity of resources during each hunting event, giving rise to the possibility of either a group sufficiently numerous so as to be able to intensively consume a large quantity of nutrients or the existence of meat surpluses, which were not immediately consumed.

The existence of surpluses and the possible secondary consumption of meat at Middle Paleolithic sites is a matter that is difficult to prove archeologically. Nevertheless, zooarcheology, along with other disciplines, can provide data that contributes to making some inferences on this matter.

According to ethnographic parallels, there are simple techniques of meat preservation which consist of smoking or drying fresh meat in the sun (Binford 1981, 1988). The technology required for these practices is compatible with that employed during the European Middle Paleolithic and therefore, with that employed at level J (existence of controlled fires, wooden structures related to the hearths, etc.). However, the analysis of the double colorations of the bones indicates that defleshing was done after the anatomic portions had been submitted to the action of fire. This contrasts with the ethnographic parallels that describe the techniques of preserving fresh meat (Binford, op. cit.). Meat, once roasted, is difficult to preserve. Therefore, it seems that the consumption of food was immediate in level J.

Nevertheless, the estimated paleoeconomic indices for level J indicate that the quantity of nutrients in Abric Romaní was very high. Such a quantity of resources would have been difficult to consume completely and immediately if the number of group members was not very high. This gives rise to the possibility that, in addition to immediate consumption, there may have been other techniques used to facilitate the preservation of certain meat surpluses (Rosell 2001). The elements required to carry out practices of this kind have been identified at Abric Romaní. In level I, for example, wooden structures related to the hearths were recovered, which were interpreted as tripods used to preserve meat (Castro-Curel and Carbonell 1995; Carbonell et al. 2002; Rosell 2001). From this perspective, the immediate consumption of meat as interpreted from the double colorations of the bones is not incompatible with the possibility that some surpluses might have been preserved.

Occupation Models and Social Articulation

As mentioned above, the faunal data from level J show a recurrence of occupations related to the annual diets of the animals in the area. These occupations occurred in the space of the shelter using relatively extensive areas. From this point of view, they repeat patterns of distribution and organization of the space. These patterns are conditioned, in some aspects, by the structural elements of the shelter: the presence of large blocks in the external zone, the shelter walls, channels and sinters. Activities related to the processing of fauna in the shelter were always carried out in the inside areas and around the hearths. The external areas were used as accumulation waste zones (toss areas) (Rando 1999). This dynamic was repeated through time and over the different occupations, in Ja as well as in Jb.

Similar phenomena have been identified at an ethnoarcheological level. According to Binford (1984a, b), at human camps activities are usually carried out inside the occupied area, whereas the outlying zones serve as areas for refuse, or for marginal or specific activities.

These repeated occupational patterns along with hunting techniques and systematic transport and processing patterns indicate that the human groups that inhabited Abric Romaní maintained traditions through time. Therefore, Abric Romaní played an important role in the course of the spatial exploitation of the human groups within this territory. The availability of resources determined the mobility of the groups in the territory, which involved a certain degree of knowledge with regard to the habits of the animals hunted. There was, therefore, an established territorial network of hominids in which the exploitation strategies of the territory and the transmission of information guaranteed the continuity and survival of the groups (Gamble 1990). This phenomenon indicates the existence of a high degree of social cohesion. In level J we observe some elements that bear witness to this social articulation and which make it possible to deduce the existence of this territorial network:

1. The seasonality of the occupations. Territorial movements in an area as diverse as that surrounding Abric Romaní are a good indicator of knowledge of the territory and the capacity of predictability of the human groups.
2. Activities related to the processing of animals around the fire. Several researchers have proposed that fire played a role as a vehicle of communication and a socializing element in human communities of the past (Carbonell and Sala 2002; Arsuaga and Martínez 1998).
3. The transport of animal resources. The system of transporting animal remains to where the rest of the group waits is governed by the need for sharing (Díez and Rosell 1998).
4. The processing patterns. The systematization of some patterns in the use of the animals is the best indication of the existence of a transmission of intergroup information and learning.

Therefore, we can deduce a "modern type" of social organization in the human communities which occupied Abric Romaní during the formation of level J. This social articulation would have manifested itself in the development of deeply rooted cultural patterns and traditions and in movements within the territory according to the seasonal availability of resources. During these movements, the shelter would have played an important referential role as a strategic place within the economical and social network of these hominids.

Conclusions

Cranial fragments and diaphyses of long bones, mainly of adult animals, characterize the faunal remains of level J. The zooarcheological study indicates that it is an assemblage of anthropic origin formed by multiple activities carried out over a relatively long period of time. The duration and time between occupations is marked by changes in the sedimentary patterns that induce the formation of at least two different sublevels (Ja and Jb).

Following a seasonal pattern, human groups occupied the shelter during the periods of the year when the most faunal resources were available in the environment, especially in spring and in autumn. These groups used hunting techniques oriented mainly towards obtaining horse meat in spring and red deer in autumn. There were occasional hunting episodes of other taxa, such as large bovids, rhinoceroses, or chamois. Individual predation seems to be the technique most commonly used, i.e. in every hunting episode one or several previously selected animals were obtained: adult or close to mature specimens.

The operational sequence of animal use follows well-established patterns, which repeat themselves systematically throughout the formation of the whole assemblage. The human groups prepared the carcasses at the acquisition site, which involved skinning, viscera removal and dismembering. This first aim of primary butchery was the transport of selected parts such as extremities and crania to Abric Romaní, which is why no remains of ribs or vertebrae are found at the site.

Once in the shelter, the extremities and the crania were roasted and later defleshed. Once the meat was removed, the bones were systematically fractured to access the marrow. The identified patterns in carrying out this activity serve a double purpose: to obtain the marrow in excellent condition and to obtain the epiphyses as a whole. These processes led to the appearance of some very well established osseous morphotypes in which the disappearance of the epiphyses is a constant (probably used as fuel). The animal processing techniques seem to be directed towards the immediate consumption of nutrients. However, the high quantity of biomass transported to the site mean that the preservation of some portions for later consumption in other places cannot be ruled out.

All this systematization of the butchery processes is related to a complex distribution of space. The activity areas are always located around the hearths, covering from the shelter wall to the zones where the large blocks fallen from the cornice have accumulated. Activity areas related to fauna have not been identified outside the shelter. However, these exterior zones contained the largest animal remains (bones larger than 10 cm). Considerable movement of bones occurred between the different activity areas, as deduced from the refits realized on the osseous remains from sublevel Ja. This indicates the occupation of relatively large spaces with the use of different areas simultaneously.

Sporadic carnivore visits occurred subsequent to the human activities. The activity of these animals is related to their intensive use of the bones abandoned by the hominids. However, the intensive use of the animal resources by the hominids meant that the activities of these prowlers did not excessively disturb the assemblage. Postdepositional processes do not disrupt the assemblage in excess, so the remains appear very close to their original position and therefore, are spatially organized.

In short, no changes in the patterns identified in the processing of animals were detected over the entire period of occupation of the two sublevels. This indicates the existence of some well-established patterns, which are the result of traditions maintained through time. From these parameters we can deduce a certain control of the territory by groups that have a considerable territorial component. Abric Romaní would have played an important role in the seasonal displacements of these groups.

This general model, developed from the zooarcheological data obtained from level J, confers a high degree of complexity in the activities of the human groups of Abric Romaní, and by extension, of the hominids of the European Middle Paleolithic, which show a high capacity for organization and planning that is comparable to that of modern humans.

Acknowledgments Special thanks to Eudald Carbonell and Manuel Vaquero for the invitation to participate in this book. This research was supported by the Ministerio de Educación y Ciencia Spanish Government Grants CGL2009-12703-C03-02 and CGL2009-7896, and by Generalitat de Catalunya Grant 2009 SGR 188.The field excavation work was supported by Departament de Cultura i Mitjans de Comunicació de la Generalitat de Catalunya, Diputació de Barcelona, Ajuntament de Capellades and Arts Gràfiques Romanyà-Valls S.A. Ruth Blasco and Maria Joana Gabucio are benificiaries of a FI Grant from Generalitat de Catalunya and financed by European Social Found. Antonio Rodríguez-Hidalgo is beneficiary of a FPI Grant from Ministry of Science and Innovation of Spanish Government. We are also grateful to Jean Philippe Brugal and Dan Adler for their critical comments on an earlier draft of this chapter.

Appendix

Annex.

Refitting of two elements from the Ja faunal assemblage (inf: infantile; juv: juvenile; ad: adult; sen: senile).

Square	N°	Ja coord. x	y	z	Size	Sp.	Age	Element	Taphonomic modifications	Distance in cm.
P51	621	5006	1555	327	Small	Unidentified		Long bone	Burned (degree 3)	4.24
P51	622	5003	1558	325	Small	Unidentified		Long bone	Burned (degree 2)	
G58	10	5704	615	529	Medium	Unidentified		Long bone		1
G58	12	5704	614	530	Medium	Unidentified		Long bone	Roots (1 side)	
H61	30	6076	782	593	Medium	Cervidae	ad	Metatarsal	Roots (1 side) Burned (degree 1)	63.06
H61	10	6047	726	585	Medium	Cervidae	ad	Metatarsal	Roots (1 side) Burned (degree 1)	
H64	1	6311	762	639	Medium	Cervidae	ad	Humerus	Roots (1 side) Burned (degree 1)	7.61
H64	2	6308	769	638	Medium	Unidentified		Flat blone	Roots (2 sides)	
J58	50	5730	921	506	Medium	Unidentified		Long bone	Roots (2 sides)	12.80
J58	52	5738	931	507	Medium	Unidentified		Long bone	Roots (2 sides)	
K50	3	4954	1000	329	Medium	Unidentified		Long bone	Roots (2 sides)	21.54
K50	8	4962	1020	345	Medium	Unidentified		Long bone	Roots (1 side)	
K54	11	5368	1027	401	Medium	Cervidae	ad	Tibia	Roots (2 sides)	13.15
K54	12	5370	1014	402	Medium	Cervidae	ad	Tibia	Roots (2 sides)	
L48	41	4704	1072	–	Medium	Unidentified		Long bone	Roots (2 sides)	98.41
L47	2	4695	1170	286	Medium	Unidentified	ad	Long bone	Roots (2 sides)	
L49	94	4860	1166	319	Medium	Cervidae	ad	Metacarpal	Roots (2 sides)	26
L49	127	4850	1190	320	Medium	Cervidae	ad	Metacarpal	Roots (2 sides)	
L56	90	5514	1160	418	Medium	Cervidae		Horn	Burned (degree 3)	3.00
L56	91	5514	1157	422	Unidentified	Unidentified		Unidentified	Burned (degree 3)	
M56	44	5536	1221	412	Medium	Cervidae	ad	Tooth		3.60
M56	45	5533	1223	412	Large	Equidae	ad	Tooth		
M55	14	5435	1214	391	Medium	Unidentified		Flat bone	Roots (1 side) Burned (degree 3)	64.49
M55	37	5467	1270	396	Medium	Unidentified		Flat bone	Roots (1 side) Burned (degree 3)	
N48	26	4701	1351	294	Medium	Unidentified		Long bone	Roots (1 side) Burned (degree 3)	8.06
N48	27	4703	1352	295	Medium	Unidentified		Long bone	Roots (1 side) Burned (degree 4)	
N48	32	4702	1387	294	Medium	Unidentified		Long bone	Burned (degree 5)	2.23
N48	33	4701	1395	293	Medium	Unidentified		Long bone	Burned (degree 5)	
N55	46	5414	1323	384	Medium	Unidentified		Long bone	Roots (2 sides)	3.16
N55	47	5411	1324	384	Medium	Unidentified		Long bone	Roots (2 sides)	
N54	15	5370	1340	376	Large	Unidentified	ad	Long bone	Burned (degree 1)	50.00
N54	158	5400	1300	374	Large	Unidentified		Long bone	Roots (2 sides)	
N50	102	4941	1338	325	Medium	Unidentified		Rib	Roots (1 side) Burned (degree 1)	24.83
N50	106	4925	1357	322	Medium	Unidentified		Rib		
N53	39	5296	1363	361	Medium	Cervidae	ad	Radius	Roots (1 side) cutmarks Burned (degree 1)	47.38
N53	60	5262	1330	358	Medium	Cervidae		Radius	Trampling Roots (1 side)	

(continued)

(continued)

Square	N°	Ja coord.			Size	Sp.	Age	Element	Taphonomic modifications	Distance in cm.
		x	y	z						
N54	163	5309	1324	367	Medium	Cervidae	ad	Femur	Burned (degree 1)	12.64
N54	170	5321	1320	372	Medium	Cervidae	ad	Femur	Roots (2 sides)	
O43	50	4275	1428	248	Medium	Unidentified		Flat bone	Roots (2 sides)	2.23
O43	78	4274	1430	247	Medium	Unidentified		Flat bone	Roots (2 sides)	
J57	5	5655	945	476	Medium	Cervidae	ad	Tibia	Roots (2 sides)	5.09
J57	6	5650	944	474	Medium	Cervidae	ad	Tibia	Roots (2 sides)	
M42	15	4178	1276	220	Medium	Cervidae	ad	Radius	Burned (degree 2) Cutmarks	37.10
M42	32	4187	1240	226	Medium	Cervidae	ad	Radius	Burned (degree 2) Cutmarks	
V50	302	4943	2119	265	Medium	Unidentified		Long bone	Burned (degree 3)	4
V50	303	4939	2119	265	Medium	Unidentified		Long bone	Burned (degree 3)	
N42	50	4163	1373	232	Medium	Unidentified		Long bone	Burned (degree 3)	57.80
N43	29	4209	1338	342	Medium	Unidentified		Long bone	Burned (degree 3)	
O47	25	4645	1488	282	Medium	Unidentified	ad	Rib	Roots (2 sides)	77.89
O47	151	4683	1420	290	Medium	Unidentified		Flat bone	Roots (2 sides)	
M45	4	4419	1214	244	Medium	Unidentified		Long bone	Roots (2 sides)	35.35
M45	71	4444	1239	251	Unidentified	Unidentified		Long bone	Roots (2 sides) Burned (degree 1) Rounded	
M50	263	4985	1248	328	Medium	Cervidae	ad	Tibia		15.29
M50	261	4982	1233	328	Large	Unidentified		Long bone		
I60	11	5940	884	570	Large/ Medium	Unidentified		Flat bone		6.08
I60	12	5941	878	570	Large/ Medium	Unidentified		Flat bone		
I60	5	5947	855	567	Large/ Medium	Unidentified		Flat bone		295.00
I63	36	6242	856	615	Large	Unidentified		Flat bone	Roots (2 sides)	
L52	18	5148	1163	348	Medium	Unidentified		Long bone	Roots (2 sides)	43.13
L52	44	5117	1193	342	Medium	Cervidae	ad	Metacarpal	Roots (2 sides)	
I61	54	6100	800		Medium	Unidentified		Long bone	Burned (degree 5) Cutmarks	230.36
G63	6	6279	655	645	Medium	Unidentified		Long bone	Burned (degree 5) Cutmarks	
N52	336	5139	1344	345	Medium	Unidentified		Long bone	Burned (degree 1) Roots (2 sides)	12.80
N52	416	5149	1352	350	Medium	Unidentified		Long bone		
N43	19	4254	1378	244	Medium	Cervidae	ad	Metacarpal	Burned (degree 4)	346.69
O46	16	4600	1400		Medium	Cervidae	ad	Metacarpal	Burned (degree 5)	
N52	392	5141	1300	345	Medium	Unidentified		Long bone	Roots (2 sides)	47.17
N52	400	5145	1347	346	Medium	Unidentified		Long bone	Roots (2 sides) Burned (degree 2)	
P51	14	5035	1545	321	Medium	Cervidae		Humerus	Burned (degree 4)	305.45
O48	76	4761	1410	300	Medium	Cervidae	ad	Humerus	Burned (degree 4)	

(continued)

(continued)

Square	N°	Ja coord.			Size	Sp.	Age	Element	Taphonomic modifications	Distance in cm.
		x	y	z						
N53	345	5289	1321	365	Medium	Unidentified		Long bone	Burned (degree 1)	49.36
N54	50	5335	1303	374	Unidentified	Unidentified		Unidentified	Burned (degree 2)	
									Cutmarks	
									Roots (2 sides)	
N49	139	4804	1337	304	Medium	Cervidae	ad	Femur	Roots (2 sides)	8.24
N49	73	4806	1345	305	Medium	Unidentified		Long bone		
G64	33	6353	655	661	Medium	Equidae	ad	Tooth	Burned (degree 3)	357.79
									Trampling	
I61	25	6035	819	592	Medium	Cervidae	ad	Metatarsal	Burned (degree 3)	
N53	297	5400	1353		Medium	Unidentified		Long bone	Roots (1 side)	32.06
									Cutmarks	
N53	313	5368	1352		Medium	Unidentified		Long bone	Roots (1 side)	
K53	1	5213	1052	388	Large	Equidae	ad	Metapodial	Roots (2 sides)	40.60
K53	5	5244	1026	382	Large	Equidae		Metapodial	Roots (2 sides)	
L49	122	4889	1145	321	Large	Equidae	ad	Tooth	Roots (2 sides)	6.40
L49	123	4893	1150	322	Large	Equidae	ad	Tooth	Roots (2 sides)	
L50	159	4973	1134	326	Large	Equidae		Tooth	Burned (degree 1)	34.78
L50	170	4940	1145	328	Mediumg	Rhinocerotidae			Burned (degree 1)	
L57	16	5684	1139	462	Large	Equidae	ad	M_2	Burned (degree 1)	59.64
									Roots (2 sides)	
L57	42	5670	1196	440	Large	Equidae	ad	Tooth	Roots (2 sides)	
									Burned (degree 1)	
L57	140	5647	1120	458	Large	Equidae	ad	M^2	Burned (degree 1)	32.38
L57	160	5660	1109	466	Large	Equidae	ad	M^3	Burned (degree 1)	
									Trampling	
L57	119	5652	1149	454	Large	Unidentified		Flat bone		17.02
L57	87	5657	1181	446	Large	Unidentified	ad	Jaw		
M55	9	5448	1275	390	Large	Unidentified		Tibia		105.51
M56	9	5550	1250	409	Large	Unidentified		Tibia		
N52	162	5103	1353	338	Large	Unidentified		Long bone	Burned (degree 2)	4.00
N52	75	5107	1353	337	Large	Unidentified	ad	Long bone	Burned (degree 2)	
N54	9	5387	1315	380	Large	Unidentified		Long bone		64.07
N54	121	5334	1351	371	Large	Unidentified	ad	Long bone		
O50	16	4935	1420	320	Unidentified	Unidentified		Unidentified	Burned (degree 3)	38.07
									Roots (1 side)	
									Trampling	
O50	53	4968	1401	324	Unidentified	Unidentified		Unidentified	Burned (degree 5)	
									Concentrated trampling	
O51	6	5024	1437	329	Large	Unidentified		Long bone	Burned (degree 3)	50.77
									Roots (1 side)	
O51	7	5067	1410		Large	Unidentified		Long bone	Burned (degree 3)	
P48	30	4705	1546	295	Large	Equidae	ad/sen	I^2		287.84
Q50	142	4984	1616	322	Large	Equidae	ad/sen	Maxiar		

(continued)

(continued)

Square	N°	Ja coord. x	y	z	Size	Sp.	Age	Element	Taphonomic modifications	Distance in cm.
R42	13	4170	1764	250	Large	Unidentified		Long bone	Burned (degree 2)	1.41
R42	14	4171	1765	251	Large	Unidentified		Long bone	Burned (degree 2)	
S42	171	4157	1828	270	Large	Unidentified		Long bone		7.21
S42	178	4163	1824	268	Large	Unidentified		Long bone	Burned (degree 2)	
N50	226	4116	1358						Burned (degree 1)	18.43
									Roots (1 side)	
N50	227	4102	1370						Roots (2 sides)	
									Percussion marks	
									Cutmarks	
									Burned (degree 1)	
M44	36	4307	1280	241	Large	Bovinae	ad	Radius/ulna	Roots (2 sides)	200.48
N45	23	4479	1383	265	Large	Bovinae	ad	Radius/ulna	Roots (2 sides)	
P51	58	5040	1533	329	Large	Equidae	ad	Humerus	Roots (2 sides)	32.65
P51	891	5069	1518	333	Medium	Unidentified		Long bone	Burned (degree 1)	
N43	49	4280	1347	247	Large	Unidentified	ad	Long bone	Burned (degree 1)	19.10
									Cutmarks	
									Roots (2 sides)	
N43	70	4293	1361	251	Large	Unidentified	ad	Long bone	Roots (2 sides)	
									Burned (degree 1)	
N44	65	4330	1371	248	Large	Equidae	ad	Tibia	Roots (2 sides)	22.67
									Burned (degree 1)	
N44	80	4313	1356	250	Large	Equidae	ad	Tibia	Burned (degree 1)	
									Cutmarks	
									Roots (2 sides)	
N50	70	4953	1374	323	Large	Unidentified		Long bone	Roots (1 side)	3.16
									Cutmarks	
									Burned (degree 2)	
N50	73	4954	1377	324	Large	Unidentified		Long bone	Roots (1 side)	
									Cutmarks	
									Burned (degree 2)	
S42	148	4156	1803	265	Unidentified	Unidentified		Unidentified	Burned (degree 4)	26.07
S42	168	4154	1829	267	Large	Unidentified		Unidentified	Burned (degree 3)	
L56	1	5549	1135	432	Middle	Unidentified		Long bone	Burned (degree 1)	258.69
									Cutmarks	
									Roots (2 sides)	
									Concentrated trampling	
N54	28	5354	1305	374	Large	Equidae	ad	Femur	Roots (2 sides)	
O48	161	4773	1404	305	Large	Equidae	ad	Jaw	Cutmarks	4.47
									Burned (degree 3)	
O48	207	4775	1408	299	Unidentified	Unidentified		Unidentified	Cutmarks	
									Burned (degree 2)	
N51	292	5060	1339	333	Large	Unidentified		Tooth	Burned (degree 5)	31.0
N51	460	5029	1334	333	Large	Unidentified		Flat bone	Burned (degree 5)	

(continued)

(continued)

Square	N°	Ja coord. x	y	z	Size	Sp.	Age	Element	Taphonomic modifications	Distance in cm.
N52	365	5125	1343	342	Unidentified	Unidentified		Unidentified	Burned (degree 3)	426.4
M56	82	5547	1283	414	Middle	Unidentified		Long bone	Burned (degree 3)	
L54	20	5356	1150	391	Middle	Unidentified		Long bone	Roots (2 sides) Percussion marks	49.4
L55	45	5402	1167		Large	Equidae		Humerus	Roots (1 side)	
M57	12	5670	1235	432	Large	Unidentified		Long bone	Rounded	468.3
I62	39	5473	810	599	Large	Unidentified	ad	Long bone	Cutmarks Burned (degree 2)	
O48	119	4755	1497	298	Large	Unidentified		Long bone	Cutmarks Burned (degree 3) Roots (1 side)	23.5
P48	20	4768	1517	296	Middle	Unidentified		Long bone	Burned (degree 3)	
P49	4	4890	1545	310	Large	Unidentified		Long bone	Roots (2 sides) Percussion marks	254-9
O47	93	4672	1413						Percussion marks	
P51	1087	5044	1508	342	Large	Bovinae	ad	Metacarpal	Roots (2 sides)	672.0
V49	85	4895	2164	275	Large	Unidentified		Metapodial	Burned (degree 1) Roots (1 side)	
N49	146	4878	1384						Burned (degree 3)	176.4
M48	83	4739	1275	320	Large	Unidentified		Long bone	Burned (degree 3) Roots (1 side)	
P48	26	4755	1520	294	Middle	Equidae	inf/juv	D^3		237.3
R46	29	4600	1700		Middle	Equidae	inf/juv	D^4		
O52	31	5118	1450	339	Unidentified	Unidentified		Long bone	Burned (degree 3)	25.6
O52	35	5143	1443	344	Unidentified	Unidentified		Long bone	Burned (degree 3)	
O52	11	5148	1425	342	Unidentified	Unidentified		Unidentified	Burned (degree 4)	1
O52	12	5148	1426	342	Unidentified	Unidentified		Unidentified	Burned (degree 4)	
N52	424	5178	1352	351	Unidentified	Unidentified		Unidentified	Burned (degree 5)	1
N52	425	5179	1352	351	Unidentified	Unidentified		Unidentified	Burned (degree 5)	
O50	249	4905	1462	321	Unidentified	Unidentified		Long bone	Burned (degree 5)	60.01
O50	277	4964	1473	330	Unidentified	Unidentified		Long bone	Burned (degree 5)	
T44	10	4348	1906	268	Unidentified	Unidentified		Unidentified	Burned (degree 3)	282.03
T44	11	4350	1908	268	Unidentified	Unidentified		Unidentified	Burned (degree 3)	

Bone refits of three or more elements from the Ja faunal assemblage

Square	N°	Ja Coord.			Orientation	Size	Taxa	Age	Element	Taphonomic modifications
		x	y	z						
L51	14	5060	1149	335	nesw	Medium	Cervidae	ad	Metacarpal	Root (2 sides)
L51	19	−5084	1140	341.5	ew	Medium	Cervidae	ad	Metacarpal	Root (2 sides)
L51	24	5073	1149	339.	ew	Medium	Cervidae	ad	Metacarpal	Root (2 sides)
L51	20	5091	1106	345.	ew	Medium	Cervidae	ad	Metatarsal	Root (2 sides)
L51	10	5083	1129	343.5	ew	Medium	Cervidae	ad	Metatarsal	Root (2 sides) Percussion notche
L52	27	5109	1122	345.5	ew	Medium	Cervidae	ad	Metatarsal	Root (2 sides)
L45	60	4428	1143	249.5	ew	Medium	Cervidae	ad	Femur	Burned (degree 3)
L45	11	4415	1194	245	nwse	Medium	Cervidae	ad	Femur	Root (2 sides)
L44	90	4391	1159	245	–	Medium	Cervidae	ad	Femur	Root (2 sides) Burned (degree 3)
L45	16	4407	1155	245	ns	Medium	Cervidae	ad	Femur	Root (2 sides) Burned (degree 1) Percussion notche
L44	67	4395	1197	249	nwse	Medium	Cervidae	ad	Femur	Root (2 sides)
O50	6	4938	1467	322.5	nesw	Medium	–	–	Long bone	Burned (degree 4)
O50	56	4944	1409	322.5	nwse	Medium	–	–	Long bone	Burned (degree 3)
O50	105	4933	1416	321	nesw	Medium	–	–	Long bone	Burned (degree 4)
H61	7	6046	601	583	ew	Medium	Cervidae	ad	Metacarpal	Root (2 sides)
H61	9	6040	620	581	nesw	Medium	Cervidae	ad	Metacarpal	Root (1 side)
H61	11	6070	651	591	nesw	Medium	Cervidae	ad	Metacarpal	Root (1 side)
N43	31	4206	1366	339	–	Medium	–	–	Long bone	Burned (degree 2)
N43	32	4207	1368	339	–	Medium	–	–	Long bone	Burned (degree 2)
N43	38	4200	1369	238	nwse	Medium	–	–	Long bone	Burned (degree 2)
N52	337	5136	1346	345	ew	Medium	Cervidae	ad	Humerus	Burned (degree 5)
N52	396	5151	1342	347.5	ew	Medium	Cervidae	ad	Humerus	Burned (degree 5)
N52	393	5145	1344	346	nwse	Medium	Cervidae	ad	Humerus	Burned (degree 5)
N52	395	5148	1347	347	ns	Medium	Cervidae	ad	Humerus	Burned (degree 5)
L48	24	4758	1178	291	ew	Medium	Cervidae	ad	Tibia	Root (2 sides)
L48	34	4743	1185	291	nwse	Medium	Cervidae	ad	Tibia	Root (2 sides)
L48	36	4750	1197	293	nesw	Medium	Cervidae	ad	Tibia	Root (2 sides)
M45	43	4469	1216	254	nesw	Large	–	–	Jaw	
M45	44	4465	1214	252	nesw	Large	–	–	Jaw	
M45	45	4460	1212	254	nesw	Large	–	–	Jaw	
O51	47	5096	1455	337.5	ns	Large	–	ad	Long bone	Burned (degree 3)
O51	30	5080	1480	339	nwse	Large	–	ad	Long bone	Burned (degree 3)
O51	36	5093	1462	337	nesw	Large	–	ad	Long bone	Burned (degree 3)
P51	1012	5017	1509	331	nwse	Large	–	–	Long bone	Burned (degree 4)
P51	1013	5016	1512	331	nwse	Large	–	–	Long bone	Burned (degree 4)
P51	1014	5017	1515	331	ns	Large	–	–	Long bone	Burned (degree 4)
N42	15	4119	1343	220	ns	Large	–	–	Long bone	Burned (degree 1) Cutmarks Root (2 sides)
N42	57	4152	1351	228	nesw	Large	–	–	Long bone	Root (2 sides)

(continued)

(continued)

Square	N°	Ja Coord. x	y	z	Orientation	Size	Taxa	Age	Element	Taphonomic modifications
N43	9	4204	1357	231	nesw	Large	–	–	Long bone	Burned (degree 1)
O42	42	4118	1458	232	nesw	Large	–	–	Long bone	Burned (degree 2)
O42	52	4121	1444	228	ew	Large	–	–	Long bone	Burned (degree 2)
O42	77	4122	1459	230	nwse	Large	–	–	Long bone	Burned (degree 2)
O42	78	4111	1471	219	nesw	Large	–	–	Long bone	Burned (degree 2)
O42	80	4117	1442	229	ew	Large	–	–	Long bone	Burned (degree 2)
O42	100	4116	1454	232	ns	Large	–	–	Long bone	Burned (degree 2)
I58	3	5773	845	516	ew	Large	Equidae	ad	Humerus	Carbonate
I58	4	5781	850	518	senw	Large	Equidae	ad	Humerus	Carbonate
I58	5	5788	847	521	–	Large	Equidae	ad	Humerus	Carbonate
N43	3	4300	1350	242	ns	Large	Equidae	ad	Humerus	Burned (degree 5) Cutmarks
N43	38	4228	1370	242	ns	Large	Equidae	ad	Humerus	Burned (degree 1) Root (2 sides)
O41	83	4098	1482	224	nesw	Large	Equidae	ad	Humerus	Impact flake. Root (1 side)
N54	58	5310	1362	361	nesw	Large	Equidae	ad	Metatarsal	Root (2 sides)
P51	461	5054	1513	334.5	nwse	Large	Equidae	ad	Metatarsal	Cutmarks
P50	83	4978	1522	323	nwse	Large	Equidae	ad	Metatarsal	Root (2 sides) Burned (degree 1)

References

Adler, D. S., Bar-Oz, G., Belfer-Cohen, A., & Bar-Yosef, O. (2006). Ahead of the game. Middle and upper Palaeolithic hunting behaviours in the Southern Caucasus. *Current Anthropology, 47*(1), 89–118.

Allué, E., Burjachs, F., García, A., López-García, J.M., Bennàsar, M., Rivals, F., Blain, H.-A., Expósito, I. & Martinell, J. (2012). Neanderthal landscapes and their home environment: flora and fauna records from level J. In E. Carbonell (Ed.), *High resolution archaeology and neanderthal behavior: Time and space in level J of Abric Romaní (Capellades, Spain)*. Dordretch:Springer.

Archer, M. (1977). Faunal remains from the excavations at Puntutjarpa Rockshelter. *Anthropological Papers of the American Museum of Natural History, 54*, 158–165.

Arsuaga, J. L., & Martínez, I. (1998). *La especie elegida*. Madrid: Temas de hoy.

Auguste, P. (1995). Chasse et charognage au Paleolithique moyen: l'apport du gisement de Biach-Saint-Vaast (Pas-de-Calais). *Bulletin de la Société Préhistorique Française, 92*, 155–167.

Azorit, M., Analla, M., Carrasco, R., Calvo, J. A., & Muñoz-Cobo, J. (2002). Teeth eruption pattern in red deer (*Cervus elaphus hispanicus*) in southern Spain. *Anales de Biología, 24*, 107–114.

Barone, R. (1976). *Anatomie comparée des mammifères domestiques*. París: Vigot Frères.

Bartram, L. E. (1997). A comparison of Kua (Botswana) and Hadza (Tanzania) bow and arrow hunting. In O. H. Donald & H. Knecht (Eds.), *Projectile technology* (pp. 321–343). New York: Plenum Press.

Bartram, L. E., & Marean, C. W. (1999). Explaining the "Klasies Pattern": Kua ethnoarcheology the Die Kelders Middle Stone Age archaeofauna, long bone fragmentation and carnivore scravaging. *Journal of Archaeological Science, 2*, 29–69.

Bartram, L. E., Kroll, E. M., & Bunn, H. T. (1991). Variability in camp structure and bone food refuse patterning at Kua San hunter-gatherer camps. In E. M. Kroll & T. D. Price (Eds.), *The interpretation of archaeological spatial patterning* (pp. 77–148). New York: Plenum Press.

Bennett, J. L. (1999). Thermal alteration of buried bone. *Journal of Archaeological Science, 26*, 1–8.

Bennett, D., & Hoffmann, R. S. (1999). Equus caballus. *Mammalian Species, 628*, 1–4.

Binford, L. R. (1978). *Nunamiut ethnoarchaeology*. New York: Academic Press.

Binford, L. R. (1981). *Bones: Ancient men and modern myths*. New York: Academic Press.

Binford, L. R. (1984a). *Faunal remains from Klassies River Mouth*. New York: Academic Press.

Binford, L. R. (1984b). Butchery, sharing, and the archaeological record. *Journal of Anthropological Archaeology, 3*, 235–237.

Binford, L. R. (1985). Human ancestors: Changing views of their behaviour. *Journal of Anthropological Archaeology, 4*, 292–397.

Binford, L. R. (1988). *En Busca del Pasado*. Barcelona: Crítica.

Binford, L. R. (1989). Etude taphonomique des restes fauniques de la Grotte Vaufrey, Couche VIII. In J. Ph. Rigaud (Ed.) *La Grotte Vaufrey: Paléoenvironnement, chronologie, activités humaines* (pp. 535–564). Paris: Mémoires de la Société Préhistorique Française 19.

Binford, L. R., Mills, M. G. L., & Stone, N. M. (1988). Hyena scavenging behavior and its implications for interpretation of faunal assemblages from FLK22 (the Zinj Floor) at Olduvai Gorge. *Journal of Anthropological Archaeology, 7*, 99–135.

Blasco, R. (2006). Estrategias de subsistencia de los homínidos del nivel XII de la Cova del Bolomor (La Valldigna, Valencia). Ms. Thesis, Universitat Rovira i Virgili.

Blasco, R., & Rosell, J. (2009). Who was the first? An experimental application of carnivore and hominid overlapping marks at the Pleistocene archaeological sites. *C. R. Palevol, 8*, 579–592.

Blasco, R., Rosell, J. and Fernández Peris, J. (2010). Several different strategies for obtaining animal resources in the late Middle Pleistocene: The case of level XII at Bolomor Cave (Valencia, Spain). *Comptes Rendus Palevol, 9*, 171–184.

Blumenschine, R. J. (1986). Carcass consumption sequence and the archaeological distinction of scavenging and hunting. *Journal of Human Evolution, 15*, 639–659.

Blumenschine, R. J. (1988). An experimental model of the timing of hominid and carnivore influence on archaeological bone assemblage. *Journal of Archaeological Science, 15*, 483–502.

Blumenschine, R. J. (1989). A landscape taphonomic model of the scale of prehistoric scavenging opportunities. *Journal of Human Evolution, 18*, 345–371.

Blumenschine, R. J., & Selvaggio, M. M. (1988). Percussion marks on bone surface as a new diagnostic of hominid behaviour. *Nature, 333*, 763–765.

Blumenschine, R. J. & Marean, C. W. (1993). A carnivore's view of archeological bone assemblage. In J. Hudson (Ed.), *From bones to behavior. Ethnoarchaelogical and experimental contributions to the interpretation of faunal remains* (pp. 273–300). Center for Archaeological Investigations, Southern Illinois University, Illinois, Occasional paper 21.

Bökonyi, S. (1972). Zoological evidence for seasonal or permanent occupation of prehistoric settlements. In P. J. Ucko, R. E. Tringham, & G. W. Dimdleby (Eds.), *Man, Settlement and Urbanism* (pp. 121–126). London: Duckworth.

Bonfield, W., & Li, C. H. (1966). Deformation and fracture of bone. *Journal of Applied Physics, 37*(2), 869–875.

Bonnichsen, R. (1979). *Pleistocene bone technology in the Beringian Refugium*. Otawa, ON: National Museum of Man.

Boyle, K. V. (2000). Reconstructing Middle Palaeolithic subsistence strategies in the south of France. *International Journal of Osteoarchaeology, 10*, 336–356.

Brain, C. K. (1969). The contribution of Namib desert Hottentots to an understanding of australopithecine bone accumulations. *Scientific Paper Namib Desert Res Stn, 39*, 13–22.

Brain, C. K. (1981). *The hunters or the hunted? An introduction to African cave taphonomy*. Chicago: University of Chicago Press.

Brantingham, J. P. (1998). Hominid-carnivore coevolution and invasion of the predatory guild. *Journal of Anthropological Archaeology, 17*, 327–353.

Bravo, P. (2001). Estudio zooarqueológico y de remontajes del subnivel Ja del Abric Romaní (Capellades, Barcelona). Ms. Thesis, Universitat Rovira i Virgili.

Bromage, T. G., & Boyde, A. (1984). Microscopic criteria for the determination of directionality of cutmarks on bone. *American Journal of Phisical Anthropology, 65*, 339–366.

Brugal, J.-Ph. (1995). Archéologie et Zoologie pour un nouveau concept, la Paléoéthnologie humaine. *Préhistoire Anthropologie Méditerranéennes, 4*, 17–26.

Brugal, J.-Ph. (1999). Étude de populations de grands Bovidés européens: intéret pour la connaissance des comportements humains au Paléolithique. In J.-Ph. Brugal, F. David, J. G. Enloe, & J. Jaubert (Eds.) *Le Bison: Gibier et Moyen de Subsistance des Hommes du Paléolithique aux Paléoindiens des Grandes Plaines* (pp. 85–103). Toulouse: Antibes, Actes du Colloque International 1995.

Brugal, J.-Ph., & Jaubert, J. (1991). Les gisements paléontologiques pléistocènes à indices de fréquentation humaine: un nouveau type de comportement de prédation? *Paléo, 3*, 15–41.

Brugal, J.-Ph., Díez-Lomana, C., Huguet, R., Michel, P. & Rosell, J. (2006). Karstic cavities, natural bone accumulations and discrete human activities in the European Palaeolithic: Some case studies. In J. A. Haws, B. S. Hockett & J.-Ph. Brugal (Eds.,) *Paleolithic Zooarchaeology in Practice* (pp. 1–12). Oxford, England: BAR International Series 1564.

Buikstra, J., & Swegle, M. (1989). Bone modification due to burning: Experimental evidence. In R. Bonnichsen & M. Sorg (Eds.), *Bone modification* (pp. 247–258). Orono, ME: Institute for Quaternary Studies, University of Maine.

Bunn, H. T. (1981). Archaeological evidence for meat-eating by Plio-Pleistocene hominids from Koobi-Fora and Olduvai Gorge. *Nature, 291*, 574–580.

Bunn, H. T. (1983). Comparative analysis of modern bone assemblages from a San hunter- gatherer camp in the Kalahari desert, Botswana, and from spotted hyena den near Nairobi, Kenya. In J. Clutton-Brock & C. Grigson (Eds.), *Animals and archaeology* (Vol. 1). *Hunters and their prey* (pp. 143–148). Oxford, England: BAR International Series 163.

Bunn, H. T. (1993). Bone assemblages at base camps: A further consideration of carcass transport and bone destruction by Hadza. In J. Hudson (Ed.), *From bones to behavior. Ethnoarchaelogical and experimental contributions to the interpretation of faunal remains* (pp. 156–168). Center for Archaeological Investigations, Occasional paper 21, Southern Illinois University.

Bunn, H. T., Bartram, L. E., & Kroll, E. M. (1988). Variability in bone assemblage formation from Hadza hunting, scavenging and carcass processing. *Journal of Anthropological Archaeology, 7*, 412–457.

Bunn, H. T., & Ezzo, J. A. (1993). Hunting and scavenging by Plio-pleistocene hominids: Nutritional contraints, archaeological patterns, and behavioural implications. *Journal of Archaeological Science, 20*, 365–398.

Butzer, K. W. (1989). *Arqueología: una Ecología del Hombre*. Barcelona: Bellaterra.

Cáceres, I., Rosell, J., & Huguet, R. (1998). Séquence d'utilisation de la biomasse animale dans le gisement de l'Abric Romaní (Barcelone, Espagne). *Quaternaire, 9*(4), 379–383.

Cáceres, I., Bennàsar, M., Huguet, R., Rosell, J., Saladié, P., Allué, E., Solé, A., Blasco, R., Campeny, G., Esteban-Nadal, M., Fernández-Laso, C., Gabucio, M.J., Ibáñez, N., Martín, P., Muñoz, L. & Rodríguez-Hidalgo, A. (2012). Taphonomy of level J of Abric Romaní. In E. Carbonell (Ed.), *High resolution archaeology and neanderthal behavior: Time and space in level J of Abric Romaní (Capellades, Spain)*. Dordretch: Springer.

Capaldo, S. D. (1995). *Inferring hominid and carnivore behavior from dual patterned archaeofaunal assemblages*. Ph. D. Dissertation, New Brunswick, New Jersey.

Capaldo, S. D. (1997). Experimental determinations of carcass processing by Plio-Pleistocene hominids and carnivores at FLK 22 (*Zinjanthropus*), Olduvai Gorge, Tanzania. *Journal of Human Evolution, 33*, 555–597.

Capaldo, S. D. (1998). Methods, marks and models for inferring hominid and carnivore behavior. *Journal of Human Evolution, 35*, 317–320.

Capaldo, S. D., & Blumenschine, R. J. (1994). A quantitative diagnosis of notches made by hammerstone percussion and carnivore gnawing on bovid long bones. *American Antiquity, 59*(4), 724–748.

Carbonell, E. (1992). Abric Romaní, nivell H: un model d'estratègia ocupacional al Plistocè superior mediterrani. *Estrat*, 5.

Carbonell, E. (2002). *Abric Romaní Nivell I. Models d'ocupació de curta durada de fa 46.000 anys a la Cinglera del Capelló*. Tarragona: Capellades Universitat Rovira i Virgili Anoia, Barcelona.

Carbonell, E., Cebrià, A., Allué, E., Cáceres, I., Castro, Z., & Díaz, R. (1996). Behavioural and organizational complexity in the Middle Palaeolithic from Abric Romaní. In E. Carbonell & M. Vaquero (Eds.), *The last Neandertals, the first anatomically modern humans. A tale about the human diversity* (pp. 385–434). Tarragona: Universitat Rovira i Virgili.

Carbonell, E., Guilbaud, M. & Mora, R. (1983). Utilización de la lógica analítica para el estudio de tecno-complejos a cantos tallados. *Cahier Noir*, 13–64.

Carbonell, E., & Rosell, J. (2004). Ocupaciones de homínidos en el Pleistoceno de la Sierra de Atapuerca. In E. Baquedano & S. Rubio (Eds.), *Miscelania en Homenaje a Emiliano Aguirre* (pp. 102–115). Alcalá de Henares: Zona Arqueológica 4. Museo Arqueológico Regional de Madrid.

Carbonell, E., & Sala, R. (2002). *Planeta humà*. Barcelona: Empuries.

Carter, R. J. (1998). Reassessment of seasonality at the Early Mesolithic site of Star Carr, Yorkshire based on radiographs of mandibular tooth development in red deer (Cervus elaphus). *Journal of Archaeological Science, 25*, 851–856.

Carter, R. J. (2005). A method to estimate the ages at death of Red deer (*Cervus elaphus*) and Roe deer (*Capreolus capreolus*) from developing mandibular dentition and its application to Mesolithic NW Europe. In D. Ruscillo (Ed.), *Recent advances in ageing and sexing animal bones* (pp. 40–61). Oxford, England: Oxbow Books.

Castro-Curel, Z., & Carbonell, E. (1995). Wood pseudomorphs from Level I at Abric Romaní (Barcelona, Spain). *Journal of Field Archaeology, 22*, 376–384.

Chase, P. G. (1986). *The hunters of Combe Grenal: Approaches to Middle Palaeolithic subsistence in Europe*. Oxford, England: BAR International Series 1.

Chase, P. G. (1988). Scavenging and hunting in the Middle Palaeolithic. In H. L. Dibble & A. Montet-White (Eds.), *Upper Pleistocene prehistory of Western Eurasia* (pp. 225–232). Philadelphia, PA: University of Pennsylvania.

Chase, P. G. (1989). How different was Middle Palaeolithic subsistence? A zooarchaeological perspective on the Middle to Upper Palaeolithic transition. In P. Mellars & C. Stringer (Eds.), *The human revolution: Behavioral and biological perspectives on the origins of modern humans* (pp. 321–327). Edinburgh: Edinburgh University Press.

Correia, P. (1997). Fire modification of bone: A review of the literature. In W. Haglund & M. Sorg (Eds.), *Forensic taphonomy the postmortem fate of human remains* (pp. 275–293). Boca Raton: CRC Press.

Costamagno, S. (1999). Taphonomie d'un aven-piège: Contribution des accumulations d'origine naturelle à l'interprétation des archéofaunes du Paléolithique Moyen. *Anthropozoologica, 29*, 13–32.

Costamagno, S., Théry-Parisot, I., Brugal, J.-Ph., & Guilbert, R. (2005). Taphonomic consequences of the use of bones as fuel. Experimental data and archaeological applications. In T. O'Connor (Ed.), *Biosphere to Lithosphere, new studies in vertebrate taphonomy* (*Proceedings of 9th conference on ICAZ, Durham, August* 02) (pp. 51–62). Durham, NC: Oxbow Books.

Costamagno, S., Théry-Parisot, I., Castel, J.-C., & Brugal, J.-Ph. (2009). Combustible ou non? Analyse multifactorielle et modèles explicatifs sur des ossements brûlés paléolithiques. In I. Théry-Parisot, S. Costamagno, & A. Henry (Eds.), *Gestion des combustibles au Paléolithique et au Mésolithique: nouveaux outils, nouvelles interprétations, Proceedings of worshop 21, XV congress UISPP, Lisboa* (pp. 65–84). Oxford, England: BAR International Series.

Cruz-Uribe, K. (1991). Distinguishing hyena from hominid bone accumulations. *Journal of Field Archaeology, 18*, 467–486.

De Graff, G. (1961). Gross effects of a primitive hearth. *South African Archaeological Bulletin, 16*, 25–26.

d'Errico, F. (2003). The invisible frontier. *A multiple species model for the origin of behavioral modernity. Evolutionary Anthropology, 12*, 188–202.

Díez, J. C. (1992). *Zooarqueología de Atapuerca (Burgos) e implicaciones paleoeconómicas del estudio tafonómico de yacimientos del Pleistoceno Medio*. Ph. D. Dissertation, Universidad Complutense.

Díez, J. C., Fernández-Jalvo, Y., Rosell, J., & Cáceres, I. (1999). Zooarchaeology and taphonomy of Aurora Stratum (Gran Dolina, Sierra de Atapuerca, Spain). *Journal of Human Evolution, 37*(3–4), 623–652.

Díez, J. C., & Rosell, J. (1998). Estrategias de subsistencia de los homínidos de la Sierra de Atapuerca. In E. Aguirre (Ed.), *Atapuerca y la Evolución Humana* (pp. 361–390). Madrid: Fundación Ramón Areces.

Domínguez-Rodrigo, M., & Pickering, T. R. (2003). Early hominid hunting and scavenging: A zooarcheological review. *Evolutionary Anthropology, 12*, 275–282.

Domínguez-Rodrigo, M., & Barba, R. (2006). New estimates of tooth mark and percussion mark frequencies at the FLK Zinj site: The carnivore-hominid-carnivore hypothesis falsified. *Journal of Human Evolution, 50*, 170–194.

Emerson, A. M. (1993). The role of body part utility in small-scale hunting under two strategies of carcass recovery. In J. Hudson (Ed.), *From bones to behavior. Ethnoarchaelogical and experimental contributions to the interpretation of faunal remains* (pp. 138–155). Carbondale, IL: Center for Archaeological Investigations, Southern Illinois University, Occasional paper 21.

Farizy, D., David, F., & Jaubert, J. (1994). *Hommes et Bisons du Paléolithique Moyen à Mauran (Haute-Garonne)*. XXXe suppl, Gallia Préhistoire. Paris: CNRS.

Fernández-Jalvo, Y., Díez, J. C., Cáceres, I., & Rosell, J. (1999). Human cannibalism in the Early Pleistocene of Europe (Gran Dolina, Sierra de Atapuerca, Burgos, Spain). *Journal of Human Evolution, 37*(3–4), 591–622.

Fernandez, P., Guadelli, J.-L., & Phose, Ph. (2006). Applying dynamics and comparing life tables for Pleistocene Equidae in anthropic (Bau de l'Aubesier, Combe-Grenal) and carnivore (Fouvent) contexts with modern feral horse populations (Akagera, Pryor Mountain). *Journal of Archaeological Science, 33*, 176–184.

Fiore, I., Gala, M., & Tagliacozzo, A. (2004). Ecology and subsistence strategies in the eastern italian Alps during the Middle Palaeolithic. *International Journal of Osteoarchaeology, 14*, 273–286.

Fischer, J. W. (1995). Bone surface modifications in Zooarchaeology. *Journal of Archaeological Method and Theory, 2*(1), 7–68.

Gamble, C. (1990). *El Poblamiento Paleolítico de Europa*. Barcelona: Crítica.

Gaudzinski, S. (1995). Wallertheim revisited: A re-analysis of the fauna from the Middle Palaeolithic site of Wallertheim (Rheinhessen/Germany). *Journal of Archaeological Science, 22*, 51–66.

Gaudzinski, S., & Roebroeks, W. (2000). Adults only. Reindeer hunting at the Middle Palaeolithic site Salzgitter Lebenstedt, Northern Germany. *Journal of Human Evolution, 38*, 497–521.

Gifford-Gonzalez, D. (1989). Ethnographic analogues for interpreting modified bones: Some cases from East Africa. In R. Bonnichsen & M. Sorg (Eds.), *Bone Modification* (pp. 179–246). Orono, ME: Institute for Quaternary Studies, University of Maine.

Gifford-Gonzalez, D. (1993). Gaps in the zooarchaeological analyses of butchery: is gender an issue? In: Hudson J. (Ed.) *From Bones to Behavior Ethnoarchaelogical and Experimental Contributions to the Interpretation of Faunal Remains* (pp. 181–199). Carbondale: Center for Archaeological Investigations, Southern Illinois University.

Gilchrist, M., & Mytum, H. (1986). Experimental archaeology and burnt animal bone from archaeological sites. *Circaea, 4*, 29–38.

Grigson, C. (1982). Sex and determination os some bones and teeth of domestic cattle: A review of the literature. In B. Wilson, C. Grigson, & S. Payne (Eds.), *Ageing and sexing animal bones archaeological sites* (pp. 7–23). Oxford, England: BAR International Series 109.

Guadelli, J. L. (1998). Détermination de l'age des chevaux fossiles et établissement des classes d'age. *Paleo, 10*, 87–93.

Hawkes, K., O'connell, J. F., & Blurton Jones, N. G. (2001). Hunting and nuclear families. Some lessons from the Hadza about men's work. *Current Anthropology, 42*(5), 681–709.

Henry, D. O. (1995). *Prehistoric cultural ecology and evolution: Insights from Southern Jordan*. New York: Plenum Press.

Hoffecker, J. F., & Cleghorn, N. (2000). Mousterian hunting patterns in the Northwestern Caucasus and the ecology of the Neanderthals. *International Journal of Osteoarchaeology, 10*, 368–378.

Hoffmann, J. L. (1992). Putting the pieces together: An introduction to refitting. In J. L. Hoffmann & J. G. Enloe (Eds.), *Putting the pieces together: An introduction to refitting* (pp. 1–20). Oxford: BAR International Series 578.

Huguet, R. (1997). Estudi Zooarqueològic de la Unitat GII del Complex Galería (Sierra de Atapuerca, Burgos). Ms. Thesis, Universitat Rovira i Virgili.

Huguet, R. (1998). Étude archéologique de l'unité GII du complexe de Galería (Sierra de Atapuerca, Burgos, Espagne). In J.-Ph. Brugal, L. Meignen, & M. Patou-Mathis (Eds.), *Économie Préhistorique: les comportements de subsistance au Paléolithique* (pp. 163–172). Sophia Antipolis: APDCA.

Huguet, R. (2007). Primeras ocupaciones humanas en la Península Ibérica: paleoeconomía en la Sierra de Atapuerca (Burgos) y en la Cuenca de Guadix-Baza (Granada) durante el Pleistoceno Inferior. Ph. D. Dissertation. Universitat Rovira i Virgili.

Huguet, R., Cáceres, I., Díez, J. C., & Rosell, J. (1999). Estudio tafonómico y zooarqueológico de los restos óseos de macromamíferos de la unidad G-II de Galería. In E. Carbonell, A. Rosas, & J. C. Díez (Eds.), *Atapuerca: Ocupaciones humanas y paleoecología del yacimiento de Galería* (pp. 245–264). Zamora: Arqueología en Castilla y León.

Isaac, G. L. (1983). Bones in contention: Competing explanations for the juxtaposition of early pleistocene artifacts and faunal remains. In J. Clutton-Brock & C. Grigson (Eds.), *Animals and archaeology, vol. 1. Hunters and their prey* (pp. 3–17). Oxford, England: BAR International Series 163.

Jaubert, J., & Brugal, J.-Ph. (1990). Contribution a l'étude du mode de vie au Paléolithique ancien et moyen: les chasseurs d'aurochs de la Borde. In J. Jaubert, M. Lorblanchet, H. Laville, R. Slott-Muller, A. Turq, & J.-Ph Brugal (Eds.), *Les Chasseurs d'Aurochs de la Borde: un Site du Paléolithique Moyen (Livernon, Lot)* (pp. 127–145). Parı́s: Documents d'Archéologie Française 27.

Jaubert, J., Kervazo, B., Bahain, J. J., Brugal, J.-Ph., Chalard, Ch., Falguères, Ch., et al. (2005). Coudoulous I (Tour-de-Faure, Lot), site du Pléistocène moyen en Quercy: bilan pluridisciplinaire. In N. Molines, M.-H. Moncel, & J.-L. Monnier (Eds.), *Données récentes sur les modalités de peuplement et sur le cadre chronostratigraphique, géologique et paléogéographique des industries du Paléolithique inférieur et moyen en Europe. Actes du Colloque International de Rennes* (pp. 227–251). Oxford, England: BAR International Series.

Johnson, W. G., Sharpe, S. E., Bullard, T. F., & Lupo, K. (2005). Characterizing a first occurrence of bison deposits in Southeastern Nevada. *Western North American Naturalist, 65*, 24–35.

Klein, R. G. (1976). The mammalian fauna of the Klasies River Mouth sites, southern Cape Province, South Africa. *South African Archaeological Bulletin, 31*, 75–98.

Klein, R. G. (1989). Why does skeletal part representation differ between smaller and larges bovids at Klasies RIver Mouth and other archaeological sites? *Journal of Archaeological Science, 26*, 1225–1234.

Klein, R. G., Cruz-Uribe, K., & Milo, R. G. (1999a). Skeletal part representation in Archaeofaunas: Comments on "Explaining the 'Klasies pattern': kua ethnoarchaeology, the Die Kelders Middle Stone Age archaeofauna, long bone fragmentation and carnivore ravaging" by Bartram & Marean. *Journal of Archaeological Science, 26*, 1225–1234.

Klein, R. G., Cruz-Uribe, K., Halkett, D., Hart, T., & Parkington, J. E. (1999b). Paleoenvironmental and human behavioral implications of the Boegoeberg 1 Late Pleistocene hyena den, Northern Cape Province, South Africa. *Quaternary Research, 52*, 393–403.

Krönneck, P., Niven, L., & Uerpmann, H. P. (2004). Middle Palaeolithic subsistence in the LoneValley (Swabian Alb, Southern Germany). *International Journal of Osteoarchaeology, 14*, 212–224.

Lam, Y. M., Chen, X., & Pearson, O. M. (1999). Intertaxonomic variability in patterns of bones density and the differential representation of bovid, cervid, equid elements en the archaeological recors. *American Antiquity, 64*, 343–362.

Lemonnier, P. (1982). La description des châines opératoires: Contribution a l'étude des systèmes techniques. *Techniques et cultures, 1*, 100–151.

Leroi-Gourhan, A. (1964). *Le Geste et la Parole. Albin Michel*. Paris: Technique et Langage.

Levine, M. A. (1982). The use of crown height measurements and eruption-wear sequences to age horse teeth. In B. Wilson, C. Grigson, & S. Payne (Eds.), *Ageing and sexing animal bones from archaeological sites* (pp. 223–250). Oxford, England: BAR International Series 109.

Lupo, K. D. (2001). Archeological Skeletal part profiles and differential transport: An ethnoarchaeological example from Hadza bone assemblages. *Journal of Anthropological Archaeology, 20*, 361–378.

Lyman, R. L. (1984). Bone density and differential survivorship of fossil classes. *Journal of Anthropological Archaeology, 3*, 259–299.

Lyman, R. L. (1985). Bone frequencies: Differential survivorship of fossil classes. *Journal of Archaeological Science, 12*, 221–236.

Lyman, R. L. (1994). *Vertebrate taphonomy*. New York: Cambridge manuals in archaeology.

Madsen, D. B., & Schmitt, D. N. (1989). Mass collecting and the Diet Breadth model: A Great Basin example. *Journal of Archaeological Science, 25*, 445–455.

Maguire, J. M., Pemberton, D., & Collett, M. H. (1980). The Makapansgat limeworks grey breccia: hominids, hyaenas, hystricids or hillwash. *Paleontologia Africana, 23*, 75–98.

Marean, C. W., & Bertino, L. (1994). Intrasite spatial analysis of bone: Subtracting the effect of secondary carnivore consumers. *American Antiquity, 59*(4), 748–768.

Marean, C. W., & Kim, S. Y. (1998). Mousterian large-mammal remains from Kobeh Cave. Behavioral implications for Neanderthals and Early Modern Humans. *Current Anthropology, 39*, 79–113.

Marean, C. W., & Spencer, L. (1991). Impact of carnivore ravaging on zooarchaeological measures of element abundance. *American Antiquity, 56*, 645–658.

Marean, C. W., Spencer, L. M., Blumenschine, R. J., & Capaldo, S. D. (1992). Captive hyena bone choice and destruction, the schlepp effect, and Olduvai Gorge archaeofaunas. *Journal of Archaeological Science, 18*, 101–121.

Mariezkurrena, K., & Altuna, J. (1983a). Contribución al conocimiento del desarrollo de la dentición y el esqueleto postcraneal de *Cervus elaphus*. *Munibe, 35*, 149–202.

Mariezkurrena, K., & Altuna, J. (1983b). Biometría y dimorfismo sexual en el esqueleto de *Cervus elaphus* würmiense, postwürmiense y actual del Cantabrico. *Munibe, 35*, 203–246.

Martínez-Moreno, J. (2005). Una aproximación zooarqueológica al estudio de los patrones de subsistencia del Paleolítico Medio Cantábrico. In R. Montes & J. A. Lasheras (Eds.), *Neandertales cantábricos, estado de la cuestión. Museo de Altamira. Monografías, 20* (pp. 209–230). Santander: Museo de Altamira.

Monchot, H. (1999). La caza del muflón (*Ovis antiqua* Pommerol, 1879) en el Pleistoceno Medio de los Pirineos: el ejemplo de la Cueva de l'Aragó (Tautavel, Francia). *Revista Española de Paleontología, 14*(1), 67–78.

Moncel, M. H., Brugal, J.-Ph., Prucca, A., & Lhomme, G. (2008). Mixed occupation during the Middle Palaeolithic: Case study of a small pit-cave-site of Les Pêcheurs (Ardèche, south-eastern France). *Journal of Anthropological Archaeology, 27*, 382–398.

Morales Pérez, J. V., Brugal, J.-Ph., Pérez Ripoll, M., Galván Santos, B., & Hernández, C. (2008). La fracturación y fragmentación de huesos largos durante el paleolítico. una aproximación teórica y metodológica. In S. Rovira Llorens, M. García-Heras, M. Gener Moret, & I. Montero Ruiz (Eds.), *Actas del VII Congresso Ibérico de Arqueometria, 8–10 de octubre 2007* (pp. 64–76). Madrid: Biomateriales.

Nicholson, R. A. (1993). A morphological investigation of burnt animal bone and an evaluation of its utility in Archaeology. *Journal of Archaeological Science, 20*, 411–428.

Noe-Nygaard, N. (1989). Man-made trace fossils on bones. *Human Evolution, 4*(6), 461–491.

Nowak, R. M. (1999). *Walker's Mammals of the World*. Baltimore, MD: Johns Hopkins University Press.

O'Connell, J. F. (1987). Alyawara site structure and its archaeological implications. *American Antiquity, 52*(1), 74–108.

O'Connell, J. F., Hawkes, K., & Blurton-Jones, N. (1988a). Hadza scavenging: Implications for Plio/Pleistocene hominid subsistence. *Current Anthropology, 29*(2), 356–363.

O'Connell, J. F., Hawkes, K., & Blurton Jones, N. (1988b). Hadza hunting, butchering, and bone transport and their archaeological implications. *Journal of Anthropological Research, 44*, 113–161.

O'Connell, J. F., Hawkes, K., & Blurton Jones, N. (1990). Reanalysis of large mammal body part transport among the Hadza. *Journal of Archaeological Science, 17*, 301–316.

O'Connell, J. F., Hawkes, K., & Blurton Jones, N. (1992). Patterns in the distribution, site structure and assemblage composition of Hadza kill-butchering sites. *Journal of Archaeological Science, 19*, 319–345.

Oliver, J. (1993). Carcass processing by the Hadza: Bone breakage from butchery to consumption. In J. Hudson (Ed.), *From bones to behavior. Ethnoarchaelogical and experimental contributions to the interpretation offaunal remains* (pp. 200–227). Carbondale, IL: Center for Archaeological Investigations, Southern Illinois University, Illinois, Occasional paper 21.

Patou-Mathis, M. (2000). Neanderthal subsistence behaviours in Europe. *International Journal of Osteoarchaeology, 10*, 379–395.

Patou-Mathis, M. (2006). Comportements de subsistence des Néandertaliens d'Europe. In B. Demansin & M. Otte (Eds.), *Neanderthals in Europe, Proceedings of the International Conference Held in the Gallo-Roman Museum in Tougereu (September 17–19th 2004)* (pp. 67–76). Liège: ERAUL 117.

Perkins, D., & Daly, P. (1968). A hunter's village in Neolithic Turkey. *Scientific American, 219*(5), 96–106.

Pickering, T. R. (2002). Reconsideration of criteria for differentiating faunal assemblages accumulated by hienas and hominids. *International Journal of Osteoarchaeology, 12*, 127–141.

Pickering, T. R., & Egeland, C. P. (2006). Experimental patterns of hammerstone percussion damage on bones: implications for inferences of carcass processing by humans. *Journal of Archaeological Science, 33*, 459–469.

Potts, R., & Shipman, P. (1981). Cut-marks made by stone tools on bones from Olduvai Gorge, Tanzania. *Nature, 291*, 577–580.

Rando, J. M. (1999). Análisis espacial del registro arqueológico del subnivel Ja. Abric Romaní (Capellades, Barcelona). Ms. Thesis, Universitat Rovira i Virgili.

Richardson, P. R. K. (1980). Carnivore damage to antelope bones and its archaeological implications. *Paleontologia Africana, 23*, 100–125.

Riglet, P. H. (1977). *Contribution à l'étude de l'âge du Cerf élaphe (Cervus elaphus L.)*. Ph. D. Dissertation, École Nationale Vétérinaire d'Alfort.

Rivals, F., Testu, A., Moigne, A. M., & de Lumley, H. (2006). The Middle Pleistocene argali (Ovis ammon antiqua) assemblages at the Caune de l'Arago (Tautavel, Pyrénés-Orientales, France): Were prehistoric hunters or carnivores responsible for their accumulation? *International Journal of Osteoarchaeology, 16*, 249–268.

Rosell, J. (1993). Impacte biologic a la base de "Gran Dolina" (Sierra de Atapuerca). Ms. Thesis, Universitat Rovira i Virgili.

Rosell, J. (1998). Les premières occupations humaines à la Sierra de Atapuerca (Burgos, Espagne). Les niveaux TDW-4 et TDW-4b. In J.-Ph. Brugal, L. Meignen, & M. Patou-Mathis (Eds.), *Économie Préhistorique: les comportements de subsistance au Paléolithique* (pp. 153–162). Sophia Antipolis: APDCA.

Rosell, J. (2001). *Patrons d'aprofitament de les biomases animals durant el Pleistocè Inferior i Mig (Sierra de Atapuerca, Burgos) i Superior (Abric Romaní, Barcelona)*. Ph. D. Dissertation, Universitat Rovira i Virgili.

Rosell, J., & Blasco, R. (2009) Home sharing: carnivores in anthropogenic assemblages of the Middle Pleistocene. *Journal of Taphonomy, 7* (4), 305-324.

Sañudo, P., Vallverdú, J. & Canals, A. (2012). Spatial patterns in level J. In E. Carbonell (Ed.), *High resolution archaeology and neanderthal behavior: Time and space in level J of Abric Romaní (Capellades, Spain)*. Dordretch: Springer.

Schiffer, M. B. (1983). Toward the identification of formation processes. *American Antiquity, 48*, 675–703.

Selvaggio, M. M. (1998). Evidence for a three-stage sequence of hominid and carnivore involvement with long bones at FLK *Zinjanthropus*, Olduvai Gorge. *Journal of Archaeological Science, 25*, 191–202.

Sergant, J., Crombe, P., & Perdaen, Y. (2006). The 'invisible' hearths: A contribution to the discernment of Mesolithic non-structured surface hearths. *Journal of Archaeological Science, 33*, 999–1007.

Shipman, P. (1981). *Life History of a Fossil. An introducction to taphonomy and paleoecology*. Cambridge, MA: Harvard University Press.

Shipman, P., & Rose, J. (1983). Early hominid hunting, butchering and carcass-processing behaviors: Approches to the fossil record. *Journal of Anthropological Archaeology, 2*, 57–98.

Shipman, P., Foster, G., & Schoeninger, M. (1984). Burnt bones and teeth: An experimental study of color, morphology, crystal structure and shrinkage. *Journal of Archaeological Science, 11*, 307–325.

Silver, I. A. (1969). La determinación de la edad en los animales domésticos. In D. Brothwell & E. Higgs (Eds.), *Ciencia en Arqueología* (pp. 229–239). México: Fondo de Cultura Económica.

Smuts, G. L. (1974). Age determination in burchell's zebra (*Equus burchelli antiquorum*) from the Kruger National Park. *Journal of the South African Wildlife Management Association, 4*, 103–115.

Spenneman, D., & Colley, S. (1989). Fire in a pit: The effects of burning on faunal remains. *Archaeozoologia, 3*, 51–64.

Steele, D. G., & Baker, B. W. (1993). Multiple predation: A definitive human hunting strategy. In J. Hudson (Ed.), *From bones to behavior. Ethnoarchaelogical and experimental contributions to the interpretation of faunal remains* (pp. 9–37). Carbondale, IL: Center for Archaeological Investigations, Southern Illinois University, Occasional paper 21.

Stiner, M. C. (1994). *Honor among thieves: A zooarchaeological study of Neandertal ecology*. Princenton, NJ: Princenton University Press.

Stiner, M. C., Kuhn, S. L., Weiner, S., & Bar-Yosef, O. (1995). Differential burning, recrystallization, and fragmentation of archaeological bones. *Journal of Archaeological Science, 22*, 223–237.

Stiner, M. C., Barkai, R., & Gopher, A. (2009). Cooperative hunting and meat sharing 400–200 kya at Qesem Cave, Israel. *Proceedings of the National Academy of Sciences, 106*(32), 13207–13212.

Texier, P.-J., Brugal, J.-Ph., Lemorini, C., & Wilson, L. (1998). Fonction d'un site du Paléolithique moyen en marge d'un territoire: l'Abri de la Combette (Bonnieux, Vaucluse). In J.-Ph. Brugal, L. Meignen, & M. Patou-Mathis (Eds.), *Économie Préhistorique: les comportements de subsistance au Paléolithique* (pp. 325–348). Sophia Antipolis: Éditions APDCA.

Théry-Parisot, I., Costamagno, S., Brugal, J.-Ph., Fosse, P., & Guilbert, R. (2005). The use of bone as fuel during the Palaeolithic, experimental study of bone combustible properties. In J. Mulville & A. K. Outram (Eds.), *The zooarchaeology of fats, oils, milk and dayring (Proceedings of the 9th ICAZ conference, Durham, August 02*, (pp. 50–59). Durham, NC: Oxbow Books.

Thieme, H. (1997). Lower Palaeolithic hunting spears from Schöningen, Germany. *Nature, 358*, 807–810.

Thieme, H. (1998). Las lanzas más antiguas del mundo: armas de caza del Paleolítico inferior de Schöningen, Alemania. In E. Carbonell, J. M. Bermúdez de Castro, J. L. Arsuaga, & X. P. Rodríguez (Eds.), *Los primeros pobladores de Europa: últimos descubrimientos y debate actual* (pp. 169–193). Burgos: Aldecoa.

Thurman, M., & Willmore, L. (1980). A replicative cremation experiment. *North American Archaeologist, 2*, 275–283.

Valensi, P. (1996). Taphonomie des grands mammifères et palethnologie à la Grotte du Lazaret (Nice, France). *Anthropozoologica, 23*, 13–28.

Valensi, P. (2000). The Archaeozoology of Lazaret Cave. *International Journal of Osteoarchaeology, 10*, 357–367.

Vallverdú, J., Allué, E., Bischoff, J. L., Cáceres, I., Carbonell, E., Cebrià, A., et al. (2005). Short human occupations in the Middle Palaeolithic level i of the Abric Romaní rock-shelter (Capellades, Barcelona, Spain). *Journal of Human Evolution, 48*, 157–174.

Vallverdú J. & Courty, M.-A. (2012). Microstratigraphic Analysis of the Level J Deposits. A dual paleoenvironmental-paleoethnographic contribution to the paleolithic archaeology at the Abric Romani site. In E. Carbonell (Ed.), *High resolution archaeology and neanderthal behavior: Time and space in level J of Abric Romaní (Capellades, Spain)*. Dordretch:Springer.

Vaquero, M. (1997). *Tecnología lítica y comportamiento humano: organización de las actividades técnicas y cambio diacrónico en el Paleolítico Medio del Abric Romaní (Capellades, Barcelona)*. Ph. D. Dissertation. Universitat Rovira i Virgili.

Vaquero, M., & Pastó, I. (2001). The definition of spatial units in Middle Palaeolithic sites: The hearth-related assemblages. *Journal of Archaeological Science, 28*, 1209–1220.

Vaquero, M., Chacón, G., Cuartero, F., García-Antón, M.D., Gómez de Soler, B. & Martínez, K. (2012). The Lithic Assemblage of Level J. In E. Carbonell (Ed.), *High resolution archaeology and neanderthal behavior: Time and space in level J of Abric Romaní (Capellades, Spain)*. Dordretch: Springer.

Villa, P., & Mahieu, E. (1991). Breakage patterns of human long bones. *Journal of Human Evolution, 21*, 27–48.

Villa, P., Castel, J.-C., Beauval, C., Bourdillat, V., & Goldberg, P. (2004). Human and carnivore sites in the European Middle and Upper Paleolithic: Similarities and differences in bone modification and fragmentation. In J.-Ph. Brugal & Ph. Fosse (Eds.), *Hommes et Carnivores au Paléolithique—Hommage à Philippe Morel* (Vol. 23(2), pp. 705–730). Genev: Revue de Paléobiologie.

Wang, X., & Martín, L. D. (1993). Natural trap cave. *Research & Exploration, 9*(4), 422–435.

White, T. D. (1992). *Prehistoric cannibalism at Mancos*. Princenton, NJ: Princeton University Press.

Yellen, J. E. (1991). Small mammals: Post-discard patterning of Kung San faunal remains. *Journal of Anthropological Archaeology, 10*, 152–192.

Chapter 9
Hearth Functioning and Forest Resource Exploitation Based on the Archeobotanical Assemblage from Level J

Ethel Allué, Dan Cabanes, Alex Solé and Robert Sala

Abstract The aim of this chapter is to present the archeobotanical record related to the hearth structures from level J from Abric Romaní. For this study we include charcoal, phytoliths and wood imprints that from different methodological approaches provide data on hearth functioning and firewood management. The results yielded by these archeobotanical assemblages point out that wood was the basic fuel used by Neanderthals for maintaining hearths and pine wood was the most appreciated. Other fuels such as grasses are also recorded and might have specific uses during the combustion processes. Finally wood imprints, even if in level J are scarce, point out the use of wood for other purposes than firewood. This study points out the complexity of Neanderthal behavior in relation to the management of fire involving fuel management and hearth functioning.

Keywords Anthracology • Phytoliths • Hearths • Wood artifacts

E. Allué (✉) · A. Solé · R. Sala
Institut Català de Paleoecologia Humana i Evolució Social (IPHES), Universitat Rovira i Virgili (URV), C/Escorxador s/n, 43003 Tarragona, Spain
e-mail: eallue@iphes.cat

A. Solé
e-mail: asole@prehistoria.urv.es

R. Sala
e-mail: robert.sala@urv.cat

D. Cabanes
Department of Structural Biology, Kimmel Center for Archaeological Science, Weizmann Institute of Science, 76100 Rehovot, Israel
e-mail: dan.cabanes@weizmann.ac.il

Introduction

The ability to light a fire, maintain it and use it as a tool is one of the greatest achievements in human history. After the development of fire technology, most prehistoric activities were carried out around hearths. Fire provided enough warmth to allow colder environments to be colonized, and its thermal energy could be exploited to cook raw food. Fire was used as a defensive or offensive weapon, for keeping away predators, or for rounding up large mammals by means of intentional wildfires for hunting purposes. Fire also helped to improve technical advances such as preparing chert for knapping and hardening the wood of spears (Oakley 1955, 1961; Perlès 1977). Fire also has a symbolic (social or cultural) value, which gave rise to a new focus for the social organization of Neanderthals (Pettitt 1997; Vaquero et al. 2004). Furthermore, studying fire remains is a way of studying the previsional cognitive capacities among the first populations to develop this technology, because the amount of fuel needed to keep a fire burning would have had to be forecasted beforehand (Heizer 1963).

The sedimentary features of the Abric Romaní infilling (Vallverdú and Courty 2012) allow us to clearly distinguish burned sediment produced during the use of fireplaces. Previous studies on this subject have shown that the activities carried out at the site were structured around hearths (Carbonell et al. 1996, 2007; Pastó et al. 2000; Vaquero and Pastó 2001; Vaquero et al. 2004). Most of these studies are based on the typological classification of the hearth established during fieldwork and the distribution of other archeological artifacts. However, in recent years new studies have been conducted based on charcoal and phytolith analyses, soil micromorphology, and travertine wood imprints (Carbonell et al. 1996; Arteaga et al. 2001; Allué 2002; Vallverdú 2002; Cabanes et al. 2007; Solé 2007). This chapter presents the results of a multidisciplinary approach to the study of Middle Paleolithic fireplaces based on the archeobotanical record of level J.

E. Carbonell i Roura (ed.), *High Resolution Archaeology and Neanderthal Behavior: Time and Space in Level J of Abric Romaní (Capellades, Spain)*, Vertebrate Paleobiology and Paleoanthropology, DOI: 10.1007/978-94-007-3922-2_9, © Springer Science+Business Media B.V. 2012

Silica phytolith analysis is a tool used to investigate the use of vegetation resources by our ancestors (Karkanas et al. 2002; Madella et al. 2002) and to reconstruct the variability of past climates (Rovner 1971; Albert et al. 2006). The physical and chemical properties of phytoliths allow them to remain preserved in a variety of environments and over different periods of time; in addition, the accumulation of silica in plant cells allows these cells to be identified on various taxonomic scales (Piperno 2006). Phytoliths pass from the living plant to the sediment when the plant is burned (they can survive at temperatures of up to 900°C) or when the plant dies and decays. Once phytoliths are in the sediment or soil, although their movement is imperceptible, they still form part of the terrestrial silica cycle (Conley 2002; Meunier et al. 2001). The identification and quantification of phytoliths in conjunction with mineralogical analysis through the Fourier Transformed Infrared Spectra (FTIR) has been used to analyze the hearths at several prehistoric sites (Albert et al. 1999, 2003; Albert and Weiner 2001; Cabanes et al. 2007; Karkanas et al. 2002; Schiegl et al. 1996, 2003, 2004). Both methods help us to understand the way in which these hearths were formed and how they were weathered by diagenetic processes (Albert et al. 1999; Weiner et al. 2002). Understanding the formation processes of the hearths allows us to make inferences about the behavior and organization of past hunter-gatherer societies.

Charcoal analysis or anthracology is an archeobotanical discipline based on the taxonomic identification of charred wood recovered from archeological or natural deposits. The aim of this discipline is to determine past vegetation formations and their evolution over time. It is also a source of data on human behavior related to the exploitation of forest resources. The value of charcoal analysis as a tool for paleoecological reconstruction has already been demonstrated by numerous authors (Vernet 1997; Figueiral and Mosbrugger 2000). Interpretations are made based on the ecological characterization of the species depending on climatic conditions and their diachronic evolution. Charcoal analysis is therefore a tool that leads to a better understanding of human behavior (Asouti and Austin 2005; Allué and García-Antón 2006). This approach is based on the consideration of firewood as an *artifact* conditioned by human choices. Both the paleoecology and human behavior aspects of the discipline should be considered in the final interpretation and depend on a meticulous and accurate sampling method. In this chapter, our analyses focus on the study of fuel, whereas the paleoecological interpretation will be developed in Allué et al. (2012). The presence of fossil wood associated with hearth structures has been exceptionally recorded in Middle Paleolithic sequences (Clark 2001; Carbonell and Castro-Curel 1992). Generally, these materials are recovered as charred and partially burned wood but also as unburned remains (Carbonell 2002). The study of these materials permits the study of firewood management and wood technology.

Materials and Methods

Wood Imprints

The Abric Romaní has yielded fossil wood remains recovered as empty cavities in the shape and the negative mold of the wood surface. These remains are empty travertine imprints of completely decayed wood (Castro-Curel and Carbonell 1995; Carbonell 2002). The sedimentary environment of Abric Romaní favors travertine formation through the precipitation of calcium carbonate onto plant structures, such as mosses, weeds, wood (trees or dead wood), leaves and pinecones. This process of travertine formation is related to the hydric activation of the rockshelter. Wood remains are preserved in anaerobic conditions, although the decay process is initiated by saprophytic organisms, bacteria and fungus (Castro-Curel and Carbonell 1995). As the wood decays, the exterior of the wood is covered in calcium carbonate. The wood totally decays leaving only an imprint in the travertine when the water process stops. These empty spaces left by the wood are called *pseudomorphs* (Castro-Curel and Carbonell 1995).

The wood imprints at Abric Romaní, as negative evidence of wood deposited in the rockshelter, were partially or totally molded during field recovery in order to obtain useful samples for study. Although this field methodology has now been improved (Solé 2007), in this chapter we describe the technique used during the excavation of level J. This technique was applied only to the largest wood imprints found in sublevel Ja, at square O47.

To obtain an accurate plastic reproduction of the wood imprints, the first step was to remove loose travertine fragment particles and microcharcoal fragments from the base of the concavity. Next, to consolidate the surfaces, the cavities were impregnated with cellulose nitrate diluted in acetone (30%) dropped from a syringe. Once the solvent had evaporated, liquid soap was poured into the hollows to facilitate the extraction of the casts. A contact coat of thinned plaster of Paris was then meticulously applied with a soft brush over the surface and, after drying, was reinforced with a wooden batten. The cavity was finally filled with a coat of thick plaster. Although easy retrieval of the hardened mold was feasible, the surrounding ground was cut into small fragments and carefully removed to allow us to successfully lift out the casts.

The wood imprints were studied using several different approaches. Spatial distribution is defined as the consideration of different depositional environments and the localization and intensity of occupations determined primarily by a focus on hearth assemblages. Our morphological study was based on the general shape and measurements of the imprints, which was primarily used to establish which part of the tree the imprint represents and the disarticulation rate of the wood. The main features we recorded when possible were length, width, thickness, cross-section morphology, outline, end shape, and the presence and shape of ramifications. The main objective of the surface examination was to detect thermal and mechanic alterations on the surface at the microscopic and macroscopic level. Modifications were determined by means of analogy with a reference collection. The micromorphological examination of the samples was conducted using a Scanning Electron Microscope (Jeol JSM 6400) at magnifications of between 0 and 200X, although in some cases it was necessary to increase magnification up to 1000X and beyond in order to study the mineralogical structure of the imprints.

Charcoal Analyses

The charcoal remains at Abric Romaní, although very well preserved, are nonetheless fragile. Water sieving could have damaged the most fragile fragments and thus prevented their taphonomic study (Allué 2002). Therefore, during the excavation each charcoal fragment of approximately 4–5 mm was extracted by hand, wrapped in aluminum foil, and appropriately labeled.

For the charcoal identification, the remains were fragmented by hand in order to obtain the three wood anatomy sections, which allowed us to describe and determine cell structure. Charcoal fragments were studied through a reflected light microscope with dark and light fields, using magnification factors of 5X, 20X, and 50X. Identifications were verified using a reference collection and various wood anatomy atlases (Schweingruber 1990).

FTIR and Phytoliths

Samples from the hearths were systematically collected during the excavation of level J (Figs. 9.1 and 9.2). From the initial assemblage of samples only 25 were selected for this study based on the characteristics of the fireplace and the amount of sediment available. The selected samples mainly contain sediments from the combustion structures (including burned sediments, reddened sediments or unaltered sediments above or below the hearths), but three independent samples were collected between the hearths, where no thermally altered sediment was clearly present. Those three samples were used as controls to differentiate the phytoliths present in the hearths from those not related to the combustion structures. The samples are identified in the text as follows: the first part of the name (until the period) indicates the square where the sample was collected, the second part of the sample name indicates the layer of the hearth where the sample was collected. Some of the samples were obtained by re-excavating soil micromorphology samples in the laboratory. Because these samples were only half impregnated, we were able to directly relate the micromorphological results with the phytoliths results.

The methods used to extract the phytoliths are mainly in accordance with those described by Madella et al. (1998) and Albert and Weiner (2001). During phytolith extraction the waste was weighed between acid treatments, which allowed us to determine the amount of carbonates, organic material, and Acid Insoluble Fraction (AIF) present in the samples by calculating the loss of mass. The margin of error for this procedure was calculated and was found to be approximately 2% for the sediments forming the Abric Romaní. The description of the phytoliths follows the ICPN (Madella et al. 2005) as closely as possible, and the number of phytoliths was calculated using the method described by Albert and Weiner (2001).

FTIR spectra from every sample were taken at the Weizmann Institute of Sciences and the Scientific and Technological Resource Services department of the Universitat Rovira i Virgili (Tarragona, Spain) using a Prospect IR Midac, with GRAM5-386TM software. FTIR were taken before and after processing the samples for phytolith extraction. The pH analyses of the sediments were performed using nine independent samples extracted from the site profile and representing sublevels Ja and Jb. These samples were collected inside the shelter, at the roofline, and outside the shelter. We used a GLP21 pHmeter and calculated pH variation using the method described by Lotti (1986).

Results

Wood Imprints: Pseudomorphs

The wood imprints recovered in level J are generally related to small segments of branches and twigs, which were recorded but not molded. There are 21 wood imprints, 20

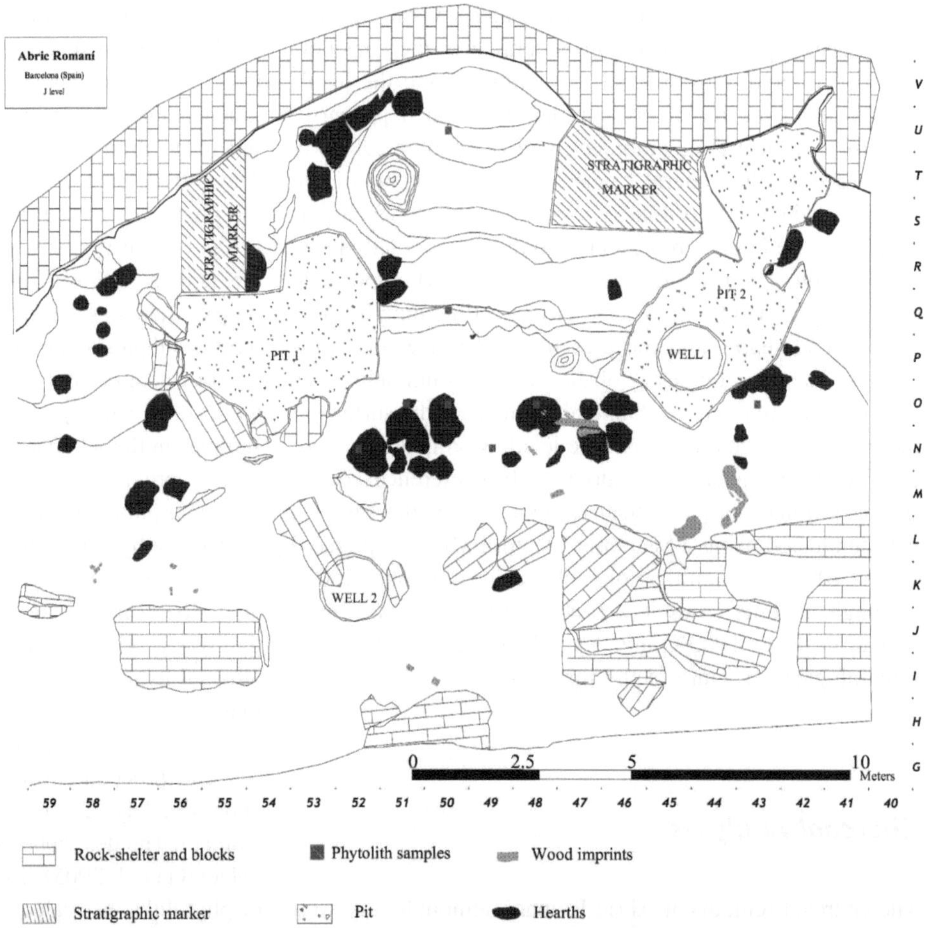

Fig. 9.1 Distribution plan of hearths, wood pseudomorphs, and phytolith samples

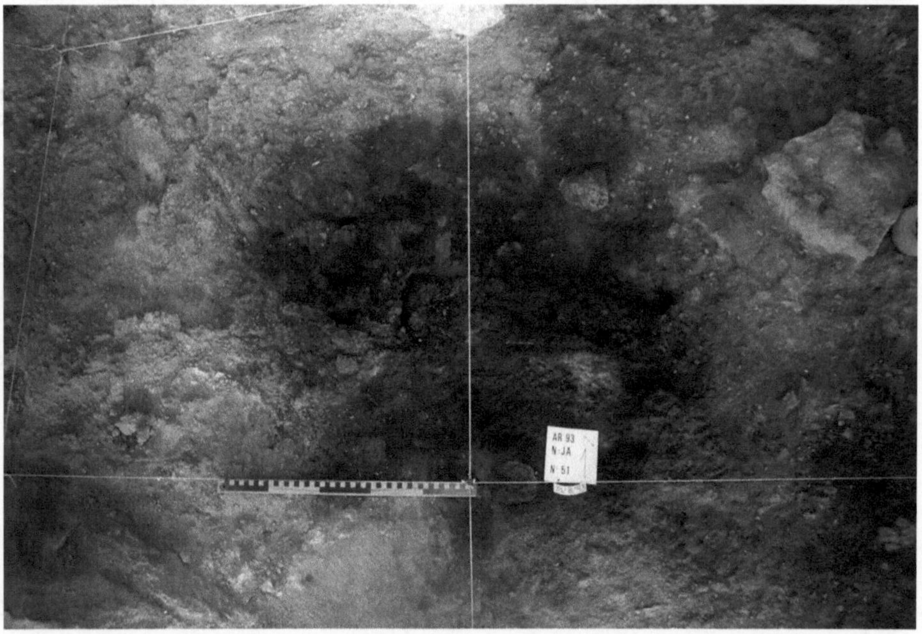

Fig. 9.2 Image of flat hearths from level J

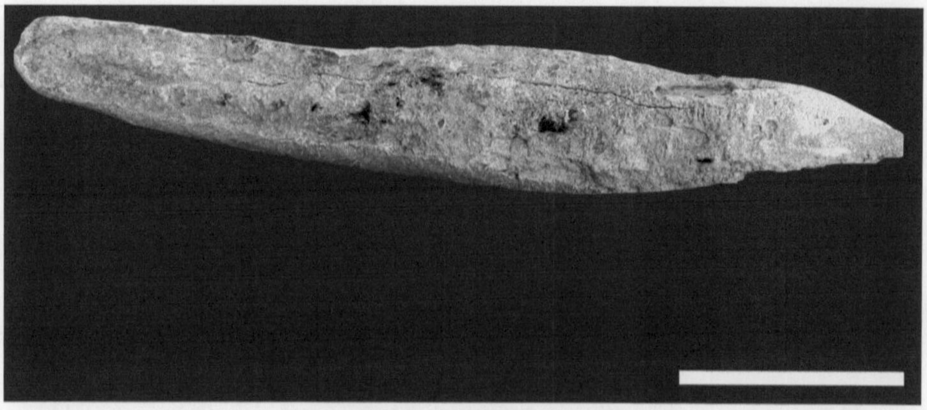

Fig. 9.3 Plaster cast "positivation" of pointed wood pseudomorphs—scale bar is 25 cm long

were recovered from sublevel Ja and one from sublevel Jb (Fig. 9.1). The average diameter of the imprints is 69 mm, and the average length 174.82 mm.

The spatial distribution of the wood imprints shows that burned fragments are located in the central part of the shelter near the hearths, whereas uncharred elements are found in the outer part of the shelter beyond the cornice fall line.

The largest wood imprint is an entirely decayed pointed wooden stick encased in carbonate sediment. Charcoal fragments preserved inside the pseudomorph allowed us to identify the wood as *Pinus sylvestris.* The dimensions and outline of the pointed pseudomorph describe an almost 100 cm-long felled straight branch that is 10 cm in diameter at the convex proximal end and that, at 80 cm in, tapers to a blunt point 2 cm in diameter at the distal end (Fig. 9.3).

Two types of grooves can be discerned in the silicone impression sample at the distal end. One type consists of a number of U-shaped 150 μm-wide grooves with parallel ridges following an oblique course with undefined endings. The other type consists of a single trapezoidal 400 μm-wide groove with a depth of 500 μm at the proximal origin and 1 mm at the center. The latter was probably made up of three incisions; the middle (400 μm wide) is slightly concave, while the U-shaped (35 μm wide) lateral striations with neat parallel ridges are polished. A transversal scratch, probably produced by use, abruptly shortens the set. Another group of diffuse, parallel striations, oriented along the longitudinal structure of the wood represent some surface grain shrinkage. U-shaped oblique striations were also found in an experimental replica of the stick, which was shaped by whittling, planing and scraping with a denticulate in an outward motion towards the point. The striations described in the archeological specimen and that also appeared in an experimental replica were clearly identified as tool marks (Fig. 9.4).

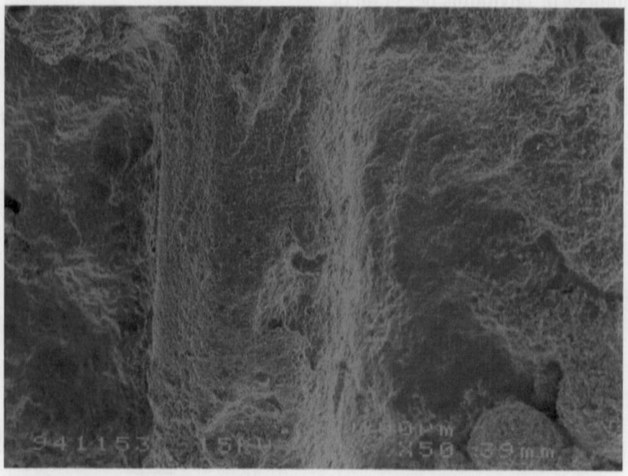

Fig. 9.4 SEM image at 50× of striae on pointed wood tip

Charcoal Analysis

The study of sublevels Ja and Jb is based on the analysis of 1,205 charcoal fragments. In both sublevels, over 90% of the charcoal fragments were identified as coniferous, and *Pinus* ssp. was the only identified taxa. The remaining undetermined conifers are probably also *Pinus* fragments, which we were not able to identify due to the small size of the fragments or to alterations in the cell structure. Table 9.1 shows the percentages of the identified taxa in sublevels Ja and Jb. Most of the charcoal remains were identified as *Pinus* type *sylvestris*, but there is also *Pinus* type *sylvestris/uncinata*, *Pinus* sp., undetermined conifers, and undetermined fragments. In addition, a single fragment of *Salix* was identified in sublevel Ja.

Conifers of the *Pinus sylvestris* type and *Pinus uncinata* type present resin ducts located on the late wood. In the

Table 9.1 Results of the charcoal analyses from sublevels Ja and Jb

	Ja		Jb	
Taxon	n° frags.	%	n° frags.	%
Pinus sylvestris type	630	71.6	217	66.8
Pinus uncinata type	21	2.4	11	3.4
Pinus sp.	61	6.9	29	8.9
Indeterminate conifer	112	12.7	47	14.5
Salix sp.	1	0.1		
Indeterminate	55	6.3	21	6.5
Number of fragments	880		325	

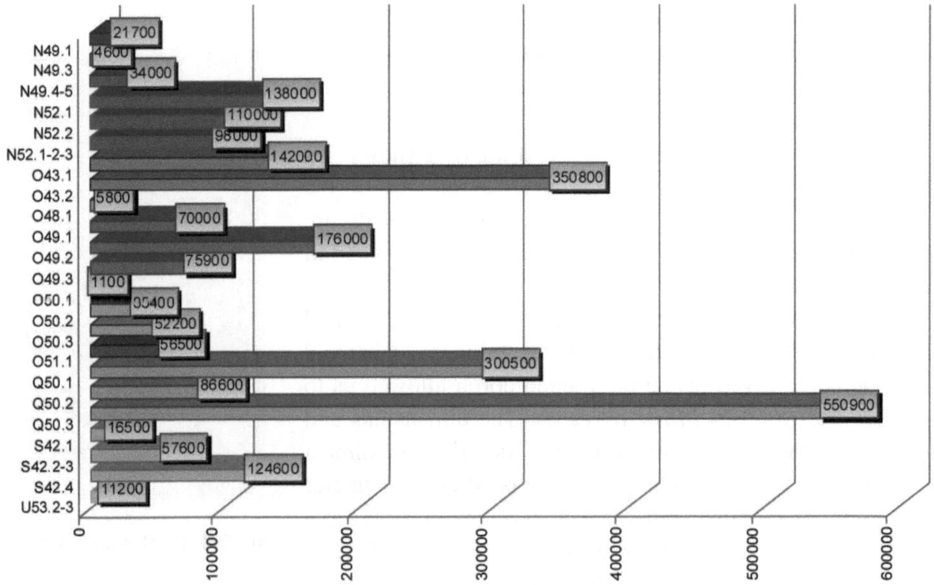

Fig. 9.5 Calculation of the number of phytoliths per gram of AIF

radial section these species show window-like cross fields and dentate transversal traqueids (Schweingruber 1990). The taxonomic category, *Pinus sylvestris* type includes *Pinus sylvestris* and *Pinus nigra*, but neither species can be distinguished on the basis of wood anatomy (Schweingruber 1990). *Pinus uncinata* type includes *Pinus sylvestris* and *Pinus uncinata* characterized by thinner growth rings, which suggest extreme environmental conditions (Heinz 1990).

Phytoliths and FTIR Analyses

Phytolith Quantification

The calculations for the number of phytoliths per gram of AIF are shown in Fig. 9.5. The samples varied widely, including within groups of samples belonging to the same combustion structure. In some samples phytoliths are virtually absent, whereas in others, such as sample Q50.3, more than half a million phytoliths per gram of AIF were detected. Note that samples extracted from the uppermost part of the fireplaces (blackened sediments, samples 3 or 4–5) usually have more phytoliths than other samples taken from the same fireplace (samples 1 or 2).

Phytolith Morphological Results

Most of the phytoliths come from wood and bark, but phytoliths from monocotyledonous plants, weathered phytoliths, indeterminate morphologies and fusiform diatoms were also identified (Fig. 9.6). There were some exceptions, such as in the case of the fireplaces in squares O50 and U53 (Fig. 9.6), and especially the sample from the hearth situated at O48, in which more than 40% of the phytoliths were determined as monocot phytoliths. All the monocot phytoliths were identified as grass phytoliths, with only two phytoliths determined as sedges, found in hearths in square U53. The anatomical origins

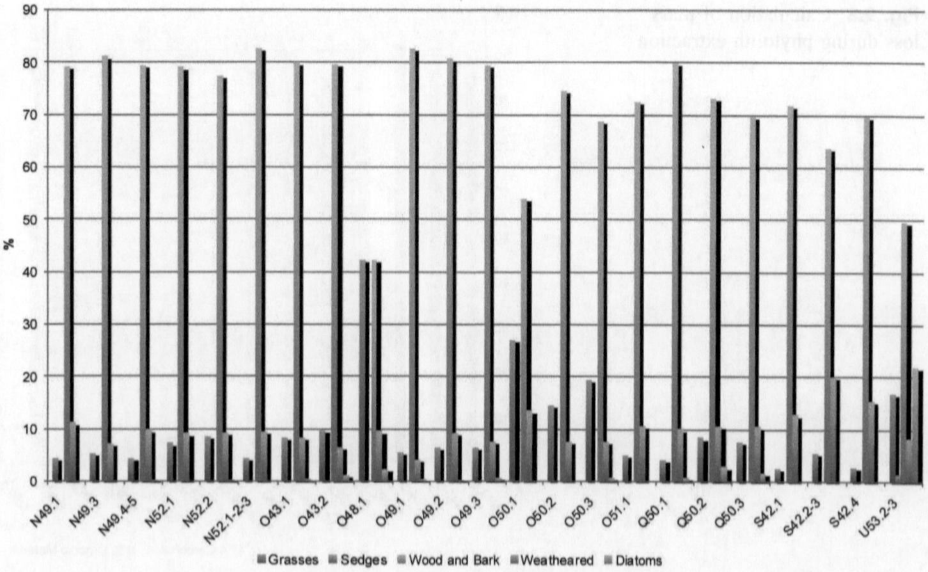

Fig. 9.6 Primary origin of the phytoliths and other elements identified in samples from sublevel Ja

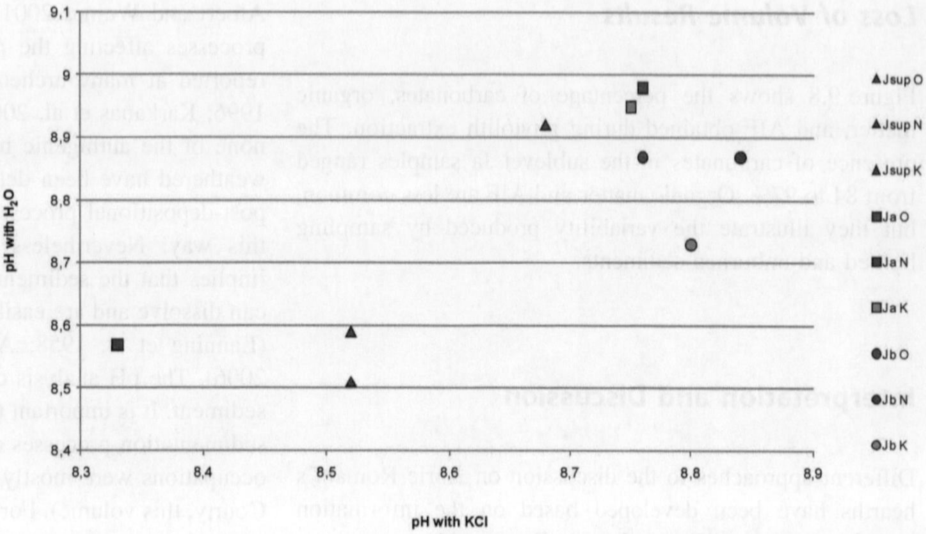

Fig. 9.7 Values for the pH analyses showing calculation with water and with KCl

of the grass phytoliths are mostly restricted to leaves and stems, however, samples from the hearths in squares O48 and O50 show a moderate percentage of phytoliths from inflorescence. These grasses were mainly festucoids, and only two phytoliths similar to chloridoid short cells were identified. Diatoms were observed in some samples, and a particularly high amount of these were found in samples U53-2.3.

FTIR Analyses

Nearly all of the samples yielded the same spectra, both for the chemically untreated samples and for the AIF sample. The initial spectrum shows a practically pure calcite, whereas the AIF spectrum is mostly formed by quartz and clay. Some AIF samples showed traces of $NaNO_3$, but this is probably due the use of nitric acid during extraction.

pH Analyses

The pH analyses revealed that the alkalinity of the sediments is considerably high, varying between values of eight and nine (Fig. 9.7). No extreme differences were detected between samples; however the samples located under the roof of the shelter were slightly more alkaline than the other samples, and sediments from sublevels Ja and Jb tended to be more alkaline than the sediments from the uppermost part of level Ja.

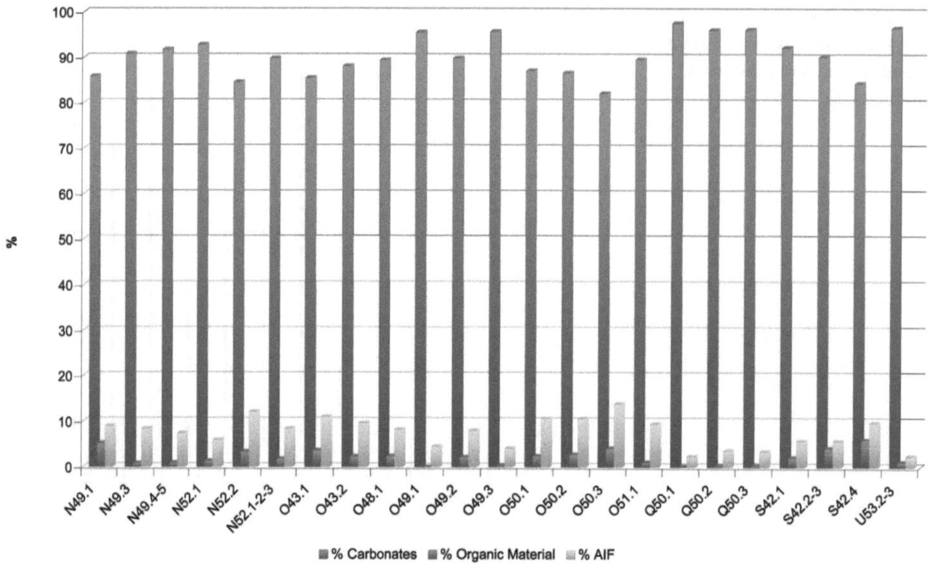

Fig. 9.8 Calculation of mass loss during phytolith extraction

Loss of Volume Results

Figure 9.8 shows the percentage of carbonates, organic matter, and AIF obtained during phytolith extraction. The presence of carbonates in the sublevel Ja samples ranged from 84 to 97%. Organic matter and AIF are less common, but they illustrate the variability produced by sampling burned and unburned sediments.

Interpretation and Discussion

Different approaches to the discussion on Abric Romaní's hearths have been developed based on the information contained in the different layers. From evidence such as spatial distribution, composition, associated artifacts, shape, etc., we can suggest patterns of hearth function (Vaquero and Pastó 2001; Vaquero et al. 2004; Carbonell et al. 2007; Vallverdú and Courty 2012). In this chapter we look at plant remains, which contribute to the understanding of hearth function in relation to firewood use and management.

The study of FTIR spectra allows us to understand the processes involved in the formation of hearths, whereas phytoliths, charcoals and wood pseudomorphs provide insight into aspects such as hearth functioning, firewood management and other plant uses. The FTIR spectra of the bulk samples used for the phytolith analyses show calcite as the main mineral forming the sediments in all cases. Calcium carbonate or calcite comes directly from the travertine forming the infilling of the site, but also from the ashes which comprise the combustion remains (Etiégni and Campbell 1991; Schiegl et al. 1996; Karkanas et al. 2000; Albert and Weiner 2001; Weiner et al. 2002). Diagenetic processes affecting the preservation of calcite have been reported at many archeological cave sites (Schiegl et al. 1996; Karkanas et al. 2000; Weiner et al. 2002); however, none of the authigenic minerals that form when calcite is weathered have been detected in level J, meaning that no post-depositional processes have affected the sediments in this way. Nevertheless, the preservation of the calcite implies that the sediments have a high pH, and phytoliths can dissolve and are easily weathered in alkaline sediments (Lanning et al. 1958; Albert et al. 1999, 2006; Piperno 2006). The pH analysis confirms the high alkalinity of the sediment. It is important to note that in level J the dominant sedimentation processes during and around the time of the occupations were mostly detritical and dry (Vallverdú and Courty, this volume). For phytoliths weathering in basic pH environments, water must be present to induce the chemical reaction. The absence of water during the deposition of the sediments from the fireplaces has partially preserved the phytolith assemblages. However, we should point out that the presence of weathered phytoliths is high in all the samples, and most of the morphologies identified show signs of weathering such as dissolution pits.

The morphology of the phytoliths does not greatly contribute to a paleoecological reconstruction, due to their general characteristics and the sampling method used (which only includes short-term deposition) and because most of the phytoliths were the product of human activities related to fire. A tentative interpretation of the results would indicate a dry climate with both forest and grass formations. However, again, we must stress that in terms of paleoecology the level J phytolith analysis contributes very little information.

Using the number of phytoliths per gram of AIF, phytolith morphologies, the amount of other microremains,

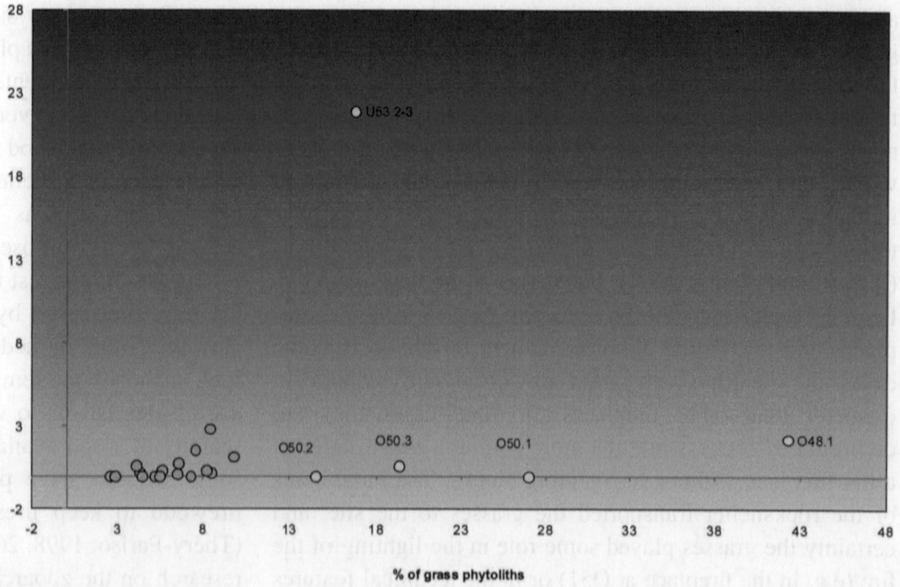

Fig. 9.9 Percentages of diatoms and grass phytoliths for the samples from sublevel Ja

and the different percentages of volume lost during phytolith extraction, we discerned four different types of hearth formation process.

The first type of formation process is called the simple hearth, also known as primary cultural microsedimentary organization (Wattez 1990). The simple hearth process consists of a fire whose remains were buried after the fuel was burned, without severe biostratinomic processes. The carbonate percentage of simple hearths decreases from the base to the top of the structure, because most of the AIF and organic remains are concentrated in the upper part of the fireplace. In addition, this type of hearth contains many more phytoliths in the uppermost sediments than in the lower ones. The fireplaces identified as simple hearth structures were those located in squares S43-42 and NO-50.

The second category of fireplace formation process is the structure with displaced sediment type. The fireplaces ascribed to these structures (hearth at N51-52) show minimal differences between the layers within the same hearth; in other words, there are roughly the same number of phytoliths and the same percentage of major sediment components in all the subsamples. The displacement of sediments is consistent with the results obtained through soil micromorphology analyses (Vallverdú 2002).

The third class of combustion-structure formation process is classified as the structure with differential inputs of firewood type. Most of the phytoliths are concentrated in the uppermost part and in the lowermost part of the structure, while the central layer has fewer phytoliths. Only one case was identified, in square N49.

Finally, the fourth type of fire is called the simple fireplace with weathering in the final phase. The percentages of carbonates, organic material or AIF are similar to the simple hearth structure. However, the phytolith number decreases in the uppermost level of the structure, and there are commonly more diatoms and grass phytoliths in the upper layer than in the lower layers of the hearth. The presence of diatoms in the sediments can be related to the initial stages of soil formation, or weathering crust. Soil formation processes usually imply the partial dissolution of the initial assemblage of phytoliths due to their recycling by the plants growing in the soil (Alexandre et al. 1997). In addition, samples taken from different soils, including from the driest places, contain an observable number of diatoms (D.C., personal observation). The fourth type of structure characterized by weathering in the final phase can also be applied to the second and third types of structure. The fireplaces with final weathering are those located at Q50 and OP42-43.

Due to the lack of more than two samples for some fireplaces, we were not able to assign those hearths to any of the above categories (fireplaces at O48, O51 and U53). However, using the type and number of phytoliths and diatoms found in those hearths, a tentative interpretation can be made.

Wood was the main fuel resource used by the Neanderthals who occupied the rockshelter, but it was not the only one. As mentioned earlier, the phytolith analyses detected grasses in different percentages. The presence of grasses and diatoms are indicative of soil formation processes (Bozarth 1993), and at Abric Romaní grasses as well as diatoms have been recovered from the sediments. Combining the percentages of grasses and diatoms, as shown in Fig. 9.9, we can distinguish two marked groups which are separate from the other samples. The first group contains practically no diatoms and differing amounts of grasses and is formed by samples taken from hearths O50, O51 and O48. The second group is formed by a single sample (U53.2.3) with a considerable number of diatoms and some grasses. It is our opinion that the presence of grasses in the first group is due to human activity, whereas the grasses and diatoms present in the second group are due to soil formation processes. As mentioned earlier, diatoms are very common in the uppermost soil horizons around the world, and the combination of grasses and diatoms is indicative of soil development (Bozarth 1993). Grass phytoliths may also be linked to firewood used by the hominids (Albert and Weiner 2001), but if this is the case, it is difficult to understand the presence of grasses in some fireplaces and the virtual absence of them in others. In some cases the hearths with grass phytoliths are smaller in diameter than other fireplaces; in other cases they are delimited by rocks, while the more common construction is a flat fireplace without surrounding blocks. The inhabitants of the rockshelter transported the grasses to the site, and certainly the grasses played some role in the lighting of the fire (e.g. in the fireplace at O51) or in the technical features of the fire.

The charcoal assemblage from sublevels Ja and Jb were found to contain the same elements, with the exception of the presence of willow in sublevel Ja. According to these data, we can conclude that the most abundant species in the immediate environment surrounding the shelter was undoubtedly pine. However, some of the pollen data collected indicate that it was not the only taxa present at the local and regional level (Allué et al. 2012). Other plant species such as junipers and riverside plants might have also grown nearby. The low variability in the charcoal assemblage therefore seems to be linked to firewood exploitation strategies and to have been reliant on supplies available in the surrounding environment.

From this perspective, we can establish firewood selection patterns that point to certain strategies. The first selection criterion is the abundance of the material in the immediate surroundings and the second is related to the presence of dead wood from trees or shrubs. The vegetation formations described in Allué et al. (2012), suggest a pine forest that probably had undergrowth shrubs such as junipers. The junipers growing in modern-day Scots pine forests are *Juniperus communis* (Blanco et al. 1998). This common juniper is a thorny-leafed species with low rate of dead wood production, whereas Scots pines produce dead wood relatively quickly. Due to meteorological conditions (wind, snow and lightning), pines produce a larger annual fall of branches than do shrubs.

Neanderthal hunter-gatherer groups occupied the shelter during short periods, so we think it likely that fallen dead branches and the lowermost dead branches of trees would have supplied enough fuel for the immediate maintenance of hearths. If we consider this as a definition of the firewood supply pattern, the species of tree in terms of fuel quality as a selection criterion loses importance. Other levels of the Abric Romaní show similar patterns of firewood exploitation strategies. Scots pine is dominant in all the levels, with one or two fragments of other species appearing only occasionally. Moreover, charcoal from this layer shows low values of altered wood due to decay processes (Allué 2002), which leads us to believe that the wood gathered was of a high quality—it was dead and dry, as was required to accomplish its purpose of hearth maintenance.

Our results suggest that firewood and some grasses were the only fuels used by the inhabitants of the rockshelter. However, other burned materials may also be considered as fuel, such as bone remains. According to some studies, the use of alternatives to wood as fuel occurs when there is a scarcity of wood available (Heizer 1963; Miller 1984), and some scholars have proposed that bones were added to firewood to keep fires burning in other European areas (Théry-Parisot 1998, 2001; Costamagno et al. 1999). Recent research on the zooarcheological remains of the Abric Romaní has looked at the importance of burned bones recovered from inside hearth structures (Rosell 2001; Cáceres et al. 2002; Rosell et al. 2012). These studies have revealed the lack of certain skeletal parts, mainly the epiphyses (Rosell et al. 2012). These gaps in the record may be interpreted as the result of the use of bones as fuel, because these parts, containing fats and spongy components, would have burned easily. Zooarcheological and taphonomy investigations suggest that bone cremation is related to marrow extraction as heating the bones decreases their elasticity thereby allowing them to be more easily fractured (Cáceres et al. 2002). Throwing bones into the hearths may also be related to cleaning activities in the occupied areas, which would also explain the presence of burned bones. However, it is important to bear in mind that the presence of unburned bones in other areas of level J suggests that they were not always thrown into the fire (Rosell et al. 2012). Furthermore, the remains of unburned wood have been identified in other levels and have been interpreted as unused firewood (Allué 2002; Vallverdú et al. 2005; Solé 2007). The presence of this unburned firewood indicates that there was no scarcity of vegetation fuel, which would imply that Neanderthals were not necessarily forced to use other fuel sources.

Ethnographic descriptions related to firewood gathering by hunter-gatherer groups suggest a preference for dry and dead branches (Heizer 1963). Firewood provisioning should not require a great expenditure of time or energy, and the use of dead branches from the forest floor or the lower branches of pine trees would save the gatherers both of these valuable resources. Some studies from other levels at Abric Romaní point out that firewood gathering strategies might be explained by an economic approach, by striking an energy balance between the investment in collecting the firewood and its efficiency as fuel (Allué 2002; Allué and

García-Antón 2006). This assertion has been used to explain dead wood collection (Shackleton and Prins 1992), and it is consistent with the presumed immediate use of resources among hunter-gatherer communities (Thery-Parisot 2001). Firewood typologies may show different collecting strategies, which are energetically balanced in terms of both investment and efficiency in relation to the abundance and distribution of resources in the environment. In terms of availability, twigs and little branches represent an abundant resource. This is in agreement with the physiological disarticulation of Scots pine which provides abundant dead wood (Martín-Closas and Gómez 2004). The wood imprint studies in levels M and I show that branches used for fuel were between 10 and 30 cm (Allué 2002; Solé 2007). In level J the preservation of wood imprints is scarce; nevertheless, the presence of the same taxonomic assemblage may suggest that the size of the branches collected was similar. Firewood gathering strategies were based on the nearest available wood. The use of wood for fuel is considered to be the exploitation of a perishable material which needs to be continuously restocked in order to prevent the fire from dying out. This implies the continuous search for firewood during the occupation without any unnecessary loss of energy.

As opposed to other layers at Abric Romaní, the occurrence of wood imprints in level J is relatively rare considering the high densities of other archeological remains. This scarcity may be related to the combination of two main factors. First, it is possible that the high occupation intensity did not produce a surplus of wood. Second, dry conditions may not have allowed for extensive wood preservation and incrustation.

Nevertheless, the evidence suggests that the pointed wooden stick was some kind of tool. This can be deduced from the shape and dimensions of the stick—which differ from those of the rest of wood assemblage—as well as its association with hearths, lithic artifacts, and bone accumulations. The only evidence from level J of the use of wood in tool-making corresponds to a pointed artifact. The archeological Paleolithic record has yielded a variety of pointed wooden artifacts. Several unambiguous wooden spears have been recovered in Schöningen, Germany (Thieme 1997, 1999), Clacton-on-Sea, UK (Oakley et al. 1977; McNabb 1989) and in Lerhingen, Germany (Movius 1950). All of these finds have a similar shape and measurement pattern: 2 m long and a maximum of 30 mm in diameter. Setting the maximum diameter next to the point would have ensured that these artifacts, used as weapons, would fly well as throwing spears, and the lightness of the proximal segment would not suggest a thrusting use. Although it seems more reasonable to relate the thick, heavy, pointed stick from the Abric Romaní site to hand-held digging activities, few and uncertain examples are given in the archeological Paleolithic record for these kinds of artifacts. Some wood items recovered in the Acheulian horizons of Kalambo Falls (Zambia) were interpreted as wooden tools (n = 36) showing a great variety of morphologies. However, these specimens had been smoothed by the flow of the stream, casting uncertainty on whether they actually showed signs of human modification. Other pointed objects have been found and one of them (Clark 2001) presents similar measurements to those of the pointed stick from Abric Romaní layer J. Some wooden digging sticks from the late Stone Age sites at Gwisho hot-springs (Zambia) are also comparable to the item we present here (Fagan and Van Noten 1966). Furthermore, wooden digging sticks are well documented in ethnographic and ethnoarcheological observations of hunter-gatherer communities (Mountford 1941; Thomson 1964; Lee 1979; Hayden 1979; Deacon 1984) showing a great variability of morphologies, measurements, functions and modes of use, although the available examples are typically more slender than this pointed branch. Finally, we must point out that the construction of dwellings and structures, which have been generally indirectly documented in the Paleolithic, would have been done with wood. The remains of these structures are visible as postholes. In conclusion, it seems wiser, due to its robustness and blunt tip, to associate the pointed stick with the distal end of a post or stake, although the possibility that it was used as a digging tool should not be ruled out.

Conclusion

The archeobotanical remains from level J from Abric Romani contribute to the understanding of hearth functioning. The results from the study of the different plant remains, charcoals, phytoliths and wood imprints provide complementary taxonomic and taphonomic data. The taxonomic analyses show that the primary firewood used was pine, whereas other types of plants, such as grasses, were probably also used during the hearth lighting process. The selection patterns for firewood exploitation were based on abundance, availability and the quality of a certain type of branch from the pine trees. The taphonomic approach, based on phytoliths and FTIR, points to different types of processes, which reflect different phases of hearth formation from lighting to post-depositional processes. This has allowed us to classify the hearths uncovered into four different categories. The study of the imprints allowed us to distinguish between firewood and worked wood, and led to the conclusion that only one imprint of those taken in level J can be classified as a manufactured, while the rest of the wood imprints are the remains of firewood. This study can be better understood in relation to other interdisciplinary approaches that also contribute to the understanding of Neanderthals' homes.

Acknowledgments The authors are responsible for the interpretation included in this chapter. We would like to express our sincere gratitude to the students, researchers and others that have contributed to the fieldwork at the Abric Romaní over the past years. We also thank the several anonymous reviewers whose comments helped to improve our manuscript.

References

Albert, R. M., & Weiner, S. (2001). Study of phytolith in prehistoric ash layers from Kebara and Tabun caves using a quantitative approach. In J. D. Meunier & F. Colin (Eds.), *Phytolith: Applications in earth sciences and human history* (pp. 251–266). Lisse: Balkema Publishers.

Albert, R. M., Lavi, O., Estroff, L., Weiner, S., Tsatskin, A., Ronen, A., et al. (1999). Mode of occupation of Tabun cave, Mt Carmel, Israel during the Mousterian period: A study of the sediments and phytoliths. *Journal of Archaeological Science, 26*, 1249–1260.

Albert, R. M., Bar-Yosef, O., Meignen, L., & Weiner, S. (2003). Quantitative phytolith study of hearths from the Natufian and middle palaeolithic levels of Hayonim cave (Galilee, Israel). *Journal of Archaeological Science, 30*, 461–480.

Albert, R. M., Bamford, M. K., & Cabanes, D. (2006). Taphonomy of phytoliths and macroplants in different soils from Olduvai Gorge (Tanzania) and the application to Plio-Pleistocene palaeoanthropological samples. *Quaternary International, 148*, 78–94.

Allué, E. (2002). Dinámica de la vegetación y explotación del combustible leñoso durante el Pleistoceno Superior y el Holoceno del Noreste de la Península Ibérica a partir del análisis antracológico. Ph.D. dissertation, Universitat Rovira i Virgili.

Allué, E., & García Antón, D. (2006). La transformación de un recurso biótico en abiótico: aspectos teóricos sobre la explotación del combustible leñoso en la prehistoria. Sociedades prehistóricas, recursos abióticos y territorio. In *III Reunión de Trabajo sobre Aprovisionamiento de Recursos Abióticos en la Prehistoria* (pp. 19–31). Loja: Fundación Ibn al-Jatib de estudios de Cooperación cultural.

Allué E., Burjachs, F., García, A., López-García, J. M., Bennàsar, M., Rivals, F., et al. (2012). Neanderthal landscapes and their home environment: Flora and fauna records from level J of Abric Romaní. In E. Carbonell (Ed.), *High Resolution Archaeology and Neanderthal Behavior: Time and Space in Level J of Abric Romaní (Capellades, Spain)*. Dordretch: Springer

Alexandre, A., Meunier, J.-D., Colin, F., & Koud, J.-M. (1997). Plant impact on the biogeochemical cycle of silicon and related weathering processes. *Geochimica et Cosmochimica Acta, 61*, 677–682.

Arteaga, I., Allué, E., Pastó, I., Vallverdú, J., & Carbonell, E. (2001). Els fogars del Paleolític Mitjà de l'Abric Romaní (Capellades, Anoia). *Cypsela, 13*, 9–19.

Asouti, E., & Austin, P. (2005). Woodland vegetation and its exploitation by past societies, based on the analysis and interpretation of archaeological wood charcoal macro-remains. *Environmental Archaeology, 10*, 1–18.

Blanco, E., Casado, M. A., Costa, M., Escribano, R., García, M., Génova, M., et al. (1998). *Los bosques ibéricos Una interpretación geobotánica*. Madrid: Editorial Planeta.

Bozarth, S. (1993). Biosilicate assemblages of boreal forest and aspen parklands. In D. M. Pearsall & D. R. Piperno (Eds.), *Current Research in phytolith analysis: Applications in archaeology and paleoecology* (pp. 95–105). Philadelphia, PA: Museum Applied Science Center for Archaeology and The University Museum of Archaeology and Anthropology, University of Pennsylvania.

Cabanes, D., Allué, E. Vallverdú, J. Cáceres, I. Vaquero, M., & Pastó, I. (2007). Hearth structure and function at level J (50kyr, BP) from Abric Romaní (Capellades, Spain): phytolith, charcoal, bones and stone-tools. In M. Madella & D. Zurro (Eds.), *Plants, people and places—recent studies in phytolith analysis* (pp. 98–106). Oxford, England: Oxbow Books.

Cáceres, I., Bravo, P., Esteban, M., Expósito, I., & Saladié, P. (2002). Fresh and heated bones breakage. An experimental approach. In M. de Renzi, M. V. Pardo, M. Belinchón, E. Peñalver, P. Montoya, & A. Márquez-Aliaga (Eds.), *Current topics on taphonomy and fossilization* (pp. 471–480). Valencia: Ajuntament de Valencia.

Carbonell, E. (2002). *Abric Romaní nivell I. Models d'ocupació de curta durada de fa 46.000 anys a la Cinglera del Capelló (Capellades Anoia, Barcelona)*. Tarragona: Universitat Rovira i Virgili.

Carbonell, E., & Castro-Curel, Z. (1992). Palaeolithic wooden artefacts from the Abric Romaní (Capellades, Barcelona, Spain). *Journal of Archaeological Science, 19*, 707–719.

Carbonell, E., Cebrià, A., Allué, E., Cáceres, I., Castro, I., Diaz, R., et al. (1996). Behavioural and organizational complexity in the middle Paleolithic from the Abric Romaní. In E. Carbonell & M. Vaquero (Eds.), *The last Neanderthals, the first anatomically modern humans: A tale about human diversity* (pp. 385–434). Tarragona: Universitat Rovira i Virgili.

Carbonell, E., Lorenzo, C., & Vallverdú, J. (2007). Centralidad especial y operative de los neanderthales. Análisis especial diacrónico de las actividades de combustion en el Abric Romaní (Anoia, Capellades, Barcelona). In E. Baquedano (Ed.), *El Universo Neandertal I* (pp. 197–219). Soria: Fundación Duques de Soria.

Castro-Curel, Z., & Carbonell, E. (1995). Wood pseudomorphs from level I at Abric Romaní, Barcelona, Spain. *Journal of Field Archaeology, 22*, 376–384.

Clark, J. D. (2001). *Kalambo falls prehistoric site III*. Cambridge, MA: Cambridge University Press.

Conley, D. J. (2002). Global biogeochemical cycles. *Terrestrial ecosystems and the global biogeochemical silica cycle, 16*(4), 1121.

Costamagno, S., Griggo, C., & Mourre, V. (1999). Approche experimentale d'un probleme taphonomique: utilisation de combustible osseux au paléolithique. *Préhistoire Européenne, 13*, 167–194.

Deacon, J. (1984). Later stone age people and ther descendants in southern Africa. In R. G. Klein (Ed.), *Southern African prehistory and paleoenvironments* (pp. 221–328). Rotterdam: Balkema.

Etiégni, L. A. G. (1991). Campbell, physical and chemical characteristics of wood ash. *Bioresource Technology, 37*, 173–178.

Fagan, B. M., & Van Noten, F. L. (1966). Wooden implements from Late Stone Age sites at Gwisho hot springs, Lochinvar, Gambia. *Proceedings of the prehistoric society, 32*, 246–261.

Figueiral, I., & Mosbrugger, V. (2000). A review of charcoal analysis as a tool for assessing quaternary and tertiary environments: achievements and limits. *Palaeogeography, Palaeoclimatology, Palaeoecology, 164*, 397–407.

Hayden, B. (1979). *Palaeolithic reflections Lithic technology and ethnographic excavation among Australian Aborigines*. Canberra, ACT: Australian Institute of Aboriginal Studies.

Heinz, C. (1990). Dynamique des végétations holocénes en mediterrannée nord-occidentale d'après l'anthracoanalyse de sites préhistoriques methodologie and paleoécologie. *Paleobiologie Continentale, XVI*, 1–212.

Heizer, R. F. (1963). Domestic fuel in primitive society. *Journal of the Royal Anthropological Institute of Great Britain and Ireland, 93*, 186–194.

Karkanas, P., Bar-Yosef, O., Goldberg, P., & Weiner, S. (2000). Diagenesis in prehistoric caves: The use of minerals that form in situ to assess the completeness of the archaeological record. *Journal of Archaeological Science, 27*, 915–929.

Karkanas, P., Rigaud, J. P., Simek, J. F., Albert, R. M., & Weiner, S. (2002). Ash bones and guano: A study of the minerals and phytoliths in the sediments of Grotte-XVI, Dordogne, France. *Journal of Archaeological Science*, *29*, 721–732.

Lanning, F. C., Ponnaiya, B. W. X., & Crumpton, C. F. (1958). The chemical nature of silica in plants. *Plant Physiology*, *33*, 339–343.

Lee, R. B. (1979). *The !Kung San*. Cambridge, MA: Cambridge University Press.

Lotti, G. (1986). *Análisis químico agrario*. Madrid: Alhambra.

Madella, M., Powers-Jones, A. H., & Jones, M. K. (1998). A simple method of extraction of opal phytolith from sediments using a non-toxic heavy liquid. *Journal of Archaeological Science*, *25*, 801–803.

Madella, M., Jones, M. K., Goldberg, P., Goren, Y., & Hovers, E. (2002). The exploitation of plant resources by Neanderthals in Amud cave (Israel): The evidence from Phytolith studies. *Journal of Archaeological Science*, *29*, 703–719.

Madella, M., Alexandre, A., Ball, T., & ICPN Working Group. (2005). International code for Phytolith nomenclature 1.0. *Annals of Botany*, *96*, 253–260.

Martín-Closas, C., & Gomez, B. (2004). Plant taphonomy and palaeoecological interpretations: A synthesis. *Geobios*, *37*, 65–88.

McNabb, J. (1989). Sticks and stones: A possible experimental solution to the question of how the Clacton spear point was made. *Proceedings of the Prehistoric Society*, *55*, 251–257.

Meunier, J. D., Alexandre, A., Colin, F., & Braun, J. J. (2001). Deciphering the dynamics of tropical soils through the study of the biogeochemical cycle of silica. *Bulletin de la Societe Geologique de France*, *172*, 533–538.

Miller, N. F. (1984). The use of dung as fuel: An ethnographic example and an archaeological application. *Paléorient*, *10*, 71–79.

Mountford, C. P. (1941). An unrecorded method of manufacturing wooden implements by simple stone tools. *Transactions of the Royal Society of South Australia*, *2*, 312–316.

Movius, H. L. (1950). A wooden spear of third Interglacial age from lower Saxony. *Southwestern Journal of Anthropology*, *6*(2), 139–142.

Oakley, K. (1955). Fire as Palaeolithic tool and weapon. *Proceedings of the Prehistoric Society*, XXI, 36–47.

Oakley, K. P. (1961a). On man's use of fire, with comments on tool-making and hunting. In S. L. Washburn (Ed.), *Social life of early man* (pp. 176–193). Chicago: Aldine.

Oakley, K. P. (1961b). On man's use of fire, with comments on tool-making and hunting. In S. L. Washburn (Ed.), *Social life of early man* (pp. 176–193). Chicago: Aldine.

Pastó, I., Allué, E., & Vallverdú, J. (2000). Mousterian hearths at Abric Romaní, Catalonia (Spain). In C. B. Stringer, R. N. E. Barton, & J. C. Finlayson (Eds.), *Neanderthals on the edge* (pp. 59–67). Oxford, England: Oxbow Books.

Perlès, C. (1977). *Préhistoire du feu*. Paris: Masson.

Pettitt, P. B. (1997). High resolution Neanderthals? Interpreting middle palaeolithic intrasite spatial data. *World Archaeology*, *29*, 208–224.

Piperno, D. R. (2006). *Phytoliths: A comprehensive guide for archaeologists and paleoecologists*. Lanham: AltaMira Press.

Rosell, J. (2001). *Patrons d'aprofitamet de les biomasses animals durant el Pleistocè inferior i Mig (Sierra de Atapuerca Burgos) i Superior (Abric Romaní, Barcelona)*. Ph. D. Dissertation, Universitat Rovira i Virgili.

Rosell, J, Blasco, R., Huguet, R., Caceres, I., Saladie, P., Rivals, F., et al. (2012). Occupational patterns and subsistence strategies in Level J of Abric Romaní. In E. Carbonell (Ed.), *High Resolution Archaeology and Neanderthal Behavior: Time and Space in Level J of Abric Romaní (Capellades, Spain)*. Dordretch: Springer.

Rovner, I. (1971). Potential of opal phytoliths for use in paleoecological reconstruction. *Quaternary Research*, *1*, 343–359.

Shackleton, C. M., & Prins, F. (1992). Charcoal analysis and the "principle of least effort": A conceptual model. *Journal of Archaeological Science*, *19*, 631–637.

Schiegl, S., Goldberg, P., BarYosef, O., & Weiner, S. (1996). Ash deposits in Hayonim and Kebara caves, Israel: Macroscopic, microscopic and mineralogical observations, and their archaeological implications. *Journal of Archaeological Science*, *23*, 763–781.

Schiegl, S., Goldberg, P., Pfretzschner, H.-U., & Conard, N. (2003). Paleolithic burned bone horizons from the Swabian Jura: Distinguishing between in situ fireplaces and dumping areas. *Geoarchaeology*, *18*, 541–565.

Schiegl, S., Stockhammer, P., Scott, C., & Wadley, L. (2004). A mineralogical and phytolith study of the middle stone age hearths in Sibudu cave, KwaZulu-Natal, South Africa. *South African Journal of Science*, *100*, 185–194.

Schweingruber, F. H. (1990). *Anatomie europäischer Hölzer ein Atlas zur Bestimmung europäischer Baum- Strauch- und Zwergstrauchhölzer Anatomy of European woods an atlas for the identification of European trees shrubs and dwarf shrubs*. Stuttgart: Verlag Paul Haupt.

Solé, A. (2007). La gestió dels recursos forestals al Paleolític mitjà a partir de les macrorestes llenyoses del nivell M de l'Abric Romaní (Capellades, Anoia). Ms. thesis, Universitat Rovira i Virgili.

Théry-Parisot, I. (1998). *Economie du combustible et Paléoecologie en contexte glaciaire et périglaciaire, Paléolithicque moyen et supérieur du sud de la France. Anthracologie, Expérimentation, Taphonomie*. Ph. D. dissertation, Université de Paris I.

Théry-Parisot, I. (2001). *Économie des combustibles au Paléolithique. Dossier de Documentation Archéologique 20*. Paris: CNRS.

Thieme, H. (1997). Lower Palaeolithic hunting spears from Germany. *Nature*, *385*, 807–810.

Thieme, H. (1999). Lower palaeolithic throwing spears and other wooden implements from Schöningen, Germany. In H. Ullrich (Ed.), *Hominid evolution: Lifestyles and survival strategies* (pp. 283–395). Gelsenkirchen: Archaea.

Thomson, D. F. (1964). Some wood and stone implements of the Bindibu tribe of central Western Australia. *Proceedings of the Prehistoric Society*, *30*, 400–422.

Vallverdú, J. (2002). *Micromorfología de las facies sedimentarias de la Sierra de Atapuerca y del nivel J del Abric Romaní. Implicaciones geoarqueológicas y paleoetnográficas*. Ph.D. dissertation, Universitat Rovira i Virgili.

Vallverdú, J. & Courty, M.-A. (2012). Microstratigraphic Analysis of the Level J Deposits. A Dual Paleoenvironmental-Paleoethnographic Contribution to the Paleolithic Archaeology at the Abric Romani Site. In E. Carbonell (Ed.), *High Resolution Archaeology and Neanderthal Behavior: Time and Space in Level J of Abric Romaní (Capellades, Spain)*. Dordretch: Springer.

Vallverdú, J., Allué, E., Bischoff, J. L., Cáceres, I., Carbonell, E., Cebrià, A., et al. (2005). Short human occupations in the middle Palaeolithic level i of the Abric Romaní rock-shelter (Capellades, Barcelona, Spain). *Journal of Human Evolution*, *48*(2), 157–174.

Vaquero, M., & Pastó, I. (2001). The definition of spatial units in middle Paleolithic sites: The hearth-related assemblages. *Journal of Archaeological Science*, *28*, 1209–1220.

Vaquero, M., Rando, J. M., & Chacón, M. G. (2004). Neanderthal spatial behavior and social structure: Hearth-related assemblages from the Abric Romaní middle Palaeolithic site. In N. J. Conard (Ed.), *Settlement dynamics of the middle Paleolithic and middle Stone age 2* (pp. 367–392). Tübingen: Kerns Verlag.

Vernet, J.-L. (1997). *L'homme et la forêt méditerranéenne de la Préhistoire à nos jours*. Paris: Ed. Errance.

Wattez, J. (1990). *Dynamique de Formation des Structures de Combustion de la Fin du Paléolithique au Néolithique Moyen*. Ph. D. dissertation, Université de Paris I.

Weiner, S., Goldberg, P., & Bar-Yosef, O. (2002). Three-dimensional distribution of minerals in the sediments of Hayonim cave, Israel: Diagenetic processes and archaeological implications. *Journal of Archaeological Science*, *29*, 1289–1308.

Part III
Conclusions

Chapter 10
Conclusions: Landscapes, Campsites, Time, and Neanderthal Behavior

Manuel Vaquero, Ethel Allué and Josep Vallverdú-Poch

Abstract The goal of this final chapter is to dicuss some issues raised by the study of Abric Romaní level J. We will focus our attention on three main questions: the adaptation of Neanderthals to their environment through subsistence strategies and settlement patterns; the characterization of Abric Romaní as a residential site; and the consequences of time resolution on assemblage formation and archeological interpretation.

Keywords Settlement patterns • Spatial organization • Resource exploitation • Time resolution • Neanderthal behavior

Level J of Abric Romaní has yielded a thought-provoking record of the behavioral strategies developed by the Neanderthals living in the Northeastern Iberian Peninsula 50 ka. The excavation of almost 290 m^2 allowed the recovery of a large quantity of archeological remains, which has made it possible to approach a wide range of behavioral domains, from the exploitation of faunal and plant resources to lithic production and tool use. Moreover, the large excavated surface has provided a spatial perspective rarely accomplished in cave and rockshelter sites. In fact, the spatial paradigm has been an essential element in the archeological research conducted at Abric Romaní and our approach to level J depends on the extensive excavation of the rockshelter. From the beginning of the current project, research at Abric Romaní has been focused on the behavior of archaic humans through the material record of living floors (Mora et al. 1988). The main goal of this research was to establish the settlement dynamics (in terms of occupation length and group size) and the spatial organization, since recent archeological research has emphasized their importance in explaining the variability of Middle Paleolithic assemblages.

This spatial point of view has been one of the hallmarks of Paleolithic research for the last 40 years, and it is strictly linked to the paleoethnographic explanations. Both the ethnographic perspective encouraged by New Archeology and the spatial research developed by Leroi-Gourhan at sites like Pincevent (Leroi-Gourhan and Brezillon 1972) contributed to putting the knowledge of hunter-gatherer ways of life on the Paleolithic archeology agenda. The ethnographic paradigm has been present in the Abric Romaní project since its beginning and most research at this site has been oriented toward the reconstruction of Neanderthal ways of life using ethnographic models. One of the most far-reaching models is that of the hunter-gatherer campsite, defined by several detailed ethnographic studies (Yellen 1977a; Binford 1978, 1987; O'Connell 1987; Fisher and Strickland 1991). This goal justified the excavation of such a large surface and is fully present in previous works (Carbonell 1992, 2002).

Abric Romaní has been considered particularly suitable for the application of the ethnographic paradigm. The accumulation of large palimpsests is less likely due to the geological formation processes that characterize most of the deposit. The archeological levels are thin, silty, sandy layers normally separated by thick strata of sterile travertine. They are therefore vertically well defined and can be considered as closed systems, since the travertine layers hamper the movement of remains between different archeological horizons. Moreover, the formation time of the archeological levels is relatively short due to the rapid sedimentation rate. These features allow the archeological

M. Vaquero (✉) · E. Allué · J. Vallverdú-Poch
Institut Català de Paleoecologia Humana i Evolució Social (IPHES), Universitat Rovira i Virgili (URV), C/Escorxador s/n, 43003, Tarragona, Spain
e-mail: manuel.vaquero@urv.cat

E. Allué
e-mail: eallue@iphes.cat

J. Vallverdú-Poch
e-mail: josep@prehistoria.urv.cat

E. Carbonell i Roura (ed.), *High Resolution Archaeology and Neanderthal Behavior: Time and Space in Level J of Abric Romaní (Capellades, Spain)*, Vertebrate Paleobiology and Paleoanthropology,
DOI: 10.1007/978-94-007-3922-2_10,

levels of Abric Romaní to be defined as high-resolution contexts, in which the blurring of spatial patterns due to occupation redundancy is less acute than in other archeological deposits. In other words, archeological time at Abric Romaní is a priori nearer to ethnographic time.

In addition, some data indicate that the archeological record and the spatial relations between remains are well preserved. In general, knapping scatters show little dispersion and refits suggest that lithic artifacts underwent limited postdepositional displacement. Bone taphonomy suggests that natural processes had relatively little impact. Few bones show rounded and polished surfaces due to water flow and when such alterations are observed, they have been produced by low energy dynamics. Evidence of reworking after burial is also scarce. Small lithic and faunal remains are clearly dominant, which also emphasizes the high level of preservation. Many refits are formed by small artifacts found close together and combustion structures are well preserved and highly visible. Other archeological items rarely preserved, such as plant remains, are also common.

However, beyond the general assessments made from the whole assemblage we should take into account the diversity of situations inferred from the taphonomic and spatial data. The archeological assemblage is not a single unit from a taphonomic point of view. Remains showing different taphonomic histories are often spatially associated. For example, bones showing evidence of water flow are located in the highest-density areas and are associated with remains that do not exhibit any alteration related to water dynamics. This is a first indicator of the temporal depth of the assemblage. There was a temporal gap between the deposition of the polished and unpolished bones. Polished bones correspond to depositional events older than those related to the non-abraded bones. This temporal incongruence is also indicated by the lithic assemblage. We observed differences in the degree of movement depending on the spatial and temporal location of the knapping events. The differential dispersion of spatially associated lithic scatters generated by different reduction sequences suggests that they have not undergone the same scattering processes and therefore originated in diachronic depositional events. This also provides a sense of temporality and suggests that the lithic assemblage is formed by remains corresponding to different depositional events and undergoing different postdepositional processes.

Variability in formation and preservation conditions is expressed both spatially and temporally. From the spatial point of view, there is a difference between the inner part of the rockshelter, close to the wall and protected by the ceiling, and the exterior area, beyond the line of large blocks. The interior sector was preferentially occupied by humans and was the place in which the hearth-related residential occupations were located. Occupation redundancy and the density of remains were higher, and postdepositional processes linked to occupation redundancy were therefore more intense. The exterior area was characterized by sporadic activities and showed a low density of remains. Some postdepositional processes, such as root damage, are more common towards the exterior areas. This spatial variability goes together with a time-related variability. The taphonomic history of the archeological remains largely depends on their temporal location in the formation sequence of the stratigraphic units. Knapping events carried out in the same areas exhibit very different scatters. Abraded bones are mixed with other remains that do not show any evidence of damage. This opens up one of the main issues that we will discuss in these pages: the temporal nature of the variability in the archeological assemblage and its implications for a behavioral interpretation. All the issues concerning archeological interpretation, from paleoenvironment to settlement patterns, are conditioned by the temporal resolution of the data.

Abric Romaní and its Landscape: Paleoenvironment, Subsistence Strategies and Settlement Patterns

Paleoenvironment played an important role in the behavioral patterns of Neanderthals, and archeological variability is often explained by changes in climatic conditions. Global climatic records from MIS 5 to 3 show rapid changes from cold to mild events (Van Andel and Davies 2003; Finlayson and Carrión 2007). However, floral and faunal records clearly differ between the northern and southern latitudes, especially on the Mediterranean peninsulas where there were significant refuge areas for plants, animals and humans (Willis et al. 2000; Finlayson and Carrión 2007; Carrión et al. 2008). High latitudes were dominated by cold fauna and mostly cold steppe-like landscapes, whereas southern latitudes were dominated by an arboreal cover with conifers and Mediterranean forests. Furthermore, due to the biogeographic diversity of the Iberian Peninsula, there are clear differences between the eastern coast and the continental highlands (Badal and Carrión 2001; Carrión et al. 2003, 2008; Finlayson and Carrión 2007).

However, paleoecological research points to the importance of time scales in the interpretation of paleoenvironmental data. A synthetic overview of the proxy data from level J provides paleoecological, climatic and environmental information from a micro to a global scale (Table 10.1). On a larger scale, the paleoecological evidence (see Allué et al. 2012) indicates a diverse local and regional environment, characterized by the coexistence of different landscapes: open areas, woodland, steep areas and river valleys. Climate

Table 10.1 Summary of environmental data from Abric Romaní level J

	Record	Microscale environment	LANDSCAPE Regional scale	Local scale	Climate
Geoarchaeogeology	Lithostratigraphy and sediments Micromorphology	Cryoturbation and temperate soil forming processes	–	–	Cold/temperate fluctuations
Palaeobotany	Charcoal	–	–	Pine woodland	Cold
	Pollen	–	Semi-open (48.3% AP) Pine woodland (39.5%) Steppe (29.2%) Prairie (13.2%)	Riverside (5.6%)	Cool (semi-cold) Mediterranean taxa (7.5%) Hygrophilous taxa (18.4%)
Palaeontology	Large mammals	–	Open habitats (45%) Woodland habitats (33%)	–	Cool (semi-cold)
	Small vertebrates	–	Woodland edges (50%) Open humid (23%) Water streams (22%)	–	Cool very humid
	Malacofauna	–	Dry steppe (37.9%) Scrubland (25.1%)	Humid areas (28.4%) Aquatic (4.7%)	Dry (41.9%) Humid (30.3%) Mediterranean taxa (2.7%)
Taphonomy	Mammals, small vertebrates and charcoal	% anthropic alteration % natural alteration	–	–	–

conditions were slightly colder than those of today, but were moderated by the proximity of the Mediterranean, as suggested by the presence of mesophilous and thermophilous taxa in the pollen record. Environmental conditions can therefore be characterized as interstadial.

However, the paleoecological interpretation is constrained by the temporal resolution of the data. A micro scale approach shows hydrological fluctuations that probably affected the habitability of the shelter. Dry periods probably enhanced the habitability of the shelter, whereas during wetter periods it would have been avoided by humans. The level J complex (see Vallverdú and Courty 2012) shows environmental changes of centennial to millennial resolution whose age seems to lie in Greenland Interstadial 13 (49–47 ka). Microstratigraphy indicates a rhythmic pattern of change in sedimentary processes, with at least four cycles of alternation between dry and wet conditions. Sublevel Jb corresponds to the second cycle, sublevel Ja to the third, and sublevel Jsup to the fourth. There is a significant association between anthropic sediments and dry lithofacies, whereas human occupation is practically non-existent during the wetter periods. Cryoturbation has been documented in the first and second cycles, affecting sublevel Jb, although taphonomy indicates a very low percentage of natural disturbances, probably due to rapid burial. The time estimate for each cycle, according to a sedimentation rate of 0.6 m/ka (Bischoff et al. 1988), suggests that the deposition of level J corresponds to a temporal gap of decades to centuries, whereas the non-deposition periods may correspond to a millennial scale.

According to the paleoecological records, Neanderthals were well adapted to a variety of environmental and landscape conditions. Their territories were varied and show different biogeographical modes. This affected adaptation and behavior patterns, indicating versatility among Neanderthals and showing their adaptive capabilities. It seems that subsistence strategies were not determined by environmental or temporal constraints. The availability, variability and predictability of resources depend on environmental conditions, so environmental changes can modify human adaptive strategies. Nevertheless, this affects these groups to different degrees in terms of decision making, selection patterns in relation to energy waste, and quality of resources. There are no significant changes in subsistence patterns throughout the Abric Romaní sequence, in spite of the abrupt environmental modifications inferred from the pollen analysis (Burjachs and Julià 1994). It therefore seems that subsistence behavior was little affected by climatic instability in the long term.

In spite of some negative constraints (e.g. the orientation of the rockshelter to the northeast), the location of Abric Romaní exhibits some features that make this place especially suitable for human settlement. First, the site is located in a strategic pathway between the interior Ebro basin and the coast: the Capellades gorge is one of the few natural passages crossing the Prelittoral Range. It was therefore a place commonly frequented by humans during their movement across the land, but probably also by the herbivore herds that constituted the main food resource of Neanderthals. In this sense, the use of the rockshelter as a stopping place during human movements seems probable. Second, Abric Romaní is located at an ecotone, which allows for easy access to different ecosystems—mountain ranges, plains and fluvial areas—allowing the exploitation of different resources. Third, the springs of the Capellades area were attractive to human settlement. However, it is not so clear that all these factors were equally important in the selection of site location. Other Mousterian sites in the margins of the Ebro basin (Roca dels Bous, Gabasa, Estret de Tragó, Fuentes de San Cristóbal) are also located in natural pathways. All these sites are on the margins of rivers that cross mountain ranges and connect the Ebro basin with other regions. At a first sight, this suggests that rockshelters in natural corridors were selected for occupation, and supports the view of Neanderthal groups as highly mobile.

However, while a variety of resources were available at these sites, not all were exploited equally by their Neanderthal inhabitants. Rather, subsistence patterns indicate a preferential choice of certain resources. In level J, the clear dominance of red deer (*Cervus elaphus*), horse (*Equus ferus*) and bovid (*Bos primigenius*) in the faunal assemblage suggests that hunting activities were focused in woodland and open areas. Chamois (*Rupicapra pyrenaica*) are rare (only eight remains, 0.6% of the identified specimens), indicating that, steep, rocky areas were infrequently exploited. Similar selectivity is also evident in the age distribution of the prey, which shows a dominance of adult individuals, although old and young animals are also represented. Finally, we have not found fish or shellfish remains indicating the exploitation of fluvial areas for subsistence. However, the common presence of fluvial cortex in some of the lithic resources brought to the shelter suggests that river valleys were frequented for lithic provisioning.

Whether this subsistence pattern is due to a selection in the strict sense of the term—choosing something in preference to another or others—or simply shows the cognitive or technological inability of Neanderthals to exploit certain species or ecosystems remains to be demonstrated. The faunal assemblage of level J reflects a generalized pattern among the Neanderthals of the Iberian Peninsula. According to this pattern, as the role of humans increases in the formation of faunal assemblages, so do the percentages of red deer, horse and bovid. These species form the bulk of the bone assemblages of most Iberian sites, especially in those that exhibit a residential component and whose faunal assemblages are clearly the product of human agency. This is the case in the Ebro basin sites mentioned above, but it is also seen in Cantabrian sites (Yravedra 2003, 2006; Castaños 2005; Martínez Moreno 2005). Mountain animals such as goat (*Capra pyrenaica*) and chamois were relatively rarely exploited in the Middle Paleolithic of Iberia. In several assemblages, such as 120 Cave, Cova Negra, Beneito Cave, Gabasa, Amalda and Esquilleu, goat remains appear to have been accumulated by carnivores, while red deer and horse were brought to the sites by humans. As carnivore input becomes more important, we see a rise in the presence of animals from mountain ecosystems. However, this pattern cannot be attributed to a behavioral handicap of Neanderthals. Although they are relatively scarce, some Middle Paleolithic assemblages are dominated by goat remains derived from human agency (Moncel et al. 1998; Adler et al. 2006).

Other northeastern sites which share the Abric Romaní's location along natural corridors also exhibit a pattern of infrequent exploitation of mountain animals. Humans preferentially selected red deer and horse in Gabasa (Blasco 1997) and Las Fuentes de San Cristóbal (Rosell et al. 2000). Mountain animals are absent from Las Fuentes de San Cristóbal and are associated with carnivore activity in Gabasa. The faunal evidence is less straightforward in Roca dels Bous and Estret de Tragó due to the poor preservation of bones (Martínez Moreno et al. 1994; Mora et al. 1992). Mountain animals are represented at these sites along with red deer and horse, but their relationship to human agency has not yet been fully ascertained. In these sites, the residential nature of the human occupation is suggested by the presence of hearths and hearth-related accumulations. Otherwise, such cave sites located closer to the shoreline or farther away from the main natural corridors (Teixoneres Cave, Arbreda Cave, 120 Cave, Gegant Cave) are characterized by a very sporadic human presence and the dominant role of carnivores in the formation of their faunal assemblages.

There is no evidence in Abric Romaní suggesting that small animals such as rabbits, fish, mollusks, or birds formed part of the diet. There is little evidence for the exploitation of small animals in the Iberian Middle Paleolithic, although some cases have been particularly well documented in recent years. In Vanguard Cave, evidence of use of a wide variety of marine resources has been found, notably among which are an accumulation of mussel (*Mytilus galloprovincialis*) shells and several seal (*Monachus monachus*) and dolphin (*Delphinus delphis*) remains, two of them showing evidence of processing by humans (Stringer et al. 2008). In addition, a broad spectrum of small animals, including rabbits, tortoises and birds, was processed and consumed by humans in the Middle Pleistocene levels of Bolomor Cave (Blasco 2008; Blasco and Fernández Peris 2009; Sanchís and Fernández Peris 2008). This suggests that not only Neanderthals, but also their ancestors, were perfectly able to exploit small game. Although they focused their subsistence strategies upon highly ranked resources, low yield species were taken when needed. This indicates that they had the capacity to engage in these behaviors, regardless of the frequency with which they did so. Why these behaviors were not more frequent remains an open question, although demographic and social explanations have recently been suggested (Stiner et al. 2000; Kuhn and Stiner 2006).

Level J has also provided information about the exploitation of plant resources. Fuel was needed in all domestic activities involving fire and wood was the best choice for use as fuel. Various types of fuel, including wood, have been identified at several Mousterian sites from different regions, demonstrating the variability of resource exploitation and fire technology (Théry-Parisot 2001; Meignen et al. 2007; Albert et al. 2007). At Abric Romaní, wood procurement was also selective. Pine (*Pinus* sp.) comprises more than 99% of the charcoals recovered at level J, despite the variability of trees and shrubs identified in the pollen analysis. However, this behavior could be strongly conditioned by the composition of the forested zones growing in the area closest to the rockshelter, since pine is also the dominant tree in pollen analysis. In this case, choice is based on the availability of pinewood, which is nearby and easy to collect. Furthermore, the taphonomic analyses of charcoal remains found at the site show that it was high quality wood and was therefore optimal firewood for maintaining hearths. However, plant resources used as food have not been detected. Plant food does not usually appear in Middle Paleolithic assemblages due to the perishability of plant remains, although the methods employed in processing plant food and the scarcity of edible plants in some areas maybe played a role in this under-representation. Nevertheless, in some cases, for example at Kebara Cave (Baruch et al. 1992; Lev et al. 2005) and Gorham's Cave (Carrión et al. 2008), the remains of edible plants have been recorded and plant food was probably part of the Neanderthal diet.

The lithics yield a more detailed picture of the patterns of movement and the size of the territory covered by Neanderthals. Some of the raw materials provide the best evidence of the exploitation of the immediate surroundings of the site. Limestone cobbles were gathered in the fluvial deposits of the Anoia River at the foot of the site, while quartz nodules were found in slope deposits close to the rockshelter. However, the nearest raw materials were not the most frequently exploited. Although chert occurs only rarely within 5 km of the site, chert artifacts make up 75.6% of the lithic remains at level J (see Vaquero et al. 2012). The recovery of some chert nodules with primary cortex from the Valldeperes formation indicates that they were gathered near the outcrops, which are located at least 25 km from Abric Romaní. The exogenous chert indicates that some resources were procured even farther away, although the precise distance is unknown. Nevertheless, the bulk of lithic resources were obtained from sources within 20 km of the site.

The preferential use of chert indicates a selective behavior similar to that observed in subsistence strategies. The most abundant raw materials in the vicinity of the site are not the most represented, because economic constraints act within the framework of previously adopted preferences. Something similar may be happening with faunal resource provisioning. The bone assemblage of level J indicates that carcass transport was conditioned by prey size. This is in agreement with the tenets of the optimal foraging theory (Winterhalder and Smith 2000). As carcass size increases, the transport of anatomical parts becomes more selective in accordance with a minimizing weight strategy. However, these economic factors influenced the array of resources previously selected by human groups. As in the case of lithics, some previous selective decisions defined the range of resources in which the cost-and-benefit principles operated.

Taking as a reference the three provisioning zones distinguished by Geneste (1988, 1989) in southwest France, the intermediate zone—ranging between 5 and 20 km—was the most significant in level J. According to Geneste, the lithic resources from this intermediate zone show the highest variability in terms of the forms in which the materials were transported to the sites. This is in agreement with the evidence from level J. Chert was transported in all the conceivable forms: complete nodules, partially shaped cores, selected flakes and retouched tools. However, to fully understand these general trends, the temporal variability in lithic provisioning during the formation of level J should be taken into account. As pointed out by Geneste (1988, p. 490) for the Grotte Vaufrey, the lithic assemblage was produced by the taphonomic mixing of different types of behavior that may have been temporally dissociated. In level J, certain provisioning strategies were more common in some occupation events or during some occupation periods, but infrequent in others.

The behavioral strategies identified by archeologists may be partly determined by assemblage resolution. The high variability of transport forms inferred from lithics is related to the event-focused approach used in this work. Identifying specific transport events allows us to recognize how materials from the same source were introduced into the site in different forms. However, this approach is not possible in faunal analysis and the maximum available resolution is determined by the categories of taxonomy and body size. As some ethnoarcheological studies (Monahan 1998) have shown, pooling together all the bones from the same species or body size can conceal the broad variability in modes of carcass transport evident at the single episode level.

Once in the rockshelter, both faunal and lithic resources were intensively exploited. The processing of faunal resources was exhaustive, including the systematic breakage of bones to extract marrow. Cut-marks indicate that all the stages of faunal processing—skinning, dismembering, eviscerating, defleshing, extraction of tendons and fat—were carried out at the site, although some of these activities are scarcely represented due to variations in the anatomical representation of prey. Furthermore, fire was an important element in food preservation and processing.

Lithic resources, especially chert nodules, also exhibit intense exploitation, although we have also documented some uneconomical events. Chert cores were reduced to exhaustion and the production of small flakes was dominant. In some instances, nearly any item was considered suitable for exploitation, as shown by the reduction of some blanks that could yield only a few small flakes. However, the intensity of core reduction is not correlated with high percentages of retouched artifacts or the presence of heavily reduced and resharpened tool types, which are normally expected when there is pressure to maximize the profitability of raw materials. Denticulates and notches are overwhelmingly dominant in level J and make up 84.6% of retouched artifacts. Most retouched artifacts—as well as most large flakes—were introduced as single artifacts from outside, in the framework of occupation events different from those during which most core reduction sequences were performed. It therefore seems that core reduction and tool retouch were dissociated behaviors affected by different constraints. Core reduction, mainly linked to residential occupations, would probably have been more affected by raw material scarcity, especially if we take into account that chert nodules were absent from the immediate surroundings of the site. On the other hand, single artifact inputs, more associated with non-residential short-term visits, would have been less constrained by raw material availability.

Abric Romaní: A Residential Site?

Settlement patterns are currently considered a major factor in explaining the variability of archeological assemblages. The first source of variability in settlement patterns is defined by the opposition between residential and non-residential sites. Unlike other aspects of settlement variability, such as occupation length, this can be easily approached through the archeological record. The first appearance of residential locations is a fundamental issue in paleoanthropological research, especially in light of the social implications of the residential function. This issue is closely linked to the debate about the first occurrences of hearths as the focus of spatial organization because hearths are an essential feature in ethnoarcheological residential campsites.

In spite of the claims about an early emergence of residential sites or home bases in the Lower or early Middle Pleistocene (Isaac 1977, 1978; Gowlet 1997; Lumley et al. 2004), they tend to be more common from the late Middle Pleistocene, coinciding with the appearance of Middle Paleolithic technocomplexes (350–300 ka). This is associated with the first clear evidence of hearths and hearth-related accumulations, suggesting the occurrence of household areas (Lumley 1969; Michel 1982, 1994; James 1989; Hallegouet et al. 1992; Fernández Peris et al. 2000; Meignen et al. 2001). These arguments have been used to suggest that the Middle Paleolithic represents a major reorganization of settlement and social systems, and thus a significant developmental stage in human evolution (Rolland 1999, 2004). According to Rolland, home bases entailed the first divergence from a primate land use pattern, which was based on a spatial dissociation between daytime and nighttime settings. Fire-making allowed for better protection against nightly carnivore threats, thereby eliminating the need to separate sleeping sites from domestic activity sites. The definition of a home base is based on subsistence—it is the place where animal resources are introduced and shared—but its primary significance is social; it is the place where socializing and knowledge transmission take place.

At any rate, residential locations were a well-established type of site during the Late Pleistocene and were a fundamental feature of Neanderthal settlement systems. Combustion structures and hearth-related accumulations are common in Middle Paleolithic sites. At least some of the occupations that formed the level J assemblage—and especially those that yielded most of the remains—were residential in nature. Different resources were transported to the rockshelter for the purpose of completing a wide range of activities that are typical of a residential context, including the production of lithic artifacts and the processing and consumption of animal resources. Other activities, like the processing of plant resources or hideworking inferred from usewear analyses, have also been documented.

Zooarcheological and taphonomic analyses indicate that humans were the main agent in the formation of the faunal assemblage. Some coprolites and bones with gnaw marks have been detected, but they correspond to carnivore scavenging of bones previously discarded by humans (Rosell et al. 2012). This establishes a clear-cut difference between Abric Romaní and several Iberian late Middle Paleolithic sites in which the contribution of carnivores to the faunal assemblage is more significant (Maroto et al. 1996; Villaverde et al. 1996; Blasco 1997; Terradas and Rueda 1998; Fernández-Jalvo and Andrews 2000; Yravedra 2006). The characteristics of the site—an open rockshelter with a very high ceiling—make it quite unsuitable for use as a carnivore den. Only in the uppermost layers, in which the appearance of the site is more markedly cave-like due to the progressive infilling of the rockshelter, is there a slight increase in carnivore bones, although the role of humans is always dominant.

However, the clearest evidence of the residential character of the Abric Romaní occupation is the role played by the hearths. More than 60 combustion structures have been identified in level J and most of the archeological remains were found in the hearth-related areas. This does not mean that all the visits to the site were organized around hearths, but the formation of hearth-related areas characterized the occupations during which most of the archeological record was generated. Hearth-related areas are the main spatial structure in all the archeological layers so far excavated and it seems clear that most activities were carried out around hearths. The densest concentrations of small bone and lithic remains—the best indicators of in situ activities—were associated with hearths. Furthermore, refits and Raw Material Units (RMUs) prove that knapping activities were preferentially carried out in these areas. The spatial distribution of bone splinters, cut-marked bones and bone refits indicates that faunal processing was also focused in the hearth-related areas. In level J, most hearths are concentrated beyond the dripline in the zone between the shelter wall and the outer line of blocks. These blocks represent an additional protection against the northern winds and made this the most sheltered and most suitable area for residential occupations.

Hearths are the best indicator of hunter-gatherer camps. Although hearths can occasionally appear in a non-residential context, like kill-butchery sites (Yellen 1977b), they are always present in residential locations. The identification of such structures is an essential requirement in characterizing an archeological assemblage as a residential camp in the ethnoarcheological sense of the term. As has been pointed out elsewhere (Vaquero and Pastó 2001; Vaquero et al. 2004), these hearth-related areas can be correlated with the household or domestic areas typical of hunter-gatherer camps (Yellen 1977a; O'Connell 1987; O'Connell et al. 1991; Fisher and Strickland 1991). In these camps, the domestic area is the location where most activities are carried out. Only some extensive or particularly messy activities are performed outside of these areas. This hearth-related behavior allows us to draw some social and cognitive inferences because hearths are essentially a social space, where most face-to-face interactions take place. It is the area where people talk while working. In fact, if somebody does not want to be disturbed by the chatting of others, the best option is to work away from the hearth (Binford 1978). This suggests that Neanderthal populations employed social and communication strategies similar to those exhibited by modern hunter-gatherers. It also suggests a social structure based on household units, whose spatial signature is similar to that of the familial units that make up modern hunter-gatherer bands.

Each domestic unit living in a campsite creates its own hearth-related domestic area and this produces a distribution pattern characterized by discrete clusters.. The best archeological correlation of this pattern can be found in some open-air Upper Paleolithic sites, such as those from the Paris basin (Leroi-Gourhan and Brézillon 1972; Olive 1988; Schmider 1992; Pigeot 2004). However, it has also been documented in some Middle Paleolithic and Middle Stone Age sites (Deacon and Deacon 1999; Henry et al. 2004; Wadley 2004) and is the dominant spatial pattern throughout the Abric Romaní sequence, characteristic of levels H, I, K and L (Vaquero et al. 2001a; Vallverdú et al. 2005; Vaquero 2008). Patterns like these are particularly visible in short-term camps, and these archeological levels were interpreted as the consequence of short occupations. As occupation length increases, the dispersion of remains due to postdepositional processes or cleaning activities and the overlapping of different domestic areas tend to blur this discrete pattern, resulting in a more continuous distribution (O'Connell 1987). In level J, we see this trend towards a continuous scatter of remains. The great number of hearths and the proximity between most of them suggests an overlapping of domestic areas. For example, the overlying level, level I, clearly showed this discrete pattern of distribution associated with hearth-related areas (Carbonell 2002). The RMU scatters and length of connection lines were shorter than in level J, and the number of remains found was also considerably lower. These data suggested a relatively short formation period for level I as well as short-term occupations (Vallverdú et al. 2005).

The importance of hearth-related areas indicates that the rockshelter had a residential function, although it seems unrealistic to assume that during the formation of level J the site was always used in the same way. The rockshelter was probably visited in different circumstances and for different purposes. The spatial pattern shown by artifacts corresponding to different modes of transport suggests that at least two basic types of rockshelter use are represented in level J: hearth-related residential occupations and short non-residential visits (Vaquero et al. 2012). Residential occupations would be associated with the introduction of bulk resources and the development of core reduction sequences. The ephemeral non-residential visits would be indicated by the abandonment of single artifacts or sets of single artifacts forming part of the personal gear transported during displacements. However, most of the remains found in this level, especially those corresponding to the main accumulations, are associated with hearths and were produced in the framework of residential occupations. Although not necessarily the most common, residential episodes are the most archeologically visible, since they tend to generate a great amount of remains. More ephemeral visits were probably common, but the remains left during such occupations were scarce and tend to be obscured by the more perceptible residential accumulations. However, the importance of these short visits cannot be dismissed, regardless of the amount of remains that can be attributed to them.

Activity areas are difficult to define in level J due to the continuous distribution of archeological items. Analogy with modern hunter-gatherers does not offer an easy solution because reoccupation of camps by hunter-gatherers is little discussed in the ethnoarcheological literature (O'Connell 1987; Kent 1991). The activity areas of level J seem to be non-specialized and waste resulting from meat, bone, plant or lithic processing was abandoned in the same locations in which those activities were performed. This is a common behavior among modern hunter-gatherers that present spatial patterns in which the functional segregation of activities is little developed (Yellen 1977a; O'Connell 1987; Fisher and Strickland 1991).

In archeological contexts, complex assemblages of activity areas are more common than individual occupational episodes. Difficulties in verifying the characteristics of activity areas using material culture are also common in present-day campsites (Gould and Yellen 1991). Level J is a palimpsest in which activity episodes are linked to the use of fire (Vaquero and Pastó 2001). At the beginning of the excavation, the scattering of combustion residues in the central zone of the rockshelter hampered the delimitation of hearths. In the internal area, combustion activity areas appear as the geological palimpsest formed by the accumulation of sublevels Ja and Jb. Microstratigraphy, taphonomy, refitting, or other evidence of cultural and natural processes must be considered when reconstructing archeological palimpsests. The numerous individual activity episodes of level J, as seen for example in the lithic production or the introduction of raw materials, correspond to a high-resolution time scale. Other lines of evidence, such as the paleoenvironmental interpretation, suggest another temporal level in which the human activities can be grouped using the temporal scale of Late Pleistocene environmental changes. This temporal record of human activities and environmental change defines the large scale geological palimpsest in level J (Bailey 2007). The recognition of spatial patterns is often distorted by the duration of activities and the reuse of activity areas. In conjunction with reoccupation, this limits the visibility of activity areas and makes the use of ethnographic analogy for interpreting the archeological record more difficult.

Ethnographic analogies are problematic given the difficulties in establishing analytical units in the level J spatial pattern. The uncertainty presented by the geological palimpsest of the internal zone and its spatial fragmentation due to previous excavations shows the limitations of the analogy posed by Henry et al. (2004). Nevertheless, the bimodal pattern in the spatial distribution of the combustion structures with respect to the shelter wall has been observed in level J (see Sañudo et al. 2012) and on other archeological levels of Abric Romaní (Carbonell et al. 2007). The sedimentary record indicates that there are different types of combustion structures in level J. The combustion structures close to the shelter wall are characterized by a very scarce cultural transformation, while those located in the central area show many cultural transformations in the form of combustion residue compaction. In the central zone there were also accumulations containing abundant charcoal and burnt travertine blocks.

The sedimentary data provides some clues on the use of space and the activity areas in level J. The microstratigraphy and the anthropic microfacies of sublevel Ja suggest a strong analogy with some open-air sites described in ethnoarcheological studies (Nicholson and Cane 1991). Combustion structures close to the wall are similar to the hearths of sleeping areas, while those in the central zone are similar to communal hearths. The burnt rocks and charcoal-rich sediments in front of the *Coveta Ripoll* and below the dripline would correspond to external refuse areas where waste from hearth cleaning was dumped, like in Thomas's exogenous model (1983). Finally, there are some empty areas in front of the *Coveta Ripoll* and along the western part of the wall. Considering the distances between the combustion structures of level I to level N, this western wall has been established as the back wall of the shelter (Carbonell et al. 2007). However, the small area of the *Coveta Ripoll* shows a use similar to that of small caves and rockshelters. Level J therefore presents both open areas, showing occupation dynamics characteristic of open-air sites, and closed areas used like small shelters and caves.

It is difficult to identify activity areas in the *Coveta Ripoll*. The spatial distribution of the combustion structures and the huge volume of remains indicate an occupational palimpsest. Differences between close-to-the-wall activity areas and the central zone have been pointed out in this context of occupational palimpsests. Identifying the spatial segregation of activities therefore depends on the development of occupational palimpsests and the thresholds of archeological visibility in site structure. The visibility of this pattern may indicate that the alterations produced by reoccupation are not so marked, and therefore a certain degree of integrity can be ensured in the assemblages of material remains from the most recent living floors (Stevenson 1991). The aspect of the organization of the residential units of level J related to maintenance activities of the domestic space, together with the bimodal pattern of fire use, shows different activities near the rockshelter wall and under the dripline. Sleeping, resting, eating and recovering were performed in the internal belt of this domestic space. In the external belt, repetitive tasks and chores were performed. This spatial segregation of behavior to accommodate non-specialized activities may have higher archeological visibility than short duration living floors, organized around a single combustion activity zone. Level J presents zones with differing archeological visibility, in which it is possible to propose:

(1) Occupation floors with a simple structure that were formed around a hearth and are highly visible through the identification of reduction sequences and lithic refits.
(2) Occupation floors with a complex structure that exhibit spatial patterns and uses similar to those documented among modern groups (Henry et al. 2004). Although these are difficult to identify through artifact distribution and refitting, they may be visible when fire use is considered as a primary cultural structure and the sedimentary record of combustion structures is well preserved.

Once the residential character of the rockshelter during several occupation events is established, we can discuss the second axis of site variability, which is related to occupation length and/or group size. In previous works, intralevel and interlevel variability was interpreted as the product of differences in occupation length (Vaquero 1999; Vaquero et al. 2001a; Vallverdú et al. 2005). This variability has previously been summarized as follows:

- Short duration occupation pattern. Isolated combustion areas measuring 4 m in length or 10–15 square meters with well delimited clusters of remains. Lithic and faunal assemblages suggest that material resources of strictly local origin are scarcely exploited, except plant resources are acquired for fire use.
- Intermediate duration occupation pattern. Interrelated combustion activity areas with well-delimited accumulations of remains and few refitted archeological remains which form a group of accumulated clusters of 10 m in length or 100–150 square meters.
- Long duration occupation pattern. Interrelated combustion activity areas with dense accumulations and poorly delimited accumulations of remains, with abundant refits, forming a group of accumulated clusters of 15 m in length occupying at least 200 square meters.

Level J was interpreted as an occupation pattern of long duration. Nevertheless, the recognition of the spatial patterns can be conditioned by the equifinality of the analytical techniques used to define duration and site size. The arguments from ethnographic analogy are the starting point for archeological interpretation, although they also show our limitations in terms of our knowledge of modern and archaic hunters-gatherers (Binford 1978). We will discuss this issue in the following section.

Temporal Patterns: Events and Structure in Assemblage Formation

These questions related to settlement patterns should not be approached without taking into consideration the temporal nature of the archeological assemblage. This temporal dimension is particularly significant if our research goal is a behavioral interpretation of the archeological record, but time resolution is also very important in dealing with other domains, like paleoenvironmental dynamics or site formation processes. Archeostratigraphy has allowed two basic subunits to be identified in this layer—sublevels Ja and Jb. These units do not correspond to occupations in the ethnographic sense, but rather they are palimpsests produced by a succession of occupation episodes. Earlier we presented evidence indicating the temporal depth of these units, such as the spatial association of remains with different taphonomic histories. However, there are other data pointing in the same direction:

- Microstratigraphic analysis (see Vallverdú and Courty 2012) has identified in some hearth deposits different combustion phases separated by horizons of natural sedimentation.
- Remains associated with carnivores—bones with toothmarks and coprolites—does not show any preferential horizon of accumulation (Rosell et al. 2012). These remains are scattered throughout the entire thickness of sublevel Ja, which suggests different visits of carnivores. If we take into account that these visits occurred after the human occupations as shown by the taphonomic data, this would indicate a succession of different human occupation events.
- Two hunting periods have been identified in sublevel Ja, autumn and winter for *Cervus elaphus* and *Bos primigenius*, and spring for *Equus ferus*. Since a continuous annual occupation seems unlikely, these would indicate that occupation events took place during different seasons. These data also suggest that different animals were targeted at different times in the annual cycle.
- There is a marked variability in the dispersion radius of different RMUs in the same areas, suggesting that they

were not equally affected by postdepositional dispersion processes (see Vaquero et al. 2012). The more scattered RMUs would correspond to the first knapping events in the formation sequence of the archeological level. The latest events—those less affected by dispersion associated with human activity—would be the most clustered. Using this criterion, we have identified a minimum of three occupation phases in the central zone of sublevel Ja.

- Lithic refits show a unidirectional pattern suggesting that the occupation of different areas was not contemporaneous. Refitting also indicates that recycling was common. The kind of recycled artifacts—nearly exhausted cores, flakes and fragments of broken cobbles—suggests that they were refuse from previous events. In some cases, the stratigraphic location of the recycled artifacts at the top of the archeological layer indicates that the recycling event took place during the latest occupations.

We have tried to establish what influence the palimpsest nature of layer J has had on assemblage variability. Specifically, we have examined whether we can identify behavioral variability that would have gone unnoticed if the assemblage had been considered a single synchronous entity. In the first place, temporal dynamics are important in evaluating the differences between level J and the other archeological layers at the site. As pointed out above, other levels—H, I, K, L and N—are characterized by discrete distribution patterns in which well-defined clusters of remains, always associated with hearths, can be recognized. The hearth-related accumulations found in these levels seem to correspond to short-term residential occupations. In addition, the absence of bidirectional links between different accumulations of the same level suggests that they were not contemporaneous, but represent diachronous occupation episodes by small groups. Level J exhibits several other differences. In addition to the considerably higher number of remains, this level shows a less clustered spatial pattern, in which the hearth-related accumulations are not so sharply defined. Moreover, lithic refitting seems to indicate an inter-connected space, although unidirectional refits are dominant. In this context, two alternative explanations can be proposed for the differences between level J and the other levels:

a. Level J is the result of longer occupations and/or larger groups.
b. Level J is characterized by a longer formation period and the number of occupation events is higher than in other layers, although the nature of these occupations is basically the same.

In previous works (Vaquero 1999; Vaquero et al. 2001), the first hypothesis was preferred and level J was considered to be the result of longer occupations. However, the fine-grained approach adopted in this work has raised some doubts as to the veracity of this picture. This discussion has important implications from the viewpoint of the variability of Middle Paleolithic settlement patterns. The presence of residential camps during the Middle Paleolithic is well established, at least from the onset of the Late Pleistocene. Several sites have provided hearth-related assemblages that can be characterized as the domestic areas typical of this kind of occupation (Meignen 1993; Rigaud et al. 1995; Barton 2000; Meignen et al. 2001; Martínez Moreno et al. 2004). However, the variability of this residential mode of occupation in terms of occupation length and group size is the subject of debate. At present, the best archeological examples of residential sites correspond to short-term events, but this may be an artifact of the archeological method. The main problem is related to the difficulty in identifying long-term occupations using archeological criteria. The possibility of explaining the archeological variability in terms of occupation length remains an open question due to the difficulties in differentiating long-term events from palimpsests, as their archeological consequences are similar.

One argument normally used to define long-term camps is the number of remains found. Archeological assemblages with large numbers and high densities of remains tend to be considered as evidence of long-term occupations (Richter 2006). However, the amount of remains depends on the number of activities and/or introduction events represented in an assemblage. Therefore, the quantity of remains is a time-dependent variable and may vary depending on the length of the period during which the assemblage was formed. A rich assemblage may be produced by long-term occupations, but it may also be a palimpsest formed by the overlapping of many short-term events. In fact, occupation redundancy can be expected in cave and rockshelter sites, which remained habitable during long periods and were probably landmarks in the cultural landscape of hunter-gatherers.

Other criteria used to identify long-term camps can be also affected by such occupational redundancy. A wide range of activities may be correlated with long-term camps, but only if it can be argued that all these activities were carried out during the same occupation, which seems difficult or impossible to demonstrate. As the number of depositional events increase, the abandonment of artifacts used in specific activities becomes more likely, including some unusual activities. As a consequence, a wider array of activities can be expected to be recognized by use-wear analyses. On the other hand, assemblages formed by few depositional events would tend to show less use-wear variability, and only the most common activities would probably be represented. In fact, use-wear patterns in these assemblages can mimic those found in "specialized" contexts. In addition, it is possible that some activities inferred from the use-wear analyses of lithic artifacts were not carried out at the site. In level J, most artifacts showing use-wear traces correspond to transported artifacts that are presumed to have had long use lives before their arrival at Abric Romaní. Most of these transported items seem unrelated to the hearth-related accumulations, and therefore the use traces identified on them do not provide information about the activities carried out during the residential occupations.

The arguments based on lithic and/or faunal variability to infer occupation length are flawed by these temporal uncertainties. All lithic categories—manuports, cores, tools, flakes from all the reduction stages—are represented in the lithic assemblage of level J, which would suggest that knapping activities were characterized by the execution of entire *chaînes opératoires* at the site. Nevertheless, this entire-reduction-sequence picture is the result of pooling together a great diversity of individual technical events (Vaquero et al. 2012). In fact, entire reduction sequences are less common than other kinds of depositional events, like the introduction and discarding of single blanks. Moreover, many *chaînes opératoires* are only represented by certain stages of the reduction sequence. Such time averaging problems also affect other criteria sometimes used to infer occupation length, like the proportions of artifact types or the percentage of carnivore damage.

The example of level J is especially relevant to interpreting intrasite spatial patterns. According to ethnoarcheological studies, camp structure may be a good indicator of occupation length, since there are some features that tend to change as stay length increases. Moreover, site structure may also be a good proxy to infer group size. Three aspects are particularly important in characterizing the residential occupations of level J:

a. The temporal relationships between different domestic areas. Contemporaneity of different areas would suggest that different domestic units were living together during the same occupation. These data can therefore be used to infer group size.
b. The different distribution of activities and the complementarity between different activity areas. In hunter-gatherer camps, most activities are carried out in hearth-related domestic areas, but some extensive or messy activities may be performed in specialized areas. The probability of those activities increases with occupation length and special activity areas tend to be more common in long-term camps.
c. The appearance of secondary disposal areas. These areas tend to be located beyond the limits of the household areas and are more common—and especially more archeologically visible—as occupation length increases.

However, level J suggests that these structural criteria also depend on temporal dynamics. As Conard and Adler (1997) have pointed out, occupational contemporaneity is often difficult to demonstrate. Refitting and RMU distribution are the best approaches to the temporal relationships between the artifacts. The absence of refits between two sets of remains does not prove that they were not contemporaneous, but we think that non-contemporaneity should be considered as the null hypothesis that we must try to disprove. At a first sight, the lithic refitting maps, especially in sublevel Ja, show several connection-lines between different areas. Along with size-sorting, this was one of the main arguments used to characterize sublevel Ja as resulting from more stable occupations than those recorded at other archeological levels (Vaquero 1999; Martínez Molina and Rando 2001). However, if we take a closer look at the character of these refits, the evidence seems less straightforward. In sublevel Jb, the two main activity areas seem to be independent and unrelated entities. The rare connections between them can be explained as the result of unintentional processes. Connections between these main clusters and other areas seem to correspond to recycling during posterior events.

Temporal patterns seem more complex in sublevel Ja. Putting to one side the refits that can be explained by unintentional factors, there are several conjoins that indicate an intentional transport of artifacts between different activity areas. However, there is no compelling evidence of robust bidirectional patterns to suggest that different activity areas were contemporaneous. The unidirectional connection patterns suggesting temporal succession are clearest. Few RMUs are distributed in more than one activity area and when this happens it seems to correspond to recycling events. Only RMU Lim-012, whose reduction was distributed in two or three different zones, may indicate contemporaneity between different areas. The bulk of the lithic evidence suggests that level J was the product of a series of small-scale events. The coexistence of different domestic units during the same occupation cannot be supported using these data.

This temporal dimension is also fundamental to approach the other two criteria used to characterize residential occupations: special activity areas and disposal areas. In other levels of the Abric Romaní sequence, like levels I and K, we have documented some extensive areas characterized by a clear dominance of bone remains and very low densities of lithics. They were interpreted as special activity areas where processing of animal carcasses was carried out. This pattern has not been found in level J. The main spatial feature of this level is the opposition between the hearth-related domestic areas, where most activities were performed, and the outer non-domestic areas, which are characterized by a very low index of activity. A special character could be proposed for these outer areas based on the use-wear data, since the artifacts showing the most developed traces—suggesting more intense use—were found outside the central domestic zones. However, these tools showing intense use-wear traces are transported artifacts that seem unrelated to the domestic events. They therefore cannot be used to suggest a functional segregation of activities during the same occupation events.

The identification of secondary refuse areas can also be conditioned by temporal dynamics. Sublevel Ja shows a spatial pattern characterized by the size sorting of both lithic and faunal remains. Small remains are clearly dominant in the inner hearth-related areas, while the frequency of large items increases towards the outside. This was interpreted as the product of the depositional processes that characterize the formation of drop and toss zones (Binford 1982). The inside would correspond to the activity area, in which there is a preferential accumulation of small remains. Large remains produced during these activities would be discarded towards the exterior, forming a low density belt defined by relatively high percentages of large items. However, the event-focused approach used in the study of the lithic assemblage indicates that there may be other processes at work in this spatial distribution. First, refits between the knapping areas and the purported dumping area are scarce, suggesting that the action of tossing large artifacts towards the outside was not common. Second, RMU distribution indicates that most large artifacts were not produced at the site, but transported as single blanks from outside. In fact, two different depositional patterns contributed to the formation of the lithic assemblage, each characterized by a different spatial distribution. On the one hand, knapping events, which produced the bulk of the lithic assemblage, are clustered in the hearth-related accumulations. On the other hand, artifacts introduced as single blanks tend to be more evenly distributed, and are well represented in the low-activity areas. Overlapping of these different depositional contexts produces a size sorting distribution that looks like that derived from refuse disposal.

Therefore, we can conclude that level J cannot be defined according to occupation length. The temporal resolution of the assemblage limits our ability to characterize the occupations beyond the distinction between residential and non-residential events. Even in a high-resolution context like Abric Romaní, it seems illusory to reach a time scale—that of occupation in the ethnographic sense—in which these questions may be answered. We can see the long-term processes acting at the time scale of the stratigraphic units (the time of the structure) and the short-term processes acting at the time scale of single activity episodes (the time of the event), but we cannot see "occupations" or segregate the archeological remains according to an ethnographic time scale.

Although this conclusion may seem a bit disappointing, this temporal dimension shines a new light on Neanderthal behavior. It allows variability levels associated with two different time scales to be discerned: the time of the event and the time of the structure (Sewell 1996; Harding 2005). Archeology is a particularly suitable domain for this kind of approach, since it allows an almost immediate access to these two temporal levels. According to Giddens (1979), structures are made up of rules and resources that are enacted in social practices. They are the underlying long-term behavioral patterns and natural conditions in which the short-term events are founded. Events are specific and contextual, singular moments embedded in circumstance, and can be considered as the actualization of structures (Beck et al. 2007). They are activities that represent adaptation to the circumstances of a particular place in a particular instant of time. The interplay between these temporal scales is essential in understanding social and cultural changes, although the causal relationships between events and structures are far from being established. On the one hand, we might consider that what happens at the time of the event is defined by the long-term structures. On the other hand, we could argue that the short-term individual engagements determine the larger-scale entities and processes (Harding 2005). According to Sewell (1996), the causes of structural changes should be looked for in the event time-scale, since events create durable ruptures between resources and schemas and therefore provide the opportunity for new frames of reference. In any case, we do not think that establishing such causal relationships is the most important endeavor at this point. For now, it is enough to identify these two time scales in the archeological record and to associate them with specific features of assemblage variability.

An event-focused approach depends on the ability to identify single activity events in the archeological assemblage. In level J, this has been possible in lithic analysis, and our perspective on this level of variability is biased towards technical activities. Other domains are less suitable for this type of temporal study, and they tend to favor the structural viewpoint of the assemblage-as-a-whole. Far from showing repetitive patterns, behavior exhibits high variability in the short term. As the resolution of analysis increases, Neanderthal behavior becomes more variable. Some data suggest that behavior was not homogeneous throughout the formation of the archeological layer. Certain behaviors are associated with some events, but not to others. For example, knapping of limestone and quartz nodules are not randomly distributed during the formation of archeological level. It is mainly concentrated in a specific area of sublevel Ja suggesting that it occurred during the same occupation event or the same formation phase. These poor quality but immediately available materials were not equally exploited throughout the level J formation sequence, but tended to be restricted to a specific time period. A similar phenomenon was also observed in level I, where the reduction of limestone and quartz cores was found to be characteristic of some accumulations (Carbonell 2002). From this perspective, characterizing a stratigraphic assemblage in behavioral terms makes little sense. It remains to be proven whether this variability at the event level is also characteristic of other behavioral realms in which we can only reach the structural level of variability due to the low resolution of the data.

Some aspects of this temporal variability can be explained by the changes brought about at the site as a consequence of human occupation. Archeological sites were dynamic entities and their appeal for human habitation changed over time, as can be seen in lithic provisioning strategies. During the beginning of the formation of level J, the rockshelter was a place devoid of lithic resources and provisioning of bulk resources in the form of entire nodules would have been more likely. One of the differences of sublevel Jb with respect to Ja is the higher percentage of RMUs introduced as complete nodules. In level Ja, some of the RMUs showing the widest scatters—probably corresponding to the first reduction sequences carried out in this sublevel—were also brought into the shelter as entire nodules. As the formation of the assemblage advanced, the site was progressively transformed into a raw material source itself and the need to introduce bulk resources would have been less likely. The dispersion radii of reduction sequences indicate that recycling events tended to take place in the last stage of the assemblage formation. Moreover, we have seen that the accumulation focused in P51 is characterized by a series of reduction sequences on highly reduced cores introduced as such into the shelter. The products of these sequences are less widely dispersed, suggesting that they were produced in the last occupations of level J. This indicates that raw material constraints changed throughout the formation of the assemblage and, consequently, the lithic provisioning strategy also changed.

Level J indicates that some of the factors normally used to explain the variability of Middle Paleolithic lithic assemblages—raw material provisioning, artifact transport, exploitation intensity—operate at the time of the event. We might wonder if this is also the case for the exploitation of floral and faunal resources, but we do not have the temporal resolution required to answer this question. However, level J also provides some insights into variability factors acting on a long-term structural time scale. These factors are defined by the characteristics that remained unchanged during the formation of the archeological assemblage in spite of the short-term adaptations observed at the event level. For example, the topography of the site determined the location of the domestic hearth-related areas in the sheltered space between the wall and the exterior line of blocks. Spatial organization was therefore conditioned by these natural factors.

Another structural trend associated with long-term behavioral patterns is the dominant exploitation of large- and middle-sized ungulates. This structural pattern persisted throughout the entire Abric Romaní sequence, regardless of the environmental changes. Other structural trends are related to technical activities, like the dominance of denticulates in the retouched assemblage and the general criteria applied in core reduction sequences. Core morphology exhibits high variability due to the use of expedient methods aimed at producing as many flakes as possible, but characterized by little concern for blank shape and size. This is a good example of the interplay between the different temporal scales of variability. The specific characteristics of each reduction sequence depend on the circumstances operating at the event level, like nodule shape or transport mode. It is at this level that the variability of core forms should be explained. However, the expedient nature of technical behavior is structural and defines the general framework in which this large variability of cores emerges. The dichotomy of expedient versus predetermined reduction methods—discoidal versus Levallois—should therefore be explained by factors acting in the long-term, but not by the adaptation to the specific constraints of each technical event. In the case of Abric Romaní, this issue may be considered from the diachronic perspective of the archeological sequence as a whole, but it is beyond the temporal limits of a single archeological layer.

Research in level J has yielded some conclusions about the usefulness of the ethnographic models we adopted as a starting point. The limits of the ethnographical explanation are mainly the consequence of the differences between the ethnographic time scale and the archeological time. The basic time scale of ethnoarcheological models, occupational time, is the most difficult to deduce from the archeological record. We have easy access to the time scale of the single activity event and the geological time scale represented by the stratigraphic unit—the structural time scale -, but the occupation time scale remains archeologically invisible. All we have are events and relationships between events, but there are serious constraints to defining these relationships in occupational terms. Through the spatial association of events we can identify activity areas and we can even expect to establish relationships between activity areas by means of refitting. Nevertheless, it seems unlikely that this chain of relationships will expand to the scale of an occupation in the ethnographic sense of the term. This does not mean that all the ethnoarcheological information is useless as a middle range framework for archeological interpretation. It implies that we must be more conscious of the consequences of time resolution in the use of such information. The ethnoarcheological evidence corresponding both to an event time scale and a structural time scale would maintain its reliability, but those ethnographic models based on the intermediate time scale of the 'occupation' are affected by these temporal problems and their usefulness should be questioned.

Level J has also provided some insights into the criteria for defining archeological assemblages. At the event time-scale, stratigraphically defined assemblages are not taphonomic, environmental or behavioral units. Therefore, stratigraphic levels cannot be behaviorally characterized, since the assemblage-as-a-whole is nothing but the addition of an unknown number of events that do not necessarily correspond to the same behavioral patterns. This explains how in a context characterized by raw material economy we have found some apparently uneconomical events. The contradictory nature of the assemblage is especially clear when we compare the core reduction sequence, aimed at maximizing the profitability of lithic resources, and the management of transported blanks, which are characterized by a comparatively wasteful behavior. In addition, this indicates that raw material economy was not ruled by structural factors. Palimpsests may be useful when we are searching for long-term trends, but they may be misleading when we try to discern the behavioral patterns that depend on short-term decisions. The definition of archeological assemblages should be adapted to the kind of issues we are dealing with.

The significance of a specific behavior is related to its frequency within a social system, not to the amount of material remains that such a behavior produces. The behavioral interpretation of palimpsests is often biased by the activity episodes that produced more remains, at the expense of other behaviors that may be more common but generate few remains. This is evident if we compare the archeological consequences of the reduction sequences carried out entirely at the site with the introduction and disposal of single blanks. The latter behavior is more common and can be considered as more significant in characterizing the technical behavior of the Neanderthals that visited Abric Romaní. However, this behavior tends to be blurred in an assemblage-as-a-whole perspective. We can only confront these problems if we focus our attention on the events and organize the archeological record according to them. The construction of an archeological discourse based on events should be considered as a challenge for future research on the evolution of human behavior.

Acknowledgments We would like to express our gratitude to all the people who have participated in the archeological works at Abric Romaní since the discovery of the site in 1909. We thank also the institutions and companies that supported the excavations and the research projects. Our special thanks go to Sally McBrearty and Eric Delson for reviewing the first draft of this chapter. This chapter also benefited from helpful comments by and discussions with the authors of the other chapters included in this book. We thank all of them.

References

Adler, D. S., Bar-Oz, G., Belfer-Cohen, A., & Bar-Yosef, O. (2006). Ahead of the game Middle and Upper Palaeolithic hunting behaviors in the southern Caucasus. *Current Anthropology, 47*(1), 89–118.

Albert, R. M., Bar-Yosef, O., & Weiner, S. (2007). Use of plant material in Kebara cave: Phytoliths and mineralogical analyses. In O. Bar-Yosef & L. Meignen (Eds.), *Kebara Cave Mt Carmel, Israel. The Middle and Upper Paleolithic archaeology* (pp. 147–162). Cambridge, MA: Peabody Museum of Archaeology and Ethnology Harvard University.

Allué, E., Burjachs, F., García, A., López-García, J. M., Bennàsar, M., Rivals, F., Blain, H.-A., Expósito, I. & Martinell, J. (2012). Neanderthal Landscapes and Their Home Environment: Flora and Fauna Records from Level J. In E. Carbonell (Ed.), *High Resolution Archaeology and Neanderthal Behavior. Time and Space in Level J of Abric Romaní* (pp.). Dordretch: Springer.

Badal, E., & Carrión, Y. (2001). Del glacial a l'interglacial: els paisatges vegetals a partir de les restes carbonitzades trobades en les coves d'Alacant. In V. Villaverde (Ed.), *De neandertals a cromanyons. L'inici del poblament humà a les terres valencianes* (pp. 21–40). València: Universitat de València.

Barton, N. (2000). Mousterian hearths and shellfish: Late Neanderthal activities on Gibraltar. In C. B. Stringer, R. N. E. Barton, & J. C. Finlayson (Eds.), *Neanderthals on the edge* (pp. 211–220). Oxford, England: Oxbow Books.

Baruch, U., Werker, E., & Bar-Yosef, O. (1992). Charred wood remains from Kebara cave, Israel: Preliminary results. *Bulletin Society Botany Fr. Actualites Botaniques, 139*, 531–538.

Bailey, G. (2007). Time perspectives, palimpsests and the archaeology of time. *Journal of Anthropological Archaeology, 26*, 198–223.

Beck, R. A., Jr., Bolender, D. J., Brown, J. A., & Earle, T. K. (2007). Eventful archaeology. The place of space in structural transformation. Current Anthropology, *48*(6), 833–860.

Binford, L. R. (1978). Dimensional analysis of behavior and site structure: Learning from an Eskimo hunting stand. *American Antiquity, 43*, 330–361.

Binford, L. R. (1982). Archaeology of place. *Journal of Anthropological Archaeology, 1*, 5–31.

Binford, L. R. (1987). Researching ambiguity: Frames of reference and site structure. In S. Kent (Ed.), *Method and theory for activity area research. An ethnoarchaeological approach* (pp. 449–512). New York: Columbia University Press.

Bischoff, J.L., Julià, R. & Mora, R. (1988). Uranium-Series Dating of the Mousterian Occupation at Abric Romani, Spain. *Nature*, 332, 68-70.

Blasco, M. F. (1997). In the pursuit of game: The mousterian cave site of Gabasa 1 in the Spanish Pyrenees. *Journal of Anthropological Research, 53*, 177–217.

Blasco, R. (2008). Human consumption of tortoises at Level IV of Bolomor Cave (Valencia, Spain). *Journal of Archaeological Science, 35*(10), 2839–2848.

Blasco, R., & Fernández Peris, J. (2009). Middle Pleistocene bird consumption at Level XI of Bolomor Cave (Valencia, Spain). *Journal of Archaeological Science, 36*(10), 2213–2223.

Burjachs, F., & Julià, R. (1994). Abrupt climatic changes during the last glaciation based on pollen analysis of the Abric Romaní, Catalonia, Spain. *Quaternary Research, 42*, 308–315.

Carbonell, E. (1992). Abric Romaní, nivell H: un model d'estratègia ocupacional al Plistocè Superior mediterrani. *Estrat, 5*, 157–308.

Carbonell, E. (2002). *Abric Romaní nivell I. Models d'ocupació de curta durada de fa 46.000 anys a la Cinglera del Capelló (Capellades Anoia, Barcelona)*. Tarragona: Universitat Rovira i Virgili.

Carbonell, E., Lorenzo, C., & Vallverdú, J. (2007). Centralidad espacial y operativa de los neanderthales. Análisis espacial diacrónico de las actividades de combustión en el Abric Romaní (Anoia, Capellades, Barcelona). In E. Baquedano (Ed.), *El universo neandertal* (pp. 197–219). Soria: Fundación duques de Soria.

Carrión, J. S., Yll, E. I., Walker, M. J., Legaz, A. J., Chain, C., & López, A. (2003). Glacial refugia of temperate, Mediterranean and Ibero-North African flora in south-eastern Spain: New evidece from cave pollen at two Neanderthal man sites. *Global Ecology and Biogeography, 12*, 119–129.

Carrión, J. S., Finlayson, C., Fernández, S., Finlayson, G., Allué, E., López-Sáez, J. A., et al. (2008). A coastal reservoir of biodiversity for Upper Pleistocene human populations: Palaeoecological investigations in Gorham's Cave (Gibraltar) in the context of the Iberian Peninsula. *Quaternary Science Reviews, 27*(23–24), 2118–2135.

Castaños, P. M. (2005). Revisión actualizada de las faunas de macromamíferos del Würm antiguo en la Región Cantábrica. In R. Montes & J. A. Lasheras (Eds.), *Neandertales cantábricos, estado de la cuestión* (pp. 201–207). Museo de Altamira, Monografías, 20. Santander: Museo de Altamira

Conard, N. J., & Adler, D. S. (1997). Lithic Reduction and Hominid Behavior in the Middle Paleolithic of the Rhineland. *Journal of Anthropological Research, 53*, 147–175.

Deacon, H. J., & Deacon, J. (1999). *Human beginnings in South Africa: Uncovering the secrets of the Stone Age.* Claremont, CA: David Philip Publishing.

Fernández-Jalvo, Y., & Andrews, P. (2000). Taphonomy of Ibex Cave, Gibraltar. In C. B. Stringer, R. N. E. Barton, & J. C. Finlayson (Eds.), *Neanderthals on the Edge* (pp. 201–213). Oxford, England: Oxbow Books.

Fernández Peris, J., Guillem, P., & Martínez Valle, R. (2000). Cova del Bolomor (Tavernes de la Valldigna-Valencia). Datos cronoestratigráficos y culturales de una secuencia del Pleistoceno Medio. In *Actas do 3° Congresso de Arqueologia Peninsular. Vol. II. Paleolítico da Península Ibérica* (pp. 81–100). Porto: ADECAP.

Finlayson, C., & Carrion, J. S. (2007). Rapid ecological turnover and its impact on Neanderthal and other human populations. *Trends in Ecology & Evolution, 22*, 213–222.

Fisher, J. W. & Strickland, H. C. (1991). Dwellings and fireplaces: Keys to Efe Pygmy campsite structure. In: C. S. Gamble & W. Boismier (Eds.), *Ethnoarchaeological approaches to mobile campsites* (pp. 215–236). Ann Arbor, MA: International Monographs in Prehistory. Ethnoarchaeological Series, 1.

Geneste, J.-M. (1988). Les industries de la Grotte Vaufrey: Technologie du debitage, economie et circulation de la matière première lithique. In J.-Ph. Rigaud (Ed.), *La Grotte Vaufrey à Cenac et Saint-Julien (Dordogne). Paléoenvironnements, chronologie et activités humaines* (pp. 441–517). Paris: Mémoires de la Société Préhistorique Française XIX.

Geneste, J.-M. (1989). Systemes d'approvisionnement en matières premièrs au Paléolithique moyen et au Paléolithique supérieur en Aquitaine. In *L'Homme de Neandertal, 8. La mutation* (pp. 61–70). Liège: ERAUL.

Giddens, A. (1979). *Central problems in social theory: Action structure, and contradiction in social analysis*. London: Macmillan.

Gould, R. A., & Yellen, J. E. (1991). Misreading the past: A reply to Binford concerning Hunter-Gatherer site structure. *Journal of Anthropological Archaeology, 16*, 283–298.

Gowlet, J. A. J. (1997). High definition archaeology: Ideas and evaluation. *World Archaeology, 29*(2), 152–171.

Hallegouet, B., Hinguant, S., Gebhardt, A., & Monnier, J.-L. (1992). Le gisement Paléolithique inférieur de Ménez-Drégan 1 (Finistère). *Premiers résultats des fouilles. Bulletin de la Société préhistorique française, 89*(3), 77–81.

Harding, J. (2005). Rethinking the great divide: Long-term structural history and the temporality of event. *Norwegian Archaeological Review, 38*(2), 88–101.

Henry, D. O., Hietala, H. J., Rosen, A. M., Demidenko, Y. E., Usik, V. I., & Armagan, T. L. (2004). Human behavioral organization in the Middle Paleolithic: Were Neanderthals different? *American Anthropologist, 106*(1), 17–31.

Isaac, G. L. (1977). *Olorgesailie Archaeological studies of a Middle Pleistocene lake basin in Kenya.* Chicago: University of Chicago Press.

Isaac, G. L. (1978). The food-sharing behavior of protohuman hominids. *Scientific American, 238*, 90–108.

Kent, S. (1991). The relationship between mobility strategies and site structure. In E. M. Kroll & T. D. Price (Eds.), *The interpretation of Archaeological spatial patterning* (pp. 33–59). New York: Plenum Press.

Kuhn, S. L., & Stiner, M. C. (2006). What's a mother to do? The division of labor among Neandertals and modern humans in Eurasia. *Current Anthropology, 47*(6), 953–980.

James, S. R. (1989). Hominid use of fire in the lower and middle pleistocene. A review of the evidence. *Current Anthropology, 30*(1), 1–26.

Leroi-Gourhan, A., & Brézillon, M. (1972). *Fouilles de Pincevent: essai d'analyse ethnographique d'un habitat magdalénien (la section 36).* Paris: Editions du CNRS.

Lev, E., Kislev, M. E., & Bar-Yosef, O. (2005). Mousterian vegetal food in Kebara Cave, Mt. Carmel. *Journal of Archaeological Science, 32*, 475–484.

de Lumley, H. (1969). Une cabane acheuléenne dans la grotte du Lazaret. *Mémoires de la Société Préhistorique Française, 7*, 1–234.

de Lumley, H., Grégoire, S., Barsky, D., Batalla, G., Bailon, S., Belda, V., et al. (2004). Habitat et mode de vie des chasseurs paléolithiques de la Caune de l'Arago (600.000–400.000 ans). *L'Anthropologie, 108*, 159–184.

Maroto, J., Soler, N., & Fullola, J. M. (1996). Cultural change between Middle and Upper Paleolithic in Catalonia. In E. Carbonell & M. Vaquero (Eds.), *The last Neandertals, the first anatomically modern humans: A tale about the human diversity* (pp. 219–250). Tarragona: Universitat Rovira i Virgili.

Martínez Moreno, J. (2005). Una aproximación zooarqueológica al estudio de los patrones de subsistencia del Paleolítico Medio Cantábrico. In R. Montes & J. A. Lasheras (Eds.), *Neandertales cantábricos, estado de la cuestión Museo de Altamira, Monografías, 20* (pp. 209–230). Santander: Museo de Altamira.

Martínez Molina, K., & Rando, J. M. (2001). Organización y funcionalidad de la producción lítica en un nivel del Paleolítico Medio del Abric Romaní. Nivel Ja (Capellades, Barcelona). *Trabajos de Prehistoria, 58*(1), 51–70.

Martínez Moreno, J., Terradas, X. & Mora, R. (1994). El jaciment del paleolític mitjà de la Roca dels Bous (Sant Llorenç de Montgai, Camarasa, Noguera). *Tribuna d'Arqueologia*, 1992–1993, 7–13.

Martínez Moreno, J., Mora, R., & de la Torre, I. (2004). Methodological Approach for Understanding Middle Palaeolithic settlement dynamics at La Roca dels Bous (Noguera, Catalunya, Northeast Spain). In N. J. Conard (Ed.), *Settlement dynamics of the Middle Paleolithic and Middle Stone Age II* (pp. 393–414). Tübingen: Kerns Verlag.

Meignen, L. (1993). Répartition spatiale des vestiges au sol de l'abri des Canalettes. In L. Meignen (Ed.), *L'abri des Canalettes. Un habitat moustérien sur les grands Causses (Nant, Aveyron). Fouilles 1980–1986* (pp. 161–197). Paris: Éditions du CNRS.

Meignen, L., Bar-Yosef, O., Goldberg, P., & Weiner, S. (2001). Le feu au Paléolithique moyen: recherches sur les structures de combustion et le statut des foyers. *L'exemple du Proche-Orient. Paléorient, 26*(2), 9–22.

Meignen, L., Goldberg, P., & Bar-Yosef, O. (2007). The hearths at Kebara cave and their role in site formation processes. In O. Bar-Yosef & L. Meignen (Eds.), *Kebara Cave Mt Carmel, Israe the Middle and Upper Paleolithic archaeology* (pp. 91–106). Cambridge, MA: Peabody Museum of Archaeology and Ethnology Harvard University.

Michel, D. (1982). Le gisement préhistorique de Port Pignot à Fermanville (Manche). *Gallia Préhistoire, 25*, 1–68.

Michel, D. (1994). *La Roche Gélétan gisement structuré du Paléolithique inférieur (Saint-Germain-des-Vaux, Manche)*. Evreux: Fondation Singer-Polignac.

Monahan, C. M. (1998). The Hadza Carcass transport debate revisited and its archaeological implications. *Journal of Archaeological Science, 25*, 405–424.

Moncel, M.-H., Patou-Mathis, M., & Otte, M. (1998). Halte de chasse au chamois au Paléolithique moyen: la couche 5 de la grotte Scladina (Sclayn, Namur, Belgique). In J.-Ph Brugal, L. Meignen, & M. Patou-Mathis (Eds.), *Economie préhistorique: les comportements de subsistance au Paléolithique* (pp. 291–308). Sophia Antipolis: Editions APDCA.

Mora, R., Carbonell, E., Cebrià, A. & Martínez, J. (1988). Els sòls d'ocupació a l'Abric Romaní (Capellades, Anoia). *Tribuna d'Arqueologia*, 1987–1988, pp. 115–123.

Mora, R., Terradas, X., Martínez Moreno, J., Pardos, E., & Pallarés, M. (1992). Primera aproximación al estudio de las ocupaciones humanas de la Cueva de l'Estret de Tragó (Os de Balaguer, Lérida). In P. Utrilla (Ed.), *Aragón/litoral mediterráneo: intercambios culturales durante la Prehistoria* (pp. 97–105). Zaragoza: Institución Fernando el Católico.

Nicholson, A., & Cane, S. (1991). Desert camps: Analysis of Australian aboriginal proto-historic campsites. In C. S. Gamble & W. A. Boismier (Eds.), *Ethnoarchaeological approaches to mobile campsites* (pp. 263–354). Ann Arbor, MA: International Monographs in Prehistory.

O'Connell, J. F. (1987). Alyawara site structure and its archaeological implications. *American Antiquity, 52*(1), 74–108.

O'Connell, J. F., Hawkes, K., & Blurton Jones, N. (1991). Distribution of refuse-producing activities at Hadza residential base camps. Implications for analyses of archaeological site structure. In E. M. Kroll & T. D. Price (Eds.), *The interpretation of archaeological spatial patterning* (pp. 61–76). New York: Plenum Press.

Olive, M. (1988). *Une habitation magdalénienne d'Étiolles: l'unité P15*. Paris: Éditions de la SPF.

Pettitt, P. B. (1997). High resolution Neanderthals? Interpreting Middle Palaeolithic intrasite spatial data. *World Archaeology, 29*(2), 208–224.

Pigeot, N. (2004). *Les derniers magdaléniens d'Étiolles. Perspectives culturelles et paléohistoriques (l'unité d'habitation Q31)*. Paris: Éditions du CNRS.

Richter, J. (2006). Neanderthals in their landscape. In B. Demarsin & M. Otte (Eds.), *Neanderthals in Europe. Proceedings of the international conference held in the Gallo-Roman museum in Tongeren (September 17–19th 2004)* (pp. 51–66). Liège: ERAUL 117.

Rigaud, J.-Ph., Simek, J. F., & Ge, Th. (1995). Mousterian fires from Grotte XVI (Dordogne, France). *Antiquity, 69*, 902–912.

Rolland, N. (1999). The Middle Palaeolithic as development stage: Evidence from technology, subsistence, settlement systems, and Hominid socio-ecology. In H. Ullrich (Ed.), *Hominid evolution. lifestyles and survival strategies* (pp. 315–334). Gelsenkirchen: Archaea.

Rolland, N. (2004). Was the emergence of home bases and domestic fire a punctuated event? A review of the Middle Pleistocene record in Eurasia. *Asian Perspectives, 43*, 248–280.

Rosell, J., Blasco, R., Huguet, R., Cáceres, I., Saladié, P., Rivals, F., et al. (2012). Occupational patterns and subsistence strategies in Level J of Abric Romaní. In E. Carbonell (Ed.), *High resolution archaeology and neanderthal behavior. Time and space in level J of Abric Romaní* (pp.). Dordretch: Springer.

Rosell, J., Huguet, R., Aimene, M., Angelucci, D. E., Canals, A., Pastó, I. et al. (2000). El yacimiento de las Fuentes de San Cristóbal (Veracruz, Huesca): un nuevo enclave del Paleolítico Medio en el Prepirineo. In *Actas do 3° Congresso de Arqueologia Peninsular. Vol. II. Paleolítico da Península Ibérica* (pp. 235–249). Porto: ADECAP.

Sanchís, A., & Fernández Peris, J. (2008). Procesado y consumo antrópico de conejo en la Cova del Bolomor (Tavernes de la Valldigna, Valencia). El nivel XVIIc (ca 350 ka). *Complutum, 19*(1), 25–46.

Sañudo, P., Vallverdú, J. & Canals, A. (2012). Spatial patterns in level J. In E. Carbonell (Ed.), *High resolution archaeology and neanderthal behavior. Time and space in level J of Abric Romaní* (pp.). Dordretch: Springer.

Schmider, B. (1992). *Marsagny, un campement des derniers chasseurs magdaléniens sur les bords de l'Yonne*. Liège: ERAUL.

Sewell, W. H., Jr. (1996). Historical events as transformations of structures: Inventing revolution at the Bastille. *Theory and Society, 25*, 841–881.

Stevenson, M. C. (1991). Beyond the formation of Hearth-Associated artifact assemblages. In E. M. Kroll & T. D. Price (Eds.), *The interpretation of archaeological spatial patterning* (pp. 269–299). New York: Plenum Press.

Stiner, M. C., Munro, N. D., & Surovell, T. A. (2000). The Tortoise and the Hare. Small-game use, the broad-spectrum revolution, and Paleolithic demography. *Current Anthropology, 41*(1), 39–73.

Stringer, C. B., Finlayson, J. C., Barton, R. N. E., Fernández-Jalvo, Y., Cáceres, I., Sabin, R. C., et al. (2008). Neanderthal exploitation of marine mammals in Gibraltar. *Proceedings of the National Academy of Sciences, 105*(38), 14319–14324.

Terradas, X., & Rueda, J. M. (1998). Grotte 120: un exemple des activités de subsistance au Paléolithique moyen dans les Pyrénées orientales. In J.-Ph Brugal, L. Meignen, & M. Patou-Mathis (Eds.), *Économie préhistorique: les comportements de subsistance au Paléolithique* (pp. 349–362). Sophia Antipolis: Éditions APDCA.

Théry-Parisot I. (2001). *Économie des combustibles au Paléolithique.* Paris: Dossier de Documentation Archéologique 20 CNRS.

Thomas, D. H. (1983). *The Archaeology of Monitor Valley: Gatcliff Shelter.* Anthropological papers, 59, 1. New York: American Museum of Natural History.

Vallverdú, J., Allué, E., Bischoff, J. L., Cáceres, I., Carbonell, E., Cebrià, A., et al. (2005). Short human occupations in the Middle Palaeolithic level i of the Abric Romaní rock-shelter (Capellades, Barcelona, Spain). *Journal of Human Evolution, 48*(2), 157–174.

Vallverdú, J. & Courty, M.-A. (2012). Microstratigraphic analysis of the level J deposits. A dual paleoenvironmental-paleoethnographic contribution to the paleolithic archaeology at the Abric Romani site. In E. Carbonell (Ed.), *High resolution archaeology and neanderthal behavior. Time and space in level J of Abric Romaní* (pp.). Dordretch: Springer.

Van Andel, T. H., & Davies, W. (2003). *Neanderthals and modern humans in the European landscape during the last glaciation. Archaeological results of the stage 3 project.* Cambridge, MA: McDonald Institute for Archaeological Research.

Vaquero, M. (1999). Intrasite spatial organization of lithic production in the Middle Palaeolithic: the evidence of the Abric Romaní (Capellades, Spain). *Antiquity, 73*, 493–504.

Vaquero, M. (2008). The history of stones: Behavioural inferences and temporal resolution of an archaeological assemblage from the

Middle Palaeolithic. *Journal of Archaeological Science, 35*, 3178–3185.

Vaquero, M., & Pastó, I. (2001). The definition of spatial units in Middle Palaeolithic sites: The hearth-related assemblages. *Journal of Archaeological Science, 28*(11), 1209–1220.

Vaquero, M., Chacón, M.G, Cuartero, F., García-Antón, Mª D., Gómez de Soler, B. & Martínez, K. (2012). The lithic assemblage of level J. In E. Carbonell (Ed.), *High resolution archaeology and neanderthal behavior. Time and space in level J of Abric Romaní* (pp.). Dordretch: Springer.

Vaquero, M., Chacón, G., Fernández, C., Martínez, K., & Rando, J. M. (2001a). Intrasite spatial patterning and transport in the Abric Romaní Middle Paleolithic Site (Capellades, Barcelona, Spain). In N. J. Conard (Ed.), *Settlement dynamics of the Middle Paleolithic and Middle Stone Age* (pp. 573–595). Tübingen: Kerns Verlag.

Vaquero, M., Vallverdú, J., Rosell, J., Pastó, I., & Allué, E. (2001b). Neandertal behaviour in the Middle Palaeolithic site of Abric Romani, Capellades, Spain. *Journal of Field Archaeology, 28*, 93–114.

Vaquero, M., Rando, J. M., & Chacón, G. (2004). Neanderthal spatial behavior and social structure: Hearth-Related assemblages from the Abric Romaní Middle Palaeolithic Site. In N. J. Conard (Ed.), *Settlement dynamics of the Middle Paleolithic and Middle Stone Age II* (pp. 367–392). Kerns Verlag: Tübingen.

Villaverde, V., Martínez-Valle, R., Guillem, P. M., & Fumanal, M. P. (1996). Mobility and the role of small game in the Middle Paleolithic of the central region of the Spanish Mediterranean: A comparison of Cova Negra with other Paleolithic deposits. In E. Carbonell & M. Vaquero (Eds.), *The Last Neandertals, the first anatomically modern humans: A tale about the human diversity* (pp. 267–288). Tarragona: Universitat Rovira i Virgili.

Wadley, L. (2004). Late Middle Stone Age spatial patterns in Rose Cottage Cave, South Africa. In N. J. Conard (Ed.), *Settlement dynamics of the Middle Paleolithic and Middle Stone Age II* (pp. 23–36). Tübingen: Kerns Verlag.

Winterhalder, B., & Smith, E. A. (2000). Analyzing adaptive strategies: Human behavioral ecology at twenty-five. *Evolutionary Anthropology, 9*(2), 51–72.

Willis, K.J., Rudner, E. & Sümegi, P. (2000). The full-glacial forests of central and southeastern Europe. *Quaternary Research, 53*(2), 203-213.

Yellen, J. E. (1977a) *Archaeological approaches to the present: Models for reconstructing the past.* New York: Academic Press.

Yellen, J. E. (1977b). Cultural patterning in faunal remains: Evidence from the !Kung Bushmen. In D. Ingersoll, J. E. Yellen, & W. Macdonald (Eds.), *Experimental archeology* (pp. 271–331). New York: Columbia University Press.

Yravedra, J. (2003). Estado de la cuestión sobre la subsistencia del Musteriense en el interior y la fachada atlántica de la Península Ibérica. *Zephyrus, 56*, 61–84.

Yravedra, J. (2006). Acumulaciones biológicas en yacimientos arqueológicos: Amalda VII y Esquilleu III-IV. *Trabajos de Prehistoria, 63*(2), 55–78.

Index

E. Carbonell i Roura (ed.), *High Resolution Archaeology and Neanderthal Behavior: Time and Space in Level J of Abric Romaní (Capellades, Spain),* Vertebrate Paleobiology and Paleoanthropology,
DOI: 10.1007/978-94-007-3922-2, © Springer Science+Business Media B.V. 2012

MIX
Papier aus verantwortungsvollen Quellen
Paper from responsible sources
FSC® C105338

If you have any concerns about our products,
you can contact us on
ProductSafety@springernature.com

In case Publisher is established outside the EU,
the EU authorized representative is:
Springer Nature Customer Service Center GmbH
Europaplatz 3, 69115 Heidelberg, Germany

Printed by Libri Plureos GmbH
in Hamburg, Germany